Arthur W. Dybdahl

Geophysical Monograph Series

Including

IUGG Volumes
Maurice Ewing Volumes
Mineral Physics Volumes

Geophysical Monograph Series

120 **GeoComplexity and the Physics of Earthquakes** *John B. Rundle, Donald L. Turcotte, and William Klein (Eds.)*

121 **The History and Dynamics of Global Plate Motions** *Mark A. Richards, Richard G. Gordon, and Rob D. van der Hilst (Eds.)*

122 **Dynamics of Fluids in Fractured Rock** *Boris Faybishenko, Paul A. Witherspoon, and Sally M. Benson (Eds.)*

123 **Atmospheric Science Across the Stratopause** *David E. Siskind, Stephen D. Eckerman, and Michael E. Summers (Eds.)*

124 **Natural Gas Hydrates: Occurrence, Distribution, and Detection** *Charles K. Paull and Willam P. Dillon (Eds.)*

125 **Space Weather** *Paul Song, Howard J. Singer, and George L. Siscoe (Eds.)*

126 **The Oceans and Rapid Climate Change: Past, Present, and Future** *Dan Seidov, Bernd J. Haupt, and Mark Maslin (Eds.)*

127 **Gas Transfer at Water Surfaces** *M. A. Donelan, W. M. Drennan, E. S. Saltzman, and R. Wanninkhof (Eds.)*

128 **Hawaiian Volcanoes: Deep Underwater Perspectives** *Eiichi Takahashi, Peter W. Lipman, Michael O. Garcia, Jiro Naka, and Shigeo Aramaki (Eds.)*

129 **Environmental Mechanics: Water, Mass and Energy Transfer in the Biosphere** *Peter A. C. Raats, David Smiles, and Arthur W. Warrick (Eds.)*

130 **Atmospheres in the Solar System: Comparative Aeronomy** *Michael Mendillo, Andrew Nagy, and J. H. Waite (Eds.)*

131 **The Ostracoda: Applications in Quaternary Research** *Jonathan A. Holmes and Allan R. Chivas (Eds.)*

132 **Mountain Building in the Uralides Pangea to the Present** *Dennis Brown, Christopher Juhlin, and Victor Puchkov (Eds.)*

133 **Earth's Low-Latitude Boundary Layer** *Patrick T. Newell and Terry Onsage (Eds.)*

134 **The North Atlantic Oscillation: Climatic Significance and Environmental Impact** *James W. Hurrell, Yochanan Kushnir, Geir Ottersen, and Martin Visbeck (Eds.)*

135 **Prediction in Geomorphology** *Peter R. Wilcock and Richard M. Iverson (Eds.)*

136 **The Central Atlantic Magmatic Province: Insights from Fragments of Pangea** *W. Hames, J. G. McHone, P. Renne, and C. Ruppel (Eds.)*

137 **Earth's Climate and Orbital Eccentricity: The Marine Isotope Stage 11 Question** *André W. Droxler, Richard Z. Poore, and Lloyd H. Burckle (Eds.)*

138 **Inside the Subduction Factory** *John Eiler (Ed.)*

139 **Volcanism and the Earth's Atmosphere** *Alan Robock and Clive Oppenheimer (Eds.)*

140 **Explosive Subaqueous Volcanism** *James D. L. White, John L. Smellie, and David A. Clague (Eds.)*

141 **Solar Variability and Its Effects on Climate** *Judit M. Pap and Peter Fox (Eds.)*

142 **Disturbances in Geospace: The Storm-Substorm Relationship** *A. Surjalal Sharma, Yohsuke Kamide, and Gurbax S. Lakhima (Eds.)*

143 **Mt. Etna: Volcano Laboratory** *Alessandro Bonaccorso, Sonia Calvari, Mauro Coltelli, Ciro Del Negro, and Susanna Falsaperla (Eds.)*

144 **The Subseafloor Biosphere at Mid-Ocean Ridges** *William S. D. Wilcock, Edward F. DeLong, Deborah S. Kelley, John A. Baross, and S. Craig Cary (Eds.)*

145 **Timescales of the Paleomagnetic Field** *James E. T. Channell, Dennis V. Kent, William Lowrie, and Joseph G. Meert (Eds.)*

146 **The Extreme Proterozoic: Geology, Geochemistry, and Climate** *Gregory S. Jenkins, Mark A. S. McMenamin, Christopher P. McKay, and Linda Sohl (Eds.)*

147 **Earth's Climate: The Ocean–Atmosphere Interaction** *Chunzai Wang, Shang-Ping Xie, and James A. Carton (Eds.)*

148 **Mid-Ocean Ridges: Hydrothermal Interactions Between the Lithosphere and Oceans** *Christopher R. German, Jian Lin, and Lindsay M. Parson (Eds.)*

149 **Continent-Ocean Interactions Within East Asian Marginal Seas** *Peter Clift, Wolfgang Kuhnt, Pinxian Wang, and Dennis Hayes (Eds.)*

150 **The State of the Planet: Frontiers and Challenges in Geophysics** *Robert Stephen John Sparks and Christopher John Hawkesworth (Eds.)*

151 **The Cenozoic Southern Ocean: Tectonics, Sedimentation, and Climate Change Between Australia and Antarctica** *Neville Exon, James P. Kennett, and Mitchell Malone (Eds.)*

152 **Sea Salt Aerosol Production: Mechanisms, Methods, Measurements, and Models** *Ernie R. Lewis and Stephen E. Schwartz*

153 **Ecosystems and Land Use Change** *Ruth S. DeFries, Gregory P. Anser, and Richard A. Houghton (Eds.)*

154 **The Rocky Mountain Region—An Evolving Lithosphere: Tectonics, Geochemistry, and Geophysics** *Karl E. Karlstrom and G. Randy Keller (Eds.)*

155 **The Inner Magnetosphere: Physics and Modeling** *Tuija I. Pulkkinen, Nikolai A. Tsyganenko, and Reiner H. W. Friedel (Eds.)*

156 **Particle Acceleration in Astrophysical Plasmas: Geospace and Beyond** *Dennis Gallagher, James Horwitz, Joseph Perez, Robert Preece, and John Quenby (Eds.)*

157 **Seismic Earth: Array Analysis of Broadband Seismograms** *Alan Levander and Guust Nolet (Eds.)*

Geophysical Monograph 158

The Nordic Seas: An Integrated Perspective

Oceanography, Climatology, Biogeochemistry, and Modeling

Helge Drange
Trond Dokken
Tore Furevik
Rüdiger Gerdes
Wolfgang Berger
Editors

American Geophysical Union
Washington, DC

Library of Congress Cataloging-in-Publication Data

The Nordic seas : an integrated perspective oceanography, climatology, biogeochemistry, and modeling / Helge Drange .. [et al.], editors.
p. cm. — (Geophysical monograph ; 158)
ISBN-13: 978-0-87590-423-8 (hardcover)
1. Oceanography—Arctic regions. 2. Ocean-atmosphere interaction—Arctic regions. 3. Climatic changes—Arctic regions. I. Drange, Helge. II. Series.

GC411N67 2005
551.46'132—dc22

2005018118

ISBN-13: 978-0-87590-423-8
ISBN: 0-87590-423-8
ISSN 0065-8448

Covers: (Front) A long phytoplankton bloom skirting the length of Norway in the Norwegian Sea (15 June 1998). (Back) Continued phytoplankton bloom off the coast of Norway (5 June 2000). Source: National Aeronautics and Space Administration (NASA) Visible Earth catalog of earth images; these two images were provided by SeaWiFS Project, NASA/Goddard Space Flight Center, and ORBIMAGE.

Printed in the United States of America.

CONTENTS

PREFACE

The world's largest positive temperature deviation from zonal mean temperatures lies within the realm of the Nordic Seas, comprising bodies of water variously referred to as the Norwegian Sea, the Iceland Sea, and the Greenland Sea. Its role as a mixing cauldron for waters entering from the North Atlantic and the Arctic Oceans, and its function as a major source of deep and abyss water, make our understanding of the Nordic Seas a crucial element in advancing the knowledge of climate dynamics in the Northern Hemisphere. In this context, its small extent (covering only 0.75% of the area of the world's oceans) and its unique location, which allows for accessibility and detailed exploration, are of special significance. The current book speaks to that significance specifically and also to assessing the region's present and future response to, and influence on, global climate change. It is the first such work since B. G. Hurdle's groundbreaking *The Nordic Seas* (published in 1986).

Certainly, there are several motivations for producing the current monograph. Climate sciences, for a number of reasons linked to developments in data gathering, analytical methods, and computational resources and to societal concerns about climate change, have been advancing rapidly during the last 20 years. An updated account of the progress on understanding the climate system in the region, therefore, is of special urgency. In that respect, this monograph marks the beginning of an era driven by a truly interdisciplinary approach to better understand the variability and stability of the climate system: Examples are given where classical oceanography, meteorology, paleoclimate observations, biogeochemistry, and numerical modeling are bridged to obtain knowledge that would be impossible to achieve based on those disciplines alone.

Another impetus for this monograph involves historical continuity: sustaining the tradition of compiling available knowledge about the Nordic Seas climate system. The eastern shores of the Nordic Seas have hosted people living off the sea for untold centuries. Knowledge of weather and the pathways of the oceans have been essential to survival, both in everyday fishing activities and for voyages of conquest and trade. Indeed, relevant insights found in print date as far back as the early 13th century. The Norse book, *The King's Mirror—Speculum Regale*, summarizes the wisdom of traders and fishermen based on the experiences and beliefs accumulated over many generations. By the 19th century, the maturation of deliberate, scientific observation brought a new era to fruition, with enormous advances made in our knowledge of the sea and of its associated climate and marine resources.

Today, compilations of state-of-the-art climate research are of particular importance for policy makers and governmental bodies to ensure proper, scientifically based, knowledge for assessing human-induced global and regional climate change issues, and for conducting optimal timing and scaling of mitigation and adaptation strategies. The Nordic Seas area, located between polar and extratropical climate regimes, is expected to be particularly vulnerable to global warming. It is appropriate, therefore, that *The Nordic Seas: An Integrated Perspective* provides a comprehensive and multidisciplinary description of the past and present climate states of the Nordic Seas, the growing realization of the importance of the Nordic Seas in the global climate system, and consideration of how the local climate system may respond to global warming. The monograph does not address how climate changes may influence people and resources in the region. It is our belief, however, that the monograph will form an essential part of the background knowledge needed to properly make such important and urgent assessments.

The monograph consists of five review papers addressing classical oceanography, large-scale atmospheric variability and its impact on ocean climate, paleoclimate observations and analyses, ocean modelling, and marine biogeochemistry. Fourteen more specialised papers, an introduction that bridges the review papers, and a paper about Vilhelm Bjerknes' achievements in weather prediction are also included.

The Editors gratefully acknowledge the efforts by the staff of the Bjerknes Centre for Climate Research, recently established by the Research Council of Norway. Their devotion made this monograph possible. A special thanks is due Elin Darelius for carrying out a substantial part of the compilation of the monograph. Guidance, advice, and technical assistance by Allan Graubard, Maxine Aldred, Dawn Seigler, and Virginia Marcum from the Books Department of AGU are also highly appreciated.

Finally, but not least, we wish to acknowledge the following colleagues for their untiring efforts in reviewing and improving the contributions to the book: Ådlandsvik, B.;

The Nordic Seas: An Integrated Perspective
Geophysical Monograph Series 158

10.1029/158GM01

Bakker, D.; Beszczynska, A.; Bjørk, S.; Bjørnsson, H.; Cheng, W.; Crucifix, M.; Cushman-Roisin, B.; Drinkwater, K.; Eiane, K.; Follows, M.; Fransson, A.; Garwood, R. W.; Gascard, J.-C.; Gulev, A.; Häkkinen, S.; Hald, M.; Hansen, B.; Hjort, C.; Hodges, K.; Holloway, G.; Høyer, J. L.; Hunke, E.; Jennings, A.; Jones, P.; Jungclaus, J.; Kaese, R.; Kara, A. B.; Karcher, M.; Koltermann, P.; Körtzinger, A.; Krahmann, G.; Løyning, T.; Marshall, J.; Matthews, J. A.; McManus, J.; McPhee, M.; Meincke, J.; Melsom, A.; Montegut, C.; Nøst, O. A.; Quadfasel, D.; Renssen, H.; Rochford, P. A.; Rögnvaldsson, S. T.; Rohling, E.; Rudels, B.; Schauer, U.; Schmidt, T.; Schott, F.; Seppä, H.; Serreze, M.; Spielhagen, R.; Stammer, D.; Steele, M.; Swift, J.; Visbeck, M.; Voelker, A.; Walsh, J.; Wanninkhof, R.; Wassman, P.; and Zhang, X.

Helge Drange
Nansen Environmental and Remote Sensing Center,
Bergen, Norway
Bjerknes Centre for Climate Research,
and Geophysical Institute,
University of Bergen, Bergen, Norway
Nansen-Zhu International Research Centre, Beijing, China

Trond Dokken
Bjerknes Centre for Climate Research, University of Bergen,
Bergen, Norway

Tore Furevik
Geophysical Institute, University of Bergen,
and Bjerknes Centre for Climate Research,
Bergen, Norway

Rüdiger Gerdes
Alfred Wegener Institute for Polar and
Marine Research, Bremerhaven, Germany

Wolfgang Berger
Scripps Institution of Oceanography,
San Diego, California, USA

The Nordic Seas: An Overview

Helge Drange[1-4], Trond Dokken[2], Tore Furevik[3,2], Rüdiger Gerdes[5], Wolfgang Berger[6], Atle Nesje[2,7], Kjell Arild Orvik[3], Øystein Skagseth[2], Ingunn Skjelvan[2,3], and Svein Østerhus[2]

The aim of this overview paper is to provide a brief synthesis of the five review papers contained in the monograph. Prevailing southwesterly winds, oceanic flow patterns, and oceanic summer heat storage make the temperatures of the Nordic Seas region 10–20°C above the mean temperature of locations at similar latitudes. The combination of the large heat import from the south plus the polar location implies that the region is prone to natural climate variations and is particularly vulnerable for external forcings. Indeed, proxy data for the Holocene epoch reveal large, high-frequency climate fluctuations, as well as long-term variations spanning the "Medieval Warm Period" and the "Little Ice Age". In phase with a strengthening of the westerly winds since the 1960s, several oceanic key variables show trends unprecedented in available instrumental records, some of which extend back 50–100 years. State-of-the-art climate models indicate that several of the changes may be linked to increased forcing of greenhouse gases and are therefore likely to be sustained or even amplified in the future. Furthermore, the marine cycling of carbon, and through that the major greenhouse gas carbon dioxide, is closely linked to the climate state of the region. The Nordic Seas region is, as one of few ocean locations, a sink for atmospheric carbon dioxide throughout the year. With the rapid developments in data acquisition, computational resources, and societal concerns for climate change and environmental issues, the review papers give an updated account of the present knowledge of the complex climate states of the Nordic Seas and how the Nordic Seas influence the climate outside the region.

1. BACKGROUND

The region north of the Greenland–Scotland Ridge (GSR) and south of the Fram Strait–Spitsbergen–northern Norway transect (e.g., Figure 1 in *Furevik and Nilsen* [this volume]), here defined as the Nordic Seas, covers about 2.5 × 10^6 km^2, or about 0.75%, of the area of the world's oceans. Despite its small extent, the region is very dynamic and diverse [*Blindheim and Østerhus*, *Furevik and Nilsen*, *Nesje et al.*, and *Skjelvan et al.*, all in this volume; *Skjoldal*, 2004, and references therein]: The topography of the sea floor is complex with shallow shelves, deep basins, mid-oceanic ridge systems, and steep slopes. The typical dynamical length scales are small, ranging from a few kilometers to some tens of kilometers. The atmosphere–ocean transfers of momentum, heat, freshwater, and gases are strong, notably during the cold winter months from November to April. Water masses

[1]Nansen Environmental and Remote Sensing Center, Bergen, Norway.
[2]Bjerknes Centre for Climate Research, University of Bergen, Bergen, Norway.
[3]Geophysical Institute, University of Bergen, Bergen, Norway.
[4]Nansen-Zhu International Research Centre, Beijing, China.
[5]Alfred Wegener Institute for Polar and Marine Research, Bremerhaven, Germany.
[6]Scripps Institution of Oceanography, San Diego, California, USA.
[7]Department of Earth Science, University of Bergen, Bergen, Norway.

The Nordic Seas: An Integrated Perspective
Geophysical Monograph Series 158

10.1029/158GM02

originating from low and high latitudes meet and interact by means of frontal mixing, deep convective mixing, subduction, and entrainment. Sea-ice is formed in the northern and western parts of the Nordic Seas in winter, whereas the region is essentially ice-free during summer. New primary production commonly equals or exceeds regenerated primary production, and fish stocks are large and the fisheries rich, particularly in the waters influenced by the warm and nutrient-rich Atlantic Water (AW). The flux of carbon per unit area from atmosphere to the ocean is among the highest in the world's oceans.

All of the above conditions within the Nordic Seas region are well appreciated and tied to its "hot spot" status as an extraordinary warm anomaly. For the present-day climate, the annual and winter mean temperatures of the central and eastern Nordic Seas are 10°C and 20°C higher, respectively, than the zonal means (Figure 1). The anomalously high temperatures are caused by three mechanisms, all of which are important [*Seager et al.*, 2002; *Rhines and Häkkinen*, 2003]: (1) prevailing westerly and southwesterly vapor-laden winds; (2) poleward transport of heat by the Gulf Stream and the North Atlantic Current system; and (3) heat released from the seasonally warmed North Atlantic mixed layer. Changes in any of the three mechanisms have the potential to significantly alter the climate in the region.

The anomalously mild climate of the Nordic Seas and the adjacent land regions has been crucial for survival at these high latitudes for untold generations [e.g., see *The King's Mirror* (*Speculum Regale*) from the early 13th century, presented in *Hellvik*, 1976]. But the region is also important in a wider context:

– For the present-day climate, about 6 Sv (1 Sv = 10^6 m^3 s^{-1}) of cold and dense water spills over the GSR [*Hansen and Østerhus*, 2000; *Blindheim and Østerhus*, this volume]. This volume transport, about one-third of the transport of water associated with the Atlantic Meridional Overturning Circulation (AMOC), drives about two-thirds of the AMOC volume transport by entraining ambient water downslope of the ridge [*Hansen et al.*, 2004].

– The intense surface forcing in the Nordic Seas has a profound impact on the hydrographic properties of the waters that enter the Atlantic Ocean either as overflow waters across the GSR or as surface waters through the Denmark Strait [*Hansen and Østerhus*, 2000; *Blindheim and Østerhus*, this volume]. It also strongly modifies the Atlantic Water that eventually ends up in the Arctic Ocean [*Furevik*, 2001; *Karcher et al.*, 2003; *Furevik and Nilsen*, this volume].

– The freshwater fluxes through the Nordic Seas are substantial. Together with the flow through the Canadian Archipelago, they represent the key routes for the oceanic limb of the hydrological cycle at high northern latitudes [*Aagaard and Carmack*, 1989; *Houghton and Visbeck*, 2002; *Peterson et al.*, 2002].

– The Northern Hemisphere atmospheric circulation is sensitive to the distribution of sea-surface temperature (SST)

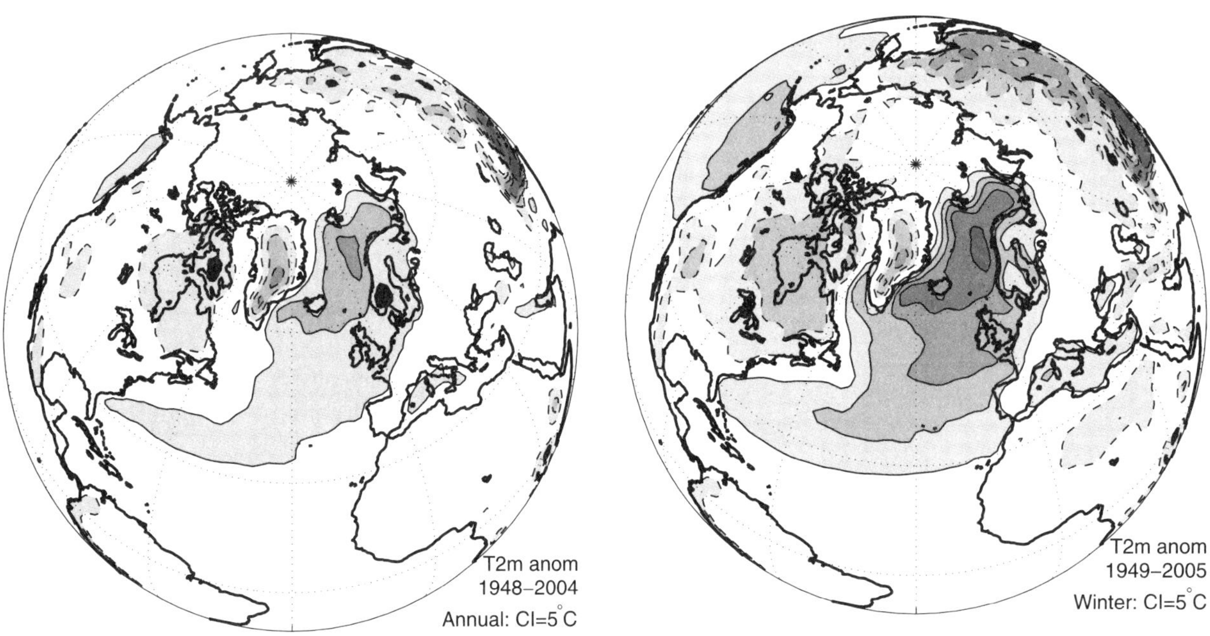

Figure 1. Deviations from zonal mean temperature at 2-m depth for annual (left) and December–February (right) means. Contours are shown for every 5°C. Solid (dashed) lines indicate positive (negative) anomalies. The figures are based on temperature data from the NCEP/NCAR reanalysis program [*Kalnay et al.*, 1996].

and sea-ice in the Nordic Seas and in the neighboring Labrador, Barents, and Kara Seas [*Deser et al.*, 2004; *Kvamstø et al.*, 2004; *Magnusdottir et al.*, 2004].

– The Nordic Seas are among the few regions of the world's oceans that take up substantial amounts of atmospheric carbon dioxide (CO_2) throughout the year. The ocean uptake of atmospheric CO_2 is large, ranging from 20 to 85 g C m^{-2} y^{-1}, which is among the highest of such fluxes in the world's oceans [*Anderson et al.*, 2000; *Takahashi et al.*, 2002; *Skjelvan et al.*, this volume].

Central questions with regard to the large-scale climate implications of the above-mentioned perspectives and with regard to the ecosystems in the Nordic Seas region are the following:

– How stable is the oceanic circulation of the Nordic Seas?

– What are the time and space characteristics of the natural variability modes?

– In which way and to which extent will global warming influence the mean climate state and the climate variability modes?

To adequately address these questions requires the use of a multidisciplinary approach in documenting available knowledge of past and present climate change and to identify and quantify the underlying dynamics and thermodynamics of the climate system. Based on such information, the fingerprints of global warming can be identified and quantified within forthcoming climate observations and model simulations. With this approach in mind, the five review chapters presented in this book have been organized to summarize available knowledge about the climate system of the Nordic Seas based on climate reconstructions from the Holocene epoch (past 11,500 years) [*Nesje et al.*], ocean observations covering the period of instrumental records (mainly the past 50 years) [*Blindheim and Østerhus*], the dynamics of the ocean response to atmospheric forcing from 1948 to the present (the period with atmospheric reanalysis products) [*Furevik and Nilsen*], observations and analyses of the cycling of inorganic carbon covering the last decades [*Skjelvan et al.*], and numerical ocean–sea-ice modeling covering the atmospheric reanalysis period [*Drange et al.*].

2. PAST AND PRESENT CLIMATE OBSERVATIONS

In *Nesje et al.* [this volume], a synthesis of temperature and precipitation reconstructions from pollen analysis, tree-line variations, chironomids (nonbiting midges), tree ring records, speleothem (chemical fingerprints of cave formations) data, glacier variations, and marine records (stable isotopes, species abundance, lithological changes) demonstrates that during most of the Holocene epoch, the climate of the Nordic Seas region was, in general, milder than the 1961–1990 mean. A climate optimum is found for the period from about 9,000 to 5,000 calendar years before the present (cal. yr BP). During this period, most of the present-day glaciers in Norway and on Svalbard melted completely, and Scots pine was found well above the present tree line in the mountain regions in Scandinavia. The annual surface temperature during this period has been estimated to be 1.8–2.0°C higher than the 1961–1990 mean [*Nesje et al.*, 2001, and this volume]. The main reason for this climate optimum was a very high summer insolation, which at 60°N exceeded the present insolation by as much as 8–10%. Furthermore, the coldest period throughout the entire Holocene was probably the "Little Ice Age", covering the time period ca. 1550–1925, in which the mean annual temperatures were approximately 1°C below the 1901–1995 mean value [*Luterbacher et al.*, 2004].

Figure 2 displays reconstructed, observed, and simulated SST variations for the period 2000 cal. yr BP to the year 2100. In Figure 2a, the deduced variability of reconstructed SST from the eastern Norwegian Sea is provided for the last two millennia. The time series show a fairly constant SST for the period preceding year 700. Thereafter, three periods with a gradual century-scale warming of 0.5–1°C are seen between the year groups 700–1000, 1200–1400, and 1500–2000. Both the first and the second warming, which make up the "Medieval Warm Period", are terminated by rather quick cooling events. The cooling at around year 1425 is particularly rapid and strong: Here the SST decreases by about 1°C in a few decades. Thus, the reconstructed SST time series clearly demonstrate that the marine climate of the Nordic Seas has experienced both century-scale warming and rapid cooling periods over the last two millennia.

The availability of high-resolution (annual- to decadal-scale resolution) climate time series covering several millennia represents a breakthrough in documenting the major variability modes in the past climate. With spatially distributed time series of a similar kind, we can, for the first time, construct the patterns and amplitudes of past climate variability models. Documentations of decadal and longer time-scale modes of climate variability are of key importance for distinguishing between natural and human-induced climate variations and for evaluating and improving climate models beyond the relatively short instrumental period.

The Nordic Seas region has a long and extensive history of instrumental observations [e.g., *Blindhem and Østerhus*, this volume]. A source of invaluable importance in this respect is the surface to abyss hydrographic observations from Ocean Weather Station Mike (OWS M) going back to 1948 [*Østerhus and Gammelsrød*, 1999]. Figure 2b presents the variations in summer SST from OWS M (thick gray line). A gradual cooling is seen for the period before 1979, followed by first a gradual and then a rapid warming. The recent warming is strong; the mean of the years 2002–2004 being 1.62°C above

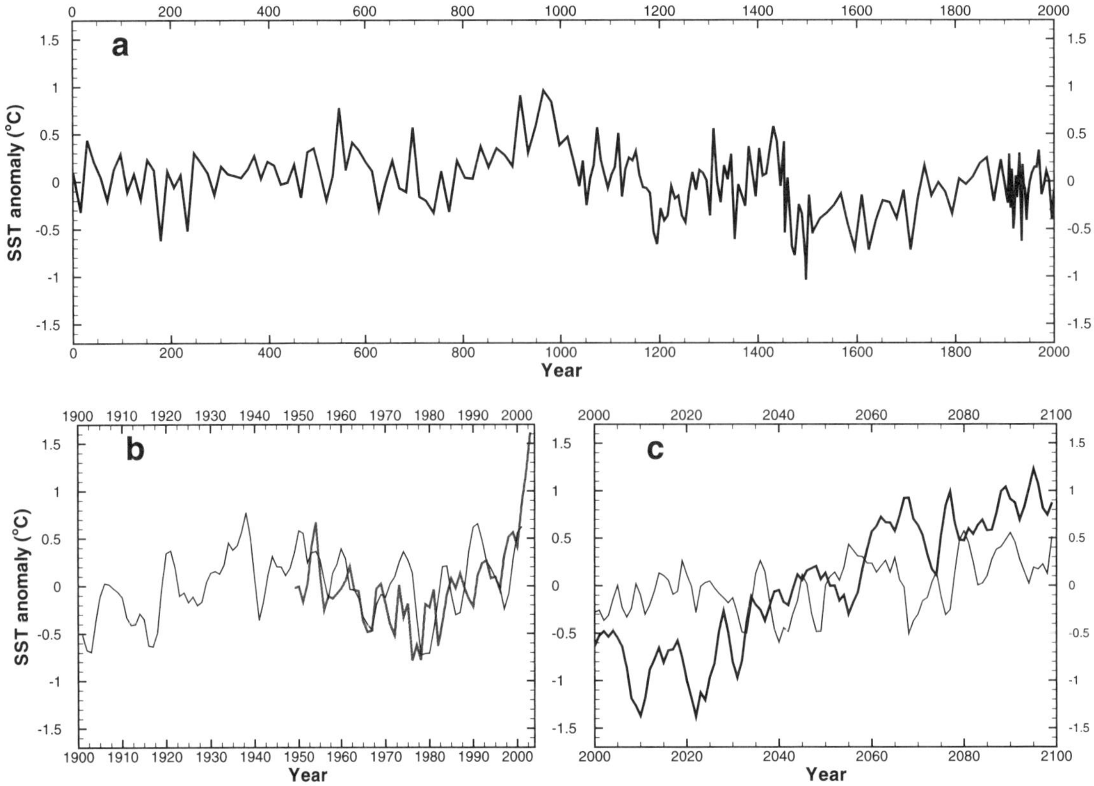

Figure 2. Temperature anomalies (°C) in the eastern and northern Nordic Seas. (a) Diatom-inferred August SST variations [*Jansen and Koc*, 2000; *Birks and Koc*, 2002; *Koc and Jansen*, 2002; *Andersen et al.*, 2004] during the last two millennia (extracted from Figure 5b in *Nesje et al.* [this volume]). (b) Observed variations in 3-yr annual mean temperature at 0–200-m depth at the Kola transect in the Barents Sea (70°30'–72°30'N, 33°30'E; thin line, updated data from *Bochkov* [1982], *Tereshchenko* [1997]) and observed variations in 3-yr mean July–September SST at OWS M (66°N,2°E; thick gray line is for unpublished data from Østerhus). (c) Simulated changes of 3-yr mean July–September SST anomalies from the eastern Norwegian Sea for 2000–2100 as based on IPCC scenarios SRES A2 (thick line) and SRES B1 (thin line) with the Bergen Climate Model [*Furevik et al.*, 2003; *Bentsen et al.*, 2004]. The horizontal model resolution in the Nordic Seas is about 300 × 150 km in the atmosphere and 80 × 80 km in the ocean. In all panels, the anomalies are relative to the mean value of the time series.

the 1948–2004 summer SST average (Figure 2b) and 0.94°C above the annual mean temperature average for the same period (both obtained by applying a 3-yr mean filter).

It is well documented that most regions at high northern latitudes were anomalously warm during the 1920s to the 1950s [e.g., *Delworth and Knudson*, 2002; *Johannessen et al.*, 2004]. Also the extent of Arctic sea-ice was less than normal during this period [*Zakharov*, 1997; *Johannessen et al.*, 2004]. On the basis of instrumentally measured air temperature from (e.g.) Stykkishólmur, Iceland (Figure 3 and *Blindheim and Østerhus* [this volume]), one can surmise that the early 20th century warm period was the warmest since the early 19th century. Figure 2b shows the 3-yr annual mean temperature from the Kola section in the Barents Sea at 0–200 m deep for 1901–2001 (thin line). The decadal-scale fluctuations in the Kola and OWS M time series follow closely. Furthermore, based on the OWS M data, it is likely that the recent warming exceeds the early 20th century warm period. While most of the early 20th century warm period can be attributed to natural climate fluctuations, at least part of the recent warming probably is the result of human activities [e.g., *Delworth and Knudson*, 2002; *Johannessen et al.*, 2004].

By comparing the temperature fluctuations along the Kola section and at OWS M (Figure 2b) with the reconstructed time series from the past 2,000 years (Figure 2a), one can argue that the recent warming is exceptional also on a millennium time perspective. This observation is, however, speculative because of uncertainties inherent in the transfer functions of the different proxies [e.g., *Mikalsen et al.*, 2001], which affect the actual interpretation of the time series [e.g., *Mann et al.*, 1999; *von Storch et al.*, 2004; *Moberg et al.*, 2005]. Bridging reconstructed and instrumental observations, a challenging

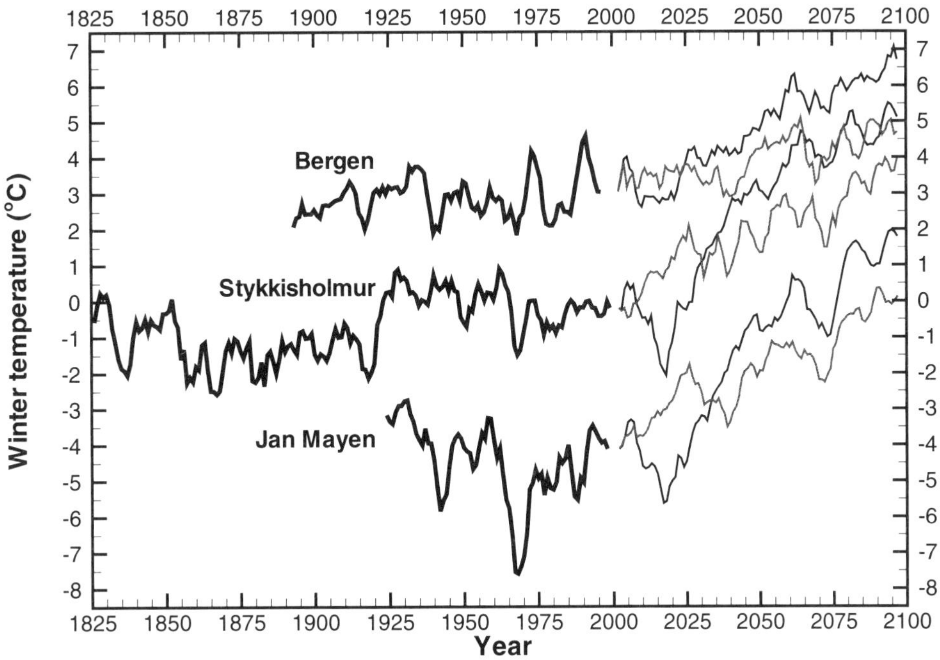

Figure 3. Observed (thick lines) and simulated (thin lines) 5-yr averaged December–April 2-m air temperature at Bergen, Norway (60°N,5°E); Stykkishólmur, Iceland (65°N,23°W); and Jan Mayen, central Nordic Seas (71°N,8°W). The two simulations are based on the IPCC SRES A2 (thin black lines) and SRES B1 (thin gray lines) with the Bergen Climate Model.

field of research, is therefore of the greatest importance for the coming years.

3. FUTURE CLIMATE AND CLIMATE MODELING

To a large degree, the evolution of the climate of the Nordic Seas depends on changes in the ocean circulation, affecting the poleward-directed heat flow. Figure 2c presents two realizations of 21st century climate from the Bergen Climate Model [*Furevik et al.*, 2004; *Bentsen et al.*, 2004], one based on the rather strong IPCC SRES A2 scenario, and one with the more modest IPCC SRES B1 scenario (for definitions of the SRES scenarios, see, e.g., *http://www.ipcc.ch*). The two model realizations show a warming of the summer SST of 2.13°C/100 yr for SRES A2 and 0.53°C/100 yr for SRES B1. The corresponding figures for the annual average SST are 2.26°C/100 yr and 0.62°C/100 yr, respectively. It is worth noting that the warming rate in SRES B1 is similar to the warming between years 1200 and 1400 (Figure 2a). The rapid and strong termination of the 1200–1400 warming period illustrates that similar cooling events may also happen in the future.

A compilation of historical temperature records and possible changes in the surface air temperatures in the Nordic Seas region, represented by Bergen, Stykkishólmur, and Jan Mayen, are presented in Figure 3. The above-mentioned early 20th century warm period is seen in the Stykkishólmur time series but is not evident in the Jan Mayen time series, thus illustrating the nonuniform climate in the region. The simulated evolution of the 21st century temperatures at these locations (thin lines in Figure 3) indicate that global warming may dominate any of the observed temperature fluctuations within a few decades.

It is important to note here that robust assessments of the development of the regional Nordic Seas climate system must be based on ensembles of integrations with thoroughly tested and evaluated climate models, not on individual model realizations such as those displayed in Figures 2c or 3. Averaging over many model systems or multimodel ensembles yields, in general, a realistic range of expectations, which is not necessarily the case for individual models or model integrations [e.g., *Cubash et al.*, 2001; *Kuzmina et al.*, 2004]. Furthermore, pulses of freshwater from the Arctic Ocean [*Dickson et al.*, 1988] and the dynamics of the warm and saline North Atlantic subtropical gyre and the cold and fresh North Atlantic subpolar gyre [*Hátún et al.*, 2005] are important for decadal-scale fluctuations of the hydrographic properties of the Nordic Seas. One must assume that such fluctuations are affected by changes in climate, including anthropogenic changes. Reliable assessment of the future Nordic Seas climate will therefore require climate models that are capable of realistically simulating the temporal and spatial characteristics of the major variability modes in the region.

4. OCEANIC CHANGES DURING RECENT DECADES

For proper descriptions of local climate features, especially changes in oceanic circulation, overflows, and convection, high spatial resolution is needed in the oceanic components of climate models. As an example, the ocean component in the climate model used in Figures 2c and 3 is about 80 km, whereas a spatial resolution of 20-40 km would generally greatly improve results [*Drange et al.*, this volume].

It is encouraging that the current generation of coupled sea-ice–ocean models is able to describe many of the key climate parameters in the Nordic Seas region [see *Drange et al.*, this volume]. The current generation of coupled sea-ice–ocean models can, for the first time, complement available ocean observations and can be used to guide forthcoming ocean observation strategies [e.g., *Hátún et al.*, this volume]. This potential for interactive research is welcome news because it will lead, over some time, to an improved representation of processes of the Nordic Seas in coupled climate models.

The inflow of Atlantic Water into the Nordic Seas is one of the key components for climate and ecology in the region, in addition to being an important factor for climate variations far outside the region. Prediction of the inflow of Atlantic Water over time scales of months to years is therefore important for many applications, ranging from basic needs of understanding the climate system to socioeconomic decisions addressing (e.g.) management of fisheries resources. On interannual time scales, *Orvik and Skagseth* [2003] demonstrated that the major northward volume transport of Atlantic Water off the Norwegian coast at 62°N responds to the North Atlantic wind stress at about 55°N with a time lag of about 15 months. Figure 4, showing a 1.5-yr extension of the analyses period presented in *Orvik and Skagseth* [2003], supports the validity of the analyses and indicates that the major branch of Atlantic Inflow is predicted to decrease until summer 2005. Figure 4 strongly suggests that the North Atlantic wind stress is of key importance for the variability of the Nordic Seas climate [see the synthesis by *Furevik and Nilsen*, this volume].

The near-surface properties of the Nordic Seas climate are intimately linked to the climate of the abyssal Nordic Seas. The hydrographic time series from OWS M are the longest continuous deep-water time series of the world's oceans. Figure 5 shows the annual mean temperature at 2,000 m deep for the period 1948–2004. The trend recapitulates the rapid warming documented up to 1996 by *Østerhus and Gammelsrød* [1999], which continues until 1998, with a weak warming thereafter. The warming trends shown in Figures 2b and 5 are caused by different mechanisms [*Blindheim and Østerhus* and *Furevik and Nilsen*, both in this volume]: Changes in the surface waters are governed mainly by the hydrographic properties of the Atlantic Inflow and the local air–sea heat fluxes, whereas changes at depth are related to the formation rate

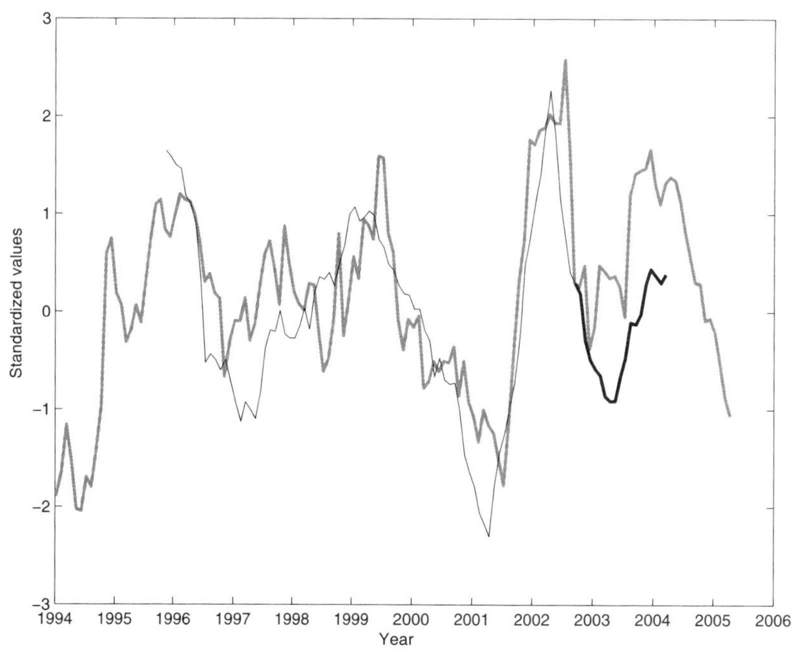

Figure 4. Time series representing the observed variability of the poleward flow of Atlantic Water at the Svinøy section off the coast of Norway at 62°N (black lines) and the North Atlantic wind stress curl at 55°N (gray line). The thick black line is updated from *Orvik and Skagseth* [2003]. In the figure, the wind stress curl precedes the flow through the Svinøy section by 15 months.

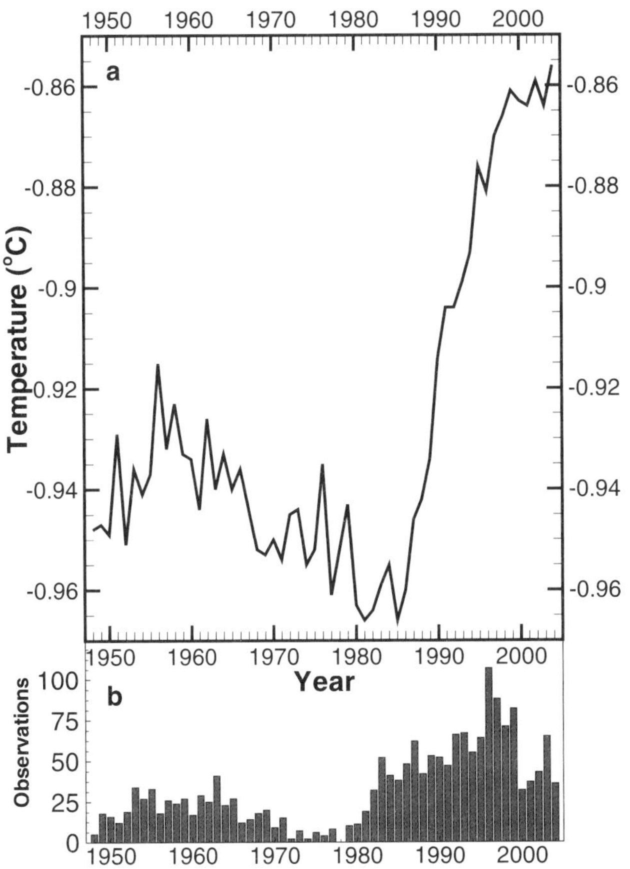

Figure 5. (a) Time series of annual mean temperature at 2,000-m depth from OWS M for the period 1948–2004 (updated from Østerhus and Gammelsrød [1999]). (b) Number of observations per year.

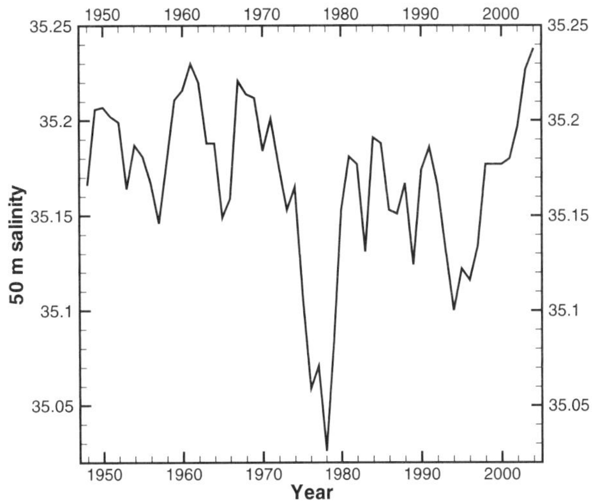

Figure 6. Time series of annual mean salinity at 50-m depth from OWS M for 1948–2004 [Østerhus, unpublished data]. The mean salinity in 2004 is 35.238.

of intermediate to deep waters, particularly in the Greenland Basin. An important task is to discover the links between these processes.

The warming shown in Figure 5 is partly linked to the gradually intensified North Atlantic westerlies since the mid-1960s [*Hurrell*, 1995; *Furevik and Nilsen*, this volume]. The increasing trend in the North Atlantic Oscillation Index between 1970 and the mid-1990s was accompanied by increasing exchange rates of volume, heat, and salt between the Atlantic and the Nordic Seas, as well as between the Nordic Seas and the Arctic Ocean. It also put a stop to the sea-ice processes important for the existence of deep convection in the Greenland Sea [*Gerdes et al.*, this volume] and thus contributed to the warming of the deep waters (Figure 5) that lack recent ventilation with waters formed at near–freezing point temperatures.

The Nordic Seas gradually freshened after the passage in the late 1960s of the freshwater burst known as the "Great Salinity Anomaly" [*Dickson et al.*, 1988]. This freshwater originated from the Arctic Ocean, propagated quickly southward towards and through the Denmark Strait in the late 1960s [*Dickson et al.*, 1998; *Haak et al.*, 2003], and possibly returned to the eastern part of the Nordic Seas in the mid-1970s (Figure 6). Recent analyses have demonstrated that essentially all of the intermediate to deep waters of the Nordic Seas and the North Atlantic have experienced a freshening over the past four decades [*Dickson et al.*, 2002; *Curry et al.*, 2003]. As in the Great Salinity Anomaly, the likely source of this freshening is a reduction in both sea-ice and liquid fresh water reservoirs in the Arctic Ocean.

Recent observations of the salinity of the Atlantic Inflow show higher values than ever before observed (Figure 6). The high salinity values are concurrent with the recent record-high SST at OWS M (Figure 2b). To what extent the exceptional warming of the Atlantic Water in recent years (Figure 2b) is linked to the interannual flow field variations of the Atlantic Inflow (Figure 4) is still an open question [e.g., *Orvik and Skagseth*, 2005].

5. INORGANIC CARBON CYCLE

An important component of the climate system involves the marine cycling of carbon and then particularly the air–sea exchange of the greenhouse gas carbon dioxide (CO_2). Global compilations of the air–sea disequilibria of CO_2 show that the northern North Atlantic and the Nordic Seas are among the most intense sinks of atmospheric CO_2 [*Takahashi et al.*, 2002]. In *Skjelvan et al.* [this volume], the physically, chemically, and biologically mediated carbon transports are quantified on the basis of available observations and modeling.

A newly established time series of the seasonal cycling of total dissolved inorganic carbon (C_T) at OWS M is displayed

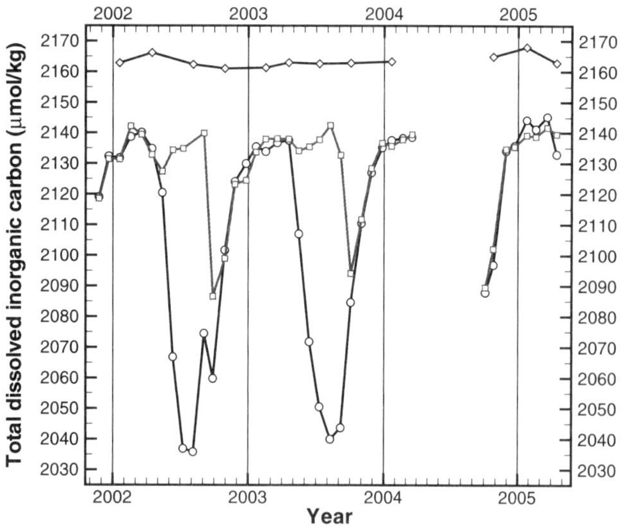

Figure 7. Illustration of the seasonal cycle of total dissolved inorganic carbon at 10-m (circles), 50-m (squares), and 2,000-m (diamonds) depths at OWS M [*Skjelvan*, unpublished data]. The curves emphasize the variations with time.

in Figure 7, which shows large fluctuations in the surface water. This variability is, to a large extent, governed by biological activity. For example, the plankton organisms start to grow in April in response to increasing insolation and stratification, which results in rapid consumption of carbon. At 50 m depth, the effect of the bloom is seen as a slight decrease in C_T followed by a slight increase attributable to respiration and remineralization. At this depth, mixing between surface waters low in C_T and deeper waters rich in C_T becomes dominant in August–September. C_T is close to constant at depths greater than 500 m. At that depth the downward transport of biologically mediated carbon is close to being balanced with advection and mixing. The presented time series is far too short to detect interannual signals, but it will be a useful tool to link observed changes in hydrography (Figures 2b and 6) and transport (Figure 4) to the air–sea exchange of CO_2. The deep-water C_T contents at selected stations from the TTO/NAS (Transient Tracers in the Ocean–North Atlantic Study) expedition in 1981 [*Brewer et al.*, 1986] have been briefly compared to deep-water C_T content at OWS M from 2003; the increase of about 6 μmol kg^{-1} during this 22-yr period indicates that the anthropogenic carbon signal has reached the Norwegian Sea abyss.

6. SUMMARY

The Nordic Seas play an important role in the dynamics of climate of the North Atlantic realm. Compared with other areas at its latitudes, this region has the strongest positive anomalies for SST and surface air temperature in the world. It is particularly important for water mass modification and formation and for air–sea interaction, and it acts as the major transport route for freshwater and heat between the North Atlantic and the Arctic Ocean. Large changes in the climate state of the region have been revealed for the Holocene epoch and have been documented with instruments for the last 50–190 years. A gradual change in the hydrography of the region has been observed since the 1960s, partly in response to large-scale changes in the atmospheric circulation. In recent years, record-high temperature and salinity values have been recorded.

How will the climate of the Nordic Seas evolve in the 21st century? We would like to know in which way and to which extent the observed changes, if continued along present trends, will influence the large-scale ocean circulation and thereby the large-scale climate system. Some indications suggest that the observed change in the climate of the region over the recent decades is at least partly a result of global warming. The perception of what is normal and what is extraordinary will thus be even more challenged in the years to come.

The majority of contributions to this book have benefited from investments from national and international programs, which have made it possible to collect relatively long observational time series from weather stations and research cruises. They have supported deployment of instruments for measuring physical and biogeochemical properties of the main water masses within the main body and through the boundaries of the Nordic Seas. It is essential for the understanding of natural climate variability modes and for proper assessment of possible changes in the climate system that key components of these campaigns continue.

The coming years will, without doubt, show many examples where climate reconstructions from the past, instrumental observations, and numerical models are used together to detect and understand causes and effects of observed changes in the climate system, and to assess how the ocean will respond to anomalous air–sea fluxes of heat, freshwater, and momentum, or how the entire climate system may change in response to, for instance, increased forcing of greenhouse gases. The review papers presented in this book form a basis for current consensus on the climate of the Nordic Seas and represent unique documentation against which forthcoming instrumental observations, paleoclimatic reconstructions, and numerical modeling will be compared, assessed, and evaluated.

Acknowledgments. The results presented here have to a large degree been supported by the projects NOClim, ProClim, RegClim, and the *Programme of supercomputing* under the Research Council of Norway. This is publication No. A98 from the Bjerknes Centre for Climate Research.

REFERENCES

Aagaard, K., E. Carmack, The role of sea ice and other fresh water in the arctic circulation. *J. Geophys. Res.*, 94, 14,485-14,498, 1989.

Andersen, C., N. Koç, A. Jennings, T. Andrews, Nonuniform response of the major surface currents of the Nordic Seas to insolation forcing: implications for the Holocene climate variability. *Paleoceanography*, 19, PA2003, doi:10.1029/2002PA000873, 2004.

Anderson, L.G., H. Drange, M. Chierichi, A. Fransson, T. Johannessen, I. Skjelvan, and F. Rey, Sesonal and annual variability in the upper Greenland Sea based on measurements and a box model. *Tellus*, 52B, 1013-1024, 2000.

Bentsen, M., H. Drange, T. Furevik, and T. Zhou, Simulated variability of the Atlantic meridional overturning circulation. *Clim. Dynam.*, doi: 10.1007/s00382-004-0397-x, 2004.

Birks, C.J.A. and N. Koç, A high-resolution diatom record of late Quaternary sea-surface temperatures and oceanographic conditions from the eastern Norwegian Sea. *Boreas*, 31, 323-344, 2002.

Bochkov, Y.A., Water temperature in the 0-200 m layer in the Kola Meridian Section in the Barents Sea, 1900-1981. *Sb. Nauch. Trud. PINRO*, 46, 113-122, 1982. In Russian.

Brewer, P.G., A.L. Bradshaw, R.T. Williams, Measurements of total carbon dioxide and alkalinity in the North Atlantic Ocean in 1981. In *The Changing Carbon Cycle: A Global Analysis*, J.R. Trabalka and D.E. Reichle (Eds.), Springer-Verlag, New York, 348-370, 1986.

Calvo, E., J. Grimalt, E. Jansen, High resolution U^{K}_{37} sea surface temperature reconstruction in the Norwegian Sea during the Holocene. *Quaternary Science Reviews*, 21, 1385-1394, 2002.

Cubash, U., G.A. Meehl, G.J. Boer, R.J. Stouffer, M. Dix, A. Noda, C.A. Senior, S. Raper, and K.S. Yap, Projections of future climate change. In *Climate Change 2001: The Scientific Basis*. J.T. Houghton, Ding Yihui, and M. Noguer (Eds.), Cambridge University Press, 2001.

Curry, R., R.R. Dickson, and I. Yashayaev, A change in the freshwater balance of the Atlantic Ocean over thepast four decades. *Nature*, 426, 826-829, doi:10.1038/nature02206, 2003.

Delworth, T.L., and T.R. Knutson, Simulation of early 20th century global warming. *Science* 287, 2246-2250, 2000.

Deser, C., G. Magnusdottir, R. Saravanan, and A. Phillips, The effects of North Atlantic SST and sea ice anomalies on the winter circulation in CCM3. part II: Direct and indirect components of the response *J. Climate*, 17, 877-889, 2004.

Dickson, B., I. Yashayaev, J. Meincke, B. Turrell, S. Dye, and J. Holfort, Rapid freshening of the deep North Atlantic Ocean over the past four decades. *Nature*, 416, 832-837, doi:10.1038/416832a, 2002.

Dickson, R.R., J. Meincke, S.-A. Malmberg, and A.J. Lee, The "Great Salinity Anomaly" in the northern North Atlantic 1968-1982. *Prog. Oceanog.*, 20, 103-151, doi:10.1016/0079-6611(88)90049-3, 1988.

Furevik, T., Annual and interannual variability of Atlantic Water temperatures in the Norwegian and Barents Seas: 1980-1996. *Deep Sea Research*, 48, 383-404, 2001.

Furevik, T., M. Bentsen, H. Drange, I.K.T.N. Kindem, G. Kvamstø, and A. Sorteberg, Description and validation of the Bergen Climate Model: ARPEGE coupled with MICOM. *Clim. Dyn.*, 21, 27-51, doi:10.1007/s00382-003-0317-5, 2003.

Haak, H., J. Jungclaus, U. Mikolajewicz, and M. Latif, Formation and propagation of great salinity anomalies. *Geophys. Res. Lett.*, 30, 1473, doi:10.1029/2003GL017065, 2003.

Hansen, B., S. Østerhus, D. Quadfasel, W.R. Turrell, Already the day after tomorrow? *Science*, 305, 953-954, 2004.

Hansen, B., and S. Østerhus, North Atlantic–Norwegian Sea Exchanges. *Prog. in Oceanography*, 45, 109-208, 2000.

Hátún, H., A.B. Sandø, H. Drange, B. Hansen, and H. Valdimarsson, Influence of the Atlantic subpolar gyre on the thermohaline circulation. *Science*, submitted.

Hellevik, A., Preface, *Kongsspegelen*, Det norske samlaget, Oslo, Norway (ISBN 82-521-0610-2), English translation [1976] by Larson, L.M., of *The King's Mirror: Speculum Regale—Konungs skuggsjá*, The American Scandinavian Foundation, New York; Humphrey Milford, London; Oxford University Press, Oxford, 1917.

Houghton, R.W., and M.H. Visbeck, Quasi-decadal salinity fluctuations in the Labrador Sea. *J. Phys. Oceanogr.*, 32, doi: 10.1175/1520-0485, 687-701, 2002.

Hurrell, J.W., Decadal trends in the North Atlantic Oscillation: Regional temperature and precipitation. *Science*, 269, 676-679, 1995.

Jansen, E., and N. Koç, Century to decadal scale records of Norwegian Sea surface temperature variations of the past 2 millennia. *PAGES/CLIVAR Newsletter*, 8, 13-14, 2000.

Johannessen, O.M., L. Bengtsson, M. Miles, S.I. Kuzmina, V. Semenov, G.V. Alekseev, A.P. Nagurny, V.F. Zakharov, L.P. Bobylev, L.H. Pettersson, K. Hasselmann, and H.P. Cattle, Arctic climate change—observed and modelled temperature and sea ice. *Tellus* Series A, 56, 328-341, 2004.

Kalnay, E., et al., The NCEP/NCAR 40-year reanalysis project. *Bull. Am. Met. Soc.*, 77, 437-471, 1996.

Karcher, M.J., R. Gerdes, F. Kauker, and C. Köberle, Arctic warming: evolution and spreading of the 1990s warm event in the Nordic Seas and the Arctic Ocean. *J. Geophys. Res.*, 108, 3034, doi: 10.1029/ 2001JC001265, 2003.

Koç, N., and E. Jansen, Holocene climate evolution of the North Atlantic Ocean and the Nordic Seas—a synthesis of new results. In Wefer, G., Berger, W.H., Behre, K.E., Jansen, E. (Eds.) *Climate and History in the North Atlantic Realm*. Springer-Verlag, Berlin, Heidelberg, 165-163, 2002.

Kuzmina, S.I., L. Bengtsson, O.M. Johannessen, H. Drange, L.P. Bobylev, and M.W. Miles, The North Atlantic Oscillation and greenhouse-gas forcing. *Geophys. Res. Lett.*, 32, L04703, doi:10.1029/2004GL021064, 2005.

Kvamstø, N.G., P. Skeie, D.B. Stephenson, Large-scale impact of localized Labrador sea-ice changes on the North Atlantic Oscillation. *Int. J. Climatol.*, 24, 603-612, 2004.

Luterbacher, J., D. Dietrich, E. Xoplaki, M. Grosjean, and H. Wanner, European seasonal and annual temperature variability, trends, and extremes since 1500. *Science*, 303, 1499-1503, 2004.

Magnusdottir, G., C. Deser, and R. Saravanan, The effects of North Atlantic SST and sea ice anomalies on the winter circulation in CCM3. part I: Main features and storm track characteristics of the response. *J. Climate*, 17, 857-876, 2004.

Mann, M.E., R.S. Bradley, M.K. Hughes, Northern Hemisphere temperatures during the past millennium: Inferences, uncertainties, and limitations. *Geophys. Res. Lett.*, 26, 759-762, 1999.

Mikalsen, G., H.P. Sejrup, and I. Aarseth, Late Holocene changes in ocean circulation and climate: Foraminiferal and isotopic evidence from Sulafjorden, western Norway. *The Holocene*, 11, 437-446, 2001.

Moberg, A., D.M. Sonechkin, K. Holmgren, N.M. Datsenko, and W. Karlén, Highly variable Northern Hemisphere temperatures reconstructed from low- and high-resolution proxy data. *Nature*, 433, 613-617, 2005.

Nesje, A., J.A. Matthews, S.O. Dahl, M.S. Berrisford, C. Andersson, Holocene glacier fluctuations of Flatebreen and winter precipitation changes in the Jostedalsbreen region, western Norway, based on glaciolacustrine records. *The Holocene*, 11, 267-280, 2001.

Orvik, K.A. and Ø. Skagseth, The impact of the wind stress curl in the North Atlantic on the Atlantic inflow to the Norwegian Sea toward the Arctic. *Geophys. Res. Lett.*, 30, 1884, doi:10.1029/2003GL017932, 2003.

Orvik, K.A. and Ø. Skagseth, The warming of the Atlantic inflow to the Norwegian Sea toward the Arctic 1995-2005: A study of temperature, volume and heat flux variations in the Norwegian Atlantic Current based on observations from moored instruments- *Geophys. Res. Lett.*, submitted.

Østerhus, S. and T. Gammelsrød, The abyss of the Nordic Seas is warming. *J. Clim.*, 12, 3297-3304, 1999.

Peterson, B.J., R.M. Holmes, J.W. McClelland, C.J. Vorosmarty, R.B. Lammers, A.I. Shiklomanov, I.A. Shiklomanov, and S. Rahmstorf, Increasing river discharge to the Arctic Ocean. *Science*, 298, 2171-2173, 2002.

Rhines, P.B. and S. Häkkinen, Is the oceanic heat transport in the North Atlantic irrelevant to the climate in Europe? *ASOF Newsletter*, 1, 13-16, 2003.

Seager, R., D.S. Battisti, J. Yin, N. Gordon, N. Naik, A.C. Clement, M.A. Cane, Is the Gulf Stream responsible for Europe's mild winters? *Q. J. R. Meteorol. Soc.*, 129, 2563-2586, 2002.

Skjoldal, H.R., (Ed.), *The Norwegian Sea Ecosystem*. Tapir Akademisk Forlag, Trondheim, Norway, 560 pp., 2004.

Stenseth, N.C., A. Mysterud, G. Ottersen, J.W. Hurrell, K.-S. Chan, and M. Lima, Ecological effects of climate fluctuations. *Science*, 297, 1292-1296, 2002.

Takahashi, T., S.C. Sutherland, C. Sweeney, A. Poisson, N. Metzl, B. Tilbrook, N. Bates, R. Wanninkhof, R.A. Feely, and C. Sabine, Global sea-air CO_2 flux based on climatological surface ocean pCO_2, and seasonal biological and temperature effects. *Deep Sea Res., II* 49, 1601-1622, 2002.

Tereshchenko, V.V., *Seasonal and year-to-year variation in temperature and salinity of the main currents along the Kola section in the Barents Sea*. Murmansk, PINRO Press. 71 pp, 1997 (in Russian).

von Storch, H., E. Zorita, J. Jones, Y. Dimitriev, F. González-Rouco, and S. Tett, Reconstructing past climate from noisy data. *Science*, 306, 679-682, 2004.

Zakharov, V.F., Sea ice in the climate system. *World Climate Research Programme/Arctic Climate System Study, WMO/TD* 782, World Meteorological Organization, Geneva, 80 pp, 1997.

W. Berger, Scripps Institution of Oceanography, University of California at San Diego, 9500 Gilman Drive, La Jolla, California 92093-0524, USA. (wberger@ucsd.edu)

T. Dokken, S. Østerhus, and Ø. Skagseth, Bjerknes Centre for Climate Research, University of Bergen, Allégt. 55, 5007 Bergen, Norway. (trond.dokken@bjerknes.uib.no; svein.osterhus@gfi.uib.no; skagseth@gfi.uib.no)

H. Drange, Nansen Environmental and Remote Sensing Center and Bjerknes Centre for Climate Research, Thormhølensgt. 47, 5006 Bergen, Norway. (helge.drange@nersc.no)

T. Furevik, Geophysical Institute and Bjerknes Centre for Climate Research, Allégt. 70, 5007 Bergen, Norway. (tore@gfi.uib.no)

R. Gerdes, Alfred Wegner Institute for Polar and Marine Research, Bussestrasse 24, 27570 Bremerhaven, Germany. (rgerdes@awibremerhaven.de)

A. Nesje, Bjerknes Centre for Climate Research and Department of Earth Science, University of Bergen, Allégt. 55, 5007 Bergen, Norway. (atle.nesje@geo.uib.no)

K.A. Orvik, Geophysical Institute, Allégt. 70, 5007 Bergen, Norway. (kjell.orvik@gfi.uib.no)

I. Skjelvan, Bjerknes Centre for Climate Research and Geophysical Institute, University of Bergen, Allégt. 55, 5007 Bergen, Norway. (Ingunn.Skjelvan@gfi.uib.no)

The Nordic Seas, Main Oceanographic Features

Johan Blindheim

Institute of Marine Research, and Bjerknes Centre for Climate Research, University of Bergen, Bergen, Norway

Svein Østerhus

Bjerknes Centre for Climate Research, and Geophysical Institute, University of Bergen, Bergen, Norway

This article presents a review of the main oceanographic features of the Nordic Seas (the Greenland, Iceland, and Norwegian Seas). It includes a short description of previous research, the bathymetry of the area, its general circulation, and the exchanges with the North Atlantic and Arctic Oceans. Water mass structure and temporal variations are discussed separately for the various basins. The most prominent variations are a cooling and freshening in the upper layers since the 1960s, particularly in the southwestern Norwegian Sea, and a rise in temperature and salinity in the deeper layers, particularly in the Greenland Sea. Some relations of these variations to ocean–atmosphere interactions are described, and the deep-water warming is ascribed to an increased influence of deep water from the Arctic Ocean and a reduced formation of deep water in the Greenland Sea. Renewal of the deep water in the Greenland Sea occurred in the late 1960s, and atmospheric observations suggest that this may have been the only renewal event in the 20th century.

INTRODUCTION

The intention with this review is to summarise some main features of the oceanography in the Nordic Seas: the Greenland, Iceland, and Norwegian Seas.

With the approximately 2600-m-deep opening through the Fram Strait, the Nordic Seas are the main connection between the North Atlantic and the Arctic Oceans. The whole area to the north of the Greenland–Scotland Ridge has therefore often been mentioned as the Arctic Mediterranean Sea, a notion first used by *Krümmel* [1879], who called the area "Das Arktische Mittelmeer".

We focus on the oceanic regions in the Nordic Seas, the deep basins and the Iceland Sea. The adjacent shelf seas, the North Sea and the Barents Sea, are not included. Neither is the Norwegian Coastal Current, which although its surface waters, under given wind conditions, may be distributed into the Norwegian and Lofoten Basins, this activity is of little importance off the shelf area. To further limit the scope of the review, we have not included results obtained from 3-dimensional numerical modelling of the Nordic Seas, as these are addressed at depth in the modelling review paper by *Drange et al.* [this volume]. For the same reason, the section on sea-ice is quite short because other chapters deal with that topic [*Kvingedal*, this volume; *Gerdes et al.*, this volume].

The Nordic Seas: An Integrated Perspective
Geophysical Monograph Series 158

10.1029/158GM03

HISTORIC BACKGROUND

Oceanographic investigations in the North Atlantic region were initiated during the second half of the 19th century. The countries in northern Europe, including Russia, were involved in this work. In Norway, efforts in practical marine investigations were initiated at about the same time. In 1857 the government requested that *A. Boech* [1871] institute investigations on the "spring herring", and in 1864 G. O. Sars was given a government grant for studies of the Lofoten cod fisheries. His results [*Sars*, 1879] were of particular importance for the development in marine biology. Among marine biologists he is well known for discovering that cod releases its brood as pelagic eggs.

Sars [1885] also participated in the first major attempt to investigate the more oceanic areas of the Nordic Seas, The Norwegian North Atlantic Expedition, during the summers of 1876–1878. Together with H. Mohn, Sars led the expedition, which yielded much basic information on the fauna in the area, its main bathymetry and, very approximately, some of its oceanographic features. *Mohn* [1887] was in charge of the oceanographic work, but due to the primitive instruments of that time, many of his results were erroneous. Theoretically, however, he was on level with his time. He applied mathematical methods for working out the oceanic circulation, and in principle, his method for computing geostrophic currents was not different from the well-known equation of *Helland-Hansen* [1905]. Helland-Hansen pointed out, though, that Mohn's method involved very laborious calculations, which, in the days before computers, was a powerful argument against its general application.

Early expeditions from different nations worked in the region and some of their observations covered marginal areas of the Nordic Seas. In particular the British "Lightning" and "Porcupine" expeditions should be mentioned. As a result of their work in the Faroe–Shetland Channel in 1868–69, Arctic and Atlantic water masses were identified for the first time. It was then observed that deep temperatures could be below 1°C near the Faroes but above 6°C at similar depths on the southern side of the channel [*Thomson*, 1873]. Wyville Thomson thought that these differences in water mass properties, within a rather limited area, might be due to topographic structures. Investigations initiated by him led to extensive soundings on the ridge system between the North Atlantic and the Nordic Seas in 1880 and 1882, particularly on its eastern part, which now is known as the Wyville Thomson Ridge [*Tizard and Murray*, 1882].

The more western parts, the Greenland–Iceland Ridge and the Iceland–Faroe Ridge, were mapped on Danish expeditions. The work on these expeditions also contributed to the oceanographic knowledge of the area. After one of the earliest expeditions, *Irminger* [1861] described the current that now bears his name, as well as the East Icelandic Current as a branch of the East Greenland Current. Some years later *Colding* [1870] found that the Irminger Current splits in the Denmark Strait, one branch continuing into North Icelandic waters, the other turning towards the southwest to flow parallel with the East Greenland Current along the southeast coast of Greenland. The currents in the region around Iceland and the Faroes were also studied more in general, and in this respect much information was obtained, especially in *Ryder's* [1901] experiments with drift bottles.

Swedish and Russian expeditions investigated the waters around Spitsbergen. As a result of their work it was established that the West Spitsbergen Current carries water of Atlantic origin into the Arctic Ocean and that some of this water deflects westward and flows southward again underneath the Polar surface waters in the East Greenland Current [*Hamberg*, 1906; *Åkerblom*, 1904]. These observations confirmed the origin of the relatively warm intermediate water observed in the East Greenland Current during the Danish Hekla expedition [*Ryder*, 1895].

The Russian investigations, which were initiated in an organised form in 1901, had main emphasis on the Murman Coast and the Barents Sea. These investigations led to the first identification of extremely saline bottom water at temperatures near the freezing point. *Knipovich* [1905] concluded that this water was formed in the Barents Sea by rejection of brine during ice formation. Such dense water formed on the shelves around the Arctic Ocean has since been identified as the principal source of its deep waters [e.g. *Midttun*, 1985; *Aagaard et al.*, 1985].

At the turn of the century, oceanographic research in Norway moved into an active period. In 1900 the Norwegian Marine and Fisheries Investigations got the specially designed research vessel "Michael Sars". Mainly on the basis of 4 years work with this vessel, *Helland-Hansen and Nansen* [1909] presented their comprehensive description of the oceanographic features in Nordic Seas. Their work described the area in such detail and to such precision that investigations during succeeding years could add little to their findings. Therefore, their description was for many years the basic source of information on the general oceanography in the area. The first observations of the cold deep water in the Greenland Sea were also obtained during this period, based on which *Nansen* [1906] presented the classical model for the formation of Greenland Sea Deep Water (GSDW).

After the active period early in the 20th century, research work in the Nordic Seas was reduced for several decades. A main reason was that the "Michael Sars" was taken over by the navy during World War I and was never handed back. A fairly large data set, however, was collected by the "Armauer Hansen" in the Norwegian Basin in 1935 and 1936 [*Mosby*, 1959; 1970]. In the central Greenland Sea, no new data were collected before the 1950s. A set of data collected during winter in the early 1950s did not show the expected homogeneous water column. Consequently *Metcalf* [1955] thought that deep convection was of little importance and suggested that cooled surface water might flow toward the bottom along gently tilted isopycnal surfaces.

The next thorough survey of the area occurred in 1958, during the International Geophysical Year, as part of the Polar Front Survey [*Dietrich*, 1969)]. Because this extensive survey also failed to observe a homogeneous water column in the Greenland Sea, *Carmack and Aagaard* [1973] proposed that the GSDW might be formed at intermediate depths through a double-diffusive process. More recent works investigating the distribution of transient tracers in the water column are, however, more in support of the convective process [e.g. *Peterson and Rooth*, 1976; *Smethie et al.*, 1986]. Evidence of cessation of deep water formation since about the early 1970s has also increased [*Schlosser et al.*, 1991; *Bönisch and Schlosser*, 1995; *Bönisch et al.*, 1997].

Several projects have focused on the Greenland Sea since 1970. In February 1971, *Malmberg* [1983] observed an almost, but not quite, homogeneous water column in the central Greenland Sea. In 1972 the GEOSECS expedition made some observations in the Nordic Seas, mainly to observe the content of transient tracers. Similar work was also the purpose of the TTO (Transient Tracers in the Ocean–North Atlantic Survey) expedition in 1981. At the same time the Working Group on Oceanic Hydrography of ICES planned a study on the formation of GSDW and collected an extensive data set, including hydro-chemical data, in the Nordic Seas in winter and spring 1982 [*Clarke et al.*, 1990]. The Greenland Sea Project also contributed to detecting the role of deep waters originating in the Arctic Ocean as a component of the deep water in the Nordic Seas [*Aagaard et al.*, 1985; *Swift and Koltermann*, 1988]. In the late 1980s attempts were made to observe the temperature field by acoustic tomography [*Worcester et al.*, 1993].

In the Fram Strait, much of the research was focused on processes in the marginal ice zone, as in the Norwegian Remote Sensing Experiment in 1979 [*Johannessen et al.*, 1983; *Svendsen et al.*, 1983] and later in the international Marginal Ice Zone Experiment in 1984 [e.g. *Johannessen et al.*, 1987; *Manley et al.*, 1987; *Quadfasel et al.*, 1987]. Convective processes in eddies along the ice edge were further studied in the Seasonal Ice Zone Experiment in 1989 [*Johannessen et al.*, 1991, *Sandven et al.*, 1999; see also *Johannessen et al.*, 1992].

An important reason for the late detection of the intrusion of Arctic Ocean Deep Water into the Nordic Seas was the late observation of the sill depth in the Fram Strait. The first reliable observations were obtained during the Polar Front Survey when Soviet vessels observed that the sill depth in the Fram Strait is about 2600 m [e.g. *Balakshin*, 1959]. Earlier a ridge was believed to run between northeastern Greenland and Spitsbergen. *Helland-Hansen and Nansen* [1909] claimed that the ridge was little known, but: "The probability is that its saddle-depth is between 700 and 800 m". Their assumption was mainly based on a comparison of the water masses in the northern Greenland Sea with those in the Arctic Ocean. Estimates of the sill depth of this ridge vary widely in later literature.

Oceanographic research in the Iceland and Norwegian Seas has to a large extent been done in relation to fisheries investigations. Annual herring surveys were initiated in the early 1950s [e.g. *Devold*, 1953], but the hydrographic observations during these surveys were mainly limited to the upper 500 m. In the summers during the 1980s, multiple-ship surveys covered the area with a dense net of CTD casts. Most stations were, however, only 1000 m deep. The Russian component of this was part of a longer series of observations initiated by PINRO, Murmansk, in the 1950s, which is still being maintained [*Borovkov and Krysov*, 1995].

The longest series of observations in the area is, however, the observations obtained from Ocean Weather Station M at 66°N, 02°E. This series goes back to 1948 and results are presented in several publications [e.g. *Mosby*, 1959; *Bøyum*, 1962; *Østerhus and Gammelsrød*, 1999].

At the southern boundary of the Norwegian Sea, the oceanographic feature that has attracted most research is the overflow of arctic water across the Greenland–Scotland Ridge. Based on observations of the Danish "Ingolf" Expedition, 1895–96, this phenomenon was first deduced by *Knudsen* [1898] and has since been dealt with by many authors. In 1960 the Hydrographical Committee of ICES coordinated a major survey of the Iceland Scotland Ridge, the International (ICES) "Overflow" Expedition, May–June 1960 [*Tait*, 1967]. A main result of the survey was that cold overflow water was found nearly everywhere on the ridge where the depth exceeded 300 m. Overflow crossing the ridge in four branches between Iceland and the Faroes was assessed at 1.1 Sv, and the overflow through the Faroe Bank Channel was assessed at 1.4 Sv. A similar survey was arranged in 1973 [e.g. *Meincke*, 1983], when the Denmark Strait also was included in the survey area [*Ross*, 1984].

In 1985 ICES initiated the ICES NANSEN (North Atlantic–Norwegian Sea Exchanges) project to study the exchanges of water, heat and other properties across the Greenland–Scotland Ridge between the Atlantic Ocean and the Nordic Seas [*Hansen and Østerhus*, 1998; 2000].

During the 1990s the Nordic WOCE project also aimed at assessing the exchanges across the Greenland–Scotland Ridge [e.g. *Østerhus et al.*, 2001; *Fogelqvist et al.*, 2003], and this research has continued in different European Community (EC) projects.

During recent years the Greenland Sea has been observed more frequently, for example, by the EC-funded ESOP projects, the results of which are referred to in the project reports [*Wadhams et al.*, editors, 1996; *Jansen and Opheim*, editors, 1999].

Several reviews of the research in the Nordic Seas have been published [*Lee*, 1963; *Trangeled*, 1974; *Hurdle*, 1986; *Hopkins*, 1991]

BATHYMETRY

The main bathymetric features of the Nordic Seas are depicted in Figure 1. The Fram Strait, with a sill depth of ~2600 m, forms the main connection with the Arctic Ocean.

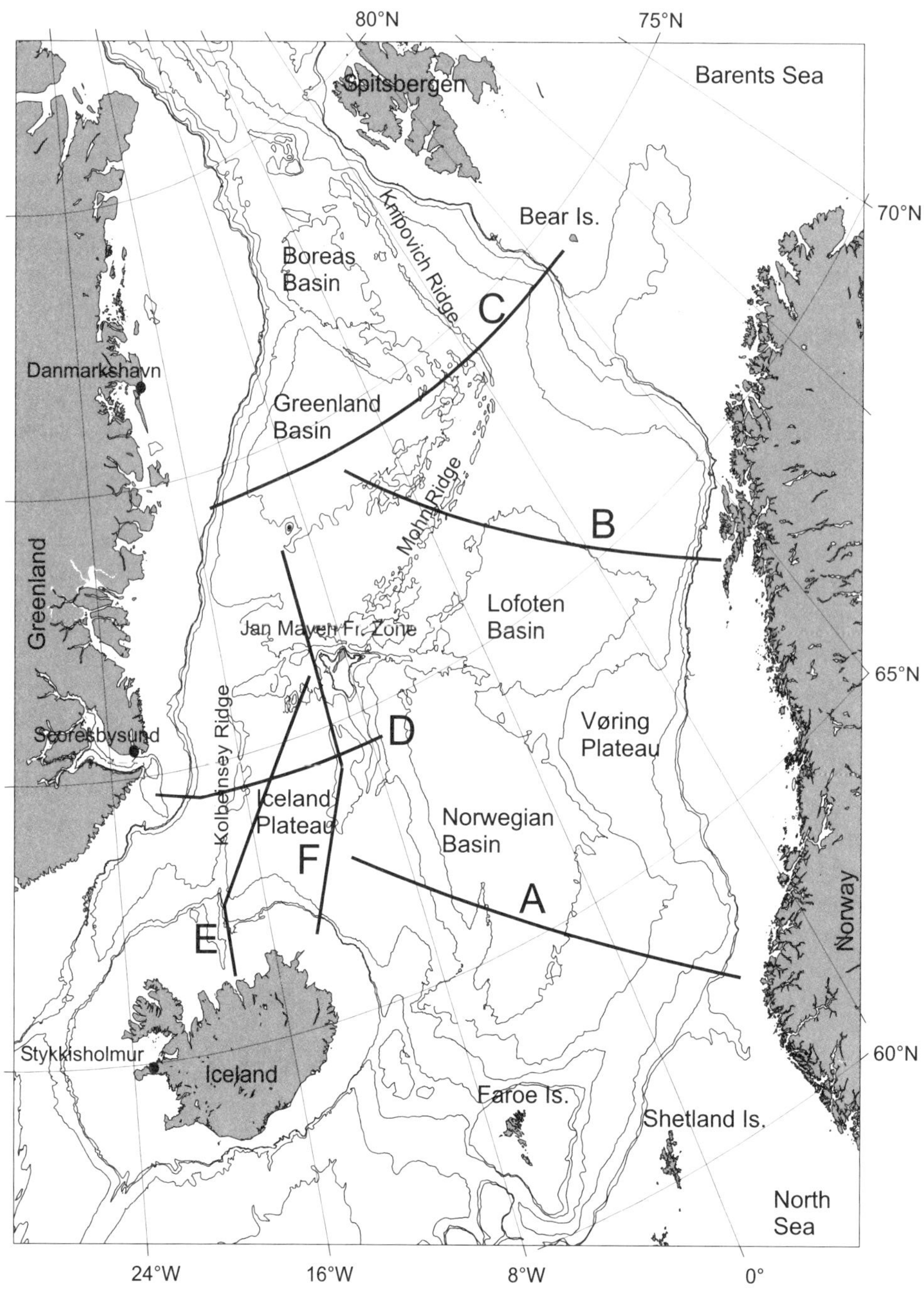

Figure 1. Main bathymetric features in the Nordic Seas, showing positions of sections that are referred to in the text.

To the south, the Greenland–Scotland Ridge, with the deepest sills located in the Faroe Bank Channel (~850 m) and in the Denmark Strait (~620 m), forms the border toward the North Atlantic. Topographic features within the Nordic Seas divide the area into the Greenland, Iceland and Norwegian Seas.

The most prominent of these features is the mid-ocean ridge. Three main parts of this ridge system are known as the Kolbeinsey Ridge, the Mohn Ridge and the Knipovich Ridge. The Kolbeinsey Ridge extends northward from the North Icelandic shelf to the latitude of Jan Mayen, where the Jan Mayen Fracture Zone cuts through it. The Mohn Ridge, between Jan Mayen and approximately 73.5°N, 8°E, has depths ranging between 1000 and 2000 m. Its rather complex topography is characterised by many isolated elevations. The Knipovich Ridge stretches from the Mohn Ridge to the Fram Strait, its shallower crests being at about 1000 m deep.

The bathymetry divides the Nordic Seas into four major basins. To the west of the mid-ocean ridge are two deep basins in the Greenland Sea. The Greenland Basin is the larger and deeper one, its floor depths ranging from 3400 to 3600 m. To the north, the Greenland Fracture Zone separates the Greenland Basin from the smaller and shallower Boreas Basin, which has depths around 3200 m.

In the Iceland Sea, to the south of the Jan Mayen Fracture Zone, the Iceland Plateau is the area between Iceland and Jan Mayen to the east of the Kolbeinsey Ridge. Along its eastern margin, which slopes into the Norwegian Basin, the Jan Mayen Ridge extends southward from Jan Mayen. West of this ridge a small basin is more than 2200 m deep.

The Norwegian Sea also has two deep basins, the Norwegian and Lofoten Basins. The former is the largest and deepest basin in the Nordic Seas, extending northward from the Iceland-Faroe Ridge and eastward from the Iceland Plateau to the Vøring Plateau and the continental slope off the Norwegian coast. Its floor lies mainly between 3200 and 3600 m, but depths exceed 3800 m in a narrow trough around 65°N, 5°W. The shallower and somewhat smaller Lofoten Basin, to the north of the Norwegian Basin and the Vøring Plateau and to the southeast of the Mohn Ridge, is about 3200 m deep.

FLUXES AND FLOW PATTERNS

The inflow from the North Atlantic Ocean, into the Nordic Seas, occurs in three locations. These inflows are the North Icelandic Irminger Current, which flows in through the eastern Denmark Strait; the Faroe Current, which enters across the Iceland–Faroe Ridge; and the Atlantic inflow, entering through the Faroe–Shetland Channel (Figure 2). The temperature and salinity (θ–S) properties of these water masses are entered in Table 1. Estimates of the mass fluxes in the various branches of Atlantic inflow to the Nordic Seas have been reported by numerous authors. Most of the estimates have, however, been based on indirect methods (e.g. geostrophic computations, budgets etc.) and vary from 2 to 12 Sv (1 Sv = 10^6 M^3 sec^{-1}) for the total flux. Only recently have fluxes based on direct measurements been obtained [*Østerhus et al.*, 2001].

In its main flow along the western side of the Reykjanes Ridge and the west coast of Iceland, the Irminger Current carries Atlantic Water that derives from the North Atlantic Current. Downstream of its bifurcation to the west of Iceland, a portion of its waters forms the North Icelandic Irminger Current. Flowing along the North Icelandic coast, this water mixes with Arctic waters deriving from the East Icelandic Current as well as coastal waters, and half-way along the coast, off Siglunes at 18.8°W, the May–June 1950–2000 average of temperature and salinity at 50 m, for the period, is 3.7°C and 34.7, respectively. Convective mixing during winter results in a water mass that *Stefánsson* [1962] has defined as North Icelandic Winter Water, identified by temperatures between 2 and 3°C and salinities of 34.85–34.90. More recently, this water mass has become fresher. Further east, the mixing with Arctic waters continues, particularly during winter and northeast of Iceland. *Read and Pollard* [1992] defined this water mass as Modified East Icelandic Water (Table 1).

Jónsson and Briem [2003] found that the Atlantic inflow to the North Icelandic Irminger Current during a 12-month period in 1999–2000 had an average volume flux of about 1 Sv. This transport represents a heat flux of approximately 25 TW (1 TW = 10^{12} W). More incomplete observations since 1985 are suggestive of considerable temporal fluctuations [*Kristmannsson*, 1998].

The Faroe Current carries Modified North Atlantic Water into the Nordic Seas. After crossing the Iceland–Faroe Ridge, it forms the Iceland–Faroe Front against the Arctic waters of the East Icelandic Current and flows eastward over the northern slope of the ridge. To the east of the Faroe plateau, some of this water recirculates into the Faroe–Shetland Channel and merges with the waters of the slope current. Most of the water continues, however, into the Norwegian Basin. When it approaches the Vøring Plateau, it turns northwest toward Jan Mayen. Near Jan Mayen, a fraction flows into the Iceland Sea, some recirculates along the western slope of the Norwegian Basin, and probably the largest portion flows northeast. Along this path the flow is parallel with the Arctic waters of the Greenland Sea, forming the Arctic Front, which to the north of Jan Mayen seems to be fixed to the Mohn Ridge.

After 4 years of current measurements, *Hansen et al.* [2003] assessed the average volume transport carried by the Faroe Current across 6°W, to the north of the Faroes, to be 3.5 Sv, which represents a heat flux of 124 TW.

The warmest and saltiest inflow to the Nordic Seas is the Northeast Atlantic Water that arrives through the Faroe–Shetland Channel (Table 1). Most of this inflow derives from the slope current to the west of the British Islands and enters mainly through the Rockall Trough. According to *Turrell et al.* [2003], this branch represents an average volume flux of 3.2 Sv and a heat flux of 127 TW.

In total, these recent assessments arrive at an Atlantic volume flux into the Nordic Seas of 7.7 Sv, the largest one being the Faroe Current. Due to the difference in temperature, the heat flux through the Faroe–Shetland Channel is, however, slightly larger than the heat carried by the Faroe Current. This results in a total heat flux from the Atlantic into the Nordic Seas of about 275 TW.

The flow that enters through the Faroe–Shetland Channel continues in the Nordic Seas as a slope current with its core off the shelf break. In the Bear Island Trough, a substantial part of its waters enters the Barents Sea, while the rest

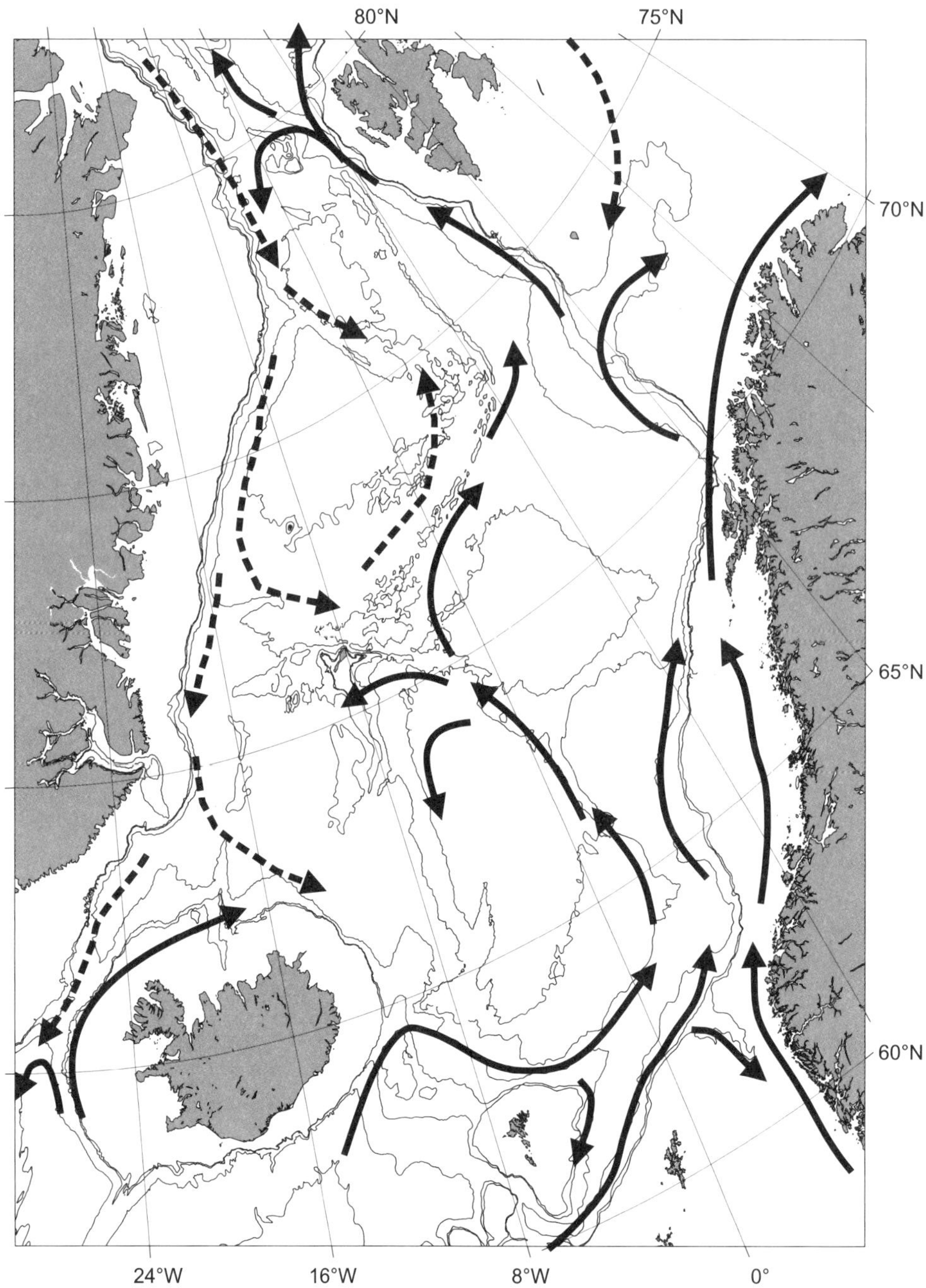

Figure 2. Schematics of the upper-layer circulation in the Nordic seas. Solid lines represent Atlantic Water. Broken lines represent Arctic and Polar waters.

continues northward in the West Spitsbergen Current, partly into the Arctic Ocean. *Rudels et al*. [1999] have described it as part of a circumpolar boundary current that flows along the shelf break around the Arctic Mediterranean.

Although the inflow of Atlantic Water to the Barents Sea is largely topographically steered, the local wind field is also of great importance for the circulation structure. The mean volume flux, averaged over 4 years of measurements (1997–2001), ranged between 1.3 Sv during summer and 1.7 Sv during winter [*Ådlandsvik and Loeng*, 1991; *Ingvaldsen et al*., 2004].

The Barents Sea is an important area for formation of dense bottom water by cooling and salt rejection during ice formation [*Knipovich*, 1905; *Nansen*, 1906; *Midttun*, 1985].

Table 1. Major Water Masses in the Nordic Seas Defined by Potential Temperature (θ) and Salinity.

Water mass	Potential temp. (°C)	Salinity
Canadian Basin Deep Water (CBDW) flows across the Lomonosov Ridge north of Greenland and enters the Nordic Seas between approximately 1500 and 2000 m over the Greenland Slope.	$-0.8 < \theta < -0.5$	>34.92
Eurasian Basin Deep Water (EBDW) occupies the deeper strata of the Eurasian Basin in the Arctic Ocean. Between approximately 2000 m and the sill depth in the Fram Strait (~2500 m) EBDW enters the Nordic Seas over the Greenland Slope.	$-0.7 < \theta < -0.9$	34.92–34.93
Greenland Sea Arctic Intermediate Water (GSAIW) is formed in the Greenland Sea by convection but is less dense than older deep water. The properties entered represent the 1990s.	$-0.9 < \theta < -0.5$	34.86–34.89
Greenland Sea Deep Water (GSDW) is a mixture of deep water formed locally by convection during cold winters, and deep waters deriving from the Arctic Ocean.	<0	34.88–34.90
Iceland Sea Arctic Intermediate Water (ISAIW) is partly formed in the northern/central Iceland Sea by winter convection and partly derives from similar intermediate waters in the Greenland Sea.	<1	34.7–34.9
Iceland Sea Deep Water (ISDW) is similar to NSDW but has slightly higher salinity due to admixture of CBDW.	~ −1	34.91–34.92
Irminger Sea Water (ISW) derives from the North Atlantic Current and is carried into the Nordic Seas by the North Icelandic Irminger Current. Properties given are those in the northern Irminger Sea.	5–7	35.05–35.10
Modified East Icelandic Water (MEIW) is an intermediate water in the southwestern Norwegian Sea characterised by waters from the East Icelandic Current and the North Icelandic Irminger Current.	1–3	34.6–34.9
Modified North Atlantic Water (MNAW) has its origin in the North Atlantic Current. It flows through the Iceland Basin and across the Iceland–Faroe Ridge into the Nordic Seas.	7.0–8.5	35.1–35.3
North East Atlantic Water (NEAW) enters the Nordic Seas through the Faroe–Shetland Channel. It derives mainly from the slope current to the west of the British Islands.	9.0–10.5	35.3–35.45
Norwegian Sea Arctic Intermediate Water (NSAIW) advects from the Iceland and Greenland Seas into the Norwegian Sea, where it occurs in a layer identified by a salinity minimum between the upper layers, composed mainly of Atlantic Water, and the deep water.	$-0.5 < \theta < 0.5$	34.7–34.9
Norwegian Sea Deep Water (NSDW) fills the Norwegian Sea beneath the NSAIW.	<0.5	34.91
Polar Intermediate Water (PIW) forms a temperature minimum underneath the upper waters of the East Greenland Current.	<0	34.4–34.7
Recirculating Atlantic Water (RAW) is water of Atlantic origin that has circulated into the East Greenland Current from the West Spitsbergen Current and the Atlantic layer in the Arctic Ocean.	0–2	34.9–35.0

Some of the dense bottom water that is formed in the eastern and central Barents Sea flows into the Norwegian Sea along the deeper northern slope of the Bear Island Trough. Measurements of the current during September–October 1978 indicated a volume flux of 0.8 Sv in this flow [*Blindheim*, 1989]. By mixing with less dense waters above, the bottom water deriving from the central and eastern Barents Sea becomes less dense along its path within the Barents Sea. In most cases it will therefore mix into intermediate depths in the Norwegian Sea. Dense bottom water formed in Storfjorden at Svalbard has a shorter route into the Norwegian Sea and is dense enough to mix into the bottom waters in the eastern Fram Strait, hence contributing to the deep water in the Arctic Ocean [*Quadfasel et al.*, 1988].

The volume exchanges through the Fram Strait at 79°N is on the order of 10 Sv in both directions, with a yearly mean net southward flow of approximately 2 to 4 Sv [*Fahrbach et al.*, 2001]. About half of the transport takes place in the deep water below 700 m, which is approximately the depth of the 0°C isotherm. The main northward transport of mass and heat takes place in the West Spitsbergen Current. *Aagaard and Greisman* [1975] report on a northward transport of 7.1 Sv, whereas *Schauer et al.* [2004] document a high variability on monthly and annual time scales, reporting values from 3 to 5 Sv. At 79°N, the East Greenland Current transports 3–4 Sv southward, including 1 Sv of Polar Water [*Foldvik et al.*, 1988; *Fahrbach et al.*, 2001]. The exchanges through the Fram Strait represent a northward net heat transport on the order of 30 TW [*Schauer et al.*, 2004].

Some Atlantic Water may deflect westward from the West Spitsbergen Current, into the northern Greenland Basin and the Boreas Basin, but most of the recirculation occurs in the Fram Strait [*Quadfasel et al.*, 1987]. Only a small part of its Atlantic Water eventually enters the Arctic Ocean, where it flows eastward along the Eurasian continental slope. The boundary current continues around the Arctic Ocean, although some of its water advects into the interior basins and forms the Atlantic Intermediate Layer. These waters exit the Arctic Ocean through the western Fram Strait, where they meet the Atlantic Water that has deflected westward from the West Spitsbergen Current [*Rudels et al.*, 1999].

These waters of Atlantic origin continue southward in the East Greenland Current, where they are known as Recirculating Atlantic Water, and flow parallel with and partly underneath the cold and fresh Polar surface waters, including sea-ice. In addition, this current also carries intermediate and deep waters from the Arctic Ocean [*Rudels et al.*, 2002]. Flowing along the coast of East Greenland, the waters of the current gradually mix with the waters of the Greenland Basin, and on topographic constraints, some of its waters divert into the basin. The first topographic feature with such effects is the Greenland Fracture Zone, which leads some of the East Greenland Current waters into the Boreas Basin [*Quadfasel and Meincke*, 1987]. Further south, in the Jan Mayen Fracture Zone, a large portion of its upper waters and all of its waters at depths greater than 1600 m flow into the Jan Mayen Current, forming the southern limb of the cyclonic gyre in the Greenland Basin (Figure 2). Further south, a portion of its upper water masses, and probably most of the bottom water, enter the East Icelandic Current and flow eastward in the Iceland Sea. The remaining part flows out of the Nordic Seas through the Denmark Strait.

Besides the inflows of warm Atlantic Water, there is an inflow of fresher and more Arctic character to the Arctic Mediterranean through the Bering Strait. *Roach et al.* [1995] measured this inflow to be close to 0.8 Sv.

These inflows have to be balanced by outflows, which take place mainly as deep overflows through the deeper sills across the Greenland–Scotland Ridge. The overflows near the bottom across the sill in the Denmark Strait and across the Iceland–Scotland Ridge are both assessed at approximately 3 Sv [e.g. *Østerhus et al.*, 2001]. Among the different views on the origin of the Denmark Strait Overflow Water, the latest version comes from *Jónsson and Valdimarson* [2004], who claim that the overflow derives from the Iceland Sea and is carried toward the sill by a jet at depths of 500–600 m over the Icelandic slope.

Further, a rather uncertain volume of surface outflow from the Arctic Mediterranean has been assessed at approximately 3 Sv, but this number is rather a balance in the budget than based on measurements. About 1.7 Sv is assumed to pass through the Canadian Archipelago [*Fissel et al.*, 1988], while the remaining 1.3 Sv are supposed to be carried by the East Greenland Current.

WATER MASS STRUCTURE

The distribution of the major water masses is illustrated in hydrographic sections across the basins as schematised in Figure 1. Dissolved oxygen and silicate are also shown in some of the sections, mainly to show the difference between the basins. In the Greenland Basin the repetitions of the section also reveal temporal variability. This is discussed jointly at the end of this chapter.

The Norwegian Basin

The Svinøy Section, in most cases along a line from 62.6°N, 04.7°E, to 64.7°N on the prime meridian, is repeated regularly to cover the southeastern Norwegian Basin in an area where the Atlantic current is relatively narrow. A prolonged section (A in Figure 1) that was worked in November 1996 is shown in Figure 3. This area was a repetition of a section worked along the same line in October 1958 as part of the IGY Polar Front Survey [*Dietrich*, 1969]. Atlantic Water with salinities between 35 and 35.26 was observed in the upper 500 m off the shelf break, its depth decreasing seaward. Its greatest horizontal extent was between 100 and 200 m, where Atlantic Water was observed to about 2°W. In the upper 100 m, however, salinities were below 34.85 west of the prime meridian, and decreased westwards to below 34.2 over the Iceland Plateau.

Water of salinities below 34.89, known as Norwegian Sea Arctic Intermediate Water, extends eastward under the Atlantic Water across the whole basin to the Norwegian slope. Only near the slope is the salinity minimum less pronounced,

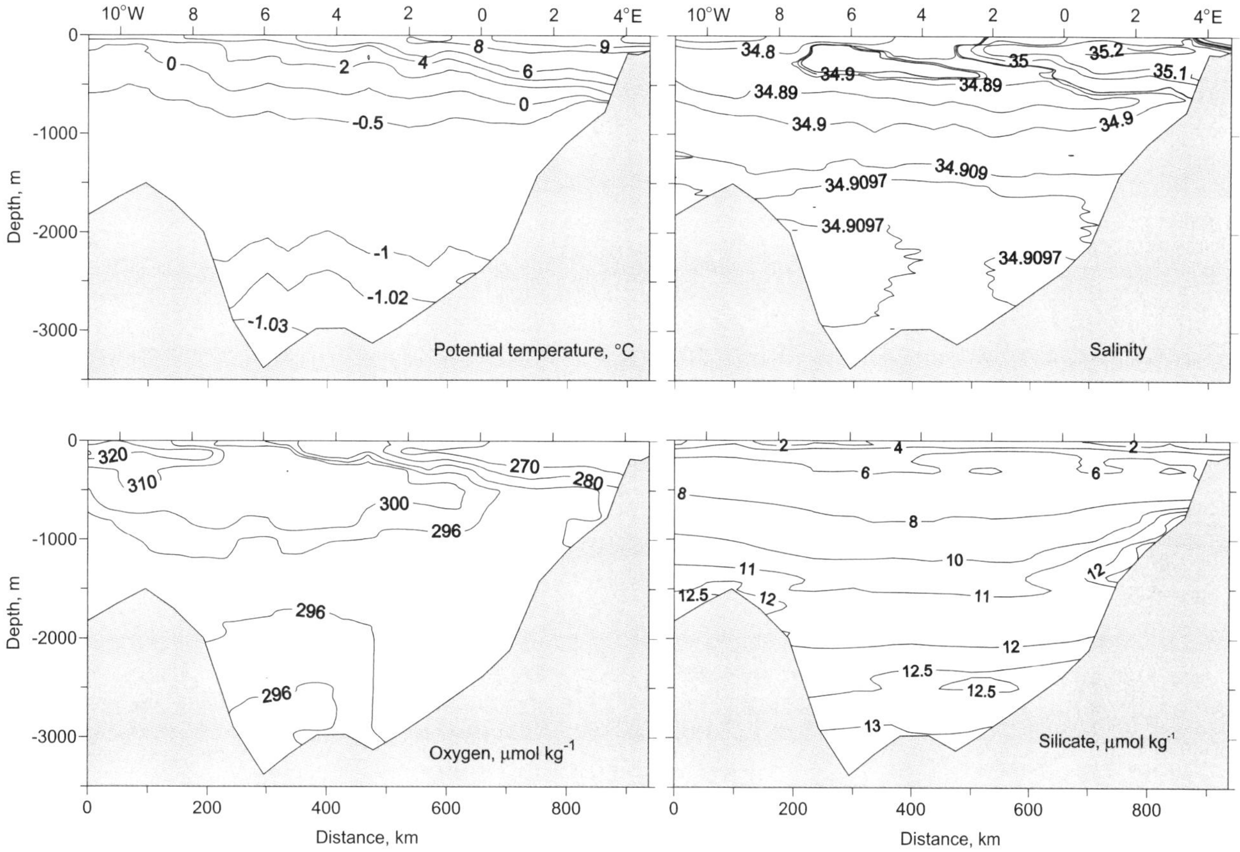

Figure 3. Potential temperature (°C), salinity, dissolved oxygen (μmol kg^{-1}), and silicate concentration (μmol kg^{-1}), across the Norwegian Basin, November 1996: A in Figure 1.

a result of more intense mixing than in the interior basin. In the zone over the upper slope, the transition between the Atlantic Water and the deep water is narrow, often less than 300 m. Repetitions of the section have shown that the depth of this transition zone varies within approximately 200 m.

Over the western slope of the basin, towards the Iceland Plateau, salinity values exceeded a maximum of 34.95 between approximately 100 and 250 m depth, underneath the fresh upper layer. This water originates from the Norwegian Atlantic Current, flows south over the western slope of the Norwegian Basin, and shows a topographically controlled cyclonic circulation in the basin. The section along the same line in 1958 had salinities in excess of 35 in this core. This circulation is also a clear feature in current maps based on surface drifters [*Poulain et al.*, 1996; *Jakobsen et al.*, 2003].

From 1958 to 2000 the temperature in the deep water increased slightly. Close to the bottom, at about 3000 m, the potential temperature rose from −1.07°C to −1.02°C. The source of this warming probably derives from the increasing fraction of Arctic Ocean Deep Water that flows in through the Greenland Sea. *Østerhus and Gammelsrød* [1999] have described a more extensive warming at Ocean Weather Station M (66°N, 2°E), where the potential temperature at 2000 m has increased by 0.1°C since the late 1980s. With an upwardly increasing delay, the warming was observed at shallower depths, at 1200 m, since 1990.

The Lofoten Basin

In June 1999, an extended Gimsøy Section (B in Figure 1) was worked across the Lofoten Basin, from the Lofoten Islands into the Greenland Basin (Figure 4). The Arctic Front was observed over the Mohn Ridge, where in the upper 500 m it was indicated by both thermal and haline isolines rising steeply against Arctic waters to the west. The salinities decreased from 35 to 34.88 over a distance of approximately 30 km. In temperature this corresponded to a decrease from 3°C to 0°C, except near the surface, where seasonal warming disturbed the frontal structure. Southeast of the front, Atlantic Water occupied the upper 600–700 m, and the 35 isohaline was associated with potential temperatures between 3°C and 4°C.

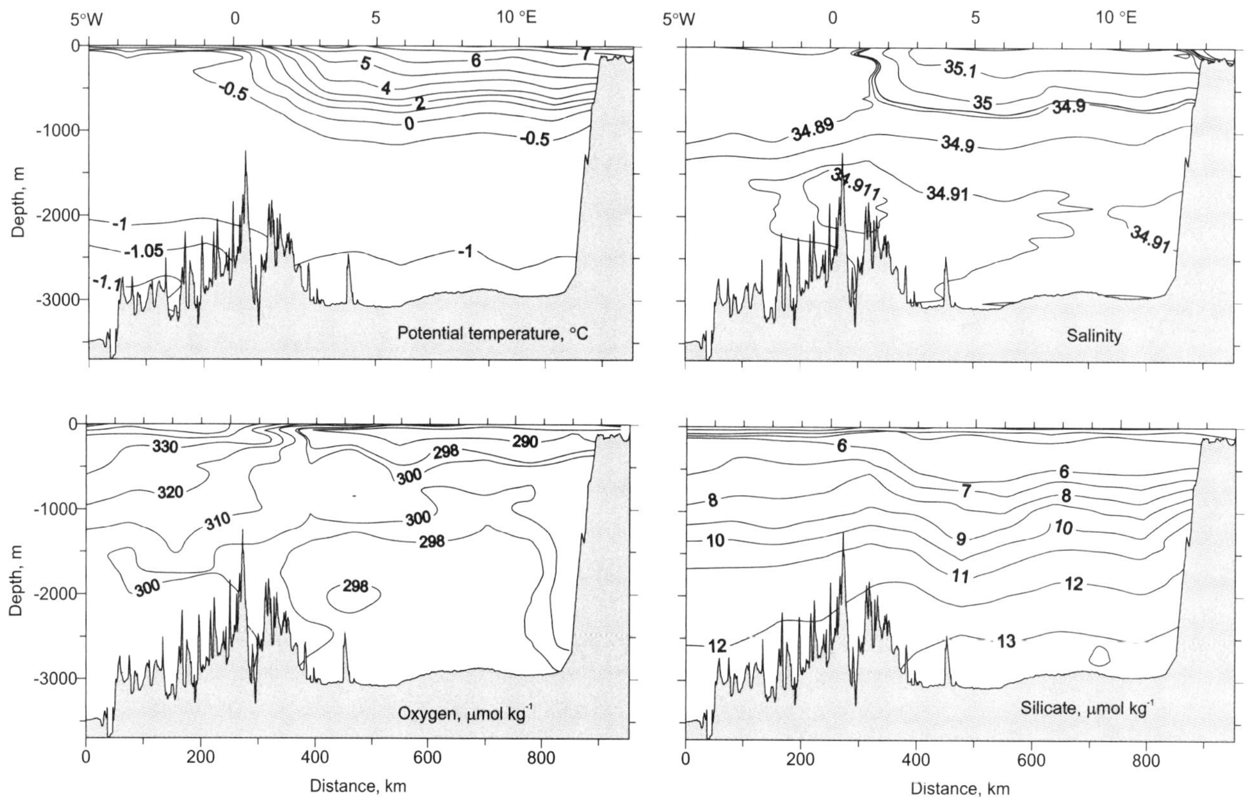

Figure 4. Potential temperature (°C), salinity, dissolved oxygen (µmol kg^{-1}), and silicate concentration (µmol kg^{-1}) in a section across the Lofoten Basin and into the Greenland Basin, June 1999: B in Figure 1.

In the Greenland Basin, to the northwest of the front, the salinities were below 34.9 to about 1200 m. As in the Norwegian Basin, this water intruded under the Atlantic Water and spread across the whole Lofoten Basin as an intermediate layer. The potential temperatures in this layer were mainly between +0.5°C and −0.5°C. The layer could also be partly identified by a slight maximum in oxygen. In the Lofoten Basin, the intermediate water occurs at greater depths than further south, normally at depths centred around 800 to 1000 m.

Deep water originating from the Arctic Ocean could be identified as a salinity maximum between 1500 and 2500 m over the Mohn Ridge. West of the peak of the ridge, the salinities ranged up to 34.913 at 1750–1800 m. The section from 1999 indicated well how this water was spreading into the basins on both sides of the ridge. To the southeast of the ridge, the bottom waters were characterised by higher silicate concentrations and lower oxygen concentrations than to the northwest of the ridge.

Measurements of potential temperature, salinity, oxygen and silicate in a section that was worked along 74.5°N in June 1999 (Figure 5) illustrate the conditions in the area along the Barents Shelf. In this section the Arctic Front was observed over the Knipovich Ridge. Because of the seasonally warmed, approximately 40-m-deep, mixed surface layer, the front was more clearly defined by salinity than by temperature. East of the front, the 300-km-wide area toward the Barents Shelf was characterised by Atlantic Water, with potential temperatures between 2°C and 7°C and salinities up to 35.14 in the upper 400–500 m. The intermediate layer of Arctic waters extended eastwards from west of the Arctic Front and was more prominent in this section than further south. The minimum salinities in the intermediate layer were slightly below 34.88 near the front and increased gradually eastward. Vigorous mixing contributed to salinities above 34.9 near the slope. The temperatures in the intermediate water were between −0.5°C and 0°C. At greater depths the maximum in salinity over the Knipovich Ridge was observed also in this section, and east of the ridge salinities were above 34.91 at depths between 1500 and 2000 m.

Time series since 1978 of temperature and salinity in the Svinøy and Gimsøy standard sections across the Norwegian Atlantic Current, and in the Sørkapp Section, along 76.33°N across the West Spitsbergen Current, are shown in Figure 6. The values are averages of the observations between at 50 and 200 m in the core of the current just off the shelf break.

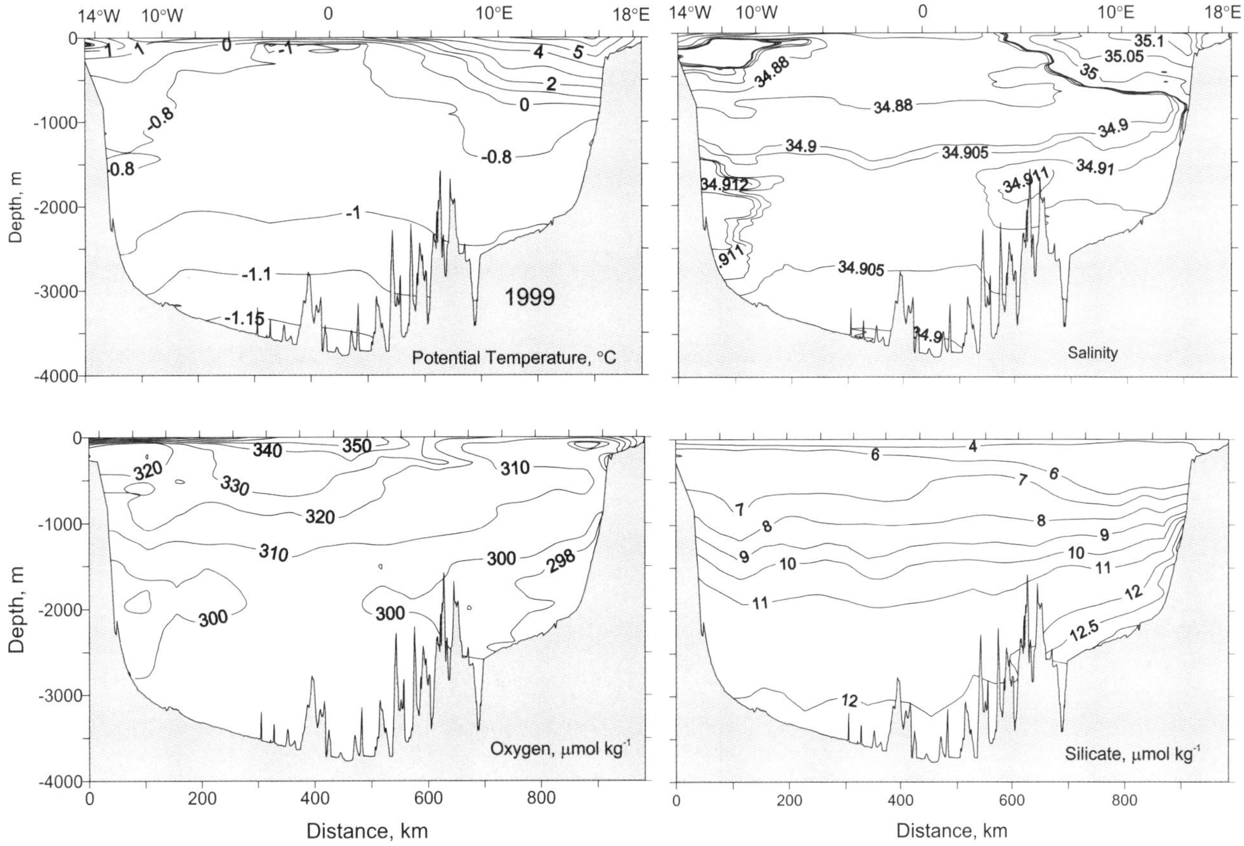

Figure 5. Potential temperature (°C), salinity, dissolved oxygen (μmol kg^{-1}), and silicate concentration (μmol kg^{-1}) along 74.5°N, between the Barents Shelf and the East Greenland Shelf in June 1999: C in Figure 1.

They represent summer conditions as the measurements are from the period late July–early September. Although the time series are still short, they suggest fluctuations with a wide spectrum of time scales. The low salinity values in the late 1970s were associated with the salinity anomaly of the 1970s [Malmberg, 1969; *Dickson et al.*, 1988]. Fluctuations on sub-decadal time scales seem to be rather regular, but there are also variations of longer duration. The trend in temperature has been toward higher values in all three sections, with an increasing trend during the period 1978–2002 of 0.6°C, 0.35°C and 0.8°C from south to north. The salinity showed a somewhat different development, decreasing from 1981 to 1996 and increasing during more recent years. The fluctuations reflect atmospheric conditions. As described by, for example, *Dickson et al.* [2000], the conditions in the Sørkapp Section in particular show a close correlation to the North Atlantic Oscillation (NAO).

Figure 7 shows the mean values from the standard sections in the temperature–salinity space. A gradual cooling and freshening of the Atlantic Water is indicated as it flows through the Norwegian Sea. Over a distance of approximately 1450 km from the position of the Svinøy Section to the Sørkapp Section, the cooling has on average been by 4.23°C and the salinity has decreased by 0.18. This cooling has been more intense between Gimsøy and Sørkapp than between Svinøy and Gimsøy, 0.0034°C km^{-1} and 0.0020°C km^{-1}, respectively. Not surprisingly, this indicates a larger heat flux to the atmosphere in the northern than in the southern Norwegian Sea. Compared with the Greenland Sea, this cooling is considerable. According to *Bøyum* [1966], the annual mean heat flux at Ocean Weather station M, averaged over the period 1948–1958, was 120 W m^{-2}. In comparison, the winter mean at Jan Mayen, averaged over the period 1957–2000, was 160 W m^{-2}. There has not been any similar difference in the freshening, however, which was 0.0001 units km^{-1} over both the distances. To what extent the freshening is due to precipitation or to admixture of less saline water from the west, and how much is from the Norwegian Coastal Current, is difficult to say. Between 50 and 200 m depth, the sum of these effects seems to be about equal to that in the Norwegian Atlantic Current along the shelf [*Mork and Blindheim*, 2003].

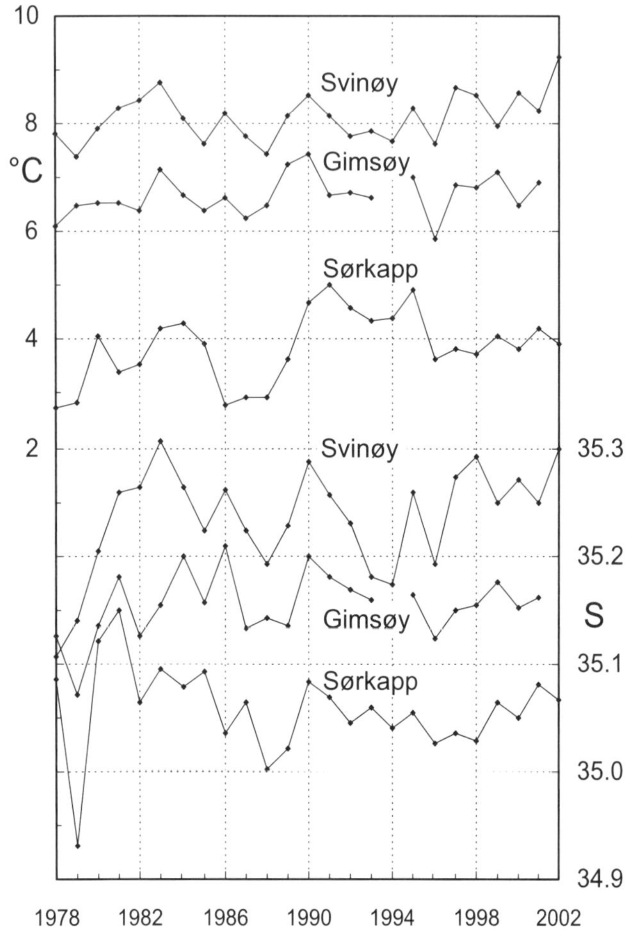

Figure 6. Temperature and salinity in the core of Atlantic Water in the standard sections Svinøy-NW, Gimsøy-NW, and Sørkapp, averaged between 50 and 200 m, July–August 1978–2002.

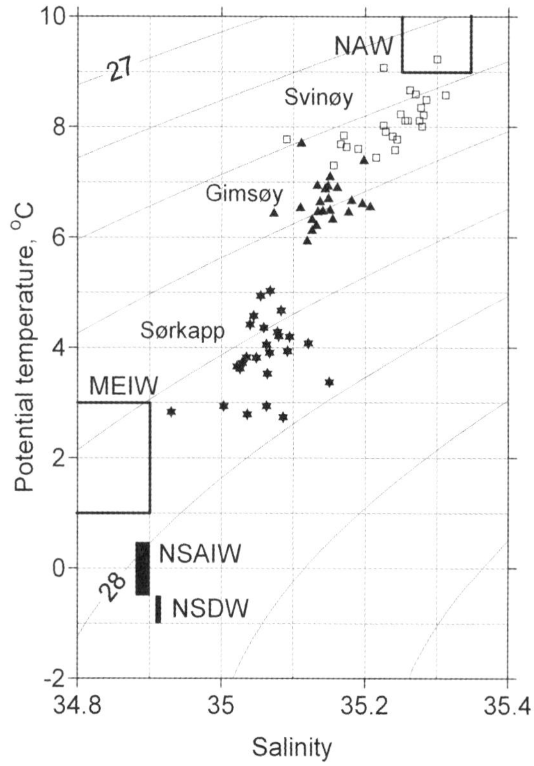

Figure 7. Temperature–salinity relationship of the mean values in the core of the Atlantic Water between 50 and 200 m, 1978–2002, in the standard sections Svinøy–NW, Gimsøy–NW, and Sørkapp. Properties of North Atlantic Water (NAW), Modified East Icelandic Water (MEIW), Norwegian Sea Arctic Intermediate Water (NSAIW) and Norwegian Sea Deep Water (NSDW) are indicated.

The Greenland Basin

The section in Figure 5 (C in Figure 1) describes the main oceanographic features in the Greenland Sea in June 1999. Favourable ice conditions allowed this section to be extended westward to the East Greenland Shelf. Near the Greenland Shelf, the surface waters were characteristic of the East Greenland Current: salinities near 33 and temperatures <0°C. A rather sharp transition was observed between these Polar surface waters and the recirculating Atlantic Water beneath and farther east. The recirculating Atlantic Water could be observed to about 7°W, approximately 175 km east of the shelf break, with maximum temperature and salinity at 100 m. Salinities and potential temperatures in the recirculating Atlantic Water ranges up to 34.95 and 1.6°C, respectively (Figure 8).

Beneath the 40-m-deep mixed surface layer, where seasonal warming had increased the temperatures to above +1.5°C, the central part of the section, between about 5°W and 3°E, showed low temperatures in the upper layers, whereas at depths of about 100 m there was a cell with temperatures below −1°C. This cold body of water, with salinities between 34.85 and 34.87 and high oxygen concentrations, seemed to be formed by the latest convective event during the previous winter or spring.

The Atlantic Intermediate Water was spread along the whole section, mainly at depths between 700 and 1100 m. In the western part of the section a salinity minimum, with salinities between 34.86 and 34.88, beneath the recirculating Atlantic Water (Figure 8). In the central Greenland Sea, colder and fresher waters at shallower depths extended to approximately 800 m. In the eastern part a salinity minimum occurred beneath the Atlantic Water, mainly with salinities between 34.89 and 34.90.

Over the Greenland Slope, the somewhat saltier and less oxygen-rich deep waters flowing in from the Arctic Ocean were clearly indicated between approximately 1500 and 3000 m. As described earlier by *Meincke et al.* [1997] and *Rudels et al.* [2002], it was possible to distinguish between

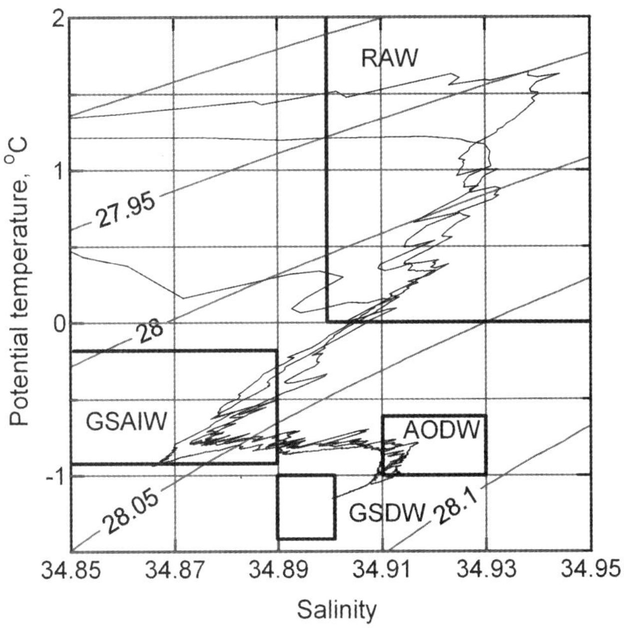

Figure 8. Temperature–salinity relationship of three CTD profiles off the east Greenland shelf (9°, 11°, and 13°W) on the section along 74.5°N in June 1999, with indications of recirculating Atlantic Water (RAW), Greenland Sea Arctic Intermediate Water (GSAIW), Arctic Ocean Deep Waters (AODW), and Greenland Sea Deep Water (GSDW).

two maxima in salinity. The upper maximum, between 1500 and 1800 m, probably originated in the Canadian Basin, and a deeper one, mainly between 2000 and 2500 m, derived from the Eurasian Basin. Both these water masses had salinities near 34.912. As described above, waters with similarly high salinities were also observed over the Knipovich Ridge on the eastern side of the basin, showing that some of the Arctic Ocean Deep Water had diverted into the Jan Mayen Current and circulated around the southern sector of the Greenland Sea Gyre.

This is further supported by a section stretching into the southwestern Greenland Basin along 10°W, shortly west of Jan Mayen (see Figure 15, presented later). To the north of Jan Mayen, the properties of the waters carried by the Jan Mayen Current indicated their origin in the East Greenland Current. The surface layer, to about 50 m deep, had the properties of Polar Water; at depths between about 100 and 500 m, the waters showed recirculating Atlantic Water characteristics. The Arctic Ocean Deep Waters were indicated by salinities in excess of 34.91 between depths of about 1100 and 2400 m. In a maximum over the slope to the north of Jan Mayen, salinities ranged up to 34.914 between 1200 and 1800 m.

If the 34.89 isohaline is chosen to indicate the limit of intermediate water toward GSDW, it may also indicate the depth of convection during the previous winter. During the first half of the 1990s the depth of this isohaline in the central Greenland Sea ranged between approximately 700 m in 1991 and 1150 m in 1995. In the years 1996–2000, when the section was occupied in spring or summer, the isohaline was observed between 1100 m and 1300 m. The vertical extent of the intermediate water may also vary seasonally due to mixing with the somewhat saltier waters below. The observations in 1996, when the section was occupied both in July and in November, are suggestive of this as the depth of the 34.89 isohaline was about 1300 m in July and 1100 m in November.

An intermediate temperature maximum, with potential temperatures between −0.85 and −0.80°C, showed gradually increasing salinities during the whole decade, from 34.885 in 1991 to 34.906 in 2000. The depth of this maximum increased from approximately 650 m in 1991 to 1400 m in 1996. Based on data up to 1996, *Budeus et al.* [1998] suggested that this might represent a downward water movement in the central Greenland Sea that might explain the observed warming of the bottom water. The more recent data show, however, that the temperature maximum has remained at depths around 1100–1300 m during the later years.

In spite of this, the layers below became progressively more characterised by Arctic Ocean Deep Waters. In 1991 these waters extended westward from the Knipovich Ridge to about 2°E, but the salinities in the central Greenland Basin were below 34.90. During the following years, however, the extent of Arctic Ocean Deep Waters gradually increased. By 1993 there was already a continuous layer with salinities above 34.9 between 2000 and 2500 m along the whole section to 7.6°W, and probably across the basin. The increase in the vertical extent of the layer with relatively warm and saline Arctic Ocean Deep Waters continued during the rest of the decade. In 1999 the salinities were higher than 34.9 from about 1500 m to near the bottom and, as seen in Figure 5, above 34.905 to about 2700 m. Also oxygen and silicate concentrations had changed considerably in 1999 compared with 1991 (*Blindheim and Rey*, 2004).

This is further demonstrated in Figure 9, which shows salinity profiles representing the central Greenland Gyre for the years 1991–2000. The profiles are averages based on two or three stations located between 74.5° and 75°N and between 1°W and 1°E. All profiles show an intermediate salinity maximum between 1500 and 2500 m, the same depth interval where the Arctic Ocean Deep Water was observed over the Greenland Slope in 1999 (Figure 5). In accordance with the spreading of the Arctic Ocean Deep Water into the interior basin, this maximum was observed at greater depths during 1991–1994 than during the later years. The increase in salinity during the decade was clear. Although the differences between consecutive years were close to or less than the observational accuracy,

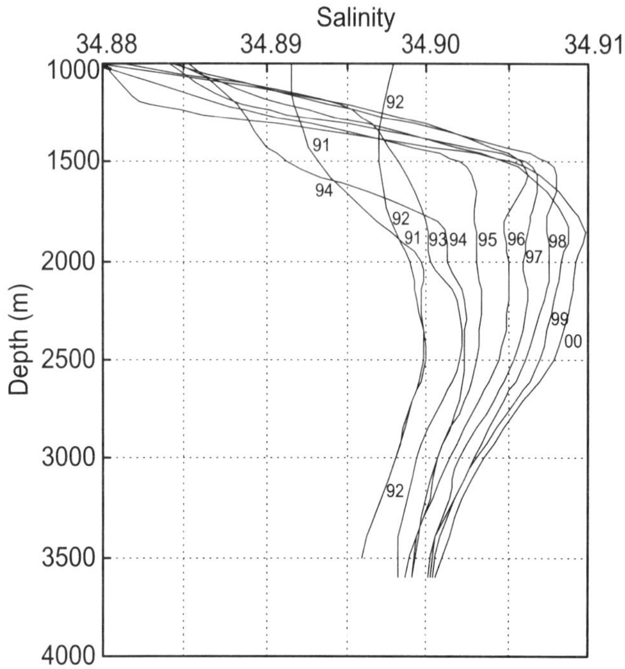

Figure 9. Profiles of mean salinity between 1000 and 3600 m deep in the central Greenland Sea, between 74.5° and 75.0°N, and between 1°E and 2°W, 1991–2000.

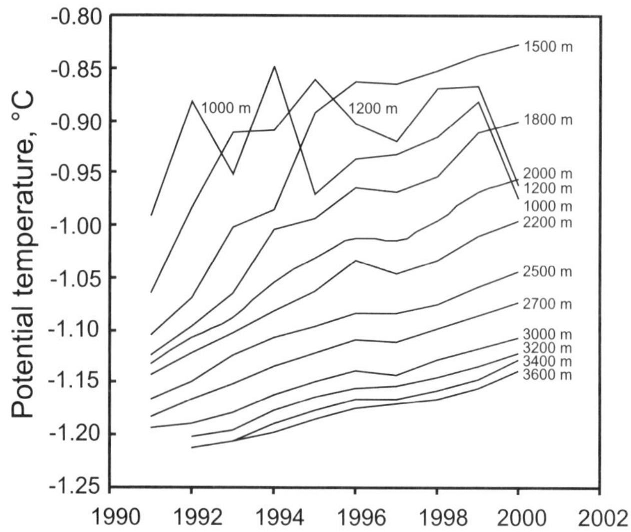

Figure 10. Potential temperature at selected depths in the central Greenland Sea during 1991–2000, based on the same stations as in Figure 9.

the total increase over the decade was evident. At 1500 m this increase was almost 0.02, decreasing to near 0.005 at 3500 m.

Figure 10 shows the temporal development from 1991 to 2000 of the potential temperature at some selected depths in the central Greenland Basin. The figure is based on the same stations as Figure 9. At depths greater than approximately 950 m there was an overall warming during the decade. At 1500 m, which was close to the depth of maximum temperature increase, the warming amounted to 0.28°C, the potential temperature reaching –0.83°C at the end of the decade. This warming decreased toward greater depths, and at 3000 m, it was 0.09°C, resulting in a potential temperature of –1.11°C in 2000.

Depending on the depth of the convection during the various winters, the series of potential temperature at 1000 and 1200 m was more irregular than the series at greater depths. The greatest temperature fluctuations occurred when the convection was relatively shallow. Hence, the 1000-m depth showed considerable cooling from 1992 to 1993 and from 1994 to 1995, whereas the layer at 1200 m was unaffected by this cooling. During the winter 1999–2000 there had been convection to depths greater than 1200 m as shown by a considerable temperature decrease at 1000 and 1200 m. Only during the period from November 1996 to May 1997 had there been a cooling that could be traced to 3000 m.

As seen in Figure 10, the cooling from 1996 to 1997 appeared as a moderate temperature decrease, possibly because the heat loss was distributed to great depths in the water column. Furthermore, the main convection probably occurred early in the winter, which is suggested by considerably lower air temperatures at Jan Mayen in December 1996–January 1997 than during the rest of the winter. This may also be the reason why the water column in the central Greenland Sea in May 1997 was more stratified than would be expected after a deep convective event. A further reason may be that lateral, isopycnal mixing may quickly disturb a homogeneous water column, particularly in the upper layers, as indicated at 1000 m. This will imply that the convective area must have been of rather small horizontal scale, and that the distance to waters with different properties had not been too great. In May 1997 a "chimney" extending to more than 2000 m depth was observed only 50 km north of the section [*Gascard et al.*, 2002]. We consider it therefore not unlikely that the observed temperature signal may be ascribed to such structures.

At the low temperatures like those of the GSDW, the haline contraction has considerably greater effect on the density of the water than the thermal expansion. However, the rise in temperature during the 1990s more than compensated for the increase in salinity, so that the density in the deeper layers of the Greenland Basin decreased. The density decrease was larger at 1500 m than at greater depths, with a stabilisation of the deeper water column as a result. Although this will have a prohibiting effect on deep convection, the average density of the deep water was reduced and the total effect is therefore in favour of overturning.

The temporal change in the GSDW is also demonstrated by the concentrations of nutrients and oxygen. From 1981 to 2000, the concentration of dissolved oxygen at 2500 m decreased from 316 to 303 μmol kg^{-1}, while the concentration of silicate increased from 10.5 to 12.0 μmol kg^{-1}. Both these trends indicate that the fraction of Arctic Ocean Deep Water in the deep Greenland Basin has been increasing (*Blindheim and Rey*, 2004).

The Iceland Sea

Few papers cover the whole Iceland Sea. *Kiilerich* [1945] included the Iceland Sea in his description of the Greenland Sea, but he had few data from its central and eastern areas. *Stefánsson* [1962] presented a thorough description of the North Icelandic waters, but his work concentrated on the waters south of 69°N. Other workers have dealt with the East Greenland Current and the Denmark Strait Overflow [e.g. *Rudels et al.*, 2002; *Strass et al.*, 1993; *Buch et al.*, 1996]. *Swift et al.* [1980] and *Swift and Aagaard* [1981] based their work on a data set that covered a wider area, which enabled them to describe most of the ice-free area of the Iceland Sea. The present description will partly be based on a data set from a fish survey in August 1988 that covered the Iceland Sea with CTD casts to 1000 m. Zonal sections were worked across the Iceland Sea at intervals of 0.5° of latitude between 68° and 71.5°N.

As described above, some of the waters of the East Greenland Current divert into the Jan Mayen Current, but most of the waters at depths shallower than 1500 m continue southward, into the Iceland Sea. Further south, approximately between 68.5° and 69.5°N, another divergence forms the East Icelandic Current. These oceanographic features divide the Iceland Sea into three regions. The largest is characterised by the waters of the East Greenland and East Icelandic Currents. In Figure 11, the area where the minimum temperature in the cold intermediate water at depths between approximately 100 and 200 m is below −1°C. South of the East Icelandic Current, the North Icelandic Irminger Current flows along the North Icelandic coast, carrying Atlantic Waters into the area. Most of these Atlantic Waters are confined to the north Icelandic shelf area without entering the central Iceland Sea [*Jónsson*, 1992; *Swift and Aagard*, 1981].

Stefánsson [1962] described the main features of the circulation in the Iceland Sea. In the eastern basin there is a cyclonic gyre centred at approximately 69.5°N, 12°W. This gyre is bordered by the East Greenland Current to the west, the East Icelandic Current to the south, and waters of Atlantic origin that flow in from the Norwegian Sea to the north of the gyre. The waters deriving from the Norwegian Sea form a somewhat warmer region in the northeastern Iceland Sea, where they characterise the properties in the upper 200–300 m of the water column. In this area the schematics in Figure 11 are based on the maximum salinities in this depth layer.

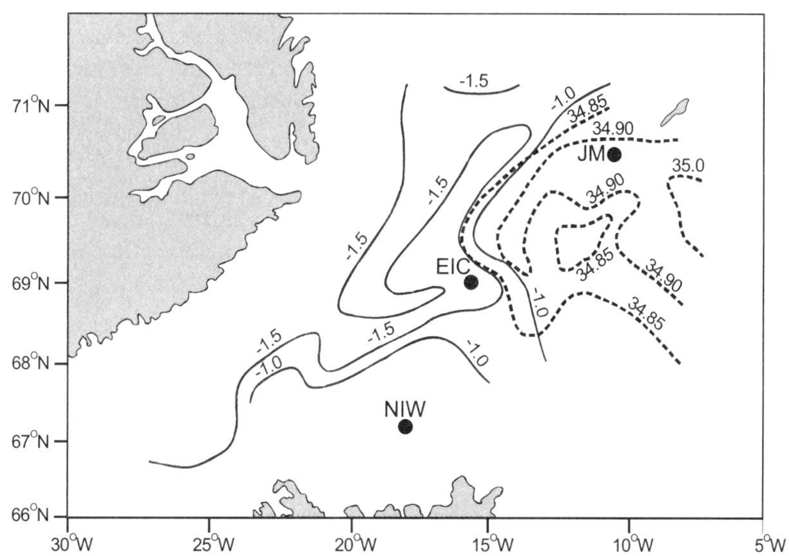

Figure 11. Schematics of three regions in the Iceland Sea. The largest area in the central and western Iceland Sea represents the East Greenland Current and the East Icelandic Current. The area is characterised by temperatures below −1°C in the cold fraction of the Iceland Sea Arctic Intermediate Water. The region to the northeast, characterised by waters flowing in from the Norwegian Sea, is bounded from the cold Arctic waters with salinity greater than 34.85 and temperatures above 0°C. The southern region also has warmer upper layers due to influence from the North Icelandic Irminger Current. Stations: EIC, East Icelandic Current; NIW, North Icelandic waters; JM, Jan Mayen.

The area is bounded from the Arctic and Polar Waters to the west and south by a salinity of 34.85. Salinities greater than 35 show the presence of Atlantic Water to the south and southeast of Jan Mayen, and a tongue of water saltier than 34.9 that turns toward south to southwest of Jan Mayen indicates the cyclonic circulation in the area.

Figure 12 shows potential temperature–salinity relationships at one station in each region. The station marked NIW (see Figure 11), is located at 67.1°N, 18.3°W, approximately 150 km off the Icelandic coast. This position is in the northern part of the region between the East Icelandic Current and Iceland that is dominated by the North Icelandic Irminger Current. The fresh surface layer on this station was relatively warm, with temperatures in the range 5–7°C. The influence of the North Icelandic Irminger Current was further indicated by a salinity maximum at potential temperatures between 2°C and 4°C. At greater depths, around 300 m, a minimum in both temperature and salinity indicated the presence of Iceland Sea Arctic Intermediate Water.

Station EIC was located within the East Icelandic Current, at 69.0°N, 15.9°W, and its features were characterised accordingly. Under a fresh, seasonally warmed surface layer with temperatures close to 3°C and salinities near 33.0 in the upper 25 m, the temperature decreased to −1.52°C at 85 m, where the salinity had increased to above 34.6 Toward greater depths, both temperature and salinity increased to a maximum around 200 m, with values of +0.6°C and 34.87, respectively. At depths greater than 300 m, temperature decreased and salinity increased to more deep water properties.

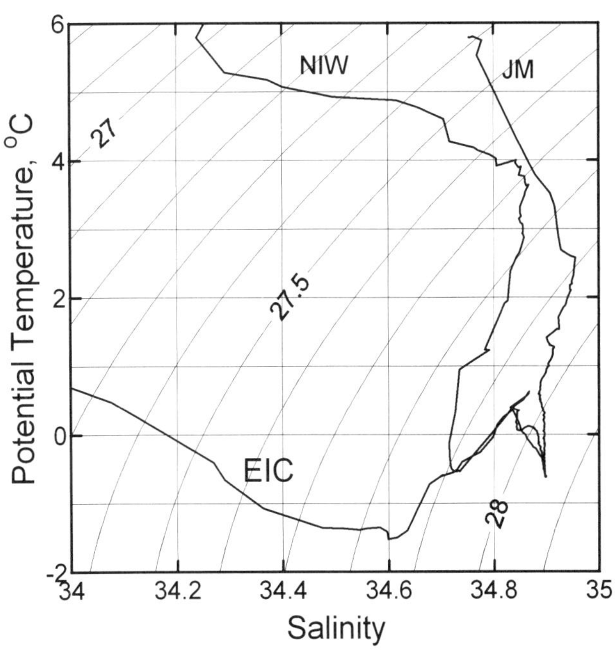

Figure 12. Potential temperature–salinity relationships in the three different regions in the Iceland Sea. Station NIW (see Figure 11) represents North Icelandic waters between the East Icelandic Current and Iceland. Station EIC is located in the East Icelandic Current. Station JM is located in the northeastern Iceland Sea, in the regime of waters flowing in from the Norwegian Basin.

Following *Rudels et al.* [2002], Polar Water is present to a depth of 50 m ($\sigma_\theta < 27.70$, $\theta < 0$; where θ is potential temperature, in °C, and σ_θ is potential density), and according to the definition of *Malmberg* [1972], Polar Intermediate Water also exists between ~50 and 100 m ($\theta < 0$, $34.4 < S < 34.7$). In the next depth layer, the temperature increased toward the maximum of temperature and salinity at depths around 200 m. This water has the properties of what *Swift and Aagaard* [1981] defined as upper Arctic Intermediate Water ($\theta < 2$, $34.7 < S < 34.9$). However, they further defined the depth layer that contains the temperature and salinity maximum as lower Arctic Intermediate Water ($0 < \theta < 3$, $S > 34.9$). At station EIC this layer clearly contained what under various names has been defined as the water of Atlantic origin that derives from the West Spitsbergen Current and the Atlantic intermediate layer in the Arctic Ocean and forms an intermediate layer in the East Greenland Current. In recent works it is described as recirculating Atlantic Water [*e.g. Rudels et al.*, 2002].

When *Swift and Aagaard* [1981] defined the intermediate maximum in temperature and salinity as lower Arctic Intermediate Water, they based the definition on observations in the eastern Iceland Sea, which showed temperatures and salinities up to + 1.3°C and 34.94, respectively. In 1988, salinities in excess of 34.9 in the intermediate maximum were observed only in the recirculating Atlantic Water in the East Greenland Current, and not in the East Icelandic Current. This is an example of difficulties with defining water masses. Using data from only one or a few years may fail to cover temporal variability. *Malmberg et al.* [1986] have demonstrated that temporal variations in the Iceland Sea may be considerable, and the data set from 1988 that is applied here supports their conclusions. According to the definition of *Swift and Aagaard* [1981] there was no lower Arctic Intermediate water in the Iceland Sea in 1988, except in the East Greenland Current. It is, however, more in agreement with the conditions described by *Jónsson* [1992], who observed that this layer can always be defined by a temperature maximum, but the maximum temperature is frequently lower than 0°C. Furthermore, there is often no salinity maximum, and the salinity at the temperature maximum is often below 34.9. The definition given by *Rudels et al.* [2002] seems more robust, although it is rather complex. Their Iceland Sea Intermediate Water includes the temperature minimum of the upper Arctic Intermediate Water, the range with increasing temperature and salinity, the temperature maximum, and the waters below this maximum with

$\sigma_{0.5} < 30.44$ kg m^{-3}. In 1988, the depth of this density was less than 1000 m only in the cyclonic gyre to the southwest of Jan Mayen, where its shallowest depth was close to 800 m.

Station JM in Figure 12 was at 70.5°N, 10.0°W, to the southwest of Jan Mayen. A fairly deep surface layer had temperatures between 5°C and 6°C, but salinities near 34.75 suggested an Arctic influence. The water flowing in from the Norwegian Sea was indicated by salinities near 34.95 and temperatures between 2°C and 3°C at depths of around 100 m. In 1988 the area with waters deriving from the Norwegian Sea extended about 100 km westward from Jan Mayen, where a rather narrow front was formed towards the Polar Waters in the East Greenland Current, which in this area extended more than 300 km eastward from the coast of Greenland.

The distribution to 1000-m depth of potential temperature and salinity along 69.5°N is shown in Figure 13. The western end of the section was located on the Greenland Shelf about 40 km from the coast, while its eastern end was on the edge of the Iceland Plateau to the southeast of Jan Mayen (D in Figure 1).

Over the Greenland Shelf a 20–30-m-deep surface layer was seasonally warmed to almost 2°C, while the salinities

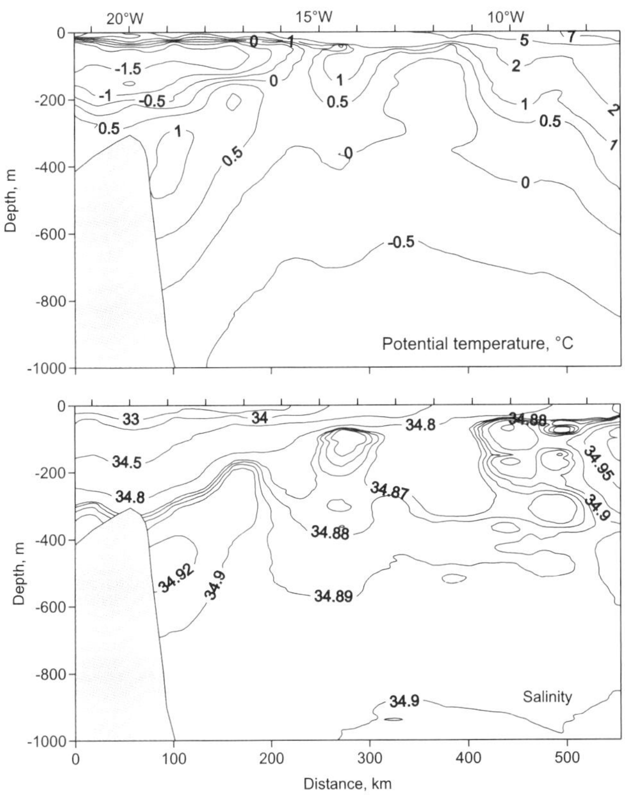

Figure 13. Potential temperature (°C) and salinity along 69.5°N in August 1988: D in Figure 1.

nearest to the coast were below 30. This surface layer was lying over a body of cold Polar Waters, extending more than 100 km eastward from the shelf edge with temperatures below −1.5°C between about 40 and 200 m along its entire extent. With salinities mostly between 34.4 and 34.7, this had the properties of Polar Intermediate Water [*Malmberg*, 1972]. The properties of this water may be maintained along the ice border during winter. The prevailing northerly winds will drive both newly formed ice and the cooled waters westward [*Johannessen et al.*, 1983], and the result at the end of the winter may have a structure similar to what is seen in Figure 13. In general, the effect of the winter cooling will reach deeper than the summer warming.

The recirculating Atlantic Water was observed just off the shelf break where it could be identified by salinities above 34.9 between 400 and 700 m depth at the slope and by temperatures up to 1.1°C around 350 m. Water warmer than +0.2°C was sloping up toward the centre of the section at about 14°W. Around this position was a 50-km-wide cell of water with temperatures up to 1.7°C and salinities slightly above 34.88 at 100 m. This was water deriving from the Norwegian Sea that was drawn into the cyclonic gyre in the eastern Iceland Sea, in the present case approximately following the 1500-m isobath. At the eastern end of the section the waters flowing in from the Norwegian Sea were identified by salinities in excess of 35.0 between 70 and 170 m.

In the centre of the gyre, relatively homogeneous water of temperatures below 0°C extended from the deep water to 100 m depth in a dome-like structure, somewhat similar to the dome in the central Greenland Sea Gyre. In the same area the salinities were between 34.80 at 50 m and 34.89 at 450 m. This might possibly be an area where winter convection could form new intermediate or deep waters. However, if the salinity of only 34.6 in the 30-m-deep surface layer remained also during the winter, convection would reach a depth of only 50 m if the surface layer cooled to the freezing point. According to *Jónsson* [1992], however, the wind-stress curl, which is strong and positive in this area during winter, gives rise to an increased cyclonic circulation and an upward Ekman pumping in the centre of the gyre. Such a circulation will promote a divergence in the centre of the gyre and a spreading of the low-salinity surface waters. The result will be a less stable surface layer, and deep convection may occur.

Figure 14 shows the distribution of potential temperature and salinity in a section across the Iceland Sea from Siglunes, North Iceland, into the southern Greenland Basin (E in Figure 1). Observations south of 68°N are from an Icelandic Siglunes Section that also was occupied in August 1988.

In the section, the relatively high temperatures in the North Icelandic Irminger Current can be seen over the Icelandic shelf. The salinities were highest between 100 and 300 m, up to 34.99. The salinities in the upper layer to about 50 m were

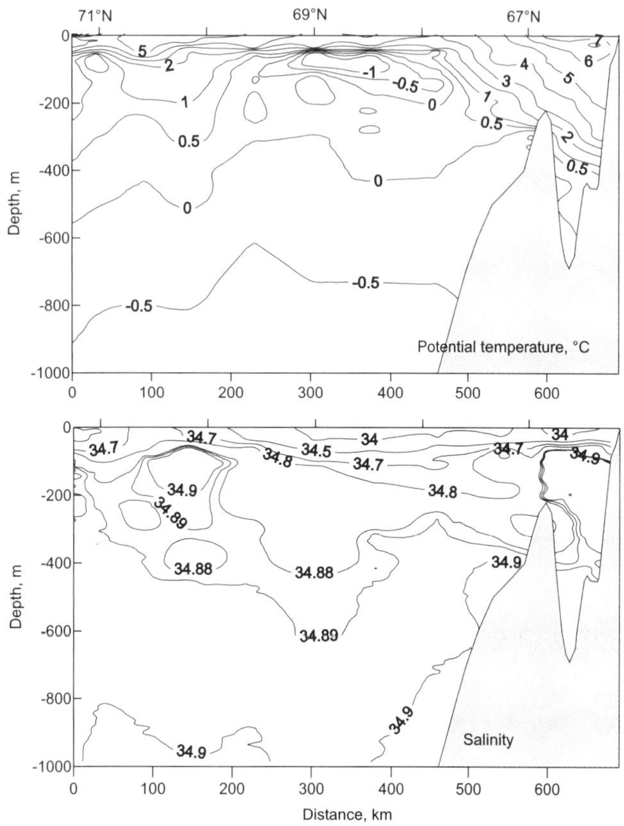

Figure 14. Potential temperature and salinity in a section from north of Siglunes, North Iceland, into the southern Greenland Sea in August 1988: E in Figure 1. Data from the Icelandic standard section are applied by courtesy of the Marine Research Institute, Reykjavik, Iceland.

below 34.5 from the Icelandic Coast to almost 70°N, and in the area of the East Icelandic Current the lowest near-surface salinities were close to 33.5. In this part of the section the surface layers were seasonally warmed to above 3°C, under which there was a sharp thermal transition layer to temperatures below −1°C, reaching −1.6°C as a minimum, between approximately 50 and 100 m depth. The coldest of these waters had salinities between 34.6 and 34.7 and may therefore be defined as Polar Intermediate Water.

North of the cold body of Polar Waters in the East Icelandic Current, conditions changed to higher temperatures, ranging between 0°C at about 500 m depth to above 5°C in the surface layer. Between 100 and 300 m deep, much of this water had salinities above 34.9, indicating its origin in the Norwegian Sea.

Most of the waters in the depth range 200–1000 m were Iceland Sea Arctic Intermediate Water, with potential temperatures below +1°C and salinities in the range 34.87–34.90. The deeper versions of the Iceland Sea Arctic Intermediate Water seem to be advected from the Greenland Sea, as also stated by *Rudels et al.* [2002]. The depth of its upper boundary layer increases from roughly 150 m at the northern end of Section E, to approximately 500 m at 69°N. When this water passes through the Iceland Sea, its properties are slightly changed by mixing. Finally it flows into the southwestern Norwegian Sea, where it is known as Norwegian Sea Arctic Intermediate Water [*Blindheim*, 1990; *Hansen and Østerhus*, 2000].

The deep waters in the Iceland Sea are not covered by the data set from 1988. It is, however, well documented that much of the upper portion of the Arctic Ocean Deep Water in the East Greenland Current flow across the Jan Mayen Fracture Zone and continue into the Iceland Sea and Denmark Strait [*Aagaard et al.*, 1991; *Buch et al.*, 1996; *Rudels et al.*, 2002]. Most of these deep waters flow into the East Iceland Current and characterise the deep waters in the Iceland Sea. This is shown in a section across the Iceland Sea from November 1994, from Langanes, northeastern Iceland, and mainly along 10°W into the Greenland Basin (Figure 15, F in Figure 1). The waters below approximately 1500 m have salinities similar to those in the core of Arctic Ocean Deep Water in the Jan Mayen Current that are shown in the same section north of the Jan Mayen plateau. These waters also form a slight salinity maximum around 1500 m in the southwestern Norwegian Basin.

Distribution of Dissolved Oxygen and Silicate

Figures 3, 4, 5, and 15 show the distribution of dissolved oxygen and silicate in the major basins within the Nordic Seas. Although we consider this information as peripheral, we include it because earlier information on the general distribution of nutrients and oxygen is rather sparse.

The source of high oxygen concentration is the central Greenland Sea, where winter ventilation, although variable, extends to the greatest depths. As a result, the oxygen concentration in the eastern Norwegian Sea is lower in the upper layers than in the intermediate water. For the same reason, the concentration of silicate is also lower in the Greenland Basin than in the other areas, particularly in the deep water.

In the deep Greenland Sea, where repeated observations are available, temporal variations are evident. During the latest two to three decades, the concentration of dissolved oxygen has been decreasing and that of silicate has been increasing. The reason for this is that the fraction of Arctic Ocean Deep Water, which has low oxygen and high silicate concentrations, has been increasing. Inflow deriving from the deep Arctic Ocean also contributes to relatively high silicate concentrations and low oxygen content in the deep waters in the Iceland Sea. This can also be traced as a minimum in oxygen at depths between approximately 1000 and 2000 m

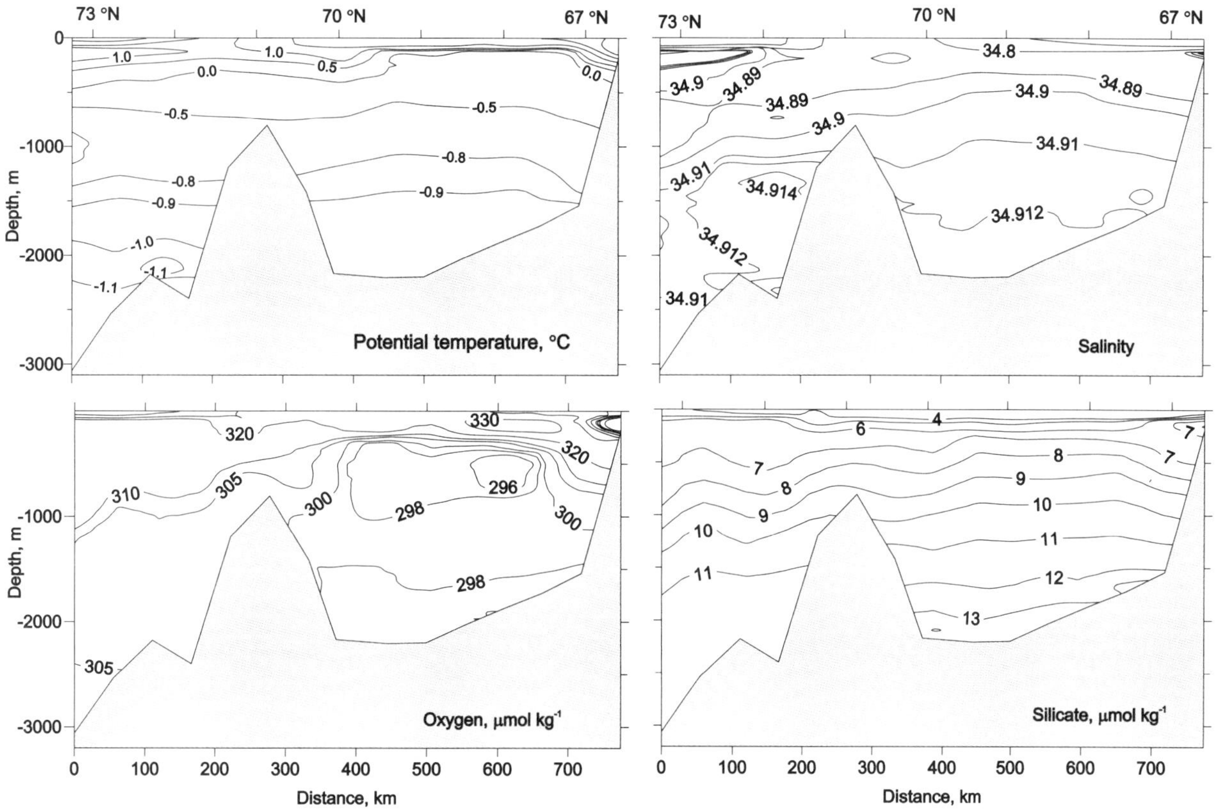

Figure 15. Potential temperature (°C), salinity, dissolved oxygen (μmol kg^{-1}), and silicate concentration (μmol kg^{-1}) in a section from Langanes, Northeast Iceland, into the Greenland Basin in November 1994: F in Figure 1.

in the western Norwegian Basin. A compilation of concentrations in the various basins is entered in Table 2.

OCEAN–ATMOSPHERE INTERACTIONS, TEMPORAL VARIATIONS

In the Nordic Seas, changes in the regional wind conditions create temporal variations in water mass distributions and properties over a wide range of time scales. It includes spatial phenomena on scales from local conditions along the coasts, to much larger systems such as the Norwegian Atlantic Current and the thermohaline circulation.

Jónsson [1991; 1992] analysed seasonal and interannual variations of wind-stress curl over the Nordic Seas and found correlation with variations of freshwater in the Iceland Sea. *Meincke et al.* [1992] expanded on this and correlated the wind-stress curl with decadal and interannual variations in the depth of the winter convection in the Greenland Sea.

The distribution and duration of the freshening in the Norwegian Sea further suggest associations with the NAO, and on a larger scale, with the Arctic Oscillation. The intensity of the westerlies and the storm tracks over the North Atlantic are connected to this system [e.g. *van Loon and Rogers*, 1978; *Hurrell*, 1995; *Rogers*, 1997]. Hence, for example, the width of the Norwegian Atlantic Current has varied in phase with the NAO. This is demonstrated in a Russian section along 65.75°N that has been repeated since the early 1960s. In the 1960s, when the NAO winter index was low, Atlantic Water was distributed across the entire Norwegian Basin. In contrast, when the NAO Index was high during the 1990s, Atlantic Waters deriving from the Norwegian Atlantic Current were hardly distributed west of the prime meridian. This difference in westward distribution amounts to more than 300 km [*Blindheim et al.*, 2000].

The effect of the NAO, or rather of the regional atmospheric circulation associated with it, was observed far north. Hence, the temperature fluctuations in the Sørkapp Section (Figure 6) is highly correlated with the NAO winter index, the correlation between their 3-year running means being 0.8 [*Blindheim et al.*, 2000; *Dickson et al.*, 2000]. Furthermore, the warming was stronger in the northern part of the Norwegian Sea than in its southern areas near the inflow—more evidence that the warming depends on the local atmospheric forcing. Simply narrowing the current when the NAO winter

Table 2. Approximate Compilation of Dissolved Oxygen and Silicate in the Norwegian, Lofoten, and Greenland Basins and on the Iceland Plateau.

Basin	Water layer	Oxygen, μmol kg^{-1}	Silicate, μmol kg^{-1}
Norwegian Basin (1996)	Upper	West: 300–310 East: <270	<6
	Intermediate	290–310	6–10
	Deep	295–296	10–13.2
Lofoten Basin (1999)	Upper	<290	<7
	Intermediate	300–310	7–10
	Deep	298–300	11–13.5
Greenland Basin (1999)	Upper	330–350	<8
	Intermediate	310–330	8–11
	Deep	305–310	11–12.5
Iceland Plateau (1994)	Upper	320–330	<6
	Intermediate	293–300	6–11
	Deep	<298	11–13.6

index is high may result in reduced cooling since a reduced surface of Atlantic Water is exposed to the atmosphere. Increased wind forcing associated with a high NAO Index also results in an increased volume transport in the slope current [*Orvik et al.*, 2001; *Orvik and Skagseth*, 2003].

An important consequence of this increased transport has been an increased heat and volume flux into the Arctic Ocean. This resulted in the volume increase and temperature rise in the boundary current and the intermediate Atlantic layer that recently has attracted much attention [*Quadfasel et al.*, 1991; *Carmack et al.*, 1995; *Swift et al.*, 1997; *Grotefendt et al.*, 1998; *Dickson et al.*, 2000; *Johannessen et al.*, 2004].

With a high NAO and a narrow Norwegian Atlantic Current, it is also likely that the circulation of Atlantic Water into the western areas of the basins is reduced. During the 1960s the westward extent of the Atlantic Water was at its largest; its westward distribution could reach the slope toward the Iceland Plateau, albeit with interannual variations. This included also the Atlantic Water that flows into the Iceland Sea south of Jan Mayen. Although there were considerable variations between years, the salinity was higher in the 1960s than during the 1990s. For the same reason it is also probable that the recirculation in the Fram Strait of Atlantic Water deriving from the West Spitsbergen Current was reduced during the same period.

Transport of Atlantic Water in the North Icelandic Irminger Current has also varied in response to the wind forcing. *Stefánsson* [1962] found correlation between the wind conditions at Stykkishólmur in May and the salinity in the coastal area east of Langanes (66.37°N, 13.58°W) in August of the same year. Later *Stefánsson and Gudmundsson* [1969] and *Ólafsson* [1999] reported on a similar correlation. *Jónsson* [1992] suggested that the mechanism by which freshwater is released from the East Greenland Current into the Iceland Sea is governed by the wind-stress curl over the area. Similarly, *Malmberg and Kristmannsson* [1992] found a positive correlation between the wind-stress curl west and north of Iceland and the water mass distribution north of Iceland 6 to 9 months later. *Blindheim and Malmberg* [2002; and this volume] investigated a century-long time series of the difference of winter (DJFM) mean sea-level pressure between Angmagssalik, East Greenland (65.60 °N, 37.63°W), and Stykkishólmur (65.08°N, 22.73°W) and correlated this to the time series of temperatures since 1950 at station S3 (66.5°N) on the Siglunes Section along 18.8°W. The two time series showed a fairly close inverse correlation, the highest correlation ($r = -0.73$) being obtained between 3-year moving averages of the temperature at 50 m depth at S3 and the winter mean sea level pressure (MSLP) gradient, 1 year delayed. This shows that the wind conditions play an important role in the bifurcation of the Irminger Current in the Denmark Strait, and consequently, in the supply of Atlantic Water to the North Icelandic Irminger Current. Although there are differences in response time, all these investigations suggest a correlation between wind conditions in the west Icelandic region and the volume of Atlantic Water carried into the North Icelandic shelf area.

This also has repercussions on the Norwegian Sea. The waters of Arctic character that are carried into the southwestern Norwegian Sea by the East Icelandic Current have three sources of salinity. These are the North Icelandic Irminger Current, the recirculating Atlantic Water in the East Greenland Current, and the waters flowing into the Iceland Sea to the south of Jan Mayen. Due to the prevailing atmospheric forcing, the supply of Atlantic Water to all these sources has been reduced since the 1960s. Thus, salinity in the southwestern Norwegian Sea has progressively reduced, as indicated by several time series [*Turrell et al.*, 1999; *Blindheim et al.*, 2000].

Another important aspect of ocean–atmosphere interaction is the transport of heat from the ocean to the atmosphere. In the Nordic Seas, the heat loss to the atmosphere, and the resulting cooling of the sea surface during the winter, may lead to deep convection. Traditionally the Greenland Sea is known as the area where the ultimate cooling may take place and where new bottom water may be formed by deep convection. This will also result in a homogenisation of the water column. Because an entirely homogeneous water column has never been observed in the Greenland Sea, some have argued that other mechanisms may lead to formation of the deep and bottom waters [*Metcalf*, 1955; *Carmack and Aagaard*, 1973]. Periods with renewal of the bottom water do, however, coincide with periods when the atmosphere over the Greenland Sea is very cold.

Aagaard [1968] compiled all relevant deep-water observations in the Greenland Sea until 1966. The data available indicated a cooling of the deep water from 1958 to 1966 but failed to show that the considerably increased cooling later in the 1960s resulted in a massive renewal of the deep and bottom waters. The low temperatures in the GSDW associated with the cooling during the 1960s was not demonstrated before *Clarke et al.* [1990] extended this examination of deep-water temperatures to cover the period from the 1950s through 1982. During the 1960s they found an annual temperature decrease of 0.005–0.010°C. After a minimum in 1972, the temperature in the GSDW again increased. The low temperatures in the deep Greenland Sea during the 1960s are also described by *Malmberg* [1983] and *Alekseev et al.* [2001].

A further extension of the time series with GSDW temperature is shown in Figure 16D, which shows that the warming has continued during the 1980s and 1990s. The temperatures plotted are averages between 2500 and 3400 m and may therefore be considered to be Greenland Sea Bottom Water since this deeper part of the basin is enclosed by the surrounding bathymetry and may be renewed only by water deriving from shallower depths. *Visbeck and Rhein* [2000] have discussed how associated diapycnal mixing may occur and attribute it to bottom boundary layer diffusivity.

Figure 16A shows normalised gradients of winter (JFM) MSLP across the Fram Strait, between Danmarkshavn, northeastern Greenland (76.77°N, 18.77°W), and Svalbard, West Spitsbergen (78.25°N, 15.47°E), as well as between Scoresbysund, East Greenland (70.48°N, 22.00°W), and Jan Mayen during the period 1950–2000. These plots of MSLP gradients show several common features. During the second half of the 1960s, both gradients show a maximum for the period 1950–2000. These strong gradients were associated with prevailing, relatively strong, northerly winds over the Greenland Sea. This wind pattern forced the fresh surface waters of the East Greenland Current toward the Greenland

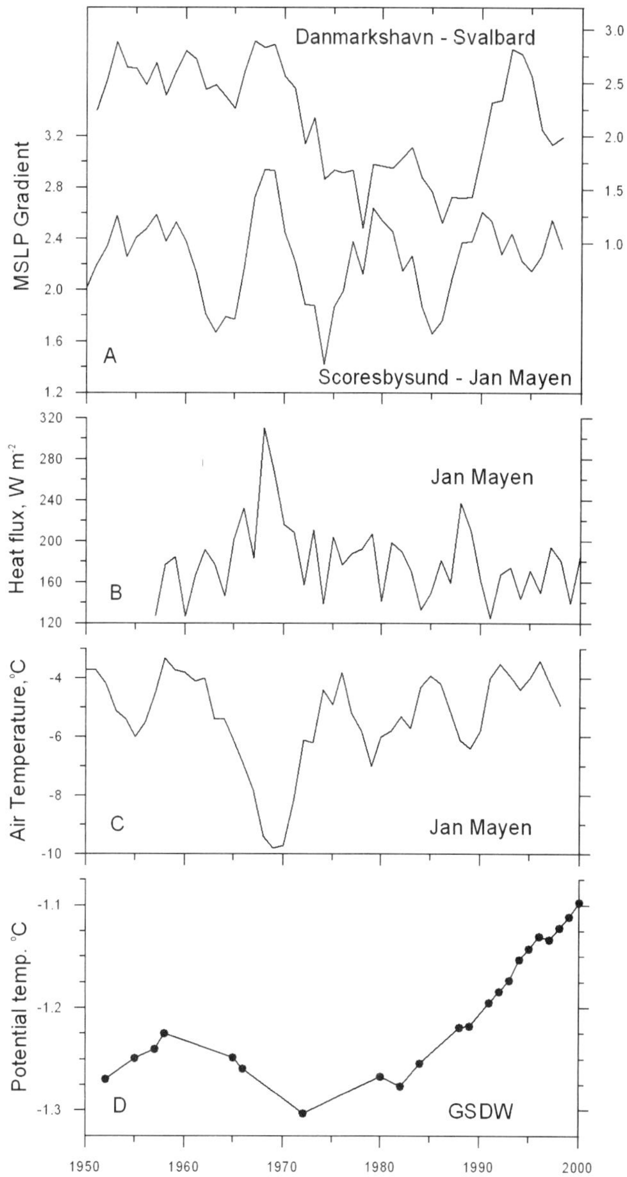

Figure 16. (A) Five-year running means of the normalised winter (JFM) mean sea-level pressure gradient between Danmarkshavn, 78.8°N, 18.8°W, and Svalbard, 78.3°N, 15.5°E (upper curve), and between Scoresbysund, 70.5°N, 22.0°W, and Jan Mayen, 70.9°N, 8.7°W (lower curve). (B) Winter means (DJFMA) of net heat flux based on observations at Jan Mayen. (C) Three-year moving averages of winter means (JFM) of air temperatures at Jan Mayen. (D) Mean temperatures for stations between 74° and 75°N, and 1°E and 1°W, averaged between 2500 and 3400 m deep in the central Greenland Basin.

Shelf at the same time as it promoted supply from the east of saltier water to the central Greenland Basin. As shown by the mean winter temperatures (DJFM) at Jan Mayen (Figure 16C) this period was also by far the coldest in the period.

Although the distance between Jan Mayen and the central Greenland Sea is almost 450 km, seasonal means of the observations at Jan Mayen seem to be representative for the Greenland Sea region. It seems therefore reasonable to conclude that the winter-means represent conditions on a regional scale. Another indication is that the time series of air temperatures observed at Stykkishólmur, West Iceland, and the time series from Jan Mayen (Figure 17) have features in common, even though they also show differences. The cold period in the late 1960s is clearly seen in both, although it was more conspicuous at Jan Mayen than at Stykkishólmur.

One of the coldest winters at Stykkishólmur during the almost 200-year period from 1823 to 2000 occurred in January 1918. The mean winter temperature (DJFMA) was then –3.90°C (JFM, –5.4°C). During the five following years, however, the winter mean increased to +0.96°C. This was the beginning of the warmest period in the time series, lasting for four decades from the early 1920s. Hence, the average winter mean for the period 1923–1963 was +0.24°C. In comparison, the average winter mean for the cold period lasting from 1850 to 1919 was –1.41°C. The winter means during the late 1960s were similar to those before 1920. The extremely low temperatures observed in the Greenland Sea in 1901, when the potential temperature at 2000 m deep was –1.4°C [*Nansen*, 1906], seems therefore to be associated with the long preceding period of cold atmosphere over the East Greenland area.

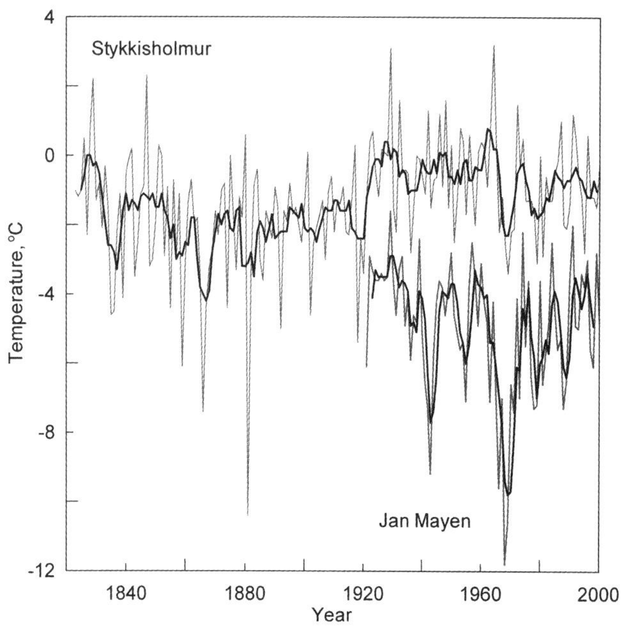

Figure 17. Five-year moving averages of winter means (JFM) of air temperature at Stykkisholmur, Iceland, and Jan Mayen. Data are from the Icelandic Meteorological Office and the Norwegian Meteorological Institute, respectively.

Figure 16B shows seasonal means (DJFMA) of total heat flux at Jan Mayen. During this period the winter mean of net heat flux varied between 125 and 311W m^{-2}. The highest heat loss from the sea surface occurred in 1968. As shown in Figure 16C, this was at the extreme of the cold period in the late 1960s.

Low salinities in the upper layers of the Greenland Sea will often prohibit deep water formation. Even after formation of heavy ice, the salinity in the layers affected by the convection may remain so low that the water will not be dense enough, even at freezing point, to sink into deeper layers. In the central Greenland Sea the relationship between the wind conditions and the surface salinity is tight, and prevailing northerly winds give rise to high upper-layer salinities [*Blindheim and Rey*, 2004]. In November of 1994 and 1995, this resulted in upper-layer salinities close to 34.9 in the central Greenland Sea. These relatively high salinities should be favourable for deep convection, but no indications of deep water formation were noted. Neither was there recorded any ice in the central Greenland Sea during these winters, the ice edge having been observed near the Greenland Slope during all winters from 1991 to 1996. The heat flux from the sea surface was simply not sufficient to cool enough water to result in deep convection. During winter 1994–1995, the December–April mean of the heat flux, based on the observations at Jan Mayen, was 170 W m^{-2}; during winter 1995–1996 it was even less, 150 W m^{-2}. In comparison, from 1965 to 1972 the average winter heat flux was 232 W m^{-2} and in 1968 it was 310 W m^{-2}.

In general, these considerations indicate that intense formation of deep water is occurring rather rarely. Probably, not much deep water was formed during the period from 1920 to the late 1950s. Similarly, very little deep water has been formed since the late 1960s.

SEA-ICE

Concentrations of sea-ice conditions in the Nordic Seas are discussed by *Kvingedal* and by *Gerdes et al.* [this volume], here we offer a brief overview. The northern and western parts of the Nordic Seas are to varying extents covered by sea-ice during the winter, usually with greatest extent in March–April. Some ice is formed locally, but to a larger extent it is efflux from the Arctic Ocean. The Fram Strait is the main gate for Arctic ice export, and the sea-ice carried by the East Greenland Current is the largest and most concentrated meridional ice flow in the World Ocean. The export of ice and liquid freshwater from the Arctic Ocean through the Fram Strait represents the major source of freshwater in the Nordic Seas [*Aagaard and Carmack*, 1989]. The amount and distribution of ice and freshwater play a crucial role in the deep convection in the Greenland and Iceland Seas. The flux of sea-ice through the Fram Strait is strongly correlated with the local wind [*Widell et al.*, 2003]. The distribution in the Nordic Seas depends on the regional pressure field and the underlying current system. Figure 18 shows the median ice border at the end of April for the periods 1963–1969 and 1989–1995, which were characterised with a minimum and a maximum, respectively, of the NAO winter index.

Explorers, whalers, sealers and fishermen have operated in the northern Nordic Seas for more than 400 years. Willem Barentsz was the first to provide information on ice conditions in the northern Barents Sea when he discovered Spitsbergen in 1596 [*de Veer*, 1609]. Observations became more frequent when whaling and sealing started early in the 17th century [*Vinje*, 2001]. In 1881 the sea-ice covered the Greenland, Iceland and Barents Seas completely and the northern part of the Norwegian Sea (Figure 18). This coincided with the lowest mean winter (JFM) temperature on record (Figure 17).

PROSPECTS

During the 20th century, the variability in the Nordic Seas has been considerable on a range of time scales. The temperature of the GSDW was 0.2°C higher in the late 1950s than in 1901. During the 1960s it decreased by 0.1°C but since then has increased, to −1.038°C in 2004. This warming results from reduced formation of GSDW and a more constant supply of deep water from the Arctic Ocean. If the situation with negligible formation of deep water in the Greenland Sea continues, it will take about 20 years before the properties of the deep and bottom waters in the Greenland Sea are practically identical to the properties of the Arctic Ocean Deep Water. Somewhat later, the Norwegian Sea Deep Water will get the same properties. It is therefore important to monitor the properties of the deep waters in the Nordic Seas. This need not involve more fieldwork than an annual single section across the major basins, but it does require the highest possible observational accuracy. New events with considerable formation of GSDW will occur during winters with a relatively cold atmosphere over the Greenland Sea, which may be indicated by winter means of air temperatures at Jan Mayen.

A different great change in the Nordic Seas is the increasing upper-layer freshening that has been observed since the 1960s, expanding eastward from the western areas. During the same period a narrowing of the Norwegian Atlantic Current has been observed [*Blindheim et al.*, 2000]. This has further been associated with higher current velocities and an increased supply of Atlantic Water to the Arctic Ocean [*Orvik et al.*, 2001]. Although this change in distribution of water masses has shown a correlation to the regional wind pattern, the freshening may not necessarily result directly from the wind stress only. It may also be due to an indirect effect that occurs as increased supply of freshwater from the Arctic Ocean. The supply seems to be increased both through the Canadian archipelago to the Labrador Sea and through the Fram Strait to the Nordic Seas [*Visbeck et al.*, 2003]. There are several possible reasons for such increased supply of freshwater from the Arctic Ocean, such as increased supply of freshwater via the Norwegian Coastal Current, increased precipitation over the basins of the Siberian rivers, and increased melting of ice, not only multiyear ice but possibly also, to a large extent, new ice. An opposite possibility may be associated with reduced deep water formation on the shelves, that may bring about an increased supply of relatively fresh waters to the mixed layer in the Arctic Ocean.

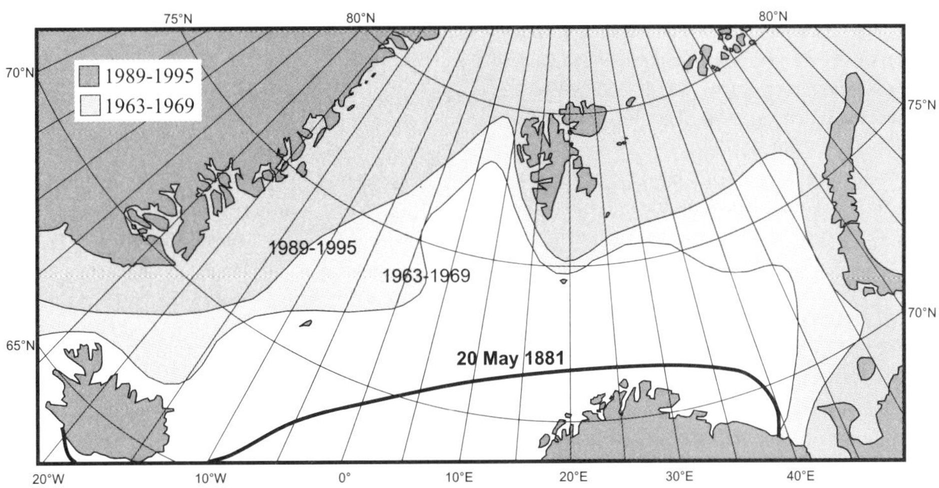

Figure 18. The sea-ice cover at the end of April for the periods 1963–1969 and 1989–1995 with a minimum and a maximum circulation index, respectively, of the North Atlantic Oscillation (NAO). Also shown is the extreme sea-ice extension from 1881.

Whatever the reasons may be, continuous monitoring of the exchanges of heat, saltwater and freshwater between the Nordic Seas and their surroundings seems important.

Acknowledgments. We would like to thank Karen Gjertsen for her skillful help in drawing the figures. Thanks are also due to the referees for their useful comments to the manuscript.

REFERENCES

Aagaard, K., Temperature variations in the Greenland Sea deep-water. *Deep-Sea Research*, 15, 281-296, 1968.

Aagaard, K., and E.C. Carmack, The role of the sea ice and other fresh water in the Arctic circulation. *Journal of Geophysical Research*, 94, C10: 14484-14498, 1989.

Aagaard, K., and P. Greisman, Towards new mass and heat budgets for the Arctic Ocean. *Journal of Geophysical Research*, 80, 3821-3827, 1985.

Aagaard, K., J.H. Swift, and E.C. Carmack, Thermohaline circulation in the Arctic Mediterranean Seas. *Journal of Geophysical Research*, 90, C3, 4833-4846, 1985.

Aagaard, K., E. Fahrbach, J. Meincke, and J.H. Swift, Saline outflow from the Arctic Ocean: Its contribution to the deep waters of the Greenland, Norwegian and Iceland seas. *Journal of Geophysical Research*, 96, 20433-20441, 1991.

Ådlandsvik, B., and H. Loeng, A study of the climate system in the Barents Sea. *Polar Research*, 10, 1, 45-49, 1991.

Åkerblom, F., Recherches océanographiques. Expedition de M.A.G. Nathorst en 1899. *Upsala Universitets Årsskrift 1903. Math. Och Naturvetenskap*, II, 80 pp., 1904.

Alekseev, G.V., O.M. Johannessen, A.A. Korbalev, V.V. Ivanov, and D.V. Kovalevsky, Interannual variability in water masses in the Greenland Sea and adjacent areas. *Polar Research* 20, 2, 201-208, 2001.

Balakshin, L.L., The water circulation and bottom contour of the northern part of the Greenland Sea. *Reprints of abstracts. International Oceanographic Congress, New York*, 430-431, 1959.

Blindheim, J., Cascading of Barents Sea bottom water into the Norwegian Sea. *Rapports et Procès-Verbaux des Reunions. Conceil International Pour L'Exploration de la Mer*, 188, 48-58. 1989.

Blindheim, J., Arctic intermediate water in the Norwegian Sea. *Deep-Sea Research*, 37, 1475-1489, 1990.

Blindheim, J., V. Borovkov, B. Hansen, S.-Aa. Malmberg, W.R. Turrell, and S. Østerhus, Upper layer cooling and freshening in the Norwegian Sea in relation to atmospheric forcing. *Deep-Sea Research I*, 47, 655-680, 2000.

Blindheim, J., and S.-Aa. Malmberg, The mean sea level Pressure (MSLP) gradient across the Denmark Strait as index for the Oceanographic conditions in North-Icelandic waters. *ICES, CM2002/O:20*, 16 pp, 2002.

Blindheim, J., and F. Rey, Water-mass formation and distribution in the Nordic Seas during the 1990s. *ICES Journal of Marine Science*, 61, 846-863, doi:10.1016/j.icesjms.2004.05.003, 2004.

Boech, A., Om silden og sildefiskerierne, Navnlig om det Norske Vaarsildfisket. Christiania, 136 pp. (in Danish), 1871.

Bönisch, G., and P. Schlosser, Deep water formation and exchange rates in the Greenland/Norwegian Seas and the Eurasian Basin of the Arctic Ocean derived from tracer balances. *Progress in Oceanography*, 35, 29-52, 1995.

Bönisch, G., J. Blindheim, J.L. Bullister, P. Schlosser, and D.W.R. Wallace, Long-term trends of temperature, salinity, density and trancient tracers in the central Greenland Sea. *Journal of geophysical Research*, 102, C8, 18553-18571, 1997.

Borovkov, V., and A. Krysov, Report from survey of distributions of Norwegian spring spawning herring and environmental conditions in the Norwegian Sea in June-July 1995. *Report from a joint meeting of Russian and Icelandic scientists in Reykjavik*, 8 pp., 1995.

Bøyum, G., A study of evaporation and heat exchange between the sea surface and the atmosphere. *Geofysiske publikasjoner, Geophysica Norwegica*, 22, 7, 15 pp., 1962.

Bøyum, G., The energy exchange between sea and atmosphere at Ocean Weather Stations M, I and A. *Geofysiske publikasjoner, Geophysica Norwegica*, 26, 7, 19 pp., 1966.

Buch, E., S.-Aa. Malmberg, and S.S. Kristmannsson, Arctic Ocean deep water masses in the western Iceland Sea. *Journal of Geophysical Research*, 101, 11965-11973, 1996.

Budeus, G., W. Schneider and G. Krause, Winter convection events and bottom water warming in the Greenland Sea. *Journal of Geophysical Research*, 103, C9, 18513-18527, 1998.

Carmack, E., and K. Aagaard, On the deep water of the Greenland Sea. *Deep-Sea Research*, 20, 687-715, 1973.

Carmack, E.C., R.W. Macdonald, R.G. Perkin, F.A. MCLaughlin, and R.J. Pearson, Evidence for warming of Atlantic water in the southern Canadian Basin of the Arctic Ocean: Results from the Larsen-93 expedition. *Geophysical Research Letters*, 22, 1061-1064, 1995.

Clarke, R.A., J.H. Swift, J.L. Reid and K.P. Koltermann, The formation of Greenland Sea Deep Water: Double diffusion or deep convection? *Deep-Sea Research*, 37, 9, 1385-1424, 1990.

Colding, A., Om strømningsforhold i almindelige ledninger og i havet. (in Danish), Copenhagen, 1870.

De Veer, G., The true and perfect description of three voyages, so strange and wonderfull that the like hath never been heard before. T. Pauier, London, 1609.

Devold, F., Herring—Norwegian Sea. *Annales Biologiques*, Copenhagen, 10, 167-171, 1953.

Dickson, R.R., J. Meincke, S.-Aa. Malmberg, and A.J. Lee (1988), The "Great Salinity Anomaly" in the northern North Atlantic 1968-1982. *Progress in Oceanography*, 20, 103-151, 1988.

Dickson, R.R., T.J. Osborn, J.W. Hurrell, J. Meincke, J. Blindheim, B. Adlandsvik, T. Vinje, G. Alekseev, and W. Maslowski, The Arctic Ocean response to the North Atlantic Oscillation. *Journal of Climate*, 13, 15, 2671-2696, 2000.

Dietrich, G., Atlas of the hydrography of the northern North Atlantic Ocean. Based on the Polar Front Survey of the International Geophysical Year winter and summer 1958. *ICES, Copenhagen*, 140 pp., 1969.

Fahrbach, E., J. Meincke, S. Østerhus, G. Rohardt, U. Schauer, V. Tverberg and J. Verduin, Direct measurements of volume transports through Fram Strait, *Polar Research*, 20, 217-22, 2001.

Fissel, D.B., J.R. Birch, H. Melling, and R.A. Lake, Non-tidal flows in the Northwest Passage. Canadian Institute of Ocean Science. Sidney, British Columbia, Canada. Tech. Rep. Hydrogr. Ocean Sci. 98, 157pp., 1988.

Fogelqvist, E., J. Blindheim, T. Tanhua, S. Østerhus, E. Buch, and F. Rey, Greenland-Scotland overflow studied by hydro-chemical multivariate analysis. *Deep-Sea Research I*, 50, 73-102, 2003.

Foldvik, A., K. Aagaard, and T. Tørresen, On the velocity field of the East Greenland Current, *Deep-Sea Research*, 35, 1335-1354, 1988.

Gascard, J.-C., A.J. Watson, M.J. Messias, K.A. Olsson, T. Johannessen, and K. Simonsen, Long-lived vortices as a mode of deep ventilation in the Greenland Sea. *Nature*, 416, 525-527, 2002.

Grotefendt, K., K. Logemann, D. Quadfasel, and S. Ronski, Is the Arctic Ocean warming? *Journal of Geophysical Research* 103, 27679-27687, 1998.

Hamberg, A. Hudrographische Arbeiten der von A.G. Nathorst geleiteten schwedischen Polarexpedition 1898. *Kongl. Svenska Vet.-Akad. Handlingar*, 41, 1. Stockholm, 1906.

Hansen, B., and S. Østerhus, editors, North Atlantic–Norwegian Sea Exchanges: The ICES NANSEN Project. *ICES Cooperative Research Report*, 225, 246 pp., 1998.

Hansen, B., and S. Østerhus, North Atlantic–Nordic Seas exchanges. *Progress in Oceanography*, 45, 109-208, 2000.

Hansen, B., S. Østerhus, H. Hátún, R. Kristiansen, and K.M.H. Larsen, The Iceland-Faroe inflow of Atlantic water to the Nordic Seas. *Progress in Oceanography*, 59, 443-474, 2003.

Helland-Hansen, B., Report on hydrographical investigations in the Faroe-Shetland Channel and the northern part of the North Sea in 1902. In *Report on Fishery and Hydrographical Investigations in the North Sea and Adjacent Waters*, edited by D.W. Thompson, pp.1-50 Fisheries Board of Scotland, Oliver and Boyd, Edinburgh, 1905.

Helland-Hansen, B., and F. Nansen, The Norwegian Sea, its physical oceanography based upon the Norwegian Researches 1900-1904. *Report on Norwegian and Marine Investigations*, 2, 2, 390 pp., 1909.

Hopkins, T.S., The Gin Sea–A synthesis of its physical oceanography and literature review 1972-1985. *Earth-Science Reviews*, 30, 175-318, 1991.

Hurdle, B.G. (editor), *The Nordic Seas*, Springer, New York., 1986.

Hurrell, J.W., Decadal trends in the North Atlantic Oscillation, regional temperatures and precipitation. *Science*, 269, 676-679, 1995.

Ingvaldsen, R., L. Asplin, and H. Loeng, The seasonal cycle in the Atlantic transport to the Barents Sea during the years 1997–2001. *Continental Shelf Research*, 24, 1015-1032, doi:10.1016/j.csr.2004.02.011, 2004.

Irminger, C., Strømninger og isdrift ved Island. (in Danish). *Tidsskrift for Søvæsen*. Copenhagen, 1861.

Jakobsen, K.P., M.H. Ribergaard, D. Quadfasel, T. Schmith, and C.W. Huges, Near-surface circulation in the northern North Atlantic as inferred from Lagrangian drifters: Variability from the mesoscale to interannual. *Journal of Geophysical Research*, 108, C8, 3251, doi:10.1029/2002JC001554, 2003.

Jansen, E., and V. Opheim (editors), European Sub Polar Ocean Programme, ESOP-2. The thermohaline circulation in the Greenland Sea. Final scientific report. European Commission Mast III Programme, 1999.

Johannessen, J.A., O.M. Johannessen, E. Svendsen, R. Shuchman, T. Manley, W.C. Campbell, E.G. Josberger, S. Sandven, J.C. Gascard, T. Olaussen, K. Davidson, and J. Van Leer, Mesoscale eddies in the Fram Strait marginal ice zone during the 1983 and 1984 Marginal Ice Zone Experiments. *Journal of Geophysical Research*, 92, C7, 6754-6772, 1987.

Johannessen, O.M., S. Sandven, and J.A. Johannessen, Eddy-related winter convection in the Boreas Basin. In Deep Convection and Deep Water Formation in the Oceans, Edited by P.C. Chu, and J.C. Gascard. *Elsevier Oceanography Series*, 57, 87-105, 1991.

Johannessen, O.M., J.A. Johannessen, J. Morison, B.A. Farrelly and E.A. Svendsen, Oceanographic conditions in thr marginal ice zone north of Svalbard in early fall 1979 with emphasis on mesoscale processes. *Journal of Geophysical Research*, 88, C5, 2755-2769. 1983.

Johannessen, O.M., W.J. Campbell, R. Shuchman, S. Sandven, P. Gloersen, J.A. Johannessen, E.G. Josberger, and P.M. Haugan, Microwave study programs of air-ice-ocean interactive processes in the seasonal ice zone of the Greenland and Barents Seas. *A.G.U. Geophysical Monograph*, 68, 261-289, 1992.

Johannessen, O.M., L. Bengtsson, M.W. Miles, S.I. Kuzmina, V.A. Semenov, G.V. Alekseev, A.P. Nagurnyi, V.F. Zakharov, L.P. Bobylev, L.H. Petterson, K. Hasselmann, and H.P. Cattle, Arctic climate change—observed and modelled temperature and sea ice variability. *Tellus*, 56A, 1-18, 2004.

Jónsson, S., Seasonal and interannual variability of wind stress curl over the Nordic Seas. *Journal of Geophysical Research*, 96 (C2), 2649-2659, 1991.

Jónsson, S., The sources of freshwater in the Iceland Sea and the mechanisms governing its interannual variability. *ICES Marine Science Symposia*, 195, 62-67, 1992.

Jónsson, S., and J. Briem, Flow of Atlantic water west of Iceland and onto the north Icelandic shelf. *ICES Marine Science Symposia*, 219, 326-328, 2003.

Jónsson, S., and H. Valdimarsson, A new path for the Denmark Strait overflow water from the Iceland Sea to Denmark Strait. *Geophysical Research Letters*, 31, L03305, doi:10.1029/2003-GL019214, 2004.

Kiilerich, A., On the hydrography of the Greenland Sea. *Meddelelser om Grønland*, 144, 2, 63 pp., Copenhagen, 1945.

Knipovich, N., Hydrologische Untersuchungen in Europäischen Eismeer. *Annalen Hydrogr*. Berlin, 33, 289-308, 1905.

Knudsen, M., Hydrografi. *Den Danske Ingolf-expedition*, København, 1, 2, 21-154, 1898.

Kristmannsson, S.S., Flow of Atlantic Water into the northern Icelandic shelf area, 1985-1989, in The ICES NANSEN project, edited by B. Hansen and S. Østerhus. *ICES Cooperative Research Report*, 225, 124-135, 1998.

Krümmel, O., *Versuch einer vergleichenden morphologie der meeresreume*. Verlag von Duncker and Humbolt, Leipzig. 110 pp., 1879.

Lee, A.J., The hydrography of the European Arctic and Subarctic Seas, in *Oceanogr. Mar. Biol. Ann. Rev.*, 1, 47-76, edited by Harold Barnes. George Allen and Unwin, London, 1963.

Malmberg, S.-Aa., Hydrographic changes in the waters between Iceland and Jan Mayen in the last decade. *Jøkull*, 19, 30-43, 1969.

Malmberg, S.-Aa., Intermediate Polar Water in the Denmark Strait—"overflow" August 1971. *ICES CM1972/C:6*, 6 pp., 1972.

Malmberg, S.-Aa., Hydrographic investigations in the Iceland and Greenland Seas in late winter 1971—"Deep Water project", *Jökull*, 33, 133-140, 1983.

Malmberg, S.-Aa., J. Ólafsson, and S.S. Kristmannsson, Hydrographic conditions in the Iceland Sea in late winter 1971, 1975 and 1982. *ICES, CM1986/C:23*, 20 pp., 1986.

Malmberg, S.-Aa., and S.S. Kristmannsson, Hydrographic conditions in Icelandic waters, 1980-1989. *ICES Marine Science Symposia*. 195, 76-92, 1992.

Manley, T.O., K.L. Hunkins, and R.D. Muench, Current regimes across the East Greenland Polar Front at 78°40′ North latitude during summer 1984. *Journal of Geophysical Research*, 92, C7, 6741-6753, 1987.

Meincke, J., The modern current regime across the Greenland-Scotland Ridge, in *Structure and development of the Greenland-Scotland Ridge, new methods and concepts*, edited by M.H.P. Bott, S. Saxov, M. Talwani and J. Thiede. *Proc. NATO Adv. Res. Inst., Padua Univ.*, 11-15, Plenum Press, N.Y. 1983.

Meincke, J., S. Jónsson, and J.H. Swift, Variability of convective conditions in the Greenland Sea. *ICES Marine Science Symposia*, 195, 32-39, 1992.

Meincke, J., B. Rudels, and H. Friedrich, The Arctic Ocean-Nordic Seas Thermohaline System. *ICES Journal of Marine Science*. 54, 283-299, 1997.

Metcalf, W.G., On the formation of bottom water in the Norwegian Basin. *Transactions, American Geophysical Union*, 36, 4, 596-600, 1955.

Midttun, L., Formation of dense bottom water in the Barents Sea. *Deep-Sea Research*, 32, 1233-1241, 1985.

Mohn, H., The North Ocean, its Depths, Temperature and Circulation. *The Norwegian North-Atlantic Expedition 1876-1878*, 2, 2, 209 pp. Grøndahl and Son, Christiania, 1887.

Mork, K.A., and J. Blindheim, Heat loss of the Norwegian Atlantic Current toward the Arctic. *ICES Marine Science Symposia*, 219, 144-149, 2003.

Mosby, H., Deep water in the Norwegian Sea. *Geofysiske Publikasjoner, Geophysica Norvegica*, 21, 3, 62 pp., 1959.

Mosby, H., Atlantic Water in the Norwegian Sea. *Geofysiske publikasjoner, Geophysica Norvegica*, 28, 1, 60 pp., 1970.

Nansen, F., Northern waters.*Videnskabs-Selskabets Skrifter, I. Matematisk-Naturvidenskabelig Klasse*, 3, 145 pp., 1906.

Ólafsson, J., Connections between oceanic conditions off N-Iceland, Lake Myvatn temperature, regional wind direction variability and the North Atlantic Oscillation. *Rit Fiskideildar*, 16, 41-57, 1999.

Orvik, K.A., Ø. Skagseth, and M. Mork, Atlantic inflow to the Nordic Seas: current structure and volume fluxes from moored current meters, VM-ADCP and SeaSoar-CTD observations, 1995-1999. *Deep-Sea Research I*, 48, 937-957, 2001.

Orvik, K.A., and Ø. Skagseth, The impact of the wind stress curl in the North Atlantic on Atlantic inflow to the Norwegian Sea toward the Arctic. *Geophysical Research Letters*, 30, 17, 1884, doi:10.1029/2003GL 017932, 2003.

Østerhus, S., and T. Gammelsrød, (1999), The abyss of the Nordic Seas is warming. *Journal of Climate*, Vol. 12, No. 11, 3297-3304, 1999.

Østerhus, S., W.R. Turrell, B. Hansen, P. Lundberg, and E. Buch, Observed transport estimates between the North Atlantic and the Arctic Mediterranean in the Iceland-Scotland region. *Polar Research*, 20 (1), 169-175, 2001.

Peterson, W.H., and C.G.H. Rooth, Formation and exchange of deep water in the Greenland and Norwegian Seas. Deep-Sea Research, Vol. 23, 273-283, 1976.

Poulain, P.-M., A. Warn-Varnas, and P. Niiler, Near-surface circulation of the Nordic Seas as measured by Lagrangian drifters. *Journal of Geophysical Research*, 101, 18237-18258, 1996.

Quadfasel, D., and J. Meincke, Note on the thermal structure of the Greenland Sea gyres. *Deep-Sea Research*, Vol. 34, No. 11, 1883-1888, 1987.

Quadfasel, D., J.-C. Gascard, and K.-P. Koltermann, Large scale oceanography in Fram Strait during the 1984 Marginal Ice Zone Experiment. *Journal of Geophysical Research*, 92, C7, 6719-6728, 1987.

Quadfasel, D., B. Rudels, and K. Kurz, Outflow of dense water from a Svalbard fjord into Fram Strait. *Deep-Sea Research*, 35. 7, 1143-1150, 1988.

Quadfasel. D., A. Sy, D. Wells, and A. Tunik (1991), Warming in the Arctic. *Nature*, 350, 385, 1991.

Read, J.F., and R.T. Pollard, Water masses in the region of the Iceland Faeroes Front. *Journal of Physical Oceanography*, 22, 1365-1378, 1992.

Roach, A.T., K. Aagaard, C.H. Pease, S.A., Salo, T. Weingartner, V. Pavlov, and M. Kulakov, Direct measurements of transport and water properties through the Bering Strait. *Journal of Geophysical Research*, 100, C9, 18443-18457, 1995.

Rogers, J.C., North Atlantic storm track variability and its association to the North Atlantic Oscillation and climate variability of northern Europe. *Journal of Climate*, 10, 1635-1647, 1997.

Ross, C.K., Temperature-salinity characteristics of the overflow in Denmark Strait during OVERFLOW `73. *Rapports et processverbaux des Réunions, Conseil international pour l'Exploration de la Mer*, 185, 111-119, 1984.

Rudels, B., H.J. Friedrich, and D. Quadfasel, The Arctic circumpolar boundary current. *Deep-Sea Research II*, 46, 1023-1062, 1999.

Rudels, B., E. Fahrbach, J. Meincke, G. Budéus, and P. Eriksson, The East Greenland Current and its contribution to the Denmark Strait overflow. *ICES Journal of Marine Science*, 59, 1133-1154, 2002.

Ryder, C., Den östgrönlandske Expedition 1891-1892, Hydrografi. *Meddelelser om Grönland*, 17. Copenhagen, 1895.

Ryder, C., Some investigations relating to the ocean currents in the sea between Norway, Scotland, and Greenland. *Nautical meteorological Annual of the Danish Meteorological Institute*, Copenhagen, 1901.

Sandven, S., O.M. Johannessen, M.M. Miles, L.H. Petterson, and K. Kloster, Barents Sea seasonal ice zone features and processes from ERS 1 synthetic aperture radar: Seasonal Ice Zone

Experiment 1992. *Journal of geophysical Research*, 104, C7, 15843-15857, 1999.

Sars, G.O., *Indberetninger til Departementet for det Indre fra Professor G. O. Sars om de af ham i Aarene 1864-1878 Anstillede Undersøgelser angaaende Saltvandsfiskeriene.* Christiania 221 pp., 1879.

Sars, G.O., *The Norwegian North-Atlantic Expedition 1876-1878, Zoology, Crustacea, I.* Grøndahl and Son, Christiania: 280 pp., 1885.

Schauer, U., E. Fahrbach, S. Østerhus, and G. Rohardt, Arctic warming through the Fram Strait—oceanic heat transport from three years of measurements. *Journal of Geophysical Research*, 109, C06026, doi: 40.1029/203JC001823, 2004.

Schlosser, P.G., G. Bönisch, M. Rhein, and R. Bayer, Reduction of deepwater formation in the Greenland Sea during the 1980s: Evidence from tracer data. *Science*, 251, 1054-1056, 1991.

Smethie, W.M., H.G. Ostlund, and H.H. Loosli, Ventilation of the deep Greenland and Norwegian seas: evidence from krypton-85, tritium, carbon-14 and argon-39. *Deep-Sea Research*, 33, 5, 675-703, 1986.

Stefánsson, U., North Icelandic waters. *Rit Fiskideildar*, 3, 269 pp., 1962.

Stefánsson, U., and G. Gudmundsson, Hydrographic conditions off the north-east coast of Iceland in relation to meteorological factors. *Tellus*, 21, 2, 245-258, 1969.

Strass, V.H., E. Fahrbach, U. Shauer, and L. Sellmann, Formation of Denmark Strait overflow water by mixing in the East Greenland Current. *Journal of Geophysical Research*, 98, C4, 6907-6919, 1993.

Svendsen, E., K. Kloster, B. Farrelly, O.M. Johannessen, J.A. Johannessen, W.J. Campbell, P. Gloersen, D. Cavalieri, and C. Mätzler. Norwegian Remote Sensing Experiment: Evaluation of the Nimbus 7 Scanning Multichannel microwave radiometer for sea ice research. *Journal of Geophysical Research*, 88, C5, 2781-2791, 1983.

Swift, J.H., K. Aagaard, and S.-Aa. Malmberg, The contribution of the Denmark Strait overflow to the deep North Atlantic. *Deep-Sea Research*, 27A, 29-42, 1980.

Swift, J.H., and K. Aagaard (1981), Seasonal transitions and water mass formation in the Iceland and Greenland Seas. *Deep-Sea Research*, 28, 1107-1129.

Swift, J.H., and K.P. Koltermann, The origin of Norwegian Sea Deep Water. *Journal of Geophysical Research*, 93, C4, 3563-3569, 1981.

Swift, J.H., E.P. Jones, K. Aagaard, E.C. Carmack, M. Hingston, R.W. Macdonald, F.A. McLaughlin, and R.G. Perkin, Waters of the Makarov and Canada basins. *Deep-Sea Research II*, 44, 1503-1529, 1997.

Tait, J.B., editor, The Iceland-Faroe Ridge international (ICES) "Overflow" expedition, May-June, 1960. An investigation of cold, deep water overspill into the north-eastern Atlantic Ocean. Papports et Procès-verbaux des Rèunions, *Conseil international pour l'Exploration de la Mer*, 157, 274 pp., 1967.

Thomson, C.W., *The depths of the sea*. An account of the general results of the dredging cruises of H.M.SS. "Porcupine" and "Lightning" during the summers of 1868 and 1870. London., 1873.

Tizard, T.H., and J. Murray, J., Exploration of the Faeroe Shetland Channel during the summer of 1880 in H.M. hired S. "Knight Errant". *Proceedings of the Royal Society*, Edinburgh, 1882.

Trangeled, S., Oceanography of the Norwegian and Greenland Seas and adjacent areas, 2, Survey of 1870-1970 literature. *SACLANT-CEN SM-4*, SACLANT Research Centre, La Spezia, 1974.

Turrell, W.R., G. Slesser, R.D. Adams, R. Payne, and P.A. Gillibrand, Decadal variability in the composition of Faroe Shetland Channel Bottom Water. *Deep-Sea Research*, 46, 1-25, 1999.

Turrell, W.R., B. Hansen, S. Hughes, and S. Østerhus, Hydrographic variability during the decade of the 1990s in the Northeast Atlantic and southern Norwegian Sea. ICES *Marine Science Symposia*, 219, 111-120, 2003.

Van Loon, H., and J.C. Rogers, The seesaw in winter temperatures between Greenland and Northern Europe. Part I: General description. *Monthly Weather Review*, 106, 296-310, 1978.

Vinje, T., Anomalies and trends of sea-ice extent and atmospheric circulation in the Nordic Seas during the period 1864-1998. *Jornal of Climate*, 14, 255-267, 2000.

Visbeck, M., and M. Rhein, Is bottom boundary layer mixing slowly ventilating Greenland Sea Deep Water? *Journal of Physical Oceanography*, 30, 215-224, 2000.

Visbeck, M., E.P. Chassignet, R.G. Curry, T.L. Delworth, R.R. Dickson, and G. Krahmann. The ocean's response to the North Atlantic Oscillation variability. The North Atlantic Oscillation: Climatic Significance and Environmental Impact. J.W. Hurrell et al. Editors. *AGU Geophysical Monograph*, 134, 113-145, 2003.

Wadhams, P., J.P. Wilkinson, and S.C.S. Wells, editors, *European subpolar ocean programme (ESOP) Sea ice–ocean interactions. Scientific Report*, 1, 247 pp., 1996.

Widell, K., S. Østerhus, and T. Gammelsrød, The sea ice flux through the Fram Strait: Results from moored instruments. *Geophysical Research Letters*, 30, 19, 1982, doi:10.1029/3003GL018119, 2003.

Worcester, P.F., J.F. Lynch, W.M.L. Morawitz, R. Pawlowicz, J.P. Sutton, B.D. Cournelle, O.M. Johannessen, W.M. Munk, W.B. Owens, R. Shuchman, and R.C. Spindel, Evolution of the large-scale temperature field in the Greenland Sea during 1988-89 from tomographic measurements. *Geophysical Research Letters*, 20, 20, 2211-2214, 1993.

Johan Blindheim, Institute of Marine Research, P.O. Box 1870 Nordnes, N-5817 Bergen, Norway. (johan.blindheim@imr.no)

Svein Østerhus, Geophysical Institute, University of Bergen, Allégaten 70, N-5007 Bergen, Norway.

Sea-Ice Extent and Variability in the Nordic Seas, 1967–2002

Børge Kvingedal

Geophysical Institute, University of Bergen, Bergen, Norway

A new data set based on weekly sea-ice concentrations, evaluated from several different data sources, reveals a marked decrease in sea-ice area in the Nordic Seas by almost 15% (4.1% per decade) during the period from January 1967 to December 2002. The sea-ice area seems to have interannual variability with a period of 9–10 years. At the present time, this variability is characterized by a negative anomaly, manifested by the record-low summer sea-ice area in 2002. All seasons undergo a reduction in sea-ice area during the period, with largest percentage decrease in autumn and smallest in spring. Variability is largest in the Greenland Sea during winter and spring and in the Barents and western Kara Seas during summer and autumn.

1. INTRODUCTION

Sea-ice in the Arctic plays an important climatic role. This includes reflecting solar radiation away from the surface, insulating warm underlying water from cold Arctic air, releasing brine during freezing periods and supplying fresh water during melting periods. Other factors such as the exchange of momentum and water vapor between the surface and the atmosphere also depend on the amount of sea-ice. Clearly, the sea-ice is a very sensitive component in the polar climate, and any changes may initiate feedback mechanisms. An example is the temperature–albedo feedback. A global temperature increase due to greenhouse gas forcing causes retreat of the ice cover. This event reduces the albedo in the polar regions with the result that more energy will be absorbed at the surface, which leads to a further temperature increase. The temperature–albedo feedback is the dominant factor as to why climate models project enhanced warming in polar regions [*IPCC*, 2001].

Knowledge of the sea-ice variability can help us understand the polar climate and climate changes. Recent studies have indicated significant reduction of Arctic sea-ice extent. Analyses of satellite passive-microwave data have detected an overall negative trend of about –3% per decade in total sea-ice extent in the Arctic since 1978 [*Parkinson et al.*, 1999; *Johannessen et al.*, 1995; *Bjørgo et al.*, 1997]. *Johannessen et al.* [1995] found a stronger negative trend of –4.3% per decade from 1987 to 1994. In a later paper, *Johannessen et al.* [1999] found a considerable reduction of about 14% in multiyear sea-ice area in winter during the period 1978 to 1998.

Regional differences in Arctic sea-ice extent have been examined by *Mysak and Manak* [1989]. They extended the 25-year data set analyzed by *Walsh and Johnson* [1979] to a 32-year period, 1953–1984, reporting monthly sea-ice concentration values. Their study of seven Arctic subregions revealed marked seasonal, regional, and interannual sea-ice variability. *Parkinson et al.* [1999] also reported large regional differences in sea-ice variability. Although the computed trend in total Arctic sea-ice extent was –2.8% per decade, the trend in the Kara and Barents Seas showed a substantial –10.5% per decade decrease. *Serreze et al.* [2003] highlighted the dramatic loss of Arctic sea-ice through analysis of the record low sea-ice extent and area in September 2002. A lack of sea-ice in the Greenland Sea contributed strongly to this minimum.

The area here referred to as the Nordic Seas comprises the Greenland, Iceland, Norwegian, Barents, and western Kara Seas, bounded by 30°W, 80°E, and 85°N. *Vinje* [2001] analyzed April and August sea-ice extent data for the Nordic Seas from 1864 to 1998. The major finding was a decrease in April sea-ice extent of ~33%. By dividing the Nordic Seas into eastern and western areas, *Vinje* [2001] found

The Nordic Seas: An Integrated Perspective
Geophysical Monograph Series 158

10.1029/158GM04

significant differences in ice extent reduction between the two areas. A reduction of ~62% in August sea-ice extent in the eastern area from 1920 to 1998 contrasts with the more modest ~22% reduction in the western areas during the same period.

The present work is an analysis of sea-ice cover in the same area analyzed by *Vinje* [2001], using parts of the same data set. In contrast to *Vinje* [2001], the present analysis uses weekly sea-ice concentration data to resolve the whole annual cycle. Both the area spanned by different ice types and total sea-ice area are analyzed. Seasonal averages based on weekly observations are also studied. The high temporal resolution of the data set, together with its length (36 years), makes it possible to get statistically reliable averages and trends. Sea-ice variability analysis for this area is performed on quality-controlled sea-ice data from January 1967 to the end of 2002.

In section 2 the sea-ice concentration data and some climatological maps are described. Time series for the different ice types are discussed in section 3. Time series of the total sea-ice area and a description of the seasonal cycle appear in section 4. Section 4 also gives correlation coefficients of year-to-year and season-to-season ice cover. Comparisons with, and discussion of, other published results follow in section 5.

2. SEA-ICE CONCENTRATION DATA

Weekly sea-ice concentration measurements have been digitalized and gridded from sea-ice charts analyzed by the Norwegian Meteorological Institute (MI) using several different data sources: Satellite images (after 1966), US ice maps (Birds Eye aircraft 1960–1971 and Joint Ice Center from 1970), visual and radar aircraft observations (Russian 1950–1970, American 1960–1970, Norwegian 1963–present), ship observations, and observations from land stations (Bjørnøya, Hopen, Isfjord Radio, and Jan Mayen since 1963). Further details can be found in *Løyning et al.* [2003]. The sea-ice concentration data are mapped onto a common grid with a polar stereographic projection and a grid cell size of 10 × 10 km. Figure 1 shows annual maximum and minimum sea-ice concentrations over the domain. On average, the maximum occurs in the first week in April (week 14), and the minimum in mid-September (week 38). (Data for these particular weeks are missing in 1976, so we used charts for the last week in March and the first week in October instead.) At the time of maximum ice-extent, some signs of the Odden feature are visible. The Odden, first reported by *Nansen* [1906], is a tongue of ice that advances rapidly and irregularly eastward into the Greenland Sea from the edge of the main pack ice [*Shuchman et al.,* 1998; *Comiso et al.,* 2001]. The Odden feature is also discussed in section 5.

Each weekly chart covers exactly the same area, making the data set temporally and spatially homogeneous. This permits identification of fluctuations on both seasonal and decadal time-scales. However, a number of factors (e.g., inaccurate concentration classification, missing data, and different observational techniques) may lead to non-uniformity of the data set.

Grid values in the original ice maps range from 0 to 10, defining the different ice types shown in Table 1. Concentration values for the different ice types in a grid cell are reported as fractional coverage, ranging from 0 out of 10 (no sea-ice) to 10 out of 10 (total ice cover). The concentration value for the ice types with grid values 1–5 are the same as MI investigators use in their ice charts. The ice types with values 6–9 are given ice concentration values defined by *Løyning et al.* [2003]. In the total area calculations discussed in section 4, each grid cell is multiplied by its mean concentration value.

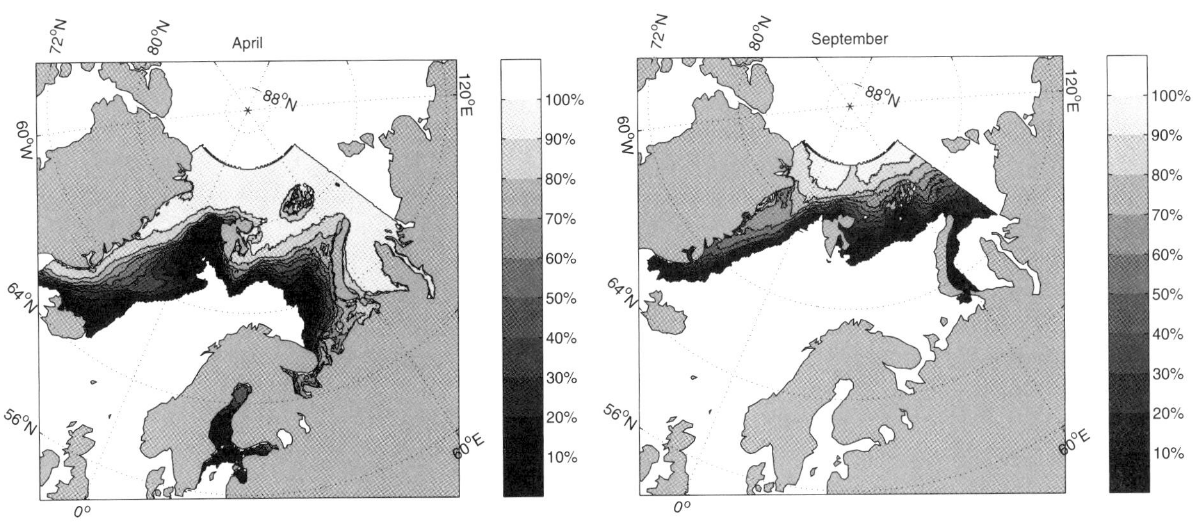

Figure 1. Mean sea-ice concentrations for early April (left) and mid-September (right).

Table 1. The Different Ice Types, Their Grid Value, Fractional Coverage, and Mean Concentration.

Grid Value	Ice Type	Fractional Coverage	Mean Concn., %
0	No ice	0/10	—
1	Fast ice	10/10	100%
2	Very close drift ice	9/10-10/10	95%
3	Close drift ice	7/10-9/10	80%
4	Open drift ice	4/10-7/10	55%
5	Very Open Drift Ice	1/10-4/10	25%
6	Young ice	>6/10	80%
7	Grease ice	>6/10	80%
8	Grease ice and old ice	<6/10	30%
9	Diverging grease ice	<6/10	30%
10	Open water	<1/10	5%

It is important to recognize the difference between sea-ice area and sea-ice extent. Sea-ice extent describes the area spanned by any ice type, regardless of ice concentration. Sea-ice area (or total sea-ice area) is the value obtained by multiplying the cell area (100 km^2) by its mean concentration. Hence, a 10 × 10 km grid cell of ice type 4 (55%) has a sea-ice extent of 100 km^2, but a sea-ice area of only 55 km^2. The sea-ice data are divided into standard calendar seasons. Each season contains an average of 13 weekly ice charts. The seasons are labeled by the year of the last month; for example, the winter season 2002 refers to the period December 2001 to February 2002.

Figure 2 shows the climatology of sea-ice extent for all four seasons. The figure gives an estimate of the probability (%) of encountering ice, no matter what type of ice. The seasonal

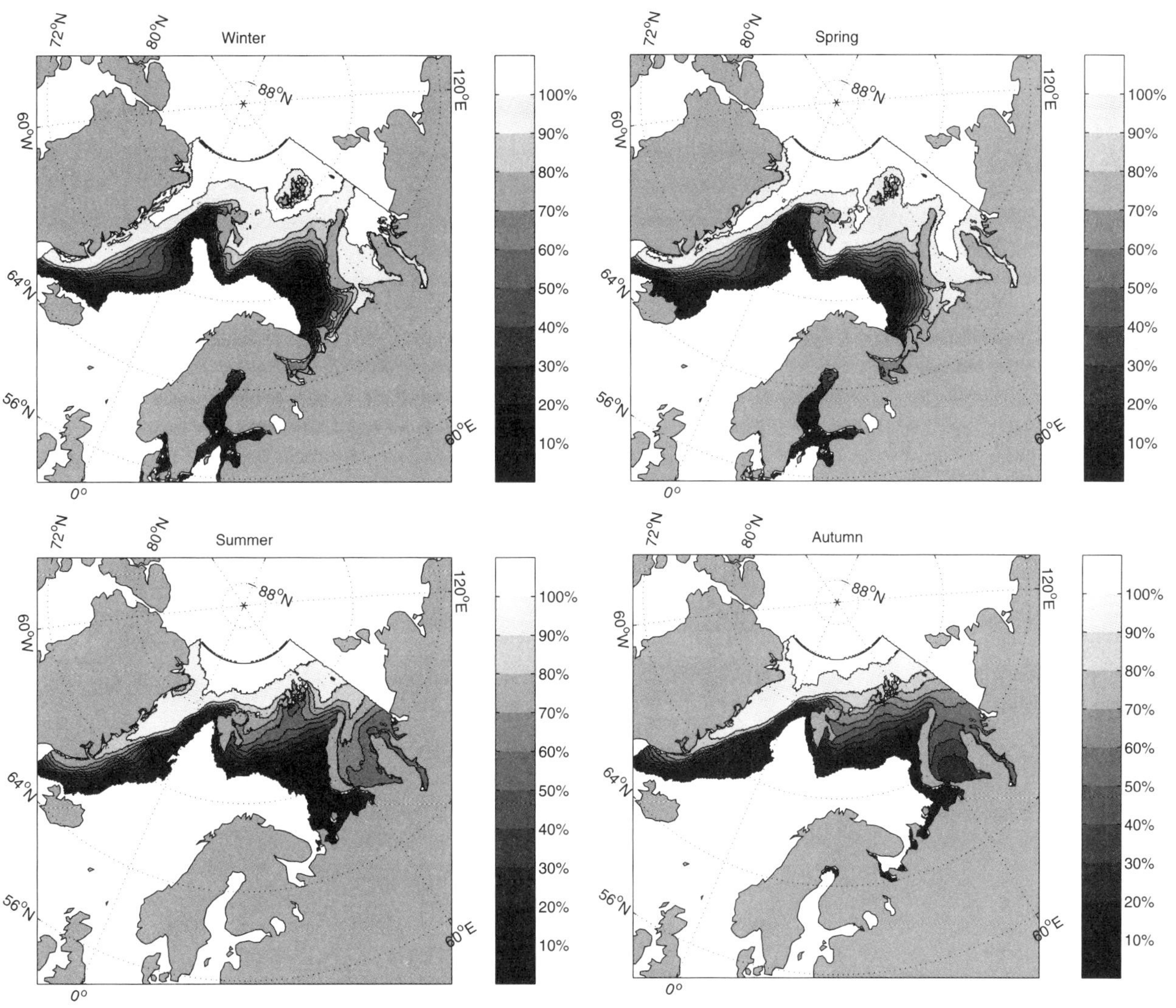

Figure 2. Seasonal means of sea-ice extent in the Nordic Seas over the period 1967–2002. The scaling shows the probability (%) of encountering ice.

climatology is constructed by first defining ice/no-ice for all grid points. Then the average of all ice charts in each season is calculated. In total, each seasonal climatology is an average of 468 ice charts, that is, 13 weeks × 36 years.

The winter and spring climatologies are rather similar, but there is greater ice area in spring. The summer and autumn climatologies also resemble each other, but with more ice in summer. In all seasons we see that the region just west of Svalbard is ice-free. This is due to the northward transport of warm water by the West Spitsbergen Current (WSC), the northward extension of the Norwegian Atlantic Current (NWAC). The warm water transported by the NWAC also explains why large parts of the Barents Sea are ice-free, despite its high latitude. Signs of ice breakup in early summer are seen south of Franz Josef Land (Figure 2, lower left). Open areas in this sector are common at this time of year.

Winter and spring show about the same interannual variability, in both the Greenland Sea and the Barents Sea. The variability in the Greenland Sea is mostly due to the fluctuating Odden feature. In summer and autumn, there is no contribution from this feature, and most of the interannual variability occurs in the Barents Sea.

3. ANNUAL CYCLE OF SEA-ICE EXTENT

The time series for the different ice types consist of 36 years with 52 weekly values. Missing weeks are interpolated from the nearest two neighboring weeks. The time development of the different ice types (Table 1) is shown in Figure 3. Note that ice type 2 was not reported until 1971. In the period 1967–1971 ice types 2 and 3 were given the same grid value (Figure 3a). Because of the difference in ice concentration between these two ice types, this practice may lead to errors in the total sea-ice area calculation. The effect on the calculations has been investigated and taken into account. Ice types 4 and 5 have been reported continuously throughout the period

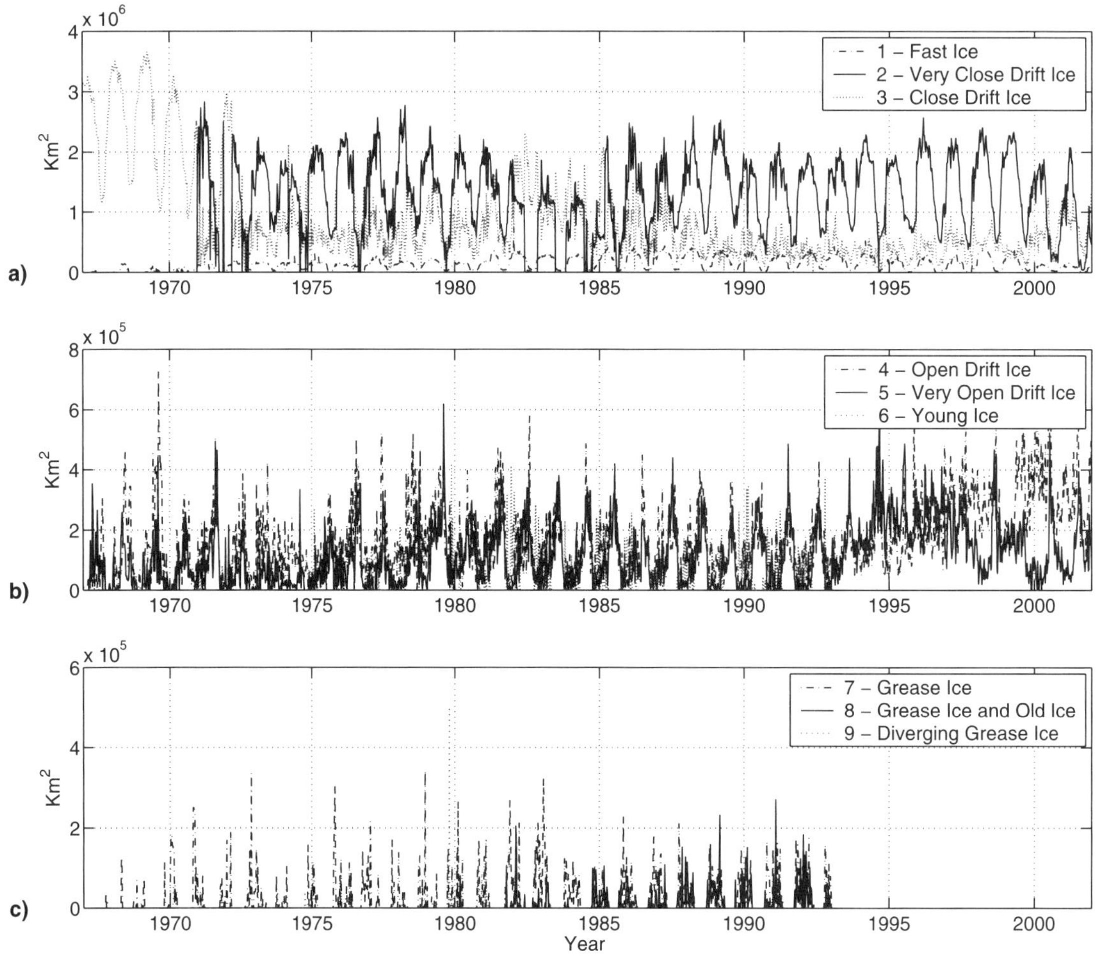

Figure 3. Time development of the different ice types that occupy the ice area (see legend). Note the different scaling of the *y*-axes.

1967–2002 (Figure 3b). Ice type 6 has been reported only sporadically during ice-growth periods. Since 1993, neither ice type 6 nor the ice types 7–9 are reported (Figure 3c).

Ice types 1 and 2 have dominant seasonal components, the Fast Ice almost vanishing every August–September (Figure 3a). On average, ice type 2 more than triples in extent from minimum in September to maximum in March–April. Ice type 3 has less pronounced seasonal variation and more high-frequency variation than the other two. It reaches minimum in the melting period (~week 38) and a less distinct maximum around February (Figure 4a).

The growth and decay of the different ice types have different characteristics during the annual cycle. Although the seasonal decline in type 2 occurs at a near constant rate of approximately 5.8×10^4 km^2 week^{-1}, the seasonal advance can be divided into two periods. The first period, from the time of ice minimum in September to mid-December, is recognized by steady and fast ice growth at a rate of 6×10^4 km^2 week^{-1}. In the second period, from January to maximum ice extent around week 13, the ice growth is slower and irregular at a rate near 3×10^4 km^2 week^{-1}. On average, the freezing period is about 6 weeks longer than the melting period.

The time of minimum and maximum extent for ice types 1–3 generally coincide. However, at the time of maximum extent of type 2, the extent of type 3 has a secondary minimum, possibly caused by freezing or convergence/divergence of sea-ice. Freezing of type 3 leads to more concentrated ice and the extent of type 2 increases. If no ice freezes or melts into type 3, the net result will be decreased extent of type 3. This means that the subsidiary peak in the extent of type 2 at around week 13 does not become evident in the total sea-ice seasonal cycle. It only "steals" ice from ice type 3, which in turn does not get renewed. At this time of year there is no net ice growth, only conversion between ice types.

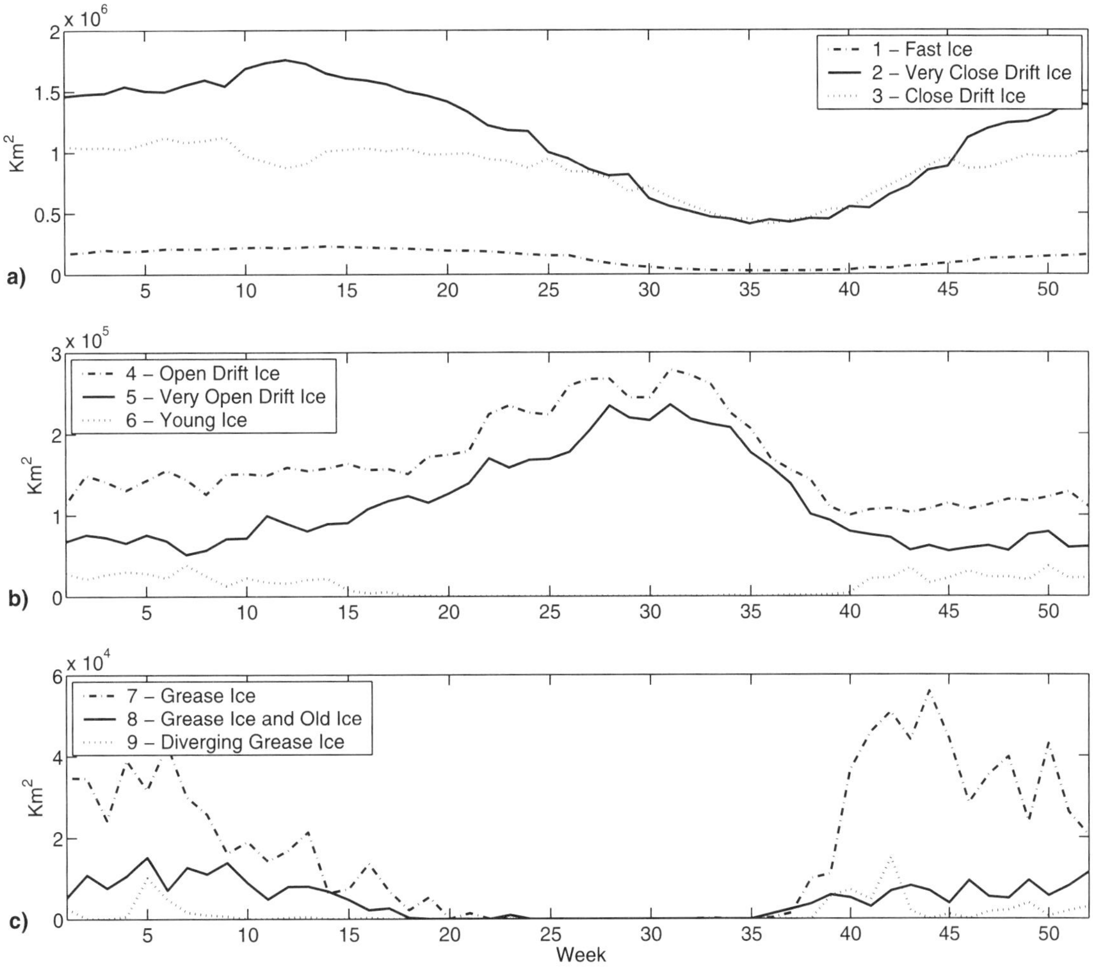

Figure 4. Mean (1967–2002) annual cycles for the ice types shown in legend.

Both ice types 4 and 5 have their maximum area during summer. While the ice types 1–3 (Figure 4a) are melting during early summer, the areas spanned by ice types 4 and 5 increase. This is due to breakup of the more concentrated ice types. Maximum extent of these two ice types occurs around week 30. In late summer, from week 32 to week 40, these ice types rapidly melt (Figure 4b). At the same time, the extent of ice type 7 increases rapidly (Figure 4c). This increase is thought to be due to breakup of the different classes of more concentrated ice types and freezing of new ice.

4. TIME SERIES OF TOTAL SEA-ICE AREA

The total sea-ice area is calculated by using the mean concentration values given in Table 1. As mentioned earlier, the number of ice classes reported were fewer before 1971. Here we have used a constant value 8×10^5 km^2 for the Close Drift Ice (annual mean after 1971). The rest of the ice extent is referred to as ice type 2. In this way we can use the ice extent as originally reported and then only estimate the compact ice area.

The total sea-ice area (solid line), the calculated trend fitted through weekly values by linear least square regression (thick solid line), and the 2-year running mean (dashed line) are shown in Figure 5. The long-term trend (1967–2002) shows a reduction in total sea-ice area of -3.28×10^5 km^2, which corresponds to a total reduction of almost −15% during the period, or −4.1% decade^{-1}. To test the significance of the trend, we constructed annual anomalies by removing the annual mean from the annual values. The autocorrelation function (not shown) of the anomalies showed no significant correlations at any lag. A standard F-test [*Pollard*, 1981] was therefore performed on the trend, testing the null hypothesis of a 0 slope. For the F-test, 34 degrees of freedom were used, two less than the number of years. The P-value (0.006) means that the null hypothesis can be rejected at a confidence level of 99%. The two trends, fitted through weekly (Figure 5) and annually (tested) values show the same reduction in sea-ice.

We see both high frequency variations (solid line in Figure 5) and interannual variability. The 2-year running mean shows four periods of above-normal ice conditions with respect to the long-term trend: the late 1960s, the late '70s extending to early '80s, the late '80s, and the late '90s. Below-normal conditions occur around the mid-'70s and '80s and the early '90s. The results also show less than normal sea-ice in the last 2 years. The frequency spectrum of the time series (not shown) suggests that the total sea-ice area oscillates at a period of approximately 10 years. The mean annual cycle of total sea-ice area was constructed by averaging the annual cycles for the 36-year period, 1967–2002

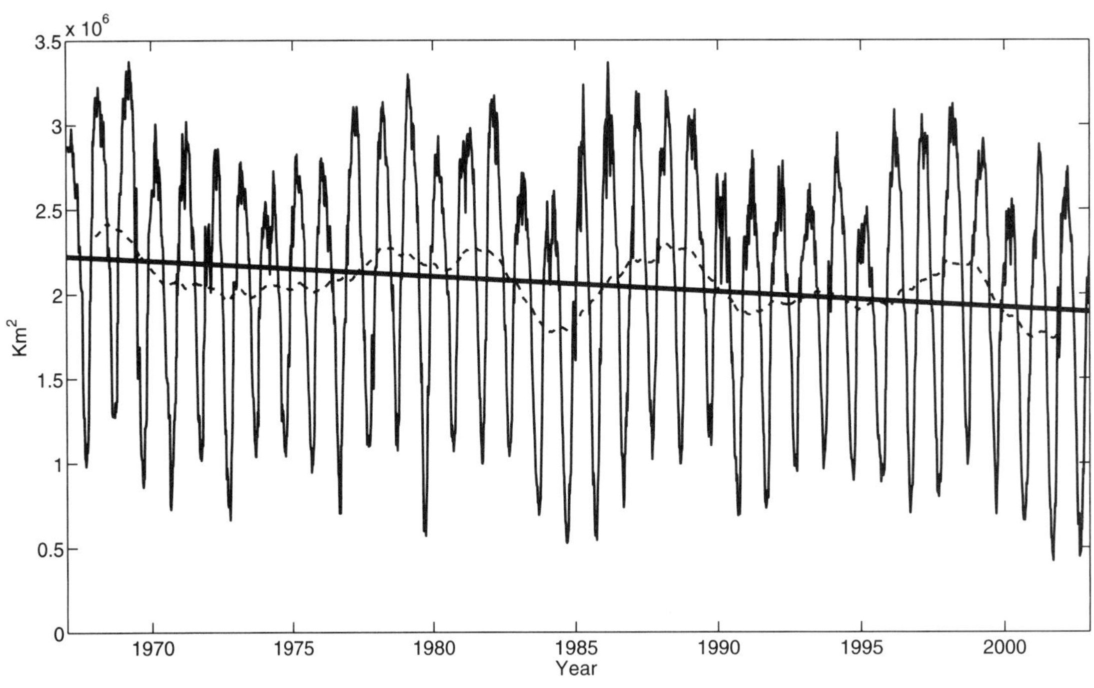

Figure 5. Total sea-ice area (solid line), long-term trend (thick solid line), and 2-year running mean (dashed line). The negative trend is statistically significant at the 99% confidence level.

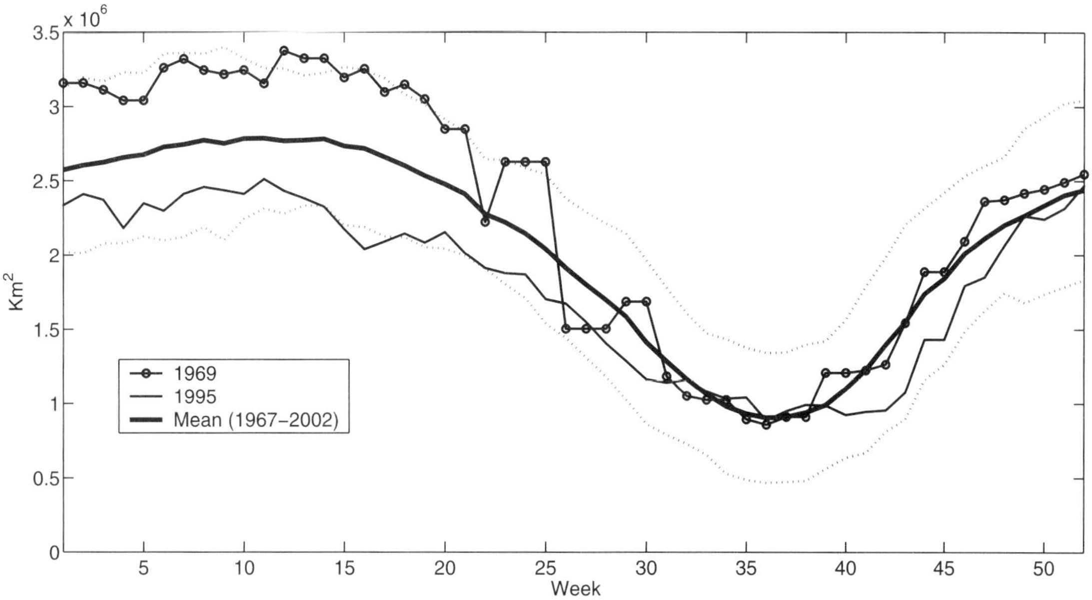

Figure 6. The mean annual cycle of total sea-ice area for 1967–2002, along with annual cycles for 1969 (with extremely large winter values) and 1995 (with extremely low winter values). Thin dashed lines show ±2 SD.

(Figure 6). The climatology shows a maximum of 2.8×10^6 km^2 in late March/early April and a minimum of 0.9×10^6 km^2 in the first week of September.

From September to December the growth is steady and rapid, adding 1×10^5 km^2 week^{-1}. Growth slows from December to March, adding only 0.3×10^5 km^2 week^{-1}. The decay period, from April to September, is most rapid in the summer season (JJA). On average, the melt period is about 8 weeks shorter than the freezing period.

To illustrate some of the interannual variability in the annual cycles, Figure 6 also shows the sea-ice areas for the years 1969 and 1995. These 2 years correspond to the maximum and minimum winter sea-ice area, respectively, over the period of study. Throughout the winter and spring, the sea-ice area in 1969 was significantly higher than in 1995. However, in late summer, the 2 years had approximately the same amount of sea-ice. In 1969 meltdown and thereafter freezing were more rapid, leading to a sharper curve around the time of minimum sea-ice. In 1995 the sea-ice had a broader and longer lasting ice-minimum period. Freezing started up about 6 weeks later, but at the end of the year the amount of sea-ice was almost the same as in 1969. As will be shown later, there is generally little correlation between winter sea-ice area and summer/autumn sea-ice area.

4.1. Seasonal Averages and Trends

Time-series and linear trends, fitted through seasonal averages, have been constructed. All four seasons show a clear reduction in total sea-ice area from 1967 to 2002 (Figure 7). In both the winter and autumn seasons, nearly 34,000 km^2 of sea-ice has been lost, yielding trends of respectively –3.4 and –5.8% decade^{-1}. In spring and summer, the melting period, the numbers are 22,000 km^2 and 24,000 km^2, respectively, giving trends of –2.2% and –4.0% decade^{-1}. Thus a lack of winter freezing seems to contribute more to the total sea-ice decrease than does increased summer melting.

A record low sea-ice area was measured in the Nordic Seas in summer 2002 (Figure 7). Summer 1990 also had little summer sea-ice. In both 1990 and 2002, however, the autumn sea-ice area was normal. All seasons in Figure 7 show interannual variability similar to that shown in Figure 5. The positive sea-ice area anomalies in the late '60s, '70s, '80s, and '90s are all shown in Figure 7.

For each season we calculated anomalies by removing the seasonal means. Autocorrelation and cross-correlation functions were then calculated from the detrended anomalies (Figure 8).

There are no strong relationships in sea-ice area from winter to winter. The 1-year lag value is statistically significant, but the correlation coefficient is only 0.40. For spring, summer, and autumn the sea-ice areas are independent of the sea-ice areas 12 months earlier, and no significant correlations are found at any lag. In spring, however, the 10- and 20-year-lagged values are close to the 95% confidence limit (Figure 8, upper right panel). This, and the 10-year-lagged winter value (Figure 8, upper left panel), suggest a 10-year cycle in winter/spring sea-ice area.

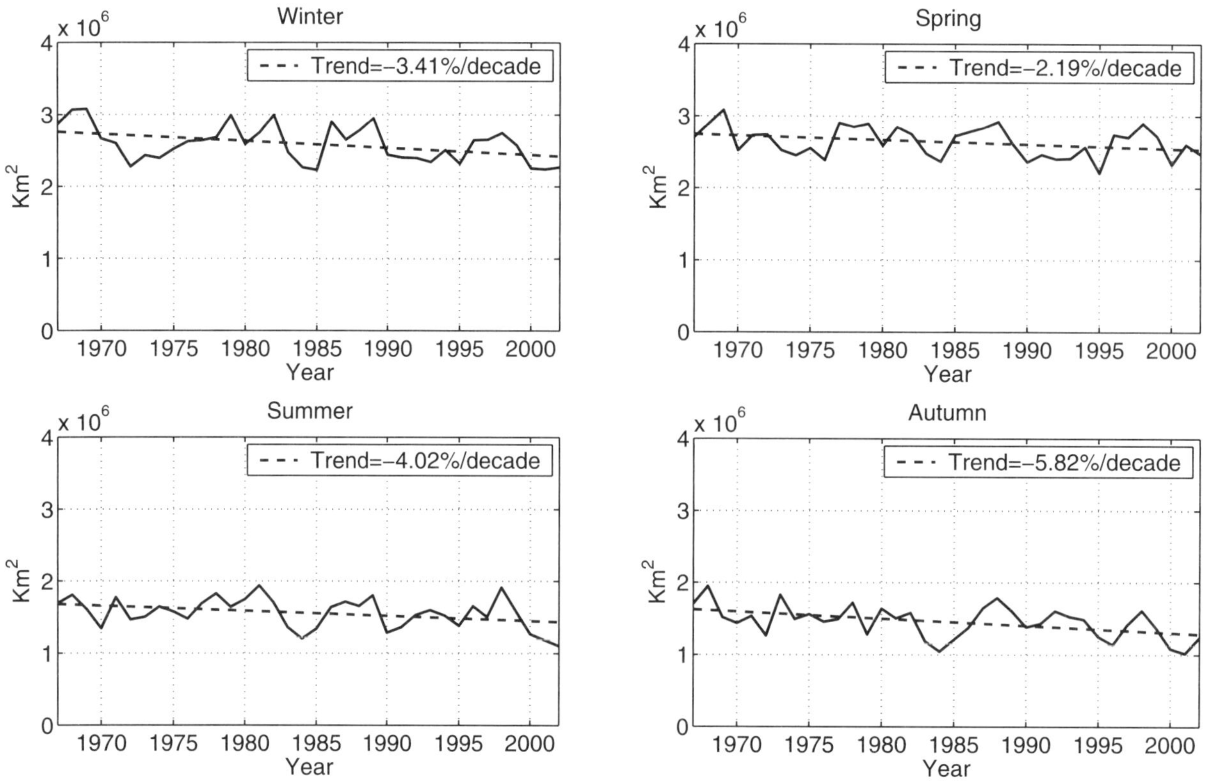

Figure 7. Time series of the winter, spring, summer, and autumn seasons, along with trends calculated by using linear least squares regression. The autumn trend is statistically significant at the 99% confidence level, summer and winter trends are significant at the 95% confidence level, and the spring trend is not significant. The trends are tested with a standard *F* test, at 34 degrees of freedom.

By studying the correlations between different seasons within a year, one can identify significant relationships (Table 2). Correlations between winter and spring, and between winter and summer, are almost the same, 0.66 and 0.65, respectively. Surprisingly, the correlation between winter and the following autumn (0.44) is higher than that between winter and the previous autumn (0.41). Spring correlates more strongly with the recent winter (0.66) than with the following summer (0.60), whereas the correlation with the following autumn is only 0.28 and not significant. The summer–autumn correlation is 0.63.

5. DISCUSSION AND CONCLUDING REMARKS

The seasonal and interannual variability of sea-ice area and extent in the Nordic Seas have been investigated for the period 1967–2002. A statistically significant decreasing trend in sea-ice area of about 4% per decade is found. This is less than half the negative trend found by *Parkinson et al.* [1999] for the same area. *Parkinson et al.* [1999] used sea-ice extent data for the shorter period 1979 to 1996. For the Greenland Sea and for the Barents and Kara Seas they found reductions in sea-ice extent of –6.1% decade^{-1} and –10.5% decade^{-1}, respectively. They found a reduction in the two regions combined of -19.7×10^3 km^2 yr^{-1}, whereas the data from 1967 to 2002 show a reduction of -9.1×10^3 km^2 yr^{-1}. In 1979 there was a decadal maximum in sea-ice area, whereas in 1996 there was a record low. This may account for the extreme negative trend found by *Parkinson et al.* [1999]. Calculating the 1979 to 1996 trend in our data, we find an 18.8×10^3 km^2 yr^{-1} reduction, in good agreement with *Parkinson et al.* [1999]. Some differences are expected because *Parkinson et al.* [1999] used sea-ice extent, whereas we used sea-ice area. Furthermore, slightly different areas are used.

Stern and Heide-Jørgensen [2003] studied trends and variability of sea-ice in Baffin Bay and Davis Strait on the west coast of Greenland for the period 1953–2001 and found a significant positive trend for the 49-year period. On comparing their results with those of *Parkinson et al.* [1999], they concluded that trends are sensitive to the period of record. The wintertime sea-ice extent recorded by *Parkinson et al.* [1999] began in 1979, a particularly low ice year in Baffin

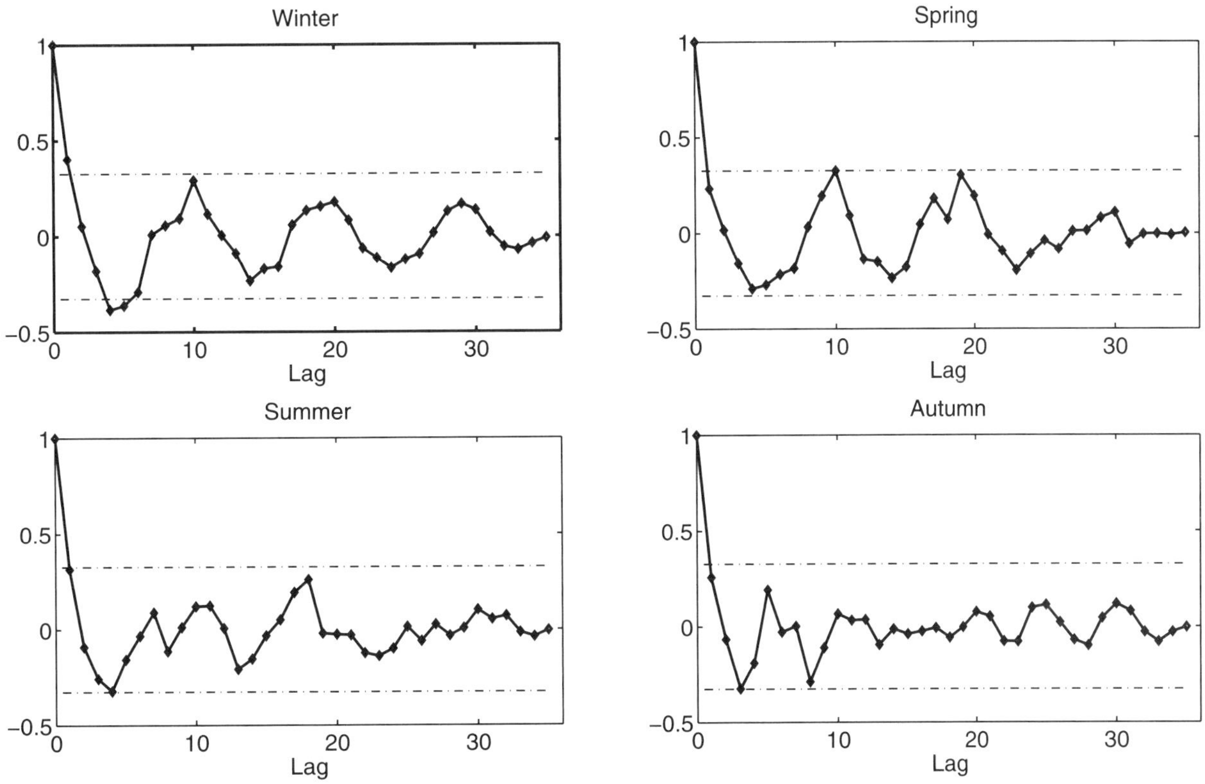

Figure 8. The autocorrelation function of seasonal anomalies for the winter (upper-left), spring (upper-right), summer (lower-left), and autumn (lower-right) seasons. The approximate 95% confidence bounds, $\pm 1{:}96/\sqrt{n}$, for an independent identically distributed process of the same length (n) are also displayed.

Table 2. Correlation Coefficients Between Sea-Ice Area Anomalies for the Different Seasons.

Season	Winter	Spring	Summer	Autumn
Winter	**1.00 (0.40)**	**0.66 (0.47)**	**0.65 (0.45)**	**0.44 (0.41)**
Spring	**0.66** (0.18)	**1.00** (0.23)	**0.60** (0.24)	0.28 (0.09)
Summer	**0.65** (0.17)	**0.60** (0.18)	**1.00** (0.30)	**0.63 (0.36)**
Autumn	**0.44** (0.14)	0.28 (0.27)	**0.63** (0.21)	**1.00** (0.26)

Statistically significant correlations (95%) are shown in bold, 1-year-lagged correlations in parentheses.

Bay, which explains the relatively large positive trend they reported over the period 1979–1996. *Stern and Heide-Jørgensen* [2003] found a cyclic period of 8 to 9 years in the wintertime sea-ice area, with low points in 1970, 1979, 1986, and 1997. All of these years except 1970 had above-normal sea-ice extent in the Nordic Seas. The dipole in ice-extent anomalies between the Nordic Seas and the Labrador Sea is well known from earlier work [*Deser et al.*, 2000; *Parkinson et al.*, 1999; *Slonosky et al.*, 1997; *Walsh and Johnson,* 1979].

The regional differences and changes in sea-ice extent are linked to the North Atlantic Oscillation (NAO) [*Hurrell*, 1995] which characterizes the strength of the westerlies between Iceland and Portugal. *Vinje* [2001] pointed out that despite the high correlations between change per year in the Lisbon-Stykkisholmur NAO index [*Furevik and Nilsen*, this volume] and the corresponding changes in the April ice extent in the Barents Sea and the Labrador Sea, –0.97 and 0.81, respectively, weaker correlations are found in the Greenland Sea (–0.6). During positive NAO periods, both the anomalous northward (warm) wind component and the positive sea surface temperature (SST) anomaly in the Nordic Seas [*Krahmann and Visbeck,* 2003] are consistent with the less than normal ice extent in the Barents Sea.

The situation in the Greenland Sea is more complex. Although the southern parts of the Greenland Sea ice cover also experience positive SST anomalies and to some extent southerly winds [*Krahmann and Visbeck,* 2003], *Kwok et al.* [2004] show that ice export from Arctic through Fram Strait increases during positive NAO periods. By neglecting years with extreme negative NAO, they found a positive correlation (0.62) between winter ice area export and the NAO index. Besides the inflow of sea-ice from the Fram Strait, Greenland Sea ice cover also consists of locally formed ice. The Odden ice tongue is a feature of the Greenland Sea that may exceed 3×10^5 km^2. This ice area, which is also a region of deep water convection [*Rudels et al.*, 1989], shows strong correlation (–0.89) with surface temperatures derived from thermal infrared sensors over the entire Odden area [*Comiso et al.*, 2001]. *Vinje et al.* [2002] addressed the importance of ocean circulation in shaping the Odden sea-ice. They suggest that sea-ice formation in the Odden area is less important for deep water convection than previously believed, whereas advection of salty water to the convection cell plays a crucial role. Furthermore, *Furevik et al.* [1998] found that the vertical stratification in the Greenland Basin is extremely weak and that even small perturbations in the surface water density can have large impacts on the convection depths and deep water formation activities. All of these factors make the air–ice–sea interactions in the Greenland Sea more complex than in the Barents Sea. The NAO index may therefore be insufficient to indicate sea-ice extent in the Greenland Sea.

The phases and strengths of observed ice extremes are in good agreement with the findings of *Mysak and Manak* [1989]. They divided the Nordic Seas into the Greenland Sea (their region 5), and the Barents and Kara seas (their region 6). Their findings can explain the less sharp curve at the time of maximum sea-ice area compared with that at the time of minimum. In their Figure 13, the annual cycle for the two regions shows that although the time of minimum sea-ice area coincides, the time of maximum sea-ice area occurs 1 month earlier in the Greenland Sea than in the Barents and Kara seas. This may lead to a smoother curve around time of maximum sea-ice area when the areas are studied as a whole.

The seasonal trends found here differ in both phase and strength from the results in *Parkinson et al.* [1999]. Differences in data coverage, area, and season definitions cannot explain the discrepancy. Whereas *Parkinson et al.* [1999] found the most negative trend in spring (–29.1 km^2 yr^{-1}) and the smallest trend in autumn (–7.8 km^2 yr^{-1}), our present analysis shows the opposite. We found the most negative trend in autumn (–9.5 km^2 yr^{-1}) and the smallest trend in spring (–6.0 km^2 yr^{-1}). Again the study period impacts the results.

Serreze et al. [2003] found that the total Arctic sea-ice extent and area in September 2002 reached their lowest values recorded since 1978. Compared with previous minimum low-extent years, the lack of sea-ice in the East Greenland Sea stands out prominently. Lack of sea-ice in this section also dominates the 1990 record-low summer sea-ice [*Maslanik et al.*, 1996].

Compared with other Arctic regions [*Parkinson et al.*, 1999], we have found a relatively strong interannual variability with a period of ~10 years. *Yi et al.* [1999] discovered approximately decadal oscillation in Arctic sea-ice concentration which agrees well with the 10-year cycle found in this study. The high variability may be caused by the extremely high frequencies of winter cyclones in and around the Nordic Seas [*Sorteberg et al.*, this volume]. The position of storm tracks is likely to influence the sea-ice coverage in both the Greenland Sea and the Barents and Kara Seas. The variability is largest during winters in the Greenland Sea, mainly because of the Odden feature, whereas the sea-ice in the Barents and Kara Seas varies most in summer and autumn. Since the beginning of the 36-year period studied, more than 20% of the total autumn sea-ice area has melted and the summer sea-ice area in 2002 reached its lowest level ever recorded.

Climate models generally agree that the sea-ice extent in the Arctic region will continue to decrease in the 21st century as the global temperatures increase. A comparison of observed sea-ice trends to control and transient integrations (with realistic forcing), reveals agreement between the observed decrease in Northern Hemisphere sea-ice extent and results from transient simulations. Future simulations project continued decreases in sea-ice extent throughout the next century [*Vinnikov et al.*, 1999; *Johannessen et al.*, 2004]. *Johannessen et al.* [2004] predicts that the Arctic summer ice cover may be reduced by 80% by the end of the 21st century. Furthermore they hypothesize that the decrease may lead to changes in pathways and spreading of meltwater in the Nordic Seas. The effects of reduced deep water formation in the Greenland Sea on the global thermohaline circulation could greatly alter the climate of the northern latitudes.

Acknowledgments. The author would like to thank Tore Furevik, Sigbjørn Grønås, and two anonymous reviewers for useful comments and suggestions.

REFERENCES

Bjørgo, E., O. Johannessen & M., Miles, Analysis of merged smmr-ssmi time series of arctic and Antarctic sea ice parameters 1978-1995, *GRL*, 24, pp. 413-416, 1997.

Comiso, J., P. Wadhams, L. Pedersen, & R. Gersten, Seasonal and interannual variability of the Odden ice tongue and a study of environmental effects, *JGR*, 106 (C5), 9093-9116, 2001.

Deser, C., J. Walsh & M. Timlin, Arctic sea ice variability in the context of recent atmospheric circulation trends, *Journal of Climate*, 13 pp. 617-633, 2000.

Furevik, T., M. Bentsen, H. Drange, J. Johannessen & A. Korablev, Temporal and spatial variability of the sea surface salinity in the Nordic seas, *JGR*, 107 (C12), 8009, 1998.

Hurrell, J., Decadal trends in the north Atlantic oscillation: regional temperatures and precipitation, *Science*, 269, pp. 676-679, 1995.

IPCC, *Climate Change 2000 - Third Assessment Report*, Cambridge University Press, 2001.

Johannessen, O., L. Bengtsson, M. Miles, S. Kuzmina, V. Semenov, G. Alekseev, A. Nagurnyi, V. Zakharov, L. Bobylev, L. Pettersson,

Hasselmann, K. & H. Cattle, Arctic climate change - observed and modeled temperature and sea ice variability, *Tellus*, 56A, 3, 2004.

Johannessen, O., M. Miles & E. Bjørgo, The Arctic's shrinking sea ice, *Nature*, 376, pp. 126-127, 1995.

Johannessen, O., E. Shalina & M. Miles, Satellite evidence for an arctic sea ice cover in transformation, *Science*, 286, pp. 1937-1939, 1999.

Krahmann, G. & M. Visbeck, Variability of the northern annular mode's signature in winter sea ice concentration, *Polar Research*, 22(1), pp. 51-57, 2003.

Kwok, R., G. Cunningham & S. Pang, Fram strait sea ice outflow, *JGR*, 109, C01009, 2004.

Løyning, T., C. Dick, H. Goodwin, O. Pavlova, T. Vinje, G. Kjærnli & T. Villinger, Acsys historical ice chart archive (1553-2002), Technical report, Norwegian polar institute, Tromsø, 2003.

Maslanik, J., M. Serreze & R. Barry, Recent decreases in arctic summer ice cover and linkages to atmospheric circulation anomalies, *GRL*, 23, 1677-1680, 1996.

Mysak, L. & D. Manak, Arctic sea-ice extent and anomalies, 1953-1984, *Atmos. Ocean*, 27, 376-405, 1989.

Nansen, F., *Norther Waters: Captain Roald Amundsen's oceanographic observations in the Arctic Seas in 1901*, Christiania, 1906.

Parkinson, C., D. Cavalieri, P. Gloersen, H. Zwally & J. Comiso, Arctic sea ice extents, areas, and trends, 1978-1996, *JGR*, 104 (C9), 20837-20856, 1999.

Pollard, J., *A Handbook of Numerical and Statistical Techniques*, Cambridge University Press, 1981.

Rudels, B., D. Quadfasel, H. Friedrich & M. Houssais, Greenland Sea convection in the winter of 1987-1988, *JGR*, 94, 3223-3227, 1989.

Serreze, M., J. Maslanik, T. Scambos, F. Fetterer, J. Stroeve, K. Knowles, C. Fowler, S. Drobot, R. Barry & T. Haran, A record minimum arctic sea ice extent and area in 2002, *GRL*, 30, No. 3, 10, 2003.

Shuchman, R., E. Josberger, C. Russel, K. Fischer, O. Johannessen, J. Johannessen & P. Gloersen, Greenland sea odden sea ice feature: Intra-annual and interannual variability, *JGR*, 103, 12709-12724, 1998.

Slonosky, V., L. Mysak & J. Derome, Linking arctic sea ice and atmospheric circulation anomalies on interannual and decadal time scales, *Atmos. Ocean*, 35, 336-366, 1997.

Stern, H. L. & M.P. Heide-Jørgensen, Trends and variability of sea ice in Baffin Bay and Davis Strait, *Polar Research*, 22(1), pp. 11-18, 2003.

Vinje, T., Anomalies and trends of sea-ice extent and atmospheric circulation in the Nordic seas during the period 1864-1998, *Journal of Climate*, 14, 255-267, 2001.

Vinje, T., T. Løyning & I. Polyakov, Effects of melting and freezing in the Greenland sea, *GRL*, 29 (23), 2129, 2002.

Vinnikov, K., A. Robock, R. Stouffer, J. Walsh, C. Parkinson, D. Cavalieri, J. Mitchell, D. Garrett & V. Zakharov, Global warming and northern hemisphere sea ice extent, *Science*, 286, pp. 1934-1937, 1999.

Walsh, J. & C. Johnson, An analysis of arctic sea ice fluctuations, 1953-1977, *JPO*, 9, pp. 580-591, 1979.

Yi, D., L. Mysak & S. Venegas, Decadal-to-interdecadal fluctuations of arctic sea-ice cover and the atmospheric circulation during 1954-1994, *Atmos. Ocean*, 37, 389-415, 1999.

Børge Kvingedal, Geophysical Institute, University of Bergen, Allegaten 70, N-5007 Bergen, Norway. (borge@gfi.uib.no)

Annual Sea Surface Height Variability in the Nordic Seas

Kjell Arne Mork

Institute of Marine Research and Bjerknes Centre for Climate Research, Norway

Øystein Skagseth

Bjerknes Centre for Climate Research, Norway

Altimeter data from May 1995 to February 2002, together with wind, heat, and fresh water fluxes and hydrographic data, are used to study the annual variability in sea surface height (SSH) in the Nordic Seas. The main contributions to the SSH are quantified with error estimates. The amplitude of the annual variation of the SSH ranges from 4 to 8 cm with largest variability in the deeper basins (Lofoten, Norwegian, and Greenland Basins) and along the Norwegian coast. The seasonal steric height can explain 1.5-4 cm of the SSH variation and arises from changes in the air–sea heat flux. However, the associated seasonal changes in the surface currents that arrive from the steric height and the heat flux are weak (a few mm/s). This is because their contributions to the SSH are nearly constant over large areas. From altimetry and hydrography, we calculated seasonal changes in bottom pressure and currents. In the Greenland and Norwegian Seas, the seasonal anomalous large-scale circulation is cyclonic during spring and anticyclonic during autumn. It is most apparent in the Norwegian Sea with a seasonal variation in speed of 1-2 cm/s. This variation can be explained by changes in the averaged wind-stress curl over the Norwegian Sea. The annual cycle of the Norwegian Atlantic slope current is also revealed. It is strongest in winter and weakest in summer with an annual amplitude in speed of 2-4 cm/s.

1. INTRODUCTION

The annual cycle is a dominant mode of variability in the world oceans. In a physical sense the ocean circulation is established as a response to wind stress and buoyancy (heat and freshwater) forcing. These forcings are associated with a profound annual cycle, and the ocean responds to these changes on various time and spatial scales, depending on the processes involved. Thus processes involved in the annual cycle might also trigger variability on other scales (e.g., climate changes) and vice versa.

The Nordic Seas: An Integrated Perspective
Geophysical Monograph Series 158

10.1029/158GM05

In a biological sense marine organisms are adapted to the annual cycle. Changes in amplitude and phase of the seasonality can therefore significantly affect the productivity and community structure of marine ecosystems. *Beare and McKenzie* [1999] found a close correspondence between the declining abundance of the marine copepod *Calanus finmarchius* in the North Sea and a decrease in the seasonal amplitude of the North Atlantic Oscillation.

There are several studies of the seasonal variability in the Nordic Seas (Norwegian, Iceland, and Greenland Seas). Investigating the wind-stress curl over the Nordic Seas, *Jonsson* [1991] found the seasonal change to be the dominant mode of variability, with maximum values in November and nearly negligible values during summer. Using hydrography data, *Mork and Blindheim* [2000] showed that the temperature at 100 m deep, in the core of the Norwegian

Atlantic Current (NwAC), south in the Norwegian Sea, changed from 7.8°C in April to 9.2°C in October. An array of current meters demonstrated to *Skagseth and Orvik* [2002] a clear winter maximum and summer minimum in the volume transport of the Norwegian Atlantic Slope Current (NwASC), explained by the variability in the along-slope component of the wind. *Skagseth* [2004] found that for time scales lasting from a few months up to a year, increased westerly wind occurs in phase with a steepening sea level slope from the northern North Atlantic into the Norwegian Sea, which also coincides with an increase in the NwASC. Despite some work on the seasonal variation of wind, hydrography, and transports, no studies have been conducted on the seasonal variation of the sea level with focus on the Nordic Seas.

The theory of the annual variability in mid-ocean regions is dealt with in a pioneering work of *Gill and Niiler* [1973]. The annual variability of sea level change can be divided into different contributions. Firstly, the response to changes in the atmospheric pressure is important (on the order of several centimeters at high latitudes) but dynamically uninteresting since water properties and currents are hardly changed. Secondly, the *barotropic* response to changes in the wind stress causes a small variation in the sea level and a corresponding change in the bottom pressure, p_b. Thirdly, there is a steric change in sea level due to expansion/contraction of the water column above the seasonal thermocline. Fourthly, there are changes connected to changes in the wind stress that involve (a) heat and salt changes in the mixed layer due to convergence and (b) Ekman pumping, which acts to displace the main seasonal thermocline (*baroclinic* response).

Considering the North Atlantic and the North Pacific at large spatial scales, on the order of 1,000 km, the main finding of *Gill and Niiler* [1973] was that the seasonal variability in the SSH is basically determined by the local heat flux, whereas advection is not particularly important. Their results are basically in accord with more recent works utilizing additional information from satellite SSH data in studies of the seasonal variability [*Ferry et al.*, 2000; *Stammer*, 1997; *Volvok and van Aken*, 2003].

The present study deals with the annual cycle of the SSH in the Nordic Seas. Figure 1 shows the area of interest. The main purpose of this study is to quantify the effects of steric height and bottom pressure to seasonal changes of the SSH, and to discuss the implications.

The paper starts with a description of the different effects on the SSH in section 2. In section 3 the data and the methods are described. The annual variation of the SSH from altimetry and steric height are presented in sections 4 and 5, respectively. In section 6 are discussed error estimates in the annual harmonic analysis. Section 7 presents the bottom pressure variation and the seasonal bottom currents.

Concluding remarks are given in section 8.

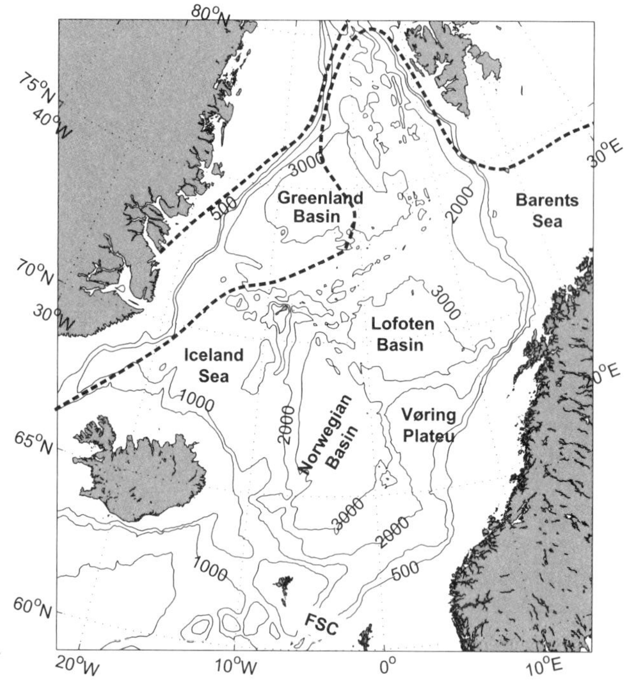

Figure 1. Bathymetry of the Nordic Seas. FSC indicates the Faeroe–Shetland Channel. The dashed lines indicate maximum and minimum climatological ice margin (ice concentration above 20%) from satellite passive-microwave observations 1978-1987 (*Gloersen et al.*, 1992).

2. THEORY OF SEASONAL VARIATION IN THE SEA LEVEL

In order to establish the framework for a discussion of the sea level variation, a careful consideration of the hydrostatic relation is required. The SSH level anomaly from the annual mean (η') can be divided into several components, as discussed in *Gill and Niller* [1973]:

$$\eta' = \eta'_s + \eta'_b + \eta'_a, \tag{1}$$

where η'_s is the steric height, η'_b is the effect of changes in the bottom pressure, and η'_a is the barometric effect on the sea level. The primes indicate anomalies from the annual mean. The steric contribution to the SSH is

$$\eta'_s = -\frac{1}{\rho_o}\int_{-H}^{0} \rho' dz \tag{2}$$

where ρ_0 is a constant characteristic density. Changes in the steric level are caused by changes in the density of the column (ρ'), implying expansion or contraction. Sea level variation due to changes in the bottom pressure (p'_b) is

$$\eta'_b = p'_b / g\rho_o \tag{3}$$

This term can be regarded as the dynamically active SSH component. A barotropic response on the ocean will produce changes in the SSH and a corresponding change in the bottom pressure [*Gill and Niiler*, 1973]. The barometric effect on the sea level is

$$\eta'_a = -\frac{p'_a}{g\rho_o} \tag{4}$$

An increase in surface air pressure (p_a) of 1 mbar corresponds to a sea level depression of approximately 1 cm. *Gill and Niiler* [1973] claimed that on a seasonal scale the barometric effect is dynamically negligible since currents are hardly changed.

The seasonal steric height (η'_s) can be split into contributions from local surface buoyancy fluxes (η'_B) and advection (η'_{ADV})

$$\eta'_s = \eta'_B + \eta'_{ADV} \tag{5}$$

The advection term includes geostrophic and Ekman transports and horizontal diffusion. We will later show that the local buoyancy flux is the dominating term. The buoyancy term comprises the local surface buoyancy flux given by the net heat (Q) and freshwater (F_{PE}) fluxes

$$\frac{\partial \eta'_B}{\partial t} = \frac{(\alpha Q)'}{\rho_o c_p} - \beta S_o F'_{PE} \tag{6}$$

where α and β are the heat and salinity expansion coefficients, respectively. Heat loss from the ocean (negative Q) cools the upper layer and results in reduction of the sea level due to contraction of the water column. A fresh water flux will have a similar effect since it reduces the upper layer density.

3. DATA AND METHODS

The altimeter data used in this study are the Collecte Localisation Satellites (CLSs) mean sea level anomalies (SLAs) gridded data set based on the merged TOPEX/POSEIDON (T/P) and ERS-1 and -2 (ERS) data sets. The data are gridded using optimal interpolation with realistic correlation function. The data set also includes a formal mapping error in percentage of signal variance. Details on the mapping of the merged T/P and ERS data are given by *Ducet et al.* [2000]. They argue that mapping errors in this region are less than or about 20% of the signal variance. The SSH data is also corrected for the inverted barometer effect, tides and dry tropospheric effects [*LeTraon and Ogor*, 1998]. As tide constituents are not resolved in the altimetry, they are alias on lower frequencies dependent on the orbit repeat periods [*Fu and Cazenave*, 2001]. The 10-day T/P orbit period have no significant tidal periods that are alias into the annual cycle. In contrast, for the 35-day ERS orbit, the tidal constituents K_1 (23.93 h) and P_1 (24.07 h) are aliased exactly into the annual cycle.

Thus, errors in the tidal models that are used for removal of tides in the ERS data effect the seasonal variation in the SLA data. That is mainly of importance north of 66°N where only ERS data exist. However, the alias effect is not significant for our study for two reasons. Firstly, in the Nordic Seas the semidiurnal tides are the dominant tidal period [*Gjevik and Straume*, 1989]. Secondly, our results show no discontinuity at 66°N that should be expected if the diurnal tides in the ERS data were aliased into the annual cycle.

The time period used in this study is from May 1995 to February 2002 with a 7-day time interval. The data have a 1/3° Mercator projection grid (i.e. 1/3° in east–west direction and weighted with cos(latitude) in north–south direction). At the equator the resolution is 37.5 km while at 70°N the resolution is 12.7 km. The altimeter data are anomalies from a 7-year mean (1993-1999). The combination of T/P and ERS data reduces the mean mapping error by a factor of more than 2 compared to T/P data only [*Ducet et al.*, 2000]. In the North Atlantic and Nordic Seas the mean error of the merged data is ~2 cm south of 66°N while it is ~3-5 cm north of 66°N. For practical reasons the altimeter data are reduced spatially. The data are gridded into 2° × 1° resolution for east–west and north–south directions, respectively. The appendix gives details on the gridding method and error estimates.

The World Ocean Atlas 2001 data set (WOA01) [*Conkright et al.*, 2002] was used to calculate the density and the steric height. These data are climatologic monthly mean values with 1° × 1° spatial resolution. The mean values are averages for the last ~30 years but most observations originate from the 1980s and 1990s. Data sets with different time spans are a general problem and are taken into consideration when they are analyzed and discussed. The WOA01 data set also includes standard errors in the monthly values of temperature and salinity. See the appendix for details on the error estimates of the steric height.

The NCEP/NCAR data sets were used for the wind, and heat and freshwater fluxes. The data sets are monthly values and have a horizontal resolution of about 1.9° in both north–south and east–west directions. Since the heat flux data do not include any errors we will just assume that there is a standard error of 15% in every value of the net atmosphere–ocean heat flux. *Godøy* [1998] investigated the ocean–atmosphere heat flux from ship observations at Ocean Weather Station "M" (66°N, 2°E) in the Norwegian Sea and estimated a standard error in the annual net heat flux of similar magnitude. With this assumption, the error in the sea level derived from the heat flux can be obtained from

Equation 6. Over ice, the data set has only ice–atmosphere heat fluxes and not ocean–ice heat fluxes. This introduces large errors in areas with ice when calculating the contribution from heat fluxes to the total steric height. This is discussed later in the paper.

The various contributions to the annual cycle are represented in terms of annual amplitude and phase by fitting the data to an annual harmonic cycle. The shape of the annual cycle is of course more complex than a harmonic cycle and fitting the data to a harmonic function will typically underestimate the annual cycle. However, the advantage of this procedure is to provide a common base for comparison of the contribution to the annual variation. In section 6, the errors in the harmonic analysis for the various contributions to the SSH are discussed. Prior to the harmonic analysis the altimeter data were detrended.

4. ANNUAL SSH VARIABILITY FROM ALTIMETRY

The annual harmonic amplitude and phase of SSH derived from the altimeter data are shown in Figure 2. Values of the amplitude lie between 4 and 8 cm, with the largest variation in the deeper basins: Greenland, Norwegian, and Lofoten Basins, and also along the Norwegian coast. Maximum sea level occurs first in the northeasterly Norwegian Sea during early September. Along the Norwegian coast and at the Scotland–Greenland Ridge the maximum occurs in early November, 2 months later than in the northeastern Norwegian Sea.

Along the eastern continental slope the phase is constant from the Faroe–Shetland Channel in the south to Spitsbergen in the north, indicating a barotropic response. This is in agreement with the results from *Orvik et al.* [2001]. From current measurements, they concluded that the inner branch of the NwASC is mainly barotropic and topographically trapped along the continental slope. Figure 2(a) shows that the amplitude along the eastern continental slope is also constant (~6 cm), at least up to the northern tip of Norway. Further north, at about 15°E, the amplitude decreases to about 4 cm. This is in the same area where the NwAC bifurcates into the northward West Spitsbergen Current and the eastward branch to the Barents Sea. Another minimum in amplitude occurs over the Vøring Plateau, at ~4 cm.

The annual harmonic cycle accounts for 40-60% of the total variance for most of the Nordic Seas (Figure 2c). In some regions it is considerable lower. Over the Vøring Plateau the minimum explained variance is only 30%. The numbers that we got for explained variance are lower than those *Isachsen et al.* [2003] obtained for the first principal component (69%) of the SSH for the Nordic Seas. This difference can be due to the different methods used. The harmonic fitting of the data, together with the lack of temporal strength in the annual cycle, will underestimate the annual variation. The spatial averaging of the altimeter data also modifies the explained variance since small and mesoscale variabilities are filtered out.

Differences in phase across the NwASC create SSH gradients and surface currents, even if the amplitudes are constant. An amplitude of 6 cm and a phase difference of 30 days at 62°-63°N gives an annual amplitude in sea level difference of 3 cm (Figure 3). The annual surface velocity amplitude of the NwASC can be calculated from the SSH. Assuming that the phase difference is taken over 126 km gives an annual amplitude of the velocity of 1.8 cm/s. Maximum SSH difference is during January and February, the same period when *Orvik et al.* [2001] observed the largest transport in the NwASC at 63°N. A sea level difference of 3 cm and a bottom depth of 400 m results in a seasonal variation of the barotropic volume transport equal to 0.9 Sv. This number is in accordance with recent estimates by *Orvik and Skagseth* [2003a]. At 69°-70°N the phase difference is larger resulting in an amplitude of the SSH difference of 7 cm (Figure 3). This gives an annual amplitude of the speed equal to 3 cm/s. The larger variability in the speed compared to that at 63°N is realistic since the continental slope is steeper there.

5. SEASONAL STERIC HEIGHT VARIABILITY

5.1. Steric Height From Hydrography

The steric height is calculated from the WOA01 data set using Equation 2. The mean northward surface velocity in the eastern Norwegian Sea, derived from the annual averaged steric height, is about 3 cm/s (not shown). The orientation of this flow has also a component toward the eastern continental margin. Through interaction with the topography this baroclinic flow is converted into, and thus feeds, the barotropic shelf edge current, the NwASC [*Orvik and Skagseth*, 2003b; and *Walin et al.*, 2003].

The annual harmonic mode of the steric height in the Nordic Seas is shown in Figure 4. The density is integrated from the surface to 600 m depth (or to the bottom when it is shallower than 600 m; a check showed only negligible differences when the integration was done to the bottom instead to 600 m). Away from boundaries, the amplitude is less than 4 cm. The large values along the East Greenland Margin are probably due to the seasonal ice melting and freshwater input from the rivers. However, when areas are ice-covered no satellite data are obtained, and it is therefore not possible to make conclusions about the difference in the steric height and altimetry at ice-covered areas (see Figure 1 for location of ice margin).

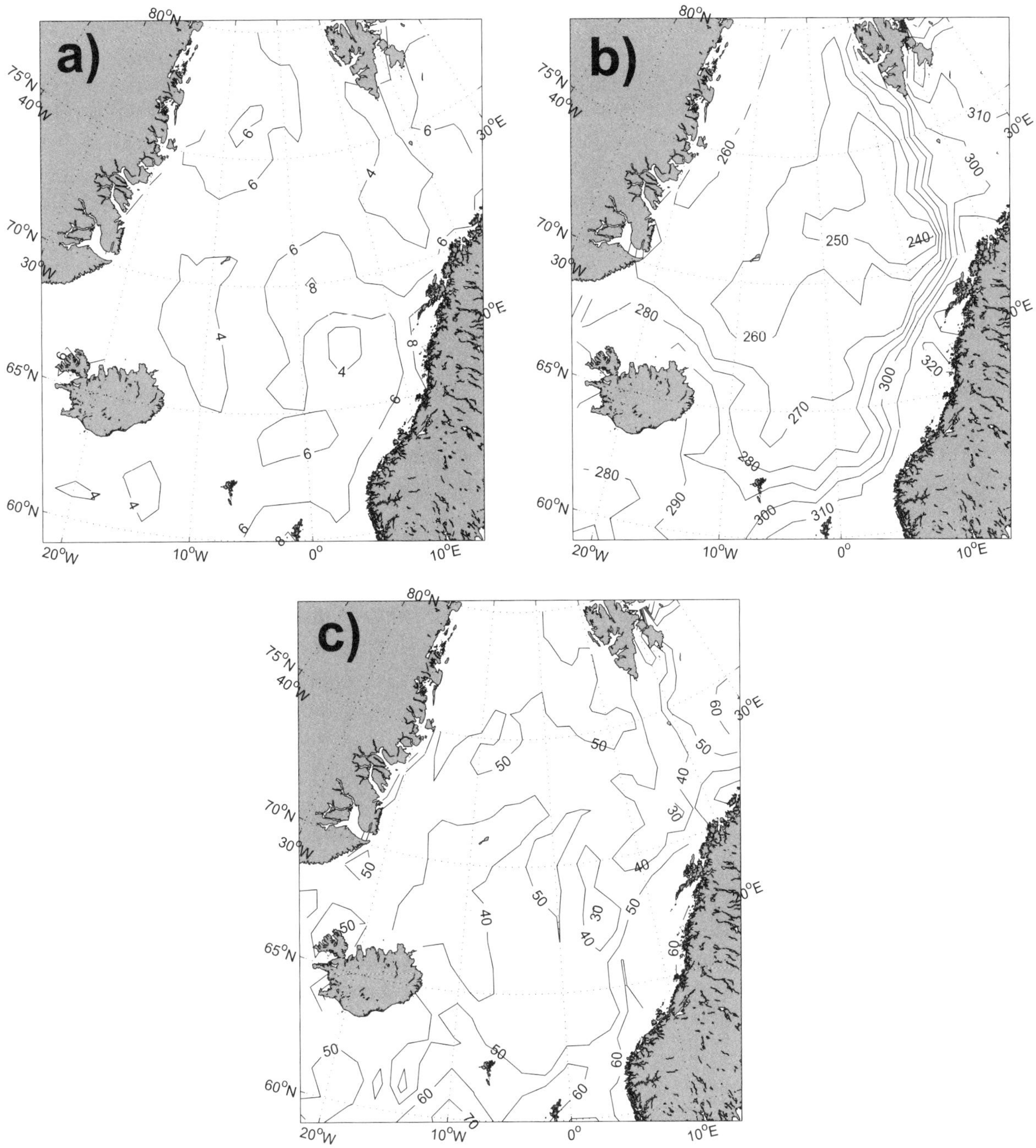

Figure 2. Harmonic analysis of SSH from altimetry(April 1995-February 2002). (a)Annual harmonic amplitude (in cm) and (b) day of year when maximum occur. Day 255 corresponds to mid-September. (c) Explained variance in percent of the annual harmonic cycle relative to the total variance.

There is also higher variability at the Norwegian continental shelf, where the water is affected by coastal water, and in the Faeroe–Shetland Channel. Lowest values are in the Iceland and Greenland Seas (~1.5 cm). The topographic influence of the Vøring Plateau is also seen. Maximum steric height occurs, for most of the region, less than 20 days later than that given from the altimetry (see Figure 2b). The latest occurrence of maximum sea level is found between northern Norway and Bear Island. There the maximum difference in phase between the altimetry and steric height is about two

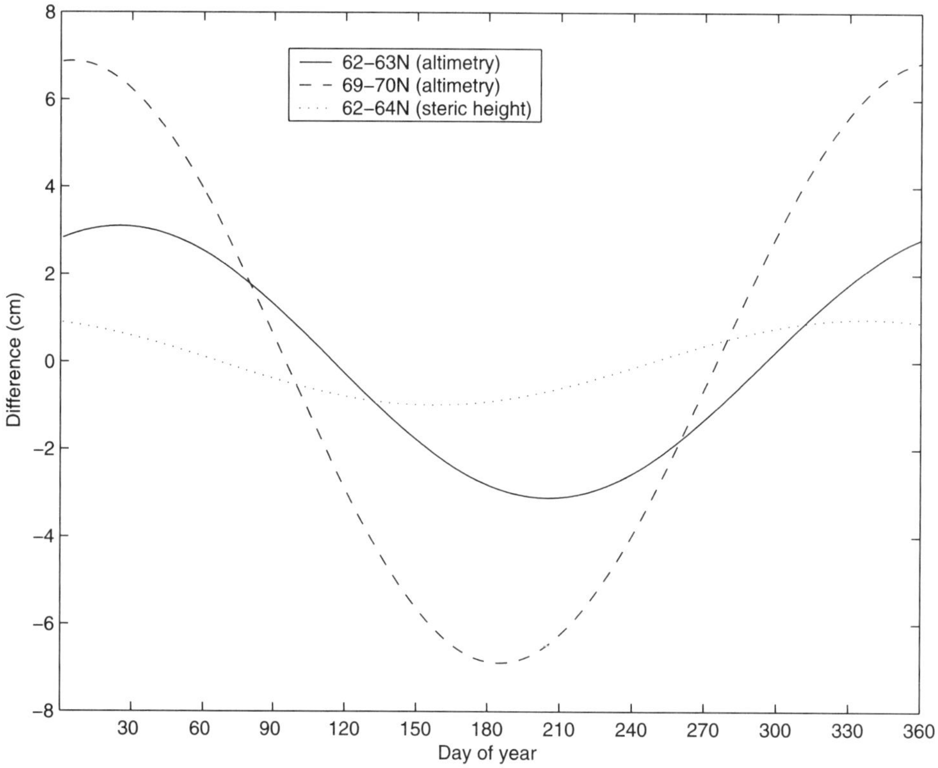

Figure 3. SSH difference (Δη) across the NwAC from the altimetry at 62°-63°N and 69°-70°N and from the steric height at 62°-64°N.

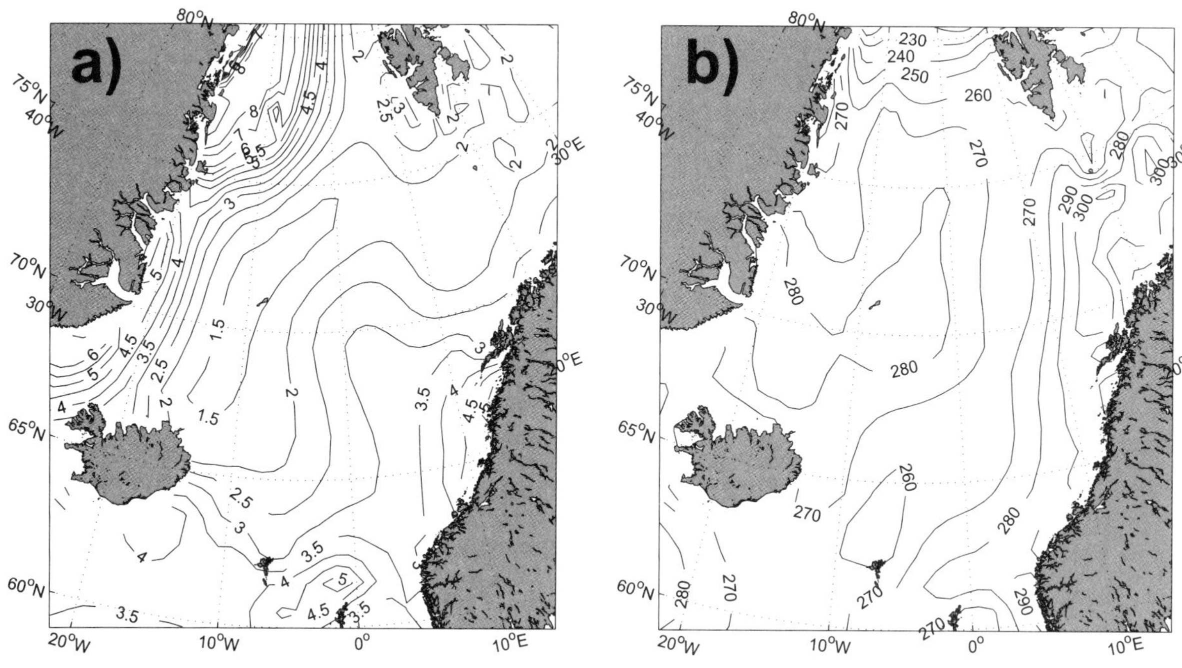

Figure 4. (a) Amplitude of the annual harmonic mode of the steric height (cm) in the Nordic Seas and (b) phase (day of year when maximum occurs). October month is the days from 270 to 300.

months. The steric height contributes ~40% of the total SLA for most of the area (compare Figures 2a and 4a).

As seen earlier, a difference in the east–west phase difference gives a seasonal variation in the transport (Figure 3). The surface currents that arise from steric height differences have an annual amplitude of 0.3 cm/s across the NwASC at 62°-64°N. Maximum northward velocity occurs in November/December, which is about 2 months different from that found by altimetry (Figure 3). The relatively low baroclinic speeds in the NwASC (compare 0.3 cm/s with 1.8 cm/s from altimetry), indicate that the current is mainly barotropic, which is in agreement with observations [*Orvik et al.*, 2001]. The difference between the total SSH change and the contribution from steric height is further investigated in section 7.

5.2. Steric Height From Local Surface Buoyancy Forcing

The contribution from the local surface buoyancy flux to the steric height was investigated using the net heat and freshwater fluxes (Equation 6). The freshwater fluxes used here are from the NCEP precipitation data set while the evaporation (E) is derived from the latent heat flux (Q_{LAT}) from $E = Q_{LAT}/L$, where L is the latent heat of vaporization. The annual ocean heat loss to the atmosphere varies from 20-40 W/m^2 in the south to about 140 W/m^2 in the north (not shown). The thermal expansion coefficient is averaged over the upper 300 m, using the WOA01 data sets. Due to large spatial variation in the mixed layer temperature the thermal expansion coefficient has large spatial differences.

The annual mean thermal expansion coefficient ranges from $0.4 \times 10^{-4}K^{-1}$ in the northwest to $1.6 \times 10^{-4}K^{-1}$ in the south, whereas the seasonal variation ranges from $0.05 \times 10^{-4}K^{-1}$ in the northwest to $0.15 \times 10^{-4}K^{-1}$ in the southeast (not shown). Thus, the coefficient varies roughly 10% in seasonal variation compared to the annual mean. The annual amplitude and phase of the heat flux contribution to the steric height are shown in Figure 5.

The largest amplitudes are on the eastern side of the Nordic Seas (~3.5 cm) and smallest at the western side (~0.5 cm). The large spatial variation in the thermal expansion coefficient explains a large part of this east–west difference. Maximum sea level derived from heat flux appears to occur at a similar time, in mid-September, for the whole region. The amplitude is very similar to that of the total steric height, derived from hydrography (compare Figures 4a and 5a). The greatest difference between the two amplitudes are in the East Greenland Current where large variation in freshwater input exists. Over ice-covered areas, heat flux variations are underestimated since the data set contains ice–atmosphere fluxes and not ocean–atmosphere fluxes. Thus, as with the altimeter data, it is difficult to make conclusions about the different results over ice-covered areas. The heat

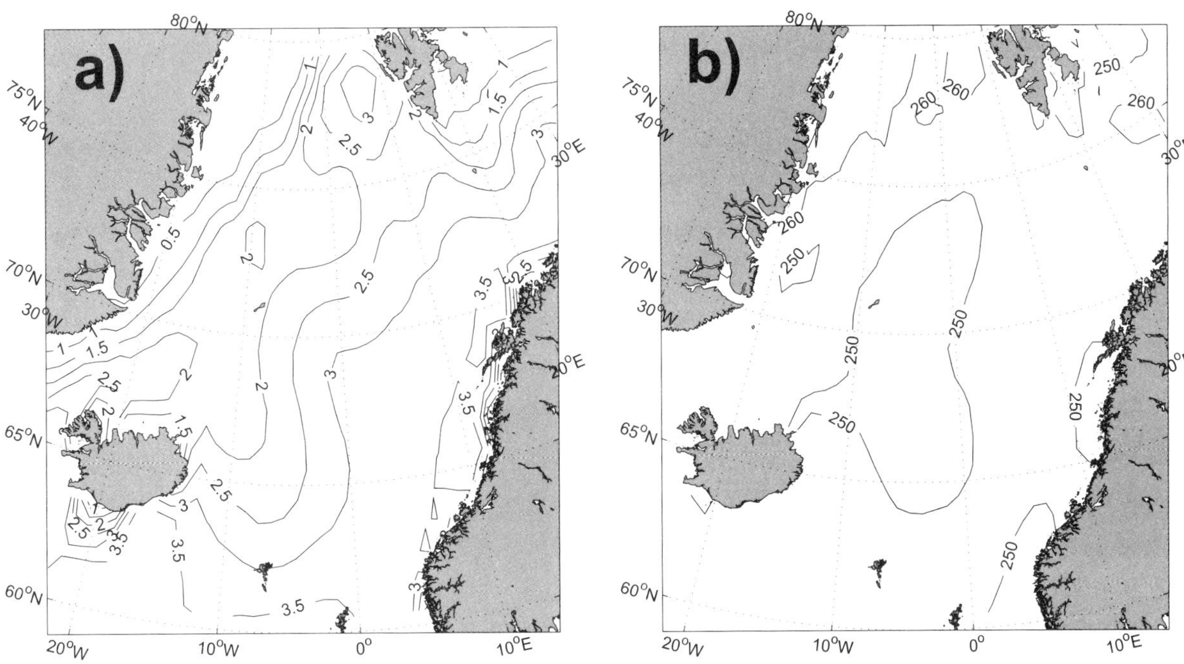

Figure 5. The net heat flux contribution to the steric height: (a) annual amplitude (in cm), (b) day of year when maximum occurs. Day 255 corresponds to mid-September.

flux contribution to the SSH has small spatial variation in phase and amplitude and thus results in associated currents that are hardly changed during the year.

The contribution of the fresh water flux to the steric height was very low (figure not shown). The annual amplitudes had typical values of less than 1 mm. Maximum values were 2-3 mm. The contributions from geostrophic advection and Ekman fluxes were also investigated and found to be small. For most of the area their contributions to the annual sea surface height variability was a few mm or less. At the northeastern coast of Greenland, the contribution from the Ekman flux can reach 1 cm maximum.

6. ERROR ESTIMATES OF AMPLITUDE AND PHASE

This section includes standard error estimates of the amplitude and phase in the harmonic analysis for SSH from altimetry, steric height, and heat flux. Two error components in amplitude and phase are derived: one that results from observation errors and one that is due to a more complex form of the annual cycle. Estimation of the first error can be found in the work of *Priestly* [1981], while a method to estimate the latter error is described by *Pawlowicz et al.* [2002]. This method is based on spectral analysis of the residual time series (after the harmonic component is removed). The errors of the harmonic coefficients are obtained by averaging the residual power spectrum over a frequency band, centered round the annual frequency. The appendix gives a more detailed description of the two error components.

The averaged amplitude and phase together with their error estimates for different regions are presented in Table 1. The errors in the harmonic analysis of the altimeter data are mainly due to the more complex form of the annual cycle and less to observation errors. These errors are between 0.7 and 1.4 cm in amplitude and between 8 and 13 days in phase. Largest errors are in the Lofoten Basin, where the amplitude also is largest. In contrast to altimetry, both steric height and heat flux have larger errors in the harmonic analysis due to errors in the observation. However, the errors derived from the residual spectrum method are questionable because those estimates are based on a low number of observations (i.e. climatologically monthly data sets). Steric height has, as for the altimetry, large errors in the Norwegian and Lofoten Basins. However, the errors are about half or less than those from altimetry.

In section 5, there was found a phase difference of ~20 days between the altimetry and the steric height. According to the error estimates of the phases, this difference may not be significant. The error estimates in Table 1 are also used in next section to give a rough estimate of the errors in the bottom currents.

7. SEASONAL VARIATION OF THE BOTTOM PRESSURE

In section 2, the different contribution to the SSH was introduced. The SSH anomaly due to changes in bottom pressure (η'_b) is calculated from Equation 1,

$$\eta'_b = \eta' - \eta'_s - \eta'_a. \tag{7}$$

The bottom pressure component can then be calculated by subtraction of the steric height (η'_s) from the altimeter data

Table 1. Overview of Averaged Amplitude and Phase With Standard Error Estimates in Four Different Regions.

Regions	Parameter	Amp	Err_{A_obs}	Err_{A_spec}	Phase	Err_{P_obs}	Err_{P_spec}
Greenland Sea	Altimetry	5.7	0.14	0.81	259	1.4	8.2
	Steric height	1.8	0.16	0.06[a]	278	4.9	1.9[a]
	Heat flux	1.3	0.06	0.06[a]	250	2.5	2.8[a]
Lofoten Basin	Altimetry	6.3	0.14	1.40	255	1.3	12.5
	Steric height	2.7	0.42	0.20[a]	270	9.3	4.5[a]
	Heat flux	2.7	0.12	0.07[a]	252	2.5	1.5[a]
Norwegian Basin	Altimetry	5.9	0.11	1.09	269	1.1	10.7
	Steric height	2.6	0.41	0.15[a]	263	9.1	3.3[a]
	Heat flux	2.5	0.11	0.04[a]	248	2.5	1.0[a]
Iceland Sea	Altimetry	4.2	0.13	0.70	263	1.7	9.8
	Steric height	1.7	0.21	0.06[a]	279	7.0	1.7[a]
	Heat flux	1.5	0.07	0.04[a]	251	2.5	1.5[a]

The regions are defined as: Greenland Sea (73°-76°N, 8°W-2°E), Lofoten Basin (69°-72°N, 0-10°E), Norwegian Basin (6°W-1°E, 63°-68°N), Iceland Sea (8°-18°W, 67.5°-71°N). The errors (Err) are errors in amplitude (A) and phase (P) from harmonic analysis estimates due to observation errors (obs) or to spectral analysis of the residual time series (spec).

[a] The estimated errors from the spectral analysis method have large uncertainties due to the few (i.e. 12) observations in the data set.

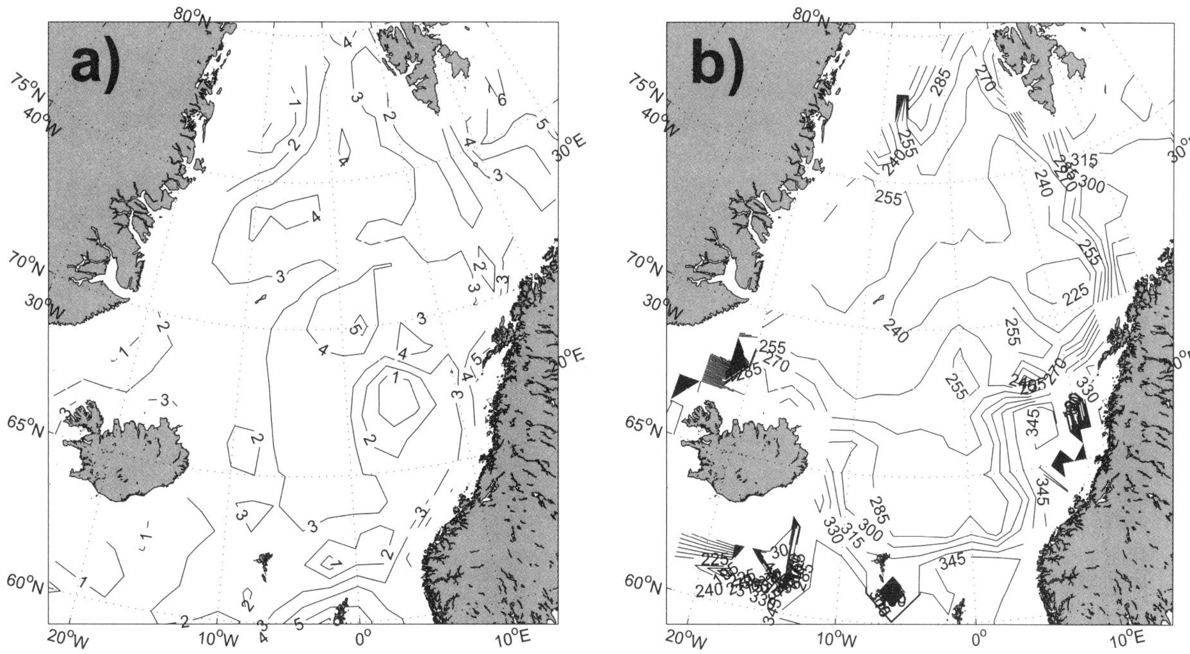

Figure 6. (a) Annual harmonic amplitude of the bottom pressure component of SSH (in cm). (b) Day of year when maximum occur. Day 255 corresponds to mid-September.

$(\eta' - \eta'_a)$. Figure 6 shows the annual amplitude and phase of the SSH derived from Equation 7. At the Vøring Plateau (about 4°E, 67°N) the amplitude is very small, less than 1 cm minimum. This indicates that changes in the SSH are mostly due to density changes in the water column. Lagrangian drifters have revealed a splitting of the northward surface currents south of the Vøring Plateau due to topographic steering [*Jacobsen et al.*, 2003]. This results in a longer residence time of the water above the Vøring Plateau and thus, a relatively larger effect of the ocean–atmosphere heat flux on the water column. The largest amplitudes are in the deep basins: Greenland, Norwegian, and Lofoten Basins.

The phase is very similar as to that of the altimetry. However, there is a phase change for most of the area of about 20 days. This difference may not be significant due to the errors in the phases (Table 1). The constant phase along the eastern continental slope, mentioned earlier, is also seen. Even if the amplitude in the NwASC is approximately half that of the altimetry only, the larger phase gradient in Figure 6b, compared with that of Figure 2b, gives about the same SSH difference (~3 cm at 62°-63°N).

The annual amplitude of the bottom velocity is calculated by using the phase and the amplitude of the bottom pressure. At 62°-63°N in the NwASC, the annual amplitude of the bottom velocity is 1.7 cm/s with maximum in January/February. This is approximately identical to that from the altimetry alone (1.8 cm/s), meaning that NwASC is mainly barotropic.

As the altimeter data are anomalies from a 7-year mean (1993-1999), bottom currents calculated from bottom pressure would be anomalies from that mean. Therefore, we have calculated seasonal bottom current anomalies from the annual mean instead. Using the annual amplitude and phase of the bottom pressure variation (see Figure 6) the bottom velocity anomalies are calculated from

$$u_b = -\frac{g}{f}\frac{\partial \eta_b}{\partial y} \tag{8a}$$

$$v_b = \frac{g}{f}\frac{\partial \eta_b}{\partial x}, \tag{8b}$$

where f is the Coriolis parameter.

From Table 1, errors in the bottom currents can also be roughly estimated. The error in the speed of the velocity (ε_v) can estimated from $\varepsilon_v = g\varepsilon_{\Delta\eta}/fL$ where L is the distance over the sea level difference ($\Delta\eta$). The error in the sea level difference ($\varepsilon_{\Delta\eta}$) is roughly $\varepsilon_{\Delta\eta} = 2^{1/2}\varepsilon_\eta$, where ε_η is the error in the sea level. Inserting realistic values ($f = 1.4 \times 10^{-4}\text{s}^{-1}$, $L = 100$ km) gives $\varepsilon_v \sim \varepsilon_\eta$. Thus, an error of 1 cm in the sea level gives an error of 1 cm/s in the speed. Error in the amplitude of the bottom pressure component is estimated from root sum squares of errors in altimetry and steric height. The estimated errors in amplitude then range from ~0.7 cm to ~1.5 cm with highest errors in the Lofoten Basin. This means that velocities with speeds of order mm/s may not be significant. These error estimates are, of course, not exact but are useful as a comparison with the derived bottom currents.

Seasonal averages of the bottom current anomalies are shown in Figure 7. The currents are mainly steered along the topography, as found by others [e.g., *Nøst and Isachsen*, 2003; *Poulain et al.*, 1996]. In the Norwegian Sea the circulation, relative to the mean, is cyclonic in the spring and anticyclonic in the autumn. The Greenland Sea also has a gyre circulation pattern but it is less apparent. Using wind stress and density fields, *Nøst and Isachsen* [2003] modeled the stationary bottom geostrophic circulation in the Nordic Seas.

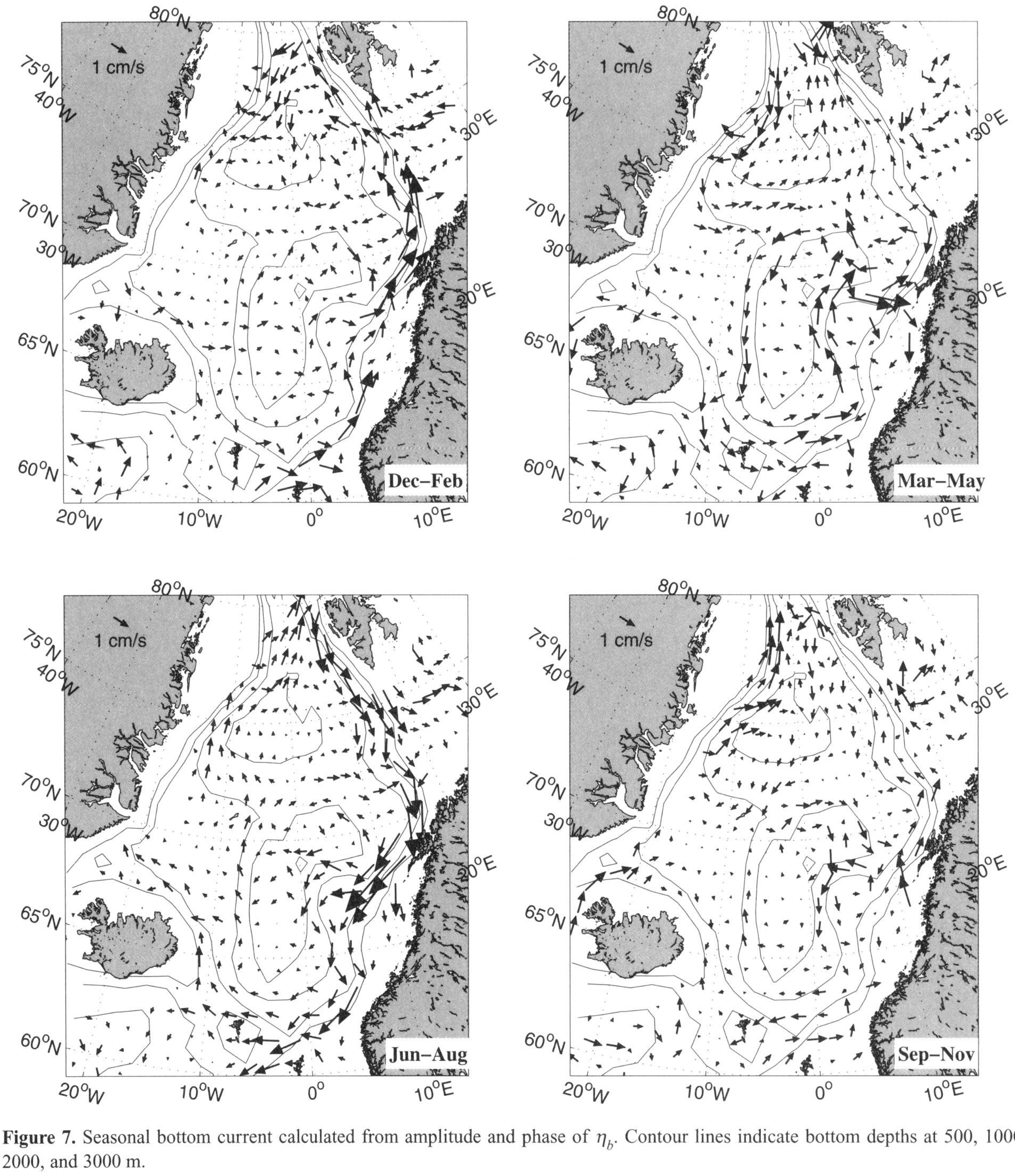

Figure 7. Seasonal bottom current calculated from amplitude and phase of η_b. Contour lines indicate bottom depths at 500, 1000, 2000, and 3000 m.

Their results gave a cyclonic circulation in the Greenland and Norwegian Seas with typical speeds of 5-10 cm/s. Our results then show that this circulation is intensified during spring and reduced during autumn. Maximum anomaly speeds in the Norwegian Sea are 1-2 cm/s.

The temporal change in the barotropic circulation around closed bottom contour lines is linked to the wind stress curl $(\nabla \times \vec{\tau})$ by

$$\frac{\partial}{\partial t}\oint_C \vec{v} \cdot d\vec{c} = \iint_A \frac{\nabla \times \vec{\tau}}{H} dA \qquad (9)$$

or

$$\frac{\partial}{\partial t}\oint_C \vec{v} \cdot d\vec{c} = \oint_C \frac{\vec{\tau}}{H} \cdot d\vec{c} \qquad (10)$$

where C is the closed contour and A is the area bounded by C. Figure 8 shows that the annual change in the velocity at a point along the 2500-m isobath coincides with the wind-stress curl, averaged over the area bounded by the 2500 m, with a difference in amplitude. We have only used the velocity at one point (7°W, 66.5°N), assuming it is a good representation (i.e. average) for all points along the 2500-m isobath in the Norwegian Sea. This simple assumption can overestimate (or underestimate) the average speed of the circulation. However, there seems to be a link between the seasonal wind forcing and the seasonal strength of the bottom circulation in the Norwegian Sea. Thus, the wind forcing speeds up the circulation ($\partial v/\partial t > 0$) from mid-September to mid-March when maximum speed is reached ($\partial v/\partial t = 0$, Figure 8).

There are several other works where links between wind forcing and circulation in the Norwegian Sea are found. Using an Ocean General Circulation Model and an analytical model, *Legutke* [1991] investigated the circulation in the Greenland and Norwegian Sea. She found coherent fluctuations between barotropic currents and wind stress integrated along similar f/H contours. With a theoretical depth integrated model and a numerical model, *Isachsen et al.* [2003] also came to a similar conclusion.

The NwASC is also revealed in Figure 7. It is strongest in the winter and weakest in the summer. Maximum anomaly speeds range from about 2 cm/s, in the south, to about 4 cm/s in the north. The time when maximum speed occurs, i.e. in the winter, is similar to what has been measured from

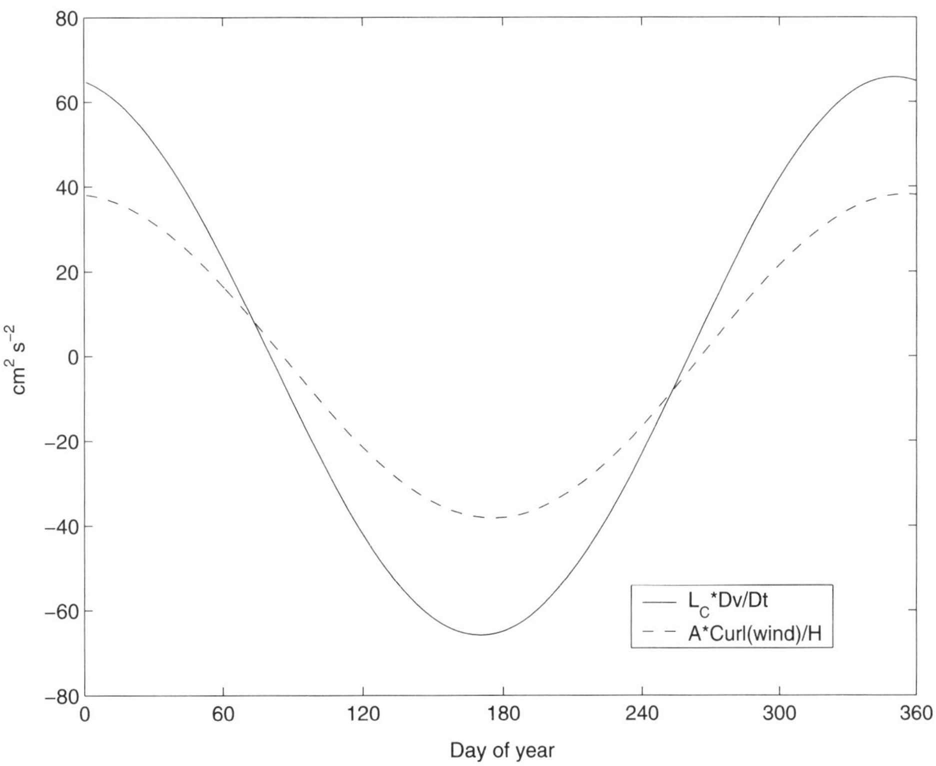

Figure 8. Temporal change in northward velocity at one point (7°W, 66.5°N) along the 2500-m isobath (solid line) and averaged wind-stress curl in the Norwegian Sea bounded by the 2500-m isobath (dashed line). L_C is the length of the contour C (i.e. the 2500-m isobath) and A is the area bounded by C. The data are harmonic fit.

current meters in the southern Norwegian Sea [*Orvik et al.*, 2001].

The inflow from the Norwegian Sea to the Barents Sea, which is mainly barotropic [*Ingvaldsen et al.*, 2002], is also greatest during winter. That is also in accordance with current meter measurements [*Ingvaldsen et al.*, 2004].

8. SUMMARY AND CONCLUDING REMARKS

In this study, we have used satellite SSH data, hydrographic data and reanalyzed atmospheric fluxes to quantify the various contributions to the annual SSH variation. The changes in the steric height and in the bottom pressure are estimated and the results are related to variations in currents and transports.

The annual amplitude of the SSH ranges from 4 to 8 cm with the largest variations in the deeper basins (Lofoten, Norwegian and Greenland Basins) and along the Norwegian coast. The steric height variation ranges from 1.5 to 4 cm for most of the region. Differing from *Volkov and van Aken* [2003], who found that the annual change in SSH is mainly due to the steric effect (60-80%), our results show that for the Nordic Seas the steric height contributes only about 40%. The main contribution to the steric height variation derives from the seasonal ocean–atmosphere heat flux variation. This is in accordance with other works [e.g. *Gill and Niller*, 1973; *Stammer*, 1997; and *Ferry et al.*, 2000]. The contributions to the steric height from geostrophic and Ekman fluxes were also investigated but were found to be small compared to that of the heat flux (less than 10%). The steric height explains a significant part (~40%) of the SSH variations but, as did *Gill and Niiler* [1973], we find that the associated currents are weak (i.e. few mm/s) due to small pressure gradients over large areas.

The annual variations in the bottom pressure, the dynamically active SSH component [*Gill and Niiler*, 1973], are largest (3-5 cm) in the Lofoten and the Greenland basins and at the Norwegian coast. Least variation is over the Vøring Plateau (~1 cm). Considering the bottom currents, some clear annual variations are identified both along the shelf break and for the interior circulation. Along the Norwegian continental shelf break the NwASC appears with nearly constant phase. The amplitude of the speed ranges from 2 cm/s to about 4 cm/s. The current is strongest in winter and weakest in summer, in agreement with observations in the southern Norwegian Sea [*Orvik et al.*, 2001]. The largest values are found at 69°N in accordance with *Poulain et al.* [1996] where the continental slope is steepest.

In the interior basins the annual variations of the currents are aligned in direction of the isobaths. The cyclonic large-scale circulation in the Norwegian Sea is strongest during spring and weakest during autumn. A similar circulation also occurs in the Greenland Sea but it is less apparent. Annual amplitude of the speed in the basin-circulation for the Norwegian Sea is about 1-2 cm/s, compared to the time-mean speed of 5-10 cm/s that *Nøst and Isachsen* [2003] calculated from an analytical model. The seasonal changes of the bottom circulation in the Norwegian Sea can be explained by the annual variability in the averaged wind stress curl in accordance with *Legutke* [1991] and *Isachsen et al.* [2003].

APPENDIX

A.1. Error Estimate of Averaged Altimeter Data in the 2°×1° Grid

The averaging of the altimeter data into 2° × 1° grid is performed by weighting (w_i) the altimeter data (ζ_i) within each grid box

$$h = \sum_{i=1}^{N} w_i \zeta_i ,$$

$$w_i = e_i^{-2} / \sum_{j=1}^{N} e_j^{-2}$$

N is the number of observations within each grid box and e_i is the standard error of the observation. The estimated error variance for the averaged value (h) is

$$\sigma_h^2 = \sum_{i=1}^{N} \sum_{j=1}^{N} w_i w_j \langle e_i e_j \rangle$$

If all the observations are uncorrelated ($\langle e_i e_j \rangle = 0,\ i \neq j$), the error variance becomes

$$\sigma_{uc}^2 = \left(\sum_{i=1}^{N} e_i^{-2} \right)^{-1}$$

Then the chosen w_i is the one that minimize the error. If instead all observations within the box are correlated, ($\langle e_i e_j \rangle = e_i^2$) the error variance becomes

$$\sigma_c^2 = N \left(\sum_{i=1}^{N} e_i^{-2} \right)^{-1} = N \sigma_{uc}^2$$

The mapping error variances in the data set are given as a ratio of the signal variance. By multiplying with the signal variance we then get the error variances (e_i^2). The mapping error in the altimeter data is a mix of both uncorrelated (measurement noise) and correlated (long-waveband) errors [*Ducet et al.*, 2000]. The white noise was computed by assuming a measurement noise of 2-cm rms for T/P and 3-cm rms for ERS. The long-wavelength error is taken as a uniform error of 2-cm rms for T/P and 4 cm rms for ERS. Since we have only the total error variance, we assume that it consists of equal parts of uncorrelated and correlated error. The standard error variance (σ_h^2) of the average value, h, then gets the form

$$\sigma_h^2 = \frac{N+1}{4}\left(\sum_{i=1}^{N} e_i^{-2}\right)^{-1}$$

A.2. Steric Height Error Estimation

The WOA01 data set contains at each grid point a mean of temperature (T_M) and salinity (S_M) together with a standard error of the mean. The estimated error in density (E_ρ) is calculated linearly by

$$E_\rho = \alpha(S_M, T_M, P)\cdot E_T + \beta(S_M, T_M, P)\cdot E_S$$

where P is the pressure and E_T and E_S are the standard error for temperature and salinity, respectively. The error variance in the steric height is then

$$\sigma_s^2 = \rho_0^{-2}\left(\sum_{i=1}^{N} E_{\rho_i}\Delta z_i\right)^2$$

Subscript i indicates an observation and N is the number of observations in the vertical. Assuming again that the density error in the vertical consists of equal parts of uncorrelated and correlated errors, the error variance then becomes

$$\sigma_s^2 = \frac{N+1}{4\rho_0^2}\sum_{i=1}^{N}(E_{\rho_i})^2(\Delta z_i)^2$$

A.3. Error Estimation in Harmonic Analysis

The time series, $y(t)$, are fit to

$$x(t) = C\cos(\omega t - \phi), \qquad \text{(A1)}$$

with the use of the least squares method. Equation A1 can be written

$$x(t) = A\cos\omega t + B\sin\omega t$$

where the amplitude (C) and phase (ϕ) are

$$C = (A^2 + B^2)^{1/2} \qquad \text{(A2)}$$

$$\phi = \tan^{-1}(B/A) \qquad \text{(A3)}$$

Errors in the observations result in errors in the harmonic coefficients. Following *Priestly* [1981], the error variance in A and B (σ_A^2, σ_B^2) due to observation errors (σ_ε) are estimated by

$$\sigma_A^2 = \sigma_B^2 = 2\sigma_\varepsilon^2/N.$$

The standard error of the amplitude and phase is estimated linearly from Equations A2 and A3 with

$$\sigma_F^2 = \left(\frac{\partial F}{\partial A}\right)^2\sigma_A^2 + \left(\frac{\partial F}{\partial B}\right)^2\sigma_B^2,$$

where F represents the amplitude (C) and the phase (P) [*Pawlowicz et al.*, 2002]. This gives the standard error variance of amplitude (σ_C^2) and phase (σ_ϕ^2)

$$\sigma_C^2 = (A^2 + B^2)^{-1}(A^2\sigma_A^2 + B^2\sigma_B^2), \qquad \text{(A4)}$$

$$\sigma_\phi^2 = (A^2 + B^2)^{-2}(A^2\sigma_B^2 + B^2\sigma_A^2). \qquad \text{(A5)}$$

When $\sigma_A^2 = \sigma_B^2 = 2\sigma_\varepsilon^2/N$, this results in

$$\sigma_C^2 = 2\sigma_\varepsilon^2/N,$$

$$\sigma_\phi^2 = 2(A^2 + B^2)^{-1}\sigma_\varepsilon^2/N.$$

In addition to the observation errors are errors due to the form of the annual cycle being more complex than a harmonic function. In the MATLAB toolbox T_TIDE, described by *Pawlowicz et al.* [2002], these errors were estimated by making a spectral estimate from the residual time series (after the harmonic component is removed). Then the power spectrum was averaged over a frequency band around the frequency of interest that gives the errors in the harmonic coefficients. They used this method for classical tidal harmonic analysis. We have used the method for the harmonic analysis of the annual cycle with a 6-month band. From the residual spectrum estimation, error variances of the coefficients $(\sigma_A^2, \sigma_B^2)_{spec}$ can be found. The errors in amplitude and phase are then estimated from Equations A4 and A5.

Acknowledgments. This work has been supported by the Bjerknes Centre for Climate Research, Norwegian Research Council project Norwegian Ocean and Climate (NOClim) and the European fifth framework programme project MAIA, EC contract EVK2-CT-1999-00008. Thanks to Henrik Søiland for valuable comments. Also, thanks to the anonymous reviewers whose comments and suggestions greatly improved the manuscript. The altimeter data (MSLA products) are supplied by the CLS Space oceanography Division, Toulouse, France. This is publication No. A66 of the Bjerknes Centre for Climate Research.

REFERENCES

Beare, D.J., and E. McKenzie (1999), Connecting ecological and physical time series: the potential role of changing seasonality, *Mar. Ecol. Prog. Ser.*, 78, 307-309.

Conkright, M.E., R.A. Locarminini, H.E. Garcia, T.D. O'Brien, T.P. Boyer, C. Stephens, and J.I. Antonov (2002), World Ocean Atlas 2001: Objective Analyses, Data Statistics, and Figures, CD-ROM Documentation. National Oceanographic Data Center, Silver Spring, MD, 17 pp.

Ducet, N., P.Y. LeTraon, and G. Reverdin (2000), Global high-resolution mapping of ocean circulation from TOPEX/Poseidon and ERS-1 and 2, *J. Geophys. Res.*, 105, C8, 19,477-19,498.

Ferry, N., G. Reverdin, and A. Oschlies (2000), Seasonal sea surface height variability in the North Atlantic Ocean, *J. Geophys. Res.*, 105, C3, 6307-6326.

Fu, L.-L. and A. Cazenave (2001), Satellite Altimetry and Earth Sciences. A handbook of Techniques and Applications, *International Geophysics Series*, Vol. 69, Academic Press, USA.

Gill, A.E, and P.P. Niller (1973), The theory of the seasonal variability in the ocean, *Deep-Sea Res.*, 20, 141-177.

Gjevik, B., and T. Straume (1989), Model simulations of the M2 and K1 tide in the Nordic Seas and the Arctic Ocean, *Tellus*, 41A, 73-96.

Gloersen, P., W.J. Campell, D.J. Cavalieri, J.C. Comiso, C.L. Parkinson, and H.J. Zwally (1992), Arctic and Antarctic Sea Ice, 1978-1987. Satellite Passive-Microwave Observations and Analysis, National Aeronautics and Space Administration, SP-511, Washington D.C.

Godøy. Ø. (1998), Climatic variability in the Nordic Seas. Dr. Scient thesis, Univ. of Bergen, Norway.

Ingvaldsen, R., H. Loeng, and L. Asplin (2002), Variability in the Atlantic inflow to the Barents Sea based on a one-year time series from moored current meters. *Cont. Shelf Res.*, 22, 505-519.

Ingvaldsen, R., L. Asplin, and H. Loeng (2004) The seasonal cycle in the Atlantic transport to the Barents Sea during the years 1997-2001, *Cont. Shelf Res.*, 24, 1015-1032.

Isachsen, P.E, J.H. LaCasce, C. Mauritzen, and S. Häkkinen (2003), Wind-driven variability of the large-scale recirculating flow in the Nordic Seas and Arctic Ocean, *J. of Phys. Ocean.*, 33, 2534-2550.

Jacobsen, P.K., M.H. Ribergaard, D. Quadfasel, T. Schmidt, and C. Hughes (2003), The near surface circulation in the Northern North Atlantic as inferred from Lagrangian drifters: variability from mesoscale to interannual, *J. Geophys. Res.*, 108(C8), 3251, doi: 10.1029/2002JC001554.

Jonsson, S. (1991), Seasonal and inter-annual variability of the wind-stress curl over the Nordic Seas, *J. Geophys. Res.*, 96, 2649-2659.

Legutke, S. (1991), A numerical investigation of the circulation in the Greenland and Norwegian Seas, *J. of Phys. Ocean.*, 21, 118-148.

LeTraon, P.Y., and F. Ogor, (1998), ERS-1/2 orbit improvement using TOPEX/Poseidon: The 2 cm challenge, *J. Geophys. Res.*, 103, 8045-8057.

Mork, K.A., and J. Blindheim (2000), Variations in the Atlantic inflow to the Nordic Seas, 1955-1996, *Deep-Sea Res. I*, 47, 1035-1057.

Nøst, O.A., and P.E. Isachsen (2003), The large-scale time-mean circulation in the Nordic Seas and Arctic Ocean estimated from simplified dynamics, *J. of Mar. Res.*, 61, 175-210.

Orvik, K.A., and P. Niiler (2002), Major pathways of Atlantic water in the northern North Atlantic and Nordic Seas toward the Arctic, *Geophys. Res. Lett.*, 29 (19), 1896, doi:10.1029/2002GL015002.

Orvik, K.A., Ø. Skagseth, and M. Mork (2001), Atlantic inflow to the Nordic Seas. Current structure and volume fluxes from moored current meters VM-ADCP and Seasoar-CTD observations 1995-1999, *Deep-Sea Res. I*, 48 (4), 937-957.

Orvik, K.A., and Ø. Skagseth (2003a), Monitoring the Norwegian Atlantic slope current using a single moored current meter, *Cont. Shelf Res.*, 23, 159-176.

Orvik, K.A., and Ø. Skagseth (2003b), The impact of the wind stress curl in the North Atlantic on the Atlantic inflow to the Norwegian Sea toward the Arctic, *Geophys. Res. Lett.*, 30(17), 1884, doi:10.1029/2003GL017932.

Pawlowicz, R., B. Beardsley, and S. Lentz (2002), Classical tidal harmonic analysis including error estimates in MATLAB using T_TIDE, *Computers and Geosciences*, 28, 929-937.

Poulain, P.-M., A. Warn-Varnas, and P.P. Niiler (1996), Near surface circulation of the Nordic seas as measured by Lagrangian drifters. *J. of Geophys. Res.*, 101(C8), 18237-18258.

Priestly, M. (1981), Spectral Analysis and Time Series, Academic Press.

Skagseth, Ø., and K.A. Orvik (2002), Identifying fluctuations in the Norwegian Atlantic Slope Current by means of Empirical Orthogonal Functions, *Cont. Shelf Res.*, 22, 547-563.

Skagseth, Ø. (2004), Monthly to annual variability of the Norwegian Atlantic slope current: connection between the northern North Atlantic and the Norwegian Sea, *Deep-Sea Res. I*, 51, 349-366.

Stammer, D. (1997), Steric and wind-induced changes in the TOPEX/Poseidon large-scale sea surface topography observations, *J. Geophys. Res.*, 102, 20.987-21.009.

Volkov, D.L. and H.M. van Aken (2003), Annual and interannual variability of sea level in the northern North Atlantic Ocean, *J. Geophys. Res.*, 108(C6), 3204, doi:10.1029/2002JC001459.

Walin, G., G. Broström, J. Nilsson, and O. Dahl (2003), Baroclinic boundary currents with downstream decreasing buoyancy; a study of an idealized Nordic Seas system, *J. of Mar. Res.*, submitted.

Kjell Arne Mork, Institute of Marine Research, Nordnesgaten 50, P.O. 1870 Nordnes, N-5817 Bergen, Norway. (kjell.arne.mork@imr.no)

Øystein Skagseth, Bjerknes Centre for Climate Research, Allégaten 55, N-5007 Bergen, Norway.

The Mean Sea Level Pressure Gradient Across the Denmark Strait as an Indicator of Conditions in the North Icelandic Irminger Current

Johan Blindheim

Institute of Marine Research, and Bjerknes Centre for Climate Research, University of Bergen, Bergen, Norway

Svend-Aage Malmberg

Marine Research Institute, Reykjavik, Iceland

Winter means of the atmospheric sea level pressure gradient across the Denmark Strait were correlated to a time series of temperatures at a repeat hydrographic station, S3, in the waters north of Iceland. These data show that variations in North Icelandic shelf waters are closely related to the gradient. The correlation between 3-year moving averages of the gradient and the temperature at S3 of the following year was 0.70. This climatic relation clearly reflected the abrupt change from Atlantic to Arctic conditions on the North Icelandic shelf in the 1960s and the variability since then.

INTRODUCTION

Areas that are of importance for the present study are the Irminger Sea and the southern Iceland Sea. In the Irminger Sea the Irminger Current carries Atlantic water, deriving from the North Atlantic current, northward along the west coast of Iceland. Off northwest Iceland, where the current meets the Greenland–Iceland Ridge, it splits into two branches. The northern one, the North Icelandic Irminger Current, turns around Horn, the northwestern corner of Iceland, to flow eastward over the North Icelandic shelf. The other branch turns southwest, and follows the east coast of Greenland toward Cape Farewell, parallel with and underneath the Arctic and Polar waters of the East Greenland Current (Figure 1).

The oceanographic conditions in the upper layers off North Iceland are characterized by these relatively warm and salty Atlantic waters of the North Icelandic Irminger Current, and the cold and relatively fresh Arctic waters in the East Icelandic Current, which diverges from the East Greenland Current to the north of the Denmark Strait. Hence, as these two quite different water masses flow eastward, they mix with each other along the shelf edge. This mixing considerably modifies the water masses, especially during winter when seasonal cooling creates convective mixing. According to *Stefánsson* [1962] this mixing during winter forms a local water mass, the North Icelandic Winter Water. This water mass merges into the East Icelandic Current as it flows into the south-western Norwegian Sea. As the current proceeds eastward over the northern slope of the Iceland–Faroe Ridge, the mixing and modification continues. In this area *Read and Pollard* [1992] defined the mixing product as Modified East Icelandic Water. Proceeding further east, these waters, originating from north of Iceland, flow into the southern Norwegian Sea and the Faroe–Shetland Channel. In this area, most of these waters occur as an intermediate Arctic water mass, but particularly during summer, its less dense versions may mix into the Atlantic waters in the surface layers of the Norwegian Atlantic Current [*Blindheim et al.*, 2000].

Based on computation of geostrophic current in the hydrographic repeat sections north of Iceland, *Stefánsson* [1962] assessed the volume flux of Atlantic Water in the North Icelandic Irminger Current to be 0.6 Sv. Later *Kristmannsson* [1998] assessed the average eastward water

The Nordic Seas: An Integrated Perspective
Geophysical Monograph Series 158

10.1029/158GM06

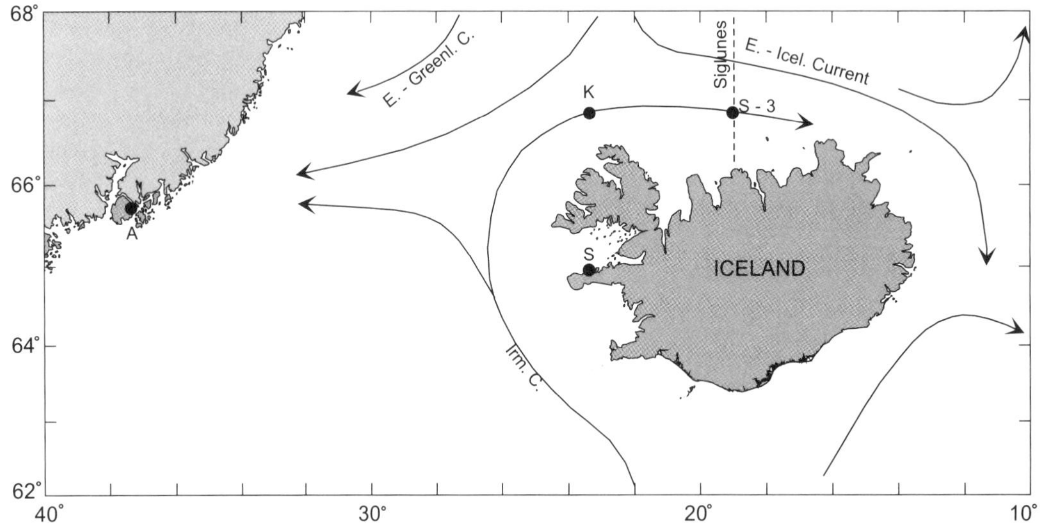

Figure 1. Schematics of the main ocean currents in Icelandic waters. Stations mentioned in the text: (A) Angmagssalik, (S) Stykkishólmur, (K) Køgur, (S-3) Station 3 on the Siglunes Section.

transport off Kögur to be 1.5 Sv. The assessments were based on a 5-year-long series, 1984-1990, of current measurements near the core of the North Icelandic Irminger Current, over a bottom depth of 250 m. The transport computations were based on only one mooring with two current meters and could have too little data to give reliable results. More recent year-long recordings (1999-2000) from an array of three moorings across the inflow revealed a transport of 0.95 Sv [*Jónsson and Briem*, 2003].

This flow of Atlantic Water into the North Icelandic region is of great importance for both oceanographic and biological conditions in the area. It has been shown that the temperature and salinity in the North Icelandic waters, as shown in the standard section off Siglunes, were drastically reduced during the late 1960s [*Malmberg*, 1969 and 1986; see also *Malmberg and Kristmannsson*, 1992]. This was the beginning of the "Great Salinity Anomaly" [*Dickson et al.*, 1988]. The biological effects in the region were drastic and have been described by many authors, for example, *Thórdardóttir* [1977]; *Ástthórsson et al.* [1983]; *Stefánsson and Ólafsson* [1991]; *Gudmundsson*, [1998]; *Ástthórsson and Gíslason*, [1998]; *Jakobsson and Østvedt* [1999].

It was shown early that the variations in oceanographic conditions north of Iceland were associated with atmospheric forcing. For example, *Stefánsson* [1962], and *Stefánsson and Gudmundsson* [1969] showed a distinct relation between the atmospheric pressure distribution in spring and the hydrographic conditions in summer. *Jónsson* [1992] suggested that the mechanism by which freshwater is released from the East Greenland Current into the Iceland Sea is governed by the mainly positive wind stress curl over the area. There is an obvious trend for large freshwater content to be associated with a low wind stress curl. Similarly, *Malmberg and Kristmannsson* [1992] found that a high positive wind stress curl over the Irminger and Iceland Seas was associated with a large supply of Atlantic Water into the shelf area to the north of Iceland, as observed 6 to 9 months later. The freshwater runoff from Iceland does not contribute much to the freshwater content in the Iceland Sea. Working on data from 1956 to 1998, *Malmberg and Jónsson* [1997] described the relationship between deep convection in the Iceland and Greenland Seas and the wind stress curl. *Ólafsson* [1999] concluded that the hydrographic regime responded to the shift between local southerly and northerly winds. He further found that the North Atlantic Oscillation (NAO) does not explain observed hydrographic variations in North Icelandic waters. In 5-year moving averages of the data from the Siglunes Section, however, *Malmberg and Valdimarsson* [2003] found some relation to the NAO.

As recently well-described in *Hurrell et al.* [2003], the NAO is a prominent mode of atmospheric variability in the North Atlantic sector. *Visbeck et al.* [2003], reviewed the ocean's response to this variability, which includes variations in circulation and properties on scales ranging from local to basin wide. Large-scale variability may also be attributed to the Arctic Oscillation [e.g. *Thompson and Wallace*, 1998].

Local variability is, however, also governed by smaller scale atmospheric features, and in the following we will briefly describe an oceanic response to variations in the Greenland High in relation to the Iceland Low. This response is shown by an inverse correlation between hydrographic conditions north of Iceland and the atmospheric mean sea level pressure (MSLP) difference between Angmagssalik,

East Greenland, and Stykkishólmur, Iceland (Figure 1). This difference may be considered as an index of the mean gradients in the field of MSLP over the Denmark Strait.

DATA

The atmospheric pressure data are from *Frich et al.* [1996] with updates. Data from Stykkishólmur on the west coast of Iceland (65°05′N, 22°44′W) and Angmagssalik on the east coast of Greenland (65°36′N, 37°38′W) are of special interest because these stations have a series of observations during the period from 1900 to 1999, albeit with some gaps in the series from Angmagssalik. Shorter series of MSLP data from two stations further north on the East Greenland coast, Scoresbysund (70°29′N, 22°00′W) and Danmarkshavn (76°46′N, 18°46′W), are also used to indicate the temporal variations in the atmospheric high pressure cell over Greenland. The oceanographic data are from the Icelandic Siglunes standard section worked in spring along 18°50′W across the northern shelf to 1000 m deep over the slope (Figure 1). The selected data are from the third station from the shore (S3) and from a selected depth (50 m), where the variability is characteristic for the region as a whole [*Malmberg and Desert*, 1999].

RESULTS AND DISCUSSION

Figure 2 shows 5-year moving averages of winter MSLP, averaged over the months December through March, from three stations on the east coast of Greenland: Angmagssalik, Scoresbysund, and Danmarkshavn. Although the series of observations are of different lengths, it is clear that they have similar large-scale variations. Further, the differences between the stations were fairly constant, and the MSLP was increasing northward along the coast. The difference in pressure during winter between Scoresbysund and Angmagssalik was on average 6.5 hPa (hectopascals), with a standard deviation of 0.7 hPa. In general, Figure 2 shows that the MSLP over Greenland increased from about 1920 to the late 1960s. Since then, the trend has been toward a lower MSLP.

Figure 3 shows that the MSLP variations in the Iceland Low, here represented by the time series from Stykkisholmur, were very similar to the fluctuations in the Greenland High, as represented by the time series from Angmagssalik. Normally the MSLP at Angmagssalik was higher than at Stykkisholmur, but this was not always the case. Particularly in the period 1947-1954, Stykkisholmur had higher MSLP than Angmagssalik. For the whole period the difference in MSLP between Angmagssalik and Stykkishólmur varied between –3.6 hPa (1948) and 6.9 hPa (1912) with an overall average of 1.82 hPa and a standard deviation of 2.32 hPa. This relation is further demonstrated by the winter MSLP difference between Angmagssalik and Stykkishólmur over the period 1900-1999 as shown in Figure 4. The gradient indicates the average wind pattern during the different winters in the Denmark Strait area; values above zero represent prevailing northerly winds.

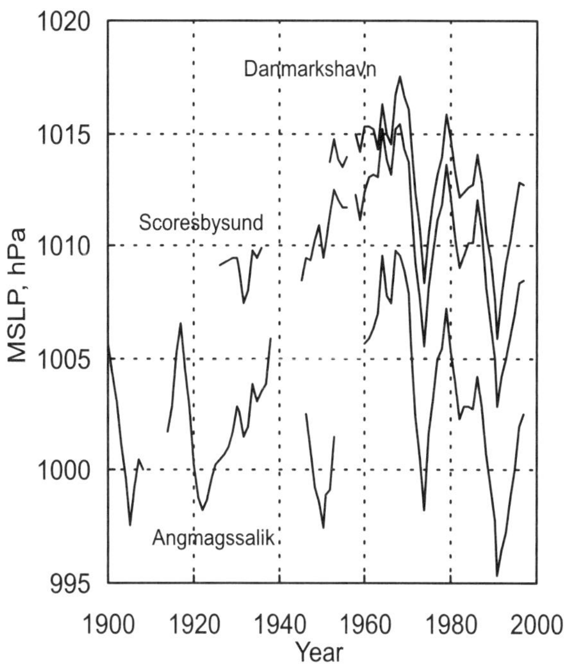

Figure 2. Five-year moving averages of atmospheric mean sea level pressure (hPa), December through March along the East Greenland Coast, at Angmagssalik since 1900, at Scoresbysund since 1926, and at Danmarkshavn since 1952.

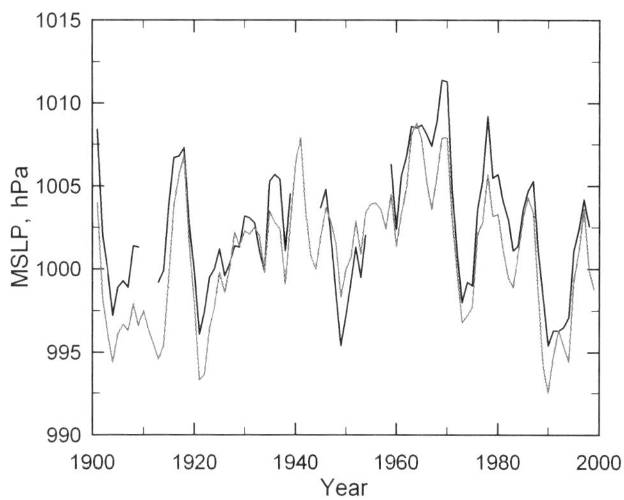

Figure 3. Three-year moving averages of mean sea level pressure (hPa), December through March (1900-1999) at Angmagssalik (graph with gaps) and Stykkishólmur.

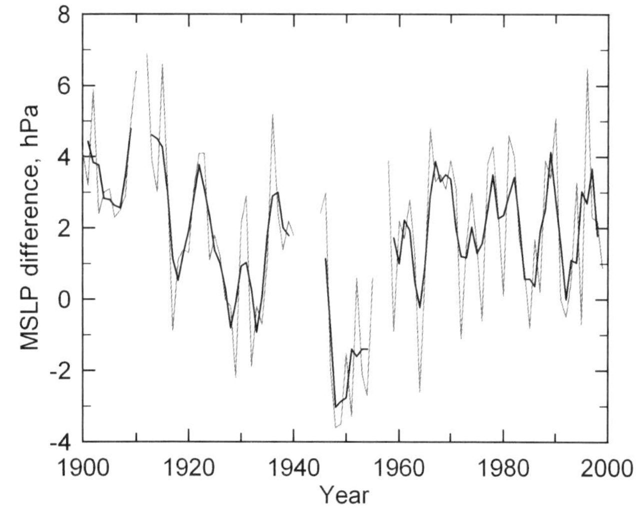

Figure 4. Mean sea level pressure difference (hPa) between Angmagssalik and Stykkisholmur with 3-year moving averages.

Although gaps in the data from Angmagssalik break the series during some intervals, Figure 4 shows that the highest winter mean pressure differences occurred around 1910, while the longest period with negative difference occurred during the late 1940s and the first half of the 1950s. These years, which formed an extreme for the whole century, fell within a period of relatively low values that lasted from about 1930 to the 1960s. These years with low frequency of northerly winds resulted in high Atlantic influence in North Icelandic waters, as also described by *Stefánsson* [1969], based on a long series of sea surface temperatures, since 1876, from the island of Grímsey near the Siglunes section. The abrupt increase in pressure difference across the Denmark Strait, from near or below zero in the early 1960s to a high of around 4 hPa at the end of the decade, had great environmental consequences, as referred to in the introduction.

The oceanographic response is demonstrated by the conditions in North Icelandic waters. The correlation coefficient during the period 1950-1999 between the inverse winter MSLP difference (each January) across the Denmark Strait and the temperature at 50 m deep at station S3 in the spring the same year was $r = 0.52$. The highest correlation, $r = 0.70$, was obtained between the 3-year moving averages of the winter MSLP gradient and 3-year averages of the S3 temperature, centered around the spring of the following year, as illustrated in Figure 5. The salinity at S3 at 50 m is also shown in Figure 5 and, not unexpectedly, is similarly correlated to the pressure difference. For comparison, the values of the difference are inverted and entered in hPa.

These results may look somewhat in conflict with earlier findings; this may not be the case, however, if the difference in spatial scale is taken into consideration. The atmospheric pressure difference between Angmagssalik and Stykkishólmur will reflect rather large-scale variations in the Irminger Sea and in the Denmark Strait. Besides reducing the volume flux of Atlantic Water in the Irminger current as a whole, prevailing northerly winds will further promote increased supply of Atlantic Water into the Irminger Current as it flows along the East Greenland slope. The wind-driven supply of Atlantic Water to the North Icelandic Irminger Current will be

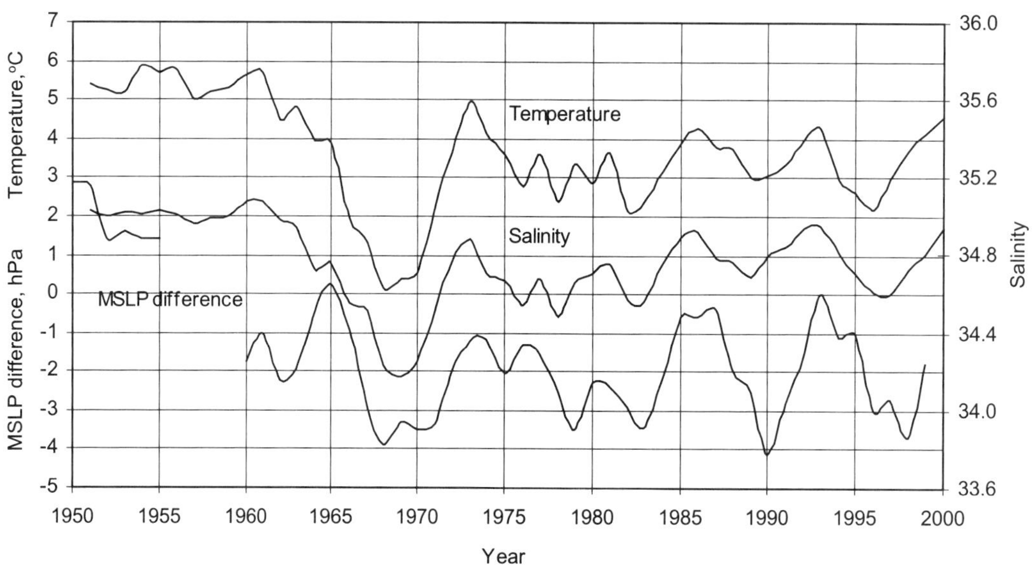

Figure 5. Three-year moving averages of the temperature (°C) and salinity at Station S3, 50 m deep, in spring 1950-2000, and inverse MSLP difference (hPa) between Angmagssalik and Stykkisholmur 1 year earlier.

reduced accordingly. It has been shown that the wind conditions have similar effects as far south and west as the former Ocean Weather Station A, which is located at 62°N,33°W in the Irminger Sea [*Blindheim*, 1968]. The delay of variations on the shelf north of Iceland in relation to the winter MSLP gradient across the Denmark Strait may, therefore, be reasonable as it has regional rather than local effects. The correlation given above is, however, only based on oceanographic data from spring, and the obtained delay may, therefore, not be exact.

On decadal time scales, using the winter MSLP gradient as a proxy, Figure 4 suggests that there was low supply of Atlantic Water to the North Icelandic area during the first two decades of the 20th century, which is in fair agreement with the results of *Stefánsson* [1969], based on the data from Grímsey. This was followed by an approximately 30-year period with larger Atlantic influence. Within this period, the late 1920s, the early 1930s, and particularly the period 1947-1954 seemed to be favored by a large supply of Atlantic Water to the shelf north of Iceland. This is also in agreement with the findings of *Stefánsson* [1962], who examined data from 1949 to 1960. In contrast, the strong northerly winds when the high pressure over Greenland peaked during the late 1960s led to years with cold conditions north of Iceland and drift ice blocking the north and east coasts of the country [*Malmberg*, 1969]. This also coincided with the period when the NAO index was at its lowest [*Hurrell*, 1995]. This intense and persistent high-pressure anomaly cell over Greenland [*Rodewald*, 1967] was the ultimate source of the cold and fresh waters of the great anomaly of the 1970s [*Dickson et al.*, 1988]

Most of the water carried by the North Icelandic Irminger Current mixes into surrounding waters, mainly in the East Icelandic Current, and flows into the southwestern Norwegian Sea, where it forms a cold tongue of Arctic character as described by *Helland-Hansen and Nansen* [1909]. The position of the Arctic Front between this water and the Atlantic waters of the Norwegian Atlantic Current may vary from year to year [*Shevchenko and Isaev*, 1985; *Monstad and Blindheim*, 1986]. Further to the east, the Arctic water mainly occurs at intermediate depths underneath the Atlantic Water in the southern Norwegian Sea and the Faroe–Shetland Channel [*Turrell et al.*, 1999]. As described by *Blindheim et al.* [2000], the narrowing of the Norwegian Atlantic Current, or in other words, the eastward shift of the Arctic Front, since the 1960s was associated with fluctuations in the NAO. Also the cooling and freshening in the upper layers of the western Norwegian Sea were ascribed to the same wind forcing. This effect shall not be ignored, but variations in the properties of the waters carried into the area by the North Icelandic Irminger Current also contribute. These characteristics, and possibly also the amount of the Arctic waters, may vary with the conditions north of Iceland. The variability in the inflow of Atlantic Water carried by the North Icelandic Irminger Current may thus be reflected in the properties of the waters flowing into the southern Norwegian Sea.

The difference between the 1950s and the 1990s in extent of water deriving from north of Iceland is demonstrated in Figure 6, which shows sections across the Norwegian Basin in the spring of 1955 and 1997. In 1955, Modified East Icelandic Water was observed only at the western end of the section, where salinities were below 34.9 in the upper 250 m, and temperatures were below 0°C except in the warmed surface layer. The volume of Atlantic Water was much larger in 1955 than in 1997. In 1955 it was distributed across the entire Norwegian Basin, to 8°W over the slope of the Iceland Plateau. In the upper 200-300 m was a continuous layer across the basin, but its vertical extent was considerably greater over the slopes, to 700 m off Norway and 400 m over the slope toward the Iceland Plateau. In this area the salinities in the upper 200 m were up to 35.12, and the Jan Mayen Front [*Read and Pollard*, 1992] between the Atlantic Water and the Modified East Icelandic Water was well developed. The accuracy of salinity determinations during the 1950s was rather low, and the salinities could be up to 0.03 too high, but even reducing the salinity by that amount would not change the observed distribution of Atlantic Water.

In 1997, Atlantic Water occurred mainly over the Norwegian slope, and it hardly extended to the prime meridian. West of this longitude the upper layers had salinities between 34.8 and 34.9, although the core of southward flowing waters, deriving from the western branch of the Norwegian Atlantic Current, was indicated by salinities between 34.9 and 34.95 at depths between 100 and 400 m. Arctic waters occupied the upper layers to the depth of almost 1000 m across the whole Norwegian Sea; in the eastern part of the section, it formed an intermediate layer between the Atlantic Water and the deep waters below. The deeper component of this water is known as Norwegian Sea Arctic Intermediate water and derives from both the Iceland and Greenland Seas [e.g., *Hansen and Østerhus*, 2000].

The section from 1955 occurred in the period when the MSLP gradient across the Denmark Strait was at a low level and conditions on the North Icelandic shelf were characterized by Atlantic Water of temperatures around 6°C and salinities up to 35.07 at S3 in spring (Figure 5). On the other hand, 1997 was within the period with high gradients and accordingly low Atlantic flux in the North Icelandic Irminger Current; corresponding temperature and salinity were 3.07°C and 34.62, respectively.

The supply of Atlantic Water via the North Icelandic Irminger Current does contribute to the difference found between the two sections in Figure 6, though it is mainly related to variability in the NAO.

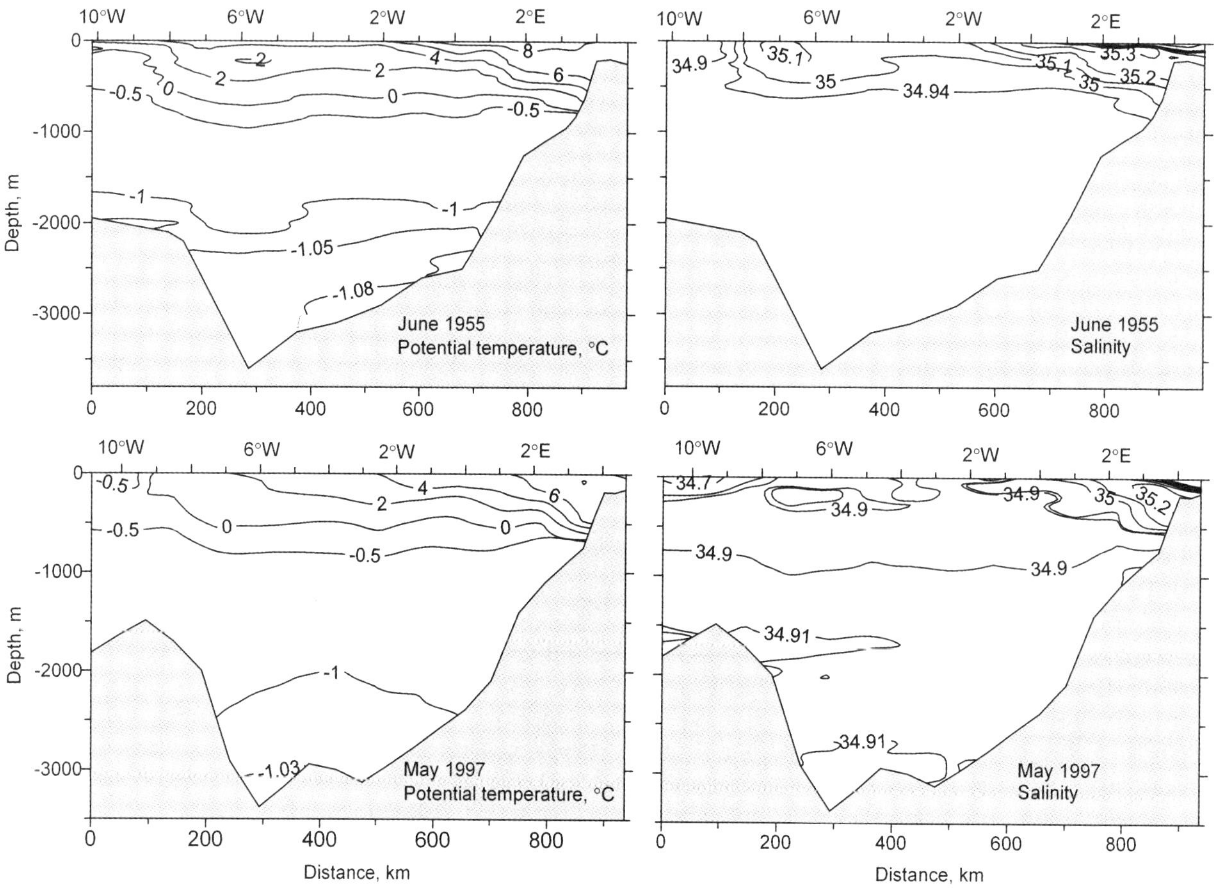

Figure 6. Potential temperature (°C) and salinity in a section across the Norwegian Basin in June 1955 (from 62°13′N, 04°53′E to 68°00′N, 10°20′W), and in a section along almost the same line in May 1997 (from 61°21′N, 04°40′E to 67°43′N, 10°53′W).

Acknowledgments. The authors extend their thanks to the three reviewers, who all provided useful criticism and advice. Thanks are also due to the Bjerknes Centre for Climate Research, for providing traveling funds.

REFERENCES

Ástthórsson, Ó.S., I. Hallgrímsson, and G.S. Jónsson, (1983) Variations in zooplankton densities in Icelandic waters in spring during the years 1961-1982. *Rit Fiskideildar* 7: 73-113.

Ástthórsson, Ó.S. and Á. Gíslason (1998) Variability in zooplankton biomass in the waters north of Iceland in relation to capelin biomass and environmental conditions. *ICES Journal of Marine Science*, 55: 808-810.

Blindheim, J., (1968). Hydrographic investigations in the Irminger Sea in the years 1954-1964. *Fiskeridirektoratets Skrifter: Serie. Havundersøkelser*. 14, 72-97.

Blindheim, J., V. Borovkov, B. Hansen, S.-Aa. Malmberg, W.R. Turrell, and S. Østerhus, (2000) Upper layer cooling and freshening in the Norwegian Sea in relation to atmospheric forcing. *Deep-Sea Research I*, 47: 655-680.

Dickson, R.R., J. Meincke, S.-Aa. Malmberg, and A.J. Lee, (1988). The "Great Salinity Anomaly" in the northern North Atlantic 1968-1982. *Progress in Oceanography*, 20: 103-151.

Frich, P., H. Alexandersson, J. Ashcroft, B. Dahlström, G.R. Demarée, A. Drebs, van Engelen, A. F. V., Førland, E. J., Hanssen-Bauer, I., Heino, R., Jónsson, T., Jonasson, K., Keegan, L., Nordli. P.Ø., Schmith, T., Steffensen, P., Tuomenvirta, H., Tveito, O. E., 1996. North Atlantic climatological dataset (NACD Version 1) - Final Report. *Danish Meteorological Institute, Scientific report*, 96-1, Copenhagen 1996, 47 pp, two diskettes with data.

Gudmundsson, K. (1998) Long-term variation in phytoplankton productivity during spring in Icelandic waters. *ICES Journal of Marine Science*, 55: 635-643.

Hansen, B. and Østerhus, S., North Atlantic-Nordic Seas exchanges. *Progress in Oceanography*, 45, 109-208, 2000.

Helland-Hansen, B. and Nansen, F. (1909) The Norwegian Sea, its physical oceanography. Based on the Norwegian researches 1900 – 1904. *Report on Norwegian Fishery and marine investigations*, 2: 390 pp.

Hurrell, J.W. (1995) Decadal trends in the North Atlantic Oscillation: Regional temperatures and precipitation. *Science* 269, 676-679.

Hurrell, J.W., Y. Kushnir, G. Ottersen, and M. Visbeck, editors (2003) *The North Atlantic Oscillation: Climatic significance and environmental impact*. Geophysical Monograph 134, American Geophysical Union, Washington DC: 279 pp.

Jakobsson, J., and O.J. Østvedt, (1999) A review of joint investigations on the distribution of herring in the Norwegian and Iceland Seas 1950-1970. *Rit Fiskideildar*, 16: 209-238.

Jónsson, S. (1992) The sources of freshwater in the Iceland Sea and the mechanisms governing its interannual variability. *ICES Marine Science Symposia*, 195: 62-67.

Jónsson, S. and J. Briem, (2003) Flow of Atlantic Water west of Iceland and onto the north Icelandic Shelf. *ICES Marine Science Symposia*, 219: 326-328.

Kristmannsson, S.S. (1998) Flow of Atlantic Water into the northern Icelandic shelf area, 1985-1989. The ICES NANSEN project. *ICES Cooperative Research Report* 225: 124-135.

Malmberg, S.-Aa. (1969) Hydrographic changes in the waters between Iceland and Jan Mayen in the last decade. *Jökull*, 19: 30-43.

Malmberg. S.-Aa. (1986) The ecological impact of the East Greenland Current on the North Icelandic Waters. S. Skreslet (editor). *The Role of Freshwater Outflow in Coastal Marine Ecosystems, NATO ASI Series, Volume G7*. Springer-Verlag Berlin Heidelberg: 389-404.

Malmberg, S.-Aa. and S.S. Kristmannsson, (1992) Hydrographic conditions in Icelandic waters, 1980-1989. *ICES Marine Science Symposia,* 195: 76-92.

Malmberg, S.-Aa. and Desert, J. (1999) Hydrographic conditions in North Icelandic waters and annual air temperature in Iceland. *ICES CM 1999/L:14*: 21 pp.

Malmberg, S.-Aa. and Jónsson, S. (1997) Timing of deep convection in the Greenland and Iceland Seas. *ICES Journal of Marine Science*, 54: 300-309.

Malmberg, S.-Aa. and Valdimarsson, H. (2003) Hydrographic conditions in Icelandic waters, 1990-1999. *ICES Marine Science Symposia*, 219: 50-60.

Monstad, T. and Blindheim, J. (1986) Relationship in distribution of blue whiting and hydrographic conditions in the Norwegian Sea during summer, 1980-85. *ICES CM 1986/H*: 54: 18 pp.

Ólafsson, J. (1999) Connections between oceanic conditions off N-Iceland, Lake Myvatn temperature, regional wind direction variability and the North Atlantic Oscillation. *Rit Fiskideildar,* 16: 41-57.

Read, J.F. and Pollard, R.T. (1992) Water masses in the region of the Iceland-Faroes Front. *Journal of Physical Oceanography*, 22: 1365-1378.

Rodewald, M. (1967) Recent variations of North Atlantic sea surface temperatures (SST) and the "type-tendencies" of the atmospheric circulation. *ICNAF Redbook 1967*, 4: 5-23.

Shevchenko, A.V. and Isaev, N.A. (1985) Peculiarities of blue whiting distribution in the Norwegian Sea in spring-summer 1983 and 1984 due to hydrographic conditions. *ICES CM1985/H*20: 11 pp.

Stefánsson, U. (1962) North Icelandic Waters. *Rit Fiskideildar*, 3(3): 269 pp.

Stefánsson, U. (1969) Temperature variations in the North Icelandic coastal area during recent decades. *Jøkull*, 19: 18-28.

Stefánsson, U. and Gudmundsson, G. (1969) Hydrographic conditions off the northeast coast of Iceland in relation to meteorological factors. *Tellus*, Vol. 21, 2: 245-258.

Stefánsson, U. and Ólafsson, J. (1991) Nutrients and fertility of Icelandic waters. *Rit Fiskideildar*, 12: 1-56.

Thompson, D.W.J., and Wallace, J.M. The Arctic Oscillation signature in the wintertime geopotential height and temperature fields. *Geophysical Research letters,* 25, 9: 1297-1300.

Thórdardóttir, Th. (1977) Primary production in North Icelandic waters in relation to recent climate changes. In M. Dunbar (Editor), *Polar Oceans. Proceedings of the Polar Oceans Conference held at McGill University, Montreal, 1974*. Arctic Institute of North America, Calgary, Canada: 655-665.

Turrell, W.R., G. Slesser, R.D. Adams, R. Payne and P.A. Gillibrand (1999), Decadal variability in the composition of Faroe Shetland Channel Bottom Water. *Deep-Sea Research*, 46, 1-25.

Visbeck, M., Chassignet, E.P., Curry, R.G. Delworth, T.L., Dickson, R.R., and Krahmann, G. (2003) The ocean's response to North Atlantic Oscillation variability. In Hurrell *et al.* editors, *The North Atlantic Oscillation: climatic significance and environmental impact, Geophysical Monograph* 134, American Geophysical Union: 113-145.

Johan Blindheim, Institute of Marine Research, P.O. Box 1870 Nordnes, N-5817 Bergen, Norway. (johan.blindheim@imr.no)

Svend-Aage Malmberg, Marine Research Institute, P.O. Box 1390, 121 Reykjavik, Iceland.

Dense-Water Production and Overflow From an Arctic Coastal Polynya in Storfjorden

Ragnheid Skogseth

The University Centre in Svalbard, Longyearbyen, Svalbard, Norway

Ilker Fer

Bjerknes Centre for Climate Research and Geophysical Institute, University of Bergen, Bergen, Norway

Peter M. Haugan

Geophysical Institute, University of Bergen, and Bjerknes Centre for Climate Research, Bergen, Norway

Storfjorden in winter is a site of active brine formation due to its recurrent polynya. The dense, brine-enriched waters fill the depressions of the fjord to its sill level and subsequently descend like a bottom gravity current, following the topography towards the shelf break. The brine formation and the overflow show strong interannual variability governed by both local factors, e.g., preconditioning, the rate of ice formation, and polynya location, and large-scale factors, e.g., Barents Sea winter ice cover and large-scale atmospheric circulation in the North Atlantic. An existing polynya and ice production model, making use of satellite images, wind data, and surface hydrography, is applied to winter 2002 and predicts large brine rejection in agreement with observations. Tuning factors for the polynya model obtained from 5 years of observations are used to extend the calculations to 33 years back in time. The ice production in 2002 is predicted to be 5.4 standard deviations away from the preceding 4 years' mean but still representative for the mean and variability of the decadal scale. The resulting strong overflow in 2002 is observed at densely spaced hydrographic stations located across its path. Significant entrainment is observed within the first 40 km of the path of the overflow, reducing the density anomaly of the overflow water by 0.18 and increasing the cross-sectional area by a factor of 2. The geostrophic shear is found to be partly responsible for the mixing. Farther downstream, entrainment is less significant. The geostrophic volume transport of the overflow watcr is approximately constant at ~0.06 Sv, in agreement with earlier estimates. The model results combined with the field data provide a picture of the link between forcing, generation of the source conditions in the basin, and the corresponding overflow.

The Nordic Seas: An Integrated Perspective
Geophysical Monograph Series 158

10.1029/158GM07

1. INTRODUCTION

On the continental shelves of the Arctic Ocean, coastal polynyas which are formed and maintained by various processes, e.g., advection of ice by off-shore winds and currents, lead to rapid, persistent, and repetitive ice formation through intense heat-exchange between the ocean and the atmosphere. Brine rejection associated with this ice formation results in brine-enriched shelf waters (BSW), which can help to maintain the cold upper halocline in the Arctic Ocean [*Aagaard et al.*, 1981]. Such dense shelf waters accumulate near the bottom and eventually spill towards the deep sea while they entrain ambient waters and change their temperature and salinity characteristics. According to *Rudels and Quadfasel* [1991], shelf convection contributes about 0.5 Sv ($1 \text{ Sv} \equiv 10^6 \text{ m}^3 \text{ s}^{-1}$) to the Arctic deep water and *Cavalieri and Martin* [1994] estimate 0.7-1.2 Sv of BSW produced by the entire Arctic coastal polynyas, both comparable to open-ocean convection in the Greenland Sea [*Smethie et al.*, 1986].

High-latitude overflows supply dense water masses to the world ocean, and are now widely accepted to be a major element of the large-scale ocean circulation. Overflows are bottom-intensified density currents and are typically much more energetic and dissipative than most other bottom currents [*Price and Baringer*, 1994]. Their relatively small spatial scales pose a central problem to ocean general circulation models. Such overflows can be better represented in large-scale ocean models when forcing, dynamics, and variability of exchange between marginal seas and the open ocean are understood through detailed process studies. Storfjorden (Figure 1), one of the few areas in the Barents Sea where BSW has been frequently observed, is estimated to supply 5-10% of the newly formed dense waters of the Arctic Ocean. Observations in Storfjorden reveal that significant quantities of dense water originate in its recurrent polynya, particularly during northeasterly winds [see, e.g., *Haarpaintner et al.*, 2001b]. Overflow of this dense water eventually sinks along Storfjordrenna and, if its density excess permits, down the continental slope into the deep Norwegian Sea and northward along the eastern slope of Fram Strait towards the Arctic Ocean, where it will appear as a positive heat anomaly due to entrainment of warmer ambient water along its path but still remain more saline than the surrounding waters [*Schauer and Fahrbach*, 1999].

In April 2002, the highest-ever recorded bottom salinity was observed in the deepest pool of Storfjorden [*Anderson et al.*, 2004]. Five months later, in September, the overflow of this source water was detected in the deep Fram Strait [*Schauer et al.*, 2003] for the first time since the only incident reported in 1986 [*Quadfasel et al.*, 1988]. Here, we apply our polynya model in an attempt to predict this large dense water production of the freezing period in 2002, and we evaluate its uniqueness with respect to long term variability conducting calculations back in time. The subsequent overflow and its mixing are further investigated by using hydrographical data acquired in August. The field data set consists of synthetic aperture radar (SAR) observations, nearby in situ meteorological data, and conventional CTD (conductivity, temperature, and depth) profiles. The outline of the work is as follows. In Section 2, a general description of the site, together with the dense-water production and overflow cycle, is presented. In Section 3 is introduced a summary of the model to estimate the ice production and brine formation. Results obtained for winter 2002 are reported and contrasted

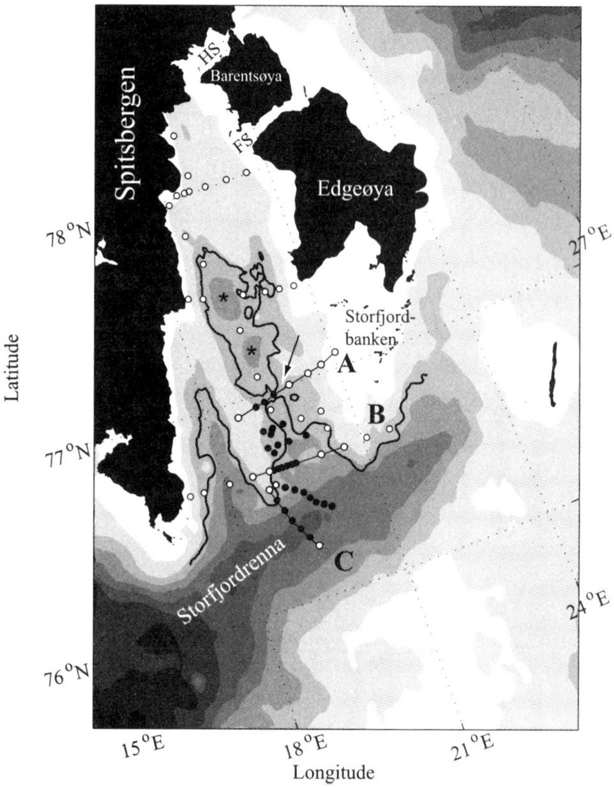

Figure 1. Storfjorden; the surrounding islands Spitsbergen, Barentsøya, and Edgeøya; the shallow bank Storfjordbanken in the southeast; and Storfjordrenna in the south. HS and FS indicate Heleysundet and Freemansundet, respectively. The 120-m-deep sill marked by an arrow is at about 77°N. The circles mark the location of the CTD stations occupied by the R/V *Håkon Mosby* during August 2002. The filled circles denote the stations where the plume was detected. Selected stations are aligned to form the sections labeled A to C. The deepest stations in Storfjorden are indicated by stars. The isobaths (ETOPO2) are drawn at 50-m intervals. Included to show the sill is the additional 125-m isobath (black contour line), derived from the 500 × 500 m resolution bathymetry presented in *Skogseth et al.* [2005].

to those of earlier years. Subsequently in Section 4, we present observations of the overflow that originated in the freezing period of 2002. Discussion and conclusions are given in Section 5.

2. GENERAL DESCRIPTION

2.1. Site

Storfjorden, located in the southeastern Svalbard Archipelago, is approximately 190 km long and 190 m deep at its maximum depth. It is enclosed by Spitsbergen to the west and north, by Barentsøya and Edgeøya to the east, the shallow bank Storfjordbanken to the southeast, and a 120-m-deep sill at about 77°N in the south (Figure 1). The basin covers an area of about 13×10^3 km^2 with a volume of 8.5×10^{11} m^3 [*Skogseth et al.*, 2005]. South of the sill, the fjord opens to Storfjordrenna, which is separated from the main part of the Barents Sea by the shallow Svalbardbanken to the south and Hopenbanken to the east. Two sounds, Heleysundet in the north and Freemansundet in the northeast, connect the basin with the northwestern Barents Sea.

2.2. Dense-Water Production and Overflow Cycle

Convective water mass transformations in Storfjorden produce dense BSW, which has been reported to descend the continental slope into deeper layers in the Fram Strait [*Quadfasel et al.*, 1988]. The cause of the high salinities is believed to be a recurrent coastal polynya situated on the lee-side of Edgeøya and Barentsøya, where ice is removed by off-shore northeasterly winds, allowing for large ice production and brine rejection. The plume of BSW overflows the 120-m-deep sill as a bottom gravity current and descends south through Storfjordrenna towards the shelf break and northward along the eastern side of Fram Strait [*Anderson et al.*, 1988; *Quadfasel et al.*, 1988]. This path is supported by the numerical model results of *Jungclaus et al.* [1995], which also suggest that another branch of the plume flows southwards, guided by the topography, into a deep trench.

The salinity of BSW, and hence the path of the overflow as well as the ultimate depth it will reach, shows strong interannual variability (Figure 2). This variability may be governed by the rate of ice formation and initial salinity of the underlying water, which mixes with the brine [*Schauer*, 1995]; the stability of the water column [*Skogseth et al.*, 2004]; and the location of the polynya [*Haarpaintner et al.*, 2001b], which may lead to higher salinities over shallow and wide shelves. Links have been found to the ice export from the Arctic Ocean into the western Barents Sea and hence to the surface salinity variability in the northwestern Barents Sea [*Maus*, 2003] as well as to the Barents Sea winter ice cover and the large-scale atmospheric circulation in the North Atlantic [*Skogseth et al.*, 2004]. Figure 2 shows a time series of the maximum observed salinity of BSW since 1981 [for data reference, see *Skogseth et al.*, 2005]. The station positions of the observations presented in Figure 2 are typically close to the deep pools inside Storfjorden, which are indicated in Figure 1. Due to missing observations in 1992 and 1994, data from moorings situated south of the sill are used. Because of mixing and entrainment in the overflow (see Section 4), the maximum salinity of BSW inside the basin is probably higher than those recorded by the moorings. The remaining observations were made between mid-April and early October. As shown later in Figure 8, the accumulated salt release increases throughout the freezing period but flattens out from mid-April. It is reasonable to assume that the salinity of BSW increases correspondingly. Therefore, April observations are representative of the BSW production throughout the freezing period. After the freezing period, the remnants of BSW in

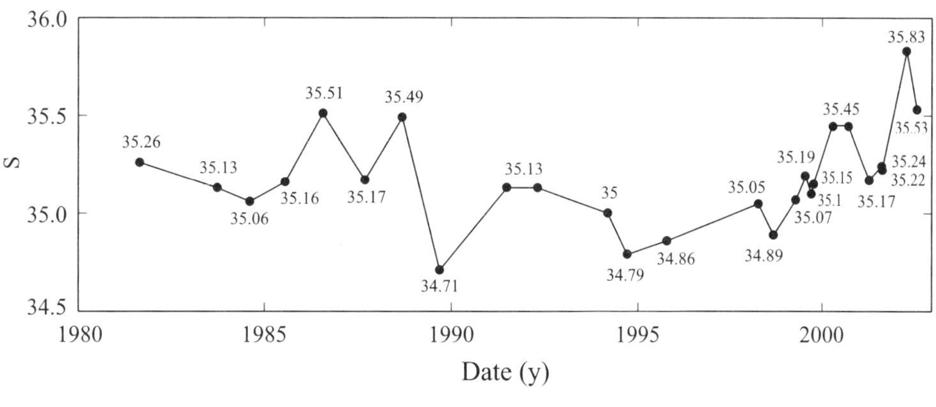

Figure 2. Time series of maximum salinity observed in the deepest part of Storfjorden. The April 2002 observation is from *Anderson et al.* [2004]; for reference to other data, see *Skogseth et al.* [2005].

the deepest layers in Storfjorden diffuse gradually due to mixing with overlying waters. Hence, the observations from late September and early October might give a lower than maximum BSW salinity for the respective years. Clearly, the temporal variability in observations influences the observed interannual variability in BSW salinity shown in Figure 2.

A cartoon of the dense water production and overflow cycle in Storfjorden is presented in Figure 3. With the onset of freezing, brine release due to ice formation produces BSW, particularly over the shelves, where polynyas allow for higher salinities. The densest water masses spill towards deeper depressions, gradually filling the basin, which increases the overall density and pushes overlying isopycnals above the sill level. A weak overflow is readily established provided that the water above the sill level is relatively denser than ambient water downstream of the sill (Figure 3a). As freezing continues, BSW accumulates in the basin reaching the sill level, and a strong steady overflow period is established when the overflow is compensated by transformation of inflow waters into BSW (Figure 3b). The overflow may reach great depths depending on its salinity anomaly. The basin can be flushed 1–11 times [*Skogseth et al.*, 2005] over each freezing period, depending on the intensity of brine production and the stability of the water column. The reported volume transport rates associated with the overflow can be summarized as follows. *Schauer* [1995] used hydrographic data and moored current meters sampling over a 1-year period in 1991/1992 to estimate a transport of 0.13 Sv of BSW during 5 months of active overflow and 0.05 Sv annually. Recently, *Fer et al.* [2003] investigated the mixing and spreading of the overflow during May 2001, using velocity and CTD profiles in the outer part of Storfjordrenna. The net volume transport out of a section close to the sill was estimated to be 0.06 Sv, which was nearly doubled in the farthest downstream section. Model results yield a transport of 0.11 Sv [*Jungclaus et al.*, 1995], comparable to the field observations. Combined results from long-term time-series observations of currents and hydrography suggest that during fall and winter near-bed currents and bottom stress can resuspend sediment in response to strong flows by surface cooling and winds [*Sternberg et al.*, 2001]. The dynamics of sediment-laden plumes are not explicitly dependent on thermohaline properties [*Fohrmann et al.*, 1998]. Such plumes propagate more rapidly, in ageostrophic balance, than a seawater plume driven by thermohaline anomalies alone. Upon reaching equilibrium density, suspended particles will be deposited, considerably reducing the density of the seawater and initiating upward convection in the water column.

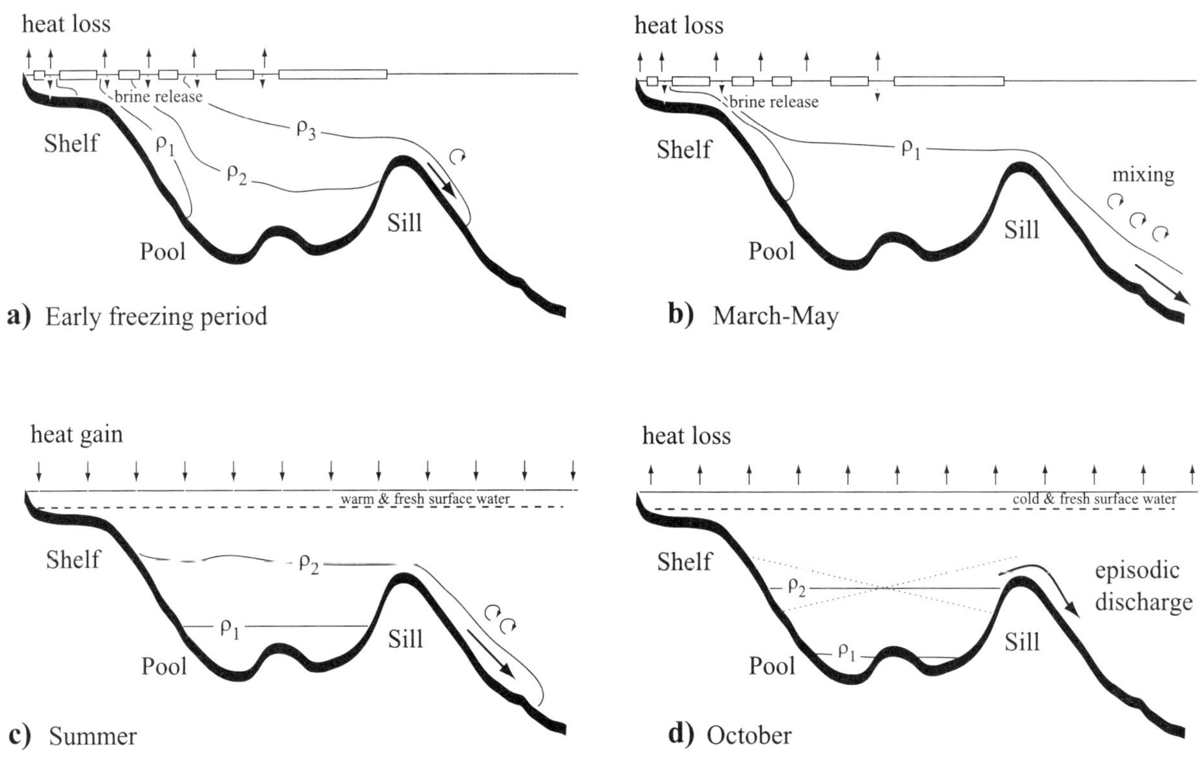

Figure 3. Schematic representation of different processes during ice formation and overflow from Storfjorden during (a) the early freezing period, (b) March-May, (c) summer, and (d) October. Isopycnals are arbitrary, with $\rho_1 > \rho_2 > \rho_3$.

After the freezing period, BSW production in the summer ceases but the overflow continues, albeit gradually weakening, presumably in a geostrophically controlled fashion [Figure 3c; see also *Skogseth et al.*, 2005]. Surface water consists of fresh melt water that is warmed by surface heating. Towards October, heat loss cools the surface layer. Dense water is trapped behind the sill, but strong wind events and seiching of the pycnocline may lead to episodic discharge (Figure 3d). The remnants of BSW in deep pools diffuse by vertical mixing, thereby finally preconditioning the basin for the next freezing period.

Variability in potential temperature, θ, and salinity, S, characteristics in the vicinity of the shelf, pool, and sill are compared in Figure 4 for a weak overflow year (1999, solid lines) and an intense overflow year (2002, dashed lines) in April, summer, and October (see Section 3.2 for an interannual comparison with respect to ice production and brine flux). Data for 1999 are taken from the Storfjorden database [*Maus*, 2003]; those for 2002 were collected during R/V *Håkon Mosby* cruises conducted in August and October. In April 2002 (not shown), CTD profiles taken from R/V *Oden* yield salinities between 34.8 (surface) and 35.8 (bottom) at the pool and between 34.75 (surface) and 35.7 (bottom) at the sill with temperatures at or near the freezing point [*Anderson et al.*, 2004]. The inflow of Atlantic Water (AW) is more pronounced in summer 2002 than in 1999. This can be due partly to the location of the Polar Front and partly to stronger compensating inflow for the strong overflow in 2002. Furthermore, an examination of the time series of the Norwegian Atlantic Current since 1995 reveals stronger northward flow of AW in 2002 than in the previous years [*Orvik and Skagseth*, 2003]. This supports the pattern in Figure 4. Brine release from ice aging and the admixture of dense shelf waters into less salty water en route to the deep pools establish the weak stratification observed in the pool region. Towards the summer, a fresh surface layer evolves from heating and melting, further enhancing the stratification. A weak isopycnal mixing with AW is apparent as a temperature maximum at middepth both in the shelf region and in the vicinity of the sill for the summer periods. The change in heat content over 75 days between October and August casts in 2002 yields 90 W m^{-2} heat gain, when integrated over the whole water column. The heat loss in the upper 50 m is 60 W m^{-2}, whereas 150 W m^{-2} is gained below 50 m from the inflow of waters of Atlantic origin. However, the shelf waters are typically at or near the freezing point before the freezing commences, typically around mid-November or early December [*Skogseth et al.*, 2004]. The heat loss needed to bring the shelf waters to the freezing point over 30 days before freezing (after October) is typically >280 W m^{-2} (or >890 W m^{-2} to bring the whole water column to the freezing point). The required large heat loss suggests an inflow of colder water to meet the preconditioning after October. This change in inflow properties can be associated with the variable position of the Polar Front. In late autumn, wind direction becomes predominantly northerly and pushes the Polar Front further southwest in Storfjordrenna, leading to an increase in ice cover and a decrease in the AW inflow.

3. ICE PRODUCTION AND BRINE FORMATION

The Storfjorden "ice-factory" is investigated by using interpreted SAR images that define the mean polynya width, its standard deviation, and area, based on pack ice boundary.

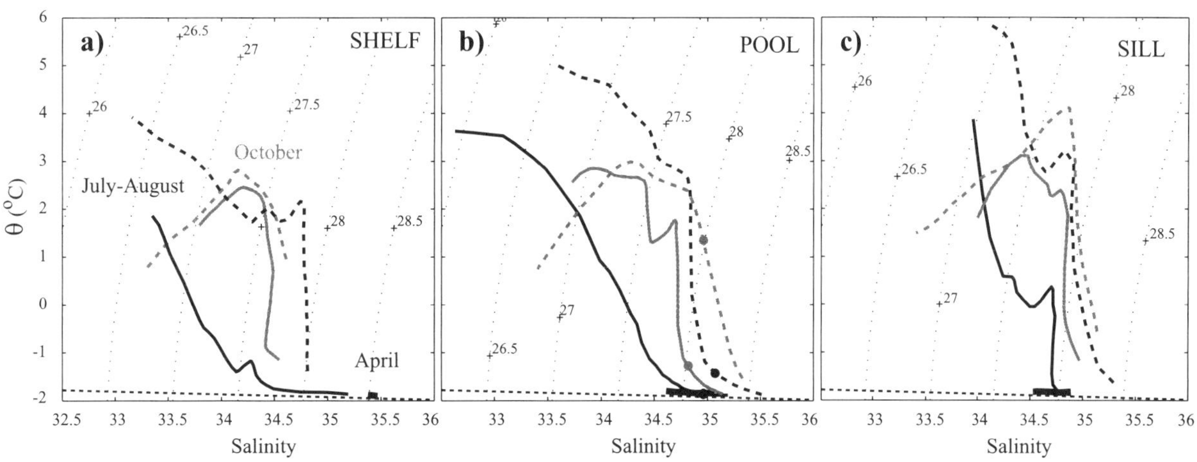

Figure 4. Potential temperature, θ, and salinity composite diagrams derived from smoothed data collected in the vicinity of (a) the shelf, (b) the pool, and (c) the sill for weak overflow (1999, solid lines) and strong overflow (2002, broken lines). Data are presented for April (thick), July-August (black), and October (gray). Contours of σ_θ are drawn at 0.5 kg m^{-3} intervals; the dotted line is the freezing point temperature. Dots in (b) mark the depth of sill level.

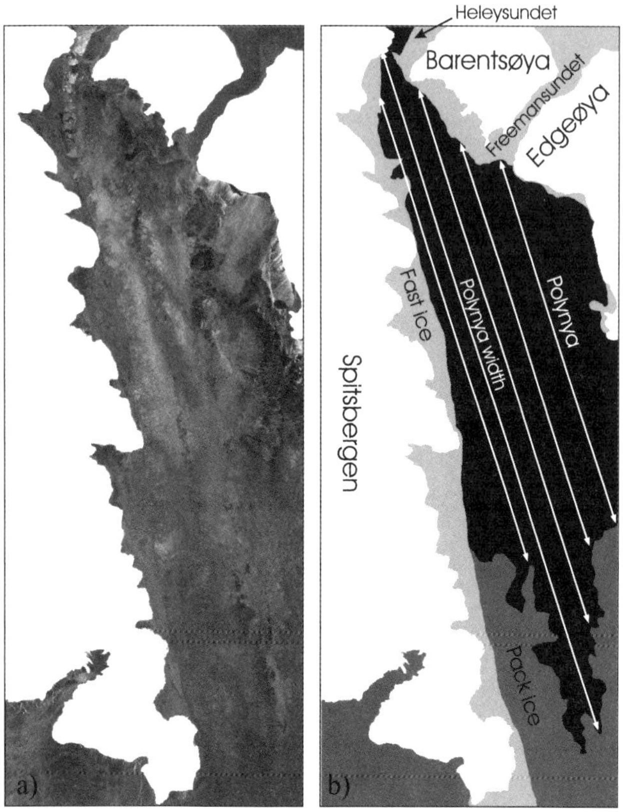

Figure 5. A representative SAR image from 10 February 2002 (original data courtesy of European Space Agency [2002], distributed by Eurimage), (a) with land mask and (b) classified into fast ice, polynya, and pack ice. The mean polynya width and its standard deviation are estimated from the polynya widths indicated by arrows.

The polynya opens during northeasterly winds and can be partly covered by thin ice. Figure 5 shows a representative, manually classified SAR image of Storfjorden on 10 February 2002. The boundary between open water and thin ice is hard to detect and is therefore parameterized in the following. During winter 2002, a total number of 12 and 26 SAR images were available for ice classification and estimation of polynya width in Storfjorden, respectively. These numbers are similar to those for the previous winters [*Skogseth et al.*, 2004].

To study the Storfjorden polynya dynamics and to determine parameters to be used in more complete models, a simple wind-driven polynya width model has been formulated [*Haarpaintner et al.*, 2001a] and re-implemented with slightly modified physics [*Skogseth et al.*, 2004]. The model provides estimates of ice and BSW production based on satellite images, wind data, and limited information of surface hydrography. The polynya width, obtained from the model on the basis of wind data from a nearby weather station, is matched to the SAR observations. Then, the tuned polynya width model and an accompanying open water width model are used as basis for ice production estimates that include contributions from open water (frazil ice), thin ice, and fast and pack ice portions of the Storfjorden basin.

A comprehensive description of the model can be found in *Skogseth et al.* [2004]; for the sake of completeness, the model is briefly reviewed in the following section. New results from the freezing period 2002 are reported and contrasted with those from earlier years in Section 3.2.

3.1. Model

The model comprises a polynya width, W_p, and an open water width, W_o, algorithm. At time $t = n$, the polynya width is found by

$$W_{p,n} = W_{p,n-1} + F(\phi_n - \phi_0)\, B_1\, U_n \cos(\phi_n - \phi_0)\, \Delta t, \quad (1)$$

where $F(\phi_n - \phi_0)$ is either an opening (OF) or closing (CF) factor (Table 1) depending on the actual wind direction, ϕ_n (OF when $-90° < \phi_n - \phi_0 < 90°$; CF otherwise). Here, $\phi_0 = -10°$ is the preferred wind direction for opening of the Storfjorden polynya, $B_1 = 0.02$ is the ice drift factor in pack ice, U_n is the wind speed, and $\Delta t = 6$ h is the time step.

Open water width in the polynya is estimated by

$$W_{o,n} = W_{o,n-1}\,(1 - \Delta h_f/h_c) + B_2\, U_n \cos(\phi_n - \phi_0)\, \Delta t \quad (2)$$

In the polynya, the ice drift factor, B_2, is increased to 0.04 to compensate for easier ice movement. The collection thickness, h_c expressed in m, is calculated from the wind speed in m s^{-1} by

$$h_c = (1.0 + 0.1U_n)/15 \quad (3)$$

The frazil ice growth during Δt, Δh_f, is found from the net heat exchange between open water and atmosphere. The latent heat of fusion of sea ice is based on the salinity of seawater and an immediate salt rejection of 69%. Growth of thin ice and fast and pack ice are derived from Stefan's law, assuming no snow on the thin ice. Thin ice is further governed by mass conservation inside the polynya. The volume of frazil ice produced at every time step is estimated from

Table 1. Opening (OF) and Closing (CF) Factors Derived From Matching Modeled and SAR-Observed Polynya Widths During the Winters 1998 to 2002.

Winter	1998	1999	2000	2001	2002	1970-2002
OF	1.0	1.8	1.0	1.3	1.5	1.0
CF	15.0	13.0	20.0	19.0	15.0	19.0

Table 2. Mean Modeled Polynya Width (W_p), Average Absolute Deviation of Modeled Polynya Width From Observed Polynya Width (ΔW_p), Modeled Polynya Area (A_p) as a Fraction of the Total Storfjorden Basin Area, Accumulated Frazil Ice (h_f) and Fast and Pack Ice (h_{fp}) Thickness, Observed Salinity of Initial (S_{01}) and Final (S_{02}) Surface Water, Surface Salinity Difference ($\Delta S_0 = S_{01} - S_{02}$) and Volume-Weighted Salinity Difference (dS) Between Spring and Preceding Autumn, Salinity of BSW (S_{bsw}), Modeled Mass of Total Produced Ice (T_{ice}), Total Released Mass of Salt (T_s), and Volume of BSW (V_{bsw}); Q_1 and Q_2 Are the Corresponding Volume Fluxes of BSW Derived Over the Freezing Periods and Annually, Respectively.

Winter	W_p (km)	ΔW_p (km)	A_p (%)	h_f (m)	h_{fp} (m)	S_{01}	S_{02}	ΔS_0	dS	S_{bsw}	T_{ice} (10^{13} kg)	T_s (10^9 kg)	V_{bsw} (10^{12} m^3)	Q_1 (Sv)	Q_2 (Sv)
1998	29	4	14.2	19.5	1.28	33.90	34.30	0.40	0.15	35.05	3.63	972.5	1.05	0.07	0.03
1999	41	28	22.2	13.1	1.03	33.40	34.60	1.20	0.65	35.18	3.35	893.1	0.86	0.06	0.03
2000	57	17	26.9	15.7	1.09	34.09	34.67	0.58	0.30	35.45	4.21	1132.8	1.11	0.07	0.04
2001	48	24	22.7	14.7	1.01	34.14	34.42	0.28	0.11	35.25	3.64	982.0	1.05	0.07	0.03
2002	76	18	33.8	18.0	1.18	34.03	34.81	0.78	0.55	35.83	5.64	1514.8	1.09	0.06	0.04

the effective open water area, where ridging and rafting processes are parameterized by assuming that 1% of the fast and pack ice area is covered with leads and 25% of the thin ice area is converted to open water due to ridging and rafting (further discussed in Section 5). The respective areas are calculated following *Skogseth et al.* [2004].

Accordingly, the volume of continuously grown ice at every time step is estimated in the fast and pack ice area and the thin ice area that are not converted to open water.

The salt release from ice production during Δt is

$$\Delta T_s = \Delta T_{ice} (P_i + P_a) S_0, \tag{4}$$

where ΔT_{ice} is the mass of total ice production (frazil and continuous), and S_0 is the salinity, in mass fraction, of the surface water, which increases from observed autumn (S_{01}) to spring (S_{02}) surface salinity (Table 2) close to the deepest stations in Storfjorden (Figure 1). Here, $P_i = 0.69$ is the fraction of salinity in the surface water that is immediately released when the ice forms and $P_a = 0.10$ is the fraction of initial salinity that is released from ice aging [*Haarpaintner et al.*, 2001a]. The ice aging is assumed to be rapid enough to occur before any ice export out of the region.

From conservation of mass and salt, the volume of BSW during Δt is given by

$$\Delta V_{bsw} = \Delta T_s / [\rho_{bsw} (S_{bsw} - S_0)]. \tag{5}$$

Here, ρ_{bsw} is the density of BSW and its salinity, S_{bsw}, is constant, equal to the observed BSW salinity at the end of each winter (Table 2).

3.2. Results

Figure 6 shows the fraction of fast ice, pack ice, and polynya (open water and thin ice) areas derived from the manually classified SAR images of Storfjorden in winters 1998 to 2002. In winter 1998, the air temperatures were relatively low and a larger portion of Storfjorden was covered with fast ice. During the following winters, the fast ice area was smaller and polynya was observed to cover a larger area with longer duration. Winter 2002 seems to have had the most frequent polynya events with the largest area and longest duration.

Observed and modeled polynya widths and corresponding ϕ_0 wind component during winter 2002 are shown in Figure 7. The error bars show the variability in the SAR-derived widths at different locations along the polynya from shore to pack ice border (Figure 5). When polynya closes more rapidly than it opens, the time derivative of the polynya width is large and negative (Figure 7). Matching iteratively the polynya widths obtained from the model with those observed in SAR images gives the model-tuned values of OF and CF listed in Table 1. For all periods, the best-fit CF is typically 10-20 times larger than the best-fit OF.

About 20% of the modeled polynya area consists of open water, which produces frazil ice that consolidates and forms thin ice. Table 2 lists the modeled mean polynya width, the average absolute deviation of modeled and observed polynya width, the fraction of modeled polynya area (open water and thin ice inside polynya), and the thickness of the accumulated frazil ice and fast and pack ice during the winters of 1998 to 2002. In accordance with observations (Figure 6), winters 1998 and 2002 have the least and the most polynya activity, respectively.

The modeled spatial fraction and ice production of the different areas during each winter are listed in Table 3. When the rafting and ridging parameterization is included, the total effective open water area roughly doubles. On average, the effective open water area occupies 10% of the Storfjorden area and accounts for about 60% of the ice production. The frazil ice production is equivalent to about 16 m y^{-1}, which is on average 12 times that of the accumulated fast and pack ice (Table 2).

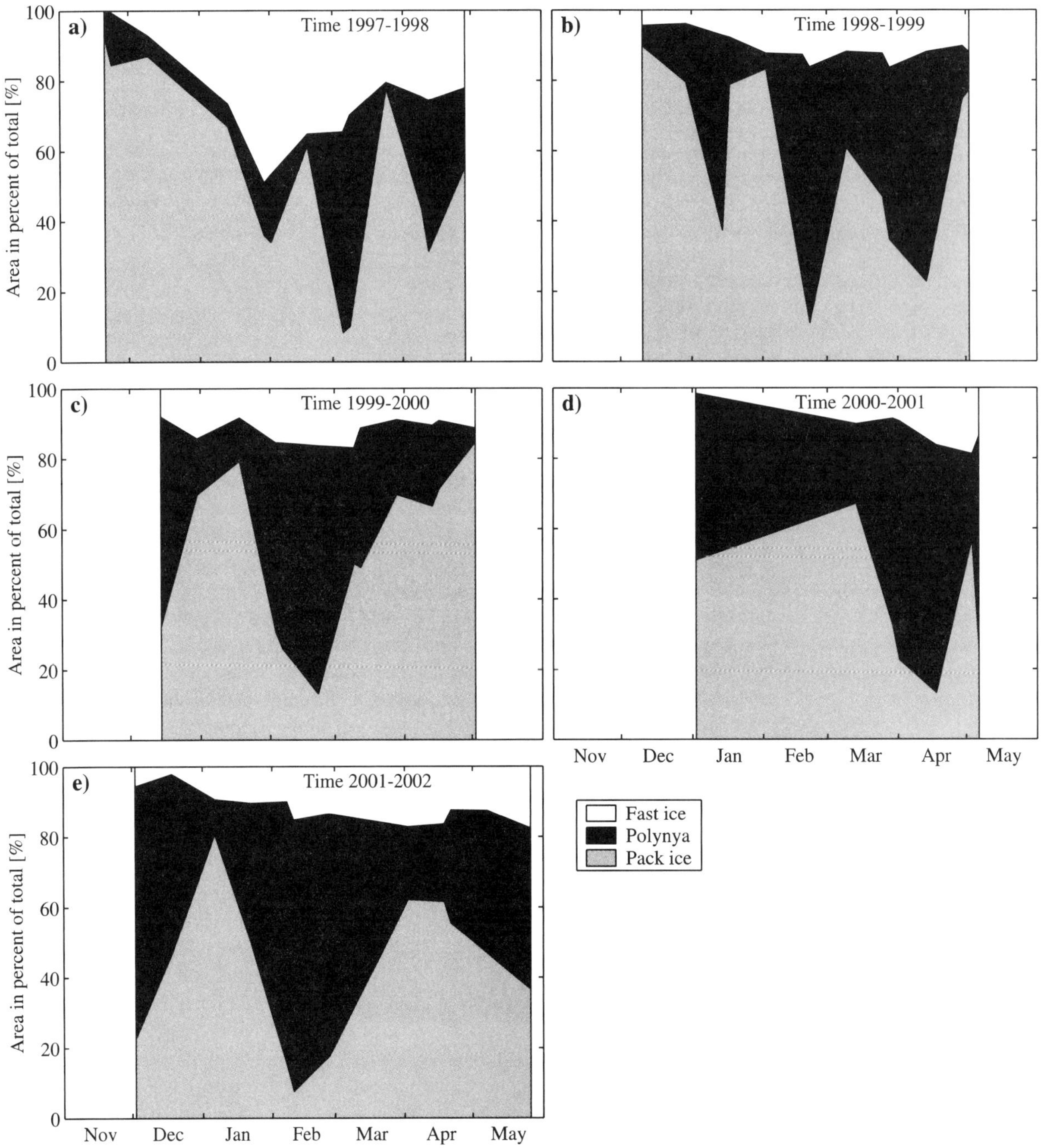

Figure 6. Fast ice, pack ice, and polynya (open water and thin ice) area as percent of total area from SAR manual classification during the freezing periods (a) 1998, (b) 1999, (c) 2000, (d) 2001, and (e) 2002.

The accumulated total salt release, T_s, in the Storfjorden basin is compared for winters 1998 to 2002 in Figure 8. Strong wind events in the direction favoring polynya opening (vertical arrows) result in enhanced and persistent salt release. Winter 2002 contains polynya events that are more frequent and last longer than in the other years (horizontal arrows). This results in significantly larger salt release, in particular for the polynya events in January and February. Note that in 2002, the mean wind speed and air temperature are moderate, 1.9 m s^{-1} and -12.6°C, respectively. The large amounts of ice formation and salt release are mainly due to the large OF (see Section 5) in addition to longer duration of strong winds in

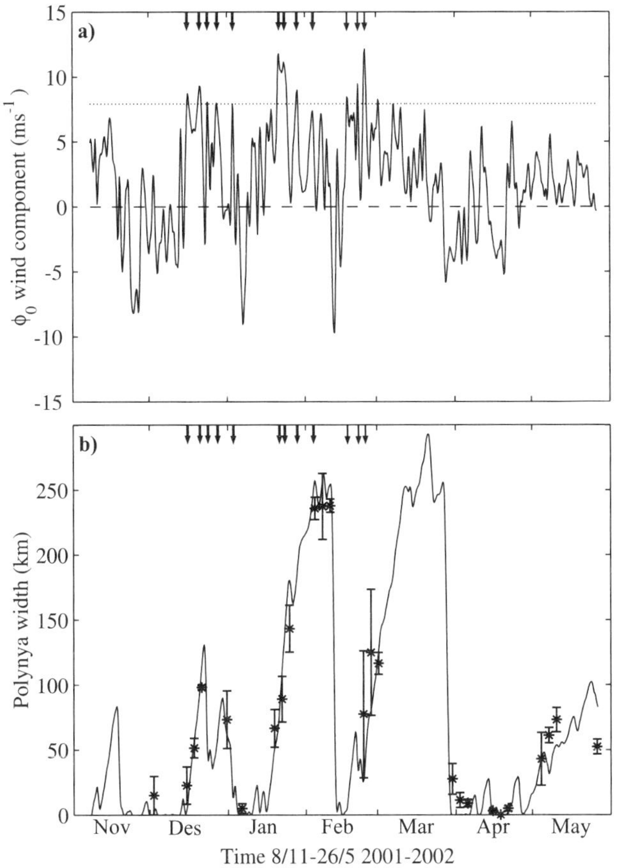

Figure 7. (a) ϕ_0 wind component and (b) polynya width obtained from the model (solid line) and from SAR images (stars and error bars) during the freezing period 2002. The error bars show the standard deviation from the mean polynya width estimated from the SAR images. The dashed line in (a) shows zero ϕ_0 wind component and the dotted line denotes twice the standard deviation (SD) from the mean during winter 2002. The arrows indicate strong wind events when the ϕ_0 wind component exceeds 2 SD.

the favorable wind direction (Figure 7) and cold air outbreaks, which give rise to large net heat flux. During the three largest polynya events (Figure 7b), the mean net heat fluxes are 357, 402, and 366 Wm^{-2}, respectively. The mean wind speed is greatly reduced by strong southwesterly winds during mid-November to mid-December, in the beginning of January, in mid-February and from the end of March to the end of April (Figure 7a).

4. MIXING OF THE OVERFLOW

The downstream evolution of the Storfjorden overflow can be characterized by an initial decrease in salinity at the early parts of the overflow where it mixes with relatively fresh East Spitsbergen water (ESW, $-1 \leq \theta \leq 0.5°C$; $34.8 \leq S \leq 34.9$); this is followed by heat gain from warm AW of comparable salinity. Leaving the sill, the dense plume will descend downslope for presumably several Rossby radii until the rotational effects are well established. It will then follow the topography as a slab of dense water through a stratified background under the influences of gravity, Coriolis acceleration, and friction. The resulting force balance will determine the descent rate (vertical displacement per unit distance of its path).

Three representative sections, with approximately even downstream separation and each spanning both edges of the overflow plume, were selected from densely spaced hydrographic stations occupied by R/V *Håkon Mosby* during 3-15 August 2002 (for station locations, see Figure 1). The fine structure from the data set was used by *Fer et al.* [2004] to estimate vertical eddy diffusivity, K_z, inferred from density overturns. Survey-averaged profiles showed enhanced mixing near the bottom in which $K_z = 10 \times 10^{-4}$ m^2 s^{-1} when averaged within the overflow. Here we will use the data to show that the geostrophic shear alone can be partly responsible for the observed levels of elevated mixing. We will further estimate the downstream development of the plume in terms of changes in density, cross-sectional area, geostrophic volume transport, and entrainment.

Table 3. Average Modeled Spatial Fraction (A) and Ice Production (V) in Open Water Area (A_o), Thin Ice Area (A_t), and Fast and Pack Ice Area (A_{fp}) From the Model for Winters 1998 to 2002 Over the Whole Storfjorden Basin (Tot). Open Water Area Contribution From Ridging and Rafting in Thin Ice and Leads in Pack Ice Is Included.

	Winter 1998 (15.11–15.05)			Winter 1999 (01.12–15.05)			Winter 2000 (10.11–15.05)			Winter 2001 (13.11–19.05)			Winter 2002 (08.11–27.05)		
	A		V	A		V	A		V	A		V	A		V
	%	%	km^3	%	%	km^3	%	%	km^3	%	%	km^3	%	%	km^3
A_o	6.6	52.4	20.3	9.6	54.3	19.4	11.3	65.7	29.5	10.4	59.3	22.9	13.1	66.4	39.9
A_t	8.4	10.9	4.3	13.4	18.1	6.5	16.3	11.9	5.3	13.1	14.6	5.7	21.4	17.6	10.5
A_{fp}	85.0	36.7	14.3	77.0	27.6	9.9	72.4	22.4	10.0	76.5	26.1	10.1	65.5	16.0	9.6
Tot	100.0	100.0	38.9	100.0	100.0	35.8	100.0	100.0	44.8	100.0	100.0	38.7	100.0	100.0	60.0

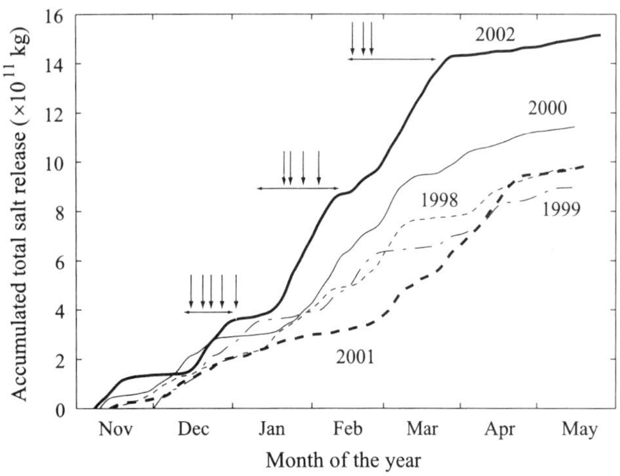

Figure 8. Time series of accumulated salt release from ice production and ice aging over the Storfjorden basin area of 13 000 km^2 during the freezing periods of 1998 (thin dashed line), 1999 (dashed-dotted line), 2000 (thin solid line), 2001 (thick dashed line), and 2002 (thick solid line). Vertical arrows mark wind events of the favorable component exceeding 2 SD (1 SD = 4.0 m s^{-1}) during 2002. Horizontal arrows show the extent of three polynya events during which the net heat loss to the atmosphere were 357, 402, and 366 W m^{-2}, respectively.

The 1-dbar-averaged CTD profiles defining a section are interpolated into a regular grid of 1-dbar bins vertically and 150 distance bins horizontally and smoothed over 6 data points. Isolines of potential density referenced to the surface (σ_θ) are shown in Figure 9 for the section along the sill (section A) and for some further downstream (sections B and C), reaching approximately 80 km. The reader is looking in the downstream direction with Spitsbergen on the right. In the following we will use the gridded data set to derive the averaged properties. The dense plume of overflow water can be identified as water with $\theta < 0°C$ and $S > 34.8$, yielding $\sigma_\theta > 27.95$ (marked by the thick isopycnal in Figure 9), consistent with *Fer et al.* [2003]. At section A, the bottom water is in excess of $\sigma_\theta = 28.5$. Because of the intense mixing with the ambient as the plume propagates, density exceeding 28.3 is no longer present at section B. However, the farthest downstream section C shows $\sigma_\theta > 28.3$ at its depression, which suggests evidence for episodes of denser overflow encountered before the survey, as noted by *Fer et al.* [2004].

The dense remnants, when included in section-averaged properties, will yield unrealistic estimates of entrainment and plume density excess and can probably account for the observed increase in salinity en route at the core of the overflow during 2001 [*Fer et al.*, 2003]. In an attempt to define the extent of the active plume water forced by its negative buoyancy, we calculate the geostrophic velocity, U_{geo}, derived from the hydrography referenced to an uneven level-of-no-motion following the top of the plume, i.e., $\sigma_\theta = 27.95$. In effect, we obtain the velocity of the plume relative to the ambient, which is the relevant velocity for entrainment parameterizations. The contours of U_{geo} are shown in Figure 9 superimposed on the density field. In section B, where the tilted isopycnals are indicative of a geostrophically balanced overflow, U_{geo} exceeds 15 cm s^{-1}. In the vicinity of the plume, where it is leaning on the sloping side, a 1 cm s^{-1} isoline appears to separate the "dynamically active" overflow from its adjacent remnant waters, as is particularly evident in section C (zero velocity isolines are not well-behaved for this purpose). Hence in the following we define the boundary of the plume as the 27.95 isopycnal emanating from the sloping side and limited by U_{geo} = 1 cm s^{-1} when the thickness of the 27.95 isopycnal (the vertical distance between the isopycnal and the sea bottom) is greater than that of the U_{geo} = 1 cm s^{-1} isoline. The plume can of course be in motion when this calculation gives zero velocity (or 1 cm s^{-1}, undistinguishable considering the associated uncertainty), but that would imply a barotropic motion. The recirculation at the deepest part of section C with the doming isopycnals can be interpreted as an anticyclonic closed cell and is indicative of previous episodes of high density overflows.

Assuming that the density anomaly of the overflow at the sill during spring was twice as large as in summer (see Section 2.2: Maximum salinity at the sill was 35.7 in April [*Anderson et al.*, 2004], compared with ~35.4 in August, Figure 3c), we can obtain an order of magnitude estimate of the time required to fill the cross-section area of O (10^6) m^2 of the remnants in section C. The maximum downslope flux per unit length of slope in a fully developed steady-state cascade is given by 0.36 $V_{nof}\, h_E$ [*Shapiro and Hill*, 1997]. Here $V_{nof} = g's/f$ is the Nof speed, i.e., the speed of along-isobath propagation in the absence of friction; $h_E = (2K/f)^{1/2}$ is the Ekman depth; s is the mean bottom slope; and K is the vertical eddy viscosity. In August, average plume density at the sill, and the mean ambient density is 28.23 and 27.9, respectively (Figure 10a), yielding a density anomaly of $\Delta\rho = 0.33$. Using the mean slope of $s \sim 3 \times 10^{-3}$, the corresponding values of V_{nof} are 14 cm s^{-1} and 7 cm s^{-1} in April and August, respectively. Assuming that the survey-averaged vertical diffusivity within the plume equals the eddy viscosity, i.e., using the Reynolds analogy $K = K_z = 10 \times 10^{-4}$ m^2 s^{-1}, and $f = 1.42 \times 10^{-4}$ s^{-1}, the Ekman thickness is 3.8 m at 77°N. The downslope transport per unit length of the slope is then 0.18 m^2 s^{-1} and can fill the depression with remnants in about 2 months. It is also noteworthy that the thickness of the overflow is greater than $2h_E$ at all sections (Figure 10b) and according to *Shapiro and Hill* [2003] this leads to a "head-up" state; i.e., the steepest sloping isopycnals are on the upslope side of the plume. Our observations are in support of their results.

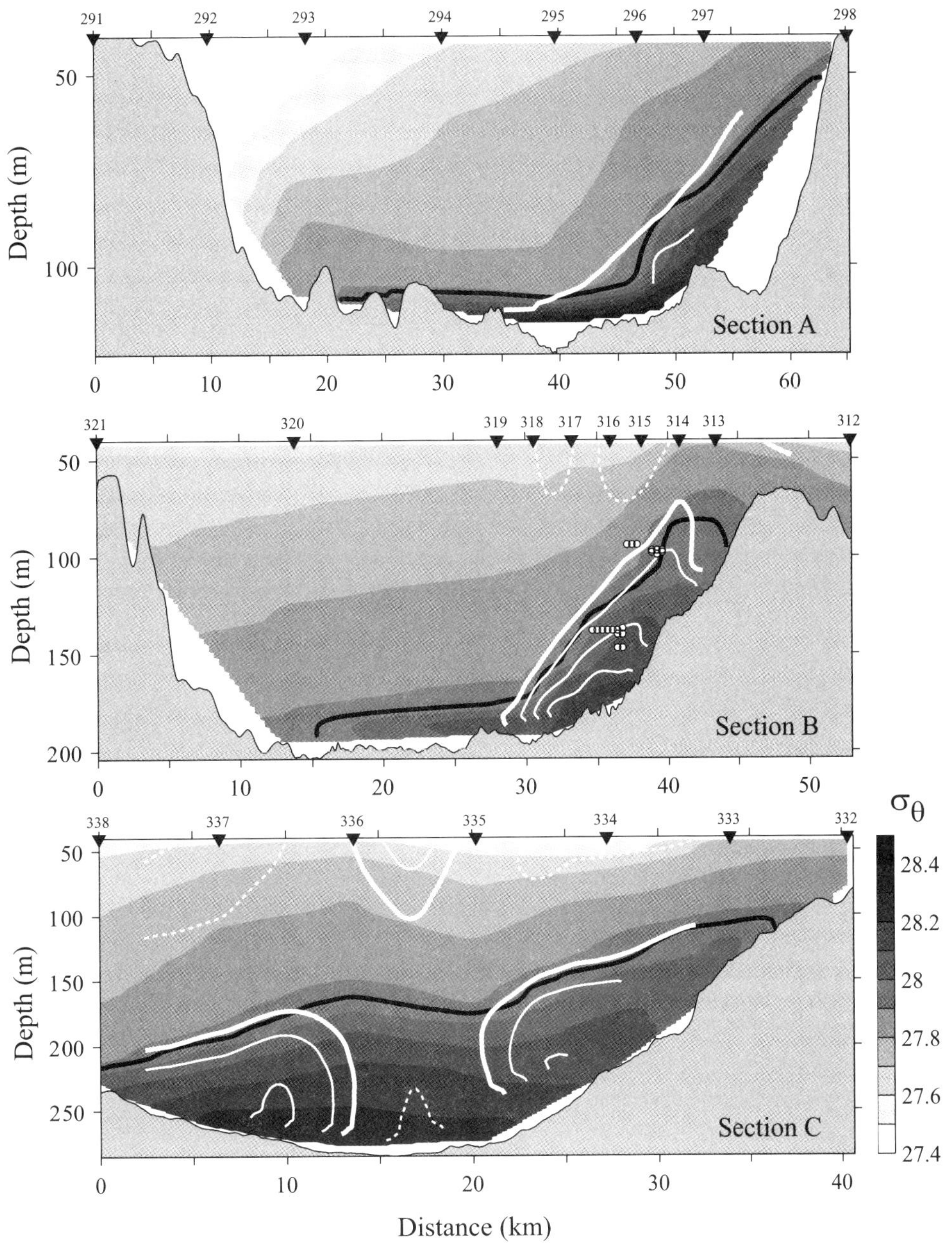

Figure 9. Distribution of σ_θ and geostrophic velocity U_{geo} (white isolines) referenced to $\sigma_\theta = 27.95$ (thick black isoline) for sections A, B, and C. Contour intervals are 0.1 kg m^{-1} for σ_θ and 5 cm s^{-1} for U_{geo}. The thick white isoline is $U_{geo} = 1$ cm s^{-1} (the zero-velocity contour is not shown). Solid contours indicate motion out of the section in the downstream direction. Upstream motion is distinguished by dashed contours. The circles in section B show the events with Ri < 1. Profiles are taken at the stations numbered, indicated by arrowheads over each panel. The 1-dbar-averaged values are interpolated into a regular grid of 1-dbar bins vertically and 150 distance bins horizontally and then smoothed over 6 data points. Bottom topography is derived from echo sounding after correcting for the sound velocity. Completing the survey including the three sections took 68 h; working each section typically took 7 h.

In a stratified flow, the gradient Richardson number, Ri (Ri = N^2/S^2, where N is the buoyancy frequency and S is the vertical shear of the horizontal current), relates the mechanical turbulence-generating forces to the stabilizing density forces. *Miles* [1961] and *Howard* [1961] independently showed that a necessary, but not sufficient, condition for shear instability in a steady stratified shear flow is Ri < 0.25, everywhere in the flow. Ri < 0.25 permits the growth of small-scale billows,

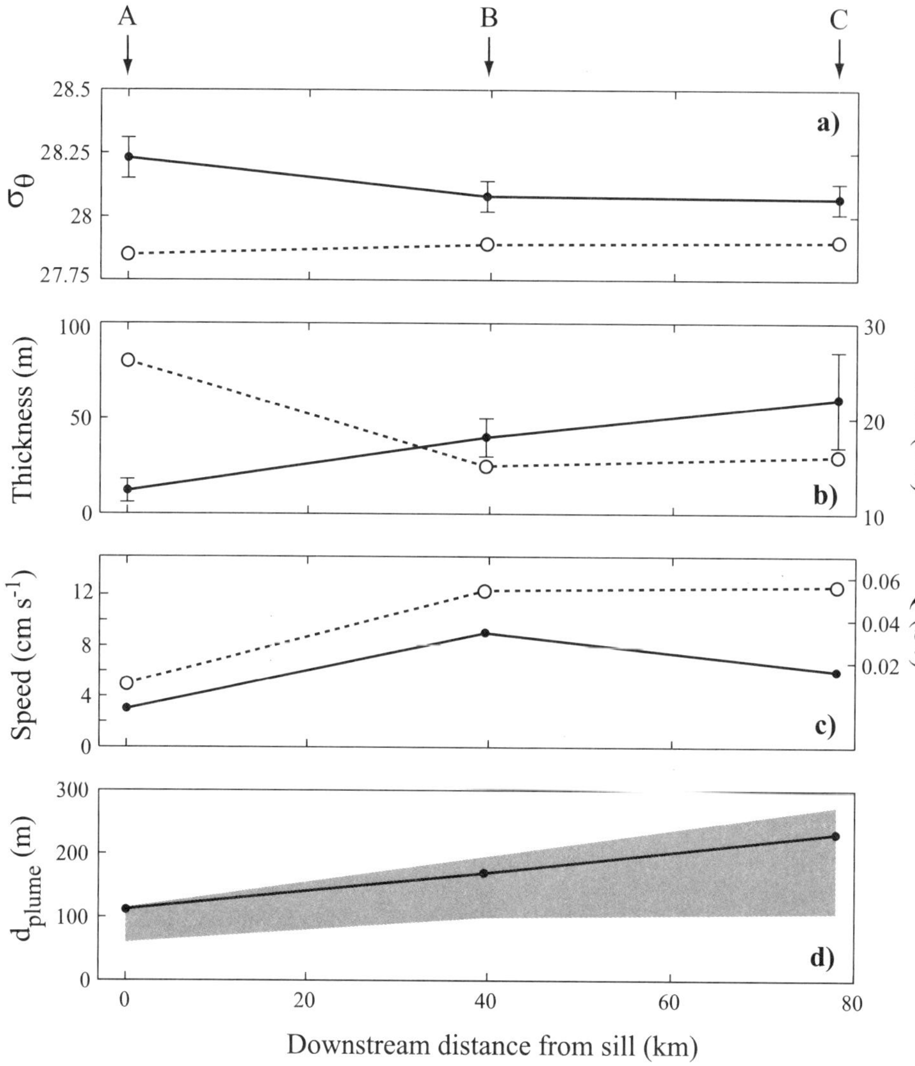

Figure 10. Downstream evolution of section-averaged (a) plume density (solid line) and ambient density (dashed line), (b) height (solid line) and width (dashed line), (c) geostrophic velocity U_{geo} (solid line) and associated volume transport Q (dashed line), and (d) density-anomaly-weighted depth, d_{plume}, associated with the plume. Sections A-C (see Figure 1) are marked by arrows above (a). The error bars show 1 SD derived for each section. The envelope in (d) marks the depth of the upper and lower edges of the plume. The horizontal axis is the downstream distance calculated from the sill following the path of the overflow.

which upon reaching sufficient amplitude collapse into turbulence [*Thorpe*, 1987]. The momentum and buoyancy is redistributed to produce a background field of higher Ri. A necessary and sufficient criterion for stability, on the other hand, is Ri > 1 [*Abarbanel et al.*, 1984]. Laboratory studies (*Rohr et al.*, 1988) and numerical simulations [*Itsweire et al.*, 1993] show evidence for turbulence in a decaying state, but still with elevated levels of dissipation, when Ri > 0.25. We have derived Ri from the geostrophic shear and the buoyancy frequency derived at 1-m vertical separations. The events with Ri < 1 are marked in Figure 9 (circles). This threshold includes active mixing events as well as decaying remnants of past mixing events. The absence of Ri < 1 does not necessarily imply absence of mixing, but raises the possibility of unresolved events by the vertical scale and the approach of this analysis. Low-Ri events are detected only in section B in the vicinity of the plume. This qualitatively agrees with the density-overturning events reported in *Fer et al.* [2004], where sections A and C were observed to yield much lower numbers of events in the vicinity of the plume, compared with section B. Furthermore, the location of Ri < 1 events in section B compares favorably with the location of the overturning events. Unstable Ri events likely drive, and those due to geostrophic shear alone can partly account for, the observed enhanced mixing (between sections A and B) and reported elevated turbulent energy dissipation rates and vertical diffusivity.

The state of the overflow, i.e., whether the plume is an accelerating event or a steady flow, can be assessed by the simple model derived by *Shapiro et al.* [2003], in which the values of the parameter R_H $(= g's/Lf^2) << 1$ represent a steady state. Here L is a relevant horizontal length scale. Using $\Delta\rho = 0.33$, $s = 3 \times 10^{-3}$ and $L = 80$ km, the distance between the sill and section C, we obtain $R_H = 6 \times 10^{-3} << 1$. Hence we expect the overflow to progress in a steady state and evaluate its downstream development by estimating its section-averaged σ_θ, height, width, geostrophic volume transport, and bottom-depth density-weighted anomaly for each section. Integrations over the cross-sectional area, A (not to be confused with the area associated with the polynya model listed in Table 2), of the overflow water are constructed according to *Girton and Sanford* [2003]. We define A as the area bounded by the boundary determined with $\sigma_\theta > 27.95$, $U_{geo} = 1$ cm s^{-1}, and the sea bottom. The section-averaged hydrographic properties σ_θ, θ, and S are calculated by integrating the vertical average of the property in each (gridded) profile across the width of the plume by using the trapezoidal method. This yields the cross-sectional area,

$$A = \iint_{area} dzdx = \int_{width} hdx, \tag{6}$$

where h is the thickness of the overflow. The overflow density is $\iint_{area} \sigma_\theta \, dzdx/A$. The geostrophic volume transport is obtained as

$$Q = \int_{width} \bar{U}_{geo} h \, dx, \tag{7}$$

where overbar denotes a layer mean for a single profile. The ambient water density is calculated as the mean within the 20 m above the plume. The density-anomaly-weighted bottom depth is

$$d_{plume} = \int_{width} \Delta\rho \, d_b \, h \, dx \, / \int_{width} \Delta\rho \, h \, dx, \tag{8}$$

where d_b is the bottom depth.

Results are summarized in Figure 10. At early parts of the overflow, there is significant entrainment of the overlying water. The height of the plume increases by a factor of approximately 4 as the width slightly decreases during adjustment for the rotational effects, increasing the cross-sectional area by a factor of two. Entrainment alone can account for this increase. Section-averaged plume salinity and θ are 35.1 and –1°C for section A and 34.95 and –0.7°C for section B, respectively. This evolution is consistent with entrainment of an equal part of ambient ESW with S = 34.8 and $\theta = -0.4$°C. Further downstream entrainment of almost equally saline AW is less significant. From section B to C, the mean temperature of the modified AW above the plume is 1°C, and the plume has to mix approximately at a rate of 5:1 to reach $\theta = -0.44$°C, the value derived for section C. The nondimensional entrainment parameter, E, which is independent of velocity estimates, equals $(h/\Delta\rho)d(\sigma_\theta)/d\zeta$ and is $O(10^{-3})$ between sections A and B and $O(10^{-4})$ between sections B and C. Here ζ is the distance along the path of the overflow. The width of the plume slightly increases as a result of the spread due to Ekman veering [*Fer et al.*, 2003]. The cross-sectional area of the plume increase by a factor of 1.5. The descent of the plume is nearly constant at a rate of 1.5 m km^{-1} (Figure 10d).

The geostrophic transport is nearly constant, ~0.06 Sv out of sections B and C. This is equal to the volume flux of BSW derived over the freezing period in 2002 (Table 2) and agrees well with the values reported in the literature (Section 2.2). The transport through section A is severely underestimated because the geostrophic adjustment is not established until farther downstream, and the section average velocity is not representative. The mean downstream velocity recorded by a moored current meter ~16 km downstream of the sill at 7 m above the bottom is 10 cm s^{-1} [*Fer et al.*, 2004]. This location is approximately 4 Rossby radii farther from the sill and we can expect a quasi-geostrophic flow. Section-averaged U_{geo} for section B, ~24 km downstream of the mooring location, is 9 cm s^{-1}. This is comparable to the current meter average and we can use the geostrophic velocity as being representative of the plume speed at sections B and C with some confidence.

5. DISCUSSION AND CONCLUSIONS

In winter 2002, relatively saline source water and a stable water column (assuming that a large surface salinity difference between spring and preceding autumn indicates a stable water column) resulted in a moderate volume of BSW estimate, despite the maximum released mass of salt (Table 2). The BSW salinity was the highest observed since 1981 and hence agrees with the modeled salt release. When compared with the four preceding winters, 2002 appears to be rather exceptional in polynya activity and corresponding ice production. The ice production (and salt release) in winter 2002 is about 1.7 standard deviations (SD) away from the 5-year 1998-2002 mean of 43.6 ± 9.7 km^3 (1099.0 [±248.1] × 10^9 kg), but 5.4 SD away from the 4-year 1998-2001 mean of 39.6 ± 3.8 km^3 (995.1 [±100.1] × 10^9 kg). For comparison, we made a 33-year time series of modeled total ice production, T_{ice}, covering winters 1970 to 2002 (Figure 11) and using constant OF and CF (Table 1) derived from a best-fit for all the winters 1998 to 2001. The average T_{ice} during this period is 39.9 ± 11.7 km^3 (±SD). Because T_{ice} is almost linearly related to the estimated total salt release, T_s, a similar fractional variability, 29%, can

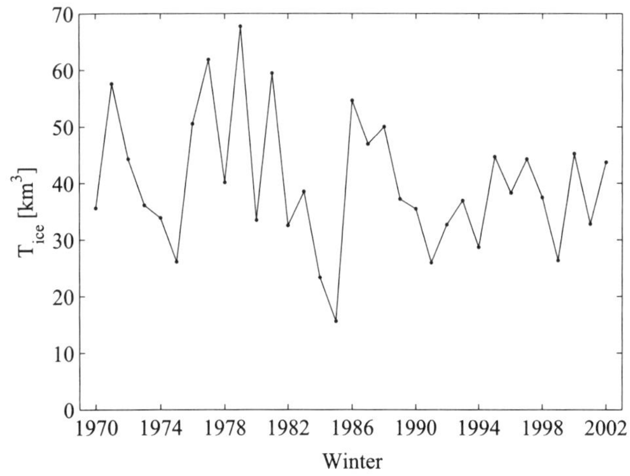

Figure 11. Time series of modeled total ice production, T_{ice}, for winters 1970 to 2002 obtained by using the best-fit OF = 1.0 and CF = 19.0 derived from matching the modeled and SAR-observed polynya widths during the winters 1998 to 2001.

be estimated for T_s. This suggests that the variability in the four preceding winters is smaller than the 33-year average (OF and CF as listed in Table 1), while the 5-year mean and variability of T_s are representative for the longer-term.

The years with T_{ice} exceeding the 33-year mean are 1971-72, 1976-79, 1981, 1986-88, 1995, 1997, 2000, and 2002, which is consistent with the BSW salinities given in Figure 2. An exception is the salinity observed in 1995; the sample to be measured was collected in mid-October, when BSW salinities are typically lower than those in the preceding spring. We note that one of the first instances when Storfjorden overflow could be followed down the continental slope into the Fram Strait was in 1986 [*Quadfasel et al.*, 1988], which appears as one of the five large ice production events in the modeled time series.

Using an open water fraction of a 1.8×10^4 km^2 Storfjorden area (1.3×10^4 km^2 in this work), *Schauer* [1995] estimated a total ice volume of 80 km^3 for winter 1992. When corrected for the area of 1.3×10^4 km^2, the corresponding ice production is 57 km^3, which is about a factor of two larger than that estimated in the 33-year time series (Figure 11). This discrepancy might be due to the use of constant OF and CF in the 33-year time series, where some winters are probably underestimated. The volume flux estimates of BSW (Table 2) are lower than 0.13 Sv (freezing period average) and 0.05 Sv (annual average), derived from moorings in winter 1992 [*Schauer*, 1995]. The mooring site is approximately 40 km downstream of the sill, and entrainment along the path of the overflow from the sill to the mooring site can account for the discrepancy between our typical volume flux estimates and those derived from moorings.

Ice production and BSW transport in Storfjorden can be compared to the values reported in the literature for other Arctic polynyas. On the mean, polynyas in Alaska and East Siberia produce an estimated ice volume of 690-850 km^3 [*Cavalieri and Martin*, 1994]. According to Cavalieri and Martin, coastal polynyas of the entire Arctic generate a mean annual brine flux of 0.7-1.2 Sv. Here, the estimated annual transports of BSW (Table 2) constitute about 3-6% of this. *Winsor and Björk* [2000], on the other hand, found an annual ice production of 300 ± 30 km^3 in the Arctic polynyas and estimated the corresponding dense water flux to be 0.2 Sv. Then, the contribution of the Storfjorden polynya is 15-20%. For the Chukchi Shelf polynyas, model studies yield ice production of 61 km^3 and dense water production of 0.05 Sv, averaged over the winters 1978-1998 [*Winsor and Chapman*, 2002]. Even though the mean ice production in Storfjorden is about 65% of that on Chukchi Shelf, the BSW flux is comparable.

The uncertainty in longer-term ice production estimates, i.e., in the 33-year time series, increases as a result of constant OF and CF. Ice production and salt release will be underestimated when the polynya opens faster and closes more slowly than assumed (need larger OF and smaller CF) and vice versa. Hence, the interannual variability in OF and CF should be included in the model to get a more reliable estimate of the variability in ice production. Ice conditions in the Barents Sea and atmospheric circulation in the North Atlantic vary interannually and influence the polynya activity in Storfjorden [*Skogseth et al.*, 2004]. This is parameterized with the model-tuned OF and CF each winter. By using constant OF and CF in the 33-year time series, the effects of the variable local ice conditions and atmospheric circulation are neglected.

The variation in OF is limited and can be explained by the variability in blocking near the mouth of the fjord caused by heavy drift ice, including multi-year ice. Blocking during opening, enhanced drag due to increased roughness and form drag during closing, and different wind regimes can partly explain the large differences between OF and CF. However, the main cause of the difference is probably rapid compaction of thin ice during closing assisted by frazil ice growth, even for weak onshore winds. As observable by SAR imagery, this has the effect of converting thin ice in the polynya into ice with pack ice characteristics and the requirement of a high closing factor.

According to *Tear et al.* [2003], two processes contribute to polynya closure: (1) Onshore drifting frazil ice continuously grows thermodynamically and piles up near the shore. This coastal boundary gradually moves offshore. (2) The consolidated ice region propagates towards the coastal boundary, which gradually expands due to piling up of frazil ice. In most of their parameter space, they find that a polynya closes faster than it opens, by a factor of 5 for two case studies north of Svalbard. They ascribe this asymmetry to initial frazil cover. Our best fit OF and CF for the winters 1998 to 2002 indicate that a polynya closes 10-20 times faster than it opens. The discrepancy between the two models can be explained by several fundamental differences. In *Tear et al.* [2003], the ratio of consolidated to frazil ice velocity is <1; in this work, we set it to 0.5 (B_1/B_2). Here, the ratio of ice velocities during opening and closing is variable, whereas *Tear et al.* [2003] use a fixed value. We apply a full surface heat budget instead of a ratio of surface heat fluxes. Furthermore, *Tear et al.* [2003] achieve a fraction of the steady-state polynya width, whereas the polynya width is variable in this work.

When rafting and ridging processes in thin ice inside the polynya and the leads in fast and pack ice are included, the total effective open water area doubles. Therefore, these processes, which can presently be included in the model in only a rather crude and ad-hoc way, are important for the total ice production. On average, the open water area occupies 10% of the Storfjorden area and accounts for about 60% of the ice production, which addresses the importance of physically acceptable representation of rafting and ridging. In winter 2002, the thin ice region was large because of the large polynya width and the relatively moderate open water width. This leads to a relatively large contribution of rafting and ridging to effective open water area compared with that in the previous years. Although this part of the model physics is still uncertain and subject to further research, its large contribution and the relatively good match with 2002 data suggest that the model would have performed poorly if this part were neglected.

Formulating theoretical models in terms of wind factors rather than ice velocities, which may vary largely, and modeling the pack ice boundary, which is more easily observed than frazil/thin ice boundary, can lead to easier comparison of observations. In future, fine-scale atmospheric model fields incorporating the effects of orography and surface conditions can be used to force two-dimensional models of polynya dynamics and thermodynamics. Fast ice and pack ice need to be distinguished from each other. Because tidal currents are strong in Storfjorden, water current stress on the ice should be included in the polynya model. Finally, future work is merited in implementing the polynya dynamics and thermodynamics in coupled atmosphere–ice–ocean grid models to simulate ice growth and brine rejection, including the effects of imposed and responding ocean circulation.

The mixing at the interface between the dense overflow plume and the ambient water is governed mainly by shear-induced internal mixing mechanisms, and salient characteristics derived from conventional CTD profiles [*Fer et al.*, 2004] can be related to elevated geostrophic shear. The influence from higher frequency variability, such as tides, is not captured with geostrophy. The magnitude of tidal velocities estimated in Storfjordrenna [*Fer et al.*, 2003] is comparable to the derived geostrophic velocity, which will raise uncertainties in snapshot calculations of volume transport. Although time-dependent fluctuations will affect the instantaneous total mass and scalar flux, their contribution to the mean fluxes can be negligible. Transport calculations through sections B and C are consistent, suggesting little tidal aliasing. To evaluate the effect of high-frequency variability on the overflow, we modeled a section (not presented) south of section B (between sections B and C) to be occupied repeatedly for three times within 30 h. The apparent variability is observed to be confined to the deepest parts of the section, where the dense water is regarded as remnants. Within the overflow, the greatest variabilities were up to 25% and 30% in geostrophic velocity and the corresponding transport estimates, respectively, whereas other section-averaged properties were constant within 8%.

The overflow observed in August is the likely source of the plume detected in deep Fram Strait in September 2002 recently reported by *Schauer et al.* [2003]. These observations credit Storfjorden as a source for ventilating the deep ocean. The mixing of the plume determines the ultimate depth where the dense water interleaves the bottom boundary and merits further studies.

Acknowledgments. This work was funded by the Norwegian Research Council, through grants 138525/410 for RS and 147493/432 for IF. The fieldwork was supported through the NOClim (Norwegian Ocean Climate) project funded by the Norwegian Research Council, grant 139815/720. We want to acknowledge Harvey Goodwin at the Norwegian Polar Institute for kindly providing us with the ERS-2 SAR data for winter 2002 and Kjell Kloster at the Nansen Environmental and Remote Sensing Center for processing the SAR data. The detailed comments by two anonymous reviewers helped to improve the paper substantially. This is publication No. A70 of the Bjerknes Centre for Climate Research.

REFERENCES

Aagaard, K., L.K. Coachman, and E. Carmack, On the halocline of the Arctic Ocean. *Deep-Sea Res.*, 28A, 529-545, 1981.

Abardanel, H.D.I., D.D. Holm, J.E. Marsden, and T. Ratiu, Richardson number criterion for the nonlinear stability of three dimensional stratified flow. *Phys. Rev. Lett.*, 52, 2352-2355, 1984.

Anderson, L.G., E.P. Jones, R. Lindegren, B. Rudels, and P.I. Sehlstedt, Nutrient regeneration in cold, high salinity bottom water of the Arctic shelves. *Continental Shelf Res.*, 8, 1345-1355, 1988.

Anderson, L.G., E. Falck, E.P. Jones, S. Jutterström, and J. Swift, Enhanced uptake of atmospheric CO_2 during freezing of seawater:

A field study in Storfjorden, Svalbard. *J. Geophys. Res.*, 109, C06004, doi:10.1029/2003JC002120, 2004.

Cavalieri, D.J., and S. Martin, The contribution of Alaskan, Siberian and Canadian coastal polynyas to the halocline layer of the Arctic Ocean. *J. Geophys. Res.*, 99, 18,343-18,362, 1994.

Fer, I., R. Skogseth, P.M. Haugan, and P. Jaccard, Observations of the Storfjorden overflow *Deep-Sea Res. I,* 50, 1283-1303, 2003.

Fer, I., R. Skogseth, and P.M. Haugan, Mixing of the Storfjorden overflow (Svalbard Archipelago) inferred from density overturns. *J. Geophys. Res.*, 109, C01005, doi:10.1029/2003JC00 1968, 2004.

Fohrmann, H., J.O. Backhaus, F. Blaume, and J. Rumohr, Sediments in bottom-arrested gravity plumes: Numerical case studies. *J. Phys. Oceanogr.*, 28, 2250-2274, 1998.

Girton, J.B., and T.B. Sanford, Descent and modification of the overflow plume in the Denmark Strait. *J. Phys. Oceanogr.*, 33, 1351-1364, 2003.

Haarpaintner, J., J.-C. Gascard, and P.M. Haugan, Ice production and brine formation in Storfjorden, Svalbard. *J. Geophys. Res.*, 106, 14,001-14,013, 2001a.

Haarpaintner, J., P.M. Haugan, and J.-C. Gascard, Interannual variability of the Storfjorden (Svalbard) ice cover and ice production observed by ERS-2 SAR. *Ann. Glaciol.*, 33, 430-436, 2001b.

Howard, L.N., Note on a paper of John W. Miles. *J. Fluid Mech.*, 10, 509-512, 1961.

Itsweire, E.C., J.R. Koseff, D.A. Briggs, and J.H. Ferziger, Turbulence in stratified shear flows: Implications for interpreting shear induced mixing in the ocean. *J. Phys. Oceanogr.*, 23, 1508-1522, 1993.

Jungclaus, J.H., J.O. Backhaus, and H. Fohrmann, Outflow of dense water from the Storfjord in Svalbard: A numerical model study. *J. Geophys. Res.*, 100, 24719-24728, 1995.

Maus, S., 2003. Interannual variability of dense shelf water salinities in the north-western Barents Sea. *Pol. Res.*, 22, 59-66.

Miles, J.W., On the stability of heterogeneous shear flows. *J. Fluid Mech.*, 10, 496-508, 1961.

Orvik, K.A., and Ø. Skagseth, The impact of the wind stress curl in the North Atlantic on the Atlantic inflow to the Norwegian Sea toward the Arctic. *Geophys. Res. Lett.*, 30, 1884, doi: 10.1029/2003GL017932, 2003.

Pease, C.H., The size of wind-driven coastal polynyas. *J. Geophys. Res.*, 92, 7049-7059, 1987.

Price, J.F., and M.O. Baringer, Outflows and deep water production by marginal seas. *Prog. Oceanogr.*, 33, 161-200, 1994.

Quadfasel, D., B. Rudels, and K. Kurz, Outflow of dense water from a Svalbard fjord into the Fram Strait. *Deep-Sea Res.*, 35, 1143-1150, 1988.

Rohr, J.J., E.C. Itsweire, K.N. Helland, and C.W. Van Atta, Growth and decay of turbulence in a stably stratified shear flow. *J. Fluid Mech.*, 187, 77-111, 1988.

Rudels, B., and D. Quadfasel, Convection and deep water formation in the Arctic Ocean - Greenland Sea system. *J. Mar. Syst.*, 2, 435-450, 1991.

Schauer, U., The release of brine-enriched shelf water from Storfjord into the Norwegian Sea. *J. Geophys. Res.*, 100, 16,015-16,028, 1995.

Schauer, U., B. Rudels, I. Fer, P.M. Haugan, R. Skogseth, G. Björk, and P. Winsor, Return of deep shelf/slope convection in the Western Barents Sea? Paper presented at *the Seventh Conference on Polar Meteorology and Oceanography and Joint Symposium on High-Latitude Climate Variations*, The American Meteorological Society, 2003.

Shapiro, G.I., and A.E. Hill, The alternative density structures of cold/saltwater pools on a sloping bottom: The role of friction. *J. Phys. Oceanogr.*, 33, 390-406, 2003.

Shapiro, G.I., and A.E. Hill, Dynamics of dense water cascades at the shelf edge. *J. Phys. Oceanogr.*, 27, 2381-2394, 1997.

Shapiro, G.I., J.M. Huthnance, and V.V. Ivanov, Dense water cascading off the continental shelf. *J. Geophys. Res.*, 108, 3390, doi: 120.1029/2002JC001610, 2003.

Skogseth, R., P.M. Haugan, and J. Haarpaintner, Ice and brine production in Storfjorden from four winters of satellite and in situ observations and modeling. *J. Geophys. Res.*, 109, C10008, doi: 10.1029/2004JC002384, 2004.

Skogseth, R., P.M. Haugan, and M. Jakobsson, Watermass transformations in Storfjorden. *Cont. Shelf Res.*, 25, 667-695, 2005.

Smethie, W.M., H.G. Ostlund, and H.H. Loosli, Ventilation of the deep Greenland and Norwegian Seas: Evidence from Krypton-85, tritium, carbon-14, and argon-39. *Deep-Sea Res. Part A*, 33, 675-703, 1986.

Sternberg, R.W., K. Aagaard, D. Cacchione, R.A. Wheatcroft, R.A. Beach, A.T. Roach, and M.A.H. Marsden, Long-term near-bed observations of velocity and hydrographic properties in the northeast Barents Sea with implications for sediment transport. *Cont. Shelf Res.*, 21, 509-529, 2001.

Tear, S., A.J. Willmott, N.R. T. Biggs, and M.A. Morales Maqueda, One-dimensional models for the closure of a coastal latent heat polynya. *J. Phys. Oceanogr.*, 33, 319-342, 2003.

Thorpe, S.A., Transitional phenomena and the development of turbulence in stratified fluids: A review. *J. Geophys. Res.*, 92, 5231-5248, 1987.

Winsor, P., and G. Björk, Polynya activity in the Arctic Ocean from 1958 to 1997. *J. Geophys. Res.*, 105, 8789-8803, 2000.

Winsor, P., and D. Chapman, Distribution and interannual variability of dense water production from coastal polynyas on the Chukchi Shelf. *J. Geophys. Res.*, 107, doi:10.1029/2001JC000- 984, 2002.

I. Fer and P.M. Haugan, Geophysical Institute, University of Bergen, Allégaten 70, N-5007, Bergen, Norway. (ilker.fer@gfi.uib.no)

R. Skogseth, The University Centre in Svalbard, Post-Box 156, 9171, Longyearbyen, Svalbard, Norway.

Pathways and Export of Greenland Sea Water

Tor Eldevik[1,2], Fiammetta Straneo[3], Anne Britt Sandø[1,2], and Tore Furevik[2,4]

The dense water that overflows the Greenland-Scotland Ridge from the Nordic Seas is a major source for the deep waters of the North Atlantic. An advective–diffusive model, with its current deduced from the data archive of a high-resolution general circulation model, has been set up to describe the spreading of Greenland Sea Water through the Nordic Seas to the overflows. A diversity of flow regimes, e.g., positive and negative North Atlantic Oscillation, can be modelled by this flexible approach. There are large differences between the simulated cases, predominantly due to the variability of the internal circulation of the Nordic Seas. The varying role played by the Jan Mayen Current is particularly striking. The model is evaluated through the observed spreading of the tracer sulphur hexafluoride, which was purposefully released in the central Greenland Sea in 1996. The model ocean compares very well with this unique field experiment.

1. INTRODUCTION

The overflows from the Nordic Seas are a major source for the deep waters of the North Atlantic and thus are important contributors to the Atlantic Meridional Overturning Circulation. Where the overflow water is formed, and how, is still a matter of debate. Proposed mechanisms involve open ocean convection, primarily in the Greenland Sea; dense water produced on the Arctic shelves during the winter [for both, see *Aagaard et al.*, 1985]; and the gradual transformation of Atlantic Water while it undergoes its cyclonic loop in the boundary current around the periphery of the Nordic Seas–Arctic Ocean [*Mauritzen*, 1996a,b].

[1]Nansen Environmental and Remote Sensing Center, Bergen, Norway.
[2]Bjerknes Centre for Climate Research, Bergen, Norway.
[3]Woods Hole Oceanographic Institute, Woods Hole, Massachusetts, USA.
[4]Geophysical Institute, University of Bergen, Norway.

The Nordic Seas: An Integrated Perspective
Geophysical Monograph Series 158

10.1029/158GM08

While drifter measurements over the past decade have provided new insight into the surface flow of the Nordic Seas [*Jakobsen et al.*, 2003; *Orvik and Niiler*, 2002; *Poulain et al.*, 1996], less is known of the intermediate and deep water flow. The general picture, based on rather sparse measurements in deep water [cf. *Hansen and Østerhus*, 2000] and diagnostic calculations using climatological hydrography and wind-stress curl as input [*Nøst and Isachsen*, 2003], reveal an intermediate and deep water flow that is generally in the same direction as the surface flow, with northward flow west of Norway and southward flow east of Greenland. Also, there is cyclonic circulation within the subbasins: the Greenland Sea, the Iceland Sea, and the Norwegian and Lofoten basins of the Norwegian Sea. Figure 1 is a synthesis of these characteristics.

The Greenland Sea is a "hot spot" for open ocean convection [*Alekseev et al.*, 2001; *Marshall and Schott*, 1999]. The spreading of intermediate and deep water from the central Greenland Sea to the Greenland–Scotland Ridge is the object of our study. It is inspired by the description by *Straneo et al.* [2003] of the spreading of Labrador Sea Water (LSW). Their advective–diffusive model is here set up for the Nordic Seas intermediate water (IW). The stationary model current is deduced from an eddy-permitting general circulation model (GCM) covering the region [*Hátún et al.*, this volume]. The four different current regimes shown in Figures 2-4 are considered: the mean over the whole GCM simulation period (1951-2000), the flows

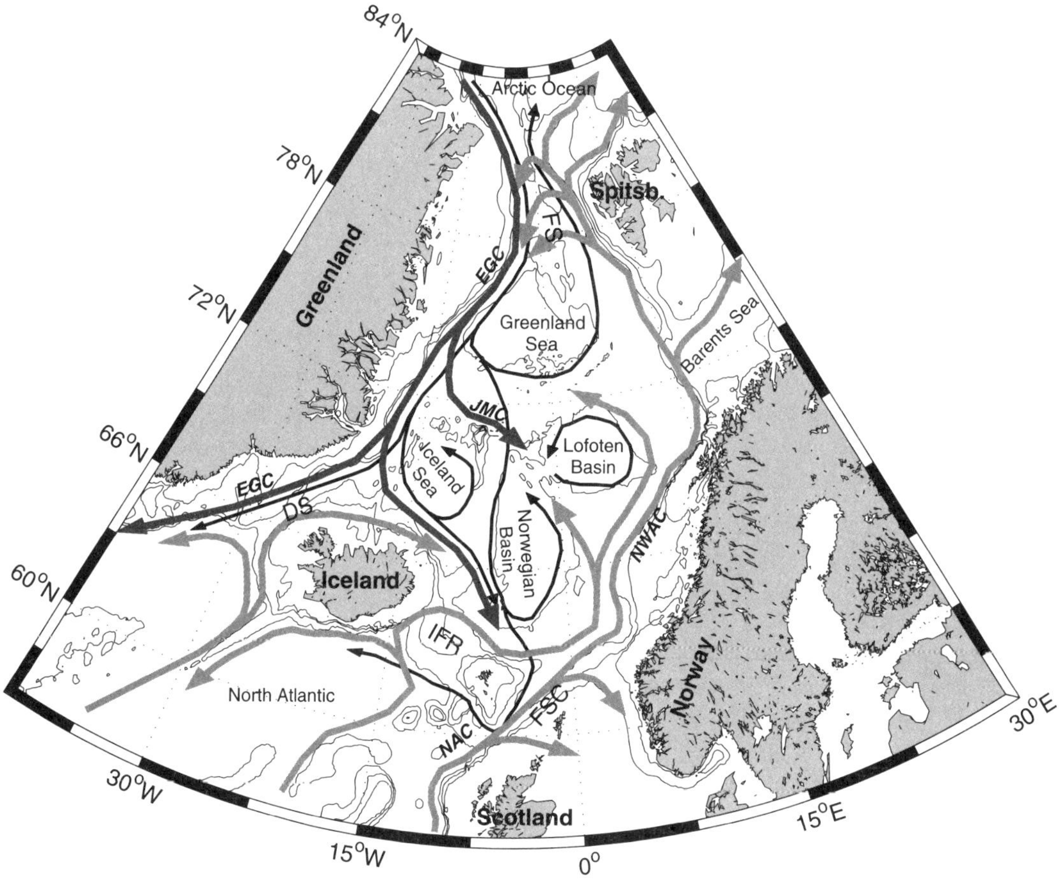

Figure 1. Sketch of the flow of the Nordic Seas. The thick arrows represent the surface flow of the Atlantic Water (light shade) and the Polar Water (dark shade), while the thin black arrows indicate the flow of intermediate and deep waters. See main text for references.

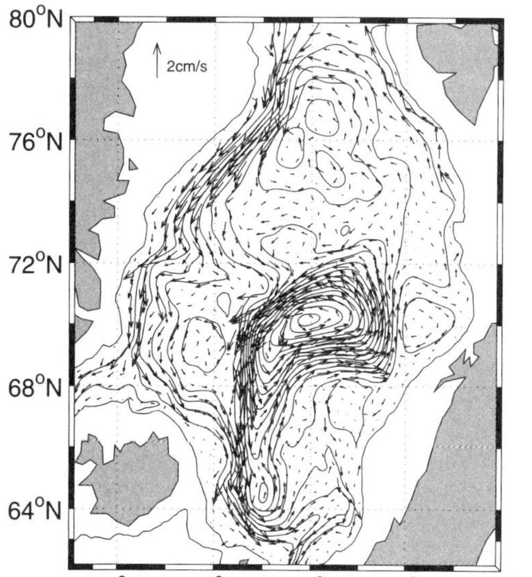

Figure 2. The mean (1951-2000) IW flow field from the GCM and the corresponding streamlines used by the advective–diffusive solver. Only a quarter of the vectors are shown. The spacing of streamlines corresponds to 0.25 Sv.

related to negative and positive phases of the North Atlantic Oscillation, and the circulation of recent years (1997-2000). In the summer of 1996 a patch of the tracer sulphur hexafluoride (SF_6) was purposefully released in the central Greenland Sea [*Watson et al.*, 1999]. The monitoring of its spreading since the release provides a unique benchmark for evaluating the model.

The goal of this study is to investigate the different pathways of the Greenland Sea Water (GSW) within the Nordic Seas and the export of GSW at the overflows to the North Atlantic. This chapter is organized as follows: The advective–diffusive model and the (proxy) current data are presented in section 2, and the model concept is illustrated by the flow and tracer spreading generated by the mean IW current. The model is evaluated in section 3, where comparison with recent SF_6 tracer data suggests that the concept is suitable for the study. The flows related to weak and strong atmospheric forcing are presented in the next section. The different model pathways and exports are compared and discussed in section 5. The importance of the Jan Mayen Current relative to the East Greenland Current and of the Faroe–Shetland Channel overflow relative to the Denmark Strait are addressed in particular. The concluding remarks of section 6 summarize the paper.

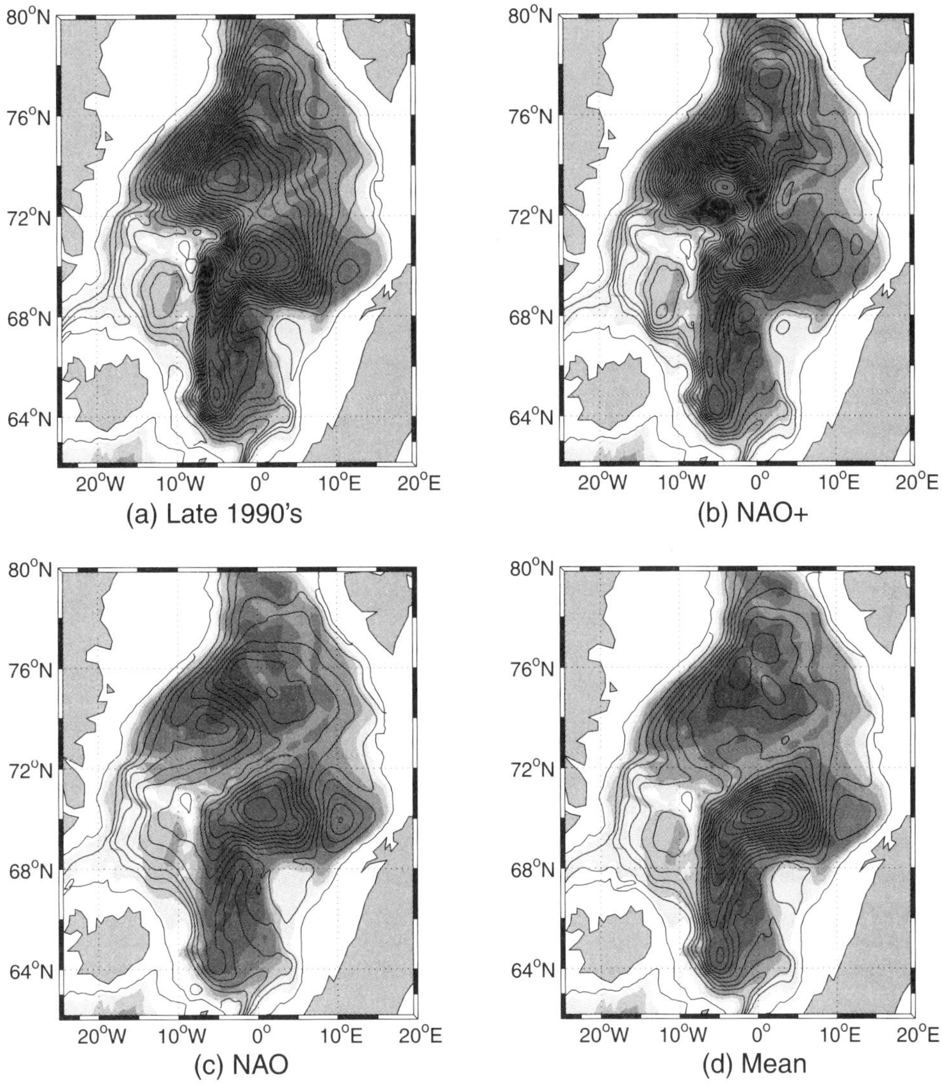

Figure 3. The IW streamfunctions for the "SF_6 years" 1997-2000 (a), and the periods of positive (1989-1992) and negative (1962-1965) NAO (b,c). The mean (1951-2000) is shown for reference in (d). The spacing of streamlines corresponds to 0.25 Sv, and the background grayscale gives the bathymetry at 500-m intervals.

2. THE MODEL OCEAN

A very simple and flexible concept is used for our study: the GSW is assumed to be a passive tracer in an advective–diffusive model ocean. The tracer is advected by a stationary horizontal current, $\mathbf{U} = \mathbf{U}(x, y)$, and mixed by an eddy diffusivity tensor, $\underset{\sim}{\kappa} = \underset{\sim}{\kappa}(x, y)$, which represents the flow at intermediate depths in the Nordic Seas. The evolution of the GSW in the basin is then governed by

$$\frac{\partial \phi}{\partial t} + \mathbf{U} \cdot \nabla \phi = \nabla \cdot (\underset{\sim}{\kappa} \cdot \nabla)\phi, \qquad (1)$$

where $\phi = \phi\,(x, y, t)$ is the concentration per unit area.

The model was originally set up by *Straneo et al.* [2003] to describe the spreading of Labrador Sea Water (LSW) in the subpolar North Atlantic. Their model flow field was deduced from float data, and hydrographic data were used to identify different scenarios of LSW generation and thereby initialize the tracer. The different pathways and timescales for LSW spreading identified in their study compare well with observations, thus supporting both the choice of the model fields and, in general, the approach used to address the spreading of the convectively formed LSW. The model also proved to be a valuable tool with which to quantify the relative importance of the different pathways.

In this study, the general approach of *Straneo et al.* [2003] is applied to the spreading of GSW with some important

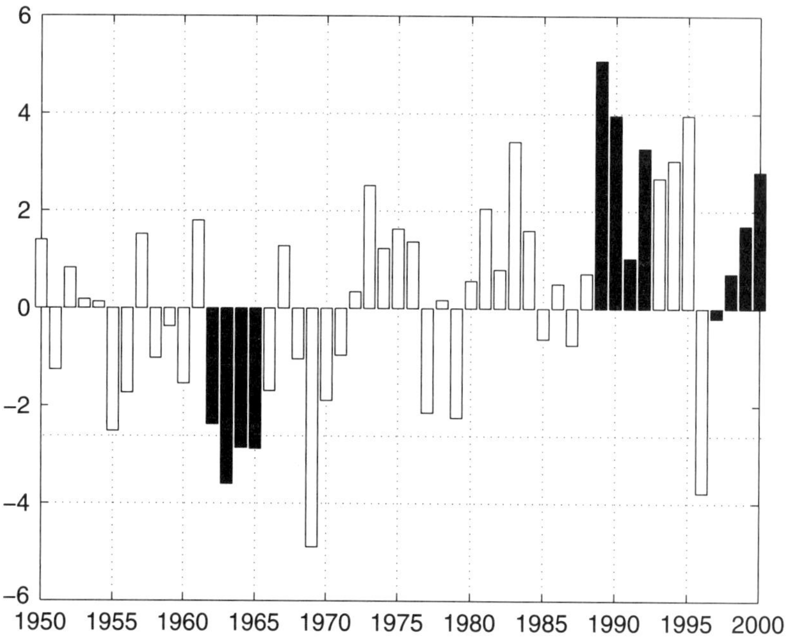

Figure 4. The winter NAO Index (values taken from http://www.cgd.ucar.edu/jhurrell/nao.html). The shading marks the three different 4-year periods used.

differences. First, the mean flow field is model-derived. This allows for a variety of mean flows to be utilized to investigate the spreading in different climate regimes (unlike *Straneo et al.*, who kept the flow field constant). Second, the ongoing SF_6 experiment conducted in the Greenland Sea provides an ideal ("real") passive tracer against which to test the validity of our approach.

2.1. The Current Data

The stationary horizontal current field and diffusivity prescribed to equation (1) should ideally be based on in situ observations (which is the case for the Labrador Sea study). The hydrography of the Nordic Seas has been sampled for more than a century [e.g., *Helland-Hansen and Nansen*, 1909; *Furevik et al.*, 2002], and drifters have in recent years revealed the flow of the surface layer in great detail [e.g., *Orvik and Niiler*, 2002]. Intermediate and deep current measurements are nevertheless sparse [*Hansen and Østerhus*, 2000; *Blindheim and Østerhus*, this volume] and can therefore not be used as a basis for our model of the intermediate waters of the Nordic Seas. We therefore have constructed proxy current data from the output of the high-resolution, synoptically forced GCM of *Hátún et al.* [this volume], which is a version of the Miami Isopycnic Coordinate Ocean Model (MICOM [*Bleck et al.*, 1992]). Their regional model simulation covers the North Atlantic Ocean and the Nordic Seas (30°N-80°N) for the years 1948-2000. The model has 25 isopycnic layers below its mixed layer and a horizontal resolution of about 20 km in the Nordic Seas region. At the lateral boundaries of the region model, the different model components are relaxed towards weekly output from a global model of the same type, with half the resolution [*Furevik et al.*, 2002].

Even with an ideally sampled GCM data archive available (model insufficiencies aside), one still has to decide on which subset to derive the velocity data from. We have chosen to average over the intermediate water (IW), herein defined to be the water between 500 and 1500 m deep. Where the ocean depth is less than 1500 m, the average velocities are weighted consistent with the reduced thickness. The choice is partly pragmatic. The advection–diffusion model describes tracer conservation by a nondivergent horizontal flow. As oceanic circulation is predominantly horizontal, the "geometrically" averaged current is close to nondivergent. The upper limit of 500 m assures that the data in general are collected well below the mixed layer. The Nordic Seas are characterized by relatively cold water below this level, which also corresponds to the typical sill depth of the Greenland–Scotland Ridge [*Hansen et al.*, 2001]. It is still sufficiently shallow to account for the overflows of the Denmark Strait (DS, 620 m deep) and the Faroe–Shetland Channel (FSC, 840 m deep). The lower limit of 1500 m exceeds the ocean depth only in the very vicinity of the overflows. The deeper flow is presumably still well accounted for: The vertical stratification and velocity shear at depth in the Nordic Seas are quite weak in hydrographic data [e.g., *Blindheim and Østerhus*, this volume;

Hopkins, 1991], as well as in the observational-based diagnostics of *Nøst and Isachsen* [2003] and in GCMs (e.g., the one at hand). Furthermore, since the 1980s convection has been ventilating the Greenland Sea only to intermediate depths [cf. *Dickson et al.*, 1996; *Watson et al.*, 1999], and the SF_6 against which the model is evaluated is found in the IW (cf. section 3).

A qualitatively different, and maybe more intuitive, alternative would be to sample an isopycnic layer (or range) representative of the wintertime ventilation in the Greenland Sea. This option was discarded because an isopycnic definition is not straightforward when linking the GSW with the overflows. The water below a few hundred meters in the central Greenland Sea has a potential density $\sigma_\theta > 28.05$ [*ESOP2*, 1999], while most of the overflowing waters in the FSC and DS are lighter than this [cf. *Fogelqvist et al.*, 2003; *Girton et al.*, 2001]. This is also the case for the waters of the GCM.

2.2. The Mean Pathways

The mean IW velocity field and streamlines from the GCM simulation period 1951-2000 (the 1948-1950 spin-up is not used) are shown in Figure 2. The GCM current is interpolated to the 10 km × 10 km grid of the advection–diffusion solver. For details on the numerical solver of equation (1), refer to *Straneo et al.* [2003]. The domain is as seen in Figure 2. Note that no streamlines cross the Iceland–Faroe Ridge, as it is shallower than 500 m. Although the IW current was defined with some care, the flow is not strictly mass-conserving. A relatively small flux correction has to be added at the open boundary to make **U** conservative. For all the simulations described herein, the correction was prescribed uniformly over the inflowing part of the Fram Strait (FS), the open boundary to the north. The overflows then remain as in the original GCM current data, and one does not risk strangling the relatively weak outflowing branch in the eastern FS. These considerations nevertheless have little practical importance. When prescribing the relaxation elsewhere, there were only marginal quantitative changes, no qualitative.

The IW streamlines of Figure 2 reproduce the general pattern from Figure 1. The cyclonic circulation covering the Norwegian and Lofoten basins is particularly pronounced in the model flow. It may possibly be too strong, but a similar pattern is present in the diagnosis of *Nøst and Isachsen* [2003]. The pattern is relatively persistent over the GCM simulation period (cf. Figure 3) and therefore should stand out in the mean.

The regional model and the corresponding coarser global versions (with grid focus on the Arctic Mediterranean) should be state of the art for GCM systems covering the Nordic Seas. The GCMs' spreading of active [*Furevik et al.*, 2002] and passive [*Gao et al.*, 2004] tracers, as well as their Atlantic–Nordic Seas exchanges [*Nilsen et al.*, 2003; *Hátún et al.*, this volume], have all been evaluated favourably against observations. The export fluxes in the case of the advection-diffusion model are 0.8 Sv (1 Sv = 10^6 m^3s^{-1}) through the DS and 1.4 Sv through the FSC. Recent overflow estimates based on observations are 0.6 Sv [*Girton et al.*, 2001] and 1.7 Sv [*Hansen and Østerhus*, 2000]. The referred DS flux is for the densest part of the overflow, $\sigma_\theta \geq 28.0$. The net model IW influx through the FS, which is by definition the sum of the DS and FSC fluxes, is 2.2 Sv.

Open ocean convection in the Nordic Seas is predominantly associated with the Greenland Sea Gyre. Although the gyre is a dynamical feature, it is normally understood to (roughly) coincide with the 3500-m isobath in the central Greenland Sea [e.g., *Bönisch et al.*, 1997; *Marshall and Schott*, 1999]. The topographical definition of the gyre agrees well with the mean circulation (cf. Figures 2 and 3d), but less so for the other cases (Figures 3a-c). This is addressed further in the discussion in section 5. For a straightforward comparison between the cases, this isobath was still used as the initial boundary of the model GSW in all the simulations except the one evaluated against SF_6 data.

The initial tracer distribution, the mean IW current, and the constant nominal diffusivity (see below) were prescribed to equation (1) to produce the GSW spreading of Figure 5. The tracer is advected and diffused from the gyre area. By year 3 (Figure 5b) it has reached the FS to the north, and two distinct branches point to the two overflows via the East Greenland Current (EGC) and the Jan Mayen Current (JMC). The EGC branch has reached the DS by year 5 (Figure 5c), where some tracer is deflected eastward to join the tracer advected by the JMC to overflow at the FSC (Figure 5d). The different straits and currents are indicated in Figure 1. The mean picture will be further characterized and compared with the other simulations in the next sections.

2.3. The Model Diffusion

The diffusive contribution to the tracer's evolution represents the action of mesoscale eddies. In regions of open ocean convection, these eddies will primarily have one of two origins: (1) the instability of the convective water mass or (2) the instability of the boundary current surrounding the convective region. The distinction between the two is important. While the former eddies are temporally and spatially tied to convection [e.g., *Jones and Marshall*, 1993, 1997], the latter are tied less directly to a distinct convective event and more to the mean gradient between the convective region and the more buoyant currents around it. Recent studies from the Labrador Sea, an open ocean convection region thought to be quite similar to the Greenland Sea

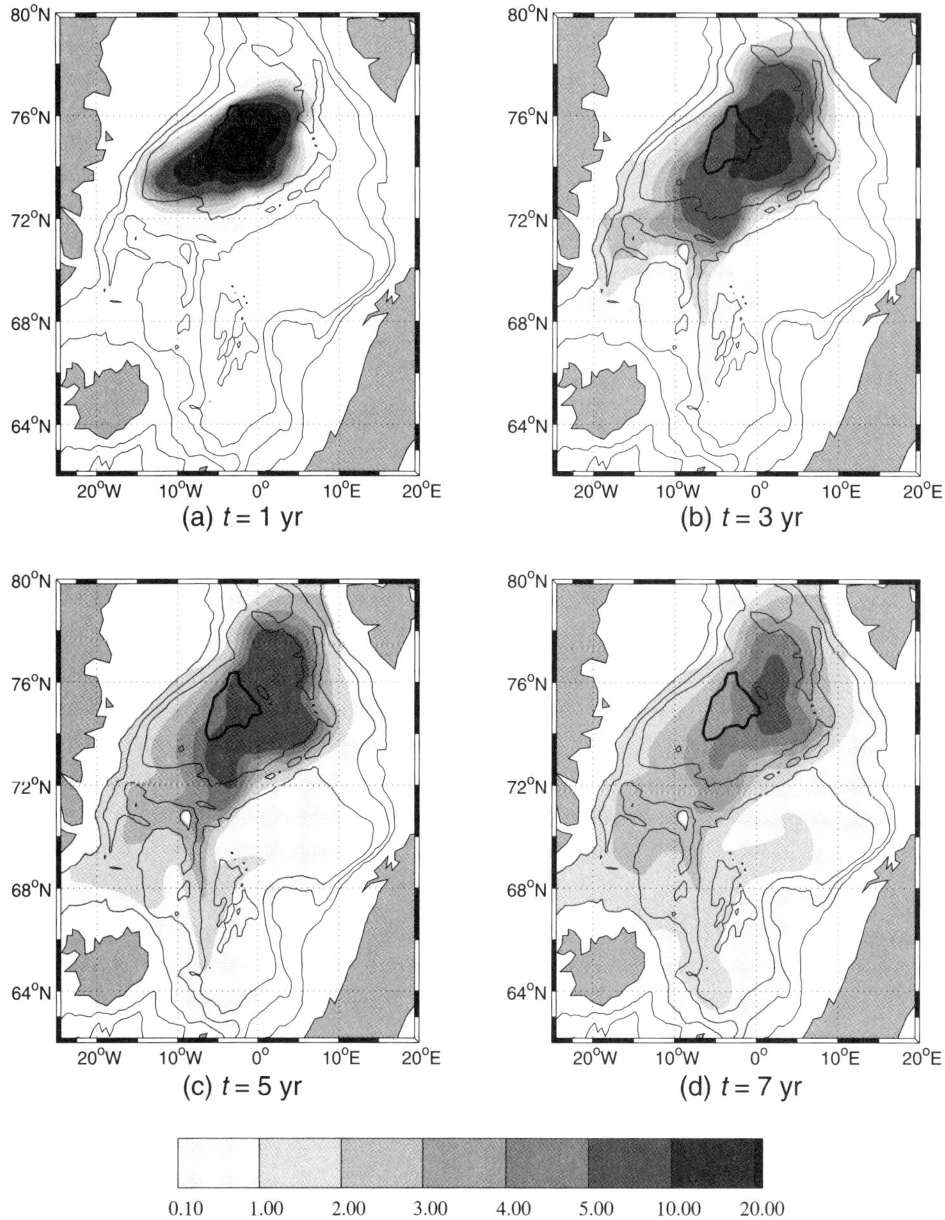

Figure 5. The spreading of tracer from the central Greenland Sea by the 1951-2000 mean IW flow. The background contours give the bathymetry at 1000-m intervals. Units in percent of the initial 100% concentration within the 3500-m isobath (thick line).

[e.g., *Marshall and Schott*, 1999], show growing evidence that spreading of the convectively formed water is a slow process dominated by the mean eddy fluxes arising from the second type of instability rather than from the seasonal, convectively generated turbulence. For example, *Lilly et al.* [1999] have shown, using mooring data, that the small-scale, high-frequency turbulence associated with convection is typically short-lived, rapidly decaying one to two months after convection. Similarly, *Straneo* [2004] showed how Labrador Sea Water export occurs mostly at a uniform rate, persisting even in years when no convection occurred (and hence is not directly tied to that year's convection). Several recent model studies also argue that the export of dense water is dominated by the continuous exchange of properties between the site of (intermittent) convection and the surrounding boundary current [*Katsman et al.*, 2004; *Spall*, 2004]. Given this, our assumption in this study (as in *Straneo et al.* [2003]) is that the lateral diffusion of the passive tracer is dominated by the mean eddy fluxes and not by the turbulence associated with convection. Effectively, this amounts to assuming a separation of timescales between the short-lived convection and the slower spreading of the dense water formed.

The eddy diffusivity is assumed to be constant in the four case studies presented here, $\underset{\sim}{\kappa} = \kappa \underset{\sim}{I}$ and is set to the nominal value $\kappa = 100\ \mathrm{m^2s^{-1}}$. From their surface-layer drifters, *Poulain et al.* [1996] estimate an eddy length scale for the Nordic Seas to be in the range 10-40 km. With $\kappa \sim u'L$ and an eddy velocity scale $u' = 1\ \mathrm{cm\ s^{-1}}$ (same as the mean flow; cf. Figure 2), an estimate of the IW diffusivity related to mesoscale eddies is then 100-400 $\mathrm{m^2s^{-1}}$. A constant value representing the basin average should then be in the very low end of this range.

Although we argue that the velocity field of the eddy-permitting GCM is an adequate substitute for lacking in situ measurements, the field is probably rather unsuited for a detailed quantification of $\underset{\sim}{\kappa}$ and its spatial distribution. The mesoscale eddies, whose mixing the eddy diffusivity represents, are mostly parameterized and not resolved by the 20-km horizontal grid of the GCM. If one nevertheless calculates the root-mean-square zonal and meridional "eddy" velocities from the weekly sampled GCM archive, using the difference between the weekly values and the 1951-2000 mean, one obtains Figure 6. The largest values are, as one would expect, associated with regions of strong gradients in the mean flow and of large variability (cf. Figure 3). The basin averages are 0.7 and 0.9 cm $\mathrm{s^{-1}}$, in the two respective directions.

For an eddy length scale of 10 km, the above velocities correspond to eddy diffusivities in units of $10^2\ \mathrm{m^2s^{-1}}$. To check the model's sensitivity to anisotropic and spatially inhomogeneous diffusion, we performed an experiment using these GCM derived eddy velocities and an eddy length scale of 10 km. A snapshot from this experiment is shown in Figure 7a. The spreading pattern differs only in details from that of the reference experiment (Figure 5c). This is consistent with *Straneo et al.* [2003], who found in their study of the Labrador Sea that the impact of the eddy diffusivity's spatial variability was quite limited. A second sensitivity experiment was done by simply doubling the reference homogeneous diffusivity to $\kappa = 200\ \mathrm{m^2s^{-1}}$. The value is probably too large to represent a basin average, but the GSW evolution in this case (Figure 7b) is still very similar to the reference. As one would expect, there are lower peak concentrations and weaker gradients.

2.4. A Note on Model Relevance

The fundamental spreading properties of the Nordic Seas flow topology is at the heart of the advection–diffusion approach. Distinct cases are easily set up within the framework, e.g., the restriction to the IW and the different four-year-mean flows of the next section. A pertinent question at

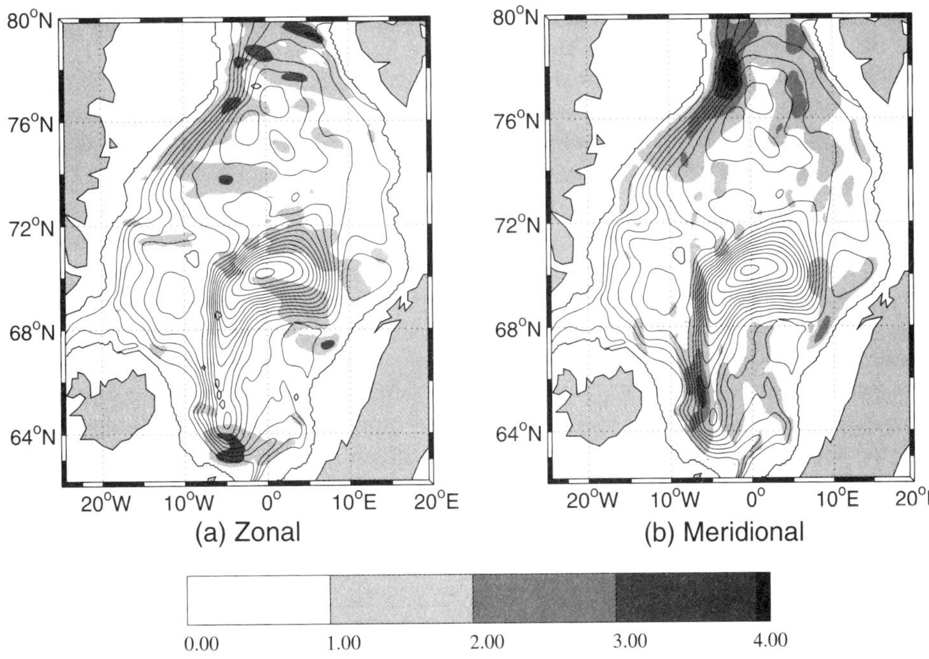

Figure 6. The IW "eddy velocities" [cm $\mathrm{s^{-1}}$] calculated from the temporal variability of the GCM flow. The streamlines of the mean flow are also shown.

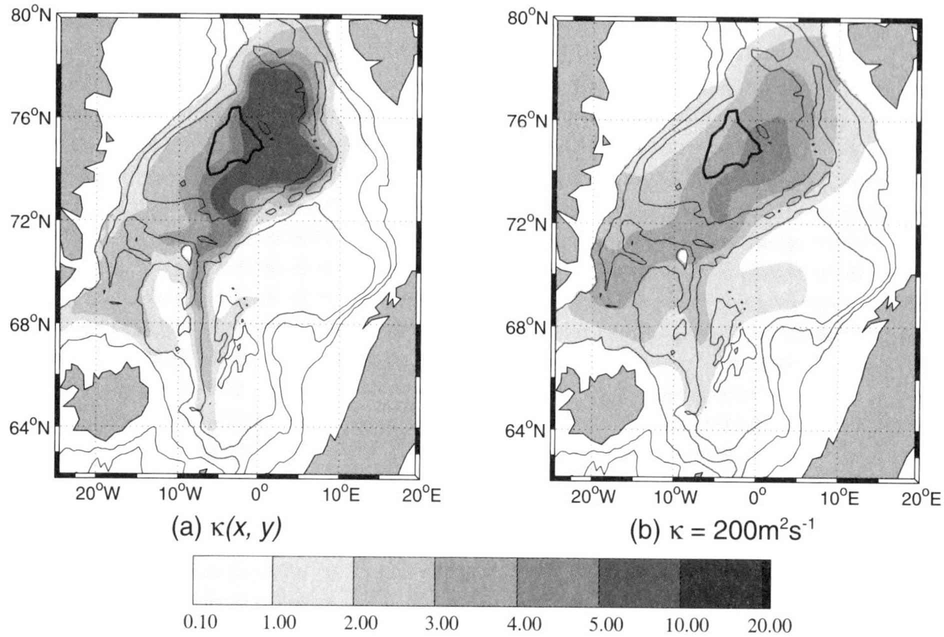

Figure 7. The spreading of tracer in the two eddy-diffusion sensitivity experiments ($t = 5$ yr).

this stage is, nevertheless, how suited is an advective-diffusive model for the present study? The model current is deduced from the data archive of a GCM. One could argue that the tracer spreading should be modelled directly by the GCM—but this is not really an option. A 10-year online tracer realisation would consume 4000 CPU-hours on a supercomputer (as of 2004). The model ocean described by equation (1), which is one way of doing off-line tracer studies based on the GCM, covers the same time span in 1 hour on a laptop computer. Furthermore, the comparison of simulated and observed SF_6 spread and export (cf. section 3) supports the use of this model in investigating the IW pathways.

3. MODEL EVALUATION: THE SF_6 EXPERIMENT

In August 1996, 320 kg of the passive tracer sulphur hexafluoride (SF_6) was released at intermediate depth in the central Greenland Sea [*ESOP2*, 1999; *Watson et al.*, 1999]. The main objective of the still-ongoing experiment is very much the same as ours: to track the GSW from its potential generation site in the gyre area along the pathways of the Nordic Seas into the neighbouring oceans. This gives important insight into the basin's present physical oceanography. More particularly, the continuous monitoring of the tracer since the release [*Messias et al.*, 2004; *Olsson et al.*, 2005] provides a unique benchmark for ocean models. The relevance of the tracer experiment to our particular model is further emphasized by the fact that the maximum concentration of SF_6 is generally measured in the IW part of the water column. Thus, as a direct evaluation of the advection–diffusion approach and of the GCM-generated model field, we present here an experiment simulating the spreading of SF_6.

The 1997-2000 mean flow field (the GCM velocity archive ends with the year 2000) is chosen as that most representative of the "SF_6-years". The streamlines are shown in Figure 3a. It does show the general IW patterns of Figures 1 and 3d, but there are features more emphasized in the late 1990s current. Note particularly that the streamlines following the JMC, diverting into the Norwegian Sea east of Jan Mayen, are more pronounced. It indicates that the JMC is more important for the export through the FSC in this case. This is confirmed by the pattern described by the model tracer and the concentrations at the overflows presented below.

For the specific comparison of model and SF_6 observations, the model tracer was initialized consistent with the August 1996 release. A snapshot of that case, corresponding to the summer of 2002, is shown in Figure 8a. The black rectangle, centered at 1.5°W, 75.25°N marks the release site. Two large-scale surveys of SF_6 were done in 2002 as parts of the EU project TRACTOR. The first was from April 20 to June 6 (*R/V Oden*) and the second from May 30 to July 1 (*R/V Knorr*). The asterisks in the figure are the hydrographic stations occupied by the *Oden* and the open circles are those occupied by the *Knorr*. The numbers 1 to 9 mark the start of the section identified by the given number. *Messias et al.* [2004] have analysed and synthesised these observations. Their SF_6 inventory (with the background subtracted) is displayed in

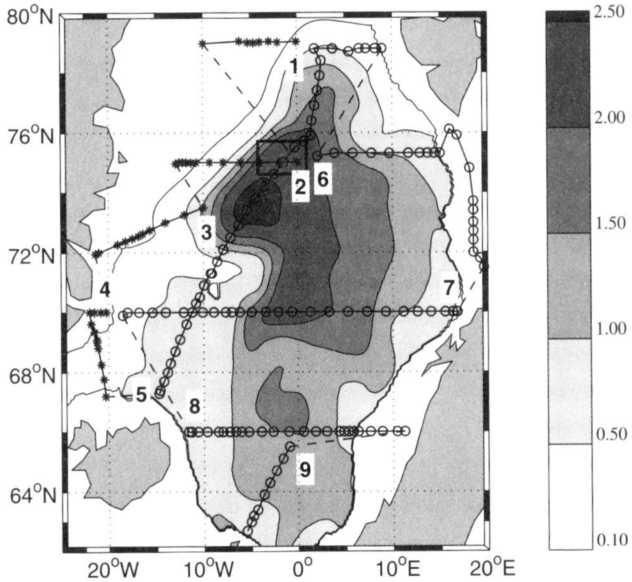

(a) Model tracer and hydrographic stations

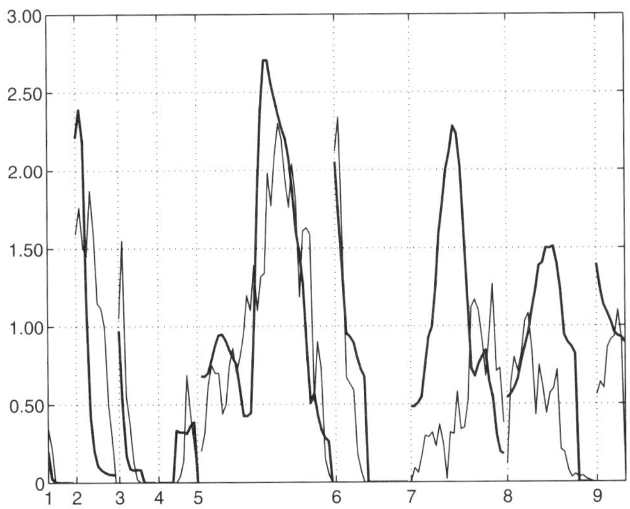

(b) Model versus observations

Figure 8. The modelled and observed SF_6 inventory (nmol m^{-2}) for the summer 2002. The top panel (a) is the modelled distribution. Also shown are the hydrographic stations occupied by *R/V Oden* (asterisks) and *R/V Knorr* (circles). The model concentration of SF_6 (thick line) is compared with the observations (thin line) at the stations in (b), where the numbers along the horizontal axis give the start of the sections as indicated in (a). The observations are taken from *Messias et al.* [2004], and these authors should be cited when reference is made to the observational data.

Figure 8b, together with the model concentration interpolated to the hydrographic stations. The stations are equally spaced along the horizontal axis, where the numbers indicate the start of the different sections. The general agreement between model and field data is very good both with respect to the magnitude and the relative distribution of the concentration. The model seems to be off in parts of sections 7 and 8, where the model has two high-concentration anomalies. One anomaly extends from the central Greenland Sea to the Lofoten Basin just crossing section 7; the other is in the eastern Norwegian Basin intersected to the south by section 8. This suggests that the model recirculation in the Lofoten and Norwegian basins (cf. Figure 3a) may be somewhat too strong; see also the discussion on this for the mean flow in subsection 2.2. The evolution of SF_6 concentration predicted by the model in the central FSC is shown in Figure 9. The time history observed by *Olsson et al.* [2005] in the Faroe Bank Channel is included. There is a remarkable agreement between the two in arrival time, build-up time, and peak concentration.

4. PHASES OF POSITIVE AND NEGATIVE NAO

The basic experiment described in subsection 2.2, using the long-term averaged GCM velocity, is meant to represent the average timescales and spreading pathways for GSW. However, the Nordic Seas are characterized by large interannual variability that is often associated with that of the North Atalantic Oscillation (NAO), the leading mode of the North Atlantic sea-level pressure pattern. The NAO represents the shift of atmospheric mass between the Icelandic Low and Azores High, and the associated index (Figure 4) is a measure for the strength of the westerlies [*Hurrell*, 1995]. During a phase of positive NAO Index, with stronger than normal westerlies, cyclonic atmospheric circulation in the Nordic Seas is enhanced, with strengthened southwesterlies in the southeast and strengthened northerlies in the west. On short time scales (~months), the direct effect of a positive NAO is to increase the wind-stress curl over the Nordic Seas. The associated divergence in the Ekman transports and sea-level changes will favour an enhanced barotropic cyclonic circulation within the Nordic Seas and sub-basins [*Furevik and Nilsen*, this volume]. On longer time scales (~years), heat and freshwater fluxes associated with the NAO can alter the thermohaline structure of the ocean and lead to slow baroclinic adjustment processes. The effect of this is much more difficult to model; it may involve propagation of Rossby waves and advection of anomalies with the mean circulation from remote regions [*Furevik and Nilsen*, this volume; *Visbeck et al.*, 2003].

Variations in the NAO, particularly the change from weak westerlies in the 1960s to extremely strong westerlies in the 1990s, have been found to correlate with changes in an abundance of physical, ecological, and even economical parameters in the North Atlantic and Nordic Seas regions [cf. *Hurrell et al.*, 2003; *Marshall et al.*, 2001]. For the Nordic

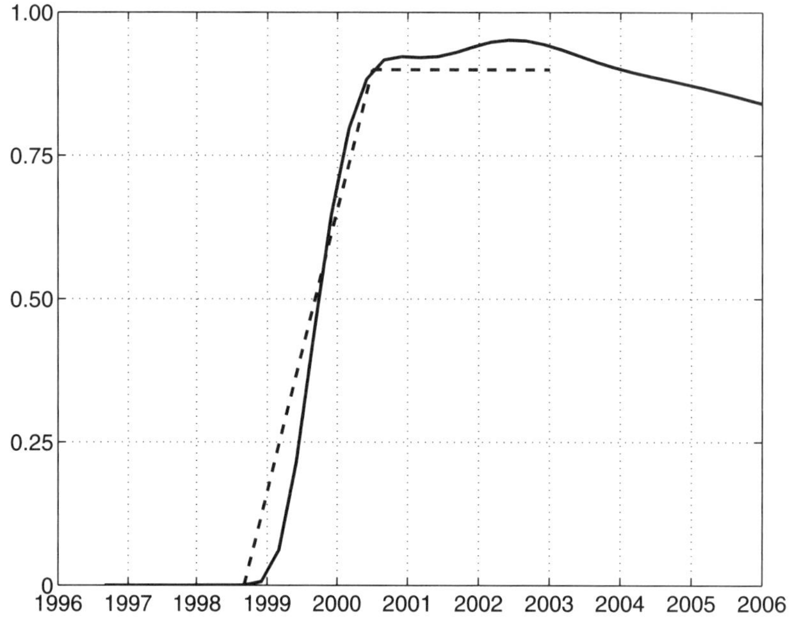

Figure 9. The model evolution of SF_6 concentration (in units of nmol m^{-2}) in the central Faroe–Shetland Channel (solid curve). The broken line is the piecewise linear approximation to observations from the Faroe Bank Channel by *Olsson et al.* [2005].

Seas area, reduced deep water formation in the Greenland Sea, increased transport of heat towards the Arctic Ocean, and general freshening of the intermediate waters, are all notifiable changes in the recent oceanographic conditions that have been related to the NAO. See *Blindheim and Østerhus* and *Furevik and Nilsen* in this volume for reviews.

The NAO Index for the time period under consideration is displayed in Figure 4. The 1960s were characterized by negative NAO, but the index has been predominantly positive since 1972 with an all-time high in 1989 (the index goes back to 1864). We construct a positive NAO case from the 1989-1992 mean of the GCM IW circulation and a negative one from the 1962-1965 mean. Within the 4-year intervals, the index is quite persistent. Also, a 4-year period in general covers the time it takes from when the tracer is released in the Greenland Sea until it arrives at the overflows (cf. Table 1 and Figure 11) and is consistent with the definition of the SF_6 case.

The intensified circulations of the subbasins and gyres of the Nordic Seas, characteristic of the strong NAO+ forcing, are clearly seen in Figure 3b. The circulation of the Greenland Sea is particularly strong in the positive phase. This is also the case for the streamlines connecting the Greenland and Norwegian Sea directly, shortcutting the GSW's journey to the FSC. The situation is quite similar to the late 1990s (Figure 3a), but the closed circulations are in general substantially stronger during NAO+. This should be expected, given that the late 1990s were characterized by a positive index of only moderate magnitude.

The energetic NAO+ case is contrasted with the expected weaker NAO– flow of 1962-1965 in Figure 3c. Compared with the others, the NAO– case is more similar to the mean flow. One difference, however, concerns the way the northern basin connects with the overflows: The JMC and the shortcut to the Norwegian Sea have practically vanished, and the meridional flow in the Norwegian Basin is much less pronounced. The different spreading of the passive tracer, initially restricted by the 3500-m isobath, for the two distinct forcing regimes is shown in Figure 10. The negative index yields slower spreading, and practically all the overflowing tracer is delivered by the EGC, as expected from the stream functions. In the positive case, the spreading is much faster and the JMC has a dominant role.

5. DISCUSSION

The variability displayed in Figure 3 is striking, both in flow strength and pathways. The JMC shortcut directly from the Greenland Sea to the Norwegian Sea clearly stands out in the late 1990s and NAO+, while it is weak or almost nonexistent in the other two. Another related feature that is highly variable is the internal circulation of the Greenland Sea. The mean case displays a Greenland Sea Gyre, which coincides roughly with the 3500-m isobath. Across the Greenland Fracture Zone to the northeast is a similar closed circulation in the Boreas Basin. This basin has been associated on a smaller scale with many of the same convective characteristics as the central Greenland Sea [cf. *Johannessen et al.*, this volume].

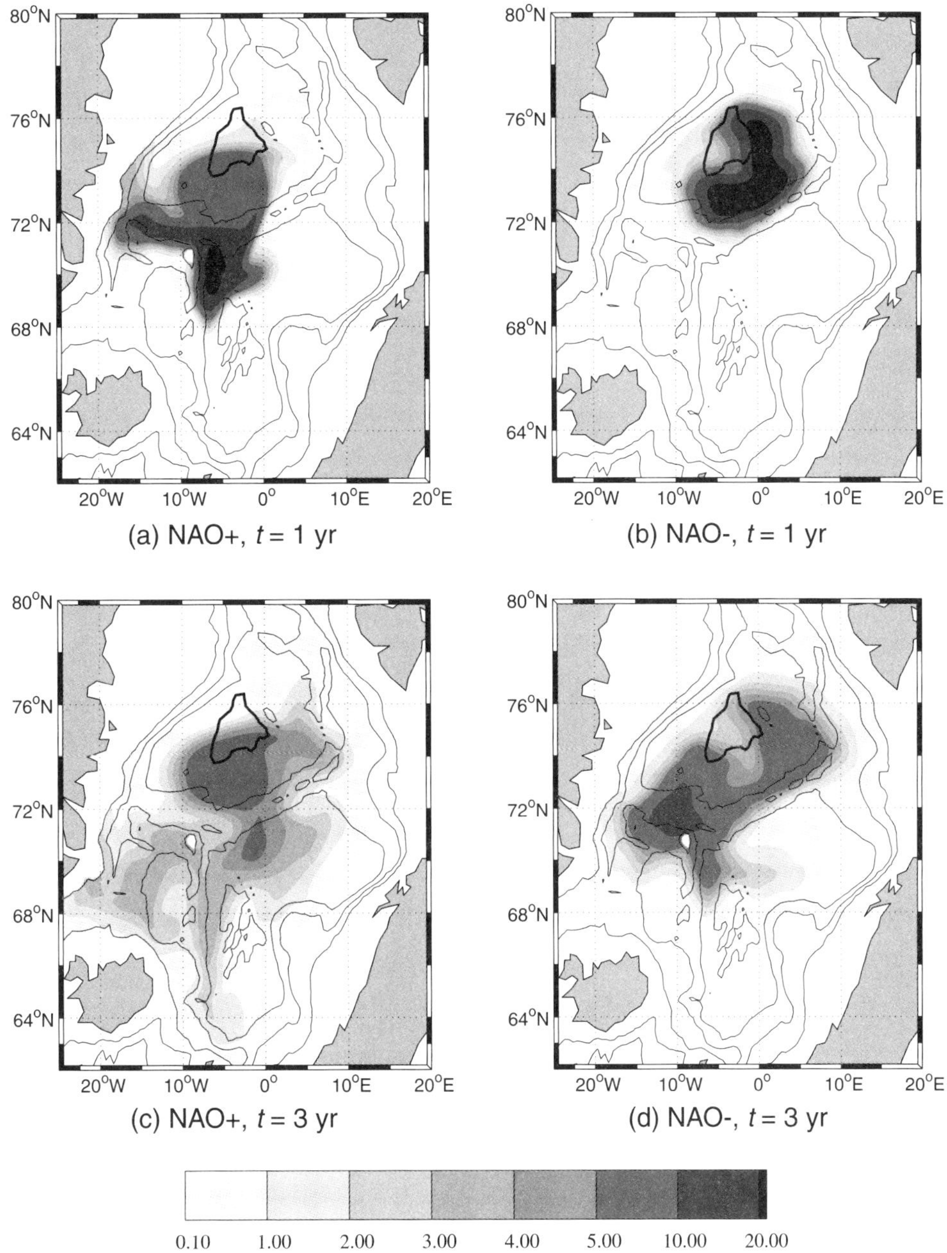

Figure 10. The spreading of GSW for positive NAO (1989-1992, left column) and negative (1962-1965, right), in years 1, 3, 5, and 7 (see also next page). Units in percent of the initial 100% concentration within the 3500-m isobath (thick line).

The negative NAO forcing moves the gyre slightly to the east. The strong wind-stress curl of the positive NAO case (and the late 1990s) characteristically spins up the circulation but also offsets it some 300 km to the south, where it is restricted by the Jan Mayen Fracture Zone and the Mohns Ridge, which separate the Greenland Basin from the Norwegian and the Lofoten basins, respectively. Such an extensive translation is not traditionally associated with the Greenland Sea Gyre. Although the GCM might be too sensitive at depth to changes in the wind forcing, we argue that these differences could be representative of the real system because the wind-stress curl is a decisive parameter in setting the gyre [e.g., *Dickson et al.*, 1996]. Also, the mean diagnostics of *Nøst and Isachsen* [2003] produce a gyre very similar to that of Figure 3d. Both gyres have relatively weak flows and are "correctly" placed over the Greenland Abyssal Plain.

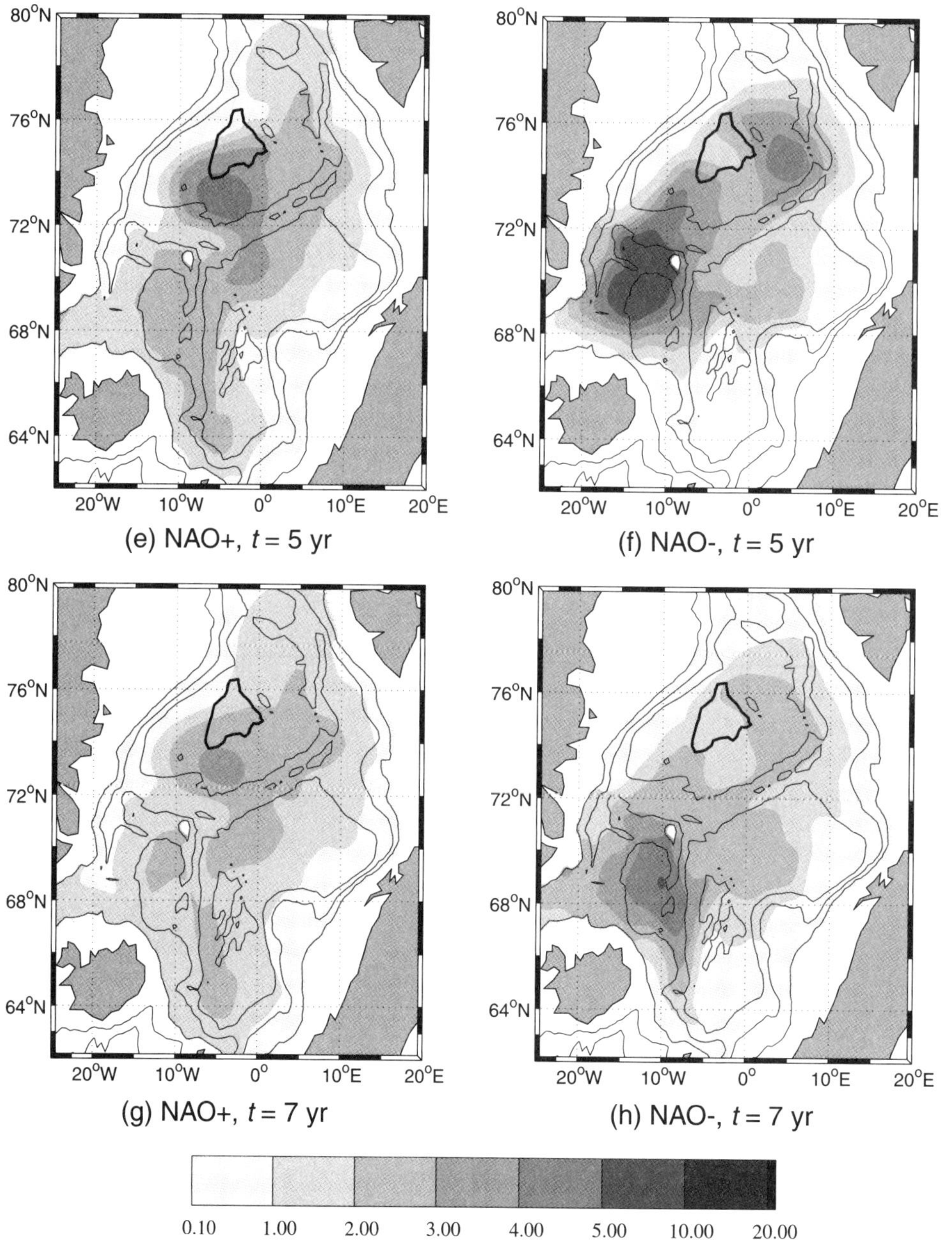

Figure 10. (Continued).

Their low inertia and the weak topographic gradients below make them vulnerable to changes in the forcing. Furthermore, hydrographic sections in the Greenland Sea are mostly taken at, or close to, 75°N [e.g., *Bönisch et al.*, 1997; *ESOP2*, 1999]. Unless a survey has sufficient three-dimensional coverage, it is often impossible to decide whether a section goes through the center or the periphery of a certain structure. For example, all the gyres of Figure 3 do intersect 75°N and all would leave the "Greenland Sea Gyre signature" of doming isopycnals there.

The evolution of GSW concentration at the overflows for all the regimes is shown in Figure 11, and the salient features are summarized in Table 1. For this comparison, we initialized the late 1990s flow like the others. The according spread of GSW was qualitatively like that of NAO+ (not shown). The arrival time at the DS is similar in all the cases. This is due to the fact that the pathway taken by the GSW exported at the DS is the EGC (e.g., Figure 5). The following build-up times are less similar, and the peak concentrations vary almost by a factor of five. This can be understood from the variable course taken by

Table 1. The arrival time (T_{arr}), the following build-up time (ΔT_{bu}), and the associated peak tracer concentration (ϕ_{max}) at the two overflows. Bold font implies the shortest times and highest peak for each case. Units are years and percent of GSW.

	DS			FSC		
	T_{arr}	ΔT_{bu}	ϕ_{max}	T_{arr}	ΔT_{bu}	ϕ_{max}
Late 1990s	2.5	2.5	0.45	**2.0**	**2.0**	**1.25**
NAO+	**2.0**	**3.0**	**1.75**	2.5	7.5	1.05
NAO–	**4.0**	**5.0**	**2.25**	6.0	≥4.0	>0.65
Mean	**3.0**	≥7.0	**1.60**	4.5	≥5.5	>0.60

the tracer to the FSC. In the negative NAO and mean cases the tracer comes via the EGC, and the part that does not leave at the DS travels southeast through the Iceland and Norwegian Seas to exit at the FSC roughly 2 years later. The setting of the positive NAO and "SF_6 years" is completely different (cf. Figure 10). The bulk of the tracer exported through the FSC is provided through the shortcut of the JMC, taking the GSW directly from the Greenland to the Norwegian Sea. Much faster arrival and a larger peak concentration at the FSC result. The concentrations in the outflow of the eastern FS are included in Figure 11 for completeness. What stands out in the GSW export to the Arctic is the mean case, followed by the NAO+. It is noteworthy that the above similarity of the mean and NAO– cases, and of the late 1990s and NAO+, is absent.

There is, maybe surprisingly, practically the same net IW flux through the model domain for the mean (2.2 Sv) and the NAO+/NAO– flows (2.3/2.2 Sv). The 2.6 Sv of the late 1990s is slightly stronger. Furthermore, the division of the influx between the overflows is the same (± 1%) for all cases, 36% going to the DS and 64% to the FSC. The large differences between the four regimes is thus not a result of varying fluxes through the open boundaries but solely of the variability of the internal circulation of the Nordic Seas. This emphasizes that a key factor for the (model) export of GSW to the North Atlantic is the role of the JMC. The JMC has previously been associated with substantial interannual variability, both from the monitoring of the "Odden" ice tongue [*Comiso et al.*, 2001], and measurements of the bottom current [*Østerhus and Gammelsrød*, 1999]. As these observations concern the surface and near-bottom layers, no quantitative comparison is attempted with the IW of the model JMC.

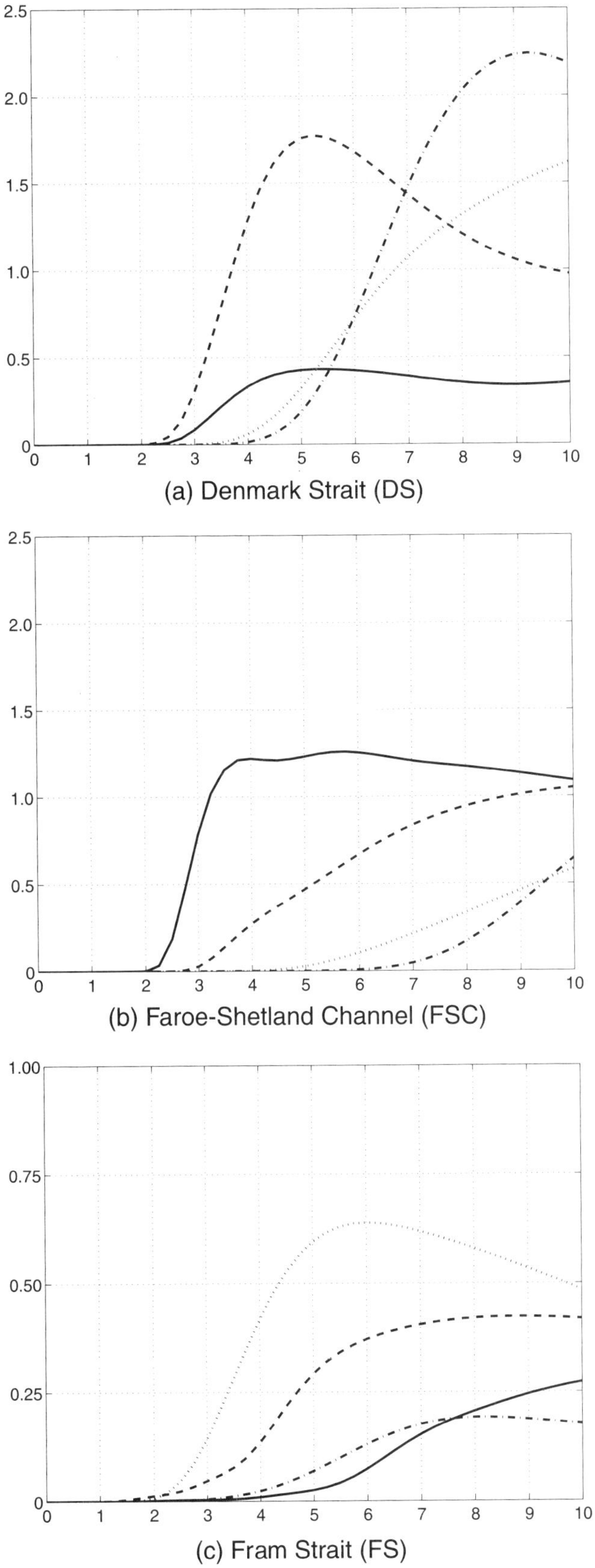

Figure 11. The evolution of tracer concentration (percent initial GSW) at the center of the two overflows and in the outflow in the eastern Fram Strait. The time is in years since release. The solid curves are the "SF_6 years" (initialized like the others) and dashed/dash-dotted is positive/negative NAO. The mean evolution is given for reference by the dotted curves. Note that the range of the vertical axis in (c) is less than that of the others.

The overflow fluxes referred to above are consistent (as they should be) with the transports in *Nilsen et al.'s* [2003] coarser version of the MICOM GCM. Their analysis suggests a basic balance between the variability of the net inflow through the FSC and the net outflow through the DS (integrating over the total water column). The GCM fluxes and the analysis are supported by what the authors can infer from relevant observations.

6. CONCLUDING REMARKS

An advective–diffusive model of the Nordic Seas has been set up to address the spreading of GSW. The model currents were taken, given a lack of in situ observations, from the data archive of a high-resolution GCM. A comparison of the model results with observations collected following the 1996 SF_6 tracer release in the central Greenland Sea lends credibility both to the velocity field from the GCM and to the approach utilized (cf. Figures 8, 9).

The variable spreading under different regimes and, in particular, under distinct phases of the NAO, are here presented. The variability between the different flow cases is striking, both in streamline patterns and in tracer distribution (cf. Figures 3, 10, 11). As reported here, the variability in tracer pathways and export reflects changes in both the pattern and the strength of the internal circulation of the Nordic Seas, not in the fluxes of the overflows. The JMC plays a key role in the different spreading. When it is weak relative to the EGC (the mean and NAO– cases), practically all of the overflowing tracer follows the latter. When the JMC is strong (the late 1990s and NAO+ cases), the pathway of the GSW to the FSC is cut short by the JMC.

NOTATION

CPU	central processing unit
DS	Denmark Strait
EGC	East Greenland Current
FS	Fram Strait
FSC	Faroe–Shetland Channel
GCM	general circulation model
GSW	Greenland Sea Water
IW	intermediate water
JMC	Jan Mayen Current
LSW	Labrador Sea Water
MICOM	Miami Isopycnic Coordinate Ocean Model
NAO	North Atlantic Oscillation
SF_6	sulphur hexafluoride

Acknowledgments. The authors thank Marie–José Messias and Anders Olsson for making the SF_6 data available through the EU TRACTOR project, and the three reviewers for their helpful comments. T.E. was supported by TRACTOR (EVK2-2000-00080) and the Norwegian Research Council project ProClim (155923/700), F.S. by NSF #0240978, A.B.S. by the research programme West-Nordic Ocean Climate of the Nordic Council of Ministers, and T.F. by the Norwegian Research Council project NOClim II (155972/720). This is contribution Nr. A69 from the Bjerknes Centre for Climate Research.

REFERENCES

Aagaard, K., J.H. Swift, and E.C. Carmack, Thermohaline circulation in the Arctic Mediterranean Seas, *J. Geophys. Res.*, 90, C3, 4833-4846, 1985.

Alekseev, G. V., O.M. Johannessen, A.A. Korablev, V.V. Ivanov, and D.M. Kovalevsky, Interannual variability of water masses in the Greenland Sea and the adjacent areas, *Polar Res.*, 20, 201-208, 2001.

Bleck, R., C. Rooth, D. Hu, and L.T. Smith, Salinity-driven thermocline transients in a wind- and thermocline-forced isopycnic coordinate model of the North Atlantic, *J. Phys. Oceanogr.*, 22, 1486-1505, 1992.

Bönisch, G., J. Blindheim, J.L. Bullister, P. Schlosser, and D.W.R. Wallace, Long-term trends of temperature, salinity, density, and transient tracers in the central Greenland Sea, *J. Geophys. Res.*, 102, C8, 18553-18571, 1997.

Comiso, J.C., P. Wadhams, L.T. Pedersen, and R.A. Gersten, Seasonal and interannual variability of the Odden ice tongue and a study of environmental effects, *J. Geophys. Res.*, 106, C5, 9093-9116, 2001.

Dickson, B., J. Lazier, J. Meincke, P. Rhines, and S.J., Long-term coordinated changes in the convective activity of the North Atlantic, *Prog. Oceanog.*, 38, 241-295, 1996.

ESOP2 Group, *The thermohaline circulation in the Greenland Sea, ESOP2/MAST3 Final scientific report*, University of Bergen, Jansen, E. and Opheim, V., Eds., 1999.

Fogelqvist, E., J. Blindheim, T. Tanhua, S. Østerhus, E. Buch, and F. Rey, Greenland-Scotland overflow studied by hydro-chemical multivariate analysis, *Deep Sea Res.* I, 50, 73-102, 2003.

Furevik, T., M. Bentsen, H. Drange, J. Johannessen, and A. Korablev, Temporal and spatial variability of the sea surface salinity in the Nordic Seas, *J. Geophys. Res.*, 107, C12, DOI 10.1029/2001JC001118, 2002.

Gao, Y., H. Drange, M. Bentsen, and O.M. Johannessen, Simulating transport of non-Chernobyl ^{137}Cs and ^{90}Sr in the North Atlantic-Arctic region, *J. Environ. Radioactivity*, 71, 1-16, 2004.

Girton, J. B., T.B. Sanford, and R.H. Käse, Synoptic sections of the Denmark Strait Overflow, *Geophys. Res. Lett.*, 28, 1619-1622, 2001.

Hansen, B., and S. Østerhus, North Atlantic–Nordic Seas exchanges, *Prog. Oceanog.*, 45, 109-208, 2000.

Hansen, B., W.R. Turrell, and S. Østerhus, Decreasing overflow from the Nordic seas into the Atlantic Ocean through the Faroe Bank channel since 1950, *Nature*, 441, 927-930, 2001.

Helland-Hansen, B., and F. Nansen, *The Norwegian Sea, its physical oceanography based upon the Norwegian Researches 1900–1904*, Report on Norwegian Fishery and Marine Investigations, Vol. 2, part 1, No. 2, Mallingske, Christiania, 390 pp., 1909.

Hopkins, T.S., The GIN Sea- A synthesis of its physical oceanography and literature review 1972-1985, *Earth Sci. Rev.*, 30, 175-318, 1991.

Hurrell, J.W., Decadal trends in the North Atlantic Oscillation: Regional temperatures and precipitation, *Science*, 269, 676-679, 1995.

Hurrell, J.W., Y. Kushnir, G. Ottersen, and M. Visbeck, editors, *The North Atlantic Oscillation: Climate Significance and Environmental Impact*, vol. 134 of *Geophysical monograph*, American Geophysical Union, 279 pp., 2003.

Jakobsen, P.K., M.H. Ribergaard, D. Quadfasel, T. Schmith, and C.W. Hughes, Near surface circulation in the northern North Atlantic as inferred from Lagrangian drifters: Variability from the mesoscale to interannual, *J. Geophys. Res.*, 108, C8, DOI 10.1029/2002JC001554, 2003.

Jones, H., and J. Marshall, Convection with rotation in a neutral ocean: a study of open-ocean deep convection, *J. Phys. Oceanogr.*, 23, 1009-1039, 1993.

Jones, H., and J. Marshall, Restratification after deep convection, *J. Phys. Oceanogr.*, 27, 2276-2287, 1997.

Katsman, C.A., M.A. Spall, and R.S. Pickart, Boundary current eddies and their role in the restratification of the Labrador Sea, *J. Phys. Oceanogr.*, 34, 1967-1983, 2004.

Lilly, J.M., P.B. Rhines, M. Visbeck, R. Davis, J.R.N. Lazier, F. Schott, and D. Farmer, Observing deep convection in the Labrador Sea during 1994/95, *J. Phys. Oceanogr.*, 29, 2065-2098, 1999.

Marshall, J., Y. Kushnir, D. Battisti, P. Chang, A. Czaja, R. Dickson, J. Hyrrell, M. McCartney, R. Saravanan, and M. Visbeck, North Atlantic climate variability: phenomena, impacts and mechanisms, *Int. J. Climatol.*, 21, 1863-1898, 2001.

Marshall, J., and F. Schott, Open-ocean convection: observations, theory, and models, *Rev. Geophys.*, 37, 1-64, 1999.

Mauritzen, C., Production of dense overflow waters feeding the North Atlantic across the Greenland-Scotland Ridge. Part 1: Evidence for a revised circulation scheme, *Deep Sea Res.* I, 43, 769-806, 1996a.

Mauritzen, C., Production of dense overflow waters feeding the North Atlantic across the Greenland-Scotland Ridge. Part 2: An inverse model, *Deep Sea Res.* I, 43, 807-835, 1996b.

Messias, M.-J., A. Watson, T. Johannessen, K.A. Olsson, W.M. Smethie, L. Anderson, S. Bacon, J.-C. Gascard, J. Olafsson, K. Oliver, J. Swift, F. Rey, K. Simonsen, J.R. Ledwell, and G. Budeus, The Greenland Sea Tracer Experiment: Formation, spreading and dispersion of intermediate waters of the Greenland Sea, In preparation, 2004.

Nilsen, J.E.Ø., Y. Gao, H. Drange, T. Furevik, and M. Bentsen, Simulated North Atlantic–Nordic Seas water mass exchanges in an isopycnic coordinate OGCM, *Geophys. Res. Lett.*, 30, DOI 10.1029/2002GL016597, 2003.

Nøst, O.A., and P.E. Isachsen, The large-scale time-mean ocean circulation in the Nordic Seas and Arctic Ocean estimated from simplified dynamics, *J. Mar. Res.*, 61, 175-210, 2003.

Olsson, K.A., E. Jeansson, L.G. Anderson, B. Hansen, T. Eldevik, R. Kristiansen, M.-J. Messias, T. Johannessen, and A.J. Watson, Intermediate water from the Greenland Sea in the Faroe Bank Channel: spreading of released sulphur hexafluoride, *Deep Sea Res.*, I, 52, 279-294, 2005.

Orvik, K. A., and P. Niiler, Major pathways of Atlantic water in the northern North Atlantic and Nordic Seas toward Arctic, *Geophys. Res. Lett.*, 29, DOI 10.1029/2002GL015002, 2002.

Østerhus, S., and T. Gammelsrød, The abyss of the Nordic Seas is warming, *J. Climate*, 12, 3297-3304, 1999.

Poulain, P.-M., A. Warn-Varnas, and P.P. Niiler, Near-surface circulation of the Nordic seas as measured by Lagrangian drifters, *J. Geophys. Res.*, 101, C8, 18237-18258, 1996.

Spall, M.A., Boundary currents and water mass transformation in marginal seas, *J. Phys. Oceanogr.*, 34, 1197-1213, 2004.

Straneo, F., The convective seasonal cycle of the Labrador Sea, Submitted to *J. Phys. Oceanogr.*, 2004.

Straneo, F., R.S. Pickart, and K. Lavender, Spreading of Labrador Sea water: an advective-diffusive study based on Lagrangian data, *Deep Sea Res.* I, 50, 701-719, 2003.

Visbeck, M., E. Chassignet, R. Curry, T. Delworth, B. Dickson, and G. Krahmann, The ocean's response to North Atlantic Oscillation variability, in *The North Atlantic Oscillation: Climate Significance and Environmental Impact*, edited by Hurrell, J.W., Y. Kushnir, G. Ottersen, and M. Visbeck, vol. 134 of Geophysical monograph, pp. 113–146, American *Geophysical Union*, 2003.

Watson, A.J., M.J. Messias, E. Fogelqvist, K.A. Van Scoy, T. Johannessen, K.I.C. Oliver, D.P. Stevens, F. Rey, K.A. Tanhua, T. Olsson, F. Carse, K. Simonsen, J. R. Ledwell, E. Jansen, D.J. Cooper, J.A. Kruepke, and E. Guilyardi, Mixing and convection in the Greenland Sea from a tracer-release experiment, *Nature*, 401, 902-904, 1999.

T. Eldevik and A.B. Sandø, Nansen Environmental and Remote Sensing Center, Thormøhlensgate 47, N-5006 Bergen, Norway. (tor.eldevik@nersc.no; anne.britt.sandoe@nersc.no)

T. Furevik, Geophysical Institute, University of Bergen, Allégaten 70, N-5007 Bergen, Norway. (Tore.Furevik@gfi.uib.no)

F. Straneo, Dept. of Physical Oceanography, Woods Hole Oceanographic Institute, MS 21, Woods Hole, MA 02543, USA. (fstraneo@whoi.edu)

Large-Scale Atmospheric Circulation Variability and Its Impacts on the Nordic Seas Ocean Climate—A Review

Tore Furevik[1]

Geophysical Institute, University of Bergen, Norway

J. Even Ø. Nilsen

Nansen Environmental and Remote Sensing Center, Bergen, Norway

The large-scale atmospheric circulation and its impacts on the Nordic Seas ocean climate are reviewed. The dominant factors for the atmospheric variability are the Icelandic low and the Azores high, determining the strength of the westerlies. From the '60s to the '90s, the atmospheric circulation shifted from record weak to record strong westerlies, and the storm tracks moved further northeast into the Nordic Seas. The reasons for this shift have most likely been forcing from the tropical ocean in combination with internal processes in the atmosphere. Associated with this low-frequency shift are changes in the atmospheric momentum, heat, and freshwater forcing of the ocean. Both local processes and advective anomalies have played active roles in the substantial changes observed in the Nordic Seas' circulation and hydrography over the same period. These include a reduction in the deep-water formation, a warming of the water going into the Arctic, and a freshening and probable reduction of the overflow water. The strengthening of the westerlies is concurrent in time with a strong increase in global mean temperatures, and we speculate that the changes in the atmospheric circulation are tied to the increased greenhouse gas forcing through a variety of forcing mechanisms. If so, the observed changes in the Nordic Seas ocean climate are likely to be amplified, and our perceptions of what constitutes normal oceanic conditions will be further challenged in the years to come.

1. INTRODUCTION

By having the North Atlantic Ocean to the south and the Arctic Ocean to the North, the Nordic Seas act as a buffer zone between the warm, saline Atlantic Water (AW) and the cold, fresh Polar Water (PW). These two water masses outline the main features of the surface water in the Nordic Seas as AW occupies the southern and eastern part of the basin and PW the northern and western part. A detailed description of the water masses, their flow patterns and water modifications, as well as the topographical features within the Nordic Seas is given by *Hansen and Østerhus* [2000] and Blindheim and Østerhus [this volume]. Only a brief outline will therefore be given here. The focus will be on the features of the surface flows that are important for the discussion of the atmospheric impacts on the oceanic circulation.

The major opening to the Arctic Ocean is the Fram Strait, but also the shelf areas of the Barents Sea provide important pathways between the Nordic Seas and the Arctic Ocean

[1]Also at Bjerknes Centre for Climate Research, Bergen, Norway.

The Nordic Seas: An Integrated Perspective
Geophysical Monograph Series 158

10.1029/158GM09

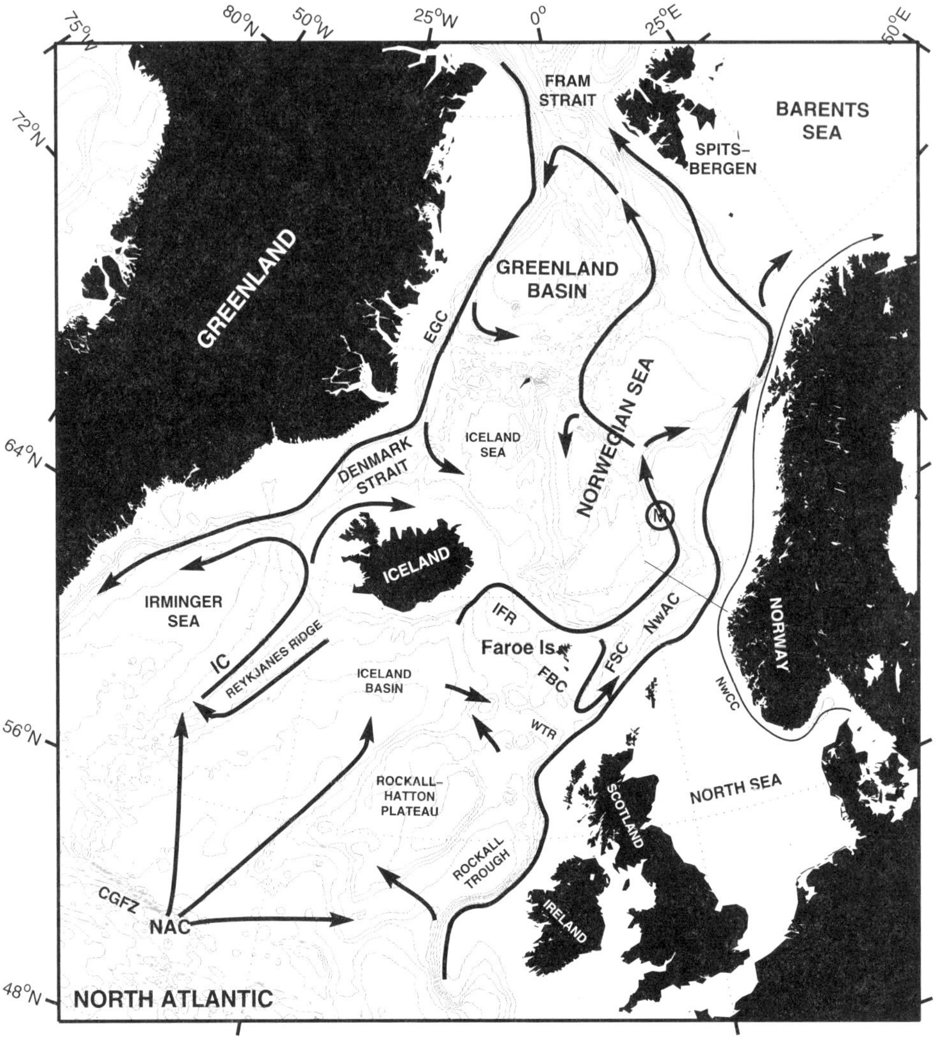

Figure 1. The northern North Atlantic and the Nordic Seas. Isobaths are drawn for every 500 m. Schematic surface currents are based on literature [*Poulain et al., 1996; Hansen and Østerhus, 2000; Bower et al., 2000; Mork and Blindheim, 2000; Holliday et al., 2000; Orvik et al., 2001; Orvik and Niiler, 2002; Jakobsen et al., 2003*]. See text for abbreviations. The circled M shows the position of Ocean Weather Station M and the thin line indicates the Svinøy Section.

(Figure 1). The main source for PW to the area is the East Greenland Current (EGC), which flows southward from the Fram Strait and along the western margin of the basin, branching off PW into both the Greenland and Iceland Seas on its way.

The Greenland–Scotland Ridge (GSR) forms the natural border to the Atlantic Ocean and provides three gateways for the exchange of water masses between the basins: The Denmark Strait between Greenland and Iceland, the Iceland–Faroe Ridge (IFR), and the Faroe Shetland Channel (FSC). Further to the south, the North Atlantic Current (NAC) starts off as a western boundary current originating from the Gulf Stream [*Bower et al.*, 2000; *Carr and Rossby*, 2001]. The NAC passes over the Charlie Gibbs Fracture Zone (CGFZ) and spreads in different paths towards the three openings of the GSR.

The AW passing over the IFR forms the Iceland Faroe Front against the PW in the central Nordic Seas and turns eastward as the Faroe Current and subsequently as the western branch of the Norwegian Atlantic Current (NwAC) [*Mork and Blindheim*, 2000; *Orvik et al.*, 2001]. Through the FSC the major source of AW is the slope current over the Scottish Slope, originating off the southwestern European coast [*Hansen and Østerhus*, 2000] and continuing northwards along the Norwegian Continental Shelf-break as the eastern branch of the NwAC [*Mork and Blindheim*, 2000; *Orvik et al.*, 2001].

Within the Nordic Seas, the strongly topographically steered eastern branch of the NwAC follows the continental slope to the Barents Sea and Fram Strait, and provides the source for the AW in the Arctic Ocean. The fate of the more meandering western branch is more disputed. Drifter measurements indicate that the flow continues into the Fram Strait [*Orvik and Niiler*, 2002], where probably most of the water recirculates south [see *Quadfasel et al.*, 1987; *Gascard et al.*, 1988].

During recent years there has been an increased focus on the many atmospheric and oceanic processes taking place in the Nordic Seas and the Arctic Ocean. There are at least four good reasons for this: Firstly, since the '60s the westerlies have strengthened, resulting in dry, cold winters in southern Europe and wet, mild winters in northern Europe [*Hurrell*, 1995]. Secondly, during the same period, the deep-water formation in the Greenland Sea has been strongly reduced, and the export of dense water to the North Atlantic seems to have weakened [*Dickson et al.*, 1996; *Hansen et al.*, 2001]. Thirdly, the Arctic Ocean has become warmer [*Carmack et al.*, 1995; *Grotefendt et al.*, 1998], and the Northern Hemisphere sea-ice cover has decreased both in extent [*Johannessen et al.*, 1999, 2004] and thickness [*Rothrock et al.*, 1999, 2003]. Fourthly, the surface and intermediate water of the Nordic Seas, the overflow water, and the deep water of the North Atlantic have all become fresher [*Blindheim et al.*, 2000; *Dickson et al.*, 2002; *Curry et al.*, 2003].

The aim of this paper is to give a review of the current knowledge of the large-scale atmospheric circulation variability over the North Atlantic, and of the role of the atmospheric forcing on the Nordic Seas ocean climate.

In Section 2 the large-scale atmospheric flow over the North Atlantic and Arctic is discussed. Focus is on the most prominent mode of sea-level pressure (SLP) variability, the North Atlantic Oscillation (NAO). The recent temporal and spatial changes in the NAO are reviewed, and proposed forcing mechanisms discussed. The variations in the atmosphere—ocean heat and momentum fluxes associated with changes in the NAO are studied in Section 3, and in Section 4 theoretical effects of the anomalous atmospheric forcing on the ocean properties in the Nordic Seas are considered. In Section 5 recent observations of the variability and trends in the ocean climate are discussed in light of anomalous forcing, and in Section 6 possible links between the changes observed in the Nordic Seas and global warming are discussed. The review is summarized and concluded in Section 7.

2. LARGE-SCALE ATMOSPHERIC CIRCULATION

The winter mean (Dec–Mar) SLP field over the Northern Hemisphere is dominated by high-pressure cells over the continents and subtropical oceans and low-pressure cells over the subpolar Atlantic and Pacific Oceans (Figure 2a). A weaker SLP maximum is the Beaufort high situated over the perennial sea-ice cover in the Canadian Basin in the western part of the Arctic Ocean. Maximum interannual variability in

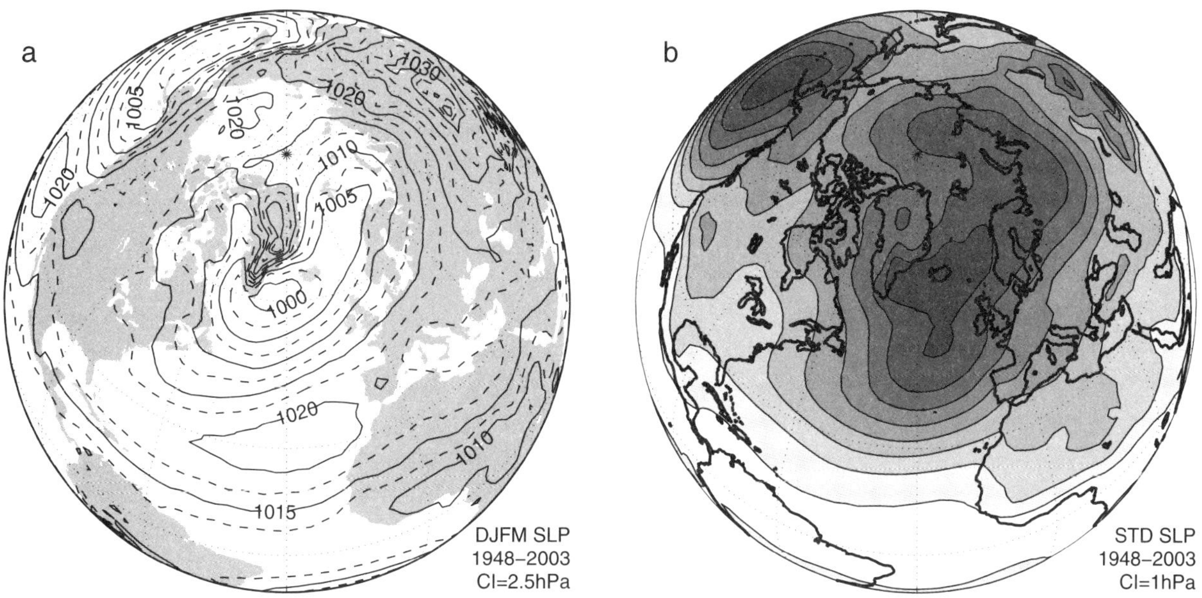

Figure 2. Winter mean (a) and year-to-year standard deviation of the winter mean SLP (b) for December through March 1949–2003 calculated from the NCEP/NCAR reanalysis data [*Kalnay et al., 1996*]. Contour intervals (CI) are given at the lower right-hand side of each plot. White colouring in (b) indicates values below 1 hPa.

winter mean SLP is found over the subpolar lows in the Atlantic and Pacific, where year-to-year standard deviation exceeds 9 hPa (Figure 2b). The Siberian high also experiences enhanced interannual variability.

2.1. Station-Based Indices of Atmospheric Variability

The strength of the westerlies between the North Atlantic subtropical high and subpolar low, known as the Azores high and the Icelandic low respectively, has a profound impact on the climate of the North European region, as well as on a wide range of physical, ecological, and social parameters in the Atlantic sector [see *Marshall et al.*, 2001b; *Hurrell et al.*, 2003, and references therein]. As a consequence, many scientists have tried to find useful indices for describing the strength of the westerlies and thus proxies for the climate of Europe. The best known are the NAO Indices, based on the SLP differences between Lisbon, Portugal, and Stykkisholmur, Iceland [*Hurrell*, 1995], or between Gibraltar and Stykkisholmur [*Jones et al.*, 1997], which succeed the Azores–Iceland indices first used by *Bjerknes* [1962] and later *Rogers* [1984]. Other indices have been constructed as proxies for more local- or regional-scale phenomena, for instance the SLP difference across the Fram Strait for the sea-ice export [e.g. *Vinje*, 2001; *Widell et al.*, 2003] or across the Barents Sea Opening to explain the inflow of AW [*Bengtson et al.*, 2004].

If the Gibraltar and Stykkisholmur SLP data are grouped by individual months, distinct annual cycles in both monthly means and monthly variances emerge (Figure 3). At Stykkisholmur, minimum SLP is in January and maximum in May, while Gibraltar is in the opposite phase, having its maximum in January and minimum in May. As a consequence, the average mean SLP difference between the two stations varies from only 2.9 hPa in May to 23.6 hPa in January, illustrating the strong seasonal dependency of the westerlies. The interannual variability in SLP varies with month, the maximum standard deviation for the two stations' monthly means being in February (9.9 and 4.0 hPa) and minimum in July (3.7 and 0.9 hPa).

For the winter months December through March, the Gibraltar and Iceland stations used to construct the Jones Index are anticorrelated at interannual and interdecadal time scales, with correlations of between –0.65 and –0.55. From May to September, correlations are small and not significant at

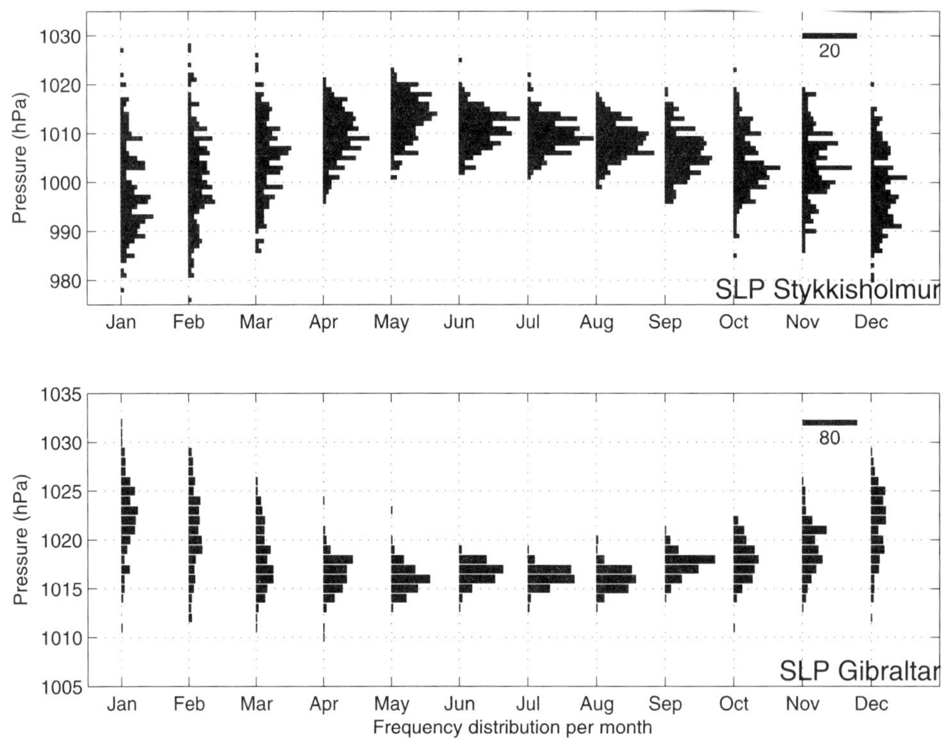

Figure 3. Frequency distribution of monthly mean SLP for Stykkisholmur (upper) and Gibraltar (lower). For a given month, the number of years with the monthly mean SLP inside each 1-hPa interval is shown. Note that the scales for SLP and for the frequency distribution (indicated in the upper right corner of each plot) differ. The plot is based on monthly mean SLP data for 1821–2000, provided by Phil Jones, Climatic Research Unit (CRU), University of East Anglia.

the 95% confidence level, mainly because the mesoscale structures are smaller and less organized during summer than in winter. As shown by *Portis et al.*, [2001] the same is true for the Lisbon and Iceland station pair used in the Hurrell Index, while the Azores and Iceland station pairs used in the Rogers (Bjerknes) Index show a significant anticorrelation throughout the year. As an alternative to the traditional index with fixed station pairs, *Portis et al.* [2001] proposed an index based on the SLP differences between the two geographical points having the highest anti-correlation, to obtain an index better at capturing the north–south SLP gradient throughout the year.

Perhaps the main reasons for the increasing interest in the NAO is the trend towards stronger westerlies that has occurred since the '60s, captured in both the *Hurrell* [1995] and *Jones et al.* [1997] indices (Figure 4a). A high (positive) NAO Index corresponds to a large-scale shift of atmospheric mass from the Icelandic low, Nordic Seas, and Arctic Ocean regions towards the Azores high (Figure 4b). Associated with a positive NAO Index, the Icelandic low extends more northeast into the Nordic Seas, where the changes in the SLP associated with a unit change in the NAO Index exceed 7 hPa. Over the Arctic Ocean, the corresponding changes range from 4 hPa near the Fram Strait to less than 1 hPa near the Bering Strait. Near the subtropical high, a unit change in the NAO Index corresponds to a 4 hPa change in SLP.

2.2. EOF-Based Indices of Atmospheric Variability

An alternative method for extracting useful indices for the strength of the large-scale atmospheric circulation is to

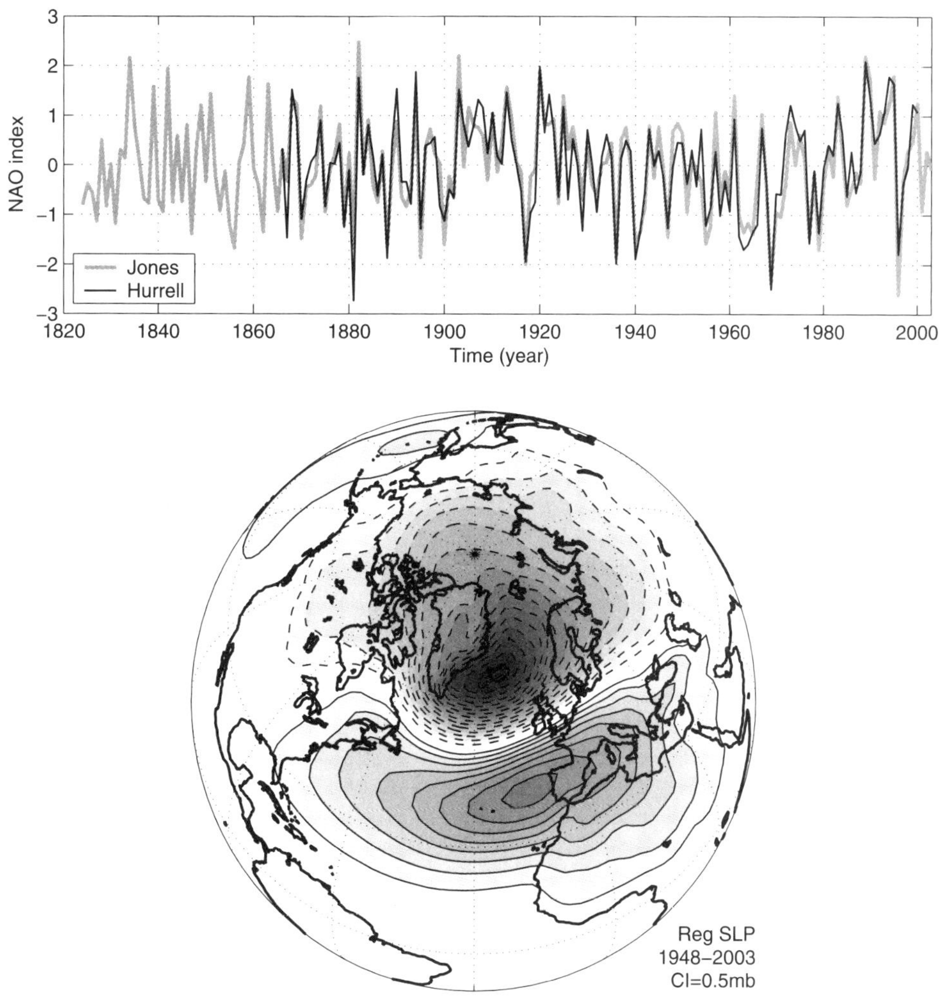

Figure 4. (Upper) The NAO winter mean (Dec–Mar) indices calculated from the Gibraltar–Stykkisholmur (Jones' Index, gray) and the Lisbon–Stykkisholmur (Hurrell's Index, black) SLP differences. (Lower) Monthly mean SLP regressed on the standardized monthly mean Jones Index for the period December 1948 to March 2003. Only winter months (Dec–Mar) are used. The values correspond to a unit change in the Jones Index. Solid lines are positive, dashed lines negative.

apply an Empirical Orthogonal Function (EOF) analysis [e.g. *von Storch*, 1995]. Typically, a singular value decomposition method is used to construct the dominant modes of variability as load functions (EOF patterns) and associated time series (principal components). The idea is to find the patterns and associated time series that explain as much as possible of the variance in the data set and, if possible, give a physical interpretation of these modes.

Figure 5 shows the leading mode of the winter SLP anomalies. The principal component defines the atmospheric index used throughout the paper, which for simplicity will be referred to as the NAO Index. In this analysis the EOF was calculated based on the December through March SLP data from the Atlantic sector only, in contrast to the November through April data from all longitudes and poleward of 20°N that *Thompson and Wallace* [1998] used to define their Arctic Oscillation Index. As pointed out by *Deser* [2000], the EOF patterns are practically identical whether the circumpolar data or only the Atlantic sector is used. For the data used here, the associated principal components have a mutual correlation of 0.99.

The EOF pattern (Figure 5b) shows a tripole structure. For positive values of the principal component, the SLP over the Arctic is lower and the SLP over the Azores high and the Aleutian low is higher than normal. This corresponds to stronger than average westerlies in the Atlantic sector, while the westerlies in the Pacific are actually reduced. The fact that the EOF is somehow annular in shape, but shows

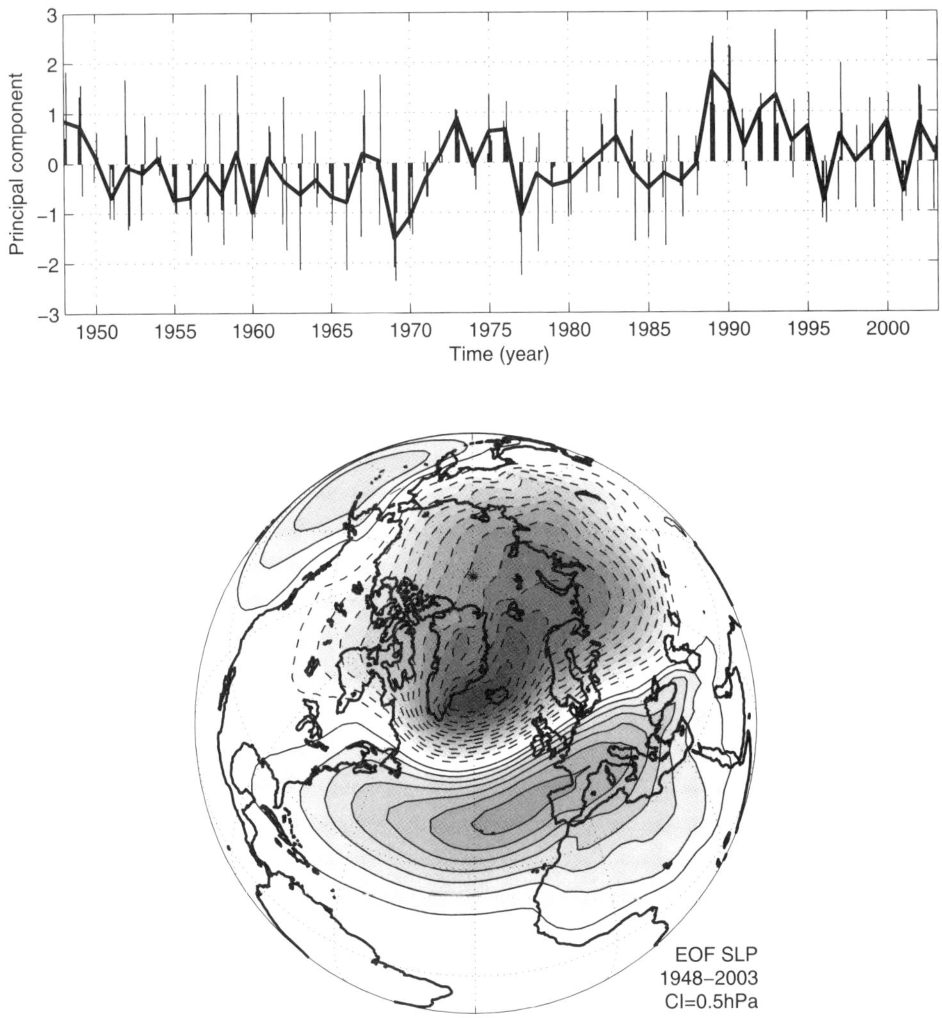

Figure 5. (Upper) The principal component of the leading mode of variability (1st EOF) of monthly mean December through March SLP, defining the NAO Index used throughout the paper. The principal component is calculated from the 1948–2003 NCEP/NCAR reanalysis data for the Atlantic sector (90°W–30°E, 20°N–80°N). The monthly values are shown as bars, and the winter means are indicated by the thick line. (Lower) Winter monthly mean SLP regressed on the principal component. Solid lines are positive, dashed lines negative.

strongest amplitudes in the Atlantic sector, is one of the reasons for the discussion on whether this is an Arctic or annular mode [*Thompson and Wallace*, 1998, 2000], or if it is physically more correct to name it the NAO [*Hurrell*, 1995; *Deser*, 2000; *Ambaum et al.*, 2001].

The EOF pattern clearly resembles the pattern found by regressing the winter mean SLP on the two-pole Jones Index (Figure 4) and is obviously a manifestation of the same feature. Nonetheless, some differences should be noted. The EOF pattern has weaker poles but at the same time a wider domain of influence, most noticeably from the Barents Sea and east into the northern parts of Siberia, and to the south of the Bering Strait. Interestingly, most of the differences are found outside the Atlantic sector used to calculate the principal component, which may indicate teleconnectivity between these areas and areas within the Atlantic sector that are not captured by the two-pole index. The differences over the Atlantic sector reflect the fact that the regression on the two-pole index will maximize the variance explained at those two particular locations, while the regression onto the principal component maximizes the variance explained over the entire Atlantic sector and thus gives a somewhat weaker pattern over the Azores high and Icelandic low.

While the two-pole indices directly give the SLP gradient and thus measure the strength of the geostrophic flow between the station pairs, EOF-based indices are based on the statistical properties of the entire SLP field in the domain analyzed. For this reason, the EOF-based indices more adequately represent the spatial domain of the mode of variability. The EOF-based indices should be treated with care, however; variability somewhere in the domain that is analyzed will influence the pattern in other areas of the domain, and vice versa. Thus, the choice of size of the domain still contains a degree of subjectivity. Furthermore, adding new data to an existing time series will change the EOF pattern and thus the principal component associated with that mode. The result is that the entire EOF-based index will change as new data are added, and results may have to be recalculated.

2.3. Temporal Changes in the Large-Scale Circulation

Since *Hurrell* [1995] demonstrated the strong influence of the NAO on the European climate, scientific interest in the subject has been escalating [see *Stephenson et al.*, 2003, for a historical review]. Due to the large impact of the large-scale atmospheric circulation on the climate in the North Atlantic region, much effort has been made to understand the causes for the NAO variability, with the ultimate goal to be able to predict the changes on seasonal to decadal time scales. A common method has been to use a statistical approach to observational data, and significant correlations are often found, for instance between the NAO and North Atlantic sea-surface properties [e.g. *Marshall et al.*, 2001b; *Czaja et al.*, 2003]. However, the generally short and inhomogeneous observational records, the fact that the NAO strongly influences the state of the ocean (see Section 4), and the indication that both the ocean surface variability and the NAO may be driven by factors outside the Atlantic sector all make it extremely difficult to assess the predictability potential by statistical methods alone.

At present, there is no general consensus about the main driver of the large-scale atmospheric variability in the North Atlantic region. At least three good candidates have been proposed. Firstly, most of the variability in the NAO seems to arise from internal processes in the troposphere, where an important mechanism is the nonlinear interactions between the time—mean flow and transient eddies, causing an essentially white spectrum [*Thompson et al.*, 2003]. Further nonlinearities or aggregation (combination of many short-lived processes) may in turn cause a reddening of the spectrum, giving rise to low-frequency behaviour similar to what has been observed in the recent decades [*Stephenson et al.*, 2000]. This is the most pessimistic explanation, because it gives no room for predictability of the NAO. However, although the bulk part of the monthly to interannual variability can be attributed to stochastic noise, this does not explain the increased variability and the strong upward trend in the last half of the 20th century [*Hurrell et al.*, 2004], nor does it explain the upward trend found in many climate projections for the 21st century [e.g. *Fyfe et al.*, 1999]. Comparing the observed trend in the NAO with the output from seven different unforced climate models—where the observed trend was outside the most extremes of the natural variability ranges simulated by the models—led *Gillet et al.* [2003] to conclude that either all the models had deficiencies in the low-frequency range or the observed trend could only be explained by external forcing. Because many of the climate models in control simulations do a fairly good job in reproducing both the spatial pattern and the strength of the interannual variability in the large-scale circulation [e.g. *Fyfe et al.*, 1999; *Stephenson and Pavan*, 2003; *Furevik et al.*, 2003], external forcing is therefore a plausible interpretation.

Secondly, observations and reanalysis data have shown that changes in the large-scale circulation may start in the stratosphere, gradually propagating downward into the troposphere [*Baldwin and Dunkerton*, 1999; *Baldwin et al.*, 2003; *Thompson et al.*, 2003]. The mechanisms, still not properly understood, may involve stratospheric refraction of upward propagating gravity waves in the troposphere [*Ambaum and Hoskins*, 2002]. Despite the lack of solid evidence, this scenario is nevertheless intriguing owing to its large potential for extended weather forecasts [*Baldwin et al.*, 2003]. Forcing from the stratosphere has also been suggested as a possible explanation for the trend towards strengthened westerlies in the latter part of the 20th century, caused by either ozone depletion in the stratosphere and associated cooling over the polar regions [e.g. *Bengtsson et al.*, 1999; *Kinden and Christiansen*, 2001], or greenhouse gas forcing, acting to warm the troposphere and cool the stratosphere [*Shindell et al.*, 1999], thus increasing the north–south temperature gradient near the tropopause [see review by *Gillet et al.*, 2003].

Thirdly, the most studied candidate for imposing low-frequency variability on the atmosphere is the ocean. The traditional approach has been to search for a statistical relationship between the underlying Atlantic sea-surface temperature (SST) and the SLP variability. An early attempt was made by *Bjerknes* [1964], who concluded that short-term variations in the SSTs were associated with local forcing from the atmosphere; longer-term variations involved changes in the oceanic circulation. Bjerknes further proposed that on long time scales, a change in the oceanic heat transport would be compensated for by an opposite-sign change in the atmospheric heat transport. Using an atmospheric general circulation model (AGCM) forced with observed SST and sea-ice anomalies, *Rodwell et al.*, [1999] reproduced most of the observed interannual and interdecadal variability in the NAO. Furthermore, when they forced the model with typical SST anomalies in the North Atlantic only, local changes in the atmosphere–ocean heat and freshwater fluxes were again found to give a thermal and geopotential structure in the atmosphere that resembled the NAO. Several other authors have found NAO-like responses to SST anomalies in the North Atlantic [e.g. *Sutton et al.*, 2000; *Marshall et al.*, 2001b; *Czaja et al.*, 2003]. However, ambiguities and the potential for predictability can be questioned because the areas where the atmosphere seems to be most sensitive to SST variability roughly coincide with the areas where the forcing from the NAO on the ocean has its maximum [*Hurrell et al.*, 2004]. A more robust result seems to be the simulated response to sea-ice anomalies in the Labrador Sea, where an increased sea-ice cover leads to a weakening of the NAO [*Kvamstø et al.*, 2004]. Similar negative feedbacks between sea-ice and the NAO (a more positive NAO phase leads to more sea-ice in the Labrador Sea) have also been reported by other groups [e.g. *Alexander et al.*, 2004; *Magnusdottir et al.*, 2004].

Recognizing that half the area of the Earth is confined to the sector between 30°S and 30°N, and that the global-scale atmospheric circulation is driven by the convection in the tropics [e.g. *Gill*, 1982], other authors have turned their attention towards these latitudes. One suggested mechanism involves the cross-equatorial SST gradient, where co-varying fluctuations in the SSTs, the trade winds, and the rainfall are known to exist [*Nobre and Shukla*, 1996]. Changes in the tropical SST and rainfall patterns could lead to changes in the extratropical Atlantic through modulation of the strength of the Hadley circulation [e.g. *Sutton et al.*, 2000]. Using an AGCM forced with observed SSTs, *Hoerling et al.* [2001] reproduced half the observed trend in the NAO since 1950 and onward. Repeating the experiment with observed SSTs only in the tropics (30°S–30°N) and annual cycle in SST elsewhere, they essentially got the same results, thus excluding the extratropics as the source for the trend in the NAO. They further argued that the main source of the NAO trend lies in the Indian Ocean and involves tropical rainfall and adiabatic heating, the drivers for the extratropical circulation. Their work is expanded in *Hurrell et al.* [2004], where four different AGCMs and a multi-model ensemble of 67 members are used. They show that the results obtained by *Hoerling et al.* [2001], including the trend in the NAO, the importance of the tropics, and in particular the SST and rainfall changes in the Indian Ocean, are reproducible in different models. *Hoerling et al.* [2004] studied the physical mechanisms linking the Indian Ocean to the North Atlantic atmospheric variability, using a series of idealized experiments where a 1°C temperature anomaly was suddenly switched on in the Indian Ocean. After less than a month, an anomalous circulation resembling the positive NAO pattern was obtained. Without being conclusive, *Hoerling et al.* [2004] speculate that the response in the Atlantic is driven primarily by eddies rather than by Rossby wave dispersion and that adjustment of the local storm track is important.

To summarize, it seems that the NAO is an inherent mode in the atmosphere onto which a variety of different forcings regress. Thus, evidence of one forcing mechanism should not exclude the others. There are reasons to believe that the changes observed in the strength of the westerlies are nonlinear combinations of stochastic noise, greenhouse gas forcing near the tropopause, forcing from the stratosphere, and forcing from the ocean, where evidence points to the Indian Ocean as a key area. Additional possible forcing mechanisms, including tropospheric aerosols, volcanoes, and solar forcing, have been discussed by *Gillet et al.* [2003].

2.4. Spatial Changes in the Large-Scale Circulation

Most of the literature about the NAO and its impacts assumes that the SLP pattern associated with the specific station-based NAO Index in use has been stationary in time. Such is, however, not the case. Among the first to observe that the correlation between the NAO and some measured or modeled quantity may break down due to changes in the spatial pattern of the NAO were *Hilmer and Jung* [2000]. In their study, they compared the NAO with modeled sea-ice export through the Fram Strait. Although they found a high correlation between the data sets in the '80s and '90s, no significant correlations were found prior to this period.

Comparing the year-to-year variability in SLP between the first and latter part of the NCEP/NCAR reanalysis period

(Figure 6a,b) shows that the variability has decreased to the west of Iceland and has slightly increased in the Nordic Seas, Barents Sea, and over the Eurasian Basin. The changes become even more pronounced when the leading modes of variability are calculated. Associated with an increase in the NAO Index an SLP decrease in the Norwegian and Barents Seas, from the first to the latter period, and an increase near the Azores high and Aleutian low (Figure 6c,d). The main consequence of this shift is that the NAO pattern becomes more zonal in the vicinity of the GSR and more meridional in the Fram Strait. Another interesting feature is the strengthening of the Pacific variability associated with the NAO, in sharp contrast to findings in the earliest reanalysis period.

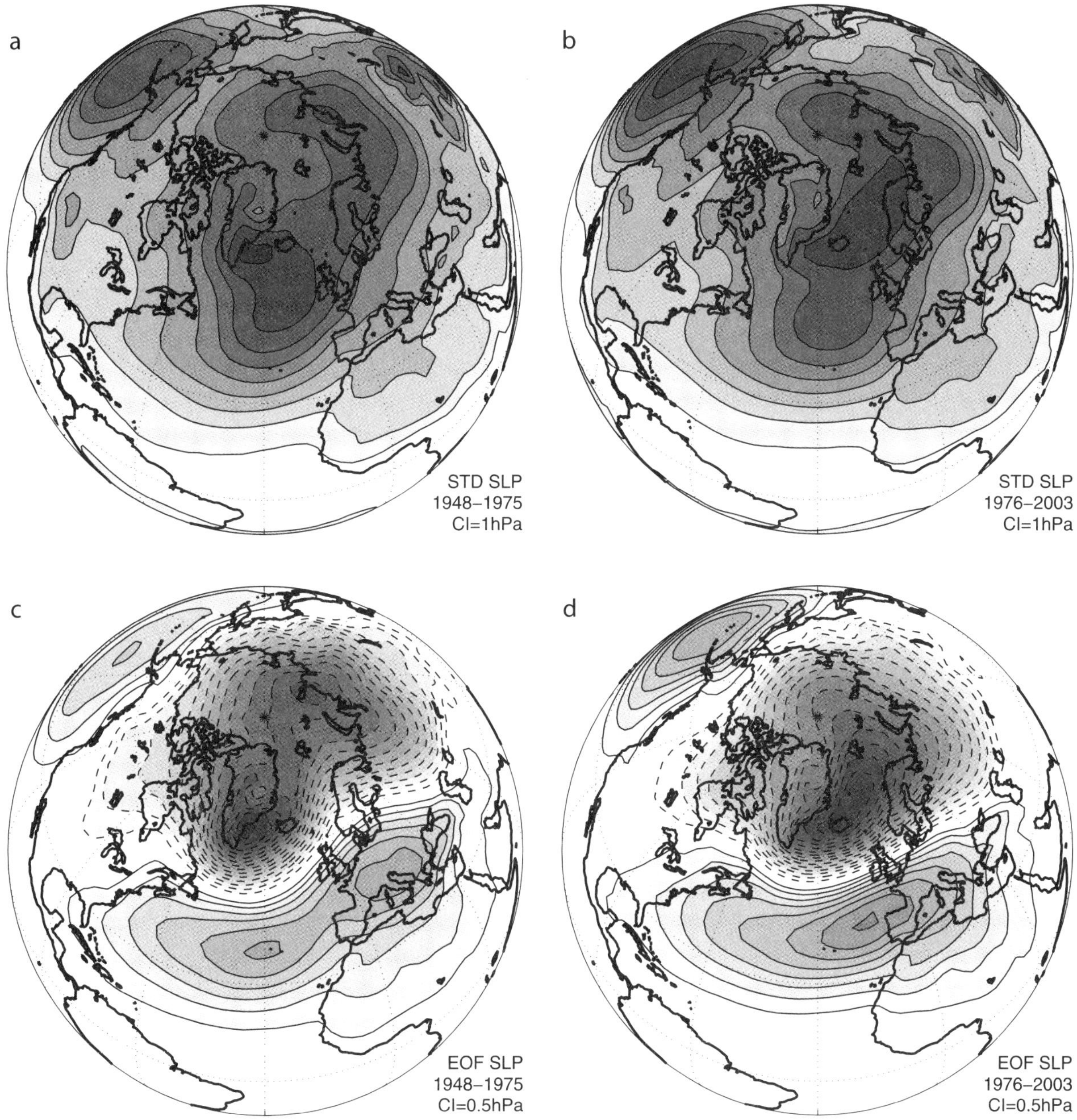

Figure 6. Standard deviation (upper panels) and the leading mode of variability (lower panels) of the winter mean SLP for the 1948–75 (left panels) and 1976–2003 (right panels). The principal components are calculated from the NCEP/NCAR reanalysis data for the Atlantic sector (90°W–30°E, 20°N–80°N). Solid lines are positive, dashed lines negative. White colouring in (a) and (b) indicates values below 1 hPa.

The reason for the apparent eastward shift in the NAO pattern is at present not clear. Sensitivity studies with atmospheric models suggest that the position of the northern centre of action may be connected to the strength of the NAO. Low index values correspond to a westward displacement from the mean position, and high index values correspond to an eastward displacement [*Peterson et al.*, 2002; *Cassou et al.*, 2004]. Other model simulations suggest that the eastward shift may be linked to an increased concentration of atmospheric CO_2 content. First pointed out by *Ulbrich and Christoph* [1999], this seems to be a tendency in most models participating in the second phase of the Coupled Model Intercomparison Project (CMIP2), as shown in a recent paper by *Kuzmina et al.* [2004].

3. ATMOSPHERE–OCEAN FLUX ANOMALIES

Associated with variability in the large-scale atmospheric circulation will be anomalies in the exchanges of momentum, heat, and freshwater between the atmosphere and the ocean. In this section we use the NCEP/NCAR reanalysis data [*Kalnay et al.*, 1996] to quantify these anomalies in terms of absolute values and deviations from the mean. Although the NCEP/NCAR data assimilation system has been kept unchanged over the reanalysis period, artificial trends or jumps may still exist in the data due to changes in the observational systems. The two major changes were the establishment of the upper-air data observational network from 1948 to the International Geophysical Year in 1957 and the introduction of satellite data in 1979 [*Kistler et al.*, 2001]. The gridded variables provided by NCEP/NCAR are divided into three classes, depending on how much they are influenced by the available observations [*Kalnay et al.*, 1996]. While output such as air temperature, pressure, and wind components are strongly constrained by the observations, and therefore considered to be the most reliable products, the variables surface heat and freshwater fluxes are determined by the model system and should therefore be used with more caution.

Although the NCEP/NCAR flux data may be strongly biased in certain areas [e.g. *Serreze and Hurst*, 2000; *Josey*, 2001], the interannual variability tends to be better correlated with independent observations [*Kistler et al.*, 2001]. Thus, using the reanalysis data in a qualitative more than a quantitative fashion, and focusing on the low-frequency impacts of the NAO, we are confident that the main results of this section are trustworthy and are not due to artificial trends in the reanalysis data.

3.1. Momentum Fluxes

The winter mean (vectorial) wind field over the Nordic Seas is dominated by the Icelandic low, giving southwesterlies over the northeastern North Atlantic, along the coast of Norway, and in the southern part of the Barents Sea (Figure 7a). Along the east coast of Greenland, northerlies dominate. The surface wind speeds for the Nordic Seas (Figure 7b) are 9-10 m s^{-1}, being greatest in the area of weakest winter-mean winds, which is along the path of the lows [see *Sorteberg et al.*, this volume].

When the wind is regressed on the NAO Index, an increase in the index is associated with strengthened southwesterly airflow towards northern Europe. Associated with a unit increase in the index is a more than 3 m s^{-1} increase in the westerly winds in the northeastern North Atlantic and a 1-2 m s^{-1} increase in the eastern part of the Nordic Seas. The impacts on the winds in the Barents Sea, Fram Strait, and along the east coast of Greenland are small.

The NAO impact on the local wind speed, reaching only 1 m s^{-1}, is much smaller than the changes in the monthly mean wind speed. The maximum is found northwest of Scotland and off the west coast of Norway. As demonstrated by *Vikebø et al.* [2003], this is also the area with the largest increase in extreme wind events throughout the reanalysis period (they defined an extreme wind event as a 6-h mean wind speed at a location that exceeds twice the mean wind speed in that area).

The monthly mean wind stress reaches 0.3 Nm^{-2} in the Denmark Strait and typically 0.1 Nm^{-2} west of Scotland and along the coast of Norway (Figure 8). Associated with a unit increase in the NAO Index is an increase in wind speed of 0.05-0.1 Nm^{-2} in the northeastern part of the North Atlantic and in the southeastern part of the Nordic Seas. A lesser increase in wind stress is found along the eastern coast of Greenland.

3.2. Heat Fluxes

From the NCEP/NCAR data we have extracted the heat fluxes divided into sensible, latent, short-wave, and long-wave radiative heat fluxes. The sensible and latent parts are shown in Figure 9a,b. Over the central parts of the Nordic Seas, the reanalysis has a sensible heat loss from the ocean, ranging from 40 to 80 Wm^{-2} during winter, and the maximum exceeding 100 Wm^{-2} along the marginal ice zone in the Barents Sea, Fram Strait, and Greenland Sea. Over the sea-ice is a downward flux of sensible heat, the surface of the ice being generally colder than the atmosphere because of the long-wave radiation from the sea-ice. The latent heat flux is the larger component in the winter heat budget, with typical values of 80-100 Wm^{-2} over most of the Nordic Seas. The maximum occurs over the pathways of the warm flow of AW within the Nordic Seas (Figure 1), which also is where the winds are strongest (Figure 7b).

If the four atmosphere–ocean heat flux components are added, the net upward winter mean heat flux is 160-180 Wm^{-2}

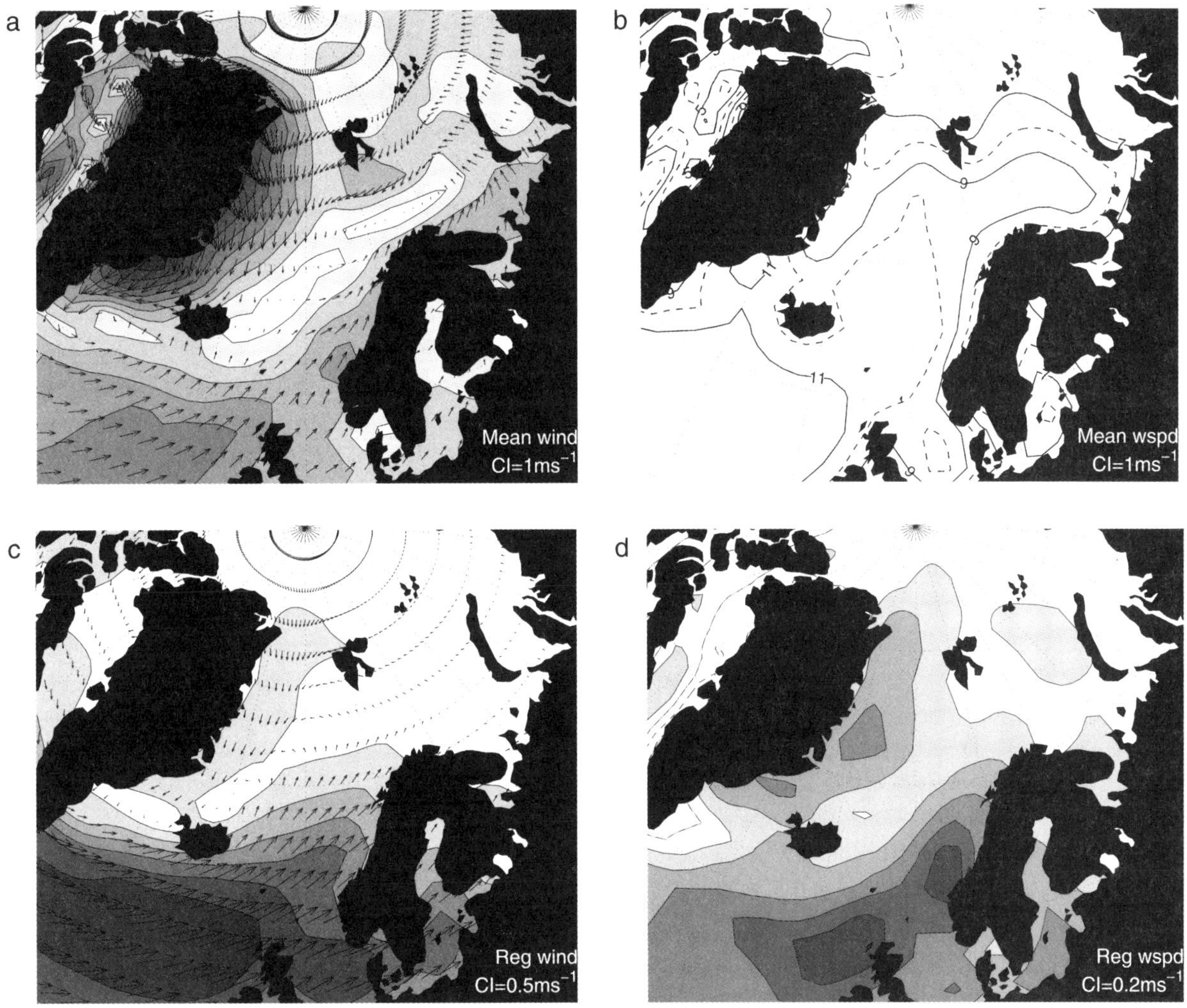

Figure 7. Winter mean wind vectors with contours for magnitude (a) and winter mean absolute wind speed (b) for the winters 1949–2003 and the regressions of these means onto the NAO Index (c,d). The values correspond to a unit increase in the index. White colouring in (a), (c), and (d) indicates values below 1, 0.5, and 0.2 m s^{-1} respectively.

over most of the region, with the maximum oceanic heat flux exceeding 300 Wm^{-2} where the relatively warm water of the West Spitsbergen Current meets the cold Arctic air in the Fram Strait. Values between 200 and 300 Wm^{-2} are also found in most of the ice-free Barents Sea, which is the subbasin within the Nordic Seas–Arctic Ocean system having the highest annual mean heat loss. In a comprehensive study, *Simonsen and Haugan* [1996] estimated that 50% of the total heat loss from the Nordic Seas occurs in the Barents Sea, totaling 136 out of 242 TW. In comparison, only 86 TW was lost from the Arctic Ocean.

The effect of the increased wind speeds associated with a more positive NAO (Figure 7d) is a strong increase in the heat loss in the northwestern North Atlantic (Figure 9d). Near the tip of Greenland, the NAO-forced heat loss is more than 40 Wm^{-2}, a 20% change from the winter mean of 200 Wm^{-2}. Both sensible and latent heat fluxes contribute to this. The maximum heat loss is found over the Labrador Sea [see also *Visbeck et al.*, 2003]. Over the Nordic Seas, the effect of a positive NAO phase is to reduce the heat loss to the atmosphere, typically on the order of 5-10 Wm^{-2}. An exception is in the marginal ice zone, where a retreat of the sea-ice cover associated with the NAO may locally more than double the heat loss to the atmosphere.

3.3. Freshwater Fluxes

According to the reanalysis data, typical evaporation rates over the Nordic Seas range from 1 mm day^{-1} near the ice-covered regions to 3 mm day^{-1} over the warm tongue of AW

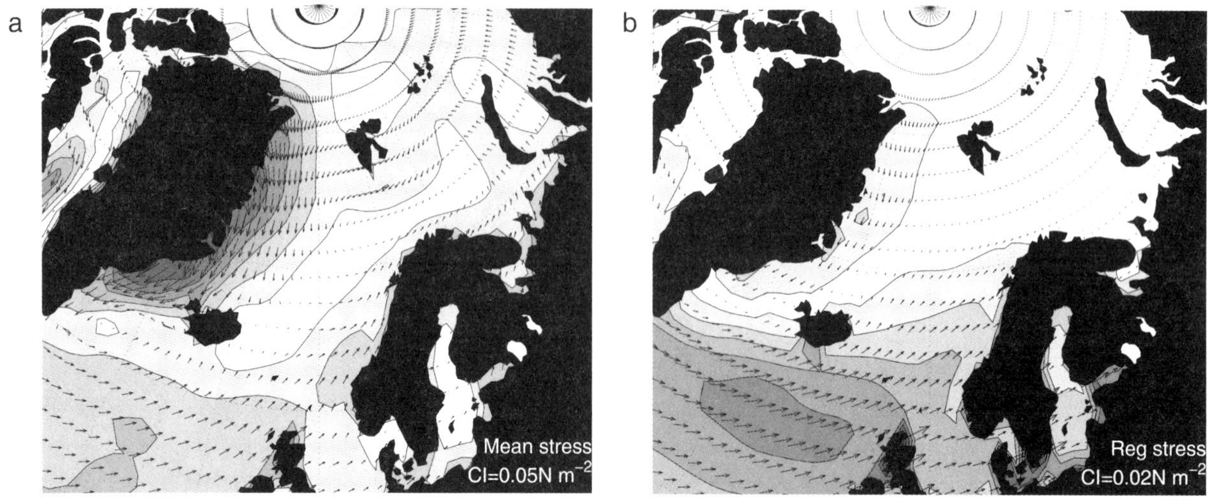

Figure 8. (a) Winter mean wind stress and (b) wind stress regressed on the NAO Index. The values correspond to a unit increase in the index. White colouring indicates values below 0.05 Nm^{-2} (a) or 0.02 Nm^{-2} (b).

(Figure 10a). Comparing this with the precipitation rates of typically 3-5 mm day^{-1} suggests a net freshwater supply to the ocean of 1-2 mm day^{-1}.

Associated with the NAO, is more evaporation to the south and west of Iceland and along the marginal ice zone east of Greenland and slightly less evaporation in the North Sea and in the Baltic Sea (Figure 10c). The NAO impact on precipitation is much stronger: Close to 1 mm day^{-1} of additional precipitation south of Iceland and along the coast of Norway is associated with a unit increase in the NAO Index (Figure 10d).

4. IMPACTS ON THE OCEAN CLIMATE

Large-scale changes in the atmosphere–ocean fluxes may impact the ocean climate through a variety of mechanisms and on a wide range of time scales, ranging from days to years. Locally, there will be a rapid response in sea-surface height, mixed layer depth, sea-surface temperature (SST), and sea-surface salinity (SSS) due to anomalies in the atmosphere–ocean momentum, heat, and freshwater fluxes. While anomalous momentum fluxes and divergences in the associated Ekman transports will change the sea-surface height and mixed layer depth directly, leading to barotropic and baroclinic adjustment processes, it will also indirectly affect the SST and SSS fields. Divergence of heat or freshwater by the Ekman transports, or diapycnal mixing with the water below the mixed layer, will add extra terms to the energy and salt balance equations.

On the longer time scales—years to decades—the Nordic Seas' response to anomalous forcing can include advection of heat or freshwater anomalies from distant regions, changes in the flow speed or position due to changes in the baroclinic structure of the upper ocean, or changes in the abyssal circulation due to buoyancy forcing in the convective regions.

In the following sections, we discuss some of the responses of the large-scale atmospheric forcing on the ocean climate. For more thorough treatments of the various aspects of the subject, readers should refer to the papers by *Anderson and Gill* [1975], *Anderson and Killworth* [1977], *Willebrand et al.* [1980], and the textbook by *Gill* [1982].

4.1. Ocean Circulation

The direct response of an anomalous wind stress $\boldsymbol{\tau}'$ acting on the ocean surface is to set up a transport in the upper waters, given by

$$\mathbf{U}_{ek} = -\frac{1}{\rho_0 f}\mathbf{k} \times \boldsymbol{\tau}' \qquad (1)$$

where $\mathbf{U}_{ek}$ is the so-called Ekman volume transport per unit length (m^{2}s^{-1}), ρ_0 is the mean density, f is the Coriolis parameter, and $\mathbf{k}$ is the vertical unit vector.

Associated with the mean wind stress over the Nordic Seas (see Figure 8), the Ekman transports show a diverging field over the central Nordic Seas, with on-shore transports towards all surrounding land areas (Figure 11). Typical estimates for the transports are on the order of 1 m^{3}s^{-1} through a section 1 m wide. Note that equation (1) is valid only for an atmospheric wind stress over an ice-free ocean. Because a sea-ice cover will significantly modify both the air–sea momentum exchanges and the associated Ekman transports, the results for the ice-covered regions should be used with caution.

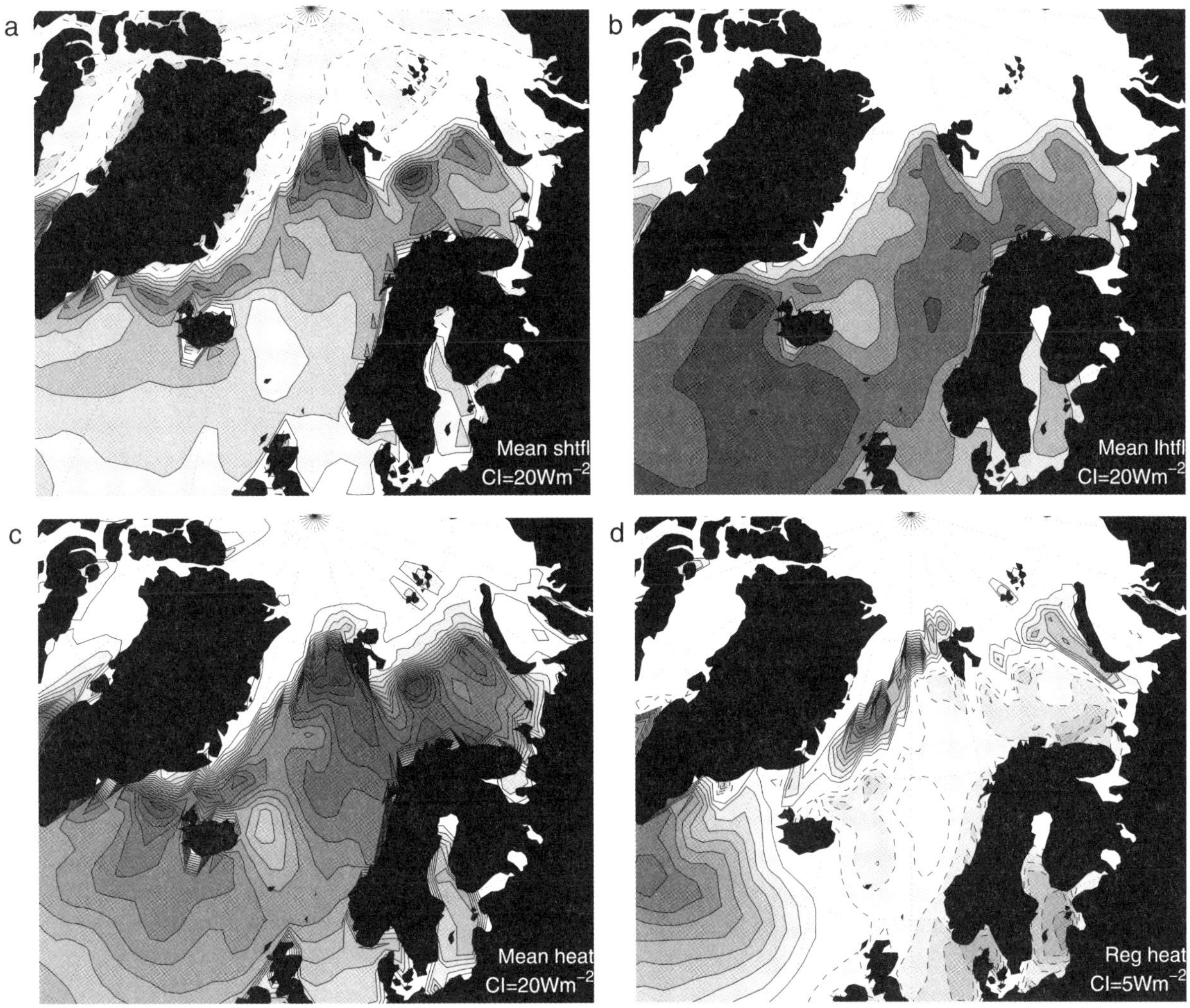

Figure 9. Winter mean sensible heat flux (a), latent heat flux (b), total heat flux (c), and total heat flux regressed on the NAO Index (d). Solid lines are positive (ocean temperatures will decrease), dashed lines negative.

As shown in Figure 8b, stronger than average westerlies are associated with an increased wind stress curl over the Nordic Seas, giving rise to strengthened Ekman transports. The region with strongest response is to the south of Iceland, where the Ekman transports (and wind stress) are nearly doubled during years when the NAO Index is one unit above average (Figure 11b).

If the wind stress was spatially uniform, the water below the mixed layer will not have been affected by the wind. However, when lateral boundaries are present or if the wind stress is spatially varying, the Ekman transports will be divergent. The result will be changes in the surface elevation and mixed layer depth and a distortion of the pressure gradient throughout the water column, producing barotropic and baroclinic flow anomalies. We will first look at the effects of side boundaries.

When the wind stress has a component τ_s parallel to a coast, linear theory gives a constant growth rate for the surface elevation η and along-shore velocity u, given by

$$\eta = \frac{\tau_s}{\rho_0 \sqrt{gH}} e^{-y/a} t \quad \text{and} \quad u = \frac{\tau_s}{\rho_0 H} e^{-y/a} t, \tag{2}$$

where g is the gravity, H is the water depth, a = $\sqrt{gH}/f$ is the Rossby radius, t is the time, and the y-axis is pointing outwards from the coast [e.g. *Gill*, 1982]. When bottom stress is included, the solutions will grow asymptotically towards equilibrium, having constant surface elevation and alongshore transports. Associated with an increase in the sea level near the coast, down-welling motion will occur in the water below the Ekman layer, changing the internal density structure in the fluid. Thus in addition to the sea-level response to

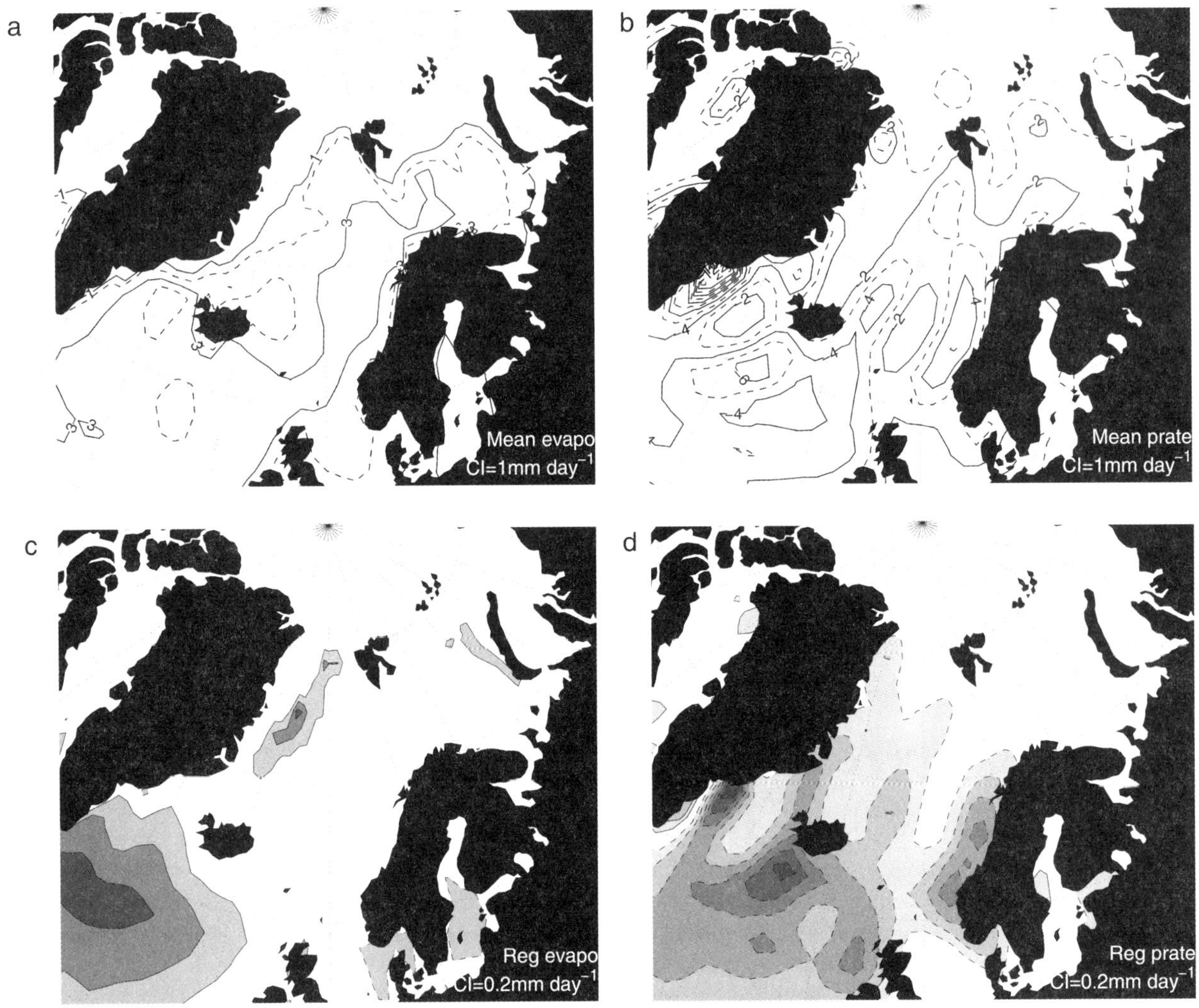

Figure 10. Winter mean evaporation (a), precipitation (b), and the evaporation (c) and precipitation (d) regressed on the NAO Index. Solid lines are positive (ocean salinities will increase), dashed lines negative.

the on-shore Ekman transport, leading to fast-propagating coastally trapped barotropic waves, modifications of the internal pressure fields will lead to much longer adjustment time scales, and slower propagating coastally trapped baroclinic waves with a typical speed of 1–2 m s^{-1}.

In the open ocean, a spatially varying wind stress will force vertical motion known as Ekman pumping, where the strength of the vertical motion, w_{ek}, is given by

$$w_{ek} = \frac{1}{\rho_0 f}\mathbf{k} \cdot \nabla \times \boldsymbol{\tau}'. \qquad (3)$$

Calculations made for the Nordic Seas area are shown in Figure 12. The dominant long-term mean features are upwelling motion (positive Ekman pumping velocities) over most of the area, with values exceeding 40 cm day^{-1} in the western part. Only the northwestern corner of the Barents Sea, where the mean wind stress shows an anticyclonic pattern (see Figure 8a), has negative Ekman pumping. The response to the enhanced wind stress associated with the leading mode of variability is positive Ekman pumping anomalies in the western part of the basin and negative anomalies to the southeast, where the anomalous westerly wind stress pattern is weakly anticyclonic (Figure 8b).

The response of the Ekman pumping in a flat-bottom ocean can be calculated using quasi-geostrophic theory [e.g. *Gill*, 1982]. An atmospheric low will introduce a positive wind stress curl to the ocean, where the barotropic response to the wind stress is a lowering of the sea level and the baroclinic response is a raising of the density surfaces below the low. Associated with the changes in the surface elevation and density distribution will be geostrophic currents

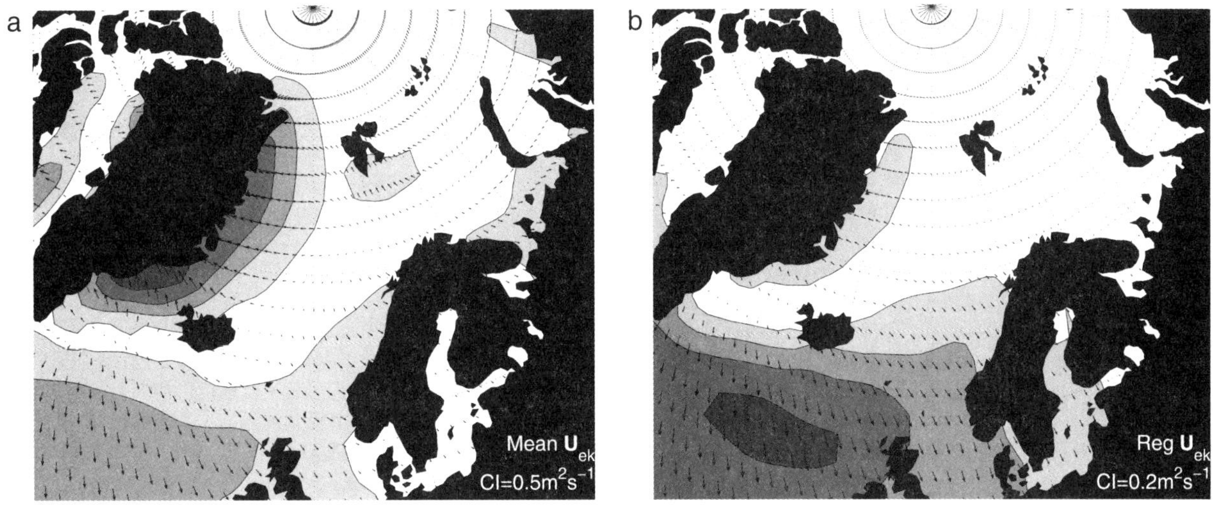

Figure 11. (a) Winter mean Ekman volume transports and (b) Ekman volume transports regressed on the NAO Index. Ekman volume transports are calculated from the NCEP/NCAR reanalysis of wind-stress fields. Effects of sea-ice are ignored. White colouring indicates values below 0.5 m^2s^{-1} (a) or 0.2 m^2s^{-1} (b).

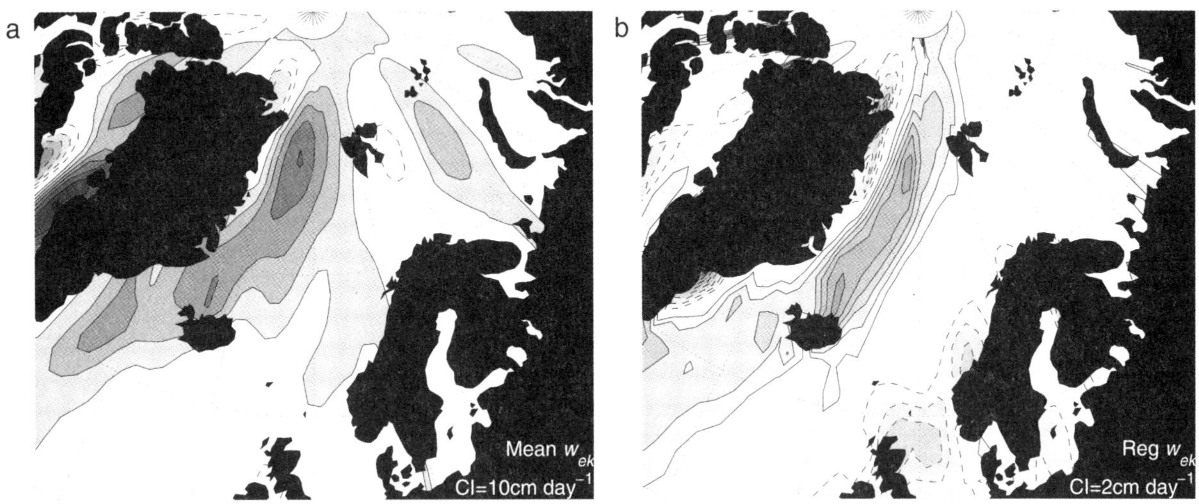

Figure 12. (a) Winter mean Ekman pumping velocities and (b) Ekman pumping velocities regressed on the NAO Index. Ekman pumping velocities are calculated from the NCEP/NCAR reanalysis of wind-stress fields. Effects of sea-ice are ignored. Solid lines indicate positive results, dashed lines negative. White colouring indicates values less than 10 cm day^{-1} (a) or 2 cm day^{-1} (b).

that increase in strength with time. If the wind stress anomaly extends to the eastern rim of the ocean, or if the background flow field can carry the signal to the eastern boundary, barotropic and baroclinic planetary waves will be radiated into the interior of the ocean [*Anderson and Gill*, 1975]. Time scales will be of the order of days for the long, nondispersive barotropic waves, and years for the baroclinic waves. After the wave front has passed, the flow becomes virtually stationary and is given by the Sverdrup balance [*Sverdrup*, 1947], which is simply the conservation of potential vorticity, given as

$$V_{sv} = \frac{1}{\rho_0 \beta} \mathbf{k} \cdot \nabla \times \boldsymbol{\tau}'. \tag{4}$$

Here V_{sv} is the meridional volume transport per unit length, and $\beta = df / dy$ is the rate of change of the Coriolis parameter with latitude.

The Sverdrup solution ignores the effects of topography and stratification. With bottom effects included, the lines of constant potential vorticity will be f/H rather than f contours, and the effect of the Ekman pumping and vortex stretching will be to produce flow across f/H contours instead of meridional flow. With stratification also present, the influence of the topography will be reduced, and the Sverdrup theory therefore provides a good approximation to the large basin-scale circulation in the world oceans [e.g. *Gill*, 1982]. It should however be noted that the baroclinic adjustment time scale in the ocean makes the time to remove the influence of topography very long [*Anderson and Killworth*, 1977]. Thus, it is not a simple task to estimate how long an anomalous wind forcing must persist to set up a quasi-stationary Sverdrup circulation in a stratified flow over topography. By calculating the response of the North Atlantic to observed wind forcing, *Willebrand et al.* [1980] found little or no coherence between the atmospheric and the oceanic response at any frequency, which they attributed to the effect of topography. The usefulness of the Sverdrup solution for the weakly stratified, strongly topographically steered flows in the Nordic Seas areas may therefore be questioned.

4.2. Heat and Freshwater Divergence

In addition to the direct effect of the anomalous heat and freshwater fluxes imposed by the atmosphere, there may also be a local warming or freshening due to heat or freshwater divergences associated with the Ekman transports.

As shown by *Marshall et al.* [2001a], the Ekman-induced contribution to the heat balance of the water column can be written

$$H_{ek} = c\rho_0 \mathbf{U}_{ek} \cdot \nabla T, \tag{5}$$

where c is the specific heat of water and T is the temperature. Only divergences in the heat transports are included in equation (5); any divergences in the Ekman-induced mass transports are neglected. A positive contribution to the water temperatures is found where warm water is advected into areas with colder water (Figure 13). This is around the northern part of Great Britain and northeast towards the western coast of Norway, where AW is advected into areas dominated by the Norwegian Coastal Current (NwCC in Figure 1) and along the marginal ice zone along the east coast of Greenland. A net cooling due to anomalous Ekman transports occurs in the northern North Atlantic and north of the Faroes, corresponding to a heat loss of more than 20 Wm^{-2}.

Following the same method as applied for the heat, one can write the Ekman transport contribution to the freshwater budget as

$$Q_{ek} = -\rho_0 \mathbf{U}_{ek} \cdot \nabla S, \tag{6}$$

where S is here the dissolved salt per mass of sea water (i.e. the dimensionless salinity value divided by 1000). The reason for the minus sign is that an Ekman transport towards higher salinities will locally freshen the water column and thus have the same effect as precipitation, which by definition is a negative freshwater flux. A positive contribution to the salinity in

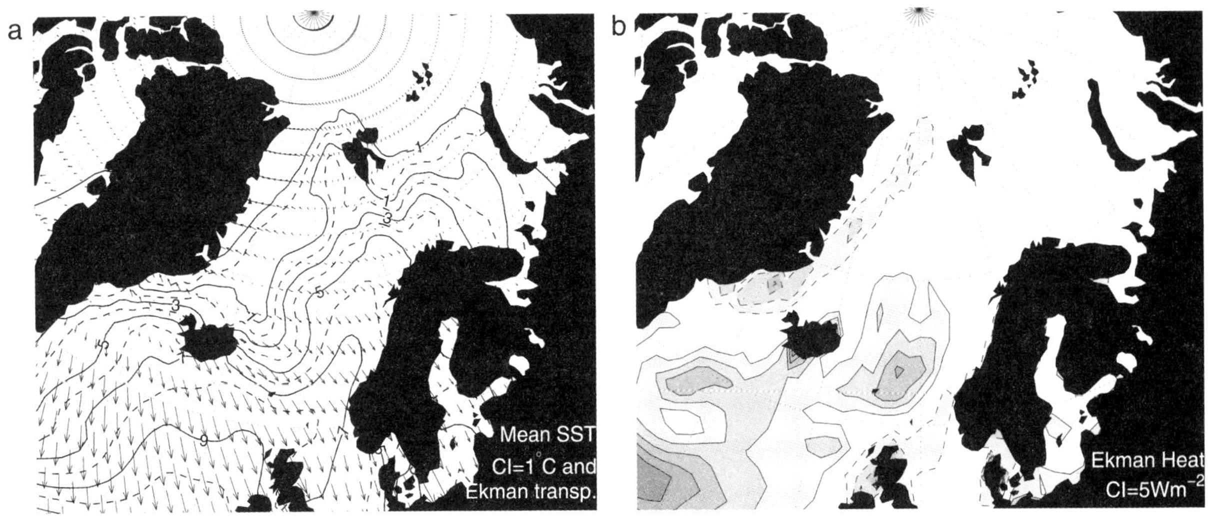

Figure 13. (a) Winter mean SST field (contours) and anomalous Ekman transports (vectors) corresponding to the NAO-related wind-stress anomalies (Figure 8b). (b) The heat divergence associated with the anomalous Ekman transports. Effects of sea-ice are ignored. Solid lines are positive (ocean temperatures will decrease), dashed lines negative. The SST data are updated from *Reynolds and Smith* [1994].

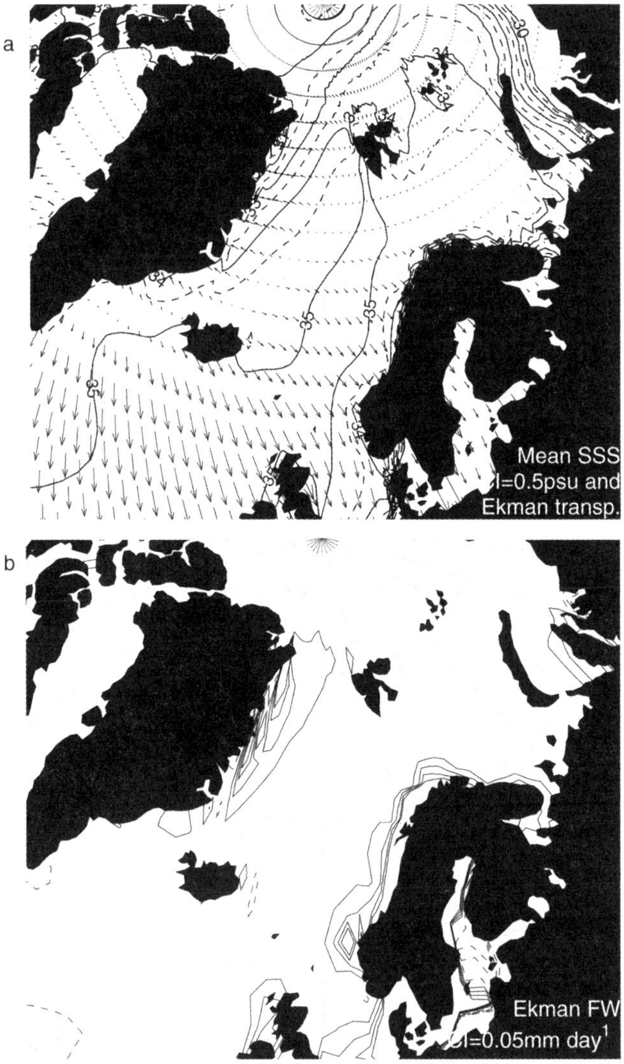

Figure 14. (a) Winter mean SSS field (contours) and anomalous Ekman transports (vectors) corresponding to the NAO-related wind-stress anomalies (Figure 8b). (b) The freshwater divergence associated with the anomalous Ekman transports. Effects of sea-ice are ignored. Solid lines indicate positive values (ocean salinities will increase), dashed lines negative. The SSS data come from the Polar Science Center Hydrographical Climatology version 2.1, updated from *Steele et al.* [2001].

the water column is found where high-salinity water is advected into areas with lower salinities (Figure 14). These areas roughly coincide with the areas having a convergence of heat, that is, to the west of Scotland, in the North Sea, along the coast of Norway, and along the east Greenland marginal ice zone. The Ekman effect here corresponds to an additional evaporation on the order of 0.1-0.2 mm day^{-1} and thus is small compared to the effects of evaporation and precipitation anomalies. The Ekman effect may, nonetheless, be important in areas with strong salinity gradients, which do not show up in smoothed climatological data sets.

In a study of the mixed layer properties in the NwAC at Ocean Weather Station M (OWSM, marked on the map in Figure 1), *Nilsen and Falck* [2004] demonstrate a clear correlation between the meridional wind component and the mixed layer salinities during summers. They find the salinities to be lowest during summers with strong northerly winds, despite the fact that these also are the summers with least precipitation, and suggest that the extra freshwater can only come from the NwCC because of the anomalous offshore Ekman transports. Thus, at least during summer, and in regions of relatively strong salinity gradients, Ekman-induced freshwater transports may be the dominant factor in establishing salinity anomalies. We have no reason to believe that it should be too different during winters, but salinity anomalies will typically be weaker then due to deeper mixed layers.

4.3. Mixed Layer Properties

As a first-order approximation, the impact on the ocean mixed layer temperature and salinity can be estimated assuming that the ocean is a passive responder to the anomalous forcing. For example, a reduction in heat loss of 10 Wm^{-2} will warm a 100-m-thick water column by 0.26 °C in 4 months (December to March), and during the same period of time an increased precipitation of 1 mm day^{-1} will freshen the same water column by 0.04.

If the SSTs are regressed on the NAO Index, the well-known tri-pole pattern emerges—colder than normal water from the equator to approximately 20°N, a band of warmer than normal water from Florida towards Europe, and a colder than normal regime in the northwestern North Atlantic [e.g. *Marshall et al.*, 2001b]. For the Nordic Seas region (Figure 15), the response is a general warming along the eastern margins, where the southwesterlies are enhanced during positive NAO years (see Figure 7c), and a cooling in the western part of the region, where the positive NAO years represent stronger northerlies and colder air over the region. Strongest warming is found in the North Sea and Baltic Sea, where the atmosphere-forced SST anomalies can exceed 1° C. In contrast, the regions to the west and south of Iceland may become several tenths of degrees colder during positive NAO years. Elsewhere, anomalies are generally small, on the order of only 0.1° C, and comparable to the simple estimate given in the introduction to this section.

In contrast to SST, which has been measured regularly by ships for more than a century and by satellites for 25 years, it is not possible to show the direct impact of the anomalous freshwater fluxes on the SSS by correlating year-to-year changes in SSS with the NAO Index. There are much less data, the NAO response in salinity is weak, and the surface

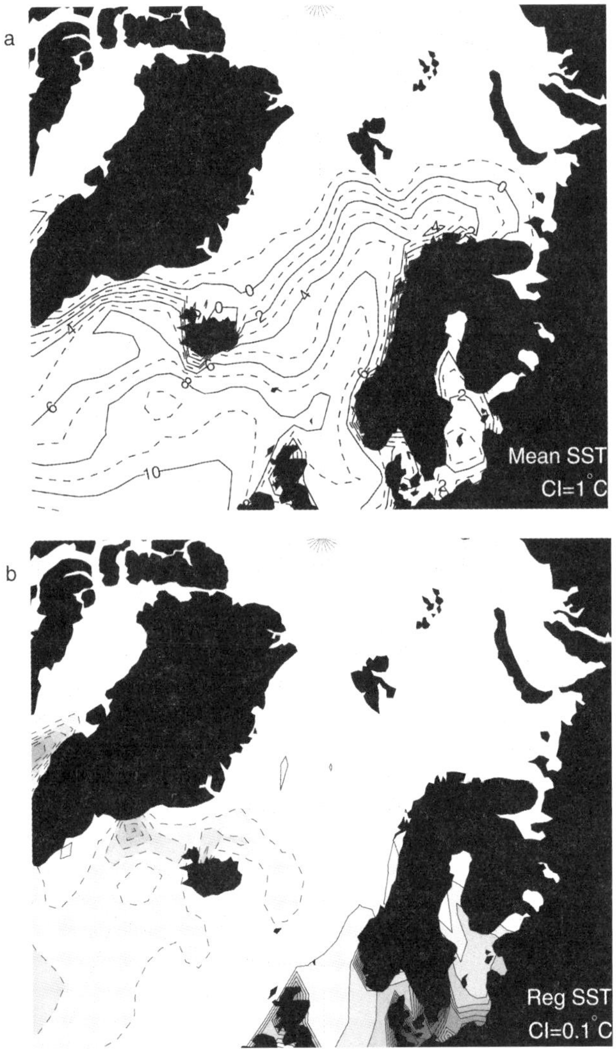

Figure 15. Winter mean SST (a) and SST regressed on the NAO Index (b). In (b) solid lines are positive, dashed lines negative.

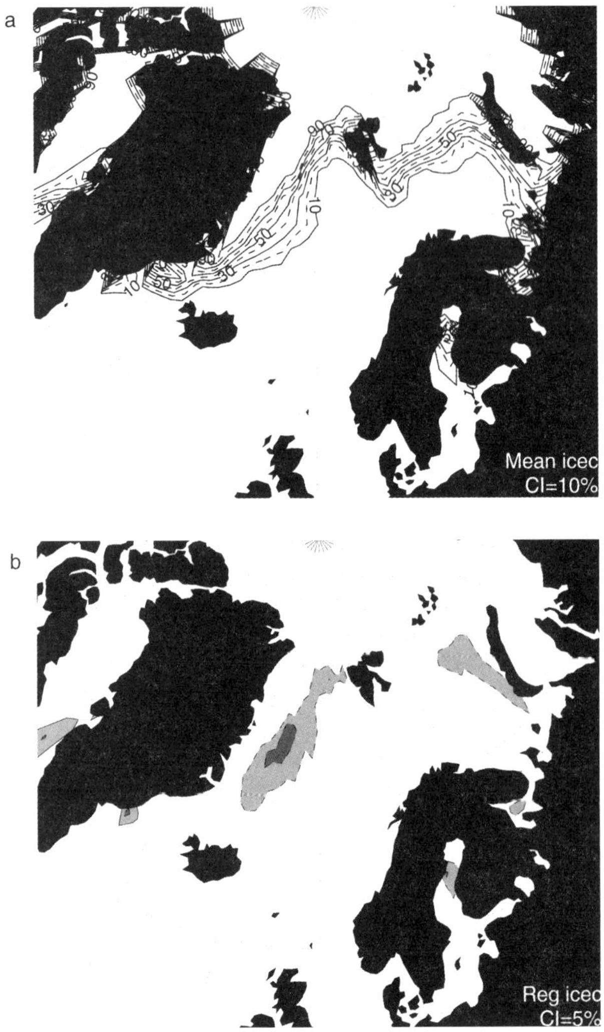

Figure 16. Winter mean sea-ice concentration (a) and sea-ice concentration regressed on the NAO Index (b). In (b) solid lines are positive, dashed lines negative.

salinity measurements are often noisy due to local precipitation events. Measurements have, nonetheless, shown a general reduction in the surface salinities for the last 3-4 decades [*Blindheim et al.*, 2000; *Furevik et al.*, 2002; *Curry et al.*, 2003], probably as a combined result of local fresh-water forcing and a changed balance between the water of Atlantic and Arctic origins.

Measurements of the extent of sea-ice have been provided by relatively good ship-based data for the last 50 years, and since 1978 spaceborne passive microwave sensors have covered the Arctic at high temporal and spatial resolutions. A regression of the sea-ice cover on the NAO Index (Figure 16) shows in general, negative values in both the Greenland and Barents Seas, and positive values in the Labrador Sea area. No detrending has been conducted here, and the positive trend in the NAO Index and the negative trends in the sea-ice data in the Nordic Seas are obviously contributing to the negative correlations in the regression values for the Nordic Seas. The sea-ice cover in the Nordic Seas has been decreasing relatively fast during the last decades, and during the record low extent in September 2002 the entire eastern coast of Greenland was ice-free for the first time on record [*Serreze et al.*, 2003]. Kvingedal [this volume] has calculated the trend in the Nordic Seas sea-ice extent for the period 1967–2002 and finds a 3.4% per decade decrease in winter sea-ice cover over the 36-year period, which is comparable to the decrease calculated for the entire Arctic since 1978 [*Johannessen et al.*, 1999].

5. OBSERVED CHANGES IN THE NORDIC SEAS OCEAN CLIMATE

As has been known for more than a century, perhaps the single most important factor for the mild and friendly climate along the northern coasts of Europe is the northern extension of the Gulf Stream. The amount of AW that enters the Nordic Seas has been estimated by indirect methods for decades and has been directly measured since 1995. Measurements are available from both the Iceland–Faroe and Faroe–Shetland passages and also from the Svinøy Section extending from the west coast of Norway (marked on the map in Figure 1).

There is today surprisingly little consensus about what is actually driving the Atlantic inflow to the Nordic Seas. Three factors seems to be present: the wind forcing; the buoyancy forcing, leading to a densification and subsequent sinking of water in the Nordic Seas and the Arctic; and the estuarine forcing, the result of the outflow of PW from the Arctic. For this last, the mechanism is that the southward flow will entrain ambient water of Atlantic origin, and because of volume conservation, there must be a compensating inflow [e.g. *Stigebrandt*, 1981].

Rapid changes in the North Atlantic, Nordic Seas, and Arctic Ocean climate have been indicated from intensified observational programmes, combining new surface, subsurface, and remotely sensed data with historical archives. We will in this section review some of the recent findings and try to assess how the changes are related to the large-scale changes that have been taking place in the atmospheric circulation over the region. For complementary discussions and references, see also *Visbeck et al.* [2003] and Blindheim and Østerhus [this volume].

5.1. Atlantic Inflow

The diverse and variable paths of the source water in the North Atlantic (Figure 1) can be expected to be susceptible to changes in the atmospheric forcing, whether from variations in the momentum input or variations in the heat- or freshwater fluxes. The main branches of AW feeding the Nordic Seas—the inner branch at the continental slope entering through the FSC, and the outer branch associated with the Arctic Front and entering over the IFR—both seem to be influenced by the large-scale circulation anomalies in the atmosphere.

In a study of surface drifters as well as altimeter data from the period 1992–98, *Flatau et al.* [2003] observed a clear southeastward shift and intensification of the current system in the subpolar North Atlantic during periods of high NAO Index. That is, the NAC flow towards the Rockall Trough (Figure 1) is intensified, ultimately giving a stronger flow of AW through the FSC. This seems to be related to findings by *Orvik and Skagseth* [2003b], who compared the volume transport in the eastern branch of the NwAC at the Svinøy Section with the wind-stress curl in the North Atlantic. Using 12-month moving averages, they found the correlation between the Svinøy transport during 7 years and the wind-stress curl in a band at 55°N to be 0.88, with a 15-month time lag. The proposed mechanism for this statistically significant correlation is Ekman pumping, which alters the density field associated with the NAC at 55°N, forcing a baroclinic Rossby wave and propagating with the baroclinic NAC until it hits the Irish–Scottish shelf 15 months later, from where fast barotropic waves transport the signal to the Svinøy Section (see Section 4.1).

Other indications of a lagged response of the NAO on the ocean circulation within the Nordic Seas also imply that baroclinic adjustment processes take place. *Blindheim et al.* [2000] found that the lateral surface extension of AW in the southeastern Norwegian Sea is negatively correlated with the NAO Index on long (>3 years) time scales and attributed this to changes in the pathways of the inflow of AW between Iceland and Scotland.

Direct measurements of current in the FSC [*Turrell et al.*, 2003] and measurements north of the Faroes capturing the water that flows over the IFR [*Hansen et al.*, 2003] show little interannual variability and a very weak seasonal cycle in the inflow of AW. This is inconsistent with the results by *Flatau et al.* [2003] and *Orvik and Skagseth* [2003b] and with the measurements at the Svinøy Section where a seasonal cycle in the eastern branch of the NAC is found [*Orvik et al.*, 2001]. The reason for this apparent discrepancy is not fully understood, but may be related to recirculation of AW and interactions between the two branches in the FSC.

Mork and Blindheim [2000] calculated the geostrophic volume transports in the two branches of the NwAC from hydrographical observations in the Svinøy Section during 1955-96. One of their results was that the two branches appeared to be in opposite phases and NAO-controlled since the summer of 1978. This out-of-phase relationship between the outer and inner branches of the NwAC is supported by a model study by *Nilsen et al.* [2003], who found a negative correlation between the flows over the IFR and through the FSC. In their model, the variability in the inflow was linked to an atmospheric pattern resembling the NAO via anomalous Ekman transports and barotropic adjustment processes.

The exact mechanisms linking the anomalous wind field to the inflow are not clear. For the period prior to the mid-'70s, stronger than normal westerlies were associated with a mean wind field that had a component from the North Atlantic Ocean and into the Nordic Seas (Figure 6c), which could support a mechanism with Ekman transport towards the Irish–Scottish shelf, and thus more flow through the FSC. After the mid-'70s, the pattern became much more zonal, and it is difficult to see how the inflow can be forced by the local wind. However, dynamical consideration (flow tends to follow f/H contours) and model results [*Nilsen et al.*, 2003] suggest there is a tight link between the inflow in the FSC and the outflow through the Denmark Strait. The northerly winds along the eastern coast of Greenland, together, with the increased southward extension of PW thus associated with high NAO Index, may inhibit inflow of AW, both in the Denmark Strait [*Blindheim et al.*, 2000] and over the IFR [*Hansen and Kristiansen*, 1994], thereby enhancing the eastward shift of the Atlantic inflow and NAO control of the flux along the eastern slope (North Atlantic/Norwegian Sea). The NAO-related variability in the inflow might thus just as well be forced by the northerly winds along the coast of Greenland and by an associated enhanced flow through the Denmark Strait, as it is forced locally at or near the inflowing region. Comparing the two periods, before and after the mid-'70s, it follows that the explanation involving enhanced outflow and a compensating inflow seems to be the most plausible for the latter period (Figure 6d). This is also the period showing highest correlation between the NAO and Fram Strait ice export [*Hilmer and Jung*, 2000], as will be further discussed in the next section.

5.2. Fram Strait Freshwater and Ice Export

The main export route of Arctic sea-ice and freshwater goes through the Fram Strait, where an estimated 2000–3000 km^3 of sea-ice and 1000 km^3 of liquid freshwater escape the Arctic every year [see *Aagaard and Carnmack*, 1989; *Vinje et al.*, 1998; *Kwok et al.*, 2004]. Variability in the strength of the export is important not only for the freshwater budget of the Arctic, but also for the downstream ocean properties and deep-water formation in the Greenland, Irminger, and Labrador Seas.

Freshwater export is a difficult parameter to measure, and most estimates are based on coupled Arctic Ocean–sea-ice models. Forcing their model with the NCEP/NCAR reanalysis data, *Zhang et al.* [2003] found that the leading atmospheric mode was the driver of many of the changes observed in the Arctic. When their Arctic Oscillation Index changed from a negative to a positive phase, the freshwater and sea-ice export through the Fram Strait increased by 12% and 56%, respectively.

The single most important factor for the ice export through the Fram Strait is, to the best of our knowledge, the SLP gradient across the strait. Using almost 3 years of data from a moored Doppler current meter in the Fram Strait, *Widell et al.* [2003] found the correlation between measured ice velocities and the SLP gradient (calculated from NCEP/NCAR reanalysis data) to be as high as 0.76 for daily means and 0.79 for monthly means.

By tracking common features in successive satellite images of sea-ice, *Kwok et al.* [2004] managed to estimate the sea-ice area flux through the Fram Strait for 1978–2002 and found a correlation of 0.62 with the wintertime NAO Index. This estimate seems to contradict the fact that the very large outflow of sea-ice and freshwater during the late '60s, which was responsible for the great salinity anomalies observed in the subpolar gyre and later along the AW pathways in the Nordic Seas [*Dickson et al.*, 1988], occurred during a period characterized by a very weak Icelandic low. However, using composites for the low-index years 1963–69, *Dickson et al.* [2000] showed that the pressure pattern also gave a northerly wind component over the Fram Strait in those years, despite the weak NAO.

Based on a hindcast run with an ocean–sea-ice model, *Hilmer and Jung* [2000] demonstrated that the correlation between the NAO and Fram Strait sea-ice export breaks down when the time series are extended back in time. Although the correlation they found was similar to that of *Kwok et al.* [2004] when analyzing the same period, the correlation dropped to almost zero for the period 1958–77. And extending their analysis to a century-scale control integration of a coupled climate model, they demonstrated that the correlation found in the last two decades was unusually high [*Jung and Hilmer*, 2001].

In a recent study, *Schmith and Hansen* [2003] used historical observations of multiyear ice from southwest Greenland for the period 1820–2001 as a proxy to deduce ice export from the Fram Strait. From their reconstructions, they found relatively large interdecadal variability, with high ice export values around 1900 and in the '60s and low values in the 1820s and in the 1930s and '40s. Because high ice export values correspond to cold SST periods in the North Atlantic, and vice versa, *Schmith and Hansen* [2003] suggested that the temperature anomalies in the North Atlantic might be a response to the ice export. They further used their observational data to redo the calculations of *Jung and Hilmer* [2001], using model data, and demonstrated the shifting influence of the NAO on the Fram Strait ice export.

The changing influence of the NAO, or more precisely, the changing representativeness of the traditional NAO Index when describing the sea-level gradient across the Fram Strait, can be seen in Figure 6c,d. During the first half of the reanalysis period, when the NAO Index was relatively low, high-index values represent a low, centered west of Iceland, with a modest SLP gradient over the Fram Strait (Figure 6c). For the period prior to the '70s, the NAO Index does not represent the wind conditions in the Fram Strait. Instead, as Figure 6c suggests, it is a better proxy for the northerly winds over the Canadian Archipelago and the Davis Strait, where *Hilmer and Jung* [2000] found the correlation between ice export and the NAO Index to be 0.63 for the early period (until 1978). In the later period, the SLP pattern changed (Figure 6d); the NAO is in a more positive phase, and the index better represents the SLP gradient across the Fram Strait. High index values now correspond to northerlies in the Fram Strait, driving more ice out of the Arctic. For the Davis Strait ice export, the NAO is not a good proxy for this latter period, as *Hilmer and Jung* [2000] calculated the correlation to be only 0.13 for the period after 1978.

5.3. Nordic Seas Circulation

As for the inflow of the AW to the Nordic Seas, it is not clear what is driving the circulation within the seas—the inertia of the inflow, local winds, or buoyancy forcing within the basins. An early attempt to study the effect of the positive wind-stress curl in the area (e.g. Figure 7) was made by *Aagaard* [1970] who digitalized hand-drawn 6-h weather maps for 1965 and calculated the monthly mean Sverdrup transports for the Nordic Seas. The annual mean wind-driven gyre transport exceeded 30 Sv in his calculations, comparable to the 35 Sv *Aagaard and Coachman* [1968] estimated from measurements at the western side of the gyre. The variations from month to month were tremendous, ranging from practically zero to almost 90 Sv.

In a recent study, *Nøst and Isachsen* [2003] combined the divergence in the Ekman transports and the observed climatological density field to calculate the geostrophic flow along f/H contours within the Nordic Seas–Arctic Ocean and got circulation patterns in close agreement with the existing observations [*Poulain et al.*, 1996; *Orvik and Niiler*, 2002; *Jakobsen et al.*, 2003]. Generally, the bottom flow and the surface flow were aligned. A noticeable exception was in the Lofoten Basin, where the model gave cyclonic bottom flow and anticyclonic surface flow, the latter agreeing with drifter measurements [*Jakobsen et al.*, 2003]. Using the assumption of a constant transport in the western branch of the NwAC, *Orvik* [2004] found the reverse deep flow to be the cause for the large deepening of the Atlantic layer in the Lofoten Basin area.

Skagseth et al. [2004] have used the combined Topex–Poseidon and ERS 1/2 weekly mean sea-level anomaly (SLA) data [*Ducet et al.*, 2000] and current meter measurements to look at coherent signals along the continental slope from the Irish–Scottish shelf towards the Fram Strait. They first compared the weekly mean geostrophic velocity derived from the SLA data with direct measurements at the Svinøy Section and found an overall correlation of 0.55 over a 7-year period. The discrepancy was mainly due to high-frequency mesoscale variability in the direct measurements that was not resolved in the altimeter data. They further found the flow along the slope to be forced by anomalies in SLP resembling the NAO pattern, with no phase lag.

Using the same dataset as *Skagseth et al.* [2004], we have calculated the phase of the annual cycle of the SLA (Figure 17a). A strong topographic control on the SLA can be seen from the bathymetry dependency of the time of annual maximum. For instance, note that the shallow areas around Iceland and on the continental shelf from Ireland to the Barents Sea and Fram Strait have SLA maximum later in the year than the central parts of the Nordic Seas and the North Atlantic. This feature cannot be fully explained by the annual cycle in heat fluxes (not shown) but must be interpreted as the combination of a pileup of water due to anomalous Ekman transports and the strong f/H control of the flow. The same conclusion can be drawn from an extension of the work by *Nøst and Isachsen* [2003], who find that the main difference between summer and winter circulation is a strengthening of the flow along the f/H contours during winters (Nøst, personal communication, 2004).

In Figure 17b the correlation between the Lisbon–Iceland NAO Index [*Hurrell*, 1995] and the SLA data is shown. While the seasonal cycle in both data sets has been removed, nothing has been done to eliminate the steric effect on the SLAs, which will contribute to the high elevations in the North Sea, along the Norwegian coast, and in the Barents Sea due to less heat loss and more precipitation (Figures 9d and 10d). Highest correlations between the SLA and the NAO Index (Figure 17b) are found west of Denmark and in the Baltics: Here correlations exceed 0.4 and a unit increase in NAO Index corresponds to a 5-cm increase in sea level (not shown). The correlation pattern is consistent with an increased cyclonic wind-stress curl over the Nordic Seas, where divergence in the Ekman transports makes a negative SLA in the central basin and a positive SLA along the surrounding coast lines. Thus, associated with stronger than normal westerlies, the northward flow of AW should be stronger than normal. This fits with measurements from the Svinøy Section, where *Orvik et al.* [2001] revealed a strong correlation between the NAO Index and the eastern branch of the NwAC, even on intra-seasonal time scales.

While surface drifters, altimeter data, and direct current meter measurements have revealed substantial information about the upper ocean circulation, much less is known about the intermediate and abyssal circulation of the Nordic Seas. Much of the knowledge we have about transport rates and transport times of the intermediate Nordic Seas, originate from the SF_6 release experiment in the central Greenland Sea in 1996 [*Watson et al.*, 1999]. Using an advective–diffusive model applied on output from a numerical model experiment, Eldevik et al. [this volume] managed to capture the main features of the observations. According to their model, the SF_6-enriched water reached the Faroe Bank overflow 2 years after the release and the Denmark Strait one year later, in good agreement with observations. The authors further

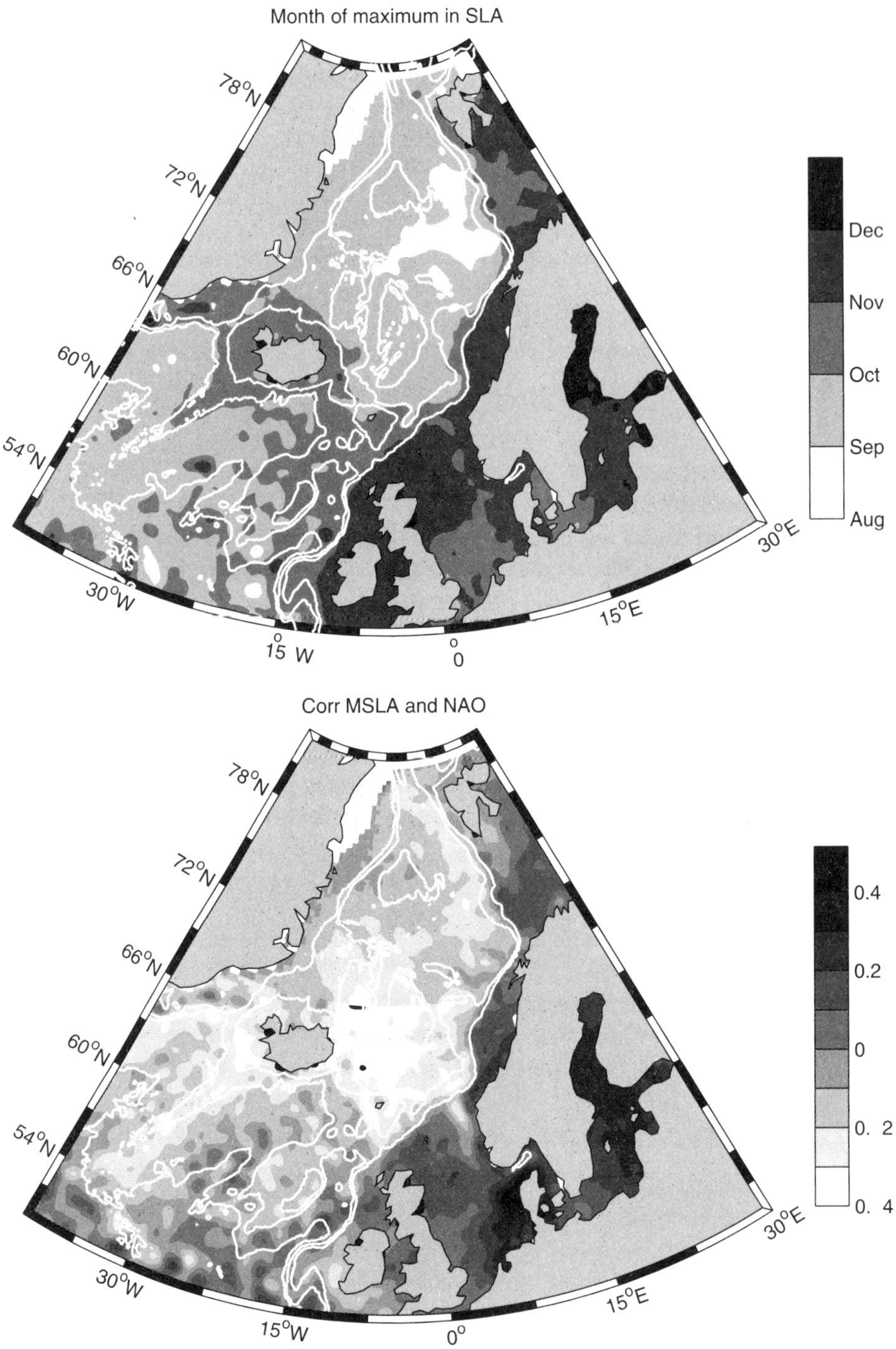

Figure 17. Combined Topex–Poseidon and ERS 1/2 sea-level anomaly (SLA). (Upper) The month of maximum climatological SLA, and (lower) the SLA correlated with the Lisbon–Stykkisholmur (Hurrell) NAO Index. The monthly NAO Index has been interpolated to weekly values, and both the NAO and the SLA data have been deseasoned prior to the correlation. The white contours indicate f/H values corresponding to 500-, 1500-, and 3000-m depths at 60°N.

point to the huge differences in the intermediate circulation patterns during the different periods in the 50-year model run and believe the differences are caused by variations in the large-scale atmospheric forcing. During the SF_6 period after September 1996, the tracer went straight to the FSC following the Jan Mayen Current. In contrast, for other periods the tracer could have taken a longer route with the East Greenland and East Icelandic Currents before reaching the Faroe Bank overflow; this flow would have used more than twice the time taken during the SF_6 experiment.

While there are indications of variations in the pathways leading to the overflow sites, there are also reports of changes in the strength of the overflow. Using OWSM data, *Hansen et al.* [2001] showed a 50-year decrease in the height of the $\sigma_\theta = 28$ density level and, by analytical arguments and comparison with direct measurements of the Faroe Bank overflow, deduced a 20% reduction in the overflow during the same period. Measurements by direct methods do not show a reduction in the inflow of AW, but such records are of shorter duration than the above-mentioned studies. For instance, 7 years of direct measurements in the eastern branch of the inflowing AW at the Svinøy Section show no indications of a reduction of the transport of AW into the area [*Orvik et al.*, 2001; *Orvik and Skagseth*, 2003a]. Combined with the above, this should imply that a reduction of the overflow could be associated with a reduced transport in the western branch that crosses the IFR. It also fits with the expected responses to the increased westerlies, that more water should move east towards the slope current off the Irish–Scottish shelf.

The results of model simulations are not conclusive, and the relatively weak changes in the thermohaline structures and deep circulation may often be difficult to distinguish from model drifts and other sources of errors. Nonetheless, hindcast simulations with an isopycnic ocean model integrated from 1948 show a reduction in the inflow of AW that quantitatively agrees with the estimates made by *Hansen et al.* [2001], and which points to the reduced inflow over the IFR as the cause [*Nilsen et al.*, 2003].

5.4. Nordic Seas Hydrography

Due to traditions going back to the pioneering works in the late 19th century [see Blindheim and Østerhus, this volume, for references], the Nordic Seas are among the most well sampled regions in the world. During the last few decades, several pronounced and far-reaching changes in the hydrography have been related to the low-frequency variability in the atmospheric circulation, with the record weak westerlies in the 1960s and the record strong westerlies in the '90s as the two extremes. We will here limit the discussion to the observed warming of the AW, the freshening of the intermediate water, and the warming of the deep water.

Furevik [2000, 2001] used both satellite-derived SST data and direct measurements to study the transport of heat anomalies from the FSC towards the Barents Sea and Arctic Ocean during the '80s and '90s (Figure 18). Two warm (W1 and W2) and one cold (C1) period were discussed. *Furevik* [2001] concluded that while the warm and cold anomalies of the '80s were advected with the AW from the FSC and through the Nordic Seas, the stronger anomaly during the early '90s was a response to reduced heat loss in the Nordic Seas. His arguments were based on the observations that firstly this heat anomaly is barely seen in the inflowing region but increases in strength towards north (Figure 18a), and secondly in the Barents Sea Opening the heat anomaly starts in the surface and gradually penetrates deeper (Figure 18b). Model studies by *Karcher et al.* [2003] mainly support this interpretation but also point to an increased inflow of AW to the Nordic Seas as a cause for the warm anomaly of the '90s.

A third factor may also be weaker transport in the western branch and stronger transport in the eastern branch of the NwAC, as indicated by an eastward [*Blindheim et al.*, 2000] and southward [*Furevik et al.*, 2002] displacement of the $S = 35$ isohaline in the southern Norwegian Sea. Thus, simple theory involving topographical steering implies that less AW should be recirculated in the Nordic Seas and furthermore suggests that an eastward displacement of the AW should force a relatively larger portion of the water through the Barents Sea. This has been shown in model runs with both idealistic [*Furevik*, 1998] and real [*Zhang et al.*, 1998] topography. Further to the north, the anomalous warm water that entered through the Fram Strait and the Barents Sea in the early '90s, and a shift to a more cyclonic atmospheric circulation over the Arctic, were apparently responsible for a significant warming and expansion of the AW layer in the Arctic Ocean [*Carmack et al.*, 1995, 1997; *Swift et al.*, 1997; *Morison et al.*, 1998; *Grotefendt et al.*, 1998; *Karcher et al.*, 2003].

Below the surface layers, the Nordic Seas have experienced a widespread freshening since the '60s, with record weak westerlies, to the '90s, when the westerlies have been the strongest measured [*Blindheim et al.*, 2000; *Dickson et al.*, 2002]. Averaged over the depth interval 400–1500 m at the OWSM, the magnitude of this freshening is 0.023, corresponding to mixing 0.7 m of freshwater into the water column. Because the depth of the freshening reaches far below the sill depths of the overflow water to the North Atlantic, the freshening signal is therefore being exported to the North Atlantic. From the 1960s to 2000 the salinity of the overflow water typically decreased by 0.01/decade [*Turrell et al.*, 1999; *Dickson et al.*, 2002]. Over a 40-year interval this corresponds to adding 1.2 m of freshwater to a 1000-m-thick water column. While sinking, the overflow water mixes

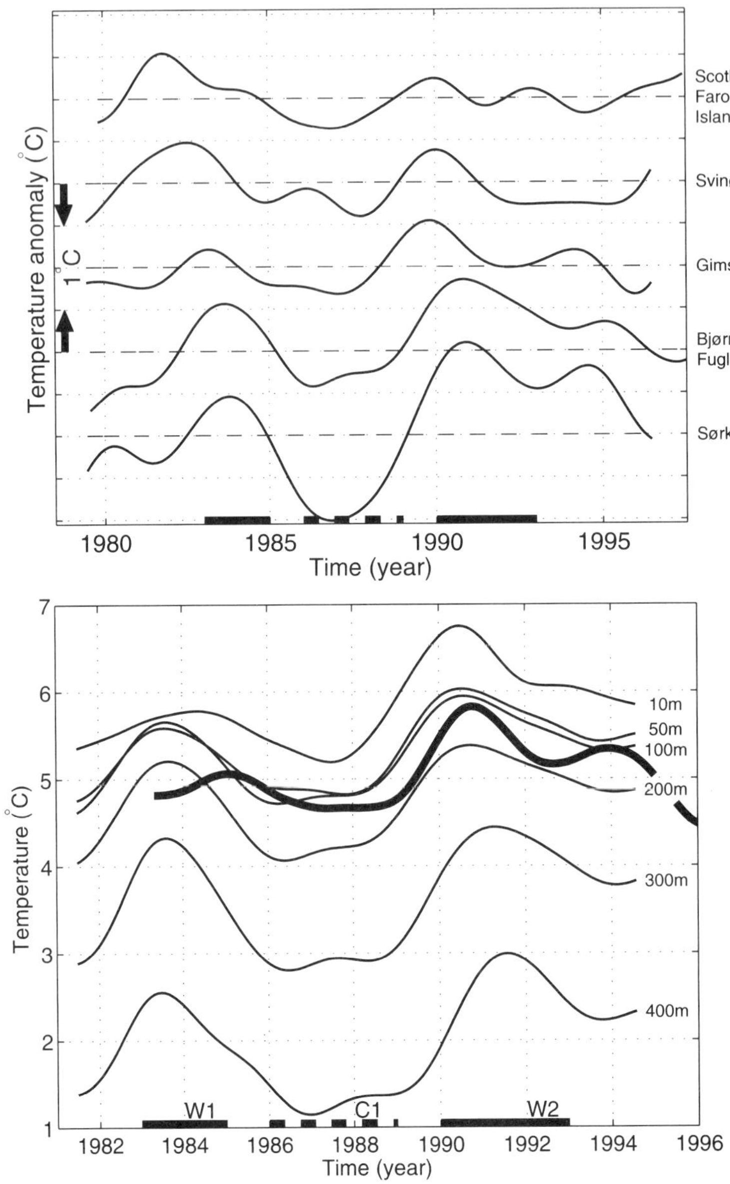

Figure 18. (Upper) Core temperatures in the flow of AW at five regular conductivity temperature depth (CTD) sections from the FSC (upper curve) to the Southcape (76°N) Section (lower curve). (Lower) Average temperatures at the surface (bold curve) and at six different depths (thin curves) in the Barents Sea Opening. The figures are adapted from *Furevik* [2001].

with ambient water, which is itself freshening, thus preserving or even amplifying the freshening signal downstream into the Labrador Sea and eventually into the deep western boundary current towards the south.

It is not clear where the excess freshwater is coming from, but several factors have been proposed, including more ice and freshwater export from the Arctic [*Vinje*, 2001; *Zhang et al.*, 2003; *Kwok et al.*, 2004], more precipitation over the region (see Figure 10d), and internal redistribution of water masses [*Blindheim et al.*, 2000] where AW is replaced by PW in the central Nordic Seas. In a recent study, *Curry et al.* [2003] used the Levitus data [e.g. *Levitus et al.*, 2000] for the Atlantic Ocean to study the changes in ocean salinity from the '60s to the '90s in a transect from 50°S to 60°N. The result of this study was a deep-reaching freshening in both the southern and northern high latitudes and more saline water occupying the upper 100m at low latitudes. This is a strong indication of an enhanced hydrological cycle, with more evaporation in the low latitudes and more precipitation in the high latitudes—which in most climate models are the predicted responses to global warming [e.g. *Cubasch et al.*, 2001].

Since the '60s, when the temperatures in the deep water of the Nordic Seas reached a minimum, the deep-water formation in the Greenland Sea has been strongly reduced or absent [*Dickson et al.*, 1996], and the Greenland Sea Bottom Water has gradually warmed [*Budéus et al.*, 1998], at least partly due to inflow of warmer water from the Eurasian Basin [*Meincke et al.*, 1997]. With a few years time lag, the warming was observed at 2000m deep at OWSM, where the signal gradually rose in the water column [*Østerhus and Gammelsrød*, 1999]. These two authors also demonstrated a reversal of the deep flow between the Greenland Sea and the Norwegian Sea and suggested this was a consequence of a reversal of the density gradient between the two basins.

6. DISCUSSION: NORDIC SEAS AND CLIMATE CHANGE

Observations have shown that during the last decades the atmospheric westerlies have strengthened, the deep-water formation in the Greenland Sea has weakened, the sea-ice extent and thickness have been strongly reduced, and the deep water of the North Atlantic has become fresher. And we may ask: Are we witnessing the impacts of global change, or are the recent observations just low-frequency manifestations of stochastic noise in the complex and still little-understood atmosphere–ice–ocean climate system?

6.1. Changes in the Surface Ocean

During the period with NCEP/NCAR reanalysis data, the strength of the westerlies has undergone a change from record weak conditions in the '60s to record strong conditions in the '90s. The changes in heat fluxes and ocean surface properties over the same time interval can be assessed by comparing the two 15-year periods 1958–72 and 1988–2002 (Figure 19). The pictures that emerge are close to what is found by simply multiplying the changes in the NAO Index by the regression patterns shown in the previous sections. In general, this includes a more cyclonic winter air flow over the Nordic Seas region (compare with Figure 5b), increased wind stress (Figure 8b), reduced heat loss to the atmosphere (Figure 9d), and increased precipitation (Figure 10d). Along the east coast of Greenland, increased northerly winds and less sea-ice have resulted in increased heat and freshwater fluxes to the atmosphere. At the ocean surface, the same period reveals a significant decrease in SST to the southwest and a warming mainly confined to the Barents Sea, the North Sea, and the Baltic Sea to the east. Furthermore, there is a substantial retreat of the sea-ice edge both in the Greenland and Barents Sea areas.

While most of the changes in the surface conditions can be explained as local responses to the variations in the atmospheric forcing, the SST and sea-ice concentration fields also indicate that changes in the ocean circulation play an active role. Two noticeable areas are to the east and west of Iceland, where there has been significant cooling despite reduced heat loss to the atmosphere. Here the AW seems to be replaced by water of polar origin, probably a result of decreased inflow of AW in the Irminger and IFR branches and perhaps also an increased outflow from the Arctic. This view is also supported by many observations and model results cited in Sections 5.1 and 5.2. A contributing factor east of Iceland may also be anomalous Ekman transports (e.g. Figure 13) associated with the strengthened atmospheric circulation (Figure 19b). The SST plot (Figure 19e) further shows the strong warming in the northern extensions of the NwAC—both in the branch flowing through the Barents Sea and in the branch going into the Fram Strait—that eventually is tied to the recent warming of the Arctic.

6.2. Changes in the Deep Ocean

Being one of the main formation areas for the deep water that ventilates world oceans, the Nordic Seas and the water mass conversion taking place there may play a pivotal role for the climate far outside its boundaries. One of the many issues that have been discussed is the role of the deep-water formation in the Nordic Seas when it comes to the strength of the NAC and the inflow of AW to the Nordic Seas and the general heat transport towards northern Europe.

Early experiments with simplified climate models indicated that we could expect a total shutdown of the AMOC in a future climate with CO_2-induced global warming [*Rahmstorf and Ganopolski*, 1999], while most state-of-art fully 3-dimensional climate models today indicate a more modest but still significant reduction in the AMOC [*Cubasch et al.*, 2001]. The main reason is a strengthened hydrological cycle, which gives increased precipitation in the high latitudes. This, in addition to generally warmer surface water, reduces the density and stabilizes the water columns in the sinking regions. In most models the reduction in the strength of the AMOC starts towards the end of the previous century.

Since the '60s, the Nordic Seas have undergone changes in hydrography that are unprecedented in instrumental records. From a regime with strong and deep-reaching Greenland Sea convection during wintertime, producing a very cold and dense bottom water, the convection is now shallow and occurs irregularly. The result has been a warming and a lowering of the densities of the Nordic Seas, a decrease in the density difference across the GSR, and a possible 20% reduction in

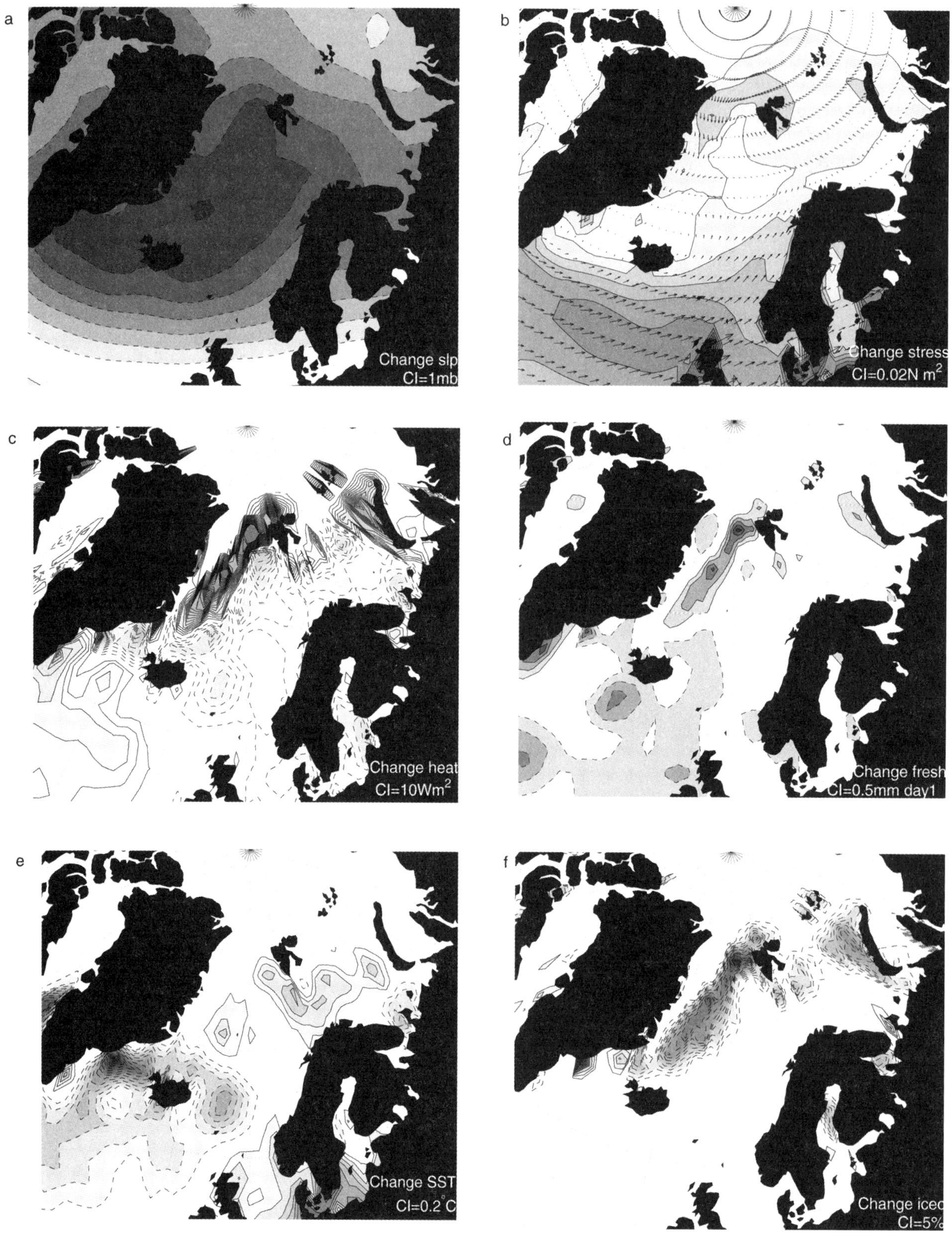

Figure 19. Change between the 15-year averages for 1958–72 and 1988–2002 for (a) SLP, (b) wind stress, (c) total heat flux, (d) total freshwater flux, (e) SST, and (f) sea-ice concentration. Ekman transports of heat and freshwater (Figures 13 and 14) are not included in (c) and (d), respectively. Solid lines are positive, dashed lines negative.

the Faroe Bank overflow. In addition, there has been a pronounced freshening of the intermediate depths of the Nordic Seas. Seen as a long-term freshening trend in the overflow water, this has resulted in a freshening of most of the deep North Atlantic. It may thus seem that the observational evidence supports the climate models when they predict a reduction in the AMOC during global warming and that such a reduction is already taking place.

Recent results, however, have led some to question this conclusion. Using the Hadley Centre's coupled climate model, *Wu et al.* [2004] found a freshening in their ensemble run that resembled the observed freshening of the North Atlantic and Nordic Seas, but without this being accompanied by a weaker AMOC. The simulated AMOC was in fact strengthened during the same period. Furthermore, the newest measurements from the Faroe Bank overflow show an increased export of dense overflow water during the last 2 years (Bogi Hansen, personal communication, 2004), calling into question the significance of the previously reported downward trend.

6.3. Nordic Seas in a Future Climate

With all the changes that have been observed in the Nordic Seas climate during the recent decades, and the growing evidence that the changes may be linked to the greenhouse gas forcing through a strengthening of the westerlies, it is tempting to speculate about the future state of the Nordic Seas.

Based on the literature reviewed in this paper, our first and strongest working hypothesis is that most of the changes we have seen in the Nordic Seas's circulation and hydrography since the '60s and onward are an inevitable response to the deepening of the Icelandic low and the associated strengthening of the westerlies [e.g. *Marshall et al.*, 2001b; *Visbeck et al.*, 2003]. Our second hypothesis is that the decadal to multidecadal variability in the atmospheric circulation is linked to the increased concentration of CO_2 and other greenhouse gases, due either to warmer tropical waters [e.g. *Hoerling et al.*, 2001] or to increased north–south temperature contrasts as a response to a warmer troposphere and a colder stratosphere [e.g. *Shindell et al.*, 1999]. Our third hypothesis is that the northeastward extension of the storm tracks, and thus the eastward displacement of the northern centre of action, is linked to the strengthening of the westerlies [e.g. *Cassou et al.*, 2004]. Our fourth hypothesis is that the temporal and spatial trends seen in the NAO in the second half of the 20th century will be sustained or even amplified during the years ahead [e.g. *Fyfe et al.*, 1999; *Ulbrich and Christoph*, 1999].

In order to assess what the consequences will be for the Nordic Seas climate, we may either rely on integrations with coarse climate models and successive dynamical downscaling with finer models, or examine what we already know about the processes that are important for the present-day Nordic Seas climate and discuss how these processes may be affected by climate change. We will here take the latter approach.

The Nordic Seas ocean climate will be determined by the local atmosphere–ocean fluxes; the volume and properties of the warm, saline water entering from the south; and the cold, fresh water entering from the north. If the long-term trend in the NAO is sustained, a reduced heat loss to the atmosphere and increased precipitation and river runoff will work to stabilize the water column and thus reduce the convection depths within the Nordic Seas. Together with the observations of the relatively strong stratification of the present-day Greenland Sea compared with that of the '60s, it is unlikely that the open-ocean deep-water formation will recover. An exception is for a scenario with many successive cold winters or with more advection of saline water into the region. Thus, we may anticipate a continuation of the warming of the deep water, the freshening of the intermediate water, and perhaps an even further reduction in the rate of overflow.

Changes in the atmospheric forcing will most likely affect all mechanisms that are maintaining the relatively stable inflow of warm AW to the Nordic Seas, and competitive effects may occur. Firstly, a reduced ice cover will mean that more water is exposed to winter cooling, which can lead to a higher production of deep water both by brine release on the shelves and by open ocean convections in new ice-free areas [e.g. *Aagaard and Carmack*, 1989]. Secondly, a strengthened hydrological cycle [*Curry et al.*, 2003] in combination with thinner ice [e.g. *Rothrock et al.*, 2003] will increase the liquid freshwater export through the Fram Strait. This can have large consequences for the inflow because, by mixing with the water below, more freshwater will strengthen the estuarine circulation and thus also strengthen the compensating inflow of AW [see *Stigebrandt*, 1981]. And thirdly, with a stronger wind-stress curl over the North Atlantic, we may expect a stronger inflow of AW to the Nordic Seas. Based on model results and observational data, the increase will likely occur through the FSC, while there may actually be a reduced inflow over the IFR [e.g. *Mork and Blindheim*, 2000; *Nilsen et al.*, 2003].

On long time scales, the local buoyancy effects of the strengthened atmospheric circulation may be reduced or perhaps canceled if the northward flow of AW carries colder and more saline water to the north. Two factors that may contribute to this are the observed enhanced cooling over the north Atlantic during positive NAO years [e.g. *Delworth and Dixon*, 2000; *Marshall et al.*, 2001b] and the increased evaporation over the tropical regions due to a spin-up of the hydrological cycle [e.g. *Curry et al.*, 2003].

To conclude, we find it likely that recent trends in the Nordic Seas climate are strongly tied to global climate change and that these changes will persist or even amplify in the decades ahead of us. We further believe that the inflow of AW into the Nordic Seas is a robust feature driven by the combined effect of deep-water formation north of the GSR, estuarine circulation effects, and forcing from the wind. The effect of global warming is likely to favor some and disfavor other mechanisms, and in total we find it unlikely that we are facing a total shutdown of the AMOC or other dramatic changes in the ocean circulation in the near future.

7. SUMMARY AND CONCLUDING REMARKS

In this paper the large-scale atmospheric circulation impacts on the Nordic Seas ocean climate have been reviewed. The Icelandic low and the Azores high, which give prevailing southwesterly winds to the southeastern part of the Nordic Seas and easterly and northeasterly winds in the northwestern part of the region, dominate the climate in the North Atlantic region. During recent decades the large-scale atmospheric circulation has undergone a low-frequency shift from record weak westerlies in the '60s to record strong westerlies in the '90s. Associated with this shift, the changes in atmospheric momentum, heat, and freshwater forcing seem to be responsible for many of the astonishingly large changes observed in the Nordic Seas' circulation and hydrography over the same period. Both local and remote processes seem to play an active role in the atmospheric forcing of the ocean.

The strengthening of the westerlies is concurrent in time with the strong increase in global mean temperatures since the '60s, which is likely linked to the increased concentrations of atmospheric greenhouse gases. The connection between global warming and strengthened westerlies is still debated. Recent dynamical considerations, however, and model experiments have pointed to tropical SSTs, particularly SSTs in the Indian Ocean, as a likely source for the energy causing the spin-up of the system. If this is the case, and if the increasing concentration of atmospheric CO_2 continues to increase the tropical temperatures, we may expect that the recent trend towards stronger westerlies will be sustained. If so, the most extreme conditions observed in the ocean climate of the North Atlantic, Nordic Seas, and the Arctic during the past decades may be only previews of the years lying ahead of us.

Acknowledgments. The study has been supported by the Research Council of Norway through the NoClim (T.F.) and RegClim (J.E.Ø.N.) projects, and by the G. C. Rieber Foundations (J.E.Ø.N.). The work was completed during a visit of T.F. at the Danish Meteorological Institute in Copenhagen, which he thanks for hospitality and good working environment. The authors also thank John Walsh and two anonymous reviewers for their constructive comments on the manuscript. The NCEP/NCAR Reanalysis data are provided by the NOAA-CIRES Climate Diagnostics Center, Boulder, Colorado, USA, from their website at http://www.cdc.noaa.gov/. This is contribution no. 68 from the Bjerknes Centre for Climate Research.

REFERENCES

Aagaard, K., Wind-driven transports in the Greenland and Norwegian seas, *Deep-Sea Res.*, 17(2), 281-291, doi: 10.1016/0011-7471(70)90021-5, 1970.

Aagaard, K., and E. Carmack, The role of sea ice and other fresh water in the arctic circulation, *J. Geophys. Res.*, 94 (C10), 14,485-14,498, 1989.

Aagaard, K., and L.K. Coachman, East Greenland Current north of Denmark Strait: Part I, *Arctic*. 21(3), 181-200, 1968.

Alexander. M.A., U.S. Bhatt, J.E. Walsh, M.S. Timlin, J.S. Miller, and J.D. Scott, The atmospheric response to realistic Arctic sea ice anomalies in an AGCM during winter, *J. Climate*, 17(5), 890-905, doi: 10.1175/1520-0442(2004)017, 2004.

Ambaum. M.H.P., and B.J. Hoskins, The NAO troposphere-stratosphere connection, *J. Climate*, 15(14), 1969-1978. 2002.

Ambaum, M.H.P., B.J. Hoskins, and D.B. Stephenson, Arctic Oscillation or North Atlantic Oscillation?. *J. Climate.*, 14(16), 3495-3507, 2001.

Anderson, D.L.T., and A.E. Gill, Spin-up of a stratified ocean, with application to upwelling. *Deep-Sea Res.*, 22(9), 583-596, 1975.

Anderson, D.L.T., and P.D. Killworth, Spin-up of a stratified ocean, with topography, *Deep-Sea Res.*, 24(8), 709-732, 1977.

Baldwin, M.P., and T.J. Dunkerton, Downward propagation of the Arctic Oscillation from the stratosphere to the troposphere. *J. Geophys. Res.*, 104 (D24), 30,937-30,946, doi:10.1029/1999JD900445, 1999.

Baldwin. M.P., D.B. Stephenson, D.W.J. Thompson. T.J. Dunkerton. A.J. Charlton, and A. O'Neill, Stratospheric memory and extended-range weather forecasts, *Science*, 301(5633), 636-640, doi: 10.1126/science. 1087143, 2003.

Bengtson, L., V.A. Semenov, and O.M. Johannessen, The early century warming in the Arctic - a possible mechanism, *J. Climate*, in press, 2004.

Bengtson, L., E. Roeckner, and M. Stendel, Why is the global warming proceeding much slower than expected?, *J. Geophys. Res.*, 104(D4), 3865-3876, doi: 10.1029/1998JD200046, 1999.

Bjerknes, J., Synoptic survey of the interaction of sea and atmosphere in the North Atlantic, *Geophys. Norv.*, 24(3), 115-145, 1962.

Bjerknes, J., Atlantic air-sea interaction, *Adv. Geophys.*, 10, 1-82, 1964.

Blindheim, J., V. Borovkov, B. Hansen, S.A. Malmberg, W.R. Turrell, and S. Østerhus, Upper layer cooling and freshening in the Norwegian Seas in relation to atmospheric forcing, *Deep-Sea Res. I*, 47(4), 655-680, doi: 10.1016/S0967-0637(9P)00070-9, 2000.

Bower, A.S., P.L. Richardson, H.D. Hunt, T. Rossby, M.D. Prater, H.-M. Zhang, S. Anderson-Fontana, P. Perez-Bruinius, and P. Lazarevich, Warm-water pathways in the Subpolar North Atlantic: An overview of the ACCE RAFOS float programme, *Int. WOCE Newsl.*, 38. 14-16, 2000.

Budéus, G., W. Schneider, and G. Krause, Winter convective events and bottom water warming in the Greenland Sea. *J. Geophys. Res.*, 103 (C9), 18,513-18,527, doi: 10.1029/98JC01563, 1998.

Carmack, E.C., R.W. Macdonald, R.G. Perkin, F.A. McLaughlin, and R.J. Pearson, Evidence for warming of Atlantic Water in the southern Canadian Basin of the Arctic Ocean: Results from the Larsen-93 expedition, *Geophys. Res. Lett.*, 22(9), 1061-1064, doi: 10.1029/95GL00808, 1995.

Carmack, E.C., *et al.*, Changes in temperature and tracer distributions within the Arctic Ocean: Results from the 1994 Arctic Ocean section, *Deep-Sea Res. II*, 44(8), 1487-1493, doi: 10.1016/S0967-0645(97)00056-8, 1997.

Carr, M., and H. Rossby, Pathways of the North Atlantic Current from surface drifters and subsurface floats, *J. Geophys. Res.*, 106(C3), 4405-4419, doi: 10.1029/2000JC900106, 2001.

Cassou, C., L. Terray, J.W. Hurrell, and C. Deser, North atlantic winter climate regimes: Spatial asymmetry, stationarity with time, and oceanic forcing, *J. Climate*, 17(5), 1055-1068, doi:10.1175/1520-0442(2004)017, 2004.

Cubasch, U., G. Meehl, G. Boer, R. Stoffer, M. Dix, A. Noda, C. Senior, S. Raper, and K. Yap, *Projections of Future Climate Change*, chap. 9, pp. 525-582, Cambridge University Press, Cambridge, United Kingdom and New York. NY, USA, 2001.

Curry, R., R.R. Dickson, and I. Yashayaev. A change in the freshwater balance of the Atlantic Ocean over the past four decades, *Nature*. 426(6968), 826-829, doi: 10.1038/nature02206, 2003.

Czaja, A., A.W. Robertson, and T. Huck, The role of Atlantic ocean atmosphere coupling in affecting North Atlantic Oscillation variability, in *The North Atlantic Oscillation: Climatic Significance and Environmental Impact, Geophysical Monograph Series*, vol. 134, edited by J. W. Hurrell, Y. Kushnir, G. Ottersen, and M. Visbeck, pp. 147-172, AGU, doi:10.1029/134GM07, 2003.

Delworth. T.L., and K.W. Dixon, Implications of the recent trend in the Arctic/North Atlantic Oscillation for the North Atlantic thermohaline circulation, *J. Climate*, 13(21), 3721-3727, doi: 10.1175/1520-0442(2000)013<3721:IOTRTI>2.0.CO;2, 2000.

Deser, C., On the teleconnectivety of the “Arctic Oscillation”, *Geophys. Res. Lett.*, 27(6), 779-782, doi: 10.1029/1999GL010945, 2000.

Dickson, B., I. Yashayaev. J. Meincke, B. Turrell, S. Dye, and J. Holfort, Rapid freshening of the deep North Atlantic Ocean over the past four decades, *Nature*, 416(6883), 832-837, doi:10.1038/416832a, 2002.

Dickson, R., J. Lazier, J. Meincke. P. Rhines, and J. Swift, Long-term coordinated changes in the convective activity of the North Atlantic, *Prog. Oceanogr.*, 38(3), 241-295, doi: 10.1016/S0079-6611(97)00002-5, 1996.

Dickson, R., T. Osborn, J. Turrell, J. Meincke, J. Blindheim, B. Adlandsvik, T. Vinje, G. Alekseev, and W. Maslowski, The Arctic Ocean response to the North Atlantic Oscillation. *J. Climate*, 13(15), 2671-2696, doi:10.1175/1520-0442(2000)013<2671:TAORTT>2.0.CO;2, 2000.

Dickson, R.R., J. Meincke, S.-A. Malmberg, and A.J. Lee, The “Great Salinity Anomaly” in the northern North Atlantic 1968-1982. *Prog. Oceanogr.*, 20(2), 103-151. doi: 10.1016/0079-6611(88)90049-3, 1988.

Ducet, N., P.Y. LeTraon, and G. Reverdin, Global high-resolution mapping of ocean circulation from Topex/Poseidon and ERS-1 and -2, *J. Geophys. Res.*, 105(C8), 19,477-19,498, doi:10.1029/2000JC900063, 2000.

Flatau, M.K., L. Talley, and P.P. Niiler, The North Atlantic Oscillation, surface current velocities, and SST changes in the Subpolar North Atlantic, *J. Climate*. 16(14), 2355-2369, doi:10.1175/2787.1, 2003.

Furevik, T., On the Atlantic Water flow in the Nordic Seas: Bifurcation and variability, Dr. scient. thesis in physical oceanography, Geophysical Institute. Univ. Bergen, Allégt. 70, 5007 Bergen, Norway, 1998.

Furevik, T., On anomalous sea surface temperatures in the Nordic Seas. *J. Climate.*, 13(5), 1044-1053. doi:10.1175/1520-0442(2000)013<1044:OASSTI>2.0.CO;2, 2000.

Furevik, T., Annual and interannual variability of Atlantic Water temperatures in the Norwegian and Barents Seas: 1980-1996, *Deep-Sea Res. I*, 48(2), 383-404, doi: 10.1016/S0967-0637(00)00050-9, 2001.

Furevik. T., M. Bentsen, H. Drange, J.A. Johannessen, and A. Korablev, Temporal and spatial variability of the sea surface salinity in the Nordic Seas, *J. Geophys. Res.*, 107(C12), 8009, doi:10.1029/2001JC001118, 2002.

Furevik, T., M. Bentsen, H. Drange, I. Kindem, N. Kvamstø, and A. Sorteberg, Description and evaluation of the Bergen Climate Model: ARPEGE coupled with MICOM, *Clim. Dynam.*, 21(1), 27-51, doi:10.1007/s00382-003-0317-5, 2003.

Fyfe, J. C., G. J. Boer, and G. M. Flato. The Arctic and Antarctic Oscillations and their projected changes under global warming, *Geophys. Res. Lett.*, 26(11), 1601-1604, doi:10.1029/1999GL900317, 1999.

Gascard, J.-C., C. Kergomard, P.-F. Jeannin, and M. Fily, Diagnostic study off the Frarn Strait Marginal Ice Zone during summer from 1983 and 1984 Marginal Ice Zone Experiment. Lagrangian observations. *J. Geophys. Res.*, 93(C4), 3613-3641, doi:10.1029/88JC01488, 1988.

Gill, A. E., *Atmosphere-Ocean Dynamics*, 662 pp., Academic Press, London, 1982.

Gillett, N.P., H. Graf, and T.J. Osborn, Climate change and the North Atlantic Oscillation, in *The North Atlantic Oscillation: Climatic. Significance and Environmental Impact, Geophysical Monograph Series*, vol. 134, edited by J. W. Hurrell, Y. Kushnir, G. Ottersen, and M. Visbeck, pp. 173-192, AGU, doi:10.l029/134GM09, 2003.

Grotefendt, K., K. Logemann, D. Quadfasel, and S. Ronski, Is the Arctic Ocean warming?, *J. Geophys. Res.*, 103(C12), 27,679-27,687, doi:10.1029/98JC02097, 1998.

Hansen, B., and R. Kristiansen, Long-term changes in the Atlantic water flowing past the Faroe Islands, *ICES CM 1994/S,* 4, 1-16, 1994.

Hansen, B., and S. Østerhus, North Atlantic-Nordic Seas exchanges, *Prog. Oceanogr.*, 45(2), 109-208, doi: 10.1016/S0079-6611(99)00052-X, 2000.

Hansen, B., W.R., Turrell, and S. Østerhus, Decreasing overflow from the Nordic Seas into the Atlantic Ocean through the Faroe Bank Channel since 1950, *Nature*, 411(6840), 927-930, doi: 10.1038/35082034, 2001.

Hansen, B., S. Østerhus, H. Hátún, R. Kristiansen, and K.M.H. Larsen, The Iceland-Faroe inflow of Atlantic water to the Nordic Seas, *Prog. Oceanogr.*, 59(4), 443-474, doi: 10.1016/j.pocean.2003.10.003, 2003.

Hilmer, M., and T. Jung, Evidence for a recent change in the link between the North Atlantic Oscillation and Arctic sea ice export, *Geophys. Res. Lett*, 27(7), 989-992, doi: 10.1029/1999GL010944, 2000.

Hoerling, M., J. Hurrell, T. Xu, G. Bates, and A. Phillips, Twentieth century north atlantic climate change, part II: Understanding the effect of indian ocean warming, *Clim. Dynam.*, 23(3-4), 391-405, doi:10.1007/s00382-004-0433-x, 2004.

Hoerling, M.P., J.W. Hurrell, and T. Y. Xu, Tropical origins for recent North Atlantic climate change, *Science*, 292(5514), 90-92. doi:10.1126/science.l058582, 2001.

Holliday, N. P., R.T. Pollard, J.F. Read, and H. Leach, Water mass properties and fluxes in the Rockall Trough, 1975-1998, *Deep-Sea Res. I*, 47(7), 1303-1332, doi: 10.1016/S0967-0637(99)00109-0, 2000.

Hurrell, J., Decadal trends in the North Atlantic Oscillation: Regional temperatures and precipitation, *Science*, 269(5224), 676-679, 1995.

Hurrell, J., M. Hoerling, A. Phillips, and T. Xu, Twentieth century north atlantic climate change. part I: Assessing determinism, *Clim. Dynam.*, doi:10.1007/s00382-004-0432-y. 2004.

Hurrell, J. W., Y. Kushnir, G. Ottersen, and M. Visbeck, An overview of the North Atlantic Oscillation, in *The North Atlantic Oscillation: Climatic Significance and Environmental Impact, Geophysical Monograph Series*, vol. 134, edited by J.W. Hurrell, Y. Kushnir, G. Ottersen, and M. Visbeck, pp. 1-36, AGU, doi:10.1029/134GM01, 2003.

Jakobsen, P., M. Ribergaard. D. Quadfasel, T. Schmith, and C. Hughes, Near-surface circulation in the northern North Atlantic as inferred from Lagrangian drifters: Variability from the mesoscale to interannual, *J. Geophys. Res.*, 108(C8), 3251, doi:10.1029/2002JC001554, 2003.

Johannessen, O.M., E.V. Shalina, and M. Miles, Satellite evidence for an Arctic sea ice cover in transformation, *Science*, 286(5446), 1937-1939, doi: 10.1126/science.286.5446.1937, 1999.

Johannessen, O.M., *et al.*, Arctic climate change: Observed and modelled temperature and sea-ice variability, *Tellus,* 56A(4), 328-341, doi:10.1111/j.l600-0870.2004.00060.x, 2004.

Jones, P.D., T. Jonsson, and D. Wheeler, Extension to the North Atlantic Oscillation using early instrumental pressure observations from Gibraltar and South-West Iceland, *Int. J. Climatol.*, 17(13), 1433-1450, doi: 10.1002/(SICI)1097-0088(19971115)17:13<1433::AID-JOC203>3.0.CO;2-P, 1997.

Josey, S., A comparison of ECMWF, NCEP-NCAR, and SOC surface heat fluxes with moored buoy measurements in the subduction region of the northeast Atlantic, *J. Climate*, 14(8), 1780-1789, doi:10.1175/1520-0442(2001)014<1780:ACOENN>2.0.CO;2, 2001.

Jung, T., and M. Hilmer, The link between the North Atlantic Oscillation and Arctic Sea ice export through Fram Strait, *J. Climate*, 14(19), 3932-3943, doi:10.1175/1520-0442(2001)014<3932:TLBTNA>2.0.CO;2, 2001.

Kalnay, E., *et al.*, The NCEP/NCAR. 40-year reanalysis project, *Bull. Am. Met. Soc.*, 77(3), 437-471, doi:10.1175/1520-0477(1996)077<0437:TNYRP>2.0.CO;2, 1996.

Karcher, M.J., R. Gerdes, F. Kauker, and C. Köberle, Arctic warming: Evolution and spreading of the 1990s warm event in the Nordic Seas and the Arctic Ocean, *J. Geophys. Res.*, 108(C2). 3034, doi:10.1029/2001JC001265, 2003.

Kindem, I.K.T., and B. Christiansen, Tropospheric response to stratospheric ozone loss, *Geophys. Res. Lett.*, 28(8), 1547-1550, doi:10.1029/2000GL012552, 2001.

Kistler, R., *et al.*, The NCEP-NCAR, 50-year reanalysis: Monthly means CD-ROM and documentation, *Bull. Am. Met. Soc.*, 82(2), 247-268, doi:10.1175/1520-0477(2001)082<0247:TNNYRM>2.3.CO;2, 2001.

Kuzmina, S.I., O.M. Johannessen, L. Bengtson, H. Drange, L.P. Bobylev, and O.G. Aniskina, The north atlantic oscillation variability and change under anthropogenic forcing, *Geophys. Res. Lett.*, submitted., 2004.

Kvamstø, N.G., P. Skeie, and D.B. Stephenson, Impact of Labrador sea-ice extent on the North Atlantic oscillation, *Int. J. Climatol.*, 24(5), 603-612, doi:10.1002/joc.l015. 2004.

Kwok, R., G.F. Cunningham, and S.S. Pang, Fram Strait sea ice outflow, *J. Geophys. Res.*, 109(C1). C01009, doi: 10.1029/2003JC001785, 2004.

Levitus, S., J.I. Antonov, T.P. Boyer, and C. Stephens, Warming of the world ocean, *Science*, 287(5461), 2225-2229, doi:10.1126/science.287.5461.2225, 2000.

Magnusdottir, G., C. Deser, and R. Saravanan, The effects of north atlantic sst and sea ice anomalies on the winter circulation in CCM3. part i: Main features and storm track characteristics of the response. *J. Climate*, 17(5), 857-876, doi:10.1175/1520-0442(2004)017, 2004.

Marshall, J., H. Johnson, and J. Goodman, A study of the interaction of the North Atlantic Oscillation with the ocean circulation. *J. Climate*, 14(7), 1399-1421, doi:10.1175/1520-0442(2001)014<1399:ASOTIO>2.0.CO;2, 2001a.

Marshall, J., Y. Kushnir, D. Battisti, P. Chang, A. Czaja, R. Dickson, M. McCartney, R. Saravanan, and M. Visbeck, North Atlantic - climate variability: Phenomena, impacts and mechanisms, *Int. J. Climatol*, 21(15), 1863-1898, doi:10.1002/joc.693, 2001b.

Meincke, J., B. Rudels, and H. Friedrich, The Arctic Ocean Nordic Seas thermohaline system, *ICES J. Mar. Sci.*, 54(3), 283-299, doi: 10.1006/jmsc. 1997.0229, 1997.

Morison, J., M. Steele, and R. Andersen, Hydrography of the upper Arctic Ocean measured from the nuclear submarine U.S.S. Pargo, *Deep-Sea Res. I*, 45(1), 15-38, doi: 10.1016/S0967-0637(97)00025-3, 1998.

Mork, K. A., and J. Blindheim, Variation in the Atlantic Inflow to the Nordic Seas, 1955-1996, *Deep-Sea Res. I*, 47(6), 1035-1057, doi: 10.1016/S0967-0637(99)00091-6, 2000.

Nilsen, J.E.Ø., and E. Falck, Variations of mixed layer depth and water properties in the Norwegian Sea during 50 years, *Prog. Oceanogr.*, submitted, 2004.

Nilsen, J.E.Ø., Y. Gao, H. Drange, T. Furevik, and M. Bentsen, Simulated North Atlantic-Nordic Seas water mass exchanges in an isopycnic coordinate OGCM, *Geophys. Res. Lett.*, 30(10), 1536, doi: 10.1029/2002GL016597, 2003.

Nobre, P., and J. Shukla, Variations of sea surface temperature, wind stress, and rainfall over the tropical Atlantic and South America, *J. Climate*, 9(10), 2464-2479, doi: 10.1175/1520-0442(1996)009 <2464:VOSSTW>2.0.CO;2, 1996.

Nøst, O.A., and P.E. Isachsen, The large-scale time-mean ocean circulation in the Nordic Seas and Arctic Ocean estimated from simplified dynamics, *J. Mar. Res.*, 61(2), 175-210, doi:10.1357/002224003322005069, 2003.

Orvik, K.A., The deepening of the Atlantic Water in the Lofoten Basin of the Norwegian Sea, demonstrated by using an active reduced gravity model, *Geophys. Res. Lett.*, 31, L01306, doi: 10.1029/2003GL018687, 2004.

Orvik, K.A., and P. Niiler, Major pathways of Atlantic water in the northern North Atlantic and Nordic Seas towards arctic, *Geophys. Res. Lett*, 29(19), 1896, doi: 10.1029/2002GL015002, 2002.

Orvik, K.A., and Ø. Skagseth, Monitoring the Norwegian Atlantic Slope Current using a single moored current meter. *Cont. Shelf Res.*, 23(2), 159-176, doi:10.1016/S0278-4343(02)00172-3, 2003a.

Orvik, K.A., and Ø. Skagseth, The impact of the wind stress curl in the North Atlantic on the Atlantic inflow to the Norwegian Sea toward the Arctic. *Geophys. Res. Lett.*, 30(17), 1884, doi:10.1029/2003GL017932, 2003b.

Orvik, K.A., Ø. Skagseth. and M. Mork, Atlantic Inflow to the Nordic Seas. Current structure and volume fluxes from moored current meters, VM-ADCP and SeaSoar-CTD observations, 1995-1999, *Deep-Sea Res. I*, 48(4), 937-957, doi: 10.1016/S0967-0637(00)00038-8, 2001.

Østerhus. S., and T. Gammelsrød. The abyss of the Nordic Seas is warming, *J. Climate*, 12(11), 3297-3304, doi: 10.1175/1520-0442(1999)012<3297:TAOTNS>2.0.CO;2, 1999.

Peterson. K.A., R.J. Greatbatch, J. Lu, H. Lin. and J. Derome, Hindcasting the NAO using diabatic forcing of a simple AGCM, *Geophys. Res. Lett*, 29(9), doi: 10.1029/2001GL014502, 2002.

Portis, D.H., J.E. Walsh, M.E. Hamly, and P.J. Lamb. Seasonality of the North Atlantic Oscillation, *J. Climate*. 14(9), 2069-2078, doi:10.1175/1520-0442(2001)014<2069:SOTNAO>2.0.CO;2, 2001.

Poulain, P.-M., A. Warn-Varnas, and P. Niiler, Near surface circulation of the Nordic Seas as measured by lagrangian drifters, *J. Geophys. Res.*, 101(C8), 18,237-18,258, doi: 10.1029/96JC00506, 1996.

Quadfasel, D., J.-C. Gascard, and K.-P. Koltermann, Large-scale oceanography in Fram Strait during the 1984 Marginal Ice Zone Experiment, *J. Geophys. Res.*, 92(C7), 6719-6728, 1987.

Rahmstorf, S., and A. Ganopolski, Long-term global warming scenarios computed with an efficient coupled climate model, *Climatic Change*, 43(2), 353-367, doi: 10.1023/A:1005474526406, 1999.

Reynolds, R., and T. Smith, Improved global sea surface temperature analyses, *J. Climate*, 7(6), 929-948, doi:710.1175/1520-0442(1994)007<0929:IGSSTA>2.0.CO;2, 1994.

Rodwell, M., D. Rowell, and C.K. Folland, Oceanic forcing of the wintertime North Atlantic Oscillation and European climate, *Nature*, 398(6725), 320-323, doi: 10.1038/18648, 1999.

Rogers, J.C., The association between the North Atlantic Oscillation and the Southern Oscillation in the northern hemisphere, *Mon. Weather Rev.*, 112(10), 1999-2015, doi: 10.1175/1520-0493(1984) 112<1999:TABTNA>2.0.CO;2, 1984.

Rothrock, D.A., Y. Yu, and G.A. Maykut, Thinning of the Arctic sea-ice cover, *Geophys. Res. Lett.*, 26(23). 3469-3472, doi:10.1029/1999GL010863, 1999.

Rothrock, D.A., J. Zhang, and Y. Yu, The Arctic ice thickness anomaly of the 1990s: A consistent view from observations and models. *J. Geophys. Res.*, 108 (C3), 3083, doi: 10.1029/2001JC001208, 2003.

Schmith. T., and C. Hansen, Fram Strait ice export during the nineteenth and twentieth centuries reconstructed from a multiyear sea ice index from southwestern Greenland, *J. Climate*. 16(16), 2782-2791, doi:10.1175/1520-0442(2003)016<2782:FSIEDT>2.0,CO;2, 2003.

Serreze, M., *et al.*, A record minimum Arctic Sea ice extent and area in 2002. *Geophys. Res. Lett.*, 30(3), doi: 10.1029/2002GL016406, 2003.

Serreze, M.C., and C.M. Hurst, Representation of mean Arctic precipitation from NCEP-NCAR, and ERA re-analysis, *J. Climate*, 13(1), 182-201, doi:10.1175/1520-0442(2000)013<0182:R.OMAPF> 2.0.CO;2, 2000.

Shindell. D.T., R.L. Miller, G.A. Schmidt, and L. Pandolfo. Simulation of recent northern winter climate trends by greenhouse-gas forcing, *Nature*. 399(6735), 452-455, doi:10.038/20905, 1999.

Simonsen. K., and P. Haugan, Heat budgets of the arctic mediterranean and sea surface heat flux parameterizations for the nordic seas, *J. Geophys. Res.*, 101 (C3), 6553-6576, doi:10.1029/95JC03305, 1996.

Skagseth, Ø., K.A. Orvik, and T. Furevik. Coherent variability of the Norwegian Atlantic Slope Current derived from TOPEX/ERS altimeter data, *Geophys. Res. Lett.*, 31(14), L14304. doi:10.1029/2004GL020057, 2004.

Steele, M., R. Morley, and W. Ermold, PHC: A global ocean hydrography with a high-quality Arctic Ocean, *J. Climate*, 14(9), 2079-2087, doi:10.1175/1520-0442(2001)014<2079:PAGOHW>2.0.CO;2, 2001.

Stephenson, D.B., and V. Pavan, The North Atlantic Oscillation in coupled climate models: A CMIP1 evaluation, *Clim. Dynam.*, 20(4), 381-399, doi:10.1007/s00382-002-0281-5, 2003.

Stephenson, D.B., V. Pavan, and R. Bojaril, Is the North Atlantic Oscillation a random walk?, *Int. J. Climatol.*, 20(1), 1-18, doi:10.1002/(SICI)1097-0088(200001)20:1<1::AID-JOC456>3.0.CO;2-P, 2000.

Stephenson, D.B., H. Wanner, S. Brönnimann, and J. Luterbacher, The history of scientific research on the North Atlantic Oscillation, in *The North Atlantic Oscillation: Climatic Significance and Environmental Impact, Geophysical Monograph Series*, vol. 134, edited by J.W. Hurrell, Y. Kushnir, G. Ottersen, and M. Visbeck, pp. 37-50, AGU, doi:10.1029/134GM02, 2003.

Stigebrandt, A., A model for the thickness and salinity of the upper layer in the Arctic Ocean and the relationship between the ice thickness and some external parameters, *J. Phys. Oceanogr.*, 11(10), 1407-1422, doi:10.1175/1520-0485(1981)011<1407:AMFTTA> 2.0.CO;2, 1981.

Sutton, R.T., S.P. Jewson, and D.P. Rowell, The elements of climate variability in the tropical Atlantic region, *J. Climate*, 13(18), 3261-3284, doi:10.1175/1520-0442(2000)013<3261:TEOCVI>2.0.CO;2,2000.

Sverdrup, H.U., Wind driven currents in a baroclinic ocean; with application to the equatorial currents of the eastern pacific, *Proceedings, National Academy of Science, Washington*, 33, 318-326, 1947.

Swift, J.H., E.P. Jones, K. Aagaard, E.C. Carmack, M. Hingston, R.W. MacDonald, F.A. McLaughlin, and R.G. Perkin, Waters of the Makarov and Canada basins, *Deep-Sea Res. II*, 44(8), 1503-1529, doi:10.1016/S0967-0645(97)00055-6, 1997.

Thompson, D.W. and J.M. Wallace, The Arctic Oscillation signature in the wintertime geopotential height and temperature fields, *Geophys. Res. Lett.*, 25(9), 1297-1300, doi:10.1029/98GL00950, 1998.

Thompson, D.W.J., and J.M. Wallace, Annular modes in the extratropical circulation. part I: Month-to-month variability, *J. Climate*, 13(5), 1000-1016, doi:10.1175/1520-0442(2000)013<1000:AMITEC>2.0.CO;2,2000.

Thompson, D.W.J., S. Lee, and M.P. Baldwin, Atmospheric processes governing the Northern Hemisphere Annular Mode/North Atlantic Oscillation, in *The North Atlantic Oscillation: Climatic Significance and Environmental Impact, Geophysical Monograph Series*, vol. 134, edited by J. W. Hurrell, Y. Kushnir, G. Ottersen, and M. Visbeck, pp. 81-112, AGU, doi:10.1029/134GM05, 2003.

Turrell, W., B. Hansen, S. Hughes, and S. Østerhus, Hydrographic variability during the decade of the 1990's in the northeast Atlantic and southern Norwegian Sea., *in Hydrobiological Variability in the ICES Area, 1990-99, ICES Mar. Sci. Symp.*, vol. 219, edited by W. Turrell, A. Lavín, K. F. Drinkwater, M. St. John, and J. Watson, pp. 111-120, ICES, 2003.

Turrell, W.R., G. Slesser, R.D. Adams, R. Payne, and P.A. Gillibrand, Decadal variability in the composition of Faroe Shetland Channel bottom water, *Deep-Sea Res. I*, 46(1), 25, doi:10.1016/S0967-0637(98)00067-3, 1999.

Ulbrich, U., and M. Christoph, A shift of the nao and increasing storm track activity over europe due to anthropogenic greenhouse gas forcing, *Clim. Dynam.*, 15(7), 551-559, doi: 10.1007/s003820050299, 1999.

Vikebø, F., T. Furevik, G. Furnes, N.G. Kvamstø, and M. Reistad, Wave height variations in the North Sea and on the Norwegian Continental Shelf 1881-1999, *Cont. Shelf Res.*, 23(3-4), 251-263, doi:10.1016/S0278-4343(02)00210-8, 2003.

Vinje, T., Fram Strait ice fluxes and atmospheric circulation: 1950-2000, *J. Climate*. 14(16), 3508-3517, doi: 10.1175/1520-0442(2001)014<3508:FSIFAA>2.0.CO;2, 2001.

Vinje, T., N. Nordlund, and Å. Kvambekk, Monitoring ice thickness in Fram Strait, *J. Geophys. Res.*, 103(C5), doi: 10.1029/97JC03360, 1998.

Visbeck, M., E.P. Chassignet, R. Curry. T. Delworth, B. Dickson, and G. Krahmann, The ocean's response to North Atlantic Oscillation variability, in *The North Atlantic Oscillation: Climatic Significance and Environmental Impact, Geophysical Monograph Series*, vol. 134, edited by J. W. Hurrell, Y. Kushnir, G. Ottersen, and M. Visbeck, pp. 113-145, AGU, doi:10.1029/134GM06, 2003.

von Storch, H., Spatial patterns: EOFs and CCA, in *Analysis of Climate Variability*, edited by H. von Storch and A. Navarra, chap. 13, Springer, 1995.

Watson, A., *et al.*, Mixing and convection in the Greenland Sea from a tracer-releaease experiment, *Nature*, 401(6756), 902-904, doi:10.1038/44807, 1999.

Widell, K., S. Østerhus, and T. Gammelsrød, Sea ice velocity in the Fram Strait monitored by moored instruments, *Geophys. Res. Lett.*, 30(19), 1982, doi: 10.1029/2003GL018119, 2003.

Willebrand, J., S. Philander, and R. Pacanowski, The oceanic response to large-scale atmospheric disturbances, *J. Phys. Oceanogr.*, 10(3). 411-429, doi:10.1175/1520-0485(1980)010<0411:TORTLS>2.0.CO;2, 1980.

Wu, P., R. Wood, and P. Stott, Does the recent freshening trend in the North Atlantic indicate a weakening therrmohaline circulation?, *Geophys. Res. Lett.*, 31(2), L02301, doi: 10.1029/2003GL018584, 2004.

Zhang. J., D.A. Rothrock. and M. Steele, Warming of the Arctic Ocean by a strengthened Atlantic inflow: Model results, *Geophys. Res. Lett.*, 25(10), 1745-1748, doi: 10.1029/98GL01299, 1998.

Zhang. X., M. Ikeda. and J.E. Walsh, Arctic sea ice and freshwater changes driven by the atmospheric leading mode in a coupled sea ice-ocean model, *J. Climate*. 16(13). 2159-2177, doi: 10.1175/2758.1, 2003.

T. Furevik, Geophysical Institute, Allégt. 70, 5007 Bergen, Norway. (tore@gfi.uib.no)

J. Even Ø. Nilsen, Nansen Environmental and Remote Sensing Center, Thormøhlensgt. 47, 5006 Bergen, Norway. (even@nersc.no)

Wintertime Nordic Seas Cyclone Variability and Its Impact on Oceanic Volume Transports Into the Nordic Seas

Asgeir Sorteberg, Nils Gunnar Kvamstø[1], and Øyvind Byrkjedal

Bjerknes Centre for Climate Research, University of Bergen, Bergen, Norway

The variability of the Nordic Seas wintertime storm tracks has been investigated using a Lagrangian approach. For the tracking algorithm, the NCEP/NCAR reanalysis has been used to identify individual wintertime storm tracks during the last 50 years. Both the number and intensity of the cyclones extending into the Nordic Seas exhibit strong interannual and decadal variability. The decadal variability was especially pronounced from the mid-1970s to the mid-1990s. On average over the 50-year period, both the cyclone intensity and number were highly correlated with the Arctic Oscillation/North Atlantic Oscillation (AO/NAO). The relationship with the two-point NAO Index broke down in the 1960s, indicating that the NAO Index should be used with caution as an index for describing the mean Nordic Seas storm track variability. In the absence of long observational time series, output from an OGCM was used to investigate the relationship between the Nordic Seas wintertime storm tracks and the ocean volume transports into and out of the Nordic Seas across the Greenland–Scotland Ridge. To the extent that the temporal variability of the modelled transports are realistic, variability in the oceanic volume fluxes over the 50-year period shows a relatively strong co-variability with both the Nordic Seas mean cyclone intensity and the number of cyclones. As noted in several studies, the relationship between the atmospheric flow variability and the variability in the ocean transports was found to be stronger from the mid-1970s and onward. Decomposing the mean Nordic Seas storm track variability into frequency and time space shows that a possible reason for this increase in co-variability during the last 25 years might be related to nonlinear effects of the increased storm track variability and possibly to the increased persistence of the storm tracks.

1. INTRODUCTION

The Nordic Seas area is one of the few locations where formation of dense and deep waters takes place. Deep water ventilation is an important component of the (Atlantic) thermohaline circulation and is also one of the controlling factors of the global sea surface temperature (SST) field of the present climate [*Hartmann*, 1994]. Locally, the open waters contribute to the anomalously warm climate that prevails in this region compared to the zonal mean [*Seager et al.*, 2002].

There are clear indications that the North Atlantic Ocean SST varies significantly with the overlying atmosphere, and it has long been recognized that large-scale fluctuations in the North Atlantic Oscillation (NAO)/Arctic Oscillation (AO) and the North Atlantic SSTs are related [*Bjerknes*, 1964].

[1]Also at Geophysical Institute, University of Bergen, Bergen, Norway.

The Nordic Seas: An Integrated Perspective
Geophysical Monograph Series 158

10.1029/158GM10

The leading mode of SST variability over the North Atlantic during winter consists of a well-known tripolar pattern [e.g. *Deser and Blackmon*, 1993; *Kushnir*, 1994]. The emergence of this pattern is consistent with the observed spatial form of the anomalous surface fluxes associated with the pattern [*Cayan*, 1992]. The strength of the correlation increases when the NAO Index leads the SST, indicating that SST is responding to atmospheric forcing on monthly time scales [*Battisti et al.*, 1995; *Delworth*, 1996; *Deser and Timlin*, 1997].

Model studies have also revealed that atmospheric variations influence the variability of, for instance, the Atlantic meridional overturning circulation (AMOC) [*Bentsen et al.*, 2004]. Again, there is a pronounced relationship between NAO/AO and the leading mode of the AMOC variability [*Bentsen et al.*, 2004]. Most of the studies investigating links between atmospheric and oceanic variability in this area represent variability in the former by the NAO/AO. The relative importance of the role of different weather phenomena in the total atmospheric forcing is less investigated. This is because the NAO is a manifestation of variability on many time scales; in the following we will investigate more closely cyclone variability and its relationship to some aspects of the variability of ocean volume flux in the North Atlantic. The synoptic low-pressure systems exert both freshwater forcing (through precipitation) and wind-stress forcing on the ocean. It is the latter that is addressed in this study.

During the last two or three decades there has been a trend towards more frequent positive anomalies of the NAO/AO Index. Storm track activity in the Northern Hemisphere is connected to not only the NAO but also to the North Pacific Oscillation (PNA) [*Lau*, 1988; *Trenberth et al.*, 1988; *Rogers*, 1990]. Positive anomalies of the NAO Index [*Rogers*, 1984; *Barnston and Livezey*, 1987; *Kushnir* 1994; *Hurrell* 1995; *Hurrell and van Loon*, 1997] are associated with a strengthening of the mid-latitude westerly flow over the North Atlantic. This strengthening manifests itself as an intensification and poleward deflection of the North Atlantic mid-latitudinal storm track. *Serreze et al.* [1993] found that the increase in number of positive NAO winters/years led to a twofold increase in the occurrence of cyclones during the period 1964-1993 in the Icelandic Low region. However, one should be aware that such an increase can be dependent on the field used to identify the cyclones and how they are counted [*Hoskins and Hodges*, 2002]. Whether the identified trends have been a result of anthropogenic climate change is being debated [*Gillett et al.*, 2003, *Kuzmina et al.*, 2005].

The purpose of this paper is twofold. First, a short description of the wintertime cyclone climatology and variability in the northern North Atlantic is given. Second, the link between the upper ocean circulation variability in this area and the variability in the North Atlantic storm tracks is investigated.

One approach of diagnosing storm track activity is to determine statistics at a set of grid points such as the variance in a frequency band associated with what are considered to be synoptic time-scales [e.g. *Blackmon*, 1976; *Blackmon et al.*, 1977]. Several properties of this approach have been given a thorough review by *Chang et al.* [2002]. *Chang et al.* [2002] and references therein also provide convincing evidence that the Atlantic storm track was weaker during the 1960s than in more recent decades. Furthermore, *Chang and Fu* [2002] have shown by means of Empirical Orthogonal Function (EOF) analysis a marked interdecadal variability in the storm track patterns. They found that the leading mode represented a simultaneous change in the intensity of the Atlantic and Pacific storm tracks. Some portion of this variability is associated with the AO and ENSO [*Chang and Fu*, 2002] variations, but when these portions are removed, a significant portion of interdecadal variability is left. The implications of such a large interdecadal variability as well as the underlying mechanism remain open [*Chang et al.*, 2002].

In assessing the storm track's importance for the ocean circulation in the Nordic Seas, we have investigated the relationship between storm track statistics and ocean volume transports across the Greenland–Scotland Ridge (GSR). Due to the lack of long time series of observed ocean volume transports across the GSR, the output from a well-tested ocean model is used as a proxy for the ocean volume transports since 1948.

In the following section, we provide a short description of the data and analysis methodology used. In section 3, we describe the climatological structure and variability of the number and intensity of the North Atlantic cyclones entering the Nordic Seas. Their relationship with the NAO/AO will be discussed in section 4, and in section 5 we investigate the relationship between the cyclones entering the Nordic Seas and the oceanic volume flow in and out of the Nordic Seas. The results are summarized and discussed in section 6.

2. METHOD AND DATA

An algorithm for feature-tracking developed by *Hodges* [1994; 1995; 1996; 1999] has been used to construct storm tracks from the NCEP/NCAR reanalysis data provided by the NOAA-CIRES Climate Diagnostics Center [*Kalnay et al.*, 1996; *Kistler et al.*, 2001]. The 850-hPa relative vorticity is used instead of the more often used mean sea level pressure (MSLP) because the former is not an extrapolated field to any large extent and is less influenced by the background flow than are fields such as MSLP and geopotential height. This method also focuses on smaller scale synoptic activity than these other fields, which has the advantage that a higher number of storm trajectories can be identified.

Relative vorticity at 850 hPa has been calculated from the 6-h wind components for the whole northern hemisphere for the wintertime (DJF) period 1948-2002. This field can be very noisy with frontal systems beginning to be resolved. Processing is therefore done at a spectral resolution of T42. The reduction in resolution associated with the T42 truncation provides some smoothing of the higher frequencies. In addition, planetary scales for total wave numbers $n \leq 7$ are removed from the field by using a filtering method discussed by *Anderson et al.* [2003]. This spectral cutoff does not make any significant change in the results since the planetary scales are relatively weak in the vorticity fields relative to those in the anomalies. Local extremes are identified in the related fields, and these are interpreted as the centres for negative and positive vorticity anomalies.

The analyses are concentrated on the positive vorticity anomalies (cyclones), the feature points associated with low-pressure systems. The feature points for two consecutive time steps are linked together by using a nearest neighbour search. A cost-function based on track smoothness in terms of changes in direction and speed [*Hodges*, 1995; 1996] is calculated for each possible ensemble of trajectories. The ensemble minimizing the cost-function is chosen to be the best fit and thereby restricts the feature to physical movement. Generalizing the technique to the spherical domain [*Hodges*, 1995] obviates the need to use projections, which may introduce systematic biases [*Zolina and Gulev*, 2002]. This allows for the analysis of global data or of large spherical regions. Semi-stationary storms (less than 10° total displacement) and short-lived storms (lasting less than 2 days) are removed.

The calculation of the statistical diagnostic fields is performed by the use of spherical nonparametric estimators from the ensembles of feature tracks [*Hodges*, 1996]. We here focus our attention on the statistical fields track density and mean intensity. Track density is scaled to a number density per month per unit area as in *Hoskins and Hodges* [2002], where a unit area is equivalent to a 5° spherical cap (~10^6 km^2). Mean intensity is computed from the size of the anomalies in relative vorticity for the feature points; the unit for this is 10^{-5} s^{-1}. Another statistical field used in the paper is the feature density, which is a measure of the occurrence of cyclones (both the number of cyclones and how long they stay in the area; thus slow-moving low-pressure systems are given more weight than the ones moving quickly through the area). A good alternative to this feature-tracking method is the use of bandpass filters. However, a limitation of bandpass filters is that although they highlight spatial patterns associated with various time scales (synoptic in this case), these patterns are not always easily related to actual weather systems [*Anderson et al.*, 2003]. This occurs partly because a narrow band emphasizes only a particular time scale whereas cyclones may have different time scales associated with their life cycles. Since one of the main aims of this study is to identify the link of low-pressure systems to the Nordic Sea volume inflow and outflow, the feature-tracking approach has been adopted.

Given the lack of long time series of ocean volume transports, we are using a recent simulation with the Nansen Center version of Miami Isopycnic Coordinate Ocean Model (MICOM; *Bleck et al.*, 1992], utilizing daily forcing from NCEP/NCAR atmospheric reanalysis as a proxy for the real transports. The MICOM version used here [*Bentsen et al.*, 2004], fully coupled to a dynamic thermodynamic sea-ice module, was driven with the NCEP/NCAR reanalysis forcing fields for the period 1948-1999; model setup and integration followed the description of the synoptic hind-cast simulations of *Furevik et al.* [2002]. No relaxation is used for temperature, whereas a diagnosed weekly resolved annual repeated restoring flux is applied for the sea surface salinity. The model has 25 vertical layers with fixed potential densities, and an uppermost mixed layer with temporally and spatially varying track density. The horizontal configuration is a local orthogonal grid mesh with one pole over North America and the other pole over the western part of Asia [*Bentsen et al.*, 1999]; grid spacing is 30 to 40 km in the North Atlantic–Nordic Seas region. The model variability has been shown to be close to the variability of the hydrographical observations and observations/estimates of Nordic Sea volume fluxes [*Nilsen et al.*, 2003; *Hátún et al.*, this volume].

3. CLIMATOLOGICAL STRUCTURE OF THE NORTH ATLANTIC CYCLONE ACTIVITY

Figure 1a and b shows the climatological distribution of the North Atlantic number of wintertime (DJF) cyclones and their mean intensity during the period 1949-2002. The well-known band with large track density from Newfoundland across the North Atlantic and into northern Europe is one of the most pronounced features in the maps. This maximum is the North Atlantic storm track, which together with smaller minima and maxima on the northern and southern flanks, respectively, constitutes a gross pattern that is apparent in any chosen quantity to represent synoptic activity [*Chang et al.*, 2002].

There are, however, differences between the many available diagnostics on a more detailed level [*Gulev et al.*, 2001; *Hoskins and Hodges*, 2002]. Such differences arise because the different analysis methods and fields used emphasize different characteristics of the storm tracks. The use of different reanalyses has been shown to introduce only negligible disagreements since these datasets are very consistent in the lower troposphere of the Northern Hemisphere (NH) [*Hodges et al.*, 2003]. Because of the many procedures for storm track

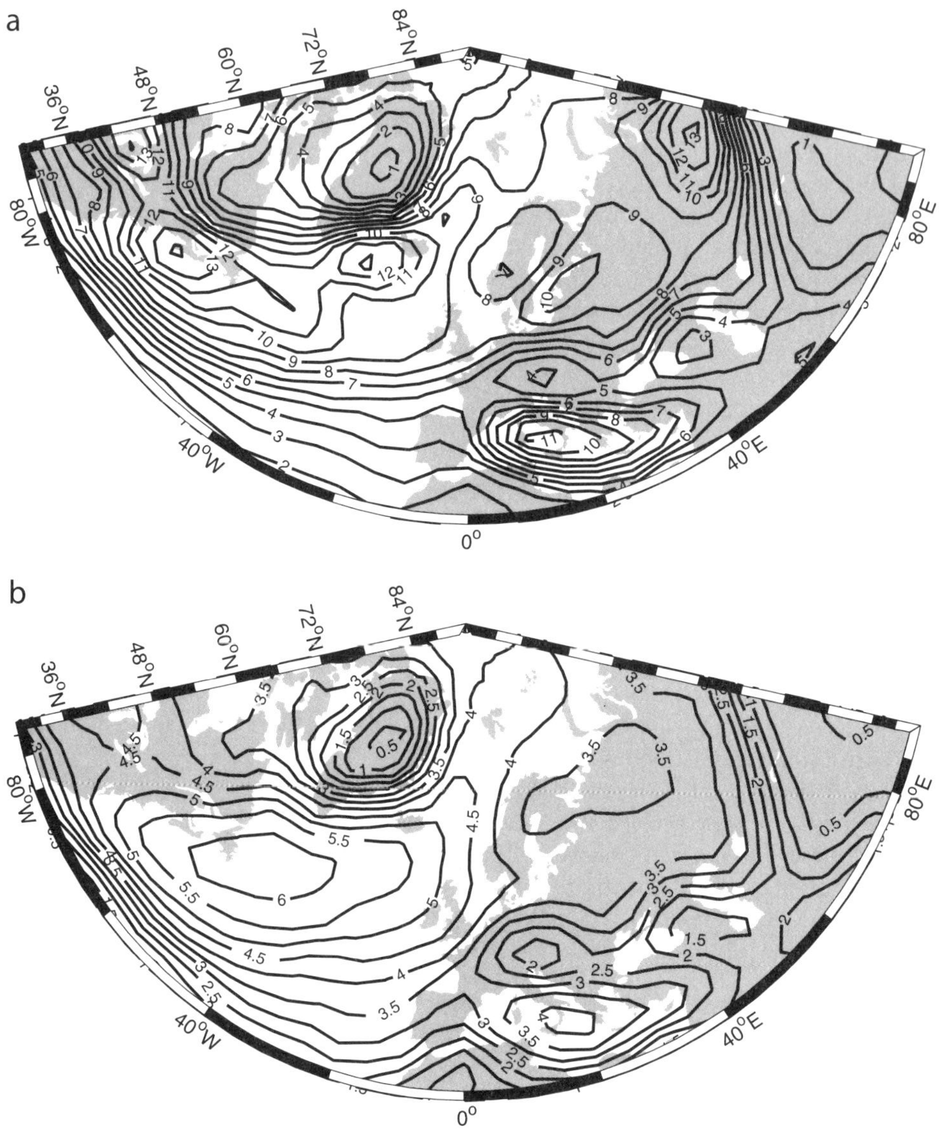

Figure 1. Mean wintertime (DJF) cyclone track density (a) and intensity (b) for the period 1949-2002 in number per month per 10^6 km^2 and 10^{-5} s^{-1}, respectively. Climatology is derived from 6-h fields of relative vorticity in 850 hPa, based on the NCEP/NCAR reanalysis.

identification [*Reitan*, 1974; *Zishka and Smith*, 1980; *Witthaker and Horn*, 1981; *Bell and Bosart*, 1989; *Serreze et al.*, 1993; *Lambert*, 1996; *Hoskins and Hodges*, 2002], a quantitative comparison is difficult. However, *Hoskins and Hodges* [2002] have, in a review of storm track perspectives, employed the same procedure as we use here, but on fields from ERA-15 (1979-1994). Their Figure 6b (track density of ξ_{850}) agrees well with our corresponding Figure 1a. A closer inspection of these figures reveals that four local maxima, located near Newfoundland, south of Greenland, east of the Baltic Sea, and in Siberia are apparent in both climatologies.

Interestingly, there are some differences with respect to the relative strength between the local maxima in the NCEP and ERA-15 reanalyses. In particular, the maximum south of Greenland is stronger than that near Newfoundland. In the case of NCEP, these two maxima have the opposite rank. A plausible explanation for this is that the most intense storm track activity has been shifted eastwards during the last part of the 1950-2000 period. An eastward shift in the North Atlantic storm track activity is also observed in *Chang et al.* [2002], who also show that this shift is associated with a general increase in storm track activity.

The distribution of storm track mean intensity (Figure 1b) has a large resemblance to the corresponding diagnostics in *Hoskins and Hodges* [2001] (their Figure 6e). This diagnostic seems to have a smoother, or larger-scale, character than the former. Also in this diagnostic we see a slight eastward shift in the North Atlantic maximum between the NCEP/NCAR and ERA-15 climatologies.

3.1. Nordic Sea Cyclone Variability

The variability in Nordic Sea cyclones and their impact on the volume transports have been investigated by constructing time series of different area-averaged cyclone parameters. The Nordic Seas area, over which we have taken the average, is here identified as 55° to 75°N and 40°E to 15°E (ocean points only). Other slightly smaller and larger areas were also tested to investigate the robustness of the findings to the chosen area. Figure 2 shows the time series of wintertime (DJF) standardized (zero mean and unit standard deviation) anomalies for the different cyclone parameters from 1949 to 2002, the 5-year low-pass-filtered variability (using a third-order Butterworth filter), and the linear trend during the time period.

Dampening thermal and pressure anomalies by radiation loss and surface drag within a few weeks should keep atmospheric fluctuations close to zero correlated from year to year if the forcings remains unchanged. However, since the atmosphere is coupled to the ocean and the (natural and anthropogenic) external forcings are not constant, we might expect a lag-correlation and slightly reddened spectra of the cyclone parameters. If, in addition, the coupling to the ocean or changes in the external forcings involve oscillatory modes, we would expect some spectral peaks, which are inconsistent with a red/white

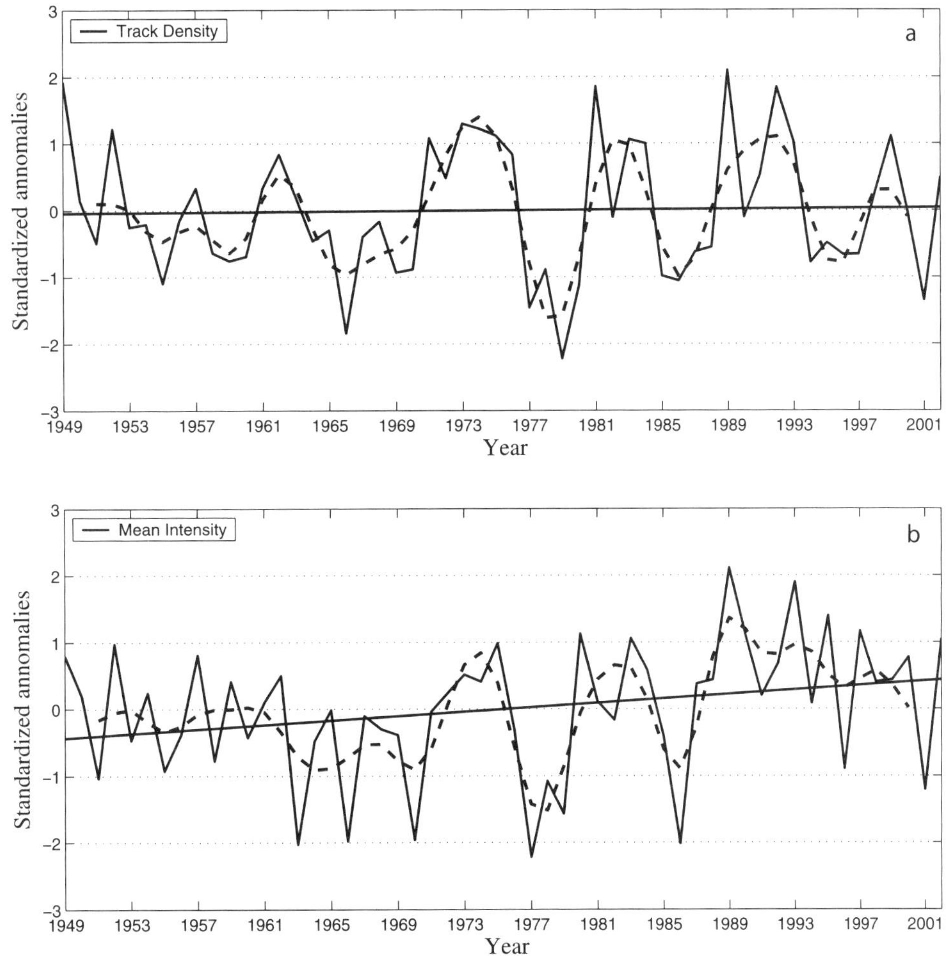

Figure 2. Wintertime (DJF) standardized (zero mean and unit standard deviation) anomalies of track density (a) and mean intensity (b) in the Nordic Sea (averages from 55° to 75°N and 40°W to 15°E, ocean points only). The thin solid line shows the basic time series, the dashed line is the 5-year low-pass-filtered time series using a third-order Butterworth filter, and the thick solid line gives the linear trend calculated using the least squares method. Values are given as deviations from the 1949-2002 mean in standard deviations (STD).

spectrum. A power spectrum of the Nordic Sea cyclone track density and intensity (not shown) indicates increased power on decadal (7-14-year) and interannual (2.5-year) time scales in both the intensity and the track density. However, the peaks are quite weak and broad, and none of them is significant at a 95% confidence level. Thus, the wintertime Nordic Sea cyclone intensity and track density variability are not significantly different from that of a red-noise process. Most of the variance, however, is on the decadal (7-14-year) and interannual (2.5-year) time scales. The variance in the cyclone track density seems to be greater on the decadal scale, whereas the intensity seems to vary somewhat more in the higher (2.5-year) frequencies. The identification of variability on decadal scales is in line with the findings of *Chang and Fu* [2002]. In their investigation of the NH wintertime upper troposphere storm tracks using the NCEP/NCAR reanalysed data, they noted significant low-frequency, decadal-scale fluctuations both in the Atlantic and Pacific regions.

While the time series may contain dominant periodic signals, these signals can vary in both amplitude and frequency over long periods of time, making the power spectra assumption of stationarity of the time series invalid (thus the strength of the peaks can become weak). To investigate the cyclone variability on interannual to decadal timescales at different times, we used a wavelet analysis. The wavelet analysis can be seen as a series of spectral analysis using bandpass filters on the time series. A comprehensive description of wavelet analysis is given by *Hubbard* [1996] and *Torrence and Compo* [1998]. For this study a Morlet wavelet is used. The wavelet analysis (Figure 3) points to the same peaks in the cyclone track density

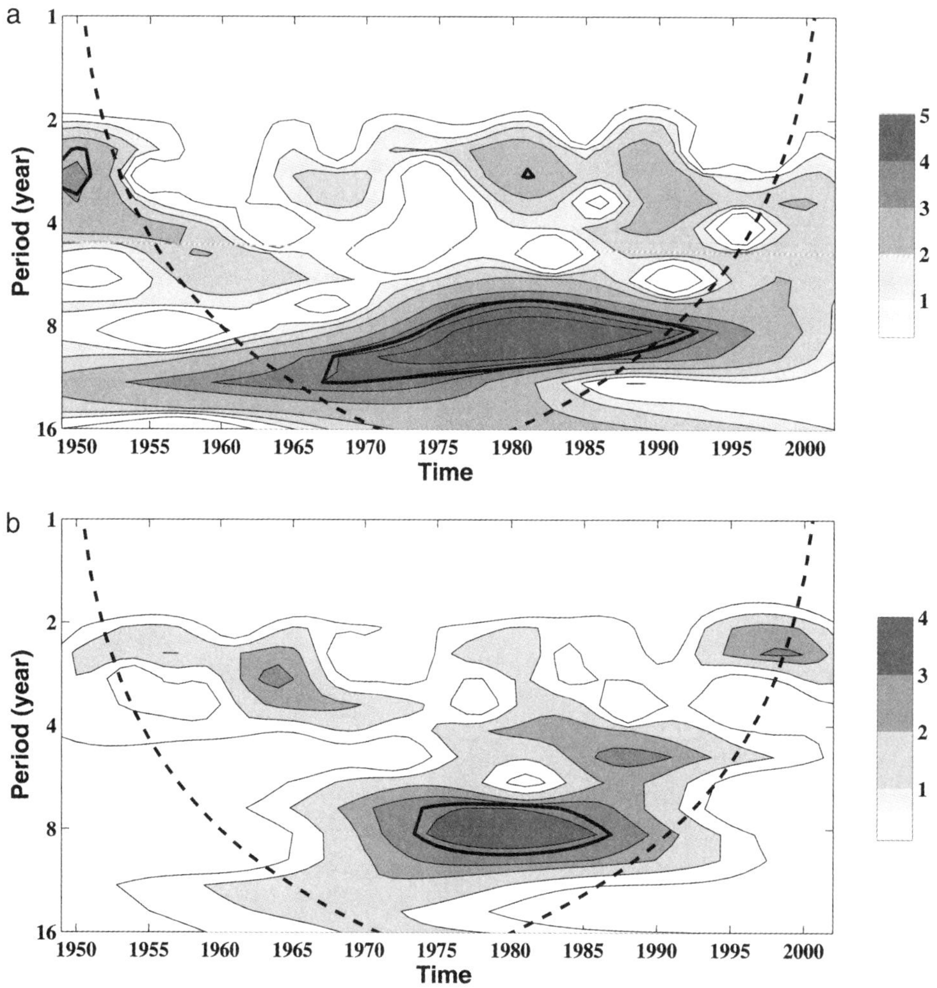

Figure 3. Wintertime (DJF) wavelet power spectrum for cyclone track density (a) and intensity (b) in the Nordic Seas area. Spectrums are calculated with a Morlet wavelet using data from 1949 to 2002. The bold contours show the 95% significance regions, using a red-noise background spectrum with a 1-year lag correlation of 0.25 and 0.0 for the cyclone track density and intensity, respectively. The cone-of-influence, where the padding with zeros reduces the variance (hence statistical significance is low), is shown as dashed lines.

and intensity spectra as in the spectrum analysis, with increased power on the interannual (2.5-year) and decadal (7-12-year) time scales (note that since the Morlet wavelet is complex, the wavelet power will combine positive and negative peaks into a single broad peak). In addition the wavelet analysis indicates that the power on the interannual and decadal scale have a significant time-dependent variability. Although the decadal peak is 95% significant, compared with a red-noise spectrum having the same 1-year lag correlation as the original time series for the 1970-1990 period, the interannual peak is not significant, except for track density in the early 1950s. In the latter half of the time series the cyclone track density has indication of more power on slightly shorter time scales (7-10 years) compared with the first half of the time series where the power is more in the 10-14-year period.

The wavelet analysis indicates that the increased power on the decadal scale in the intensity appears simultaneously with the increased power on the decadal scale in the cyclone track density. This suggests that the two parameters might be linked on this time scale, while on shorter time scales the variability is less connected. Thus, at least on decadal time scales, decades with a large number of cyclones could often experience more intense cyclones. Several of the Nordic Sea cyclone parameters experienced a maximum in the late 1980s to early 1990s. Figure 3 indicates that this was partly due to the increased variability on decadal time scales during this time. In more detail the increased variability of cyclone intensity and track density were seen in 1980-1990 (Figure 2), mainly due to increased power on the decadal time scale (some on the 5-year scale in the early 1990s) for intensity and on both the interannual and decadal time scales for track density. Interestingly, *Chang and Fu* [2002] noted a transition in the upper troposphere storm track intensity during the early 1970s. They found the same (but slightly weaker) signal in the radiosondes. Looking at the wavelet analysis for the Nordic Sea cyclone variability, we suspect the increased power on the decadal scale after 1970 may be linked to the reported transition.

4. RELATIONSHIP BETWEEN NAO/AO AND THE NORDIC SEA CYCLONE PARAMETERS

The AO is a large-scale atmospheric mode of climate variability characterized by the leading EOF in MSLP (20-90°N) [*Thompson and Wallace*, 1998; 2000). The AO is made from the MSLP without any filtering and encompasses meteorological phenomena on different temporal and spatial scales that interact and co-vary to make up the leading EOF. The NAO, on the other hand, is originally a two-point index based on the difference in MSLP between the Atlantic Subtropical High (Azores, Lisbon, or Gibraltar) and the Subpolar Low (Iceland) [*Jones et al.*, 1997]. The two indexes are highly correlated (~0.8) during wintertime. Bearing in mind that vorticity conservation leads to an upscale energy cascade in barotropic fluids [e.g. *Pedlosky,* 1987], that large-scale patterns may evolve from small-scale eddies, and that positive eddy-zonal feedbacks exist [*Lorenz and Hartmann*, 2003], it is no surprise that the temporal variability in the cyclone parameters in the Nordic Seas area are strongly correlated with the AO (Table 1). This strong relationship holds for several of the cyclone variables including the number of cyclones (0.74), their intensity (0.79), and their speed (0.70 for the longitudinal velocity component and 0.69 for the latitudinal component). Note that the correlations are valid only for the period where data are available and should not be generalised.

To investigate the differences in correlations between the cyclone parameters and AO/NAO with time, we calculated the correlations by using a 10-year moving window. Figure 4 shows how the correlation between AO/NAO and the Nordic Seas cyclone track density varies with time. Most Nordic Seas cyclone parameters show reduced or vanishing co-variability with AO/NAO during the 1960s and especially high co-variability from the mid-1970s to the mid-1990s. For at least some Nordic Sea cyclone parameters, the correlation with NAO/AO appears to be increasing with increasing cyclone variability. By comparing the wavelet analysis in Figure 3 with the correlations in Figure 4, there are indications of stronger correlation with NAO/AO during times of strong variability. Whether this is a robust feature is not clear, and the existence of a plausible physical explanation has to be further investigated.

The increase in correlation between NAO/AO and the Nordic Sea cyclone activity take place at the same time as, according to several studies, a shift in the location of the NAO/AO "action centres" [e.g. *Hilmer and Jung*, 2000; *Peterson et al.*, 2003]. This shift in the NAO centres of action represents a change in

Table 1. Wintertime (DJF) Correlation Between Average Cyclone Parameters Over The Nordic Seas Area (55°N to 75°N and 40°W to 15°E, ocean points only) and NAO/AO.

Variable	NAO	AO
Mean intensity	0.72*	0.79*
Track density	0.65*	0.75*
Feature density	0.64*	0.56*
Mean speed	0.39*	0.65*
Speed, longitudinal component	0.49*	0.70*
Speed, latitudinal component	0.54*	0.69*
Mean growth/decay rate	−0.40*	−0.23
Mean lifetime	0.11	0.01

Time periods: NAO, 1949-2000; AO, 1949-2001. NAO is based on Gibraltar/Iceland pressure difference [*Jones et al.*, 1997] and AO is based on *Thompson and Wallace* [1998].

*Significant at 99%.

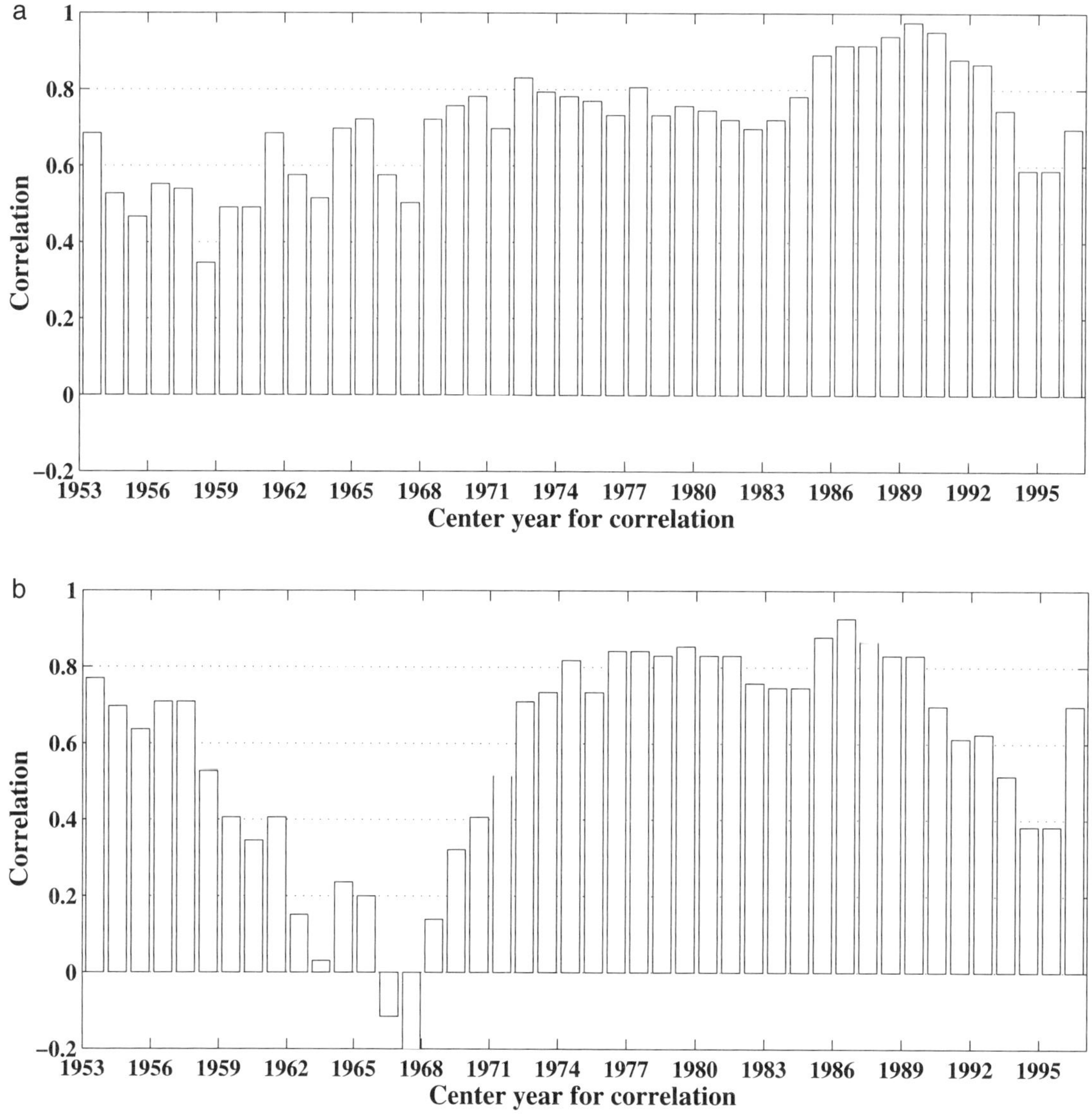

Figure 4. Ten-year moving correlations between wintertime (DJF) Nordic Seas cyclone track density and AO (a) and NAO (b).

the areas having large MSLP variability and corresponds well with increased Nordic Seas wintertime storm track variability. Since the Nordic Seas cyclonic storm track variability constitutes an important part of the principal component–based NAO/AO pattern, a change in the variability of Nordic Sea cyclonic storm tracks may indeed be manifested as a change in the location of the centres of action in a the NAO/AO pattern. A more detailed description and discussion of this shift in centres of action and its implications for the Nordic Sea ocean variability are given in *Furevik and Nilsen* (this issue).

Interestingly, AO in general seems to be slightly better correlated with the cyclone statistics than is the two-point NAO Index. This slightly lower correlation between several of the Nordic Seas cyclone parameters and NAO is mainly attributable to low correlations with NAO in the 1960s. It is not clear if this is a result of the more hemispheric character of the AO Index or the fact that the two-point NAO Index is not as good as the principal component–based AO in picking up the general North Atlantic MSLP variability in this time period. Correlation between Nordic Sea cyclone statistics and a principal component–based NAO (www.cpc.noaa.gov) was also evaluated (not shown), which in general showed weaker correlations with the Nordic Sea cyclone variability but its behaviour in time was similar to that of the two-point NAO Index, indicating that the AO might be the most robust measure of the Nordic Sea storm track variability.

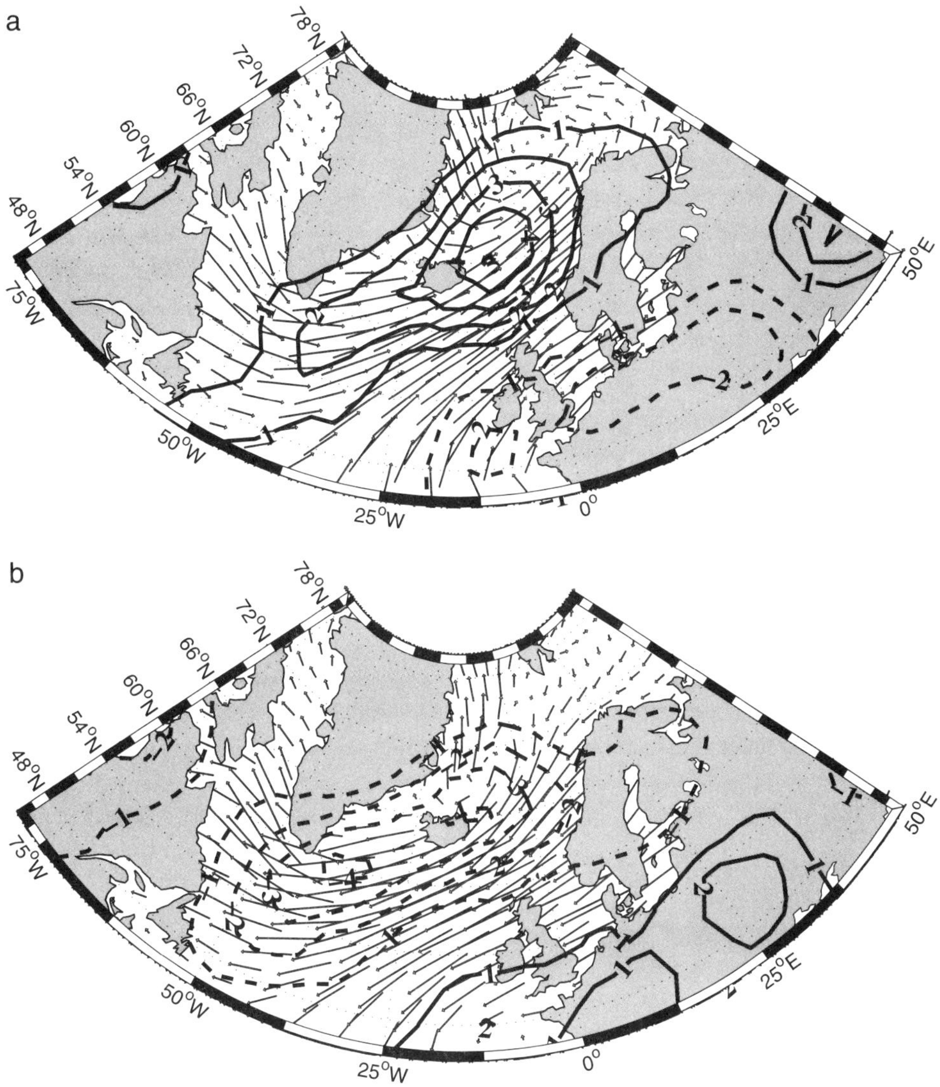

Figure 5. Composites of wintertime (DJF) mean anomalies in cyclone track density [number $(\text{month}*5°)^{-1}$] and anomalies in 10-m wind speed for years with DJF AO Index greater than 1 STD (9 years; a) or less than −1 STD (10 years; b).

Figure 5 displays the departure from climatology of the mean number of cyclones in wintertime (DJF) Nordic Seas for years in which the AO index exceeded ±1 standard deviation (STD). Also displayed are anomalies in the 10-m wind speed (NCEP/NCAR reanalysis). During years of high AO, the number of cyclones is increased in the whole Nordic Seas area, with a northeastward shift in the storm tracks (compare with the climatology in Figure 1a).

The mean number of cyclones for years with AO above (below) 1 STD is on the order of 30-50% higher (20-30% lower) in the area northeast of Iceland. In addition to an increase in the number of cyclones, the mean intensity in the Nordic Seas is also increased (decreased) during years with a high (low) AO Index [5-15% during years with AO above (below) 1 (−1) standard deviation]. In central Europe the reduction in number of cyclones is 25-35%, whereas the intensity is reduced by as much as 30-50% during years with AO above 1 STD. The mean wintertime wind anomalies strongly reflect the changes in the number of cyclones. During years of high AO, the increase in cyclones gives strong south-westerly wind anomalies to the east of Iceland. Years of low AO show a similar, but opposite, pattern of strong northeasterly wind anomalies. East of Greenland are northerly wind anomalies during high-AO years and southerly during low ones. Another striking feature is that in high-AO years, the cyclones in the Nordic Seas and especially northwest of the British Islands have on average a stronger northward and eastward velocity component (not shown). Thus the speed of the low-pressure systems is higher, and more of the lows west of the British Islands move into the Nordic Sea area.

5. NORTH ATLANTIC CYCLONE VARIABILITY AND OCEANIC VOLUME FLOW INTO THE NORDIC SEA

Recent work has emphasized the relationship between atmospheric circulation and ocean volume transport across the GSR [*Bacon*, 1998; *Furevik*, 1998; *Blindheim et al.*, 2000; *Dickson et al.*, 2000; *Nilsen et al.*, 2003; *Orvik and Skagseth*, 2003]. These studies have investigated the NAO, the wind stress, and the curl of the wind stress as measures of the atmospheric forcing on the ocean.

In this section we attempt to separate the influence of the wintertime synoptic low-pressure systems on the transport across the GSR to quantify the role of these atmospheric systems and to add knowledge as to which atmospheric scales and processes might be important for the variability in the ocean volume transports. Hence, we use the cyclone dataset to investigate the relationship between the cyclone variability and the variability of the ocean volume transport in the three sections along the GSR (see Figure 6). We focus on the relationship between simultaneous co-variability of the cyclone and ocean transport variability. This relationship is believed to be dominated by near instantaneous barotropic adjustment of the ocean flow to the wind-stress anomalies related to the cyclone variability. However, at times where the time series are serially correlated, the correlations may also be influenced by slower processes related to the curl of the wind stress, which induces Ekman pumping [*Orvik and Skagseth*, 2003].

As already mentioned, we used output from the Nansen Center version of MICOM (section 2) to obtain time series of the volume transports across each of the three sections along the GSR (Figure 6). For each section, inflow, outflow, and net transports denote the net northward transport, the net southward transport, and the net northward minus the net southward transport, respectively.

5.1. *Influence of the Mean Nordic Sea Cyclone Variability*

As a first approach, we investigate the relationship between the area-averaged Nordic Seas cyclone track density and mean intensity described in section 3, and the inflow, outflow, and net (inflow minus outflow) volume transport in

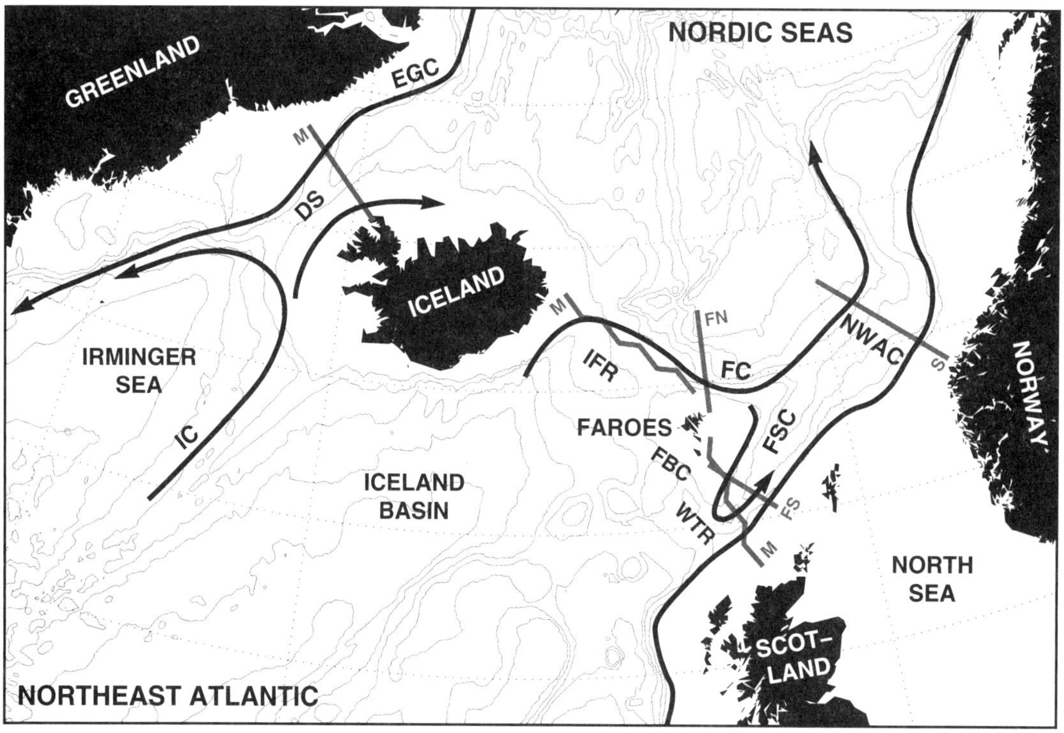

Figure 6. Schematic picture of surface currents. The sections used for calculating the oceanic inflow (northward) and outflow (southward) are indicated with an M. FSC: Faroe–Shetland Channel, DS: Denmark Strait, IFR: Iceland–Faroe Ridge.

the three GSR sections. Spearman rank correlations are used instead of the more conventional linear correlations because of the sensitivity of the linear correlation to outliers. Also the assumption of a linear relationship between the ocean volume transports and the atmospheric variables describing the statistics of the low-pressure systems becomes less crucial when using the Spearman rank correlation. The Spearman rank correlations will identify any monotonic relationships, not only the strictly linear ones [see *Wilks*, 1995]. The search for a monotonic rather than a linear relationship is also motivated by the fact that the wind effect on the ocean is given by the wind stress, which may not be linearly related to the cyclone parameters. The time series has been detrended before correlations are calculated to emphasize the co-variability on interannual to decadal scales.

Table 2 shows the Spearman rank correlations between the intensity and track density of wintertime Nordic Seas area cyclones versus the simulated wintertime (DJF) inflow, outflow, and net volume flow. The interannual variations in the Denmark Strait and the Faroe–Shetland Channel seem most influenced by the average Nordic Seas cyclone parameters, showing significant correlations both with the intensity and number of cyclones. Also added is the correlation with the cyclone feature density, which is a measure of the occurrence of cyclones (both the number of cyclones and how long they stay in the area); that is, slow-moving low-pressure systems are given more weight than the ones moving quickly through the area. Not surprisingly, the correlation with the feature density is slightly stronger than for the track density, which only counts the number of cyclones and not how long they stay in the area. Both the Denmark Strait and the Faroe–Shetland Channel show a slightly stronger co-variability with the number of cyclones than with cyclone intensity.

Table 2. Spearman Rank Correlations (Detrended Time Series) Between Parameters Describing the Mean Wintertime (DJF) Cyclone Activity in the Nordic Seas Area and Mean Wintertime (DJF) Nordic Sea Inflow, Outflow, and Net Flow as Provided by the MICOM Model for the Period 1949-1999.

Variable	Iceland–Faroe	Faroe–Shetland	Denmark Strait
		INFLOW	
Mean intensity	0.12	0.39*	–0.37*
Track density	–0.01	0.48*	–0.55*
Feature density	–0.16	0.53*	–0.62*
		OUTFLOW	
Mean intensity	–0.23	–0.34**	0.53*
Track density	–0.28**	–0.48*	0.53*
Feature density	–0.35**	–0.51*	0.59*
		NET FLOW	
Mean intensity	0.26	0.38*	–0.53*
Track density	0.12	0.49*	–0.58*
Feature density	0.01	0.55*	–0.64*

*Significant at 99%, **significant at 95%. Inflow and outflow denotes water masses flowing into (northward) and out (southward) of the Nordic Seas, respectively; net flow is the difference between the inflow and outflow.

For the Denmark Strait the fact that the relationship between cyclone variability and both inflow (northward currents of Atlantic water on the Icelandic side, predominantly at the surface) and outflow (southward currents located beneath the inflowing Atlantic water and at the surface towards Greenland) is fairly strong indicates close relationships with the atmospheric forcing. It is also interesting to note that the Faroe–Shetland outflow, which consists mainly of deep overflow, correlates just as well as cyclone variability correlates with inflow, which is mainly in the upper ocean. There are at least two plausible explanations for the coupling between the atmospheric (surface) forcing and the outflows: The correlation may arise purely due to the relationship between the parts of the model outflow that are near the surface layer, or the deeper part of the outflow may directly or indirectly feed or integrate the atmospheric forcing. The first interpretation is valid for the upper polar waters in the Denmark Strait but is not obvious for the (on average) deep Faroe–Shetland outflow. The second interpretation is possible during times where there is some persistence (positive lag-correlation), or at least absence of negative lag-correlation, in the cyclone variability. The correlation coefficient then reflects both a fast (Ekman) response and the fact that the cyclone variability is serially correlated with the variability at previous seasons or years. If the latter is the case, we should expect a higher correlation between the Faroe–Shetland outflow and cyclone variability during years when there is stronger decadal cyclone variability. This is indeed the case; by calculating the correlation between the outflow and the Nordic Seas track density, we determined the correlation was –0.22 during the period with low decadal activity (1950-1970), increasing to –0.76 during the period with strong decadal activity (1975-1990).

Due to correlation between the intensity and number of cyclones (track density), the total effect of the variability in the mean Nordic Sea number of cyclones and cyclone intensity on the volume fluxes is not the sum of the variances explained by the intensity and track density individually. Thus, a simple way of indicating the total effect of intensity and track density on the volume fluxes is to use linear multiple regression. This estimate will give an approximation of the linear relationship between wintertime cyclones and the wintertime volume flux variability. We have used the method of least squares, where the mean of the errors is zero and the errors have zero correlation with each of the predictors. Adding more cyclone parameters, such as the mean speed of the cyclones or their lifespan, to the regression gave a marginally better fit to the MICOM-simulated volume transports; for simplicity, however, we have used only the number of cyclones and their intensity in the regression calculation.

Thus the net volume fluxes are estimated as: Net variability in volume flux = a * variability in Nordic Sea cyclone intensity + b * variability in Nordic Sea cyclone track density, where a and b are the regression coefficients indicating how many Sv (1 Sv = 10^6 m^3s^{-1}) deviation from the mean flow results from 1 STD changes in the wintertime Nordic Sea cyclone intensity and track density, respectively. However, given the correlation between the cyclone track density and the intensity, any separate interpretation of the regression coefficients a and b should be done with care.

Figure 7 shows that a portion of the interannual variance in the net wintertime volume flow through the Denmark Strait and the Faroe–Shetland Channel is linearly related to the mean cyclone variability in the Nordic Sea area (39% and 28% of the variance co-vary, respectively). Thus, under the assumption that the relationship between the Nordic Sea cyclone variability and variability in ocean volume transport is linear, a 1 STD deviation in both the track density and intensity should give a change of 0.7-0.8 Sv in the net volume flow through the Denmark Strait and the Faroe–Shetland

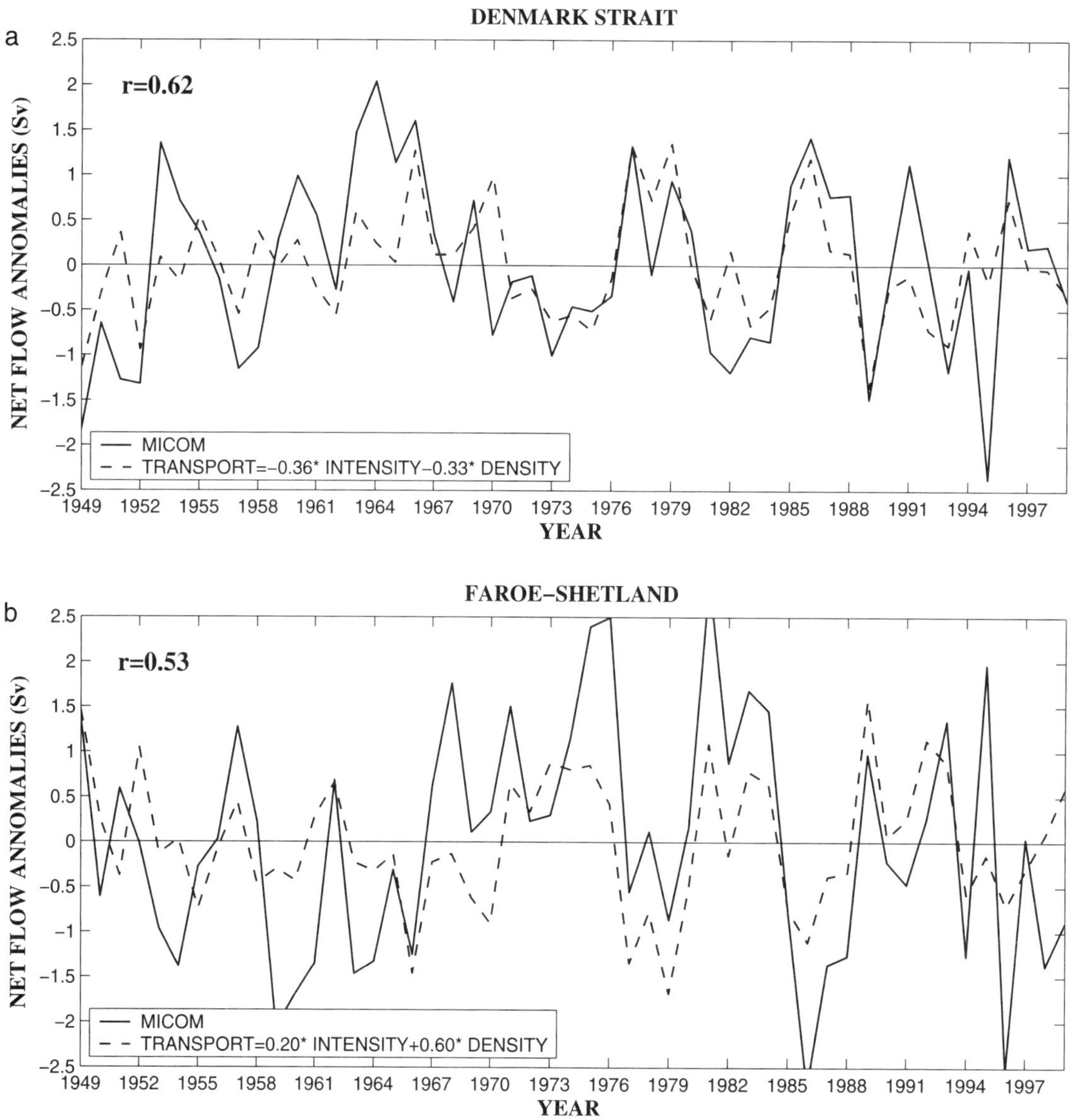

Figure 7. Wintertime (DJF) anomalies (detrended) of the net volume flow (STD) in the Denmark Strait (a) and the Faroe–Shetland Channel (b) from the MICOM model (solid) and the multiple linear regressions based on the DJF Nordic Seas cyclone track density and intensity (dashed). Note: Volume fluxes are detrended before the regression is calculated. *r* denotes the linear correlation between the MICOM-simulated net volume transport and the regression estimate.

Channel. On the other hand, the variability over the Iceland–Faroe Ridge seems to be more influenced by multi-decadal variations that are not reflected in the area-averaged Nordic Sea cyclone parameters.

5.2. The Appearance and Disappearance of Atmosphere–Ocean Relationships

According to several studies, oceanic variables that have been significantly correlated with NAO from the mid-1970s and onwards show no significant correlation before this period [*Jung and Hilmer*, 2000; *Lu and Greatbatch*, 2002]. As seen in Figure 4, NAO shows a stronger correlation with the synoptic low-pressure systems during the latter period. The question arises, does the wintertime synoptic low-pressure variability give a stronger correlation with the volume transports into the Nordic Seas in periods where the correlation towards the NAO is absent? To examine this, we have plotted the correlation within a 16-year moving window between the simulated net flow over the Faroe–Shetland Channel and the Nordic Seas number of cyclones (track density) variability. Figure 8 indicates that the wintertime net oceanic volume transports into the Nordic Seas are significantly less correlated to the wintertime cyclone variability during the early part of the period in the same manner, as was the case with the NAO (the same picture emerged when we calculated the moving correlation between the storm tracks and the net flow over the Denmark Strait). This indicates that not only changes in the ability of the NAO to represent the Nordic Sea cyclone variability make the correlation between NAO and oceanic parameters break down before the mid-1970s but also a lesser impact of Nordic Sea cyclone variability on the oceanic volume transports during this period. This reduced correlation between the oceanic volume transports into and out of the Nordic Seas and the Nordic Seas cyclone variability in the early period may be explained in two ways: through a nonlinear relationship between the cyclonic variability (here measured by tracking the extremes in the curl of the wind) and the ocean volume fluxes, or through a change in the persistence of the storm tracks.

The first mechanism is independent of the time scale of the cyclone variability and is related only to the fact that strong low-pressure variability might have a nonlinear impact on the variability of the wind stress, since the wind stress is approximately proportional to the square of the wind speed, which in turn would lead to a barotropic ocean adjustment. How a change in the lag-correlation (persistence) of the cyclone parameters may influence the effect of the cyclone variability on the ocean circulation may be explained by the possibility of a forcing with a lag-correlation (or at least a lack of strong negative lag-correlation) could impose a forcing of the same sign on the ocean for a longer time and therefore make the fast barotropic and slower response to Ekman pumping work together. Investigation of the 1-year lag-correlation of the Nordic Sea cyclone intensity and track density show that the lag-correlation have indeed changed with time (Table 3). This change in 1-year lag-correlations is linked to variability

Table 3. One-Year Lag Correlation for Wintertime (DJF) Cyclone Intensity and Track Density at Different Time Periods.

Parameter	Year		
	1950–1965	1965–1995	1949–2002
Track intensity	–0.50	0.14	0.00
Track density	0.12	0.35	0.25

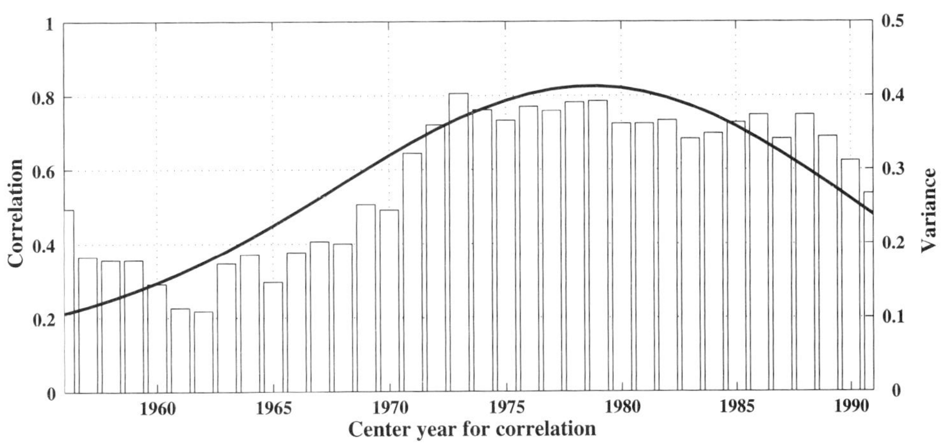

Figure 8. Bars: 16-year moving correlations between wintertime (DJF) area-averaged Nordic Sea track density and net oceanic volume transports through the Faroe–Shetland Channel. Solid line: moving 16-year average of the decadal (7-12-year) variance of the track density, calculated by using wavelet analysis (see section 3).

in the variance on the interannual and decadal scales (see Figure 3). The tendency for a more positive lag-correlation in the mid-1970s to mid-1990s is due to the increased power on decadal scales (Figure 3), which counteracted the tendency for a negative 1-year lag-correlation imposed by the interannual variability. When there is strong variability both on the interannual and the decadal scales, the resulting lag-correlation is close to zero.

Figure 8 indicates that changes in co-variability between ocean volume transport and Nordic Seas cyclones is related to the strength of the decadal cyclone variability, the relationship being strong (correlations of 0.7–0.8 and −0.7 to −0.8, with the net flow through the Faroe-Shetland Channel and Denmark Strait, respectively) in time periods with strong decadal variability. Thus, we cannot rule out the possibility that the persistence in the storm tracks may be a supplementary effect of the nonlinear response to wind stress. One might speculate that the strong Nordic Seas cyclone activity in the 1970s to 1990s explains the reported strong correlations between NAO and oceanic variables in these decades as well as the breakdown in correlation during earlier periods, when the Nordic Seas cyclone activity was weaker.

5.3. North Atlantic Cyclone Patterns During High- and Low-Volume Transports Into the Nordic Seas

In addition to the relationship between the wintertime mean Nordic Sea cyclone variability and the wintertime ocean volume transport determined in Table 2 and Figure 7, we wanted to get a spatial picture of the North Atlantic cyclone variability during years with strong anomalies in the oceanic transports across GSR. As investigated in section 5.2, the co-variability between the cyclone and oceanic volume flow variability seems highest during strong deviations in cyclone behaviour from the mean state. This justifies the use of composites (means over years with anomalies higher than 1 STD and lower than −1 STD) to display the spatial pattern of variability in the North Atlantic cyclones during years with high and low oceanic inflow and outflow through the different sections.

5.3.1. Relationships between North Atlantic cyclone variability and flow through the Denmark Strait and the Faroe-Shetland Channel. Figure 9 shows a composite of the number of cyclones (track density) during high and low inflow through the Denmark Strait. As indicated by the correlations (Table 2) and in the linear regression analysis (Figure 7), both the variability in inflow (Figure 9a and b) and in outflow (not shown) are related to the cyclone variability in the Nordic seas. Positive anomalies exceeding 1 STD in the inflow (increased flow northward) are linked to a 10-30% reduction in the number of low-pressure systems in the North Atlantic with the biggest reduction in the Icelandic area. The intensity of the low-pressure systems is reduced in the same areas (not shown). Due to this reduction in cyclone track density and intensity near Iceland, the area east of Greenland has southerly wind anomalies (i.e. reduced northerly winds). This could increase the inflow through the Denmark Strait into the Nordic Seas. The situation is opposite during low inflow, with increased cyclone number and intensity in the Icelandic area giving less inflow through the Denmark Strait. In addition, years with low inflow are associated with years with few cyclones in the Labrador Seas, which further increases the northerly wind anomalies east of Greenland. The composites of outflow anomalies depict the same type of pattern as the inflow composites. Interestingly, the composites for the Faroe–Shetland Channel inflow (Figure 10) and outflow (not shown) are similar to the composites for the Denmark Strait described above, the cyclone activity in the area close to Iceland being important for both the inflow and outflow. Thus the variability in the Nordic Sea wintertime cyclones may explain the strong anticorrelation found by *Nilsen et al.* [2003] between the net volume fluxes in the Denmark Strait and the Faroe–Shetland Channel (−0.72 for the wintertime transports and −0.74 for the annual 3-year smoothed data, defining net as the difference between inflow and outflow from the Nordic Seas).

5.3.2. Relationships between North Atlantic cyclone variability and flow through the Iceland–Faroe Ridge. Table 2 and Figure 7 indicate that the interannual variability in the wintertime volume flow across the Iceland–Faroe Ridge is only weakly connected to the area-averaged variability of wintertime cyclones in the Nordic Seas. However, as seen in Figures 11 and 12, the composites of high inflow and high and low outflow indicate that the anomalies in the Iceland–Faroe Ridge volume transport might be related to the cyclone variability in the vicinity of the Nordic Seas that is not picked up by the average Nordic Sea cyclone variability investigated in section 5.1. Low inflow seems connected to high low-pressure activity in the Kategat (Figure 11) with an increase in the number of cyclones entering the Kategat of 20-30% and an increase in the intensity of 10% during years with a inflow anomalies lower than −1 STD. This induces northwesterly wind anomalies (reduced southwesterly winds) in the Iceland–Faroe region, which may be responsible for the reduced inflow. On the other hand, increased outflow (Figure 12a) seems related to an increase in the number of cyclones in the area between northern Norway and Svalbard and fewer cyclones in the area west of British Islands (20% increase/reduction in the number of cyclones, respectively, for years where the Iceland–Faroe outflow deviates by 1 STD from the mean). Both of these changes will give northwesterly wind anomalies in the Iceland–Faroe area. Low outflow

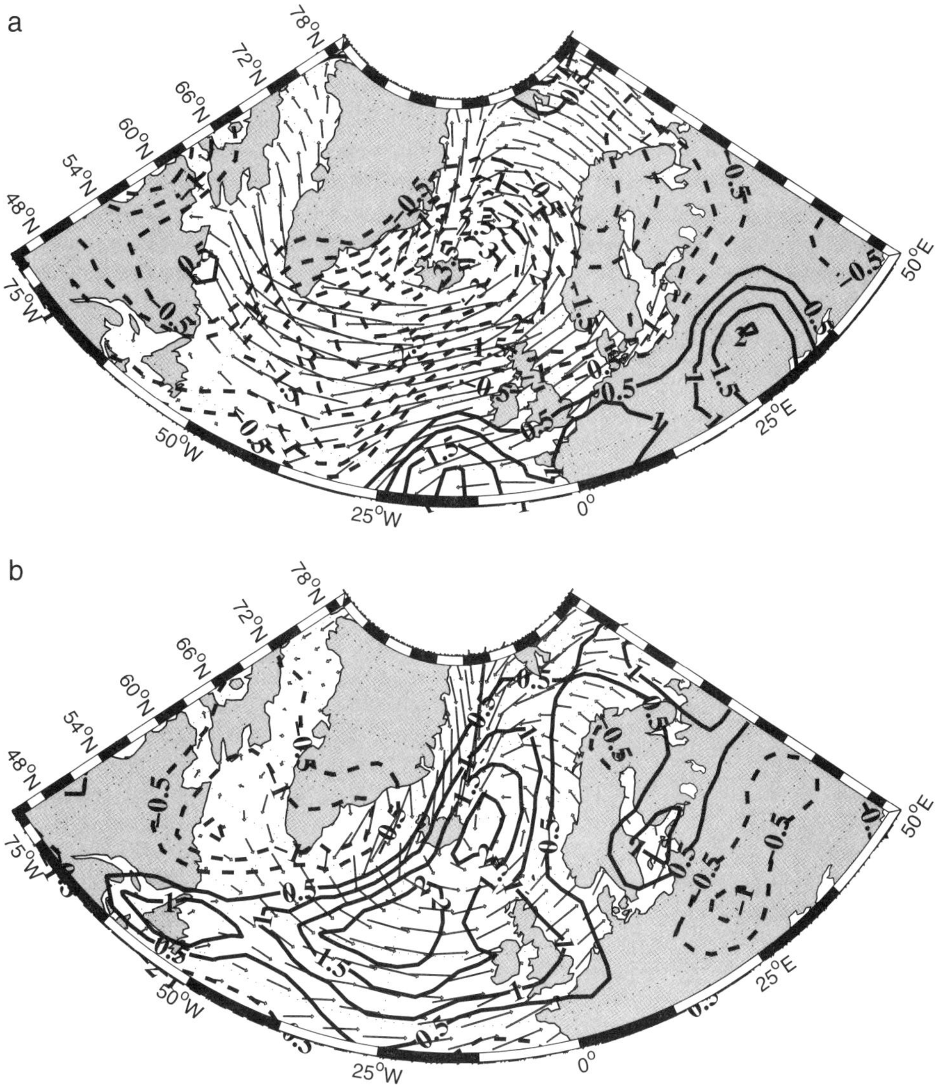

Figure 9. Composites of wintertime (DJF) mean anomalies in number of cyclones (track density) and anomalies in 10-m wind speed are displayed for years in which the DJF Denmark Strait inflow is greater than 1 STD (9 years; a) or less than −1 STD (7 years; b).

(Figure 12b), on the other hand, seems more related to the increase in number of cyclones in the Icelandic area. This might explain the small but significant correlation of −0.28 (Table 2) between the mean Nordic Seas cyclone track density and the outflow.

As shown, there is no instant connection between the Iceland–Faroe Ridge inflow and the Nordic Seas cyclone activity (Table 2), so the area-averaged Nordic Seas cyclone variability cannot explain the strong negative correlation (−0.8) in the annual 3-year-smoothed net flow between the Faroe–Shetland Channel and the Iceland–Faroe Ridge reported in *Nilsen et al.* [2003]. However, a possible contribution to this correlation might be the cyclonic activity in the Kategat area, which reduces the inflow across the Iceland–Faroe Ridge and pushes more water into the Faroe–Shetland branch. Such action would tend to anticorrelate the wintertime inflow in the two branches (correlation of −0.49) as indicated in *Nilsen et al.* [2003].

6. SUMMARY AND CONCLUSIONS

A feature-tracking algorithm of wintertime storm tracks has been used to investigate the relationship between wintertime cyclones and ocean transports in the Nordic Seas for

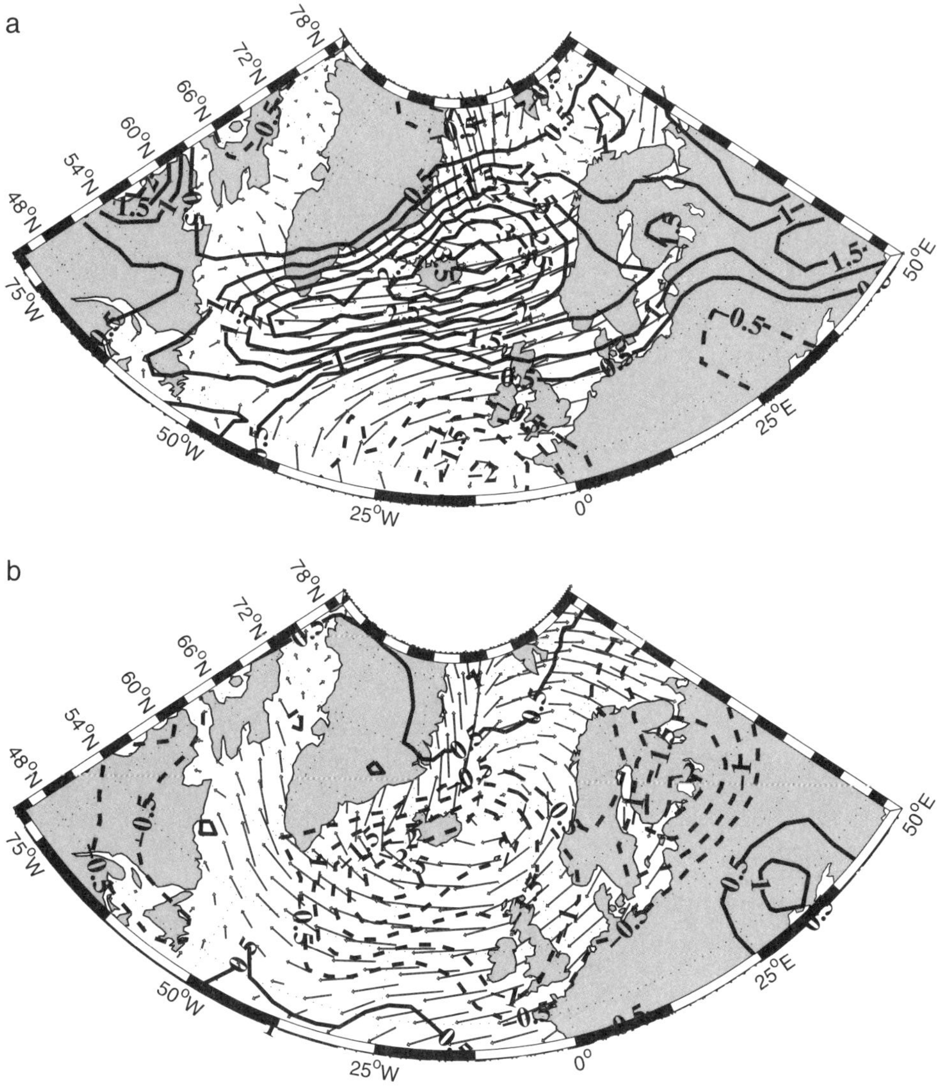

Figure 10. Composites of wintertime (DJF) mean anomalies in number of cyclones (track density) and anomalies in 10-wind speed for years with DJF Faroe–Shetland inflow greater than 1 STD (10 years; a) or less than –1 STD (10 years; b).

the period 1949-2002. The number and intensity of the mean Nordic Sea cyclones show interannual (2.5-year) and decadal (7-12-year) fluctuations with pronounced peaks in the early 1970s, '80s, and '90s. Both the number and intensity of the Nordic Seas cyclone variability are linked to the NAO and AO, the correlation being slightly stronger with the AO due to a reduced NAO correlation in the 1960s. The strong relationship between the Nordic Sea cyclones and AO/NAO confirms that the Nordic Seas cyclone variability is a major part of both the AO and NAO pattern. However, the breakdown of the correlation with NAO in the 1960s indicates that the NAO Index should be used with caution as an index for the wintertime Nordic Seas storm tracks. The NAO Index is best seen as an index containing information of variability in sea-level pressure on all spatial scales ranging from planetary to synoptic scales and also including the effect of variability in high-pressure systems. We have noted that the eastward shift in the NAO centres of action coincides with the period of increased decadal variability of the Nordic Sea wintertime storm track variability, indicating a possible link between the two phenomena.

Due to the lack of long observational time series of oceanic volume transport across the GSR we used a simulation with the Nansen Center version of the ocean model MICOM forced with daily atmospheric NCEP/NCAR reanalysis fields as a proxy for the real volume transports. Our results are

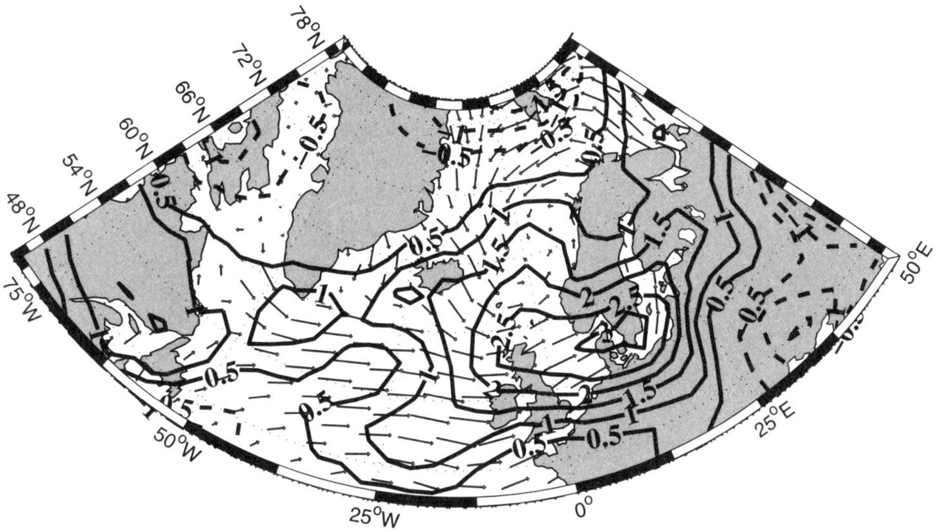

Figure 11. Composites of wintertime (DJF) mean anomalies in number of cyclones (track density) and anomalies in 10-m wind speed for years with DJF Iceland-Faroe inflow less than −1 STD (8 years).

therefore heavily dependent on the ability of this ocean model to reproduce the main features of the variability of the real volume transports. The wintertime variability in volume transports through the Faroe–Shetland Channel and the Denmark Strait was significantly correlated to the mean Nordic Seas cyclone variability (both number and intensity). This relationship with the wintertime cyclone variability, especially with the cyclones close to Iceland, might contribute to the strong anticorrelation between the annual net flow into the Nordic Seas through the Faroe–Shetland Channel and the Denmark Strait reported in *Nilsen et al.* [2003]. High (low) cyclone activity near Iceland induces strong southwesterly (northeasterly) wind anomalies in the Faroe–Shetland area simultaneously with strong northerly (southerly) wind anomalies in the Denmark Strait region.

The variability in the net flow across the Iceland–Faroe Ridge shows less influence of the mean Nordic Sea cyclone variability. However, years with low inflow over the Iceland–Faroe Ridge appear to be related to higher cyclone activity in the Kategat, giving northwesterly wind anomalies in the Iceland–Faroe region.

Nilsen et al. [2003] reported a strong anticorrelation between the annual net flow into the Nordic Seas in the Faroe–Shetland Channel and over the Iceland–Faroe Ridge. The wintertime cyclone activity in Kategat might have contributed to this when periods of strong cyclone activity in Kategat reduced the inflow across the Iceland–Faroe Ridge and pushed more water into the Faroe–Shetland branch. Apparently years with strong outflow (southward flow) over the Iceland–Faroe Ridge are associated with years having stronger than normal low-pressure activity in the area between northern Norway and Svalbard and weaker than normal activity west of the British Islands, both of which will contribute to northwesterly wind anomalies in the Iceland–Faroe area.

As for NAO, the correlation between the mean Nordic Sea cyclone activity and oceanic volume transports in and out of the region was strongest from the mid-1970s into the 1990s. Two mechanisms have been proposed in this respect. The first is the increase in Nordic Seas cyclone variability during this period, which could lead to a nonlinear impact (the wind stress is approximately proportional to the square of the wind speed) of the cyclone variability on the oceanic volume transports. This could make the co-variability between the variability in the cyclones and the oceanic volume transports more apparent during years with strong cyclone variability. A second effect might be that stronger persistence in the storm track due to increased variability on decadal scales might impose a forcing of the same sign on the ocean for a longer time and therefore make the fast barotropic and slower baroclinic ocean adjustments work together. The analysis in section 5.2 and the result of *Orvik and Skagseth* [2003], who reported a lagged response of 15 months between the North Atlantic wind-stress curl at about 55°N and the Norwegian Atlantic Current, indicate that this effect may be supplementary to the first one.

Note that the presented analysis considers only the seasonal co-variability of wintertime oceanic volume transport anomalies and the variability in wintertime synoptic low-pressure systems. An extension of this analysis will be to examine the ocean–atmosphere relationships in other seasons. Studies indicate that other relationships, including both

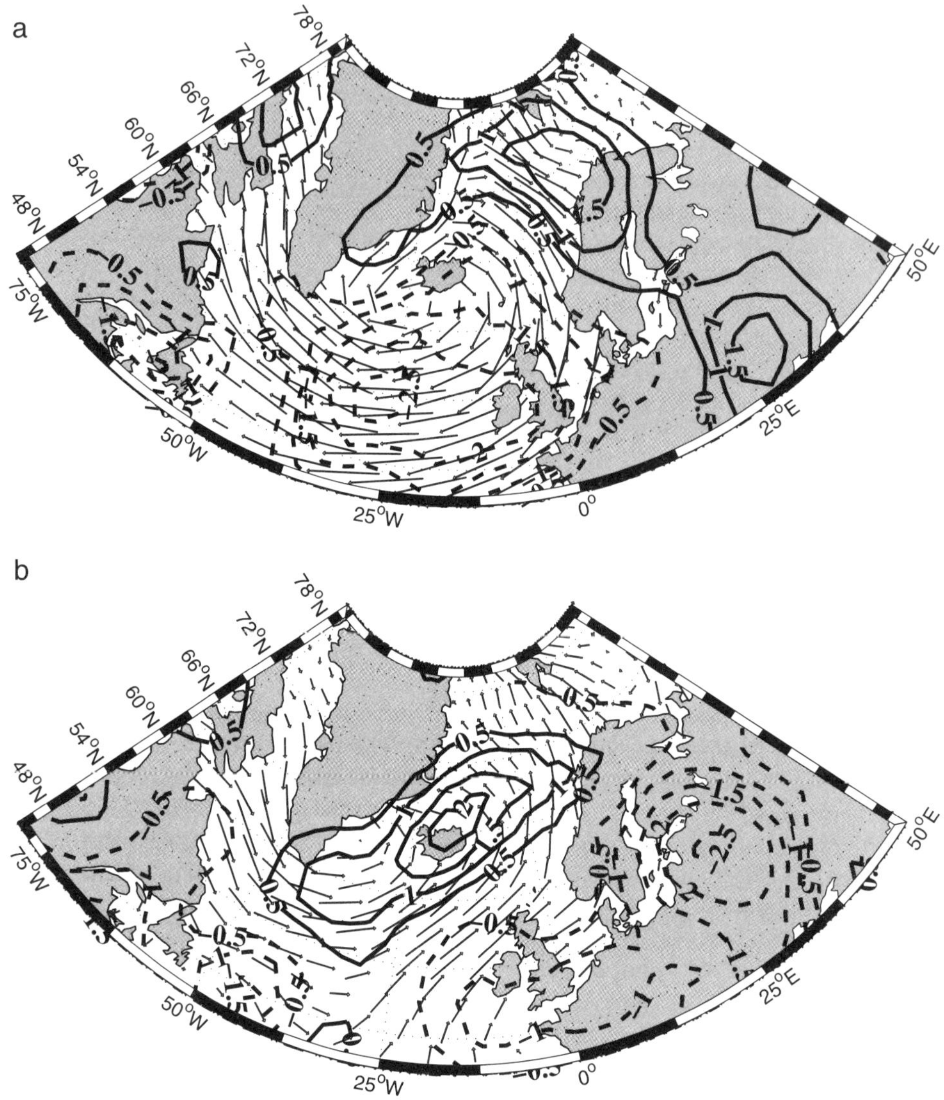

Figure 12. Composites of wintertime (DJF) mean anomalies in number of cyclones (track density) and anomalies in 10-m wind speed for years with DJF Iceland–Faroe outflow greater than 1 STD (8 years; a) or less than −1 STD (7 years; b).

the hydrological cycle and more large-scale ocean–atmosphere interactions [e.g.; *Blindheim et al.*, 2000; *Mork and Blindheim*, 2000; *Nilsen et al.*, 2003; *Visbeck et al.*, 2003], are important both for the strength of the mean volume transport and for the long-term variability. Some of these aspects are treated in *Furevik and Nilsen* (this issue).

Acknowledgments. The authors thank Dr. Kevin Hodges for providing the tracking algorithm; Dr. Jan Even Øye Nilsen for providing the ocean data, Figure 6, and valuable suggestions; and Professor Helge Drange and three anonymous reviewers for their insightful comments. This work was funded by the Norwegian Research Council under the SFF and ECOBE contracts. This is publication No. A87 from the Bjerknes Centre for Climate Research.

REFERENCES

Anderson, D., K.I. Hodges, and B.J. Hoskins, Sensitivity of Feature-Based Analysis Methods of Storm Tracks to the Form of Background Field Removal. *Monthly Weather Review*, 131, 565-573, 2003.

Bacon, S, Decadal variability in the outflow from the Nordic Seas to the deep Atlantic Ocean. *Nature*, 394, 871-874, 1998.

Barnston, A.G., and R.E. Livezey, Classification, seasonality and persistence of low-frequency atmospheric circulation patterns. *Monthly Weather Review*, 115, 1083-1126, 1987.

Battisti, D.S., U.S. Bhatt, and M.A. Alexander, A modeling study of the interannual variability in the wintertime North Atlantic Ocean. *Journal of Climate*, 8, 3067-3083, 1995.

Bell, G.D., and L.H. Bosart, A 15-year climatology of Northern Henisphere 500 mb closed cyclone and anticyclone centres. *Monthly Weather Review*, 117, 2142-2162, 1989.

Bentsen, M., Drange, H., Furevik, T., and Zhou, T., Variability of the Atlantic meridional overturning circulation in an isopycnic coordinate OGCM. *Clim. Dyn.,* in press, (2004).

Bentsen, M., *et al.*: Coordinate transformation on a sphere using conformal mapping, *Monthly Weather Review*, 127, 2733-2740, 1999.

Bleck, R., C. Rooth, D. Hu, and L.T. Smith, Salinity-driven thermohaline transients in a wind- and thermohaline-forced isopycnic coordinate model of the North Atlantic, *Journal of Physical Oceanography*, 22, 1486-1515, 1992.

Blackmon, M.L., A climatological spectral study of the 500 mb geopotential height of the Northern Hemisphere. *Journal of Atmospheric Science*, 33, 1607-1623, 1976.

Blackmon, M.L., J.M. Wallace, N.-C. Lau, and S. Mullen, An observational study of the Northern Hemisphere wintertime circulation. *Journal of Atmospheric Science*, 34, 1040-1053, 1977.

Blindheim, J., Borovkov, V., Hansen, B., Malmberg, S.-A., Turrell, W., and Østerhus, S., Upper layer cooling and freshening in the Norwegian Sea in relation to atmospheric forcing. *Deep-Sea Research*, 47, 655-680, 2000.

Cayan, D., Latent and sensible heat-flux anomalies over the northern oceans—The connection to monthly atmospheric circulation, *Journal of Climate*, 5, 354-369, 1992.

Chang, E., and Y. Fu, Interdecadal variations in the Northern Hemisphere winter storm track intensity, *Journal of Climate*, 15, 642-658, 2002.

Chang, E., S. Lee, K.L. Swanson, Storm track dynamics, *Journal of Climate*, 15, 2163-2183, 2002.

Delworth, T.L., North Atlantic interannual variability in a coupled ocean-atmosphere model. *Journal of Climate*, 9, 2356-2375, 1996.

Deser, C., and M. L. Blackmon, Surface climate variations over the North Atlantic Ocean during winter: 1900-1993. *Journal of Climate*, 6, 1743-1753, 1993.

Deser, C., and M.S. Timlin, Atmosphere-ocean interaction on weekly time scales in the North Atlantic and Pacific. *Journal of Climate*, 10, 393-408, 1997.

Dickson, R.R., T. Osborne, J. Hurrell, J. Meincke, J. Blindheim, B. Aadlandsvik, T. Vinje, G. Alekseev, and W. Maslowski, The Arctic response to the North Atlantic Oscillation. *Journal of Climate*, 13, 2671-2696, 2000.

Furevik, T., On the Atlantic Flow in the Nordic Seas, bifurcation and variability. Ph.D. thesis. Geophysical Institute, University of Bergen, 1998.

Furevik, T., M. Bentsen, H. Drange, J.A. Johannessen, and A. Korablev, Temporal and spatial variability of the sea surface salinity in the Nordic Seas, *Journal of Geophysical Research*, 107, 2002.

Geng, Q., and M. Sugi, Variability of the North Atlantic Cyclone activity in winter analysed from NCEP-NCAR Reanalysis data. *Journal of Climate*, 14, 18, 3863-3873, 2001.

Gillett, N.P., Graf, H.F., and Osborn, T.J., Climate change and the NAO. In: The North Atlantic Oscillation. Ed. J.W. Hurrell. *AGU Geophysical Monograph Series,* 134, ISBN 0-87590-994-9, pp. 193-210, 2003.

Gulev, S.K., O. Zolina, and S. Grigoriev. Extratropical cyclone variability in the Northern Henisphere winter from the NCEP/NCAR reanalysis data, *Climate Dynamics*, 17, 795-809, 2001.

Hodges, K.I. A general method for tracking analysis and its application to meteorological data. *Monthly Weather Review*, 122, 2573-2586, 1994.

Hodges, K.I. Feature Tracking on the Unit Sphere. *Monthly Weather Review*, 123, 3458-3465, 1995.

Hodges, K.I. Spherical Nonparametric Estimators Applied to the UGAMP Model Integration for AMIP. *Monthly Weather Review*, 124, 2914-2932, 1996.

Hodges, K.I.: Adaptive Constraints for Feature Tracking. *Monthly Weather Review*, 127, 1362-1373, 1999.

Hodges K.I., B.J. Hoskins, J. Boyle, and C. Thorncroft A Comparison of Recent Reanalysis Datasets Using Objective Feature Tracking: Storm Tracks and Tropical Easterly Waves, *Monthly Weather Review*, 131, 2012-2037, 2003.

Hoskins, B.J., and K.I. Hodges New Perspectives on the Northern Hemisphere Winter Storm-Tracks, *Journal of Atmospheric Science*, 59, 1041-1061, 2002.

Hubbard, B.B. *The World According to Wavelets: The Story of a Mathematical Technique in the Making.* Ed. A.K. Peters. (ISBN: 1-56881-072-5), 1996.

Hurrell, J.W. Decadal trends in the North Atlantic Oscillation: regional temperatures and precipitation, *Science,* 269, 676-679, 1995.

Hurrell, J.W., and H. van Loon Decadal variations in climate associated with the North Atlantic Oscillation, *Climate Change*, 36, 301-326, 1997.

Jones, P., T. Jonsson, and D. Wheeler Extension to the North Atlantic Oscillation using early instrumental pressure observations from Gibraktar and South-West Iceland. *International Journal of Climatology*, 17, 1433-1450, 1997.

Kalnay, E., *et al.* The NCEP/NCAR 40-year Reanalysis Project, *Bulletin of American Meteorological Society*, 77, 437-471, 1996.

Kistler, R., E. Kalnay, W. Collins, S. Saha, G. White, J. Woollen, M. Chelliah, W. Ebisuzaki, M. Kanamitsu, V. Kousky, H. van den Dool, R. Jenne, and M. Fiorino The NCEP-NCAR 50-year reanalysis: Monthly means CD-ROM and documentation, *Bulletin of American Meteorological Society*, 82, 247-267, 2001.

Kushnir, Y. Interdecadal variations in the North Atlantic sea surface temperature and associated atmospheric conditions, *Journal of Climate*, 7, 142-157, 1994.

Kuzmina, S.I., L. Bengtsson, O.M. Johannessen, H. Drange, L.P. Bobylev, and M.W. Miles: The North Atlantic Oscillation and greenhouse-gas forcing, *Geophysical Research Letters*, in press, 2005.

Lambert, S. Intense extratropical Northern Hemisphere winter cyclone events, *Journal of Geophysical Research*, 101, 21319-21325, 1996.

Lau, N.-C. Variability of the observed midlatitude storm tracks in relation to low-frequency changes in the circulation pattern, *Journal of Atmospheric Sciences*, 45, 2718-2743, 1988.

Lorenz, D.J., and D.L. Hartmann Eddy-Zonal Flow Feedback in the Northern Hemisphere winter. *Journal of Climate,* 16(8), 1212-1227, 2003.

Mork, K.A., and J. Blindheim: Variations in the Atlantic inflow to the Nordic Seas 1955-1995. *Deep-Sea Research, I,* 47, 1035-1057, 2000.

Orvik, K.A., and Ø. Skagseth: The impact of the wind stress curl on the North Atlantic inflow to the Norwegian Sea toward the Arctic. *Geophysical Research Letters*, 30, No. 17, 1884, 2003.

Nilsen, J.E.Ø., Y. Gao, H. Drange, T. Furevik, and M. Bentsen: Simulated North Atlantic-Nordic Seas water masses exchanges in an isopycnal coordinate OGCM. *Geophysical Research Letters*, 30, No. 10, 1536, 2003.

Pedlosky, J.: *Geophysical Fluid Dynamics*. Springer-Verlag, New York, 2nd ed., 1987.

Peterson, K.A., J. Lu, and R.J. Greatbatch, Evidence of nonlinear dynamics in the eastward shift of the NAO, *Geophysical Research Letters*, 30(2), 10.1029/2002GL015585, 2003.

Reitan, C.H.: Frequency of cyclones and cyclogenesis for North America, 1951–1970, *Monthly Weather Review*, 102, 861-868, 1974.

Rogers, J.C.: The association between the North Atlantic Oscillation and the South Atlantic Oscillation in the Northern Hemisphere. *Monthly Weather Review*, 112, 1999-2115, 1984.

Rogers, J.C.: Patterns of Low Frequency Monthly Sea Level Pressure Variability (1988-1986) and Associated Wave Cyclone Frequencies, *Journal of Climate*, 3, 1364-1379, 1990.

Seager, R., D.S. Battisti, J. Yin, N. Gordon, N.H. Naik, A.C. Clement, and M.A. Cane: Is the Gulf Stream responsible for Europe's mild winters? *Quarterly Journal of the Royal Meteorological Society*, 128, 2563-2586, 2002.

Serreze, M.C., J.E. Box, R.G. Barry, and J.E. Walsh: Characteristics of Arctic synoptic activity, 1952-1989, *Metrological and Atmospheric Physics*, 51, 147-164, 1993.

Thomson, D.J.: Spectrum estimation and harmonic analysis, *Proc. IEEE*, 70, 1055-1096, 1982.

Thompson, D.W.J., and J.M. Wallace: The Arctic Oscillation signature in the wintertime geopotential height and temperature fields. *Geophysical Research Letters*, 25, 1297-1300, 1998.

Thompson, D.W.J., and M. Wallace: Annular Modes in the Extratropical Circulation. Part I: Month-to-Month Variability. *Journal of Climate*, 13, No. 5, 1000-1016, 2000.

Torrence, C., and G.P. Compo: A Practical Guide to Wavelet Analysis. *Bulletin of American Meteorological Society*, 79, 61-78, 1998.

Trenberth, K.E., G.W. Branstator, D. Karoly, A. Kumar, N.-C. Lau, and C. Ropelewski: Progress during TOGA in understanding and modeling global teleconnections associated with tropical sea surface temperatures, *Journal of Geophysical Research*, 103, 14291-14324, 1988.

Visbeck, M., E.P. Chassignet, R.G. Curry, T.L. Delworth, R.R. Dickson, and G. Krahmann: The ocean's response to the NAO. In: The North Atlantic Oscillation. (Ed. J. Hurrell) *Geophysical Monograph Series*, 134. Washington, DC: American Geophysical Union, 2003.

Witthaker, L.M., and L.H. Horn: Geographical and seasonal distribution of North American cyclogenesis, 1958-1977, *Monthly Weather Review*, 109, 2312-2322, 1981.

Wilks, D.S., 1995: Statistical Methods in the Atmospheric Sciences. *International Geophysics Series, Vol. 59.* Academic Press.

Zishka, K.M., and P.J. Smith: The climatology of cyclones and anticyclones over North America and surrounding ocean environments for January and July 1950-1977, *Monthly Weather Review*, 108, 387-401, 1980.

Zolina, O., and S.K. Gulev: Improving the accuracy of mapping cyclone numbers and frequencies. *Monthly Weather Review*, 130, 748-759, 2002.

Øyvind Byrkjedal and Asgeir Sorteberg, Bjerknes Centre for Climate Research, University of Bergen, Allegaten 55, N-5007 Bergen, Norway. (oyvind.byrkjedal@bjerknes.uib.no; asgeir.sorteberg @bjerknes.uib.no)

Nils Gunnar Kvamstø, Geophysical Institute, University of Bergen, Allegaten 70, N-5007 Bergen, Norway. (nilsg@gfi.uib.no)

A Review of the Inorganic Carbon Cycle of the Nordic Seas and Barents Sea

Ingunn Skjelvan[1,2], Are Olsen[1,2], Leif G. Anderson[3], Richard G.J. Bellerby[1], Eva Falck[2], Yoshie Kasajima[1], Caroline Kivimäe[2], Abdirahman Omar[1,2], Francisco Rey[4], K. Anders Olsson[1], Truls Johannessen[1,2], and Christoph Heinze[2,1]

Studies of the inorganic carbon cycle have been performed in the Nordic Seas and the Barents Sea since the 1980s. Here we present a review over current knowledge on carbon transport between the different reservoirs in the area, and the transformation of inorganic carbon to organic matter. The carbon transport and transformation are closely related to other biogeochemical processes in the region, for instance air–sea gas exchange and primary production. Horizontal transport of carbon through the Nordic Seas (about 9 Pg C yr^{-1}) is orders of magnitude larger than transports due to air–sea gas exchange and deep mixing. The Nordic Seas and the Barents Sea are sinks for atmospheric carbon dioxide throughout the year (in the range of 20-85 g C m^{-2} yr^{-1}, for the different regions). However, the strength of the sink appears to be decreasing, at least for the Nordic Seas. The carbon transport due to deep-water formation has been estimated to 0.12 Pg C for the year 1994 in the Greenland Sea; however, this will vary with location, duration, and magnitude of deep mixing, showing large fluctuations. Carbon is also sequestered in the Nordic Seas and the Barents Sea due to biological activity, and the export production out of the surface layers in this area is in the range of 15-75 g C m^{-2} yr^{-1}. Rising atmospheric carbon dioxide levels and changes in climate will to a varying degree affect the different processes, and this is briefly addressed in the paper.

1. INTRODUCTION

The Nordic Seas (Greenland Sea, Norwegian Sea, and Iceland Sea), as well as the Barents Sea, are active areas for carbon sequestration from the atmosphere as well as for formation of intermediate to deep waters. At present the atmosphere receives an increasing amount of carbon dioxide (CO_2) from human activities like burning of fossil fuel and deforestation, which is reflected in an annual increase of atmospheric CO_2 of about 1.4 μatm [*Keeling and Whorf*, 2004]. This perturbation of the natural carbon cycle might have adverse effects on climate [*IPCC*, Chapter 3, 2001].

The atmospheric content of carbon in the form of carbon dioxide is about 750 Pg (Pg = 10^{15} g), which is about 2/3 of the carbon stored in the surface ocean and 1/50 of the amount stored in the deep ocean. The ocean has capacity to store an enormous amount of carbon, and it is estimated that on time scales of thousands of years about 90% of the man-made or anthropogenic carbon input to the atmosphere will be sequestered in the ocean [*Bolin and Eriksson*, 1959; *Archer et al.*, 1998]. However, the bottleneck in this process is the slow overturning of the ocean, which restricts

[1] Bjerknes Centre for Climate Research, University of Bergen, Bergen, Norway.
[2] Geophysical Institute, University of Bergen, Bergen, Norway.
[3] Department of Chemistry, Göteborg University, Göteborg, Sweden.
[4] Institute of Marine Research, Bergen, Norway.

The Nordic Seas: An Integrated Perspective
Geophysical Monograph Series 158

10.1029/158GM11

the exchange between upper and deeper layers [*Bolin and Eriksson*, 1959]. With respect to carbon dioxide (CO_2), the surface layer equilibrates with the atmosphere in about 1 year, while the mixing time between upper and deep ocean is approximately 1000 years [*Broecker*, 1974], and this clearly illustrates that mixing between surface and deep ocean limits the rate of carbon storage in the ocean.

The north–south conveyor belt of carbon is, in a simplistic way, expressed as follows: Carbon flows northwards with the surface currents towards the Nordic Seas and Barents Sea, and on its way the water cools, which favours the introduction of carbon from the atmosphere into the surface waters. At higher latitudes, like the Greenland Sea, the cold and dense surface water mixes to larger depths, feeding the deep ocean with carbon and oxygen-rich water. The deep water flows towards the south, and on its way, a gradual upward movement occurs across the global oceans. Carbon is eventually released to the atmosphere at low latitudes. The circulation is complemented by sinking of organic matter to the deep water as a result of biological productivity in the surface ocean. On short time scales, the introduction of anthropogenic CO_2 into the atmosphere will not change the circulation of carbon described above, but the amount of carbon transported between areas and reservoirs will be perturbed.

This paper reviews the current knowledge regarding how the Nordic Seas and the Barents Sea act as carbon sinks. We have focused on several processes. Carbon is transported horizontally between areas and vertically from surface waters to larger depths by physical processes. In addition, atmospheric CO_2 enters the surface ocean as a function of wind speed and the air–sea carbon dioxide concentration gradient. Further, during the biologically productive season, carbon is transformed into biological matter, which is re-cycled in the upper water or transported towards larger depths. This latter process will remove carbon from the atmosphere–sea surface cycle for centuries. The term export production is used to quantify the amount of carbon removed from the surface layer by sinking organic material, and different approaches to estimate this term are presented and compared. Finally, the different processes are discussed in the context of rising atmospheric carbon dioxide levels and changing climate.

2. CHEMISTRY OF DISSOLVED INORGANIC CARBON

Dissolved inorganic carbon (C_T) appears in the surface water as carbon dioxide ($CO_2(aq)$), carbonic acid (H_2CO_3), hydrogen carbonate ion (HCO_3^-), and carbonate ion (CO_3^{2-}). In seawater with a typical pH of 8.1, about 90% of C_T is present as hydrogen carbonate and 9% as carbonate, which leaves less than 1% to dissolved carbon dioxide and carbonic acid. The distribution reflects the capacity of the seawater to act as a buffer for CO_2, which is expressed in the following reaction,

$$CO_2(aq) + H_2O + CO_3^{2-} \Leftrightarrow 2HCO_3^- \qquad (1)$$

Changes in pH control the distribution of inorganic carbon between the different species in the ocean. A slightly more acidic ocean, as a result of large uptake of anthropogenic CO_2 from the atmosphere, will be reflected in a reduced oceanic buffer capacity. This will reduce the availability of carbonate ions for further oceanic CO_2 uptake.

Total alkalinity (A_T) is closely connected to the buffer capacity, and this quantity expresses the capacity of the water to take up CO_2; i.e. high A_T values reflect seawater with large uptake potential [*Stumm and Morgan*, 1996].

3. HORIZONTAL EXCHANGE OF CARBON WITH SURROUNDING SEAS

The Nordic Seas experience a large exchange of waters with surrounding seas. The major inflow comes from south with the Atlantic water. Much of this originally warm and salty water returns to south after being modified within the Nordic Seas, while a fraction continues into the Arctic Ocean, both through the Fram Strait and the Barents Sea (Figure 1a and b). In the north, colder and fresher water from the Arctic Ocean enters the Nordic Seas, and to a large extent this water follows the East Greenland continental slope southwards as the East Greenland Current (EGC) and continues into the North Atlantic Ocean. Cold deep water also exits the Arctic Ocean into the Nordic Seas, mainly following the East Greenland continental slope, but significant amounts turn eastwards and flow into the deep Nordic Seas. The EGC entrains waters from the Greenland and Iceland Seas during the flow southwards, thus making the water that flows into the North Atlantic a mixture of a number of different source waters.

This exchange of water with surrounding seas will of course also transport dissolved carbon (both inorganic and organic), contributing to the carbon budget of the Nordic Seas. Dissolved organic carbon (DOC) concentrations in seawater of the Arctic Ocean and Nordic Seas are typically between 50 and 100 μmol kg^{-1} [e.g. *Anderson*, 2002], while C_T concentrations are largely a function of salinity and varies between 2000 and 2200 μmol kg^{-1} [e.g. *Anderson et al.*, 1998]. Furthermore both the DOC and the C_T concentrations are affected by biological activity, where primary production consumes C_T and produces DOC, while respiration has the opposite effects. Consequently, any attempt to compute the transport of carbon into and out of the Nordic Seas will be heavily impacted by the season as well as the volume fluxes.

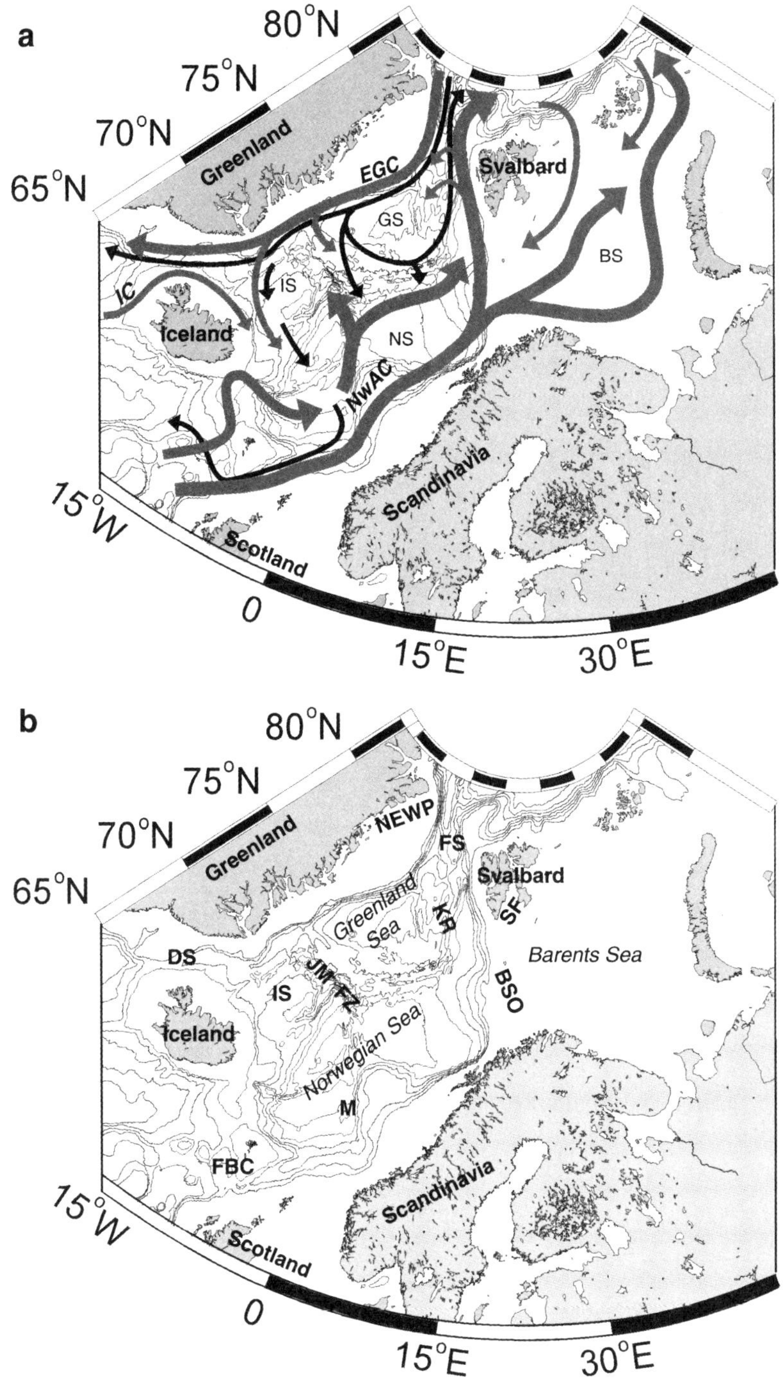

Figure 1. (a) Nordic Seas and Barents Sea with dominating currents. Gray arrows denote surface water currents and black arrows denote deep-water currents. GS: Greenland Sea, NS: Norwegian Sea, BS: Barents Sea, IS: Iceland Sea, NwAC: Norwegian Atlantic Current, EGC: East Greenland Current, IC: Irminger Current. (b) Nordic Seas and Barents Sea with geographical features. BSO: Barents Sea Opening, SF: Storfjorden, FS: Fram Strait, NEWP: Northeast Water Polynya, KR: Knipovich Ridge, JMFZ: Jan Mayen Fracture Zone, M: Ocean Weather Station Mike, IS: Iceland Sea, DS: Denmark Strait, FBC: Faroe Bank Channel.

Carbon budget computations for the Nordic Seas and the Arctic Ocean have been performed with the objective to assess the air–sea flux of CO_2 from the difference between the in- and out-fluxes [*Lundberg and Haugan*, 1996; *Anderson et al.*, 1998]. Such computations are much more robust than those of the absolute horizontal fluxes of carbon into and out of the Nordic Seas, because the former is based on a balance in the volume fluxes, which tend to compensate for some of the uncertainties in the individual volume fluxes. This is based on the fact that the C_T concentration is closely coupled to salinity in the high-salinity waters (where the largest uncertainties in the volume fluxes exist) but less so in the low-salinity upper waters (where the volume fluxes are relatively well known).

The major deviation from the salinity–C_T relationship in the waters of the Nordic Seas occurs for the low-salinity waters of the EGC. These have a signature of sea-ice melt water and river runoff, where the latter has high relative C_T concentration as a result of decay of organic matter in the drainage basins. The rivers also supply terrigenous DOC to the Arctic, resulting in elevated concentrations of both DOC and C_T in the EGC, where the former can be traced by lignin content and carbon-isotope ratios [*Opsahl et al.*, 1999].

Table 1 gives an overview of the major fluxes and carbon characteristics of the water flowing through the Nordic Seas. To put some perspective on the horizontal transport of carbon, the volume fluxes have to be considered. *Hansen and Østerhus* [2000] estimated that 8 Sv (1 Sv = 10^6 m^3 s^{-1}) enters the Nordic Seas from the North Atlantic. Further north, about 5 Sv enters the Arctic Ocean from the Nordic Seas through the Fram Strait and the Barents Sea [*Mauritzen*, 1996]. As the water cools on its way northwards, the uptake of CO_2 from the atmosphere increases caused by increased solubility. Consequently, the mean C_T concentration is higher in the seawater that flows into the Arctic Ocean than in that entering the Nordic Seas in the south. Typical winter C_T concentrations in the waters entering the Nordic Seas from the south are 2120 μmol kg^{-1} [*Falck and Anderson*, in press, 2004] and in the waters exiting to the Arctic Ocean 2140 μmol kg^{-1} [*Kaltin and Anderson*, in press, 2004]. A back of an envelope calculation gives corresponding carbon transports of 6.4 and 4.0 Pg C yr^{-1}, respectively. To put these amounts in perspective, the air–sea flux of CO_2 within the Nordic Seas has been estimated as ~0.09 Pg C yr^{-1} from budget computations (combination of the estimates from *Lundberg and Haugan* [1996] and *Anderson et al.* [1998]); compared to this, the horizontal fluxes dominate the carbon transport of the Nordic Seas by one to two orders of magnitudes.

Calculating the transport of carbon from the Arctic Ocean to the Nordic Seas is a greater challenge, as there is a considerable salinity gradient in the EGC at Fram Strait, which also has an impact on the C_T concentration. The vertical variability in salinity, temperature, and C_T in the top 500 m of the EGC at latitude of 81.5°N in the late winter (May 9, 2002) is illustrated in Figure 2. The volume transport to the south in the EGC should closely match that of the volume flowing north from the Nordic Seas to the Arctic Ocean due to conservation of mass. The only other in- and outfluxes are those through the Bering Strait and river runoff, and through the Canadian Arctic Archipelago, respectively. Hence, the uncertainty is how much of the Bering Sea inflow does not exit through the Canadian Arctic Archipelago. The normalized C_T profile in Figure 2 shows that the excess C_T in the low-salinity surface water contributes to a carbon transport to the south that is larger than the northward transport, and this is mainly attributed to the addition by river runoff. To put some numbers on the carbon flux into the Nordic Seas from the Arctic Ocean, volume estimates from *Mauritzen* [1996] and typical carbon values from Figure 2 are used. Back of an envelope calculation gives carbon fluxes of about 3.1 Pg C yr^{-1}.

Locally in the Nordic Seas, the transport of C_T into and out of the Greenland Sea has been calculated by combining measured C_T and modelled water transports using the Miami

Table 1. Horizontal Fluxes Into and Out of the Nordic Seas.

	From	Transport through	Volume flow (Sv)	Transport of carbon (Pg yr^{-1})	Typical C_T values (μmol kg^{-1})
INTO	South	Greenland Scotland Ridge	8[H&Ø]	6.4	2120[F&A]
Nordic Seas	North	Fram Strait (surface)	1.6[M]	1.3	2090[A]
		Fram Strait (deep)	2.2[M]	1.8	2140[A]
OUT OF	South	Denmark Strait (surface)	–1.3[H&Ø]	–1	2140[S]
Nordic Seas		Greenland Scotland Gap (deep)	–5.6[H&Ø]	–4.6	2155[O&A]
	North	Fram Strait and Barents Sea	–5[M]	–4.0	2140[K&A]
Total			~0	–0.1	

A is *Anderson* [unpublished data], F&A is *Falck and Anderson* [in press, 2004], H&Ø is *Hansen and Østerhus* [2000], K&A is *Kaltin and Anderson* [in press, 2004], M is *Mauritzen* [1996], O&A is *Olsen and Anderson* [unpublished data], S is *Skjelvan et al.* [1999b]. Note that transports of carbon in Pg are back-of-an-envelope calculations.

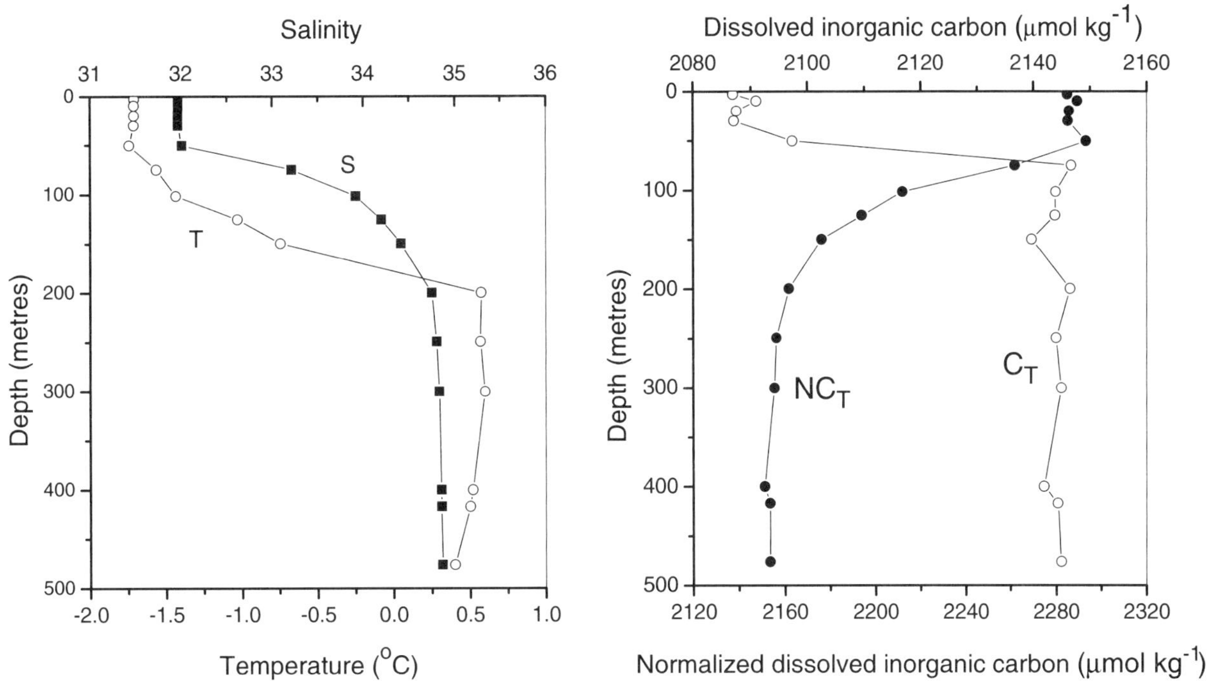

Figure 2. Depth profiles of salinity and temperature (left) and C_T as well as C_T normalized to a salinity of 35 (right) in the top 500 m of the EGC at latitude of 81.5°N in May 9, 2002.

Isopycnic Coordinate Ocean Model [*Bleck et al.*, 1992; *Chierici et al.*, 1999]. The main transport of carbon into the Greenland Sea occurs through Fram Strait in the surface mixed layer (SML) and the deep layer, and from the Atlantic in the SML, whereas the main transport out is in the SML crossing the Jan Mayen Fracture Zone and across the Knipovich Ridge (Figure 1b). The carbon transport calculation reveals that during wintertime more carbon is transported out of the Greenland Sea than into the basin, corresponding to an oceanic CO_2-uptake from the atmosphere of 0.024 ± 0.006 Pg C yr^{-1}. This amount represents 20-30% of the total CO_2-uptake in the Nordic Seas.

Hansen and Østerhus [2000] estimate the overflow across the Greenland Scotland Ridge to be 5.6 Sv, prior to entrainment of water in the North Atlantic. The overflow is equally divided between the Iceland–Scotland Ridge and the Denmark Strait, and the main sources of this overflow water are the Arctic Ocean, the Greenland Sea, and recirculating Atlantic water. Measurements from the Faroe Bank Channel and the Denmark Strait indicate that the inorganic carbon concentration of the overflow water is 2150 to 2160 μmol kg^{-1} [*A. Olsen* and *L. Anderson*, unpublished data], which gives an annual flux of about 4.6 Pg C with the Nordic Seas deep water to the North Atlantic.

The horizontal carbon fluxes through the Nordic Seas are summarized in Table 1. We stress that these fluxes have to be taken with care, since there are large uncertainties in the volume transport. However, the discrepancy between carbon transports into and out of the Nordic Seas of approximately 0.1 Pg yr^{-1}, which is attributed to air–sea gas uptake, expresses the orders of magnitude difference between the horizontally transported carbon and the fluxes of carbon due to other processes. There is not enough information available to accurately assess the uncertainties in the volume transport, but the largest are most likely in the in- and outflow through the Fram Strait. These uncertainties could easily be up to 25%, and in this context the uncertainties in C_T become negligible.

4. AIR–SEA GAS EXCHANGE

The carbon transport between atmosphere and ocean is controlled by wind speed and ΔpCO_2, the latter being the difference between atmospheric and sea-surface partial pressure of CO_2. In the following, the term fugacity, *f*, is used for the partial pressure, which expresses the non-ideal nature of CO_2 and which differs from pCO_2 by less than 0.5%. Whereas the atmospheric concentration of CO_2 is slowly varying in time and space, the sea-surface fCO_2, and by that the ΔfCO_2, is largely influenced by sea-surface temperature (SST), ocean advection, and mixing processes, and biological primary production. It has been shown, at least for the Greenland Sea, that about half of the air–sea flux is driven by heat exchange with the atmosphere and half by biological production [*Anderson et al.*, 1999]. Heat loss to the atmosphere when surface water moves northwards will decrease its temperature and hence increase its gas solubility. Consequently sea-surface fCO_2 decreases with increasing latitude, which will induce an uptake of CO_2 from the atmosphere. Non-calcareous biological primary production also decreases sea-surface fCO_2 and stimulates uptake from the atmosphere (see also section 6.3). Variability in fCO_2 of up to 100 μatm has been observed during spring in the surface waters of the Greenland Sea in areas with similar hydrographical conditions but at different stages with respect to spring bloom development [*Skjelvan et al.*, 1999a].

The air–sea flux of CO_2 can be estimated according to:

$$F_{CO_2} = k_w K_0 \Delta fCO_2 \tag{2}$$

where K_0 is the solubility of CO_2 and k_w is the gas transfer coefficient. The solubility depends on the salinity and temperature of the water [*Weiss*, 1974], and ΔfCO_2 can be determined from measurements. However, the gas transfer coefficient is more of a challenge. Parameterization of the gas transfer coefficient is a never-ending subject for debate, but mostly a relationship with wind speed is used. Depending on which relationship is used, the carbon flux estimates might vary by a factor of two. All the estimates presented here are based on the relationship of *Wanninkhof* [1992] (his equations 1 and 3), which expresses the gas transfer as a function of the square of the friction wind speed (i.e., the 10-m wind speed). The wind speed in the Nordic Seas shows huge fluctuations, which has a large impact on the air–sea flux of CO_2.

The Nordic Seas and Barents Sea surface water is in general undersaturated with respect to CO_2 throughout the year, and these areas act as sinks for atmospheric CO_2, the Greenland Sea being the largest sink due to the high fCO_2 gradients across the air–sea interface [*Skjelvan et al.*, 1999a,b; *Olsen et al.*, 2003a].

4.1. Estimates of the Air–Sea Flux

Several approaches have been used to estimate air–sea fluxes in the Nordic Seas and the Barents Sea. The most straightforward approach is the one based on measured ΔfCO_2 data; thus, the challenge of this method is of course the scarcity of data. In the Nordic Seas several estimates of the air–sea flux are based on such direct measurements, and Table 2 summarizes the estimates presented in the following paragraphs.

Skjelvan et al. [1999a] calculated a carbon flux into the Greenland and Norwegian Seas of 67–85 and 36 g C $m^{-2} yr^{-1}$, respectively, based on data from the years 1995 and 1996–1997. (The fluxes are recalculated with respect to the ones published by weighting the monthly flux values differently. Winter, spring, summer, and fall fluxes are assumed to last for 2 months each, which leaves 4 months between the seasons where linear interpolation of the fluxes is used). *Anderson et al.* [2000a] used all available carbonate system data in the Greenland Sea to compute the annual surface water fCO_2 cycle, giving a carbon flux of 53 g C m^{-2} yr^{-1} into the ocean. For the Norwegian Sea, *Falck and Anderson* [in press, 2004] calculated a flux of CO_2 from the atmosphere into the surface water at Ocean Weather Station M (OWSM) of 32 g C m^{-2} yr^{-1}, based on observed surface water fCO_2 data from 1991–1994. Air–sea flux of carbon into the Iceland Sea was estimated by *Skjelvan et al.* [1999b], who calculated the oceanic uptake over the years 1993 to 1997 to be 69 g C m^{-2} yr^{-1}.

Olsen et al. [2003a] presented a different approach, where an empirical relationship between SST and fCO_2 valid for

Table 2. Air–Sea Fluxes of Carbon Into the Surface Water of Greenland, Norwegian, Barents, Iceland Seas, Storfjorden Polynya, and Northeast Water Polynya and Characteristics of These Areas. Fluxes are annual estimates if nothing else is noted.

	Greenland Sea	Norwegian Sea	Barents Sea	Iceland Sea	Polynyas
Area (–10^{12} m^2)	0.81[S&H]	1.39[S&H]	1.4[Sa]	0.51[S&H]	0.0013**
Ice cover	Temporary (Dec-Apr)	None	Temporary (Dec-July)	Temporary (drift ice)	Temporary
Air–sea gas exchange (g C m^{-2})	53[A] 67-85[S1]* 40-50[Ol] 54[H]	36[S1]* 20[Ol] 32[F&A]	98[Om1] 29[Om1] (in PS) 29[K] (in PS) 44[Fr]	69[S2] 40-50[Ol]	65[Om2] (winter)** 5[Fa] (summer)***

A is *Anderson et al.* [2000a], Fr is *Fransson et al.* [2001], Fa is *Falck* [1999], F&A is *Falck and Anderson* [in press, 2004], H is *Hood et al.* [1999], K is *Kaltin et al.* [2002], Ol is *Olsen et al.* [2003a], Om1 is *A.Omar* [personal communication], Om2 is *Omar et al.* [this issue], S1 is *Skjelvan et al.* [1999a], S2 is *Skjelvan et al.* [1999b], Sa is *Sakshaug et al.* [1992], S&H is *Simonsen and Haugan* [1996].

*Estimate is recalculated with respect to the one previously published. **Estimate is from a coastal polynya in the Storfjorden at Svalbard. *** Estimate is from the Northeast Water Polynya at the shelf east of Greenland. NA, not available. PS, productive season.

winter months (October–March) was determined. This relationship was used to extrapolate fCO_2 data in both time and space in the Nordic Seas and the northern North Atlantic (Figure 3a). Figure 3b shows fluxes into the Greenland, Norwegian, and Iceland Sea in the order of 40–50, about 20, and 40–50 g C m^{-2} yr^{-1}, respectively. Summarised over the whole area, the air–sea carbon flux during winter totalled 0.081 Pg C (see Table 3). This number is considerable lower than the one by *Takahashi et al.* [2002], who presented an air–sea estimate from the North Atlantic north of 50°N (in revised form at http://www.ldeo.columbia.edu/CO2) of 0.31 Pg yr^{-1}. The discrepancy is most likely due to the different ways of projecting data onto a single year. *Takahashi et al.* [2002] assumed that there is no anthropogenic increase in the sea water fCO_2 of the North Atlantic, while *Olsen et al.* [2003a] accounted for such an increase and sized it to be at the same rate as the atmospheric increase in fCO_2.

The approach of *Olsen et al.* [2003a] represented an attempt to overcome the lack of fCO_2 time series measurements in the Nordic Seas. *Olsen et al.* [2003a] employed their relationship on objectively analysed fields of bias-corrected satellite SST data [*Reynolds and Smith*, 1994], thus determining fCO_2 fields covering winter months from 1981 through to 2001. Interannual variations of the Nordic Seas fCO_2 were determined to typically be around ±0.5% to 2%, corresponding to ±1.5 to 6 µatm (Figure 3c). Furthermore they determined interannual wintertime flux variations to be typically around 10-20% over large areas of the Nordic Seas

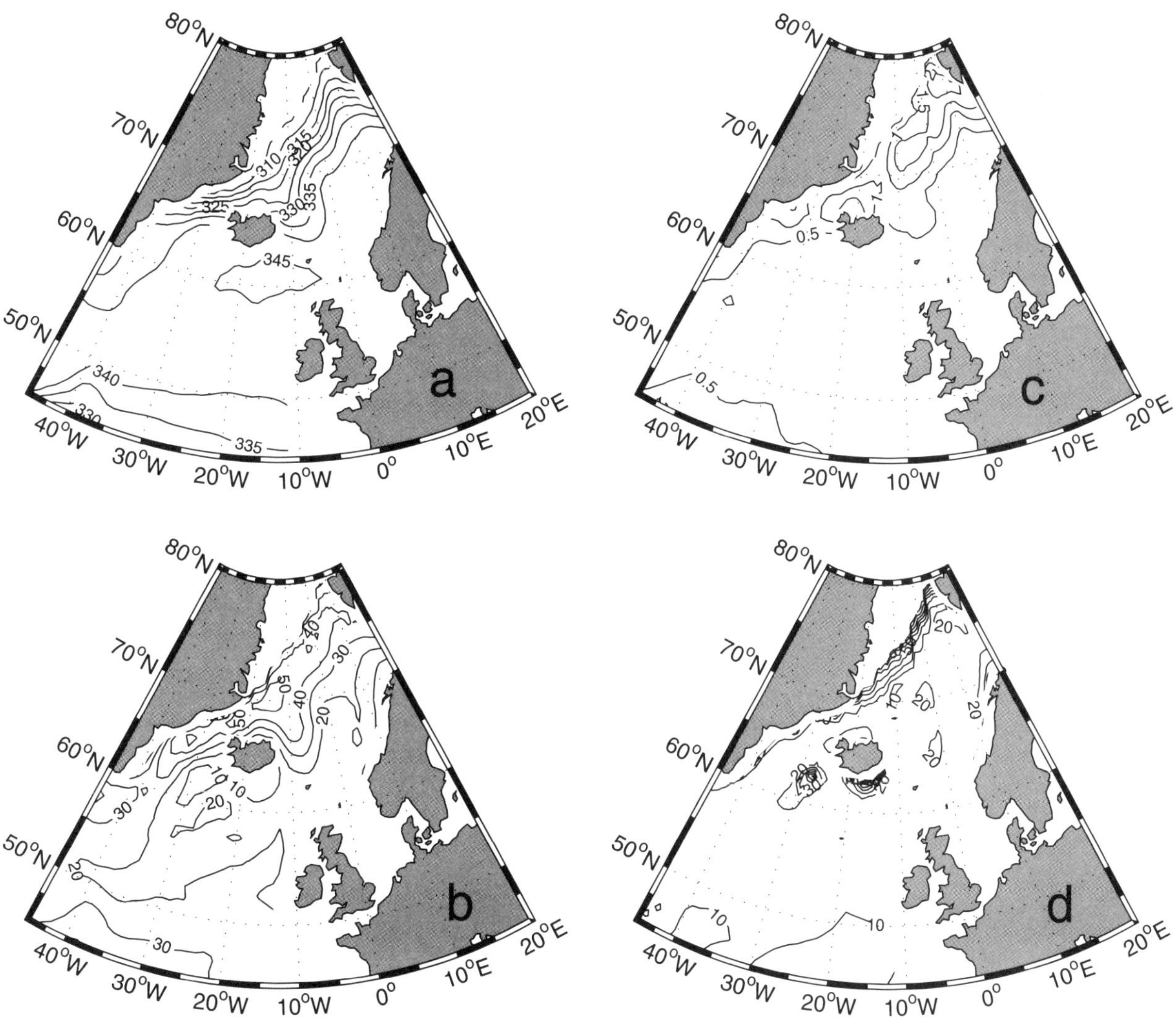

Figure 3. Averaged fCO_2 values in µatm (a), air–sea flux of CO_2 in g C m^{-2} yr^{-1} during winter season (October to March) (b), interannual standard deviation in per cent for fCO_2 (c), and interannual standard deviation in per cent for the CO_2 flux (d). The figure is reprinted from *Olsen et al.* [2003a] with permission from Elsevier.

Table 3. Air–Sea Fluxes of Carbon (in Pg) Into the Nordic Seas (Greenland Sea, Norwegian Sea, and Iceland Sea). The flux values are annual if nothing else is noted.

Estimated from	Air–sea flux (Pg C)	Comments
Lundberg [1994]	0.08	Ice extent of 37% assumed for the Nordic Seas
Skjelvan et al. [1999a, 1999b]	0.09 + 0.01	Ice extent of 37% assumed for the Nordic Seas
Takahashi et al. [2002]	0.31	North of 50°N, Arctic and Barents Sea included
Olsen et al. [2003a]	0.081	North of 45°N, Arctic and Barents Sea not included, realistic sea-ice extent, October to March.

(Figure 3d). They concluded that factors that control interannual flux variations, such as sea-surface fCO_2, wind speed, and atmospheric fCO_2 (a consequence of variations of the atmospheric pressure at sea level), seemed to be of about equal importance in the Nordic Seas. The very high interannual flux variations in the Greenland Sea (Figure 3d) are a consequence of interannual variations of ice-cover extent.

Hood et al. [1999] used a combined method of drifting buoys fCO_2 data and a SST–fCO_2 relationship to determine air–sea flux of carbon into the Greenland Sea. Their estimate of 54 g C m^{-2} yr^{-1} into the sea is comparable with the ones presented above.

Considering the whole Nordic Seas, *Skjelvan et al.* [1999b] estimated a CO_2 uptake of 65 g C m^{-2} yr^{-1}, from summation of flux estimates for the different seas, weighted by area. This estimate is larger than the one by *Lundberg* [1994], who calculated a flux of CO_2 into the surface ocean of 49 g C m^{-2} yr^{-1}, based upon fCO_2, C_T, alkalinity and pH measurements in the area.

Taking into account the ice-free area of the Nordic Seas, which is assumed to be 63% of the total area [*Lundberg*, 1994], the annual carbon sink of these seas is estimated to be 0.09 ± 0.01 Pg C. This is in good agreement with *Lundberg* [1994], who calculated an annual uptake of atmospheric CO_2 of 0.08 Pg C (Table 3). From this we can estimate an annual carbon uptake from the atmosphere of 0.03 Pg for the Greenland Sea, which covers an area of about 1/3 of the total Nordic Seas. This latter calculation does not take into account the fact that the Greenland Sea experiences a heavy ice-extent compared to the Norwegian and Iceland Seas. In spite of this, the estimate for the Greenland Sea is similar to the one by *Chierici et al.* [1999], of 0.024 ± 0.006 Pg C yr^{-1}, based on modelled water transports and measured C_T values.

For the Barents Sea, *A. Omar* [personal communication] used an empirical relationship to estimate sea-surface fCO_2 from apparent oxygen utilization (AOU), SST, salinity, and phosphate. He found air–sea fluxes into the surface water of the Atlantic sector in the Barents Sea of 29 g C m^{-2} from late winter to early summer, which is the most productive season. If the surface water were to be exposed to the atmosphere throughout the year, an annual uptake from the atmosphere of 98 g C m^{-2} yr^{-1} would be estimated. However, the Barents Sea is partly ice covered, which will reduce this estimate.

The above Barents Sea estimate is similar to the one by *Kaltin et al.* [2002], who used changes in the carbonate system in addition to salinity and nitrate and conservation of mass to estimate the uptake of atmospheric carbon into the upper 100 m of the Atlantic water in the Barents Sea. They computed the air–sea uptake to be 29 g C m^{-2} from late winter to early summer. According to *Fransson et al.* [2001] 44 g C m^{-2} yr^{-1} is introduced from the atmosphere into the Atlantic water of the Barents Sea, based on comparison of relative deficits of different chemical constituents, integrated over the upper 150 m. This latter estimate must be considered as the average uptake that occurs in the Atlantic water during its residence in the Barents Sea, which is assumed to be about 1 year.

Omar et al. [this issue] estimated the wintertime flux of atmospheric carbon into a coastal polynya in Storfjorden off Svalbard to be 65 g C m^{-2}, based on inorganic carbon and hydrographic data collected during late summer 2001 and spring 2002. Biological processes were accounted for in their estimate, which was about 13 times higher than for a non-polynya (i.e., ice-covered) region in the same area. This illustrates that Arctic polynyas may be important areas for air–sea CO_2 exchange in the wintertime and should be included in carbon budgets of the Arctic region.

The Northeast Water Polynya (NEWP) is found above the continental shelf off northeast Greenland and is a recurrent feature in this area during summer. The major part of the surface water in this polynya is of Pacific origin [*Falck*, 2001], with a decreasing Pacific influence downward in the water column. *Falck* [1999] has calculated fCO_2 values for this polynya from C_T and alkalinity data collected during the summer of 1993. The calculated CO_2-uptake in the surface layer for the ice-free period (2 to 3 months) gave a mean value of the order of 5 g C m^{-2}, ranging from 2.4 to 7.2 g C m^{-2} for individual stations. The differences seen in CO_2-uptake between the individual stations were mainly due to ice conditions, i.e., how long time a given station had been ice-free. The low fluxes reported from the NEWP can be attributed to the low wind speeds in this region during summer.

Table 2 summarises the air–sea fluxes reported above. All flux estimates are directed into the sea and represent a large range, in general between 20 and 85 g C m^{-2} yr^{-1} for the different seas. The uncertainties are in general not included in this review, since they are estimated by variable methods in the different references. However, if we assume a general error of 10-20% in the air–sea flux estimates reported here, the average uptake rates should be most reliable.

Little is known on the interannual variations of air–sea CO_2 exchange in the Nordic and Barents Seas, and more than anything else this reflects the scarcity of fCO_2 data from these regions. *Skjelvan et al.* [1999a] compared fCO_2 data obtained in the Greenland Sea on a cruise in February-March 1995 with data obtained in February-March 1997. The fCO_2 was 20-30 (~5-10%) μatm lower in 1997 than in 1995. The decrease in fCO_2 between these two cruises appeared to be a consequence of lower SST and more extensive ice cover in 1997.

The estimated fluxes presented so far express the transport of the sum of natural and anthropogenic carbon between atmosphere and ocean. When making budget computations to assess the air–sea flux of CO_2, the contribution by anthropogenic CO_2 also has to be considered. This is largely a function of the time since the water mass was in contact with the atmosphere but is also affected by the sea-ice cover that the water might have experienced at the surface, which can hamper exchange with the atmosphere. The issue of how much anthropogenic CO_2 that is taken up from the atmosphere at a given location depends on the degree of equilibrium between the surface water and the atmosphere, as well as whether this degree of equilibrium has changed from preindustrial times to present. Assuming that the ocean circulation and biological activity of the surface waters have not changed since preindustrial times (which no evidence indicates), the degree of saturation of the surface waters with respect to CO_2 (ΔfCO_2) has not changed either. If this is accepted, we can compute the contribution that anthropogenic CO_2 has on the total C_T concentration, applying the present-day ΔfCO_2. However, as sea-ice hampers air–sea gas exchange, any change in sea-ice cover will affect the degree of saturation and thus also the contribution of anthropogenic CO_2 on the total C_T concentration. Changes in the sea-ice cover have been observed during the last 30 years and this effect therefore has to be examined.

4.2. Negative Feedback on the Uptake of Atmospheric Carbon

Takahashi et al. [2002] make the assumption that as a result of deep mixing at high latitudes the CO_2 properties of surface water at those latitudes remain unchanged, which means that there is no increase in surface water fCO_2 over the years. Since the content of atmospheric CO_2 is increasing by about 1.4 μatm yr^{-1} due to anthropogenic CO_2 input [*Keeling and Whorf*, 2004], we would also expect the ΔfCO_2 to increase so that there is an increasing uptake of anthropogenic CO_2 from the atmosphere in the Nordic Seas, which is also indicated by a majority of numerical models [*Orr et al.*, 2001]. However, it has been shown by *Omar et al.* [2003] and *Lefèvre et al.* [2004], based on an extensive amount of data from the Barents Sea and the North Atlantic, that surface ocean fCO_2 in these regions is increasing more or less in parallel to the atmospheric CO_2. *Omar et al.* [2003] calculated an annual fCO_2 increase of 1.3 μatm in the Barents Sea surface water based on data from cruises in 2000-2001 and 1967. *Lefèvre et al.* [2004] used a large data set from 1981 to 1998 to estimate the annual surface water fCO_2 increase to be 1.8 μatm. Unpublished surface fCO_2 data from M/S *Nuka Arctica* in the North Atlantic show an annual increase in surface fCO_2 of 1.6 μatm. Thus it appears that the fCO_2 of the northern North Atlantic and Nordic and Barents Seas is increasing at a similar rate as the atmospheric CO_2, and from this we conclude that the major part of the Nordic Seas does not, at least for the last few decades, have the potential for extra uptake of anthropogenic CO_2 from the atmosphere.

A possible explanation for this observation is the increased transport of carbon into the Nordic Seas and Barents Sea from the south that follows from the reduced out-gassing of CO_2 at lower latitudes in comparison with preindustrial times. This is verified by *Anderson and Olsen* [2002a], who show, using a simple advective model, that lower latitudes have the largest uptake of anthropogenic CO_2 from the atmosphere, while higher latitudes have a smaller uptake or even are a source of anthropogenic CO_2 to the atmosphere (Figure 4). The reduced out-gassing at lower latitudes is caused by a higher atmospheric CO_2 concentration at present than in previous times, which results in a smaller ΔfCO_2. Consequently the buffer capacity of the northward-flowing surface water is reduced and so is the uptake of CO_2 in the north. This is in concert with the explanation of *Wallace* [2001]. He showed, by using an idealized model, that the amount of CO_2 that must be transported across the air–sea interface to maintain equilibrium between surface and atmosphere is reduced by 15-20 μmol kg^{-1} for the present ocean compared with the preindustrial ocean. This is attributed to the decreased buffer capacity of seawater at higher levels of fCO_2. Consequently, the net transfer of CO_2 across the air–sea interface is smaller today than in the preindustrial era, despite higher present atmospheric CO_2 levels. This explanation suggests that the Nordic Seas is a transit area for anthropogenic CO_2 on its way to the deep ocean, rather than an area of large uptake of anthropogenic CO_2 from the atmosphere. By deep mixing at high latitudes, the anthropogenic CO_2 is transported away from the atmosphere and sequestered for long times.

There are of course uncertainties connected to the above estimates for surface-water fCO_2 increases. The estimate of

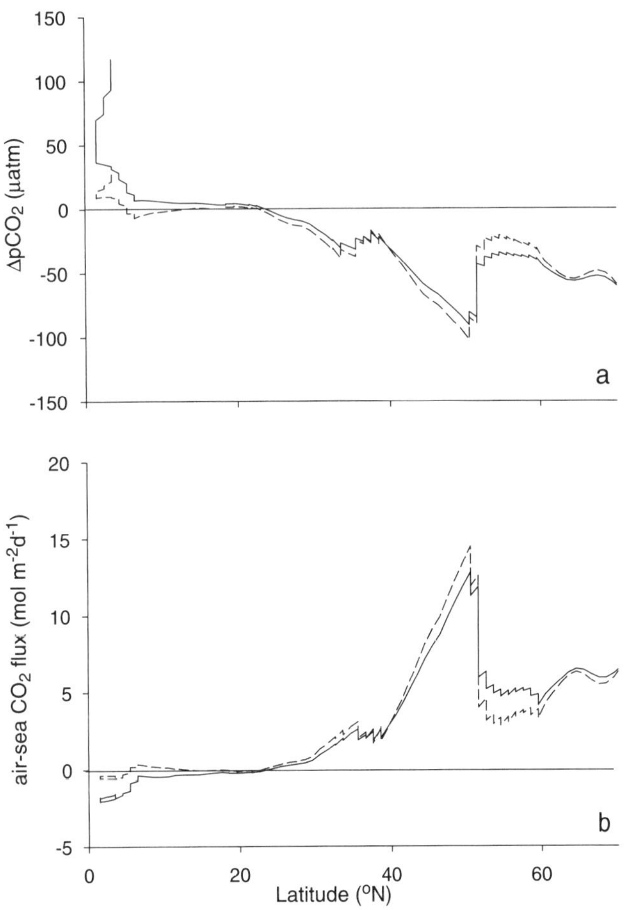

Figure 4. ΔpCO_2 (atmosphere – ocean) (a) and air–sea CO_2 flux (b) as a function of latitude along a North Atlantic flow path as computed by *Anderson and Olsen* [2002]. The solid and dashed lines show results when calculations were done using a preindustrial atmospheric CO_2 concentration (280 ppm) and 2002 atmospheric CO_2 concentrations (370 ppm), respectively.

Omar et al. [2003] was corrected for biological effects and temperature changes throughout the year, which makes the estimate of annual fCO_2 increase in the Barents Sea surface water a robust estimate for anthropogenic CO_2 input. The results from *Lefèvre et al.* [2004], however, are not corrected for such effects, and a fraction of the increase can be attributed to changes in SST and biology.

5. TRANSPORT OF CARBON DUE TO PHYSICAL PROCESSES IN THE OCEAN

Deep mixing has a large potential for transporting carbon from the surface to the deep ocean. The Greenland Sea is one of the most important regions for deep open ocean mixing in the world [e.g. *Helland-Hansen and Nansen*, 1909; *Rudels*, 1989], and the ventilated water moves southwards as a part of the meridional overturning circulation.

A cyclonic circulation plays an important role for deep mixing. It raises the pycnocline and brings high-density water closer to the surface, preconditioning for deep convection. Surface water is cooled and becomes dense in winter, and the water column turns over in the central gyre [*Killworth*, 1983]. The cyclonic circulation in the Greenland Sea gyre is composed of the northward-flowing Norwegian Atlantic Current and the southward-flowing EGC, the latter carrying cold and fresh water out of the Arctic Ocean. The central Greenland Sea is usually ice-free, and strong cooling increases the water density and carbon solubility. Deep mixing takes place through overturning of the fluid column in numerous plumes, which have a horizontal scale on the order of less than 1 km and with vertical velocities of up to 10 cm s^{-1} [*Schott and Leaman*, 1991].

Long-lived anticyclonic eddies with a diameter of about 10 km have also been observed in the Greenland Sea [*Gascard et al.*, 2002]; these eddies have a large vertical extension (about 2000 m), which is relevant to the deep-mixing depth. The development of these features is not known in detail; however, this is another potential mechanism of vertical transport of carbon. This is verified from measurements of chemical tracers in a similar eddy detected in the central Greenland Sea in September 2003. These measurements revealed that surface water is one of the contributors to the core water of the eddy [*Y. Kasajima*, personal communication].

According to *Schlosser et al.* [1991], the magnitude of the deep-water formation in the Greenland Sea has decreased since the early 1980s, and consequently the transport of carbon has probably decreased. From a ventilation model applied by *Anderson et al.* [2000b], an annual renewal by deep-water formation of about 1% appeared below 1500 m, which corresponds to a mean ventilation of 0.17 Sv. Based on this ventilation, *Anderson et al.* [2000b] calculated a storage of anthropogenic CO_2 of 0.0024 Pg C for the year 1994. Assuming that the anthropogenic C_T concentration in the surface water (36 µmol kg^{-1} according to *Anderson et al.* [2000b]) is about 2% of the natural surface-water C_T, this gives an estimate of total carbon sequestered by deep-water formation in the Greenland Sea of 0.12 Pg C in 1994.

On a smaller scale, vertical mixing brings carbon and nutrients back to the surface layer when the mixed layer deepens during fall and winter. This entrainment into the surface layer of water from below has been examined with a simple box model [*Anderson et al.*, 2000a; *Falck and Anderson*, in press, 2004; *Olsen et al.*, 2002], which initially was used to determine the export production out of a given depth layer (see section 6.1). Carbon entrainment into the surface layer from below is calculated to be 11 and 30 g C m^{-2} yr^{-1} for the Greenland and Norwegian Seas, respectively. For the Iceland

Sea the entrainment is set to be 20 g C m^{-2} yr^{-1}. According to *Jakobsson* [2002], the shelf areas of the Greenland, Norwegian, and Iceland Seas represents about 40%, 30%, and 45% of the total areas, respectively. Based on the assumption that the entrainment is reduced to 50% of the above values in the shelf areas, the transport of carbon due to entrainment into the surface layer is estimated as 0.004, 0.04, and 0.005 Pg C yr^{-1}, for the Greenland, Norwegian, and Iceland Seas, respectively. The former estimate is based on an annual ice cover of 44% [*Anderson et al.*, 2000a], and the latter region is assumed to have an annual ice cover of 40%. For the Nordic Seas as a whole this sums to about 0.05 Pg C yr^{-1} entrained into the surface water. In the Barents Sea the carbon entrainment from below is assumed to be negligible, similar to the negligible oxygen entrainment estimate in this area from *Olsen et al.* [2002].

In shelf areas of the Arctic Ocean, formation of sea-ice during winter will produce high-density brine-enriched shelf water (BSW) [e.g. *Quadfasel et al.*, 1988]. Guided by the topography, plumes of the BSW flow towards the Arctic Ocean and Nordic Seas, where they can reach great depths, depending on the initial density of the plumes and the degree of en route mixing with the surrounding water masses. BSW of Barents Sea origin has been traced all the way into great depths in the Fram Strait [*Quadfasel et al.*, 1988]. The formation and outflow of BSW therefore represent a mechanism that transports atmospheric carbon into the deep ocean. *Omar et al.* [this issue] estimated a wintertime flux of CO_2 from the atmosphere to the sea surface of $(0.15 \pm 0.05) \cdot 10^{-3}$ Pg C, which accompanies the formation of BSW in this region. This calculation was based on a typical annual production of the BSW and an observed excess carbon in the BSW, which has been attributed to wintertime air–sea gas exchange that primarily occurs in a coastal polynya in Storfjorden.

During times of changing climate, the deep-water formation can be perturbed and thereby the vertical transport of carbon. *Aagaard and Carmack* [1989], investigating the impact of fresh water input to the Greenland Sea, reported that a doubling of the input of fresh water to the convective gyre would shut down the deep mixing in this area. Such a fresh water input could come from changes in the input from precipitation over land and river runoff into the Arctic or from transport of ice into the Greenland Sea gyre due to changing meteorological conditions.

Model simulations presented in *Johannessen et al.* [2004] predict that during this century the Arctic Ocean might become nearly ice-free during summer, while only moderate reductions will be seen in wintertime sea-ice cover. Hence the seasonal variability in sea-ice extent will be much larger than at present. How this will affect the carbon fluxes has not yet been studied, but one can speculate that due to an enhanced formation of new ice during winter, the brine formation will increase, which in turn will increase the amount of carbon transported with BSW. On the other hand, the production of BSW and consequently the transport of carbon can eventually cease if global warming continues to the extreme that sea-ice formation is heavily reduced or stops.

6. BIOLOGICALLY MEDIATED TRANSPORT OF CARBON

The surface-ocean C_T concentration varies with the season as a result of physical and biological processes, which is also the case for oxygen and nutrients. Primary production consumes inorganic carbon and nutrients, and secondary processes like grazing by zooplankton will continue the transformation of carbon into dissolved organic matter and particulate debris that sinks towards the bottom. The dramatic increase in biological activity during spring and summer is a result of the increased solar radiation, the shallowing of the upper mixed layer, and the establishing of a seasonal pycnocline [*Sverdrup*, 1953; *Rey*, 2004]. In fall and winter the major part of the organic matter left in the water column regenerates, and dissolved inorganic carbon and nutrients are reintroduced into the surface water concurrent, the mixed layer deepening due to cooling of the surface water.

Large regional differences are seen in preconditioning and characteristics of the bloom events. In the Barents Sea, which is partly ice-covered during winter, a highly stratified marginal ice zone (MIZ) develops as the ice melts. The MIZ sweeps over the region as the ice recedes northwards, and a concurrent melt-water bloom quickly depletes the surface-layer nutrients. Strong vertical density gradients prevent mixing in of new nutrients from below [*Rey and Loeng*, 1985]. In non-ice-covered regions of the Barents Sea, however, stratification develops due to heat input, and vertical density gradients are smaller than in the MIZ. This allows the reintroduction of nutrients into the SML so that a higher annual production can be sustained [*Slagstad and Wassmann*, 1996; *Rey et al.*, 1987].

Along the coast of Norway, the spring bloom starts when water from the Norwegian Coastal Current spreads out as a wedge over the eastern part of the Norwegian Atlantic Current [*Rey*, 1981; *Olsen et al.*, 2003b]. The recurrent springtime spreading of the coastal water (CW) is a consequence of offshore Ekman drift caused by the prevalence of northerly winds along the Norwegian coast during summer [*Gade* 1986; *Sætre et al.*, 1988]. These winds are directed by the offshore increase in atmospheric pressure at sea level created by the differential heating of ocean and landmasses. *Olsen et al.* [2003b] analysed temperature, salinity, and nutrient data obtained at the entrance to the Barents Sea. In near-shore areas (within ~220 km), the spring bloom is triggered by offshore deflection of CW, and heat input controls the further shoaling over summer. Further offshore (220-330 km), the spring bloom starts following the thermal stratification, and the development of the bloom through summer depends on the extent of spreading of CW.

In the open-ocean region of the Norwegian Sea, stratification due to heating of the water triggers the spring bloom [*Sverdrup*, 1953; *Rey*, 2004]. In these regions events of strong wind may interrupt the build-up of surface stratification and prolong the spring bloom development [*Sakshaug et al.*, 1995]. In the post-bloom phase, episodic mixing events may bring in new nutrients that refuel the bloom.

According to *Redfield et al.* [1963], phosphate, nitrate, carbon, and oxygen are utilized and released in fixed proportions during formation and mineralization of organic matter; P:N:C:O_2 equals 1:16:106:–138, which is denoted the RKR ratios. However, there is a debate about whether the classical RKR ratios apply under all circumstances [e.g. *Sambrotto et al.*, 1993; *Anderson and Sarmiento*, 1994], and (e.g.) *Rey* [2004] has estimated RKR ratios for the Norwegian Sea and adjacent ocean areas of 1:16:122:–180. *Klausmeier et al.* [2004] have recently discussed the question on what decides this ratio.

6.1. Export Production in Open Ocean and Shelf Regions

In a carbon budgeting perspective it is relevant to quantify the amount of carbon removed from the surface layer due to biological activity. According to *Eppley* [1989], the term *export production* should be used for the organic matter, which is not remineralized in the surface water but rather sinks out of the euphotic zone or the seasonal upper-mixed layer. The fraction of the organic matter that sinks out of the euphotic zone depends on factors like mixing and entrainment, intensity and timing of the phytoplankton bloom, and grazing pressure by zooplankton. There exist several other production definitions, which for instance describe the amount of carbon transformed into organic matter based on nitrate supply from outside the surface layer. We have chosen to denote the production estimates presented below as export production, based on the assumption that on an annual basis, nutrients lost due to export of organic material are usually replaced by mixing of deeper water into the surface layer during winter [e.g. *Eppley and Peterson*, 1979; *Platt et al.*, 1989]. The similarity of the presented estimates indicates that this might be a proper assumption. However, we are aware of the ongoing debate about comparing different production terms [e.g. *Laws*, 1991].

Several approaches have been used to study the biological production in the Nordic Seas. *Anderson et al.* [2000a], *Skjelvan et al.* [2001], and *Falck and Anderson* [in press, 2004] have all used a similar box model where oceanic carbon and oxygen fluxes were calculated based on observed deficits of carbon, oxygen, and nutrients in the upper layer (Figure 5). The model approach was based upon the assumption that the only source of nutrients was from below in the water column, with vertical mixing as the transport route. RKR ratios [*Redfield et al.*, 1963] for the *Anderson et al.* [2000a], *Falck and Anderson* [in press, 2004], and *Skjelvan et al.* [2001] estimates were C:N equals 7.5, P:N:C:O_2 equals 1:16:106:–138, and P:N:C:O_2 equals 1:16:106:–138, respectively. The depth ranges for the export estimates were 150, 100, and 300 m for *Anderson et al.* [2000a], *Falck and Anderson* [in press, 2004], and *Skjelvan et al.* [2001], respectively. Fluxes of carbon out of the surface layers due to biological production were determined, and the results are summarized in Table 4. The calculations show that the export production in the Greenland and Norwegian Sea is relatively similar; 15-35 g C m^{-2} yr^{-1}, despite the seasonal differences observed in these two seas. The Norwegian Sea experiences a heavy grazing pressure early in the season, which prevents accumulation of phytoplankton stock while nutrient depletion is delayed compared to what is seen in the Greenland Sea [*Bodungen et al.*, 1995]. In the latter sea, the nutrients are depleted relatively early in the season. Organic matter builds up, and almost all of it is recycled due to a winter mixed layer that reaches depths of up to 1000 m [*Noji et al.*, 2000]. Such a deep mixed layer effectively increases the

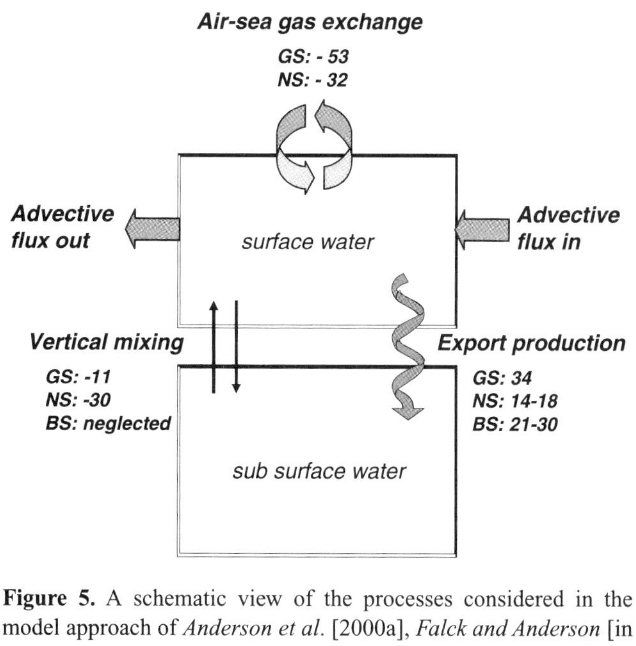

Figure 5. A schematic view of the processes considered in the model approach of *Anderson et al.* [2000a], *Falck and Anderson* [in press, 2004], *Olsen et al.* [2002], and *Skjelvan et al.* [2001], including the carbon fluxes (in g C m^{-2} yr^{-1}) determined from the model. The carbon flux due to air–sea gas exchange in the Barents Sea was not estimated in these studies. GS: Greenland Sea, NS: Norwegian Sea, BS: Barents Sea. Negative signs indicate flux of carbon into the surface water.

Table 4. Export Production and Sedimentation in the Greenland, Norwegian, Barents, and Icelandic Seas and in Polynyas. The estimates are annual estimates if nothing else is noted.

	Greenland Sea	Norwegian Sea	Barents Sea	Iceland Sea	Polynyas
Export production ($g C m^{-2}$)	34[A2] 27[B] 52-62[R] (no remineralization)	24-32[S3] 14-18[F&A] 21-29[B] 39[F&G]	21-30[O]	75[G]	8[A1] (summer) * 9[A1] (winter)* 15-65[F] (3 summer months)**
Sedimentation ($g C m^{-2}$)	1.9[N] (upper layer) 1.2[N] (deep)	NA	NA	NA	NA

A1 is *Anderson et al.* [1988], A2 is *Anderson et al.* [2000a], B is *Bodungen et al.* [1995], F is *Falck* [1999], F&A is *Falck and Anderson* [in press, 2004], F&G is *Falck and Gade* [1999], G is *Gudmundsson* [1998], N is *Noji et al.* [2000], O is *Olsen et al.* [2002], R is *Rey et al.* [2000], and S3 is *Skjelvan et al.* [2001]. *Estimate is from a polynya in the Storfjorden at Svalbard. ** Estimate is from the Northeast Water Polynya at the shelf east of Greenland. NA, not available.

depth below which carbon export for sedimentation can occur. This was verified by sediment traps deployed at 2-300 and 2000 m deep in the Greenland Sea during the years 1993-1995, which showed sedimentation rates of 1.9 $g C m^{-2} yr^{-1}$ (3.39 $g C m^{-2} yr^{-1}$ when including accumulation of DOC in the upper 200 m and migration of zooplankton below 200 m) and 1.2 $g C m^{-2} yr^{-1}$, respectively [*Noji et al.*, 2000].

A similar model approach was used for the Barents Sea. *Olsen et al.* [2002] used oxygen and nutrient data, RKR ratios of P:N:C:O_2 of 1:16:106:–138, and a depth range of 150 m to calculate an export production in the Atlantic sector of the Barents Sea of 21-30 $g C m^{-2} yr^{-1}$ (Table 4). The different numbers reflect warm and cold years, which will be discussed in section 6.2. This Barents Sea estimate is in concert with the estimates from the Greenland and Norwegian Seas.

Anderson et al. [1988] examined the biologically mediated carbon transport by BSW in the Barents Sea and later summarized this in *Anderson and Jones* [1991]. During summer, biological drawdown of seawater fCO_2 drives a flux of CO_2 into the water column. When the productive season is over, the organic matter from dead organisms sinks and quickly reaches the shallow surface sediment, where it starts to decay. Some of the carbon produced during the decay process is added to the overlying BSW, which leaves the shelf as a bottom current towards the Arctic Ocean. Using data from Storfjorden, *Anderson et al.* [1988] determined that the addition of carbon into the BSW by the decay of organic matter occurs at a rate of 8 $g C m^{-2} yr^{-1}$. This estimate was reached through examining the relationships between increasing C_T and phosphate and decreasing oxygen towards depth (with increasing salinity) in Storfjorden, at a station from which they measured a RKR ratio close to the classical ratio [*Redfield et al.*, 1963]. A similar uptake was suggested for the winter season of 2001-2002, when an increase in C_T corresponding to about 9 $g C m^{-2}$ relative to the concentration in the top 50 m was observed in late April 2002 [*Anderson et al.*, 2004]. The surface water was undersaturated with respect to fCO_2 and the observed excess was suggested to be a result of enhanced air–sea exchange of CO_2 caused by sea-ice formation, with an efficient exchange through the surface film during the ice-crystal formation and the rapid transport of the high-salinity brine out of the surface layer. The high uptake rate is possible only in a region where the sea-ice that is formed is transported away by wind, as in the polynya of Storfjorden.

In the Northeast Water Polynya, export production rates as reported in the literature are in the range 0.17-0.72 $g C m^{-2} d^{-1}$, depending on the method of calculation [e.g. *Wallace et al.*, 1995; *Falck*, 1999]. Assuming an ice-free period of 3 months, this amounts to 15-65 $g C m^{-2} yr^{-1}$. However, the number of months the polynya is open varies between years, and this will influence the export production.

The above estimates for production in the Greenland, Norwegian, and Barents Seas are in the middle of the range compared to other estimates; *Bodungen et al.* [1995] estimated export production by ^{15}N tracer technique to be 27 and 21-29 $g C m^{-2} yr^{-1}$ in the Greenland Sea and Norwegian Sea, respectively. *Falck and Gade* [1999] applied an oxygen budget for the Nordic Seas, and for Ocean Weather Station Mike (OWSM) in the Norwegian Sea they found an export production of 39 $g C m^{-2} yr^{-1}$. For the Greenland Sea, *Rey et al.* [2000] calculated an export production of 52-62 $g C m^{-2} yr^{-1}$ based on nitrate removal during summer season. The latter estimate does not account for remineralization during fall and is therefore expected to be higher than the export production calculated by e.g *Anderson et al.* [2000a].

The Iceland Sea represents a highly productive region of the Nordic Seas. *Gudmundsson* [1998] determined a primary productivity of 4.3 $mg C m^{-3} h^{-1}$ in the region north of Iceland. Assuming an euphotic zone of 20 m, a productive season of 2.5 months, and export of approximately 50% of the primary production out of the euphotic zone, this amounts to an export production of about 75 $g C m^{-2} yr^{-1}$.

To get an estimate of the biologically mediated carbon export from each of the ocean areas discussed in this review, we need to extrapolate the export estimates over larger regions than they have been constrained to. This is certainly an exercise with which one can put a large question mark, since blooming events are patchy both in time and space. However, if we assume, based on values in Table 4, mean annual export production estimates in the Greenland, Norwegian, Iceland, and Barents Sea of 30, 25, 75, and 25 g C m^{-2}, respectively, and also assume that approximately half of each of the basin areas is exposed to biological production, export production from the Greenland, Norwegian, Iceland, and Barents Seas can be estimated to 0.012, 0.017, 0.019, and 0.018 Pg C yr^{-1}, respectively. The latter assumption also takes care of the fact that some of the regions are temporarily ice-covered. Summarizing over the whole area of the Nordic Seas (Greenland, Norwegian, and Iceland Seas), an annual export production of about 0.05 Pg C is estimated. These estimates have to be taken with care, and the uncertainties are large. However, the order of magnitude of the estimates is in concert with the carbon transports estimated from air–sea gas exchange and deep mixing.

6.2. Climate Effects on Export Production

The Barents Sea is probably the most well studied region of the ocean areas covered by this review with respect to marine biological response to climate variations. This response is a consequence of the large interannual fluctuations in ice-cover extent related to the properties of the Atlantic Water feeding the Barents Sea [*Loeng*, 1991]. In cold years the ice reaches—and can extend or drift—across the Polar Front, the boundary between the Polar and Atlantic regimes. In these years Barents Sea biological productivity is less than in warm years, when the ice cover does not extend as far south. Model results by *Slagstad and Wassmann* [1996] also showed decrease of vertical carbon export in cold years, but these were obtained using a fixed meso-zooplankton module, thus excluding variable zooplankton grazing. Observations from the Barents Sea show that stabilisation of the water column due to heating of the water leads to slowly developing blooms, which favours a tight match between phyto- and zooplankton development. Blooms that start due to melt-water input in the MIZ, on the other hand, develop at a greater rate. This results in a mismatch between phyto- and zooplankton development and consequently a low grazing pressure appears, which leads to massive sedimentation of phytoplankton as depletion sets in [*Rey et al*, 1987; *Rey and Skjoldal*, 1987].

In the middle part of the Barents Sea Opening (220-330 km off the coast of northern Norway), the bloom development is tightly connected to the atmospheric pressure gradients over the area [*Olsen et al.*, 2003b]. In years with a low pressure in the north (75°N, 20°E) and a high pressure in the south (70°N, 20°E), the pressure gradient will induce prevailing westerly winds and the spreading of CW into the region is limited. The SML remains quite deep throughout summer, and the bloom appears to follow a typical Atlantic pattern and is unable to fully utilise available nitrate. Extensive grazing probably terminates the bloom. In years with a high pressure in the north and low pressure in the south, however, easterly winds prevail, and the Ekman drift will be directed to the north so that the wedge of CW reaches out into the region of the Barents Sea Opening. This CW is low in nutrients since production has occurred for quite some time, and, additionally, the strong vertical density gradients will efficiently cut off communication between the SML and the deeper, nutrient-rich waters. The bloom is terminated due to nutrient limitation in these years, and annual primary production is probably reduced.

Vertical export of carbon is tightly coupled to the balance between phyto- and zooplankton during the spring bloom [*Wassmann*, 1998]. A time lag between phytoplankton blooms and the increase of major grazers, such as copepods, results in large and intense blooms that are able to deplete the nutrients in the SML. As depletion sets in, aggregates of fast-sinking phytoplankton form so that organic carbon is effectively transported to larger depths and sediments. A close match between producers and consumers, on the other hand, intensifies recycling in the euphotic zone. A rapidly developing spring bloom will favour imbalance between phyto- and zooplankton due to the reproductive time lag of the latter, while grazing should be higher and sedimentation lower when a bloom develops slowly and not too early in the season [*Sakshaug and Slagstad*, 1992].

As the timing and development of the spring bloom in the Nordic Seas and Barents Sea are closely related to climatic factors, *viz.* sea-ice distribution, heat input, pressure gradient between land and ocean masses, and wind strength, climate change will clearly change the relative importance of sedimentation and recirculation in the marine food chain. Such changes in the biological pump may have a significant impact on oceanic carbon storage. Data from the Greenland Sea show that only 10% of the new production available for export found its way to layers deeper than 200 m in the mid-1990s; the rest was remineralised in the upper ocean and thus reduced the potential for further CO_2 absorption from the atmosphere [*Miller et al.*, 1999; *Noji et al.*, 1999; *Rey et al.*, 2000]. Evidently the potential for added carbon storage through decreased recycling of organic matter is substantial.

6.3. Acidification of the Ocean

The pH decrease, as an effect of the invasion of carbon dioxide from the atmosphere to the surface ocean from 1750 to the present day, has been estimated to be about 0.1 pH units for the North Atlantic [*Wallace*, 2001]. General ocean circulation models predict a maximum reduction of surface pH up to 0.7 pH units relative to present levels, based on extrapolated anthropogenic CO_2 emissions [*Caldeira and Wickett*, 2003]. Even over reasonably short time scales it is expected that the surface pH will decrease significantly; the projected surface-layer pH levels at the end of this century are up to 0.35 pH units lower than present [*Wallace*, 2001; *Bellerby et al.*, this issue]. The degree of pH decrease exhibits large regional variability within the Nordic Seas with the waters influenced by polar waters showing a greater reduction in seawater pH than the waters of dominantly North Atlantic origin [see *Bellerby et al.*, this issue].

The biological feedback to oceanic CO_2 increase is difficult to characterise due to differing response of planktonic species to changes in the marine CO_2 system. Calcareous plankton species such as *Emiliana huxleyi* and *Gephyrocapsa oceanica* exhibit lower calcification rates and malformed coccoliths when grown under low pH levels [*Riebesell et al.*, 2000]. Such a slowdown in calcification on regional scales would reduce the present seasonal decrease in carbonate alkalinity in surface waters, and as calcification increases seawater fCO_2, this would be a negative feedback to increasing atmospheric CO_2 levels by increasing the potential of the solubility pump. Changes in calcification and preservation records have been shown in the distributions of sedimentary carbonates in the Nordic Seas by *Helmke and Bausch* [2002], who further hypothesise that much of the glacial decrease in the atmospheric CO_2 may be attributed to the changes in the alkalinity of the oceans.

Mesocosm studies on the effect of high initial ambient CO_2 (750 μatm) on coccolithophore assemblages have shown an increase in particulate organic carbon production [*Zondervan et al.*, 2002] relative to the present-day atmospheric CO_2 levels. However, evidence is conflicting that increasing CO_2 concentrations will increase productivity in the ocean. *Tortell et al.* [1997] showed that the growth of plankton assemblages dominated by large diatoms (using chlorophyll *a* as a proxy for growth) was not enhanced by increased CO_2 availability and concluded that organisms that utilise carbonic anhydrase to facilitate direct utilisation of HCO_3^- are not affected by ambient CO_2 levels and so would not benefit from enhanced CO_2 concentrations. Conversely, *Hein and Sand-Jensen* [1997] have shown that increasing the CO_2 concentration in natural oligotrophic phytoplankton communities, including diatoms, did have a positive effect on primary productivity. In a wider perspective, one can speculate on eventual effects for organisms at higher levels in the food chain. Any prediction is difficult, however; according to *Fosså* [2002], changes in the productivity and distribution of marine plankton assemblages will have a consequence for food chain dynamics and ultimately fish stocks through the availability and quality of food sources.

7. SUMMARY OF TRANSPORTS

Figure 6 summarizes the carbon fluxes connected to the Nordic Seas, with Figure 6a focusing on fluxes into and out of the Nordic Seas and Figure 6b focusing on the fluxes within the basin. The horizontal carbon transport through the Nordic Seas, about 9.5 Pg C yr^{-1}, is larger by orders of magnitude than the transport due to air–sea gas exchange (0.09 Pg C yr^{-1}), export production (0.05 Pg C yr^{-1}), deep mixing (0.12 Pg C yr^{-1}), and entrainment into the surface layer from below (0.05 Pg C yr^{-1}).

Uncertainties of variable sizes are connected to these estimates. Precise absolute estimates of the horizontal carbon fluxes are hard to achieve due to the large uncertainties (up to 25%) in the volume fluxes, and alternative ways of determining these are recommended, like calculations that require balance in the volume fluxes in and out of the Nordic Seas.

As regards the air-sea gas exchange estimate, this is probably the estimate in this review that is based upon the most extensive dataset, however, uncertainties of ±10-20% are discussed in the different references.

The presented estimate for export production is associated with the largest uncertainty in this review. The estimate suffers

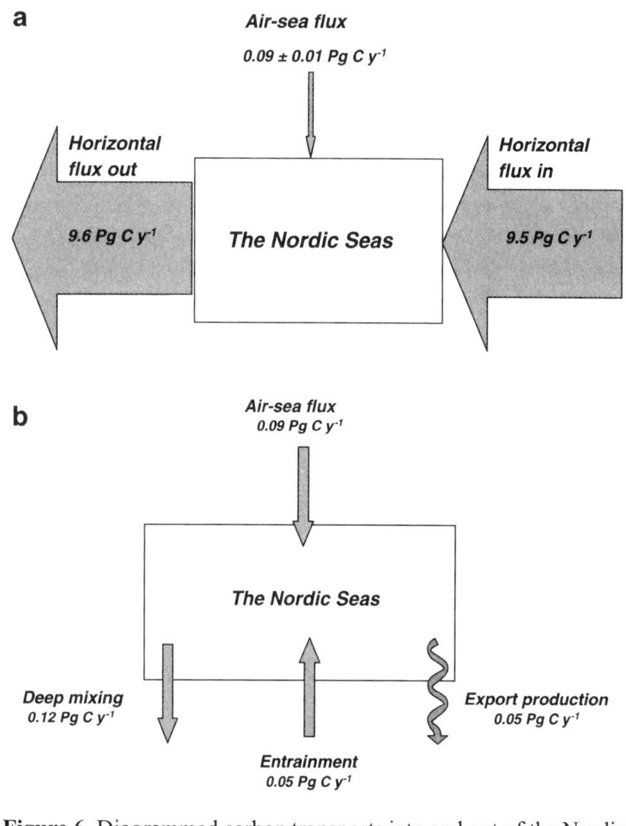

Figure 6. Diagrammed carbon transports into and out of the Nordic Seas (a) and within the Nordic Seas (b).

from the fact that we have extrapolated a spatially and temporally fluctuating biological production to large areas, and the export production estimate could easily vary by a factor of two. This would influence a budget where carbon transports due to air–sea gas exchange and vertical mixing are considered but be of minor importance when the large horizontal carbon transports are taken into account.

The estimate for carbon sequestered by deep-water formation, 0.12 Pg C, is constrained to the Greenland Sea during a specific year. However, this amount is adopted as a carbon to deep-water transport estimate for the Nordic Seas as a whole, since the Greenland Sea is assumed to be the dominating region for deep mixing in the Nordic Seas [*Swift and Aagaard*, 1981]. The uncertainty of this estimate is ±25%, and the carbon transport will be influenced by the high interannual and interdecadal variation in the amount of deep-water formation [e.g. *Bönisch et al.*, 1997]. Intermediate water is also formed in the Iceland Sea, and this will increase our estimate [*Malmberg and Jónsson*, 1997].

Carbon entrainment into the surface water of the Nordic Seas from below has been determined from a simple box model initially used to elucidate the export production from a certain depth layer. Uncertainty in the entrainment is estimated to be on the order of 20%, primarily based on deviations in the RKR number used.

It is important to note that the horizontal flux represents a carbon transport in both the surface layer and the deep layer of the Nordic Seas. The region is important for carbon transport from the surface ocean to intermediate and deep layers with an annual export of carbon through the deeper parts of the Greenland Scotland Ridge of about 6 Pg. As regards transport routes into the region, about 0.09 Pg C yr^{-1} enters the region from the atmosphere and approximately 9 Pg is transported into the region in the surface layers (Table 1). These numbers show that the Nordic Seas is an important transit region for CO_2.

Our estimates for the surface to deep-ocean carbon transport account for only 0.17 Pg C (0.05 due to export production and 0.12 associated with deep mixing in the Greenland Sea). The deep-mixing estimate is limited to depths greater than 1500 m in the Greenland Sea, and more work is needed in order to fully understand the carbon transport associated with the large-scale sinking of water masses in the Nordic Seas.

For the Barents Sea, no estimate for the horizontal transport of carbon is made, since we are in lack of observations. The carbon flux due to export production is estimated to be 0.018 Pg C yr^{-1}. As stressed above, this amount has to be taken with care, and the estimate could easily, as described for the Nordic Seas, vary by a factor of two. The discrepancy in air–sea gas exchange estimates from the Barents Sea is large, and an averaged uptake of 50 g C m^{-2} yr^{-1} is used to estimate an annual uptake of 0.07 Pg C for an ice-free Barents Sea, which is clearly an overestimation. The uptake has to be coupled to an annual mean ice extent to achieve a more reasonable estimate. Uncertainties of ±10–20% are connected to the air–sea gas exchange estimates from the Barents Sea. For Storfjorden in the Barents Sea, the formation of BSW, which represents a way of transporting carbon towards deeper layers, is accompanied by an uptake of 0.00015 Pg C from the atmosphere. An extrapolation of this number depends on ice extent and total area of polynyas in the Barents Sea. Finally, entrainment from below is assumed to be negligible according to *Olsen et al.* [2002].

8. FINAL COMMENTS

Observational studies of the Nordic Seas and the Barents Sea are challenging. The region is dynamic and complex with water masses of very different origins, from the warm and salty Atlantic water entering the area from the south to the cold and fresh East Greenland Current flowing southwards at the western boundary of the Nordic Seas.

Process studies are of crucial importance when it comes to increasing our knowledge on carbon cycling. Over the year, air–sea CO_2 exchange is directed into the sea surface of the Nordic Sea and the Barents Sea. However, the size of this carbon sink appears to be decreasing, which is in contradiction to global ocean model results. What causes this discrepancy is not fully understood, but it might be attributed to a too weak northward heat transport and consequently a too weak overturning in the models [*Völker et al.*, 2002], or to the fact that the models are typically forced with time-invariant atmospheric fields. Further investigation concerning these matters is required.

The biologically mediated carbon transport probably represents the largest gap in our knowledge. To quantify the amount of carbon removed from the surface layer, the term export production has been used. We realize that, indeed, this term depends on the context it is used in. Surprisingly similar estimates for export production have been found for the Nordic Seas and the Barents Sea. However, the blooming events are patchy, which makes it difficult to extrapolate export production estimates over large areas.

In the context of a changing climate we do expect perturbations in the carbon cycle, which has been discussed in the sections above. For example, different species in a planktonic community have been shown to respond differently to an increased atmospheric CO_2 level, which in turn might influence the export production. The quantification of this perturbation, however, is a complicated task, and will be a subject of study for years to come.

Acknowledgment. We would like to express our gratitude to all the persons who have spent their time collecting and analysing data. Without their contribution this work would not have been possible. This publication is Nr A77 from the Bjerknes Centre for Climate Research.

REFERENCES

Aagaard, K. and E.C. Carmack, The role of sea ice and other fresh water in the arctic circulation, *J. Geophys. Res.*, 94, pp. 14485-14498, 1989.

Anderson, L.A. and J.L. Sarmiento, Redfield ratios of remineralization determined by nutrient data analysis, *Glob. Biogeoc. Cycl.*, 8, pp. 65-88, 1994.

Anderson, L.G., E.P. Jones, R. Lindegren and B. Rudels, Nutrient regeneration in cold, high salinity bottom water of Arctic shelves, *Cont. shelf Res.*, 8, pp. 1345-1355, 1988.

Anderson, L.G. and E.P. Jones, The transport of CO_2 into Arctic and Antarctic Seas: similarities and differences in the driving processes, *J. Mar. Sys.*, 2, pp. 81-95, 1991.

Anderson, L.G., K. Olsson and M. Chierici, A Carbon budget for the Arctic Ocean, *Glob. Biogeoc. Cycl.*, 12, pp. 455-465, 1998.

Anderson, L.G., K.Y. Børsheim, M. Chierici, H. Drange, E. Falck, A. Fransson, E.M. Hood, T. Johannessen, L. Merlivat, T.T. Noji, J. Ólafsson, F. Rey and I. Skjelvan, The carbon cycle in the Greenland Sea. *in The Thermohaline Circulation in The Greenland Sea—Final scientific report of ESOP-2 project*, EC MAST III Programme, Contract MAS3-CT95-0015, 1999.

Anderson, L.G., H. Drange, M. Chierici, A. Fransson, T. Johannessen, I. Skjelvan and F. Rey, Annual carbon fluxes in the upper Greenland Sea based on measurements and a box-model approach. *Tellus*, 52B (3), pp. 1013-1024, 2000a.

Anderson, L.G., M. Chierici, E. Fogelqvist and T. Johannessen, Flux of anthropogenic carbon into the deep Greenland Sea, *J. Geophys. Res.*, 105, pp. 14339-14345, 2000b.

Anderson, L.G. and A. Olsen, Air–sea flux of anthropogenic carbon dioxide in the North Atlantic, *Geophys Res Letters*, 29, pp. 16.1-16.4, 2002.

Anderson, L.G., DOC in the Arctic Ocean, *in Biogeochemistry of Marine Dissolved Organic Matter*, edited by D. A. Hansel and C.A. Carlson, Academic Press, pp. 665-683, 2002.

Anderson, L.G., E. Falck, E.P. Jones, S. Jutterström and J.H. Swift, Enhanced uptake of atmospheric CO_2 during freezing of seawater: A field study in Storfjorden, Svalbard, *J. Geophys. Res.*, 109, C06004, doi:10.1029/2003JC002120, 2004.

Archer, D., H. Kheshgi and E. Maier-Reimer, Dynamics of fossil fuel CO2 neutralization by marine CaCO3, *Glob. Biogeoc. Cycl.*, 12, pp. 259-276, 1998.

Bellerby, R.G.J., A. Olsen, T. Furevik and L.G. Anderson, Response of the surface CO_2 system in the Nordic Seas and northern North Atlantic to climate change, this issue, 2004.

Bleck, R., C. Rooth, D. Hu, and L. T. Smith, Salinity-driven thermocline transients in a wind- and thermohaline-forced isopycnic coordinate model of the North Atlantic, *J. Phys. Oceanogr.*, 22, 1486-1505, 1992.

Bodungen, B.V., A. Anita, E. Bauerfeind, O. Haupt, W. Koeve, E. Machado, I. Peeken, R. Peinert, S. Reitmeier, C. Thomsen, M. Voss, M. Wunsch, U. Zeller and B. Zeitzschel, Pelagic processes and vertical flux of particles: an overview of a long-term comparative study in the Norwegian Sea and Greenland Sea, *Geol. Rundsch.*, 84, pp. 11-27, 1995.

Bolin, B. and E. Eriksson, Changes in the carbon dioxide content of the atmosphere and sea due to fossil fuel combustion, *in The atmosphere and the sea in motion, Rossby Memorial Volume*, edited by B. Bolin, Rockefeller Inst., New York, pp. 130-142, 1959.

Bönisch, G., J. Blindheim, J.L. Bullister, P. Schlosser, and D.W.R. Wallace, Long-term trends of temperature, salinity, density, and transient tracers in the central Greenland Sea, *J. Geophys. Res.*, 102 (C8), pp. 18553-18571, 1997.

Broecker, W.S., *Chemical oceanography*. Harcourt Brace Jovanovic, Inc., pp. 1-214, 1974.

Caldeira, K. and M.E. Wickett, Anthropogenic carbon and ocean pH, *Nature*, 425, pp. 325-325, 2003.

Chierici, M., H. Drange, L.G. Anderson and T. Johannessen, Inorganic carbon fluxes between the Greenland Sea and the surrounding areas, calculated by applying model results. *J. Mar. Sys.*, 22, pp. 295-309, 1999.

Eppley, R.W. and B.J. Peterson, Particulate organic matter flux and planktonic new production in the deep ocean, *Nature*, 282, pp. 677-680, 1979.

Eppley, R.W., New production: History, methods, problems, *in Productivity of the ocean: Present and past*, edited by W.H. Berger, V.S. Smetacek and G. Wefer, John Wiley & Sons, pp. 85-97, 1989.

Falck, E. and H.G. Gade, Net community production and oxygen fluxes in the Nordic Seas based on O_2 budget calculations, *Glob. Biogeochem. Cycl.*, 13 (4), pp. 1117-1126, 1999.

Falck, E., On the behaviour of oxygen in the euphotic zone of the Nordic Seas and on carbon and nutrients in the Northeast Water Polynya, Ph-D thesis, University of Bergen, Norway, 1999.

Falck, E., Contribution of waters of Atlantic and Pacific origin in the Northeast Water Polynya, *Polar Research*, 20, pp. 193-200, 2001.

Falck, E. and L.G. Anderson, The dynamics of the carbon cycle in the surface water of the Norwegian Sea. *Mar. Chem.*, in press, 2004.

Fosså, J.H. 2002. Havets miljø 2002. *Fisken og havet*, sørnr. 2-2002 (with summaries in English), 2002.

Fransson, A., M. Chierici, L.G. Anderson, I. Bussmann, G. Kattner and E.P. Jones, J.H. Swift, The importance of shelf processes for the modification of chemical constituents in the waters of the Eurasian Artic Ocean: Implication for carbon fluxes, *Cont. Shelf Res.* 21, pp. 225-242, 2001.

Gade, H.G. Features of fjord and ocean circulation, *in The Nordic Seas*, edited by B.G. Hurdle, Springer, New York, pp. 183-189, 1986.

Gascard J.-C., A.J. Watson, M.-J. Messias, K.A. Olsson, T. Johannessen and K. Simonsen, Long-lived vortices as a mode of deep ventilation in the Greenland Sea, *Nature*, 416 (6880), pp. 525-527, 2002.

Gudmundsson, K., Long-term variation in phytoplankton productivity during spring in Icelandic waters, *ICES Journal of Marine Science*, 55, pp. 635-643, 1998.

Hansen, B. and S. Østerhus, North Atlantic - Nordic Seas exchanges, *Progress in Oceanography* 45 (2), pp. 109-208, 2000.

Hein, M. and K. Sand-Jensen, CO_2 Increases Oceanic Primary Production, *Nature*, 388(6642), pp. 526-527, 1997.

Helland-Hansen, B. and F. Nansen, The Norwegian Sea. Its physical oceanography based upon the Norwegian researchers 1900-1904. *Rep. Norw. Fish. Mar. Invest., Rep. Technol. Res.* 2(Pt. 1, No2): pp. 1-390, 1909.

Helmke, J.P. and H.A. Bauch, Glacial-interglacial carbonate preservation records in the Nordic Seas, *Global and Planetary Change*, 33, pp. 15-28, 2002.

Hood, E.M., L. Merlivat, and T. Johannessen, Variations of fCO_2 and air-sea flux of CO_2 in the Greenland Sea gyre using high-frequency time series data from CARIOCA drift buoys, *J. Geophys. Res.*, 104 (C9), pp. 20571-20583, 1999.

IPCC, Climate change 2001: The scientific basis. Contribution of working group I to the Third Assessment Report of the Intergovernmental Panel on Climate Change. Available online at http://www.ipcc.ch/), 2001.

Johannessen, O.M., L. Bengtsson, M.W. Miles, S.I. Kuzmina, V.A. Semenov, G.V. Alekseev, A.P. Nagurnyi, V.F. Zakharov, L. Bobylev, L. H. Pettersson, K. Hasselmann and H.P. Cattle, Arctic climate change: observed and modelled temperature and sea-ice variability, *Tellus,* 56A (4), 328-341, 2004.

Kaltin, S., L.G. Anderson, M. Chierici, A. Fransson and K. Olsson, Uptake of atmospheric carbon dioxide in the Barents Sea, *J. Mar. Syst*., 38, pp. 31- 45, 2002.

Kaltin, S., and L.G. Anderson, Uptake of atmospheric carbon dioxide in Arctic shelf seas: evaluation of the relative importance of processes that influence pCO_2 in water transported over the Bering-Chukchi Sea shelf. *Mar. Chem*., in press, 2004.

Keeling, C.D. and T.P. Whorf, Atmospheric CO_2 records from sites in the SIO air sampling network, *in Trends: A Compendium of Data on Global Change*. Carbon Dioxide Information Analysis Center, Oak Ridge National Laboratory, U.S. Department of Energy, Oak Ridge, Tenn., U.S.A, 2004.

Killworth, P.D., Deep convection in the world ocean, *Rev. Geophys*., 21(1), pp. 1-26, 1983.

Klausmeier, C.A., E. Litchmann, T. Daufresne and S.A. Levin, Optimal nitrogen-to-phosphorus stoichiometry of phytoplankton, *Nature*, 429, pp. 171-174, 2004.

Laws, E.A., Photosynthetic quotient, new production and net community production in the open ocean, *Deep-Sea Res*., 38, pp. 143-167, 1991.

Lefèvre, N., A.J. Watson, A. Olsen, A. Rios, F.F. Pérez, and T. Johannessen, A decrease in the sink for atmospheric CO2 in the North Atlantic, *Geophys Res Letters*, 31, pp. 1-4, 2004.

Loeng, H., Features of the physical oceanographic conditions of the Barents Sea, *in Proceedings of the Pro Mare Symposium of Polar Marine Ecology* edited by E. Sakshaug, C.C.E. Hopkins and N.A. Øritsland, Trondheim, 12-16 May 1991. *Polar Research* 10, pp. 5-18, 1991.

Lundberg, L., CO_2 air-sea exchange in the Nordic Seas. An attempt to make an estimate based on data, *Oceanologica Acta*, 17, 159-175, 1994.

Lundberg, L. and P.M. Haugan, A Nordic Seas-Arctic Ocean carbon budget from volume flow and inorganic carbon data. *Glob. Biogeochem. Cycl*., 10, pp. 493-510, 1996.

Malmberg, S.-A., and S. Jónsson, Timing of deep convection in the Greenland and Iceland Seas, *ICES J. Mar. Sci*., *54* (3), pp. 300-309, 1997.

Mauritzen, C., Production of dense overflow waters feeding the North Atlantic across the Greenland-Scotland Ridge. Part 2: An inverse model. *Deep-Sea Res. I*, 43, pp 807-835. 1996.

Miller, L.A., M. Chierici, T. Johannessen, T.T. Noji, F. Rey, and I. Skjelvan, Seasonal dissolved inorganic carbon variations in the Greenland Sea and implications for atmospheric CO_2 exchange, *Deep-Sea Res II,* 46, pp. 1473-1496, 1999.

Noji, T.T., F. Rey, L.A. Miller, K.Y. Børsheim and J. Urban-Rich, Fate of biogenic carbon in the upper 200 m of the central Greenland Sea, *Deep-Sea Res. II*, 46, pp. 1497-1509, 1999.

Noji T.T., L.A. Miller, I. Skjelvan, E. Falck, K.Y. Børsheim, F. Rey, J. Urban-Rich and T. Johannessen, Constraints on carbon drawdown and export in the Greenland Sea. *in The Northern North Atlantic - a changing environment* edited by P. Schäfer, W. Ritzrau, M. Schl‚ter and J. Thiede, Springer Verlag, pp. 39-52, 2000.

Olsen, A., L.G. Anderson and T. Johannessen, The impact of climate variations on fluxes of oxygen in the Barents Sea, *Cont. Shelf Res*., 22, pp. 1117-1128, 2002.

Olsen, A., R.G.J. Bellerby, T. Johannessen, A.M. Omar and I. Skjelvan, Interannual variability in the wintertime air-sea flux of carbon dioxide in the northern North Atlantic, 1981-2001, *Deep Sea Res*. I, 50, pp. 1323-1338, 2003a.

Olsen, A., T. Johannessen and F. Rey, On the nature of the factors that control spring bloom development at the entrance to the Barents Sea and interannual variability, *Sarsia*, 88, pp. 379-393, 2003b.

Omar, A., T. Johannessen, S. Kaltin and A. Olsen, Anthropogenic increase of oceanic *p*CO2 in the Barents Sea surface water, *J Geophys Res*., 108, pp. 18.1-18.8, 2003.

Omar, A., T. Johannessen, R. Bellerby, A. Olsen, L.G. Anderson and C. Kivimäe, Sea ice and brine formation in Storfjorden: implications for the Arctic winter time air-sea CO2 flux, this issue, 2004.

Orr, J.C, E. Maier-Reimer, U. Mikolajewicz, P. Monfray, J.L. Sarmiento, J.R. Toggweiler, N.K. Taylor, J. Palmer, N. Gruber, C.L. Sabine, C. Le Quéré, R.M. Key, and J. Boutin, Estimates of anthropogenic carbon uptake from four three-dimensional global ocean models, *Glob. Biogeochem. Cyc*., 15, pp. 43-60, 2001.

Platt, T., W.G. Harrison, M.R. Lewis, W.K.W Li, S. Sathyendranath, R.E. Smith and A.F. Vezina, Biological production of the oceans: the case for a consensus, *Mar. Ecol. Prog. Ser*., 52, pp. 77-89, 1989.

Quadfasel, D., B. Rudels, and K. Kurz, Outflow of dense water from a Svalbard fjord into the Fram Strait, *Deep-Sea Res*. 35, pp. 1143-1150, 1988.

Redfield, A.C., B.H. Ketchum and F.A. Richards, The influence of organisms on the composition of seawater, *in The Sea*, edited by M.N. Hill, John Wiley, New York, pp. 26-77, 1963.

Rey, F. The development of the spring phytoplankton outburst at selected sites off the Norwegian coast. *in The Norwegian Coastal Current*, vol. II, edited by R. Sætre and M. Mork, University of Bergen, Bergen, pp. 649-680, 1981.

Rey, F. and H. Loeng, The influence of ice and hydrographic conditions on the development of phytoplankton in the Barents Sea, *in Marine Biology of Polar Regions and Effects of Stress on Marine Organisms*, edited by J.S. Gray and M.E. Christiansen, John Wiley & Sons Ltd, pp. 49-63, 1985.

Rey, F. and H.R. Skjoldal, Consumption of silic acid below the euphotic zone by sedimenting diatom blooms in the Barents Sea, *Mar. Ecol. Prog. Ser., 36*, pp. 307-312, 1987.

Rey, F., H.R. Skjoldal and D. Slagstad, Primary production in relation to climatic changes in the Barents Sea, *in The effect of oceanographic conditions in distribution and population dynamics of commercial fish stocks in the Barents Sea, proceedings of the third Soviet-Norwegian Symposium, Murmansk, 26-18 May 1986*, edited by H. Loeng, Institute of Marine Research, Bergen, pp. 29-46, 1987.

Rey, F., T.T. Noji and L.A. Miller, Seasonal phytoplankton development and new production in the central Greenland Sea, *Sarsia, 85*, pp. 329-344, 2000.

Rey, F., Phytoplankton: The grass of the sea, *in The Norwegian Sea Ecosystem*, edited by H.R. Skjoldal, Tapir, Trondheim, Norway, pp. 93-132, 2004.

Reynolds, R. and T. Smith, Improved global sea surface temperature analyses, *J. Clim.*, 7, pp. 929-948, 1994.

Riebesell, U., I. Zondervan, B. Rost, P.D. Tortell, R.E Zeebe and F.M.M. Morel, Reduced calcification of marine plankton in response to increased atmospheric CO_2, *Nature*, 407, pp. 364-367, 2000.

Rudels, B., Haline convection in the Greenland Sea, *Deep-Sea Res.*, 37 (9): pp. 1491-1511, 1990.

Sætre, R., J. Aure and R. Ljøen, Wind effects on the lateral extension of the Norwegian Coastal Current, *Cont. Shelf. Res.*, 8, pp. 239-253, 1988.

Sakshaug, E. and D. Slagstad, Sea-ice and wind: Effects on primary productivity in the Barents Sea, *Atmosphere-Ocean*, 30, pp. 579-591, 1992.

Sakshaug, E., A. Bjørge, B. Gulliksen, H. Loeng and F. Mehlum, Økosystemer i Barentshavet, PRO MARE; Norwegian Research Program for Marine Arctic Ecology, 1992.

Sakshaug, E., F. Rey and D. Slagstad, Wind forcing of marine primary production in the northern atmospheric low-pressure belt. *in Ecology of Fjords and Coastal Waters*, edited by H.R.Skjoldal, C. Hopkins, K.E. Erikstad, H.P. Leinaas, Elsevier, Amsterdam, pp. 15-25, 1995.

Sambrotto, R.N., G. Savidge, C. Robinson, P. Boyd, T. Takahashi, D.M. Karl, C. Langdon, D. Chipman, J. Marra and L. Codispoti, Elevated consumption of carbon relative to nitrogen in the surface ocean, *Nature*, 363, pp. 248-250, 1993.

Schlosser, P., G. Bönisch, M. Rhein and R. Bayer, Reduction of deepwater formation in the Greenland Sea during the 1980s: Evidence from tracer data, *Science*, 251, pp. 1054-1056, 1991.

Schott, F. and K.D. Leaman, Observations with moored acoustic Doppler current profiles in the convection regime in the Golfed u Lion, *J. Phys. Oceanogr.*, 21, pp. 558-574, 1991.

Shindell, D.T., Simulation of recent northern winter climate trends by greenhouse forcing, *Nature*, 399, pp. 452-455, 1999.

Simonsen, K. and P.M. Haugan, Heat budgets of the Arctic Mediterranean and sea surface heat flux parametrizations for the Nordic Seas. *J. Geophys. Res.* 101, 6553-6576, 1996.

Skjelvan, I., T. Johannessen and L.A. Miller, Interannual variability of fCO_2 in the Greenland and Norwegian Seas, *Tellus*, 51B, pp. 477-489, 1999a.

Skjelvan, I., M. Chierici and J. Ólafsson, Horizontal distributions of NC_T, NA_T and fCO_2 in the Nordic Seas, *in Carbon and oxygen fluxes in the Nordic Seas*, edited by I. Skjelvan, Ph-D thesis, University of Bergen, Norway, 1999b.

Skjelvan, I., E. Falck, L.G. Anderson and F. Rey, Oxygen fluxes in the Norwegian Atlantic Current, *Mar. Chem.*, 73 (3-4), pp. 291-303, 2001.

Slagstad, D. and P. Wassmann, Climate change and carbon flux in the Barents Sea: 3-D simulations of ice-distribution, primary production and vertical export of particulate organic carbon, *Mem. Natl. Inst. Polar. Res., Spec. Issue* 51, pp. 119-141, 1996.

Stumm, W. and J.J. Morgan, *Aquatic Chemistry (3rd ed.)*. Wiley & Sons, New York, pp. 1022, 1996.

Sverdrup, H.U., On conditions for the vernal blooming of phytoplankton, *Journal du Conceil Permanent International pour l'Exploration de la Mer,* 18, pp. 287-295, 1953.

Swift, J.H. and K. Aagaard, Seasonal transitions and water mass formation in the Iceland and Greenland seas, *Deep-Sea Res. A*, *28A* (10), pp. 1107-1129, 1981.

Takahashi, T., S. C. Sutherland, C. Sweeney, A. Poisson, N. Metzl, B. Tilbrook, N. Bates, R. Wanninkhof, R. A. Feely and C. Sabine, Global sea-air CO_2 flux based on climatological surface ocean pCO_2, and seasonal biological and temperature effects, *Deep Sea Res. II*, 49, pp. 1601-1622, 2002.

Tortell, P.D., J.R. Reinfelder and F.M.M Morel, Active uptake of bicarbonate by diatoms, *Nature*, 390, pp. 243–244, 1997.

Völker, C., D.W.R. Wallace and D.A. Wolf-Gladrow, On the role of heat fluxes in the uptake of anthropogenic carbon in the North Atlantic, *Glob. Biogeochem. Cycl.*, 16 (4), 1138, doi:10.1029/2002GB001897, 2002.

Wallace, D.W.R., P.J. Minnett and T.S. Hopkins, Nutrients, oxygen, and inferred new production in the Northeast Water Polynya, 1992. *J. Geophys.Res.*, 100, pp. 4323-4340, 1995.

Wallace, D.W.R., Storage and transport of excess CO_2 in the oceans: The JGOFS/WOCE global CO_2 survey, *in Ocean circulation and climate: observing and modelling the global ocean*, edited by G. Siedler, J. Church and J. Gould, Academic Press, pp. 489-521, 2001.

Wanninkhof, R., Relationship between wind speed and gas exchange over the ocean, *J. Geophys. Res.*, 97(C5), pp. 7373-7382, 1992.

Wassmann, P., Retention versus export food chains: processes controlling sinking loss from marine pelagic systems, *Hydrobiologia*, 363, pp. 29-57, 1998.

Weiss, R.F., Carbon dioxide in water and seawater: the solubility of a nonideal gas, *Mar. Chem.*, 2, 201-215, 1974.

Zondervan I., B. Rost and U. Riebesell, Effect of CO_2 concentration on the PIC/POC ratio in the coccolithophore *Emiliania huxleyi* grown under light-limited conditions and different daylengths, *J. Exp. Mar. Biol. Eco.*, 272, pp. 55-70, 2002.

L.G. Anderson, Department of Chemistry, Göteborg University, 41296 Göteborg, Sweden.

R.G.J. Bellerby, C. Heinze, T. Johannessen, Y. Kasajima, A. Olsen, K.A. Olsson, A. Omar, and I. Skjelvan, Bjerknes Centre for Climate Research, University of Bergen, Allegaten 55, NO-5007 Bergen, Norway. (ingunn.skjelvan@gfi.uib.no)

E. Falck and C. Kivimäe, Geophysical Institute, University of Bergen, Allegaten 70, NO-5007 Bergen, Norway.

F. Rey, Institute of Marine Research, Postbox 1870 Nordnes, NO-5817 Bergen, Norway.

Sea-Ice and Brine Formation in Storfjorden: Implications for the Arctic Wintertime Air–Sea CO_2 Flux

Abdirahman Omar[1,2], Truls Johannessen[1,2], Richard G.J. Bellerby[1], Are Olsen[1,2], Leif G. Anderson[3], and Caroline Kivimäe[2]

We investigated the air–sea CO_2 flux associated with the formation of brine-enriched shelf water (BSW) in Storfjorden: a water mass formed by sea-ice formation and brine rejection during winter. We used inorganic carbon and auxiliary hydrographic data collected in August 2001 and April 2002. After accounting for the effects of biological and other processes, we found that the formation of BSW is accompanied by an uptake of atmospheric CO_2. This uptake accounted for a carbon enrichment of 17 ± 4 μmol kg^{-1} in the BSW. About 2/3 of the uptake occurred during winter in a coastal polynya for which a wintertime CO_2 flux (calculated as C) was 65 ± 40 g m^{-2}. The flux was about 13 times lower (≈ 5 g m^{-2}) for the nonpolynya regions of Storfjorden. The computed areal fluxes were extrapolated to polynyas (and leads) and seasonally ice–covered regions in the Arctic. Based on this extrapolation, Arctic coastal polynyas were found to account for a winter time CO_2 uptake of 2.3 ($\pm$1.4) $\times 10^{12}$ g yr^{-1}. The carbon taken up in coastal polynyas is most likely transported into the deep ocean through brine-driven deep water formation. For polynyas in the central Arctic and for seasonally ice-covered regions, a seaward wintertime air–sea CO_2 flux of 43 ($\pm$22) $\times 10^{12}$ g yr^{-1} was computed. Further, it was found that this flux can increase to 124 ($\pm$55) $\times 10^{12}$ g yr^{-1} by about year 2100, if the extent of seasonal ice increases by 43% (as modeled in *Johannessen et al.,* 2002) and central Arctic polynyas become 10 times larger than present. However, the importance of the computed flux with respect to sequestration of atmospheric CO_2 is not clear, due to limited knowledge on the depth penetration of the brine water associated with the formation of new ice in central Arctic polynyas and seasonally ice-covered regions.

1. INTRODUCTION

The Arctic Ocean (AO) could become nearly ice-free during summer by the end of this century. Simulation results from coupled atmosphere-ice-ocean models have indicated a moderate reduction in winter sea ice extent and a drastic decrease for the summer, implying a greater seasonal variability of the Arctic ice cover [*Johannessen et al.,* 2002].

A disappearance of the permanent ice cover is expected to enhance carbon fluxes into the AO. Enhanced carbon fluxes are thought to result from greater CO_2 solubility, due to freshening of the surface water, and through greater primary production, due to better light conditions [*Anderson and Kaltin,* 2001]. However, the effect of changes in the extent of

[1]Bjerknes Centre for Climate Research, University of Bergen, Bergen, Norway.
[2]Geophysical Institute, University of Bergen, Bergen, Norway.
[3]Department of Analytical and Marine Chemistry, Göteborg University, Göteborg, Sweden.

The Nordic Seas: An Integrated Perspective
Geophysical Monograph Series 158

10.1029/158GM12

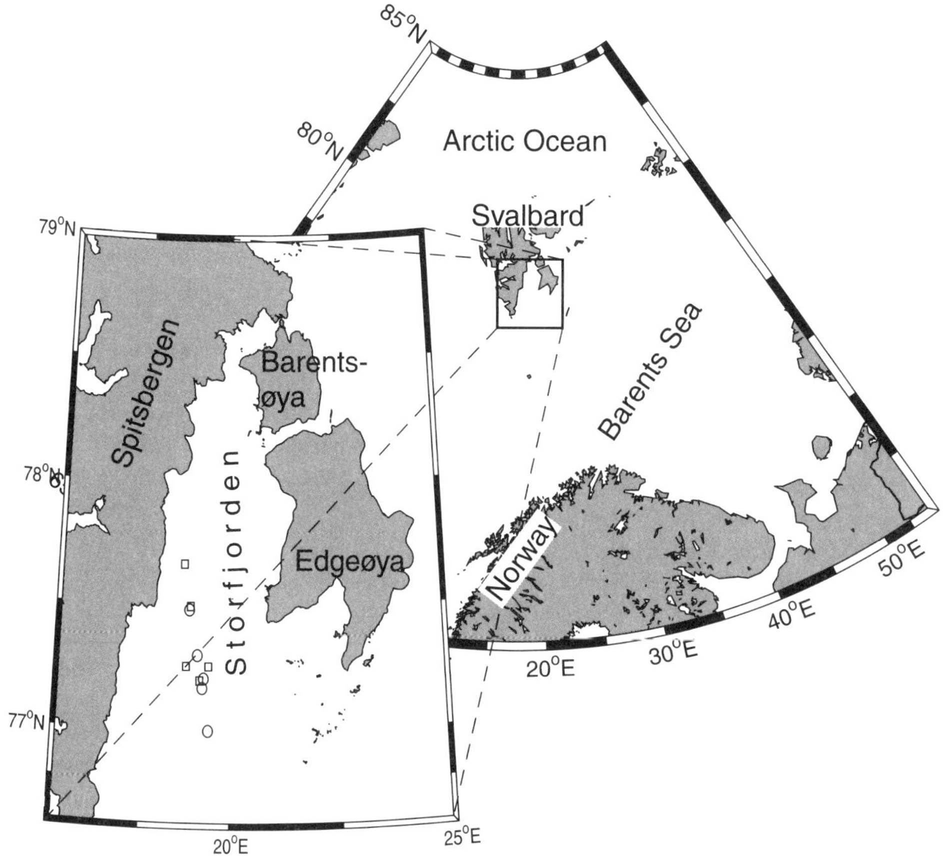

Figure 1. Map of Storfjorden, showing the stations visited during *Håkon Mosby* 2001 (squares) and *Oden* 2002 (diamonds) cruises.

seasonal ice on the carbon fluxes into the Arctic has not been studied. CO_2 flux and seasonal ice formation are related through the high-density water that forms through sea-ice formation and subsequent brine rejection. Such high-density waters supply the AO's deep and intermediate waters [e.g. *Jones et al.,* 2001]. Further, there is an important conduit of CO_2 to the AO through biological and transport processes [*Anderson and Jones,* 1991].

At present, the areal extent of seasonal ice in the Arctic is about 7×10^{12} m^2. Polynyas and leads (henceforth referred to as polynyas) are the major sites where extensive new ice formation and brine rejection occur. In the central Arctic, polynyas occupy about 1% of the winter ice extent, but they produce similar amounts of ice as that produced below multiyear ice [e.g. *Skogseth & Haugan,* 2003; *Gow and Tucker,* 1990]. Along the coastal margin, Arctic coastal polynyas produce large quantities of ice and high-density brine-enriched water [e.g. *Cavalieri and Martin* 1994]. The area of polynyas will probably increase in the future following the aforementioned greater seasonal variation of the arctic ice cover.

The present study investigates air–sea CO_2 flux during sea-ice formation and brine rejection, by using data collected from Storfjorden (Figure 1). This is a fjord situated in the western Barents Sea, between the islands of Spitsbergen, Barentsøya, and Edgeøya in the Svalbard Archipelago. Ice formation and subsequent brine rejection during winter produce brine-enriched shelf water (BSW), which is defined by $S \geq 35$ and $T < -1.5°C$ [*Loeng,* 1991; referred to there as bottom water (BW)]. Further, most of the BSW production occurs in a coastal polynya situated at the lee-side of Edgeøya and Barentsøya [*Skogseth and Haugan,* 2003; *Haarpaintner et al.,* 2001], where offshore winds enhance ice production. Data on total dissolved inorganic carbon (C_T), total alkalinity (A_T), dissolved oxygen (O_2), temperature (T), and salinity (S) were collected from Storfjorden in August 2001 and April 2002. By evaluating the processes governing C_T concentration in BSW, the air–sea CO_2 flux associated with sea-ice formation and brine release over winter is determined. In particular, the flux is determined separately for the polynya and nonpolynya regions of the fjord. The computed carbon fluxes are then extrapolated to polynyas and seasonally ice-covered regions of the entire Arctic. Finally, the effect of changes in seasonal ice coverage and polynya extent on wintertime air–sea CO_2 flux in the Arctic is estimated.

2. DATA AND METHODS

2.1. Data and Analytical Methods

Data used in this work originate from two surveys (from the *Håkon Mosby* in 2001 and from the *Oden* in 2002) conducted in Storfjorden. Figure 1 shows the stations occupied during these surveys. Table 1 shows which parameters were measured during each cruise. Water samples were drawn from Niskin bottles mounted on a rosette with a CTD and analyzed onboard within 24 hours.

C_T was determined by gas extraction of acidified water samples followed by coulometric titration with a SOMMA system [*Johnson et al.*, 1985; 1987]. The precision was ±4 and ±1 μmol kg^{-1} during *Håkon Mosby* 2001 and *Oden* 2002, respectively. The accuracy in C_T measurements was set by running Certified Reference Material (CRM) supplied by Andrew Dickson at the Scripps Institute of Oceanography (San Diego, California).

A_T was determined by titrating samples with 0.1 M HCl as described by *Haraldson et al.* [1997]. The precision was determined to be ±5 and ±1 μmol kg^{-1} during *Håkon Mosby* 2001 and *Oden* 2002, respectively. The accuracy was set in the same way as for C_T.

O_2 was determined by using the Winkler method with visual detection of the titration end point during *Håkon Mosby* 2001 and with automated detection during *Oden* 2002.

2.2. Calculation Methods

Depth variations in the concentration of C_T are due to (i) biological activity, i.e., the formation/decay of organic matter and formation/dissolution of calcium carbonate, (ii) mixing of water masses, (iii) dilution/evaporation, (iv) brine rejection, and (v) air–sea CO_2 exchange. We corrected the measured C_T data for the effects of (i)–(iv) so we could evaluate the contribution from the air–sea flux of CO_2. The corrected C_T (nC_T) can be written as:

$$nC_T = (C_T - R_{C:O} \times AOU - 0.5 \times (nA_T - k) - b) \times \frac{34.17}{s} + b \qquad (1)$$

where $R_{C:O}$ is the Redfield ratio of remineralization for carbon to oxygen, AOU is the apparent oxygen utilization, nA_T is the alkalinity corrected for the formation/decay of organic matter and then adjusted to a constant salinity of 34.17, and k and b are constants. The term $-R_{C:O} \times$ AOU accounts for C_T changes attributable to organic matter, while the term $-0.5 \times (nA_T - k)$ accounts for the changes due to the presence of calcium carbonate.

AOU was calculated after *Najjar and Keeling* [1997] according to:

$$AOU = (O_{2meas} - O_{2sat}) + \left(1 - \frac{p}{1013.25}\right) O_{2sat} \qquad (2)$$

where O_2 is the oxygen concentration and the subscripts "sat" and "meas" refer to saturation and measured, respectively. The saturation concentration of O_2 is a function of atmospheric pressure (p). Hence, the second term of the right-hand side of Eq 2 is used to normalize our AOU values to 1 atm. For p, we used NCEP/NCAR reanalysis data of mean annual atmospheric pressure [*Kalnay et al.,* 1996] obtained from the NOAA CIRES Climate Diagnostics Center at http://www.cdc.noaa.gov. We assumed that AOU values calculated from Eq 2 quantified the amount of O_2 produced (AOU > 0) or consumed (AOU < 0) during the formation or decay of organic matter. Then we computed the corresponding changes in C_T from the product of AOU × $R_{C:O}$. We subtracted these results from the measured C_T to account for C_T changes that are due to formation/decay of organic matter (Eq 1).

Alkalinity is influenced by the same processes as for C_T except air–sea CO_2 exchange. The formation/decay of organic matter increases/decreases A_T slightly. For our area of investigation, adding the concurrent changes in nitrate to A_T will account for the above alkalinity changes [e.g. *Brewer et al.,* 1975]. We calculated the nitrate change from the product AOU × $R_{N:O}$, where $R_{N:O}$ (= −16/138) is the Redfield ratio of remineralization for nitrogen to oxygen. The results were added to the measured alkalinity. The resulting corrected alkalinity is henceforth referred to as *corr*A_T. We adjusted *corr*A_T to a constant salinity of 34.17 to account for the changes due to mixing, dilution/evaporation, and brine rejection. For this purpose, we followed the procedure outlined by *Friis et al.* [2003], which better accounts for the effect of runoff, upwelling, and lateral mixing than does the traditional procedure. We first determined the intercept (b) of the linear relationship between *corr*A_T and salinity and then adjusted

Table 1. Cruise Dates and Chemical Sampling From Storfjorden.

Cruise	Date (season)	S	T	C_T	A_T	O_2
Håkon Mosby 2001	Aug 23-25 (summer)	✓	✓	✓	✓	✓
Oden 2002	Apr 26-29 (winter)	✓	✓	✓	✓	✓

S = salinity, T = temperature, C_T = total dissolved inorganic carbon, A_T = total alkalinity, O_2 = dissolved oxygen.

all data of $corr\mathrm{A_T}$ to 34.17 salinity to determine $n\mathrm{A_T}$, according to:

$$n\mathrm{A_T} = \frac{(corr\mathrm{A_T} - \mathrm{b}) \times 34.17}{\mathrm{S}} + \mathrm{b} \qquad (3)$$

The intercept b was determined separately for samples influenced by brine rejection (identified by $T \leq 0°C$) and for the rest of the dataset (with $T > 0°C$). For the latter samples, mixing between modified Atlantic water (with $S \approx 34.85$ and $corr\mathrm{A_T} \approx 2300$ μmol/kg) and fresher surface water (with $S < 33.5$ and $corr\mathrm{A_T} < 2230$ μmol/kg) produced a linear relationship between $corr\mathrm{A_T}$ and salinity with a slope and intercept of a_1 and b_1. A different linear relationship was found for samples influenced by brine rejection, and its coefficients were such that $a_2 > a_1$ and $b_2 \ll b_1$. Further, the two relationships intersected at S = 34.17. At this salinity, the value of $corr\mathrm{A_T}$ was equal to k (see Eq 1), which is why we chose to adjust our data to S = 34.17. The above implies that, for each sample, any nonzero residual obtained from $(n\mathrm{A_T} - \mathrm{k})$ results from the formation/dissolution of $CaCO_3$. Further, the corresponding change in $\mathrm{C_T}$ is one-half of this residual. Thus, we included the term $-0.5 \times (n\mathrm{A_T} - \mathrm{k})$ in Eq 1 to account for this change.

Once $\mathrm{C_T}$ values were corrected for the effect of biological processes, i.e., once $\mathrm{C_T} - \mathrm{R_{C:O}} \times \mathrm{AOU} - 0.5 \times (n\mathrm{A_T} - \mathrm{k})$ was determined, the results were then adjusted to 34.17 salinity, to correct for changes due to (ii)–(iv) above. Specifically, we assumed that the freshwater alkalinity end-member (i.e., b) was entirely due to bicarbonate ions. Therefore, the same values of b could be used for $\mathrm{C_T}$. Hence, by replacing $corr\mathrm{A_T}$ in the right hand side of Eq 3 with $\mathrm{C_T} - \mathrm{R_{C:O}} \times \mathrm{AOU} - 0.5 \times (n\mathrm{A_T} - \mathrm{k})$, we obtained the right hand side of Eq 1.

For each cruise, we constructed a composite profile for each of the variables T, S, AOU, $n\mathrm{A_T}$, and $n\mathrm{C_T}$ by taking the 10-m depth-averages of data acquired from different stations. These profiles reveal depth variations during each cruise; comparison between two profiles for the same parameter but from different cruises reveals the seasonal (summer–winter) variation of that particular parameter.

2.3. Uncertainties

As depicted in Eqs 1 and 3, $n\mathrm{C_T}$ and $n\mathrm{A_T}$ are functions of measured and calculated quantities. The uncertainties of the former type of quantities are given by their analytical imprecision while the uncertainties of the latter are given by their standard deviations. We calculated the total uncertainties in $n\mathrm{C_T}$ and $n\mathrm{A_T}$ that are due to the above individual uncertainties. For this purpose, we neglected the individual uncertainties in S and AOU because they were negligible compared to the individual uncertainties associated with the other parameters used for the calculation. Hence, for each pair of values of S and AOU, $n\mathrm{C_T}$ would be described by a linear function of $\mathrm{C_T}$, $\mathrm{R_{C:O}}$, $n\mathrm{A_T}$, k, and b, with proportionality constants of (34.17/S), (34.17/S) × AOU, (34.17/2S), (34.17/2S), and (1 – 34.17/S), respectively (see Eq 1). Further, we denoted the uncertainties associated with $\mathrm{C_T}$, $\mathrm{R_{C:O}}$, $n\mathrm{A_T}$, k, and b as σ_{C_T}, $\sigma_{R_{C:O}}$, σ_{nA_T}, σ_k, and σ_b, respectively. The total uncertainty in nCT was then computed according to (e.g.) *Bentley* [1988]:

$$\sigma_{nC_T} = \left(\left(\frac{34.17}{S} \times \sigma_{C_T}\right)^2 + \left(\frac{34.17}{S} \times \mathrm{AOU} \times \sigma_{R_{C:O}}\right)^2 + \left(\frac{34.17}{2S}\sigma_{nA_T}\right)^2 + \left(\frac{34.17}{2S} \times \sigma_k\right)^2 + \left(1 - \frac{34.17}{S}\right)^2 \times \sigma_b^2 \right)^{1/2} \qquad (4)$$

where by using similar arguments, σ_{nA_T} was calculated according to:

$$\sigma_{nA_T} = \left(\left(\frac{34.17}{S} \times \sigma_{A_T}\right)^2 + \left(\frac{34.17}{S} \times \mathrm{AOU} \times \sigma_{R_{N:O}}\right)^2 + \left(1 - \frac{34.17}{S}\right)^2 \times \sigma_b^2 \right)^{1/2} \qquad (5)$$

Table 2 lists the values of the individual uncertainties (i.e. values of σ_{C_T}, $\sigma_{R_{C:O}}$, σ_{nA_T}, σ_k, σ_b and $\sigma_{R_{N:O}}$ Values of $\pm 0.15 \times \mathrm{R_{C:O}}$ and $\pm 0.15 \times \mathrm{R_{N:O}}$ were choosen for $\sigma_{R_{C:O}}$ and $\sigma_{R_{N:O}}$.These particular values were chosen in order to account for the wide range of values that are often reported in the literature for $\mathrm{R_{C:O}}$ and $\mathrm{R_{N:O}}$, which typically differ up to 15% [e.g. *Redfield et al.* [1963]; *Anderson et al.* [1988,

Table 2. Values of Individual Uncertainties Used for Computation of Eqs 4 and 5.

Parameter	Value [μmol kg^{-1}]
σ_{C_T}	±4 for *Håkon Mosby* 2001
	±1 for *Oden* 2002
σ_{A_T}	±5 for *Håkon Mosby* 2001
	±1 for *Oden* 2002
$\sigma_{R_{C:O}}$	$\pm 0.15 \times \mathrm{R_{C:O}}$
$\sigma_{R_{N:O}}$	$\pm 0.15 \times \mathrm{R_{N:O}}$
σ_k (see Table 3)	±4 for samples with $T \leq 0°C$
	±3 for samples with $T > 0°C$
σ_b (see b in Table 3)	±42 for samples with $T \leq 0°C$
	±46 for samples with $T > 0°C$

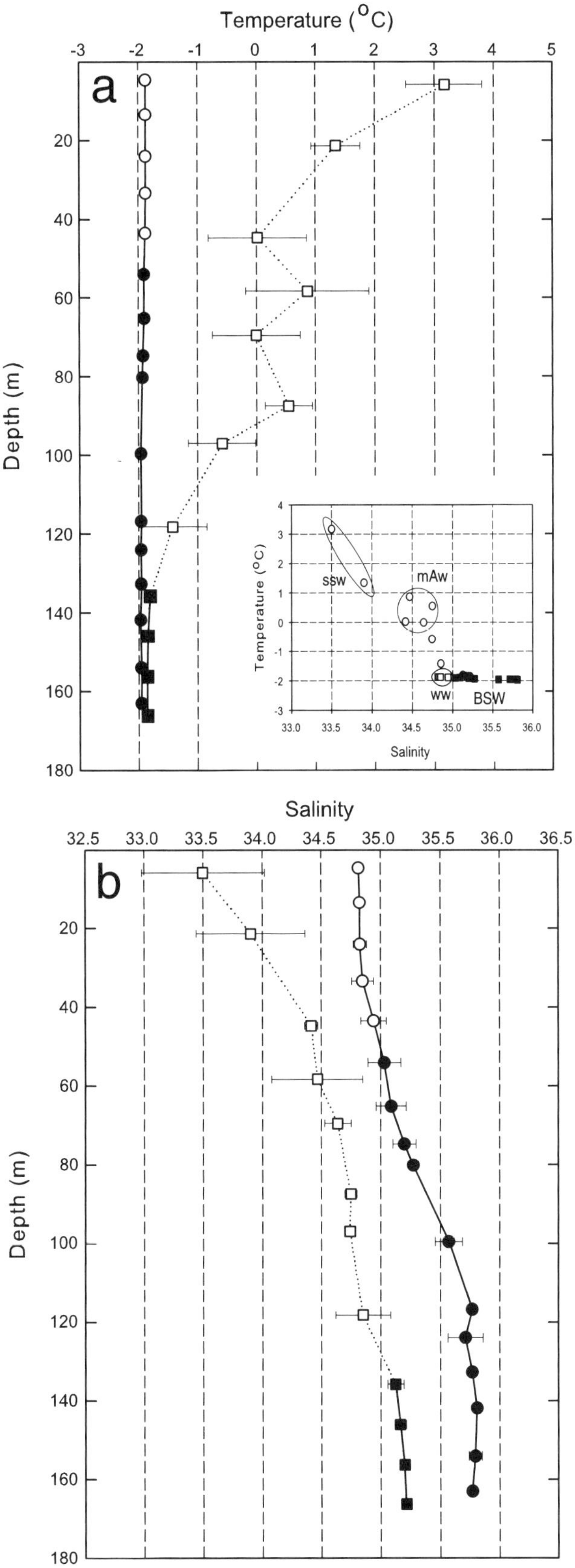

and references therein]; *Anderson and Sarmiento* [1994]; *Körtzinger et al.* [2001]].

The total uncertainty associated with each depth-averaged data point of nC_T and nA_T was calculated according to:

$$\sigma_{nC_T}(\text{mean}) = \frac{\sigma_{nC_T}}{\sqrt{n}} \text{ and } \sigma_{nA_T}(\text{mean}) = \frac{\sigma_{nA_T}}{\sqrt{n}} \tag{6}$$

where n is the number of data points on which the averages were based. Note that according to Eqs 4, 5, and 6 greater magnitudes of measurement imprecision and/or of AOU results in higher values of σ_{nC_r} (mean) and σ_{nA_T} (mean). For the averages of T, S, and AOU we chose to present the associated standard deviations, rather than the much smaller total uncertainties which result from measurement imprecision. These standard deviations are measures of the spatial variability. The corresponding spatial variability in nC_T and nA_T were generally less than σ_{nC_T} (mean) and σ_{nA_T} (mean).The latter can, therefore, be viewed both as measures of the total uncertainty and as measures of the spatial variability.

3. RESULTS

Temperature and salinity profiles are shown in Figs 2a and 2b, the former with an inset that depicts the T and S definitions of the water masses of interest. A fresher (S < 34) and warmer (T > 1°C) water resides at the surface during summer. This water mass, which has been diluted by meltwater and runoff from Svalbard and heated by solar radiation, will henceforth be referred as to summer surface water (ssw). Temperature decreases while salinity increases with depth at the base of ssw, a consequence of mixing with the underlying modified Atlantic water (mAw) that occupies depths between 45 and 90 m. Beneath the mAw, temperature decreases to near freezing while salinity increases to over 35. This is a result of mixing with BSW, which resides at depths below 135 m. During winter, the whole water column is at or near freezing (Figure 2a). A fresher layer with S ≈ 34.83 occupies the upper 35 m of the fjord (Fig 2b), henceforth referred to as winter water (ww). Below the ww layer, salinity increases almost linearly with depth to ≈ 35.8 at around 115 m. A nearly constant salinity of ≈ 35.8 is observed below the 115-m depth.

Figure 2. Depth profiles of (a) temperature and (b) salinity. Squares denote the summer data acquired during *Hkon Mosby* 2001.Circles denote the winter data acquired during *Oden* 2002. Inset in (a): T and S definitions of the water masses of interest, i.e., brine-enriched shelf water (BSW), modified Atlantic water (mAw), summer surface water (ssw), and winter water (ww). Note that filled symbols indicate data of BSW. Error bars depict the spatial variability (see section 2.3 of the main text).

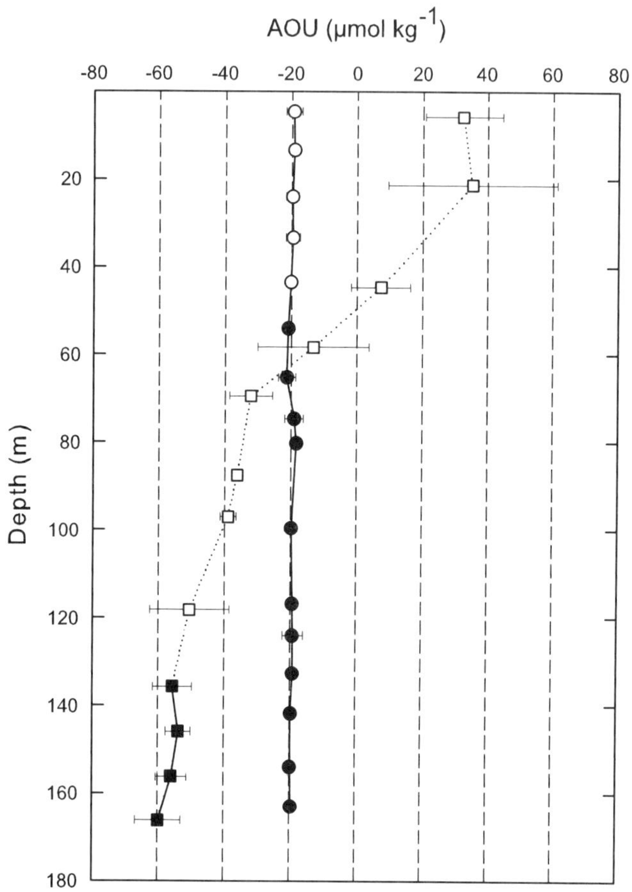

Figure 3. Depth profiles of AOU (apparent oxygen utilization). Symbols and error bars as in Figure 2.

Note that the BSW had about 0.6 higher salinity in 2002 compared to 2001 (Figure 2b). This indicates that the BSW observed in summer 2001 was effectively flushed during winter 2001/2002.

AOU values were >0 in the upper 50 m and <0 for the rest of the water column during summer (Figure 3). This means that O_2 was produced in the upper 50 m as a result of primary production, but was consumed due to decay of organic matter in the rest of the water column. During winter, on the other hand, a signature of decay of organic matter (AOU < 0) was observed throughout the water column.

Values of the coefficients and associated statistics for the two relationships between *corr*A_T and S (section 2.2) are given in Table 3.

nA_T is constant at 2264 ± 4 μmolkg^{-1} (Figure 4), irrespective of season or depth. This means that dissolution/formation of calcium carbonate is a less significant process in Storfjorden. We also note the total uncertainties (±5.1 μmol kg^{-1}) of the summer data are higher those in winter (±1.7 μmol kg^{-1}). This is partly due to better measurement

Table 3. Values of Coefficients and Statistical Parameters for the Two Relationships Between *corr*A_T and Salinity.

Parameter	For samples with	
	T > 0°C	T ≤ 0°C
a	55.38 ± 1.30	65.74 ± 1.20
b	372 ± 46	18 ± 42
R^2	0.99	0.96
Number of samples	15	97
Standard error of the fit (σ_k)	±3	±4

imprecision (summer: ±4; winter: ±1 μmol kg^{-1}) and partly due to greater magnitudes of AOU (summer: 30 to –60; winter: ≈ –20 μmol kg^{-1}) (see also section 2.3 and Figure 3).

Carbon contribution from air–sea exchange produces both seasonal and depth variations of nC_T (Figure 5). During winter, nC_T shows a two-layer structure; relatively constant values of 2101 ± 1.6 μmol kg^{-1} are observed in the upper 45 m,

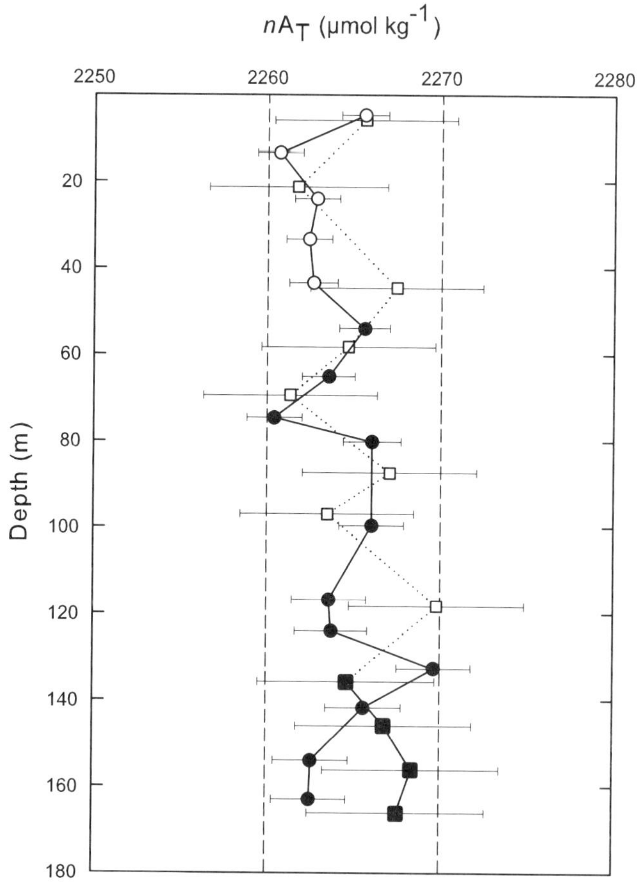

Figure 4. Depth profiles of nA_T (total alkalinity, corrected for the effect of formation/decay of organic matter, and adjusted to a constant salinity of 34.17). Error bars depict the total uncertainty (see section 2.3 of the main text). Symbols as in Figure 2.

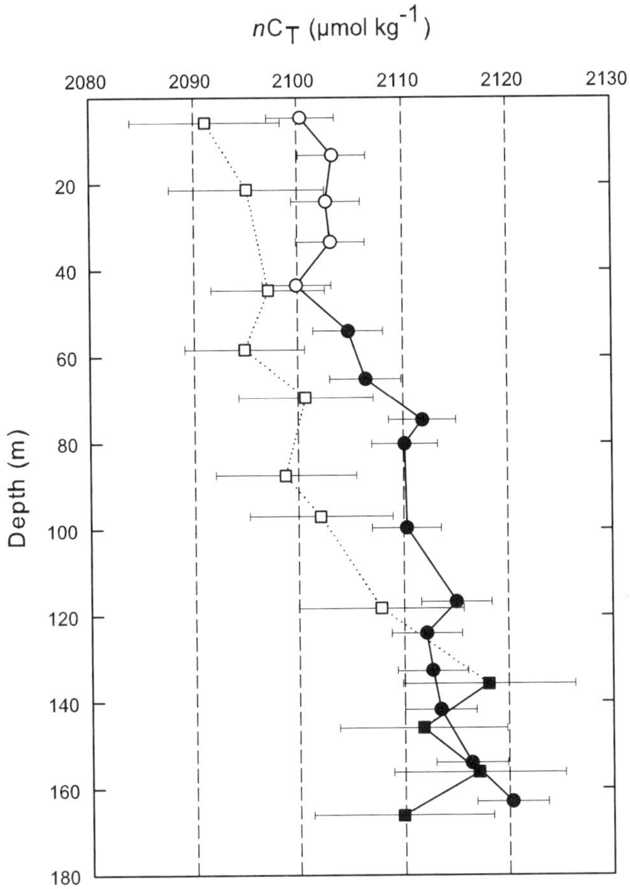

Figure 5. Depth profiles of nC_T (total dissolved inorganic carbon, corrected for the effects of processes other than air–sea CO_2 exchange, adjusted to a constant salinity of 34.17). Error bars depict the total uncertainty (see section 2.3 of the main text). Symbols and lines as in Figure 2.

2113 ± 3.2 µmol kg^{-1} in BSW. These two layers are divided by a gradient of values between 2105 and 2106 µmol kg^{-1}. The water forming this gradient is also BSW, but its nC_T values have been reduced because of mixing with the overlying water. A similar two-layer structure is evident in the summer nC_T profile: values of 2096 ± 3.3 µmol kg^{-1} observed in the upper 90 m and 2114 ± 4 µmol kg^{-1} in BSW. A strong gradient, with values between around 2100 and 2110 µmol kg^{-1}, divides these two layers. The total uncertainties for the summer data are higher (~±7.3 µmol kg^{-1}) than in winter (±3.3 µmol kg^{-1}), for the same reasons as given earlier for nA_T.

From the above, it is clear that essentially the same nC_T values were observed for BSW in the summer of 2001 and the winter of 2002. This is as anticipated, because BSW forms during winter only and no seasonal variability is expected. However, this observation also means that there were no significant year-to-year variations of BSW nC_T between 2001 and 2002. Conversely, a seasonal variability is evident for the water that occupied the upper parts of the fjord. nC_T values observed in the upper 90 m during summer were on average 5 ± 3.7 µmol kg^{-1} lower than those observed in the upper 45 m during winter.

4. DISCUSSION

4.1. Degree of Oxygen Saturation

Different magnitudes of AOU could be calculated from the O_2 data, depending on the degree of O_2-saturation assumed. The profiles shown on Figure 3 were computed by assuming 100% saturation (section 2.2). Any O_2-saturation lower than 100% would result in higher AOU values (Eq 2). For instance, if we used a 94.5% saturation, which was observed in the upper 45 m during winter, then we would obtain winter AOU values of ≈ 0 µmol kg^{-1}; for the summer profile, the transition from AOU < 0 to AOU > 0 would deepen from 50 m to 60 m (Figure 3). Also, both summer and winter nC_T profiles (Figure 5) would be shifted towards higher (15–16 µmol kg^{-1}) values. However, the magnitudes of neither depth gradients nor seasonal variations in nC_T would change. That is, the absolute values of the computed nC_T are sensitive to the choice of degree of O_2-saturation, but the relative seasonal and depth variations of nC_T are not. Therefore, nC_T changes due to air–sea CO_2 flux are correctly quantified irrespective of the degree of O_2-saturation chosen.

4.2. Uptake of Atmospheric CO_2 During the Formation of Sea-Ice and BSW

The quantification of atmospheric CO_2 uptake concurrent with the production of BSW is easier to demonstrate after the following conceptual description of the formation of this water mass. We conceptualize that heat loss to the atmosphere during fall, and brine rejection during early winter, transform ssw and mAw into ww. The ww is under-saturated with CO_2 (verified from the data used in this study, not shown). Thus, a flux of CO_2 moves into the water column in the absence of ice cover. As the winter progresses, sea-ice formation and subsequent brine rejection takes place primarily in the polynya but also under the pack ice. While air–sea CO_2 exchange is hampered in the ice-covered areas of the fjord, it continues to take place in the polynya, and the brines formed there can approach CO_2 saturation. On sinking, however, the rejected brines entrain ambient waters of intermediate depth with lower carbon concentration. The admixture forms BSW, which is not fully saturated with CO_2. "New" ww, which was ice-covered and has a lower concentration of carbon, advects or upwells into the polynya and replaces the brines that sank. As this continues throughout the winter, the deeper part of the fjord is gradually filled with high-CO_2 BSW, while the surface of the fjord contains low-CO_2 ww (Figure 5). When a sufficient volume of BSW has been produced, it drains over the sill and flows to the open ocean as a bottom current, carrying the excess carbon with it.

Simultaneously with the transformation of ssw and mAw into ww, uptake of atmospheric CO_2 increases the nC_T of ssw by 5 ± 3.7 μmol kg^{-1} (from 2096 ± 3.3 to 2101 ± 1.6) (Figure 5). Further CO_2 uptake increases nC_T of ww by 12 ± 3.6 μmol kg^{-1} (from 2101 ± 1.6 to 2113 ± 3) (Figure 5), as this water is transformed into BSW. The latter carbon enrichment is henceforth denoted as $\Delta C_{T\text{-win}}$. We believe this enrichment takes place mainly in the polynya (see below). However, other parts of the fjord may also contribute through a one-time uptake, before the ice-cover becomes an effective barrier for air–sea gas exchange. In total, the ssw and mAw take up 17 ± 4 μmol kg^{-1} [= 5 (±3.7) + 12 (±3.6)] during their transformation into BSW. This carbon enrichment is henceforth denoted as $\Delta C_{T\text{-tot}}$.

There is a theoretical foundation for a total uptake of 17 ± 4 μmol kg^{-1}. For each sample, the increase in C_T (ΔC_T) that is expected from a temperature decrease (ΔT) can be determined according to *Watson et al.* [1995]:

$$\Delta C_T = \frac{0.0423}{RF} \times C_T \times \Delta T \quad (7)$$

where RF is the Revelle Factor. We determined this factor according to the procedure given in *Takahashi et al.* [1993] and obtained a value of 16.5. ΔT was determined by assuming that all waters were cooled from their initial temperatures to −1.9°C (the mean temperature observed in BSW). Then, using Eq 7, we obtained a mean ΔC_T of 15 ± 2 μmol kg^{-1} for ssw and mAw, which is in agreement with the observed value.

The cooling-induced uptake is not, however, instantaneous. The nC_T in ww is around 5 μmol kg^{-1} higher than that in ssw and mAw (Figure 5), only one-third of the 15 μmol kg^{-1} that would be obtained by using Eq 7. We believe the rest of the uptake happens during a prolonged residence time in the polynya. This would be especially true if sea-ice formation enhances air–sea exchange, as was suggested in a recent study by *Anderson et al.* [2004]. Hence, nonpolynya parts of the fjord may also contribute to $\Delta C_{T\text{-win}}$ under the process of formation of the ice cover. In any case, our nC_T data (Figure 5) and the above calculation show that the excess carbon observed in BSW ($\Delta C_{T\text{-tot}}$) is the same amount that would be expected from the observed temperature decrease.

We combined $\Delta C_{T\text{-tot}}$ with typical annual mean volume of BSW (V_{BSW}) to compute the corresponding total carbon uptake from the atmosphere (CU_{tot}), in grams of C per year, according to:

$$CU_{tot} = \Delta C_{T\text{-tot}} \times V_{BSW} \quad (8)$$

We determined CU_{tot} to be 0.21 (±0.06) × 10^{12} g yr^{-1}, assuming V_{BSW} = 9.9 (±1.7) × 10^{11} m^3, as modeled for the winters 1997/1998–2000/2001 by *Skogseth and Haugan* [2003].

4.3. Extrapolation of Storfjorden Wintertime Carbon Fluxes to the Entire Arctic

Sea-ice formation and production of brine-enriched water are common processes on the Arctic shelves [e.g. *Martin and Cavalieri,* 1989] and in the Arctic Ocean in general. Therefore, in this subsection, we estimate the wintertime CO_2 uptake that takes place simultaneously with ice and brine production in Arctic coastal polynyas. We also estimate the wintertime air–sea CO_2 flux that accompanies ice formation in central Arctic polynyas and in seasonally ice-covered regions. For the latter calculations, we use both present and future projections of the extent of seasonal ice and central Arctic polynyas. Note that wintertime air–sea CO_2 flux in Arctic coastal polynyas is referred to as "wintertime carbon uptake" because the word "uptake" reflects the fact that, in Arctic coastal polynyas, atmospheric carbon is most likely transported into the deep ocean through air–sea flux and subsequent formation of deep water. On the other hand, wintertime carbon flux associated with ice formation in the rest of the Arctic is referred to as "wintertime air–sea CO_2 flux", emphasizing that this is a seaward CO_2 flux across the air–sea interface and that knowledge about the likely fate of the atmospheric carbon involved is limited (i.e. it may return to the atmosphere in summer).

Table 4 shows values and sources of the different parameters used in the following calculations.

We combined $\Delta C_{T\text{-win}}$ and V_{BSW} to compute the winter carbon uptake (CU_{win}) for Storfjorden by using Eq 9.

$$CU_{win} = \Delta C_{T\text{-win}} \times V_{BSW} \quad (9)$$

During the winters of 1997/1998–2000/2001 the polynya occupied 10% of Storfjorden area but accounted for 58% of the total ice production [*Skogseth and Haugan*, 2003]. We assume that a similar partitioning also prevailed during the winter 2001/2002 and that a certain fraction (f) of V_{BSW} was produced in the polynya. Then, the winter uptake attributable to the polynya (CU^{p}_{win}) is given by:

$$CU^{p}_{win} = f \times \Delta C_{T\text{-win}} \times V_{BSW} \quad (10)$$

whereas the part of CU_{win} attributable to the nonpolynya regions of the fjord (CU^{np}_{win}) is given by:

$$CU^{np}_{win} = (1 - f) \times \Delta C_{T\text{-win}} \times V_{BSW} \quad (11)$$

The average winter CO_2 fluxes for polynya and nonpolynya regions of Storfjorden (F^P and F^{nP}) [C, g m^{-2}] were then

Table 4. Values and Sources of Different Parameters Used for Calculation of the Entries in Table 5.

Parameter	Value	Reference
Excess C_T ($\Delta C_{T\text{-win}}$) in BSW, due to uptake in winter	0.15 ± 0.044 g m^{-3}	This study
Area of Storfjorden	13×10^9 m^2	*Skogseth and Haugan,* 2003
Area of polynya in Storfjorden	1.3×10^9 m^2	*Skogseth and Haugan,* 2003
Volume of BSW (V_{BSW}) formed in Storfjorden in 1998-2001	$(9.8 \pm 1.8) \times 10^{11}$ m^{-3}	*Skogseth and Haugan,* 2003
Fraction (f) of V_{BSW} that form in Storfjorden polynya	0.58 ± 0.29	Assumption
Winter ice extent in the Arctic	14×10^{12} m^2	*Johannessen et al.,* 2002
Area of Arctic coastal polynyas	3.5×10^{10} m^2	*Winsor and Björk,* 2000
Area of central Arctic polynyas	11×10^{10} m^2	*Gow and Tucker,* 1990
Extent of seasonal ice in the Arctic	7×10^{12} m^2	*Johannessen et al.,* 2002
Winter ice extent in future* Arctic	12×10^{12} m^2	*Johannessen et al.,* 2002
Area of central Arctic polynyas in the future [in % of the above]	1, 10, or 20	Assumption
Extent of seasonal ice in future Arctic	10×10^{12} m^2	*Johannessen et al.,* 2002

*Around year 2100.

computed by dividing the outputs of Eqs 10 & 11 by the corresponding areas (Table 4). The first two entries of Table 5 show the results of these calculations when the value of f chosen was 0.58 ± 0.29. No information on f was available to us, but we assumed a one-to-one correspondence in the amount of ice and brine formed per unit area. Accordingly, 58% of V_{BSW} would be formed in the polynya, since 58% of the total ice production occurred there. Further, to acknowledge that this partitioning is not well constrained, we allowed the value of f to vary as much as 50%. As is evident from Table 5, the average winter CO_2 flux computed for Storfjorden polynya is about 13 times higher than the rest of the fjord. However, polynya and nonpolynya regions of Storfjorden contributed nearly equally to the total winter uptake (U_{win}) (Table 5, third entry), the area of the nonpolynya being 10 times larger than the polynyas.

Wintertime carbon uptake for Arctic coastal polynyas was determined from the product of F^p (Table 5) and the area of the polynyas (Table 4). The latter was inferred, from Tables 2 and 3 of *Winsor and Björk* [2000]. The result is shown in the fourth entry of Table 5.

We postulate that there are, analogous to Storfjorden, wintertime air–sea CO_2 fluxes occurring in central Arctic polynyas and in seasonally ice-covered regions. That is, air–sea CO_2 flux occurs in the central Arctic polynyas throughout winter, while a one-time air–sea CO_2 flux occurs in the seasonally ice-covered region, just before the ice cover becomes an effective barrier to air–sea CO_2 exchange. However, unlike the case for Arctic coastal polynyas, the carbon taken up through this process may not be transferred into the deep ocean. The reason is that deepwater formation in the Arctic takes place primarily on the shelves [e.g. *Jones,* 2001]. Nevertheless, due to the vastness of

Table 5. Computed Wintertime Air–Sea CO_2 Fluxes for Polynya and Nonpolynya Regions of Storfjorden and for Present and Future Arctic.

Parameter	Value
For Storfjorden	
Areal winter flux for the polynya region (F^P) [in g m^{-2}]	65 ± 40
Areal winter flux for the non-polynya region (F^{np}) [in g m^{-2}]	5.2 ± 3.2
Total winter uptake (CU_{win}) [in 10^{12} g yr^{-1}]	0.15 ± 0.05
For contemporary Arctic [*in* 10^{12} g yr^{-1}]	
Winter uptake for Arctic coastal polynyas	2.3 ± 1.4
Wintertime air–sea CO_2 flux for central Arctic polynyas	6.8 ± 4.2
Wintertime air–sea CO_2 flux for seasonally ice-covered Arctic	36 ± 22
Total winter flux for the Arctic, excluding coastal polynyas	43 ± 22
For future Arctic, excluding coastal polynyas* [*in* 10^{12} g yr^{-1}]	
If polynyas occupy 1% of future Arctic winter ice extent	59 ± 32
If polynyas occupy 10% of future Arctic winter ice extent	124 ± 55
If polynyas occupy 20% of future Arctic winter ice extent	196 ± 98

*Around year 2100.

the areas involved, it is important to consider the wintertime air–sea CO_2 flux associated with ice formation in central Arctic polynyas and in seasonally ice-covered regions. For this purpose, we assumed F^P and F^{nP} (Table 5) represented polynyas and seasonally ice-covered regions, respectively. Then we multiplied F^P and F^{nP} by the corresponding areas (Table 4). The obtained values for the individual fluxes and their sum are shown in entries 5–7 of Table 5. Due to the set of assumptions made during the calculations, the computed fluxes are probably subject to errors that are even larger than the uncertainties listed on Table 5. For instance, the surface waters of the Arctic are not CO_2-undersaturated everywhere during winter [e.g. *Kelley*, 1968]. Thus, in some regions the CO_2 flux may be directed to the atmosphere rather than into the ocean. The above-determined air–sea CO_2 fluxes should, therefore, be taken as first-order estimates. Nevertheless, we expect the fluxes to be in the right order of magnitude. In that case, the computed total wintertime air–sea CO_2 flux for the Arctic is comparable to that for the northern North Atlantic [*Olsen et al.*, 2003]. Hence, we suggest that there is a regionally significant wintertime air–sea CO_2 flux associated with ice formation in the Arctic.

Model results show that the extent of seasonal ice in the Arctic will increase to about 10×10^{12} m^2 [*Johannessen et al.*, 2002; their Figure 7] around year 2100. Further, observations have shown that a thinning of the ice [*Rothrock et al.*, 1999, 2003; *Johannessen et al.*, 2002] will accompany its decrease. Because of this thinning, the extent of leads and polynyas will probably increase in the future. At present, the area occupied by leads and polynyas increases from 1% of the winter ice cover to 10–20% during summer [*Gow and Tucker*, 1990]. Quantitative information on how this may change in the future was not available to us. Therefore, we calculated the total (for central Arctic polynyas + seasonally ice covered regions) wintertime air–sea CO_2 fluxes for three different future scenarios, depending on the extent of polynyas by the end of this century; For all three cases, we used a seasonal ice extent of 10×10^{12} m^2 and assumed the polynyas would occupy 1%, 10%, and 20% of the future winter ice-cover (Table 4). The results of these computations are shown in the last three entries of Table 5. As can be inferred from Tables 4 and 5, wintertime CO_2 (C) flux into the Arctic can increase by 80 ($\pm$ 60) $\times 10^{12}$ g yr^{-1}, if seasonal ice extent increases by 43% and central Arctic polynyas become 10 times larger than present.

5. CONCLUSIONS

Cold and high-salinity BSW forms in Storfjorden during winter. This is due to extensive sea-ice production, which primarily takes place in a coastal polynya. We have investigated air–sea CO_2 flux associated with this process by analyzing inorganic carbon and auxiliary hydrographic data collected in August 2001 and April 2002. Uptake of atmospheric carbon occurs simultaneously with the transformation of the source waters into BSW. This uptake accounted for a carbon enrichment of 17 $\pm$ 4 μmol kg^{-1}. The observed enrichment could be computed theoretically by using the observed temperature decrease. However, further analysis revealed that only about one-third of the carbon enrichment occurred in parallel with the heat loss, indicating that the cooling-induced CO_2 uptake was not instantaneous. The remaining two-thirds of the uptake has been suggested to occur later in the polynya. The areal wintertime CO_2 flux of 65 $\pm$ 40 g m^{-2} we computed for the polynya was about 13 times lower ($\approx$ 5 g m^{-2}) for the nonpolynya parts of the fjord.

Extrapolating the fluxes computed for Storfjorden to the entire Arctic gave a total CO_2 uptake of Arctic coastal polynyas of 2.3 ($\pm$1.4) $\times 10^{12}$ g yr^{-1}. This carbon is most likely sequestered from the surface–ocean–atmosphere system through brine-driven deep-water formation in Arctic coastal polynyas. For central Arctic polynyas and for seasonally ice-covered regions, we estimated a wintertime air–sea CO_2 flux of 43 ($\pm$22) $\times 10^{12}$ g yr^{-1} C. This is a regionally significant flux, about half of the wintertime flux into the northern North Atlantic [*Olsen et al.*, 2003]. Further, by the end of this century, this air–sea CO_2 flux can increase to 124 ($\pm$55) $\times 10^{12}$ g yr^{-1}, if seasonal ice extent increases by 43% [*Johannessen et al.*, 2002; their Figure 7] and central Arctic polynyas become 10 times larger than present. With respect to the sequestration of atmospheric carbon, the importance of the computed wintertime air–sea CO_2 flux depends on the penetration of brine–driven convection in the involved regions.

Acknowledgments. This work has been partly supported by the Norwegian Research Council under project no. 122470/720, and partly by the European Commission through the project "Tracer and Circulation in the Nordic Seas Region" (TRACTOR) under contract no. EVK2-2000-00080. This is publication no. A50 of the Bjerknes Centre for Climate Research.

REFERENCES

Anderson, L.G, E. Falck, E.P. Jones, S. Jutterström, and J.H. Swift, Enhanced uptake of atmospheric CO_2 during freezing of seawater: A field study in Storfjorden, Svalbard. *J. Geophys. Res.*, 109, C06004, doi:10.1029/2003JC002120, 2004.

Anderson, L.G. and S. Kaltin, Carbon fluxes in the Arctic Ocean–potential impact by climate change. *Polar Res.*, 20, 225-232, 2001.

Anderson, L.A. and J.L., Sarmiento. Redfield ratios of remineralization determined by nutrient data analysis. *Global Biogeochem. Cy.*, 8, 65-80, 1994.

Anderson, L.G. and E.P. Jones, The transport of CO2 into Arctic and Antarctic Seas: similarities and differences in the driving processes. *J. Marine Syst.*, 2, 81-95, 1991.

Anderson, L.G., P.E. Jones, R. Lindegren, B. Rudels, Nutrient regeneration in cold, high salinity bottom water of Arctic shelves *Cont. Shelf Res.*, 8, 1345-1355, 1988.

Bentley, J.P. *Principles of measurement systems* (2nd ed.). Longman scientific & technical, UK. 503 pp. 1988.

Brewer, P.G., G.T.F. Wong, M.P. Bacon, and D.W. Spencer, An ocean alkalinity problem? *Earth Planet. Sc. Lett.,* 26, 81-87, 1975.

Cavalieri, D.J. and S. Martin, The contribution of Alaskan, Siberian, and Canadian coastal polynyas to the cold halocline layer of the Arctic Ocean. *J. Geophys. Res.*, 99 (C9), 18343-18362, 1994.

Friis, K., A. Kortzinger, and D.W.R, Wallace, Salinity normalization of marine inorganic carbon chemistry data. *Geophys. Res. Lett.*, 30, NO. 2, 1085, doi:10.1029/2002GL015898, 2003.

Gow, A.J. and W.B. Tucker, Sea ice in the polar regions, *in Polar Oceanography, edited by* W.O. Smith, pp 47-122, Academic Press, Inc, London, 1990.

Haarpaintner, J., J. Gascard, and P.M. Haugan, Ice production and brine formation in Storfjorden, Svalbard. *J. Geophys. Res.*, 106 (C7), 14001-14013, 2001.

Haraldson, C., L.G. Anderson, M. Hasselov, S. Hulth, and K. Olsson, Rapid, high-precision potentiometric titration of alkalinity in the ocean and sediment pore waters. *Deep-Sea Res., I,* 44, 2031-2044, 1997.

Johannessen, O.M., L. Bengtsson, M.W. Miles, S.I. Kuzmina, V.A. Semenov, G.V. Alekseev, A.P. Nagurnyi, V.F. Zakharov, L. Bobylev, L.H. Pettersson, K. Hasselmann and H.P. Cattle, Arctic climate change—observed and modeled temperature and sea-ice variability. *Tellus,* 56A, 328-341, 2004.

Johnson, K.M., A.E. King, and J.M. Sieburth, Coulometric TCO_2 analyses for marine studies; an introduction, *Mar. Chem.*, 16, 61-82, 1985.

Johnson, K.M., J.M. Sieburth, P.J. Williams, and L. Brandstrom, Coulometric total carbon dioxide analysis for marine studies: automation and calibration, *Mar. Chem.*, 21, 117-133, 1987.

Jones, E. P., Circulation in the Arctic Ocean. *Polar Res.*, 20, 139-146, 2001.

Kalnay, E., M. Kanamitsu, R. Kistler, W. Collins, D. Deaven, L. Gandin, M. Iredell, S. Saha, G. White, J. Woollen, Y. Zhu, M. Chelliah, W. Ebisuzki, W. Higgins, J. Janowiak, K.C. Mo, C. Ropelewski, A. Leetmaa, R. Reynolds, and R. Jenne, 1996. The NCEP/NCAR 40-year Reanalysis Project. Bull. Am. Met. Soc. 77, 438-471.

Kelley, J., Carbon dioxide in the seawater under the Arctic Ice. *Nature* 218, 862-865, 1968.

Körtzinger, A., J.I. Hedges, and P.D. Quay, Redfield ratios revisited: Removing the biasing effect of anthropogenic CO2. *Limnol. Oceanogr.*, 46, 964-970, 2001.

Martin, S. and D. Cavalieri, Contribution of the Siberian polynyas to the Arctic Ocean intermediate and deep water. *J. Geophys. Res.*, 94, 12725-12738, 1989.

Najjar, G. and R. Keeling, Analysis of the mean annual cycle of dissolved oxygen anomaly in the world ocean. *J. Mar. Res.*, 55, 117-151, 1997.

Olsen, A., R.G.J. Bellerby, T. Johannessen, A.M. Omar, and I. Skjelvan, Interannual variability in the wintertime air-sea flux of carbon dioxide in the northern North Atlantic. *Deep Sea Res. I,* 50, 1323–1338, 2003.

Redfield, A.C., B.H. Ketchum, and F.A. Richards, The influence of organisms on the composition of seawater, *in The Sea,* edited by M. N. Hill, pp 26-77, John Wiley, New York, 1963.

Rothrock, D.A., Y. Yu, and G.A. Maykut, Thinning of the Arctic Sea-Ice Cover. *Geophys. Res. Lett.,* 26, 3469- 3472, 1999.

Skogseth, R. and P. Haugan, Ice and brine production in Storfjorden from four winters of satellite and in situ observation, *in Dense water production processes in Storfjorden (PhD thesis) by* Skogseth R., University of Bergen, Norway, 2003.

Takahashi, T., J. Olafsson, J.G. Goddard, D.W. Chipman, and S.C. Sutherland, Seasonal variation of CO_2 and nutrients in the high-latitude surface oceans: a comparative study. *Glob. Biogeochem. Cy.*, *7*, 843-878, 1993.

Watson, A., P.D. Nightingale, and D.J. Cooper, Modelling atmosphere-ocean CO_2 transfer. *Phil. Trans. R. Soc. lond.* B, 125-132, 1995.

Winsor, P. and G. Björk, Polynya activity in the Arctic Ocean from 1958 to 1997. *J. Geophys. Res.*, 105 (C4), 8789-8803, 2000.

L.G. Anderson, Department of Analytical and Marine Chemistry, Göteborg University, 41296 Göteborg, Sweden.

R.G.J. Bellerby, Bjerknes Centre for Climate Research, University of Bergen, Allegaten 55, N-5007 Bergen, Norway.

T. Johannessen, A. Olsen, and A. Omar, Bjerknes Centre for Climate Research, University of Bergen, Allegaten 55, N-5007 Bergen, Norway, and Geophysical Institute, University of Bergen, Allégaten 70, 5007 Bergen, Norway. (abdirahman.omar@bjerknes.uib.no)

Caroline Kivimäe, Geophysical Institute, University of Bergen, Allégaten 70, 5007 Bergen, Norway.

Response of the Surface Ocean CO_2 System in the Nordic Seas and Northern North Atlantic to Climate Change

Richard G.J. Bellerby[1,2], Are Olsen[1,2], Tore Furevik[1,2], and Leif G. Anderson[3]

Using mixed-layer salinity and temperature output from the Bergen Climate Model and empirical relationships between sea surface temperature and surface water CO_2 fugacity, and between sea surface salinity and total alkalinity, we predict modifications in the winter and springtime CO_2 system condition in the Nordic Seas (Greenland, Norwegian, and Iceland Seas) and the northern North Atlantic, using both contemporary and 2× present-day atmospheric CO_2 simulations. The model shows that over the next 70 years there will be only very small changes in total alkalinity with the influence of the North Atlantic Current responsible for increases of between 6 and 14 μmol.kg^{-1} in the northern North Atlantic and Norwegian Sea. The East Greenland Current and the Iceland Sea show small reductions in mixed-layer salinity and statistically insignificant reductions in total alkalinity. The total dissolved inorganic carbon (C_T) content in the mixed layer increases between the two simulations. However, the increase is not uniform, with a strong south to north decrease in the content of C_T (88 to 76 μmol.kg^{-1}), showing that the surface waters of the Nordic Seas will continue to be a partial return source of anthropogenic carbon, previously taken up to the south, to the atmosphere in agreement with recent findings. Mixed-layer pH_T is shown to decrease by 0.335 pH units in the western Greenland Sea and East Greenland Current. A minimum decrease of 0.285 pH units is seen in the North Atlantic. The greatest uncertainty within the model is the extent of CO_2 disequilibria between the atmosphere and the ocean in the future scenario. The Bergen Climate Model shows large changes in mixed-layer depth, and the implications of this on changes in carbon air–sea equilibria and future changes in seawater pH_T are discussed.

1. INTRODUCTION

A consequence of the anthropogenic release of carbon dioxide into the atmosphere is believed to be significant perturbations of climate, with globally averaged surface temperatures projected to increase by 1.4 to 5.8°C over the period 1990 to 2100 [*IPCC*, 2001]. Due to the strong impact of the North Atlantic/Arctic Oscillation (NAO) [*Shindell et al.*, 1999] on the northern high latitudes [*Hurrel*, 1995] and the difficulties encountered when trying to simulate the present and future states of this climate mode, large differences are found when different climate projections for the Arctic and sub-Arctic regions are compared [*Räisänen*, 2001]. Nevertheless, most models agree that the warming

[1]Bjerknes Centre for Climate Research, University of Bergen, Bergen, Norway.
[2]Geophysical Institute, University of Bergen, Allégaten 70, Bergen, Norway.
[3]Department of Chemistry, Göteborg University, Göteborg, Sweden.

The Nordic Seas: An Integrated Perspective
Geophysical Monograph Series 158

10.1029/158GM13

will be most extensive over the northern high latitudes, which together with ice melting and increased precipitation will reduce the upper ocean density, probably the main reason for the reduction in the ocean thermohaline circulation found in many climate models [*IPCC*, 2001]. Thus we may expect that the Nordic Seas, in particular, will undergo significant increases in air and ocean temperatures, and sea-ice melting over the Arctic Ocean and increases in precipitation over the northern latitudes will tend to reduce the salinity in the area. Conversely, as the North Atlantic and Nordic Seas are strongly influenced by the state of the NAO, any tendency towards a more positive state of the NAO may increase the influence of the North Atlantic on the eastern Nordic Seas through a strengthening of the North Atlantic Current [*Blindheim et al.*, 2000; *Mork and Blindheim*, 2000; *Orvik et al.*, 2001; *Bersch*, 2002].

A further result of anthropogenic CO_2 release is an increase in the inorganic carbon concentration in the ocean due to oceanic CO_2 uptake. There is a build-up of CO_2 in the surface ocean due to the slow connection with intermediate and deep waters. The CO_2 system in the surface oceans is controlled on decadal scales by the surface ocean thermohaline properties and the atmospheric concentration of CO_2. The salinity is, generally, strongly correlated with the total alkalinity, which in concert with the thermal control on CO_2 solubility determines the characteristics of the CO_2 system in the surface ocean. Through changes in the carbonic acid system, the ocean's buffer capacity, or ability to absorb CO_2, is reduced, causing a further positive feedback on atmospheric anthropogenic CO_2 increase. Significantly, pH of the surface ocean will decrease considerably, which may have consequences for ecosystem performance and may incur regime shifts. Understanding the response of the ocean to increases in carbon loading from the atmosphere is important in determining its future uptake capacity. The result of this is more accurate feedback information to incorporate into the emission-based CO_2 scenarios used in climate predictions [i.e. *IPCC*, 2001]. Existing climate models incorporating ocean CO_2 generally lack regional detail, which is important in determining local effects. Latitudinally averaged modelling [e.g. *Joos et al.*, 1999; *Plattner et al.*, 2001] is not sufficient to determine the basin-scale ocean response necessary to focus on the importance of dominant local processes. Regional-scale scenarios of climate change have been suggested to be an important first step in assessing climate impact, reducing the downscaling generalities associated with many global climate models [*Hewitson*, 2003].

To focus locally on the Nordic Seas, an excellent definition and description of this region is found in *Hansen and Østerhuis* [2000], a first-order sensitivity study combining regional modelling output of hydrographic change with empirical CO_2 parameter proxies to estimate the surface ocean response to climate change during the 21st century. The study also includes the northern North Atlantic and the waters of the Fram Strait due to their strong influence on the Nordic Seas. The study concentrates on the late winter period, as empirical understanding of CO_2 proxies is limited to this period. Because this season also represents the starting point for biological production, the study thus sets the carbon chemical baseline from which the spring bloom will commence.

2. METHODS

2.1 Output From the Bergen Climate Model

The Bergen Climate Model (BCM) consists of ARPEGE/IFS [*Déqué et al.*, 1994] coupled to MICOM [*Bleck et al.*, 1992], using the coupling tool OASIS [*Terray et al.*, 1995]. In the present version, a linear T_L63 grid (horizontal resolution about 2.8° × 2.8° degrees) with 31 vertical levels is used for the atmosphere, and a resolution of 2.4° × 2.4° with 24 vertical layers is used for the ocean. The ocean resolution increases towards the equator and towards the two model poles, which are situated over Antarctica and Siberia. Flux adjustment has been applied for the heat and fresh water fluxes, where seasonal varying flux adjustment terms are calculated from the heat and fresh water that had to be added to or subtracted from the model to keep the surface temperature and salinity close to climatology in a control simulation (constant CO_2). A detailed description and evaluation of the BCM system are given in *Furevik et al.* [2003].

A twin experiment consisting of control and perturbation integrations has been conducted. In the 300-year control integration, atmospheric concentrations of greenhouse gasses and aerosol particles are kept at the present day-values, while in the perturbation integration, the atmospheric CO_2 concentration is increased by 1% each year, i.e. doubling after about 70 years. The experiments are described in *Furevik et al.* [2003] and *Sorteberg and Kvamstø* [2005]. For the present paper, mixed-layer salinity and temperature output from the model (Figure 1) was taken as 20-year averages from the control and perturbation experiments between model years 60 to 80. For the control climate, the atmospheric CO_2 concentration was fixed at the 1997 level, while for the perturbation integration, it was between 82% and 122% above this level. Since the ocean model grid in BCM is irregular, the data were interpolated onto a regular 2° × 1° longitude × latitude grid prior to the analysis.

2.2. Inorganic Carbon Model

The inorganic CO_2 system can be determined by knowing two of the four major CO_2 system parameters: pH_T, inorganic

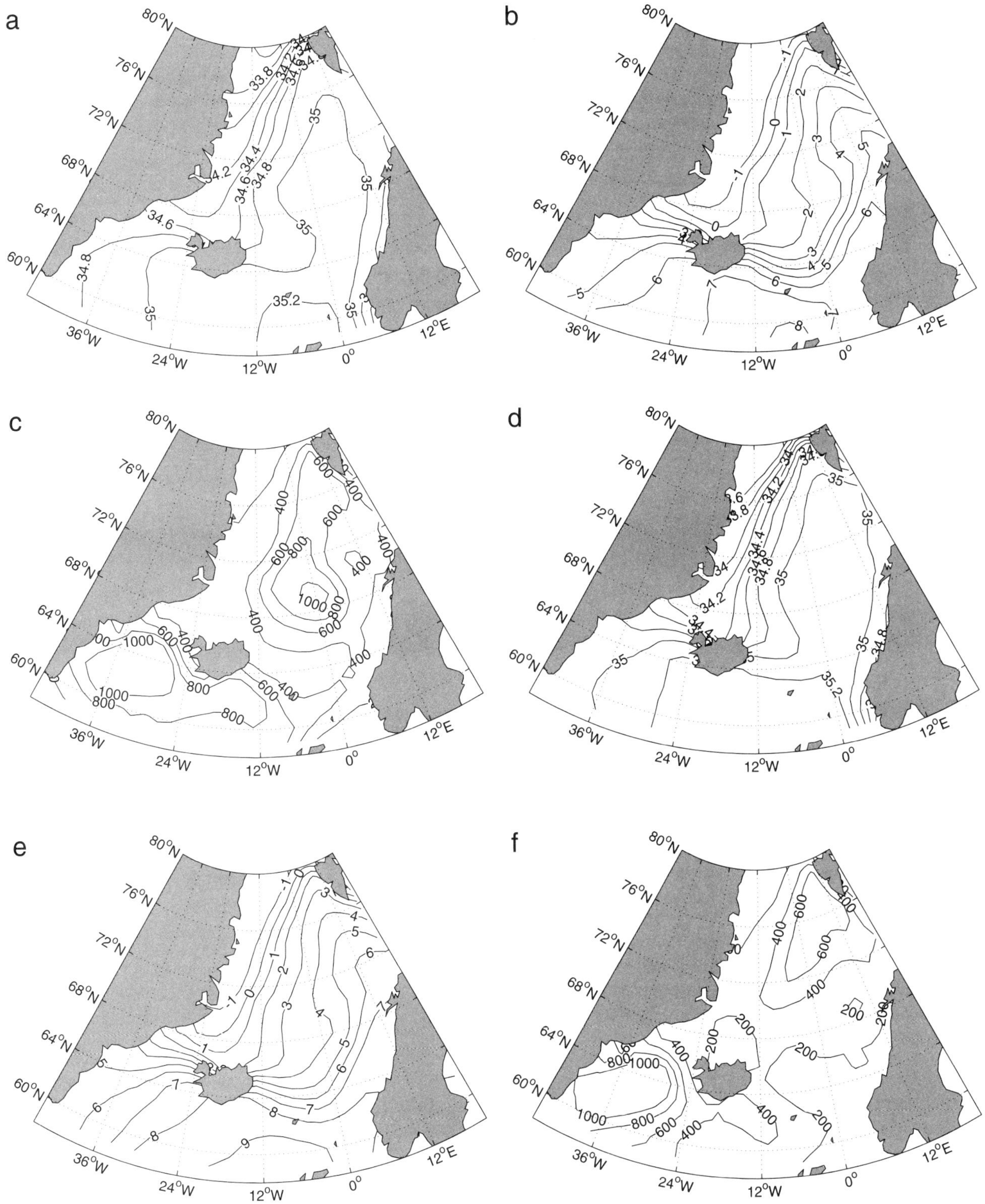

Figure 1. Hydrographical data from the Bergen Climate Model output used in this work. The output is for the control run (a–c) and future climate scenario (d–f). The figures show mixed layer salinity (a, d), mixed-layer temperature (°C) (b, e), and mixed-layer depth (m) (c, f).

carbon (C_T), total alkalinity (TAlk), and surface water CO_2 fugality (fCO_2). The latter pair can be estimated using empirical relationships between thermohaline and CO_2 system variables determined for the Nordic Seas regions.

Following *Olsen et al.* [2003], the surface wintertime fCO_2 in the Nordic Seas, normalised to 1995 and 5°C, is represented by the relationship

$$fCO_{2\text{-}95N}^{sw} = 391.13 - 8.71 \times SST - 0.36 \times SST^2 + 0.011 \times SST^3 \quad (1)$$

where SST is the sea surface temperature, and converted to in situ temperature using the isochemical relationship according to *Takahashi et al.* [1993].

The influence of the Arctic Ocean on the TAlk of the Nordic Seas negates the use of the single, linear North Atlantic–specific alkalinity relationship suggested by *Millero et al.* [1998]. Waters originating from the Arctic have a higher specific alkalinity intercept than waters of North Atlantic origin. Therefore, TAlk was estimated from the combined unpublished relationships from the 2002 Nordic Seas expeditions aboard the *RV*s *Knorr* and *Oden* [Richard Bellerby and Leif Anderson, respectively, both unpublished] (Figure 2) resulting in two salinity regime–dependent relationships:

$$\text{TAlk} = 66.96.\text{S} - 36.803 \quad (2)$$

($r^2 = 0.81$) (S > 34.6);

and

$$\text{TAlk} = 17.24.\text{S} + 1683.3 \quad (3)$$

($r^2 = 0.65$) (S < 34.6)

The TAlk and salinity values used in both equations are from the top 150 m of the ocean; the relationships shown in Figure 2 include measurements from all regions, including ice-covered and coastal stations. The two slopes in Equations 2 and 3 represent the North Atlantic and Arctic waters, respectively. The relationship in Equation 3 has a very similar slope to that defined by *Fransson et al.* [2001] for the Arctic Ocean.

The thermohaline output from the BCM was combined with Equations (1) to (3), and the CO_2 system was defined using an inorganic carbon speciation model [*Lewis and Wallace*, 1998]

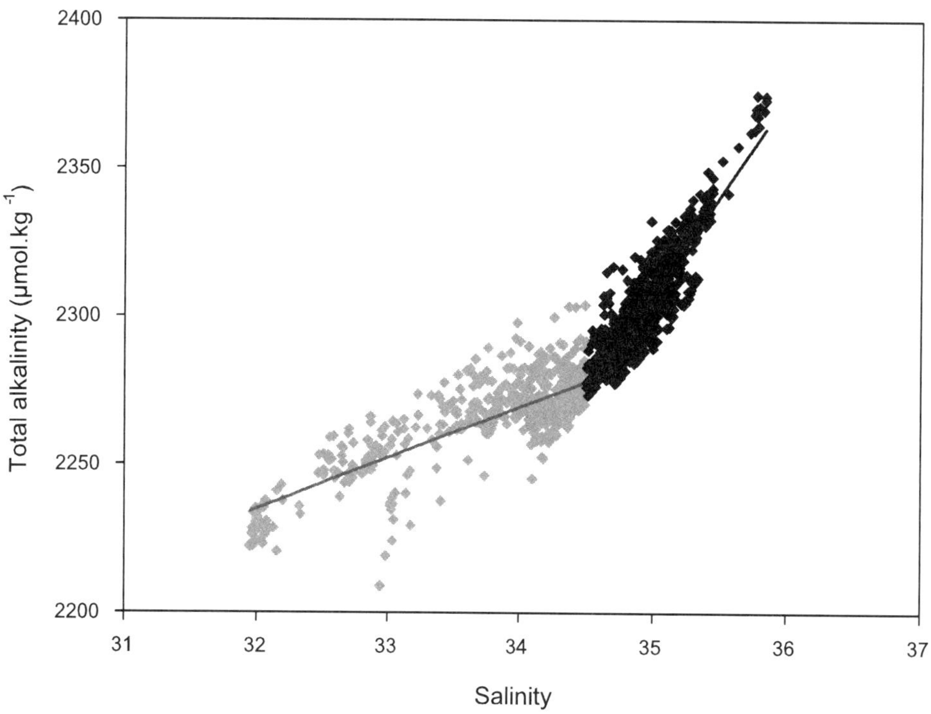

Figure 2. The empirical relationships between total alkalinity (μmol.kg^{-1}) and salinity used in Equations 2 and 3 to model the total alkalinity of the surface waters. The association hinges at a salinity of approximately 34.5 with the observed slopes given by: total alkalinity (gray symbols) = 17.24*salinity + 1683.3 (S < 34.5, $r^2 = 0.65$, $n = 588$) and total alkalinity (black symbols) = 66.98* salinity – 36.803 (S > 34.5, $r^2 = 0.81$, $n = 1871$).

on the total hydrogen ion scale (H_T) using the *Roy et al.* [1993, 1994] dissociation constants.

For the future scenario, the model uses the fit of Equation (1) with the future SST values, and the atmospheric fCO_2 is increased by 371 μatm, which corresponds to the difference between the 1995 (364 μatm) and the average concentration used in the model integration (735 μatm.)

3. RESULTS

The sea surface control fCO_2 field in Figure 3a agrees closely with the measured winter and spring values of *Skjelvan et al.* [1999] for the Greenland and Norwegian Seas and with the spring values of *Hood et al.* [1999] for the western Greenland Sea. The model sea surface control fCO_2 output is about 5 μatm lower than that empirically modelled for springtime by *Olsen et al.* [2003]. Although the use of flux adjustment keeps the SST in the control integration close to the *Reynolds and Smith* [1994] climatology, the same dataset used by *Olsen et al.* [2003], there is an underestimation of approximately 0.1°C in SST in the BCM in the Nordic Seas [*Furevik et al.*, 2003], which would account for much of the difference in sea surface fCO_2 between this model and the *Olsen et al.* [2003] model. The remaining difference may be due to small errors in representing the surface salinity field in the model, which we did not test in this work (the *Olsen et al.* [2003] study does not use salinity). The sea surface fCO_2 in the future integration (Figure 3b) maintains the general structure in Figure 3a with higher values evident in the North Atlantic and eastern Nordic Seas and lowest values to the west of the Greenland Sea and Iceland Sea.

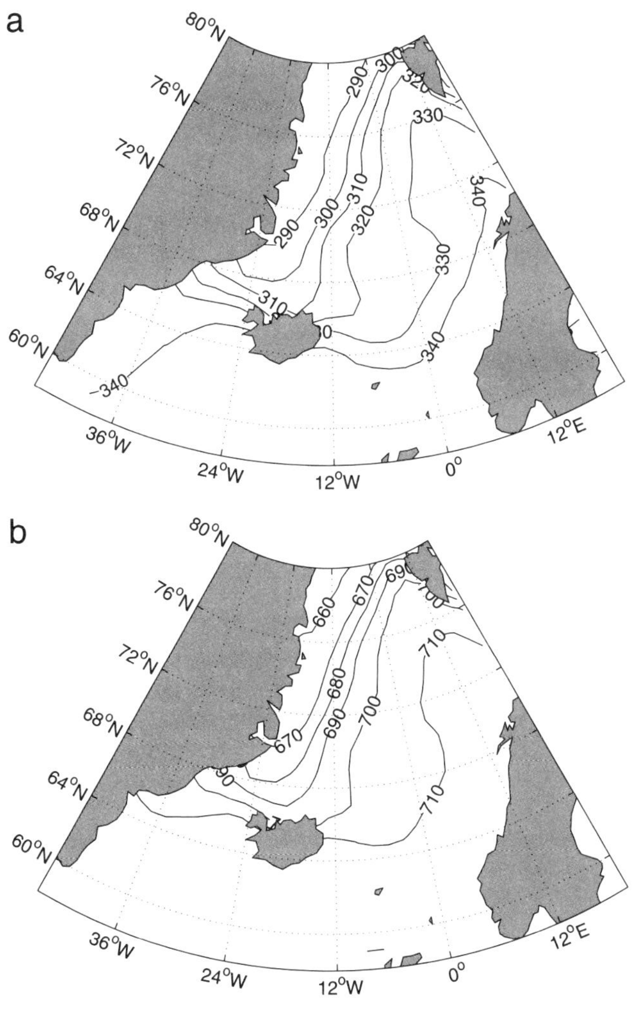

Figure 3. Modelled fCO_2 (μatm) for the control (a) and the future (b) simulations.

The modest salinity increases in the region are reflected in small changes in surface TAlk (Figure 4a). The largest (8-14 $\mu mol.kg^{-1}$) and only statistically significant changes in TAlk are associated with the largest increases in salinity (>0.2) predicted in North Atlantic, south of 64°N. All other TAlk changes are insignificant compared to the variation in the relationships derived from Figure 2. Between the control integration to the future scenario, the Nordic Seas surface waters have undergone a large increase in the C_T content due to the invasion of carbon from the atmosphere (Figure 4b). The largest increase is in the North Atlantic (80-88 μmol kg^{-1}). The Norwegian and Greenland Seas show an almost monotonic C_T increase (78-80 μmol kg^{-1}), and the waters of the East Greenland Current (EGC) and Iceland Sea have the lowest increase, 70 μmol kg^{-1}.

The reduction in pH_T has a more longitudinal structure in the Nordic Seas than was evident in total alkalinity and total carbon dioxide (Figure 4c). The largest decreases are shown to be in the Western Greenland Sea and the EGC with maximum decreases of 0.335 pH units. Minimum decreases of 0.285 pH units are seen in the North Atlantic.

4. DISCUSSION

4.1. CO_2 System Modification

The reduction in salinity witnessed in the EGC is the result of increased precipitation over the Siberian plateau in the BCM. The contemporary influence of the carbonaceous continental runoff [*Fransson et al.*, 2001] is signified by the high intercept in Equation 3. A comparison of Figures 1a and 1b shows that reductions in seawater salinity are limited to the EGC and the Iceland Sea, the increase in the northward influence of the North Atlantic Current acting to increase salinity (and thus potentially TAlk) over the central and eastern Nordic Seas. The increase in salinity and TAlk in the North Atlantic and (potentially) eastern Nordic Seas reflects,

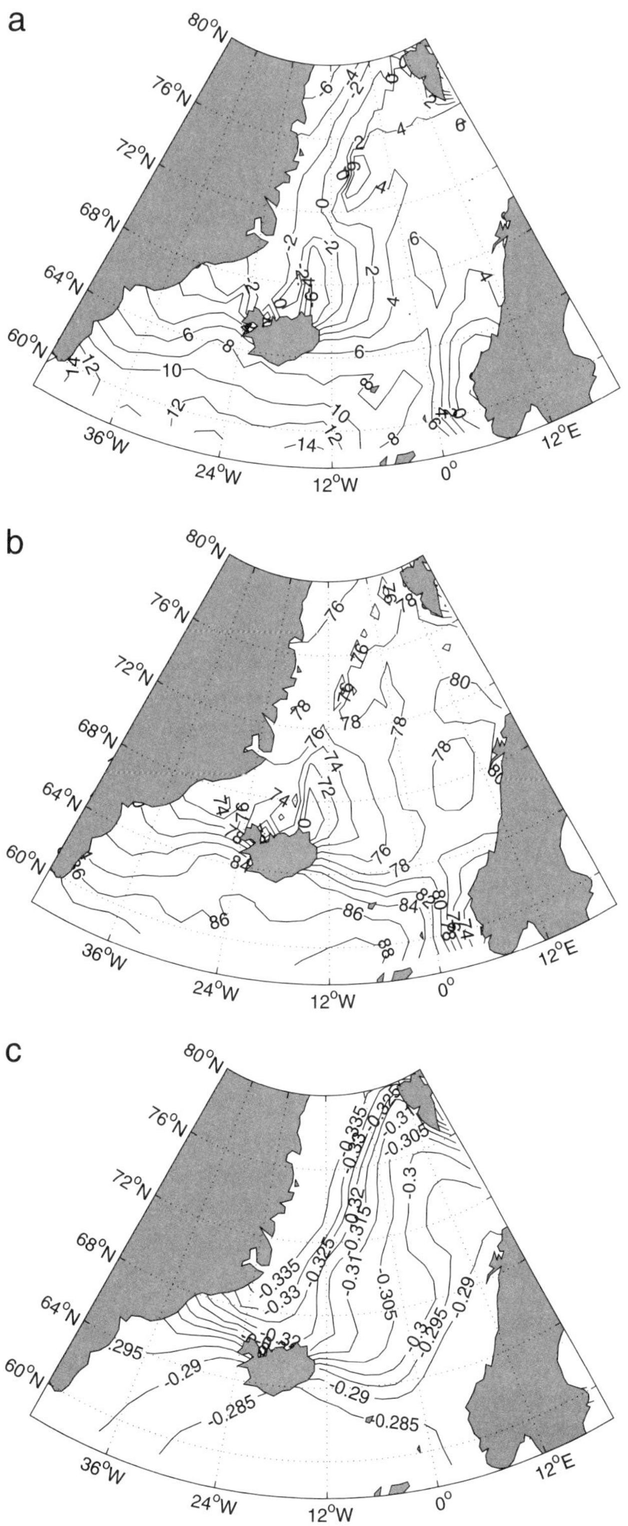

Figure 4. Changes in total alkalinity (μmol.kg^{-1}) (a), total carbon dioxide (μmol.kg^{-1}) (b), and pH_T between the future and control simulations.

respectively, the influence of subtropical waters on the northern North Atlantic and a northward concurrent extension into the North Atlantic Current influenced by an increasingly positive trend in model NAO and associated wind-forcing.

One major unknown in the model is the use of the modified fCO_2–SST relationship (Equation 1) in the future scenario. It is very uncertain how the wintertime surface fCO_2 will respond to the changes in the mixed-layer depth (MLD) predicted by the BCM. Our assumption that the slope of the fCO_2–SST relationship in Equation 1 will hold is based on previous studies of the North Atlantic and Nordic Seas [*Anderson and Olsen*, 2002; *Omar et al.*, 2003]. The difference in $\Delta fCO_2^{\text{air-sea}}$ between preindustrial and present times in the latitudinal band 60° to 72 °N of the north-flowing Atlantic Water, as computed by a conceptual advective model [*Anderson and Olsen*, 2002], was 2.7 ± 1.9 μatm ($n = 13$). The model computation is based on the same wind field and surface MLD, indicating that the $\Delta fCO_2^{\text{air-sea}}$ is determined by the air–sea flux of CO_2 if the climatic conditions are constant. This computation is supported by *Omar et al.* [2003], who have shown that in the Barents Sea the surface fCO_2 has remained in phase with the increasing atmospheric CO_2 concentration between 1967 and 2001.

The BCM shows a substantial regional variability in the MLD. The northern North Atlantic and southern Nordic Seas exhibit a reduction of up to 800 m in winter MLD (Figure 1c and 1f), whilst the MLD in the western North Atlantic and northern Nordic Sea regions remains similar to the control run. Shallowing of the mixed layer could enable much greater CO_2 equilibrium between the atmosphere and the surface ocean than is witnessed in the control. However, the tightness of the fCO_2–SST relationship in *Olsen et al.* [2003] illustrates the dominant control of SST on the surface fCO_2. As the BCM implicitly accounts for thermal exchange modifications associated with changes in the MLD, it may be expected that the future fCO_2 scenario (Figure 3b) will include the effects of MLD variation.

Increases in C_T between the two scenarios reflect the absorption of CO_2 from the atmosphere. Most of the anthropogenic carbon in the North Atlantic and Nordic Seas is thought to be advected from further south [*Anderson and Olsen*, 2002]. However, the increase is not geographically uniform. A general decrease in C_T from south to north shows that the majority of the Nordic Seas will be a significant return source of anthropogenic carbon from the ocean to the atmosphere. This result, hypothesised by *Wallace* [2001] and later shown in a conceptual advective model of *Anderson and Olsen* [2002], is due in main to the decreased buffer capacity of waters at higher latitudes due to their high C_T content and low temperatures.

Assuming that the projected disequilibria between the atmosphere and surface ocean holds, we regard our estimate to be the lower bound of the future C_T content of the surface mixed layer. The likely increase in the C_T concentration will not be enough to compensate for the decrease in MLD, resulting in a lower total uptake of atmospheric CO_2. However, there remains an increase in C_T after the partial atmospheric return discussed above. If the water of the surface mixed layer is the source of ventilating deep waters in the Nordic Seas that overflow to the North Atlantic, the result will be increased sequestration of anthropogenic CO_2 if the ventilation rate does not change. The latter might not be likely, but this aspect points to the many facets of the issue.

4.2. Sensitivity Analysis

In order to estimate the potential mixed layer C_T and pH_T response due to variations in the equilibrium efficiency of the solubility pump, the calculations of the future CO_2 system were re-done using Equation 1 and a future increase of 371 ± 50 μatm (Figure 5). The reanalyses show geographically uniform differences in C_T and pH_T for both scenarios compared to the standard future scenario (Figure 4). However, it is more likely that there will be regionally uneven changes in air–sea equilibrium associated with MLD changes. A prominent example of this would be a prediction that the Norwegian Sea would become much more acidic in the control run due to regionally increased C_T content in the shallow mixed layer compared to the remaining Nordic Seas. Such small changes in ocean pH will regulate the chemical perturbation on plankton and marine ecosystems (e.g. *Riebesell et al.* [2001]), highlighting the requirement for an improved first-order understanding of the feedback between atmospheric carbon increase and ocean carbon chemistry.

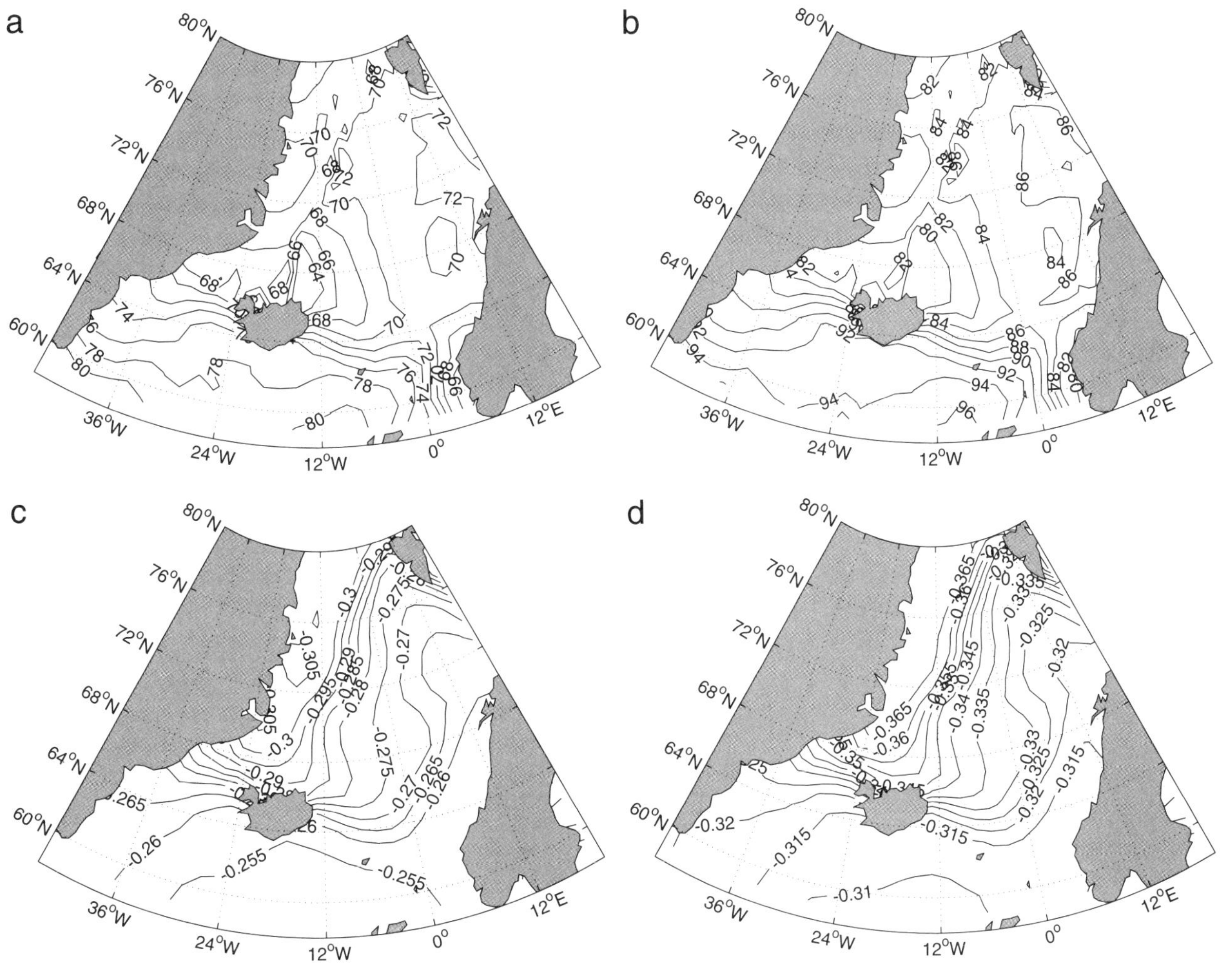

Figure 5. Results from a sensitivity analysis of the model to changes in the equilibration efficiency of the surface ocean with the atmospheric CO_2 concentration. The result from Equation 1 in the future scenario analysis was allowed to deviate by ±50 μatm. The figures show the perturbations in total carbon dioxide (μmol.kg^{-1}) (a, b) and pH_T (c, d) for the minimum and maximum fCO_2 adjustment, respectively, from the standard equilibrium fCO_2 result shown in Figure 3b.

More work is required to understand the control of mixed-layer hydrography on carbon chemistry in the ocean, in particular the response of fCO_2. The chemical impact by pH on biological production needs to be further quantified, and biological feedback models linked to the ocean–atmosphere carbon cycle will further enhance the performance and reliability of global models such as the BCM.

Acknowledgments. The authors would like to thank Mats Bentsen and the BCM group for the work done with the climate simulations. This work has been supported by the Commission of the European Union under contracts EVK2-2000-00080 (Tracers and circulation in the Nordic Seas Region (TRACTOR)) and EVK2-CT-2000-00088 (Carbon Variability Studies by Ships Of Opportunity (CAVASSOO)) and by the Norwegian Research Council contract 151784/432. This is publication Nr. A 90 from the Bjerknes Centre for Climate Research.

REFERENCES

Anderson, L.G. and A. Olsen, 2002. Air-Sea flux of anthropogenic carbon dioxide in the North Atlantic, *Geophys. Res. Lett.*, 29 (17), 10.1029/2002 GL014820.

Bersch M., North Atlantic oscillation-induced changes in the upper layer circulation in the northern North Atlantic Ocean, *J. Geophys. Res.*, *107(C10)*, 3156, doi:10.1029/2001JC111901, 2002.

Bleck, R., C. Rooth, D. Hu, and L.T. Smith, Salinity-driven thermohaline transients in a wind- and thermohaline-forced isopycnic coordinate model of the North Atlantic *J. Phys. Oceanogr.* 22, 1486-1515 (1992).

Blindheim J., V. Borovkov, B. Hansen, S.-Aa. Malmberg, W.R. Turrell, and S. Østerhus, Upper layer cooling and freshening in the Norwegian Sea in relation to atmospheric forcing, *Deep-Sea Res. I*, 47, 655-680, 2000.

Déqué, M., C. Dreveton, A. Braun, and D. Cariolle, The ARPEGE/IFS atmosphere model: A contribution to the French community climate modelling, *Clim. Dyn.*, 10, 249-266 (1994).

Fransson A., M. Chierici, L.G. Anderson, I. Bussman, G. Kattner, E.P. Jones, and J.H. Swift, The importance of shelf proccesses for the modification of chemical constituents in the waters of the Eurasian Arctic Ocean: implication for carbon fluxes, *Cont. Shelf Res.*, 21, 225-242, 2001.

Furevik, T., M. Bentsen, H. Drange, I.K.T. Kindem, N.G. Kvamstø, and A. Sorteberg, (2003): Description and Validation of the Bergen Climate Model: ARPEGE coupled with MICOM, *Clim. Dyn*, 21, 27-51.

Hansen, B., and S. Østerhuis, 2000. North Atlantic-Nordic Seas exchanges. *Progress in Oceanography*, 45(2), 109-208.

Hurrell, J.W., 1995: Decadal trends in the North Atlantic Oscillation regional temperatures and precipitation. *Science*, 269, 676-679.

Hewitson B., 2003. Devoping perturbations for climate change impact assessments. *EOS*, 84(35), 337 and 341.

Hood E.M., L. Merlivat, and T. Johannessen, 1999. Variations of fCO_2 and air-sea flux of CO_2 in the Greenland Sea gyre using high frequency time series data from CARIOCA drift buoys, *J. Geophys. Res.*, 104 (C9), 20,571-20,583.

IPCC, 2001: Climate Change 2001: *The Scientific Basis. Contribution of Working Group I to the Third Assessment Report of the Intergovernmental Panel on Climate Change* [Houghton J.T., Ding Y., Griggs D.J., Noguer M., van der Linden P.J., Dai X., Maskell K., and Johnson C.A. (eds)]. Cambridge University Press, Cambridge, United Kingdom and New York, NY, USA, 881 pp.

Joos F., G.-K. Plattner, T.F. Stocker, O. Marchal, and A. Schmittner, Global warming and marines carbon cycle feedbacks on future atmospheric CO_2, *Science*, 284, 464-467, 1999.

Lewis, E., and D.W.R. Wallace, Program developed for CO2 system calculations, ORNL/CDIAC-105, Carbon Dioxide Information Analysis Centre, Oak Ridge National Laboratory, US Department of Energy, Oak Ridge, TN, 1998.

Millero, F.J., K. Lee, and M. Roche, Distribution of alkalinity in the surface waters of the major oceans, *Mar. Chem.*, 45, 111-130, 1998.

Mork, K.A., and J. Blindheim, Variations in the Atlantic inflow to the Nordic Seas, *Deep-Sea Res. I*, 47, 1035-1057, 2000.

Olsen A., R.G.J. Bellerby, T. Johannessen, A.M. Omar, and I. Skjelvan, 2003. Interannual variability in the wintertime air-sea flux of carbon dioxide in the northern North Atlantic, 1981-2001, *Deep-Sea Res. I*, 50 (10-11), 1323-1338.

Omar A., T. Johannessen, S. Kaltin, and A. Olsen, 2003 The anthropogenic increase of oceanic pCO_2 in the Barents Sea surface water. *J. Geophys. Res.* 108(C12) 3338 doi: 10.1029/2002JC001628.

Orvik, K.A., Ø. Skagseth, and M. Mork, Atlantic inflow to the Nordic Seas: current structure and volume fluxes from moored current meters, VM-ADCP and Seasoar-CTD observations, 1995-1999, *Deep-Sea Res. I*, 48, 937-957, 2001.

Plattner, G.-K., F. Joos, T.F. Stocker, and O. Marchal, Feedback mechanisms and sensitivities of ocean carbon uptake under global warming, *Tellus*, *53B*, 564-592, 2001.

Räisänen, J. 2001. CO_2-induced climate change in CMIP2 experiments: Quantification of agreement and role of internal variability. *J. Climate* 14, 2088-2104.

Riebesell, U., I. Zondervan, B. Rost, P.D. Tortell, R.E. Zeebe, F.M.M. Morel, 2000. Reduced calcification of marine plankton in response to increased atmospheric CO_2, *Nature,* 407, 364-367.

Reynolds, R., and T. Smith, 1994. Improved global sea surface temperature analyses. Journal of Climate 7, 929-948.

Roy R.N, L.N. Roy, K.M. Vogel, C. Porter-Moore, T. Pearson, C.E. Good, F.J. Millero and D.M. Campbell, (1993) The dissociation constants of carbonic acid in seawater at salinities 5 to 45 and temperatures 0 to 45°C. *Marine Chemistry*, 44, 249-267.

Roy, R.N., L.N. Roy, K.M. Vogel, C. Porter-Moore, T. Pearson, C.E. Good, F.J. Millero, and D.M. Campbell, (1994) Erratum for: The dissociation constants of carbonic acid in seawater at salinities 5 to 45 and temperatures 0 to 45°C. *Marine Chemistry*, 45, 337.

Shindell, D.T., R.L. Miller, G.A. Schmidt, and L. Pandolfo, 1999. Simulation of recent northern winter climate trends by greenhouse-gas forcing. *Nature* 399, 452-455.

Skjelvan I., T. Johannessen, and L.A. Miller, Interannual variability of fCO_2 in the Greenland and Norwegian Seas, *Tellus*, *51B*, 477-489, 1999.

Sorteberg, A., and N.G. Kvamstø, 2005. The effect of Internal Variability on Anthropogenic Climate Projections RegClim General Technical Report No. 8.

Takahashi, T., J. Olafsson, J.G. Goddard, D.W. Chipman, S.C. Sutherland, 1993. Seasonal variation of CO_2 and nutrients in the high latitude surface oceans: a comparative study. *Global Biogeochemical Cycles*, 7, 843-878.

Terray, L. *et al*. Climatology and interannual variability simulated by the ARPEGE-OPA model, *Clim. Dyn.,* 11, 487-505 (1995).

Wallace, D.W.R., 2001. Storage and transport of excess CO_2 in the oceans: The JGOFS/WOCE global CO_2 survey, in *Ocean circulation and climate: observing and modelling the global ocean,* Edited by G. Siedler, J. Church and J. Gould.

L.G. Anderson, Department of Chemistry, Göteborg University, S-412 96 Göteborg, Sweden. (leifand@chem.gu.se).

R.G.J. Bellerby, Bjerknes Centre for Climate Research, University of Bergen, Allégaten 55, N-5007, Bergen, Norway. (Richard.Bellerby@bjerknes.uib.no).

T. Furevik and A. Olsen, Geophysical Institute, University of Bergen, Allégaten 70, N-5007, Bergen, Norway. (Tore.Furevik@gfi.uib.no; Are.Olsen@gfi.uib.no).

Ocean General Circulation Modelling of the Nordic Seas

Helge Drange[1-4], Rüdiger Gerdes[5], Yongqi Gao[1,2,4], Michael Karcher[5,6], Frank Kauker[5,6], and Mats Bentsen[1,2]

The complexity of the state-of-the-art Ocean General Circulation Models (OGCMs) has increased and the quality of the model systems has improved considerably over the last decades. The improvement is caused by a variety of factors ranging from improved representation of key physical and dynamical processes, parallel development of at least three classes of OGCM systems, accurate and cost-effective numerical schemes, an unprecedented increase in computational resources, and the availability of synoptic, multidecadal atmospheric forcing fields. The implications of these improvements are that the present generation of OGCMs can, for the first time, complement available ocean observations and be used to guide forthcoming ocean observation strategies. OGCMs are also extensively used as laboratories for assessing cause-relationships for observed changes in the marine climate system, and to assess how the ocean system may change in response to, for instance, anomalous air–sea fluxes of heat, freshwater, and momentum. The Nordic Seas are a particularly challenging region for OGCMs because of characteristic length scales of only a few to about 10 km, a variety of complex and interrelated ocean processes, and extreme air–sea fluxes. This paper gives an overview of the status of the prognostic modelling of the Nordic Seas marine climate system. To exemplify the status, we present output from two widely different state-of-the-art OGCM systems. We also address processes that are still inadequately described in the current generation of OGCMs, thus providing guidelines for the future development of model systems particularly tailored for the Nordic Seas region.

1. INTRODUCTION

Numerical modelling of the climate of the Nordic Seas and the adjacent waters is an important method to assess and predict the influence of the region on the Atlantic Meridional Overturning Circulation (AMOC), the climate of Arctic Ocean, the cycling of freshwater, and the ocean productivity. The field has made considerable progress in recent years due to a larger number of contributing modelling groups, the introduction of greater spatial resolution, and improved representation of important processes. Increasingly realistic atmospheric forcing data have been employed although their reliability in the Nordic Seas has still to be assessed. The combination of realistic forcing and the existence of high-quality in situ observations and continuous time series at key locations of the oceanic circulation have lead to validated models. Still, many problems remain, caused by the complexity of

[1]Nansen Environmental and Remote Sensing Center, Bergen, Norway.
[2]Bjerknes Centre for Climate Research, University of Bergen, Bergen, Norway.
[3]Geophysical Institute, University of Bergen, Bergen, Norway.
[4]Nansen-Zhu International Research Centre, Beijing, China.
[5]Alfred Wegener Institute for Polar and Marine Research, Bremerhaven, Germany.
[6]O.A.Sys, Ocean Atmosphere Systems GbR, Hamburg, Germany.

The Nordic Seas: An Integrated Perspective
Geophysical Monograph Series 158

10.1029/158GM14

the topography; the small, short spatial and temporal scales of the main processes in the region; and the inherent coupling between the Nordic Seas and the entire Atlantic and Arctic Oceans. As in all oceans at high latitudes, baroclinic adjustment processes mediated by baroclinic Rossby waves are very slow. This has consequences for the ocean circulation response to atmospheric variability on seasonal to interannual time scales. The response here is more barotropic and more guided by topography than in midlatitude oceans.

Numerical ocean–sea-ice models address a broad range of scientific questions regarding the development of the circulation and water mass distribution in the Nordic Seas. They aid the interpretation of observations that are by their nature sparse in space and time [e.g., *Hátún et al.*, this issue]. Usually, results of hindcast calculations are used for this purpose where a realistic time history of data on atmospheric forcing is prescribed. This approach can, in principle, be used to assess the characteristics of decadal variability modes, compared with (e.g.) trends in forced greenhouse gases. Available atmospheric forcing fields cover the period since 1948 [e.g., *Kalnay et al.*, 1996], meaning that the hindcast approach is currently somewhat restricted to decadal-scale studies.

Combined with observational data, models can provide a detailed state description of the system and its temporal development. The state description is valuable in itself but can also be utilized to initialise coupled ocean–sea-ice–atmosphere models. This approach is of key importance for conducting reliable seasonal to decadal climate predictions [e.g., *Collins et al.*, 2005]. Another important branch of modelling is numerical experimentation that acts as a substitute for physical experiments that are usually not feasible (nor desirable) with the climate system. Idealized experiments and scenario calculations belong to this category.

No dedicated review exists on the sea-ice–ocean modelling of the Nordic Seas. However, a description of mesoscale sea-ice–ocean models and sea-ice–ocean process studies in general is covered by *Häkkinen* [1991], and *Hopkins* [1991], *Stevens* [1991], and *Heburn et al.* [1995] introduced and reviewed relevant modelling of the Nordic Seas until the mid 1990s. The most relevant model review is by *Mellor and Häkkinen* [1994], addressing the development of coupled ice–ocean models with particular focus on the Arctic Ocean.

To limit and focus the scope of the paper, this review is heavily based on the output from two state-of-the-art OGCMs, both of which are driven by realistic, daily atmospheric forcing fields for the period 1948-2002. The objectives are to explicitly illustrate similarities and differences between the two models, to address the degree of realism in the simulated fields based on comparison with observations, and by that to demonstrate strengths and weaknesses of the present generation of OGCMs applied to the Nordic Seas region. In this respect the paper is more than a review because it presents the first multiple-model comparison for the Nordic Seas.

The two model systems used differ substantially by construction: The Alfred Wegener Institute (AWI) North Atlantic/Arctic Ocean–Sea-Ice Model (NAOSIM) uses levels of fixed depth as the vertical coordinate and, as such, is a representative of the classical family of OGCMs, whereas the Nansen Center (NERSC) model, which is a derivative of the Miami Isopycnic Coordinate Ocean Model (MICOM), uses surfaces of constant density as the vertical coordinate. Also the applied model domain, the spin-up procedure, the horizontal and vertical grid resolution, the details of the atmospheric forcing, the mixed-layer dynamics, and several of the model parameters differ (see Appendix) implying that the results presented should not be interpreted as the output from a planned project for model intercomparison.

The review starts with a brief overview of the available model systems and gives some examples of how the various model systems have been used to study the marine climate system of the Nordic Seas. Next, a comparison between MICOM and NAOSIM is presented., in which the mean state and the variability of the volume fluxes into and out of the Nordic Seas, the horizontal circulation pattern, the seasonal sea surface salinity (SSS) and sea surface temperature (SST) fields, and the thickness of the upper mixed layer and the sea-ice extent are considered. Comparisons are also made between observed values of several of these quantities and the values predicted in each model. Discussion of the particular challenges for the ocean modelling of the Nordic Seas is followed with concluding remarks.

2. PROGNOSTIC OCEAN MODELLING

Numerical ocean model systems can in general be grouped into three main categories; diagnostic, prognostic, and data-assimilation systems. In diagnostic modelling, the state of the ocean is directly derived from observations [e.g., *Engedahl et al.*, 1988]. In prognostic modelling, the topic of this review, the evolution of the model system is governed by time- and space-dependent continuity and momentum equations, an equation of state, and prescribed atmospheric forcing fields. Data-assimilation systems are built in response to the fact that both observations and the prognostic model systems are imperfect: The data-assimilation system therefore feeds available observations into the prognostic model so that the model state remains close to the observed state throughout the integration [e.g., *Bennett*, 1992; *Evensen*, 2003; *Stammer et al.*, 2003a,b]. For climate research, data-assimilation systems are of fundamental importance for seasonal to decadal climate prediction assessments because the initial state is of key importance for the evolution of the climate system [e.g., *Collins et al.*, 2005].

Prognostic model systems are fully governed by the primitive equations [e.g., *Müller and Willebrand*, 1989] of coupled time- and space-dependent momentum and continuity equations and by an equation of state. The primitive equations are regarded as good approximations, provided that vertical motion is much smaller than horizontal motion and that the fluid layer depth is small compared to the radius of the sphere. Once the prognostic model is started from an initial state, it will compute, time-step by time-step and grid-point by grid-point, a complete set of internally consistent dynamic and thermodynamic fields. The air–sea boundary conditions are prescribed fields of wind stress, heat, and freshwater fluxes. These fields represent the atmospheric forcing of the ocean–sea-ice system. The most common atmospheric forcing products are monthly mean climatological fields, and daily variable fields available from (e.g.) the NCAR/NCEP (http://www.cdc.noaa.gov/cdc/reanalysis/) and ECMWF (http://www.ecmwf.int/research/era/) reanalyses projects. By its nature, a prognostic model is the commonly used model system to study the dynamics of past, present, and future atmosphere and ocean climate states.

Prognostic 3-dimensional ocean modelling extends back to the pioneering works of *Bryan* [1969] and *Cox* [1984]. This model system, now known as the Modular Ocean Model (MOM), was extended to the polar regions by *Semtner* [1976a,b]. The model system has been continuously developed through improvements of the code, the physical parameterizations, and the numerics. The latest state of MOM is documented in *Griffies et al.* [2004]. The MOM system and derivations thereof are without comparison the most frequently and widely used OGCM system of today. NAOSIM, for example, is a derivative of the MOM system.

The Bryan–Cox (or MOM) model system is also known as a level or geopotential coordinate OGCM because the vertical discretisation is based on layers at fixed depths. There are two alternative formulations for the vertical discretisation in OGCMs: In one, the vertical coordinate follows topography, and this system hence is known as terrain-following (or sigma) coordinate OGCM. The Princeton Ocean Model (POM) developed by *Blumberg and Mellor* [1987] was the first of these models. A derivative of this system has been applied to the North Atlantic Ocean, the Nordic Seas, and the Arctic Ocean by, in particular, Sirpa Häkkinen [*Häkkinen and Mellor*, 1992; *Häkkinen*, 1995, 1999, 2002; *Mauritzen and Häkkinen*, 1999].

A third approach is to treat surfaces of constant potential density, or isopycnals, as the vertical coordinate. Such a model system is known as an isopycnic coordinate OGCM. The motivation behind the latter choice of vertical discretisation is that the ocean transport and mixing are mainly directed along surfaces of constant density. The Ocean isoPYCnic (OPYC) model, developed by *Oberhuber* [1993], played a pioneering role in the use of isopycnal models for high latitude. For instance, *Aukrust and Oberhuber* [1995] used an Atlantic-Arctic configuration with a grid focussed in the Nordic Seas while *Holland et al.* [1996] and *Karcher and Oberhuber* [2002] applied OPYC to study the mixed layer circulation and the exchanges of different water masses between the Arctic and the subarctic seas, respectively. Today, however, MICOM [*Bleck et al.*, 1992] is the only widely used isopycnic coordinate model system. The NERSC model used in this review is a derivative of MICOM.

The geopotential, the terrain-following, and the isopycnic coordinate OGCMs have all inherent advantages and weaknesses. As a consequence of this, hybrid coordinate OGCMs have been developed. The most well-known hybrid OGCM is the HYbrid Coordinate Ocean Model (HYCOM; *Bleck* [2002]) in which the open stratified ocean is treated with isopycnic coordinates, the shallow coastal regions are treated with terrain-following coordinates, and the upper mixed layer is treated with geopotential coordinates. In the limit of infinite numerical resources and hence infinite grid resolution—the basic features of the various OGCM concepts will converge into one consistent model system [*Semtner*, 1995].

For a thorough review of prognostic ocean modelling in general, see *Griffies et al.* [2000]. Of high relevance are also the results of several model intercomparison projects. One such project for the North Atlantic that includes geopotential, terrain-following, and isopycnic-coordinate OGCMs is documented by *DYNAMO Group* [1997], *Böning et al.* [2001], *New et al.* [2001], and *Willebrand et al.* [2001]. Another is the Arctic Ocean Model Intercomparison Project (AOMIP) results presented in *Proshutinsky et al.* [2001], *Steiner et al.* [2004], and *Uotila et al.* [2005]. More information about AOMIP is available at http://fish.cims.nyu.edu/project_aomip/overview.html.

3. MODEL–MODEL AND OBSERVATION–MODEL COMPARISONS

This section is split into two parts: The first part addresses similarities and differences between the MICOM and NAOSIM systems. Such a comparison is useful as it points towards robust model responses (in the sense that the models respond similarly to the applied forcing, indicating that the model physics is adequately represented) and model uncertainties (indicating where the model systems need to improve). The second part contains a comparison between simulated and observed fields—the only way to assess the degree of realism of the models. Both model–model and observation–model analyses quickly become extensive [e.g., *Karcher et al.*, 2003; *Steiner et al.*, 2004; *Hátún et al.*, this issue], so this review has been restricted to a display and a brief discussion of some of the key large-scale features of the Nordic Seas.

The fields to be addressed are the following: the annual mean circulation at 150-m depth, the annual mean barotropic (or vertically integrated) circulation, the long-term mean and temporal variability of the volume transport, the winter SST and SSS fields, the mean and maximum thickness of the mixed-layer in March and the mean, minimum, and maximum sea-ice concentrations in March. Short descriptions of the two model systems are given in the appendix.

3.1. Simulated Circulation in the Nordic Seas Region

The most fundamental bulk properties of the marine climate system of the Nordic Seas are the volume, heat, and salt fluxes into and out of the region. The poleward flow of Atlantic Water transports significant amounts of heat, salt, carbon, and plant nutrients into the region and further into the Arctic Ocean [e.g., *Hansen and Østerhus*, 2000; *Blindheim and Østerhus*, this issue; *Skjelvan et al.*, this issue], whereas the southward flowing cold and fresh Polar Ocean waters off the coast of Greenland and the dense Greenland–Scotland Ridge (GSR) overflow waters are of key importance for the hydrographic, and probably also the dynamic state, of the North Atlantic climate system [e.g., *Dickson et al.*, 2002; *Curry et al.*, 2003; *Hansen et al.*, 2004].

Figure 1 shows the observed surface circulation from the 1990s based on analyses of drifter trajectories [*Jakobsen et al.*, 2003] and the mean simulated circulation at 150-m

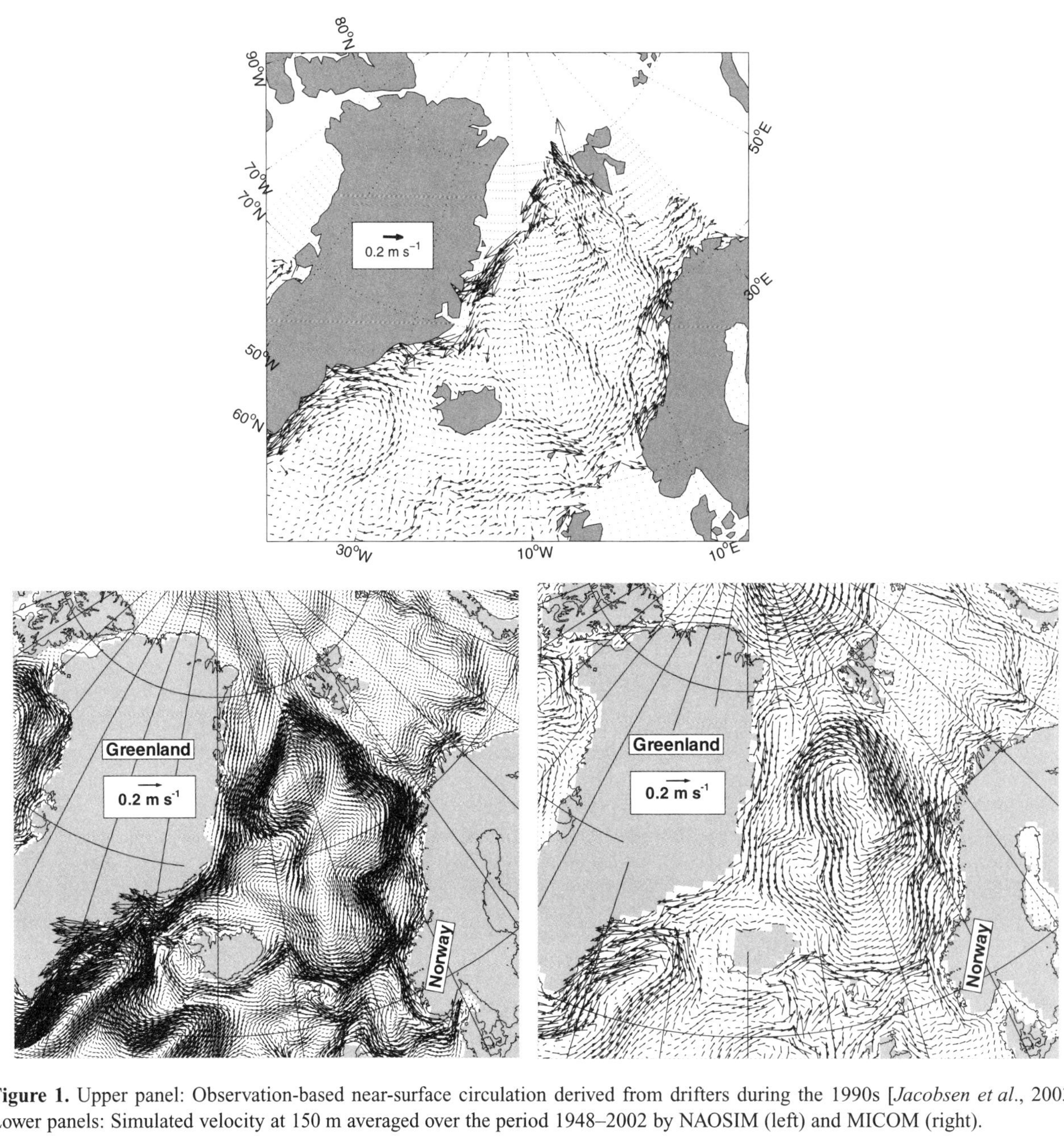

Figure 1. Upper panel: Observation-based near-surface circulation derived from drifters during the 1990s [*Jacobsen et al.*, 2003]. Lower panels: Simulated velocity at 150 m averaged over the period 1948–2002 by NAOSIM (left) and MICOM (right).

depth for the period 1948–2002. The model systems show, in accordance with the drifters, that the inflow of Atlantic Water to the Nordic Seas takes place just east of Iceland, southeast of the Faroes, and along the Scotland slope, with a minor inflow component located on the eastern side of the Denmark Strait (the Irminger Current). Furthermore, the Atlantic Water crossing the Iceland–Scotland Ridge flows into the North Sea and continues close to the coast of Norway or in an outer branch following the outer Norwegian continental slope. At about 70°N, the Atlantic Water splits into two branches, one meandering into the Barents Sea close to northern Norway, another branch heading towards the Fram Strait. The simulated circulation fields show a general agreement with the drifter field for all of these features.

However, the circulation pattern in the Fram Strait region differs between the models, NAOSIM results being closest to the drifter data. NAOSIM produces a two-current system with northward-flowing water on the Spitzbergen side and southward-flowing water on the Greenland side of the strait, whereas almost all the northward-flowing water apparently recirculates in the strait in the MICOM simulation. Further south, the Polar Water follows the coast of Greenland towards the Denmark Strait. One branch of the Polar Water continues through the Denmark Strait, whereas the other branch flows eastward north of Iceland as the East Icelandic Current. The East Icelandic Current is also fed by the northward flowing Irminger Current. Both models simulate the two-directional flow in the Denmark Strait. In the Nordic Seas interior, however, both models fail to simulate the topographically steered, northwestward-directed flow structure seen in the central basin in the drifter data. The reason for this failure is not known but might be linked to using too smooth a topography in the models. The internal circulation is rather similar between the models, but NAOSIM shows sharper gradients and more details. A profound difference, however, is the location of the Greenland Gyre. which is more to the west in NAOSIM. As will be shown in section 3.5, the central Greenland Sea gyre is too intensive and is located slightly too far to the west in NAOSIM, whereas it is too diffusive and is located too far to the east in MICOM. The latter difference also implies that NAOSIM produces a more prominent eastward-directed current north of Jan Mayen.

It is apparent from both Figures 1 and 2 that NAOSIM produces stronger gradients and more details than MICOM. This probably results from the essentially doubled horizontal resolution in NAOSIM (28 km vs. 40 km) but can also be attributed to the quite different characteristics of the wintertime mixing in the region and the somewhat different extent of sea-ice (see section 3.5).

Figure 2 shows the barotropic (or vertically integrated) flow field. Many of the features from the flow at 150 m are present in the barotropic field, indicating both the weak stratification and the importance of the topography (see Figure 1 in *Blindheim and Østerhus* [this issue]) in guiding the circulation, including the surface flow, in the region. The controlling role of topography on the surface circulation has long been known but was first demonstrated in OGCMs of the Nordic Seas by *Legutke* [1991] and later confirmed by *Aukrust and Oberhuber* [1995].

3.2. Simulated and Observed Volume Fluxes in the Nordic Seas Region

More quantitative results are obtained by examining the volume transports into and out of the region, i.e., the northward, southward, and net flow through the Denmark Strait,

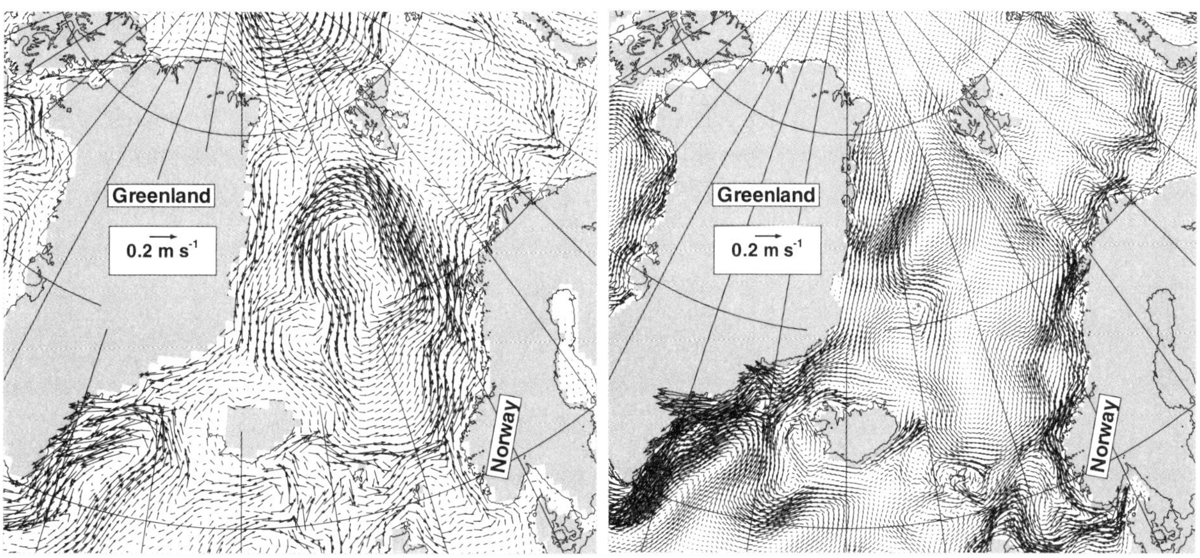

Figure 2. Simulated barotropic velocity field averaged over the period 1948–2002 by NAOSIM (left) and MICOM (right).

across the Iceland–Faroe Ridge, between the Faroe Islands and Scotland, through the Barents Opening and through the Fram Strait. In addition, there is a small component of flow through the English Channel.

Among numerous observation-based estimates of the volume, heat, and freshwater transports through the Nordic Seas, some unfortunately vary by a factor two to three [e.g., *Simonsen and Haugan*, 1996; *Hansen and Østerhus*, 2000; *Blindhein and Østerhus*, this issue]. Because accurate observations from moored current meters have become available since only around 1994 [*Østerhus et al.*, 2005], we have split the present discussion into two parts: the full integration period from 1948 to the present, essentially without reliable observational guidelines, and the period since 1994 with direct observations from current meters.

We are aware, however, that even with accurate current-meter observations, the derivation of volume, heat, and salt fluxes is far from solved. Lateral movements of strong fronts, eddy activity, and spatial gaps in the coverage (because of, e.g., instrument loss or failure) often hamper a reliable estimate of these fluxes. *Schauer et al.* [2004] recently discussed this problem for flux estimates from observations in the Fram Strait. They estimate the error as being on the order of the net volume fluxes through the strait. Because the fluxes are variable in time and because usually there is no synoptic cover of all straits with observations, the possibility of closure of missing information from one strait with observations from the others is eliminated.

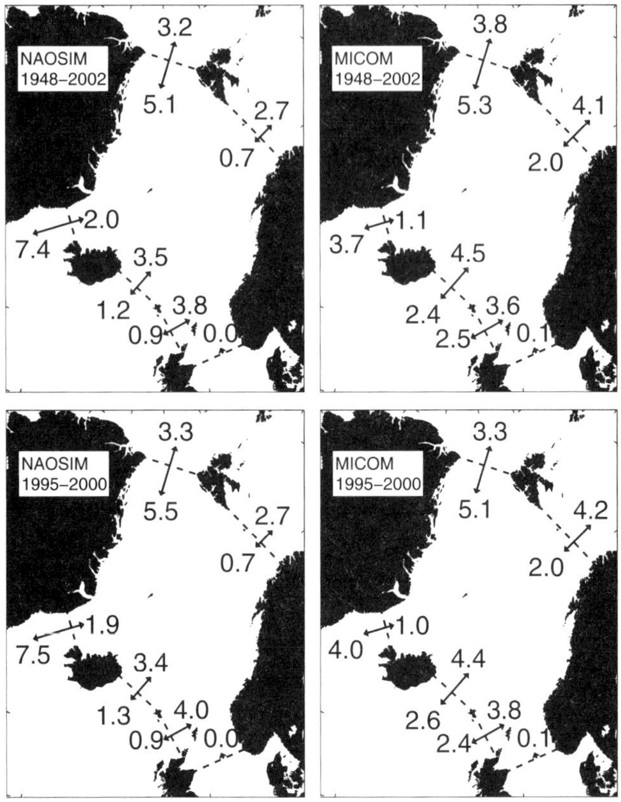

Figure 3. Simulated mean northward and southward volume fluxes (Sv) for 1948–2002 (upper panels) and 1995–1999 (lower panels). Left panels: NAOSIM simulations; right panels, MICOM simulations. Throughout the manuscript, the transports from MICOM are calculated online, i.e., time step by time step, whereas weekly output of velocity, temperature, and salinity are used to diagnose the transports from NAOSIM.

The simulated net volume transports into and out of the Nordic Seas for the period 1948–2002 are shown in Figure 3. It follows that the total amount of water flowing northward across GSR is 9.3 Sv in NAOSIM and 9.2 Sv in MICOM, whereas the corresponding fluxes into the Arctic Ocean are 5.9 Sv and 7.9 Sv, respectively. For the net southward flow, the numbers are 9.5 Sv (NAOSIM) and 8.6 Sv (MICOM) across GSR, and 5.8 Sv (NAOSIM) and 7.3 Sv (MICOM) for the two Arctic sections. The flow through the English Channel is small, about 0.1 Sv in the two models.

The above volume fluxes, and even more the individual fluxes for the sections in Figure 3 and Table 1, show some differences between the two model systems. Several reasons account for these differences. First of all, the version of MICOM used in this review is global in extent, meaning that any residual flow of water through the Nordic Seas region, averaged over a few days or longer, is balanced by volume transports through the Bering Strait and the Canadian Archipelago. In the MICOM model, the net northward flow is 0.6 Sv through the Nordic Seas (Figure 3 and Table 1). This residual is balanced by a poleward flow of 1.0 Sv through the Bering Strait and a net southward flow of 1.6 Sv through the Canadian Archipelago. In contrast, NAOSIM indicates near-zero residual flow through the Nordic Seas, with a corresponding near-zero net flow through the Canadian Archipelago because of the closed Bering Strait. Therefore, volume transport differences of about 0.5 Sv can be expected between the two model systems, at least for some of the transport routes.

Secondly, even OGCMs that are forced—and by that to quite a degree constrained—by prescribed atmospheric momentum, heat, and freshwater fluxes, as is the case for MICOM and NAOSIM, have a certain degree of internal variability. This internal variability originates from the essentially unknown initial state of the ocean. Differences in the initial ocean state, particularly hydrographic differences in the weakly stratified regions in the Nordic Seas and in the North Atlantic subpolar gyre along with the flux of freshwater through the Fram Strait, have the potential to moderate the ocean circulation in the region on time scales from years to decades [e.g., *Haak et al.*, 2003; *Bentsen et al.*, 2004].

Repeated model simulations with identical model systems but different ocean initial states indicate that the major volume fluxes in the Nordic Seas may differ from a few tenths of Sv

Table 1. Simulated Northward, Southward, and Net Northward (Northward – Southward) Volume (Sv); Heat Flow (TW); and Freshwater Flux (0.01 Sv) Through the Denmark Strait (DS), Across the Iceland–Faroe Ridge (IFR), Between the Faroes and Shetland (FS), Across the Barents Opening (BO), and Through the Fram Strait (FS) in MICOM and NAOSIM.

Section	**DS**		**IF**		**FS**		**BO**		**FS**	
					Northward volume transports (Sv)					
	F_V	σ_V	F_V	σ_V	F_V	σ_V	F_V	σ_V	F_V	σ_V
NAOSIM	**2.0**	.13	**3.5**	.33	**3.8**	.45	**2.7**	.34	**3.2**	0.49
MICOM	**1.1**	.16	**4.5**	.42	**3.6**	.46	**4.1**	.33	**3.8**	0.63
					Southward volume transports (Sv)					
	F_V	σ_V	F_V	σ_V	F_V	σ_V	F_V	σ_V	F_V	σ_V
NAOSIM	**7.4**	.49	**1.2**	.21	**0.9**	.07	**0.7**	.13	**5.1**	0.60
MICOM	**3.7**	.48	**2.4**	.32	**2.5**	.26	**2.0**	.37	**5.3**	0.62
					Net (northward – southward) volume transports (Sv)					
	F_V	σ_V	F_V	σ_V	F_V	σ_V	F_V	σ_V	F_V	σ_V
NAOSIM	**–5.4**	.46	**2.3**	.36	**2.9**	.48	**2.0**	.40	**–1.9**	0.33
MICOM	**–2.7**	.49	**2.0**	.37	**1.2**	.69	**2.1**	.32	**–1.5**	0.36
					Northward heat transports (TW)					
	F_H	σ_H	F_H	σ_H	F_H	σ_H	F_H	σ_H	F_H	σ_H
NAOSIM	**26**	3.5	**101**	10.8	**137**	16.1	**49**	8.7	**28**	6.7
MICOM	**26**	3.4	**106**	16.1	**136**	15.6	**86**	8.6	**25**	2.8
					Southward heat transports (TW)					
	F_H	σ_H	F_H	σ_H	F_H	σ_H	F_H	σ_H	F_H	σ_H
NAOSIM	**22**	7.2	**29**	5.3	**14**	1.5	**5**	1.5	**–5**	2.7
MICOM	**7**	1.2	**37**	6.5	**24**	5.5	**21**	4.3	**24**	5.2
					Net (northward – southward) heat transports (TW)					
	F_H	σ_H	F_H	σ_H	F_H	σ_H	F_H	σ_H	F_H	σ_H
NAOSIM	**4**	6.1	**72**	11.6	**123**	16.5	**44**	8.5	**33**	6.6
MICOM	**19**	3.8	**69**	12.1	**112**	19.4	**65**	6.8	**1**	4.2
					Northward salt transports (0.01 Sv)					
	F_S	σ_S	F_S	σ_S	F_S	σ_S	F_S	σ_S	F_S	σ_S
NAOSIM	**–0.6**	0.3	**3.6**	0.4	**5.1**	0.6	**1.2**	0.3	**–0.2**	0.4
MICOM	**–0.4**	0.3	**3.4**	0.8	**4.8**	0.7	**0.2**	0.5	**2.0**	1.6
					Southward salt transports (0.01 Sv)					
	F_S	σ_S	F_S	σ_S	F_S	σ_S	F_S	σ_S	F_S	σ_S
NAOSIM	**–9.2**	1.8	**1.1**	0.2	**0.8**	0.1	**–0.1**	0.2	**–6.7**	1.6
MICOM	**–4.4**	1.0	**0.9**	0.5	**1.6**	0.3	**–0.2**	0.3	**–8.1**	2.6
					Net (northward – southward) salt transports (0.01 Sv)					
	F_S	σ_S	F_S	σ_S	F_S	σ_S	F_S	σ_S	F_S	σ_S
NAOSIM	**8.6**	1.6	**2.5**	0.4	**4.4**	0.6	**1.3**	0.2	**6.5**	1.5
MICOM	**4.0**	0.8	**2.5**	0.5	**3.2**	0.8	**0.4**	0.4	**10.2**	3.0

The heat flows are given relative to T = 0°C and the freshwater fluxes relative to S = 34.8. Negative heat and freshwater fluxes means that the heat and freshwater transports (relative to T = 0°C and S = 34.8) are directed opposite to the flow. Because of rounding-off errors, the volume transport budgets balance within 0.1–0.2 Sv. σ_V, standard deviation.

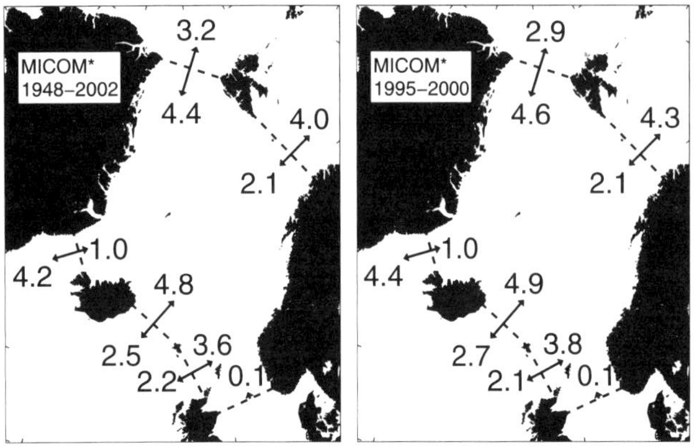

Figure 4. As in Figure 3, but for a second realisation with MICOM, labelled MICOM*. MICOM* is initialised with the ocean state of MOCOM at the end of 2002 and is then integrated from 1948 and onwards.

to about 0.5 Sv. This is illustrated in Figure 4, showing the net volume transports given by MICOM for an integration continuing from the ocean state at the end of the MICOM realisation shown in Figures 1–3. Significant differences may be noticed, e.g., in the Fram Strait, where the long-term mean northward flow in the second integration is reduced from 3.8 to 3.2 Sv, and the southward flow from 5.3 to 4.4 Sv. The second MICOM integration (MICOM*) is included here only to illustrate the effects different ocean initial states may have on the simulated ocean climate; it will not be used further.

To sum up: The combined effect of regional versus global model domain, and internal variability of any OGCM system based on the essentially unknown ocean initial state, may produce volume transport differences of about 0.5 Sv (based on the presented comparison between the NAOSIM, MICOM, and MICOM* realisations; differences may be larger if more model systems or realisations are included in the analyses). Consequently, volume transport differences exceeding, say, 1 Sv, cannot be simply explained by differences in the model domain (i.e., global versus regional domains) or the unknown initial state of the ocean. In such a case, model differences can be attributed only to the intrinsic properties of the models—such as horizontal and vertical model resolution, formulation and parameterization of resolved and unresolved ocean processes, and the numerical implementation of the governing equations.

3.3. Comparison Between Observed and Simulated Mean Volume Transports

The simulated volume transports in Figure 3 can be compared with available observation-based transport estimates. As already mentioned, reliable velocity measurements are available only from 1994 onwards. A compilation of available

Table 2. Observed [*Østerhus et al.*, 2005] and Simulated Properties of the Atlantic Inflow to the Nordic Seas (heat flow is relative to 0°C).

Atlantic inflow	Period		Average fluxes	
			Volume	Heat
			Sv	TW
Denmark Strait	1994–2000	**Observed**	**0.75**	**19**
		NAOSIM	1.9	26
		MICOM	0.8	24
Iceland–Faroe	Jun 1997–Jun 2001	**Observed**	**3.5**	**124**
		NAOSIM	3.4	98
		MICOM	4.3	95
Faroe–Scotland	1994-2000	**Observed**	**3.2**	**149**
		NAOSIM	4.0	147
		MICOM	3.7	137

literature values, together with simulated values for the same time period as for the observations, is presented in Table 2. For the period October 1994 to August 2000, the observation-based estimate of the northward flow of Atlantic Water in the Denmark Strait is 0.75 Sv in *Østerhus et al.* [2005]. The respective MICOM value is 0.8 Sv, whereas that of NAOSIM is greater, 1.9 Sv. For the Iceland–Scotland openings, both MICOM and NAOSIM are close to the observed transports, but with the MICOM transport across the Iceland– Faroe Ridge and the NAOSIM transport across the Faroe–Scotland opening, 0.8 Sv too high. Note, however, that the estimates for the Atlantic Water inflow across the Iceland–Faroe ridge range from 3.3 Sv [*Hansen and Østerhus*, 2000] to 3.8 Sv [*Østerhus et al.*, 2005], depending on the method and the time interval.

A striking difference between the two models is apparent in the southward transports through Denmark Strait with 7.5 Sv for NAOSIM and 4.0 Sv for MICOM. This is partly balanced by the southward flow across the Iceland–Scotland ridges of 2.2 Sv for NAOSIM in contrast to 5.0 Sv for MICOM. Observationally based estimates for the deepest southward flows across the Iceland–Scotland ridges are 2.2–2.9 Sv with large uncertainties [*Hansen and Østerhus*, 2000]. For the Denmark Strait an observationally based estimate of 3 Sv for deep overflow to the south is rather well accepted [*Blindheim and Østerhus*, this issue]; in contrast, the shallow outflow with the EGC is very uncertain. *Hansen and Østerhus* [2000] calculated the combined outflow at the surface through Denmark Strait and the Canadian Archipelago to be 3 Sv, based on the residual from estimates for all other flows in and out of the Nordic Seas.

For the Barents Opening, most recent estimates are based on current-meter observations from the major inflow section between Bear Island and Norway from 1997 to 2001. *Ingvaldsen et al.* [2004] calculated from these observations 1.5 Sv for the inflowing Atlantic Water and added 0.5–1 Sv for the Norwegian Coastal Current, yielding a total of 2–2.5 Sv. This compares with an inflow of 2.7 Sv for NAOSIM and 4.2 Sv for MICOM in the 1995–2000 period. For the Fram Strait, flow estimates are uncertain because of the very strong recirculation in the strait.

When evaluating the differences of model and observationally based estimates, we have to consider that several of the water masses that pass the Nordic Sea openings are subject to intense recirculation at the openings, which complicates the interpretation of the transports at fixed sections: This is especially true for the Fram Strait, the Denmark Strait, and the Faroe–Scotland gap. Moreover, the fact that different observational estimates are based on different definitions for the passing water masses may lead to confusion. Here we use vertically integrated total water column transports for the models without performing a detailed comparison along the various definitions of water mass. Such complications are discussed in (e.g.) *Nilsen et al.* [2003].

In conclusion, the correspondence between the observed and simulated volume transports is encouraging, particularly for the inflow of Atlantic Water between Iceland and Scotland. Also the simulated heat transports (Table 2) are consistent with the observation-based values.

3.4. Model–Model Comparison of the Interannual Variability of the Volume Transports

The major variability modes of the Nordic Seas marine climate system can be explored by examining the variability of the amount and properties of water flowing into and out of the region. Irrespective of the actual match or mismatch between the simulated mean transports through the different sections shown in Figure 3, one would expect that the simulated flow *anomalies* should show similarities over the integration period. The correspondence should be particularly clear for quantities that are directly or to a large degree driven by the applied atmospheric forcing fields, and for processes that are properly resolved by the models. Or conversely, simulated quantities that show a high degree of covariance over time are probably forced, in a direct way, by the applied atmospheric fields, and hence the governing ocean dynamics is appropriate.

Figure 5 shows the simulated northward and southward volume transport anomalies for both model systems for the five open ocean sections in Figure 3. The figure caption includes the linear correlation of the transport anomalies between MICOM and NAOSIM. It follows from Figure 5 that the variability in the northward flow across the Denmark Strait is weak (standard deviation of 0.16 Sv and 0.13 Sv for MICOM and NAOSIM, respectively), with no correlation between the two simulated time series. The situation is opposite for the northward flow anomalies between Iceland and Scotland and across the Barents Opening. Here the mean standard deviations of the two models are 0.37 Sv (Iceland–Faroe), 0.45 Sv (Iceland–Scotland), and 0.33 Sv (Barents Opening), and the given correlations are significant. It is interesting, and a positive result, that two widely different model systems produce so consistent anomalies in flow through these three sections for such a long time period. The correlation between the two model systems breaks down for the Fram Strait, however, which may not be too surprising given the differences in the circulation in the region (cf. Figures 1–2).

The southward volume transport anomalies are shown in the right panels of Figure 5. The two model systems produce quite consistent transport anomalies in the Denmark Strait, across the Iceland–Faroe Ridge, between the Faroes and Scotland, and through the Barents Opening, but not for the Fram Strait.

In general, the magnitude of the volume transport anomalies is comparable between the two model systems, although MICOM tends to produce slightly larger anomalies than NAOSIM.

The extent to which the temporal evolution of the simulated anomalies in Figure 5 is realistic can be assessed only by comparison with continuous and high-quality current-meter observations (see the discussion in section 3.2). Figure 6 shows, for example, observed and simulated transports for the northward flow in the Faroe–Shetland Channel for the period October 1998 to December 2002. Consistent with Table 2, both MICOM and NAOSIM are, on average, close to the observed northward volume flux. The monthly mean variability is, however, larger in the observation-based time series. Subtracting the mean values of the time series and normalising the resulting anomalies, produces a close match. This finding illustrates the potential for combining observations and prognostic model systems to better understand the variability modes of the marine climate system in the region and shows that available model systems have the potential to be used to extend, in time and space, available observations (see, e.g., *Hátún et al.* [this issue]).

3.5. Simulated Thermodynamic Surface Properties

In the following, a model–model comparison addresses the thermodynamic properties of the Nordic Seas. We start with SSS or, equivalently, the upper ocean fresh water content. Away from the ice edge where strong salt fluxes might occur, SSS is far less influenced by the atmospheric forcing than SST. Therefore, salinity can be considered a tracer for the transport and mixing of the major water masses in the region.

The upper panels in Figure 7 display the March distribution of SSS in MICOM and NAOSIM. Because of the differences in resolution, NAOSIM describes more details and accepts sharper gradients than MICOM. This is especially pronounced in the EGC that is visible as a fresh and cold

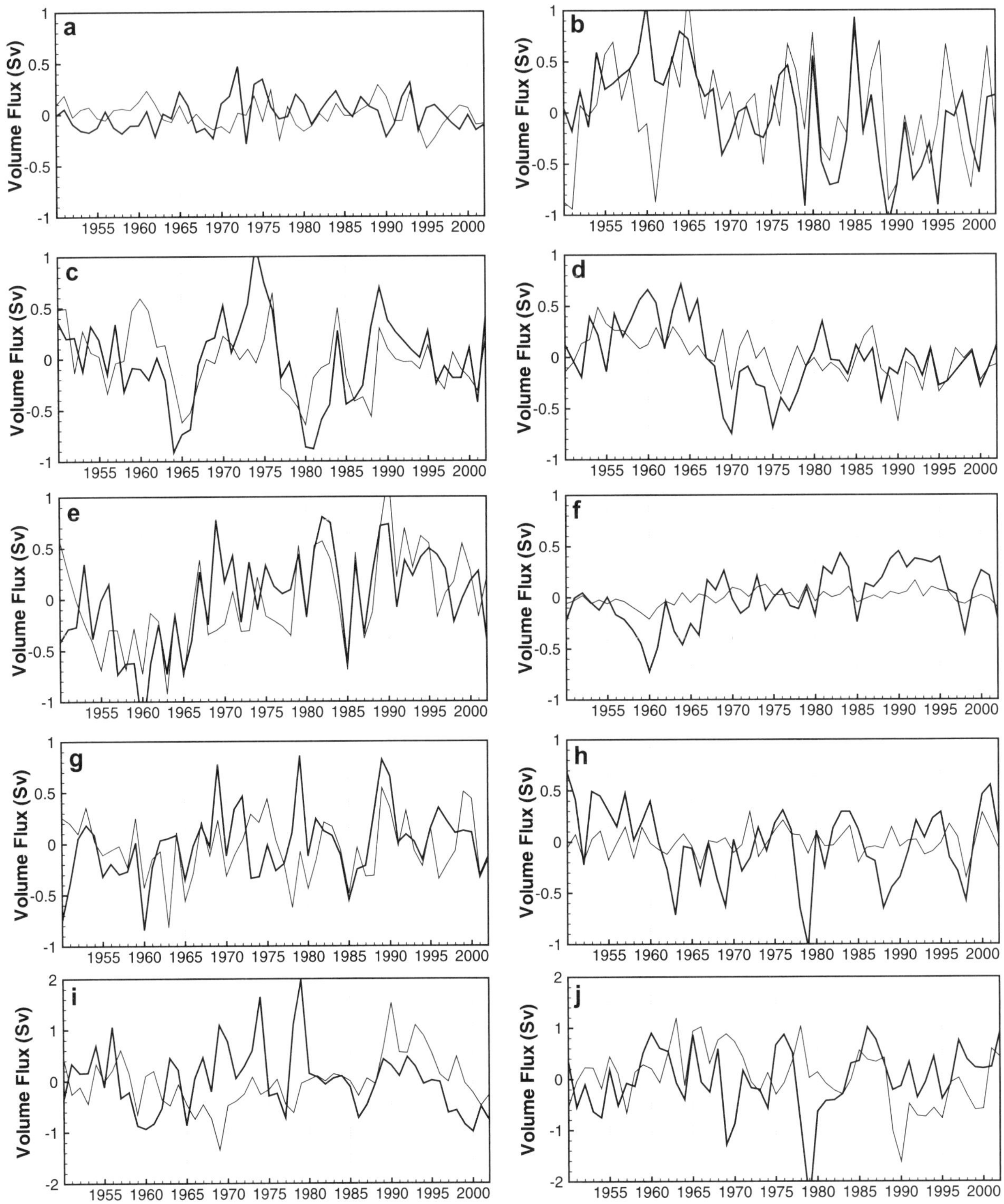

Figure 5. Simulated annual mean volume transport anomalies (Sv) for NAOSIM (thin lines) and MICOM (thick lines). Left (right) column shows anomalies in net northward (southward) transports. The panels represent (a,b) the Denmark Strait (r = 0.05/**0.46**), (c,d) the Iceland–Faroe Ridge (r = **0.51**/**0.45**), (e,f) the Faroe–Scotland section (r = **0.59**/**0.63**), (g,h) the Barents Opening (r = **0.28**/**0.35**), and (i,j) the Fram Strait (r = –0.05/0.04). The numbers in parentheses give the linear correlation coefficient between the two model time series for the northward/southward volume fluxes, respectively, the bold numbers indicating 95% or higher significance. The sections are identified in Figure 3.

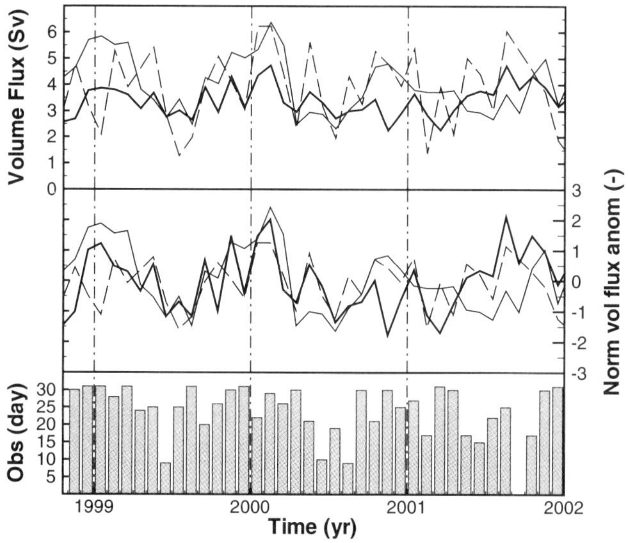

Figure 6. Observed and simulated northward volume transport through the Faroe–Shetland Channel. Upper panel: Observed transport (Sv) based on current-meter measurements [*Østerhus et al.*, 2005] in dashed line, simulated NAOSIM (MICOM) transport (Sv) in thin (thick) solid line. Middle panel: Corresponding volume flux anomalies, normalized with respect to the standard deviation of the respective time series. Lower panel: The number of days per month when observations were made. The correlation between the observed time series and MICOM is 0.43, significant at the 99% confidence level. The corresponding figure for NAOSIM is 0.26 at the 90% confidence level (serial correlations are included in the analysis).

boundary layer all the way to the southern tip of Greenland in NAOSIM, whereas MICOM produces a front between the polar waters of the EGC and the subpolar Atlantic Water aligned with the Greenland–Iceland Ridge. The northward-flowing Atlantic Water is evident off the coast of Norway. NAOSIM is fresher than MICOM in the central and eastern part of the Nordic Seas.

The March SST is quite similar between the two models in the east (lower panels in Figure 7), but with lower surface temperatures in MICOM in the central part of the Nordic Seas, and with a very cold region in the central Greenland Sea in NAOSIM.

Observed and simulated vertical distributions of temperature in July 1999 along 75°N, and thus crossing the central Greenland Sea basin, are provided in Figure 8. The observations clearly show the warm poleward-flowing Atlantic Water towards east, extending below 500 m eastward of 9°E. At the surface, waters exceeding 6°C are found in the upper 50 m eastward of about 3°E. In the central Greenland Basin near the prime meridian, temperatures below 0°C are found below about 50 m deep. In the west, the return Atlantic Water flows along the Greenland continental slope, whereas cold Polar Water constitutes the main water mass on the Greenland shelf.

Both the poleward-flowing Atlantic Water and the return Atlantic Water are clearly seen in the simulations (lower panels in Figure 8). The poleward-flowing Atlantic Water is, however, too narrow in both models, particularly in MICOM. The mixed-layer temperature in MICOM is close to the observed temperature, whereas in NAOSIM the surface layer is too warm. Towards the west, NAOSIM produces too strong and prominent return Atlantic Water. Both models capture the cold Polar Water on the Greenland shelf. It is encouraging, despite the mentioned differences, that both model systems are able to produce realistic subsurface temperature distributions after a total integration time of about 100 years (spin-up plus the reanalysis integration; see Appendix).

Figure 9 displays the simulated mean and extreme sea-ice concentrations for the period 1948–2002. The corresponding observed sea-ice edge position is given for the period 1978–2002. The Is Odden—the sea-ice structure extending into the Greenland Sea in the upper left panel of Figure 9—is very pronounced in NAOSIM, as documented in the mean sea-ice concentration over the integration period. Sea-ice formation and transport in the Is Odden are important processes for deep convection in NAOSIM [*Gerdes et al.*, this volume]. The NAOSIM maximum sea-ice extent seems to exceed the observed sea ice extent even in the extremely cold year of 1881 [*Blindheim and Østerhus*, this issue]. One should keep in mind, however, that the extrema of ice extent shown in Figure 9 are derived from time series at individual grid points throughout the integration period 1948–2002, and that the map does not correspond to a state the model has actually occupied during the integration. For instance, high sea-ice cover in the Barents Sea and in the Greenland Sea virtually never occur simultaneously. A detailed comparison of observed sea-ice variability with that simulated by NAOSIM is given in *Kauker et al.* [2003].

The combination of winter SSS and SST governs the thickness of the upper, well-mixed layer in the ocean. MICOM and NAOSIM treat the mixed layer differently, with an explicit mixed-layer model in MICOM, based on the *Gaspar* [1988] bulk representation, whereas NAOSIM does not use an explicit mixed-layer parameterization. A deep winter mixed layer is a result of convective mixing caused by unstable stratification. For the diagnostic, the mixed-layer thickness in NAOSIM is based on the depth at which the density is 0.02 kg m^{-3} higher than the surface density. It is therefore difficult to perform a one-to-one comparison of the two mixed-layer fields. Nevertheless, the structure of deep mixing in March is quite similar, with rather deep mixing south of GSR, along the path of the Atlantic Water in the

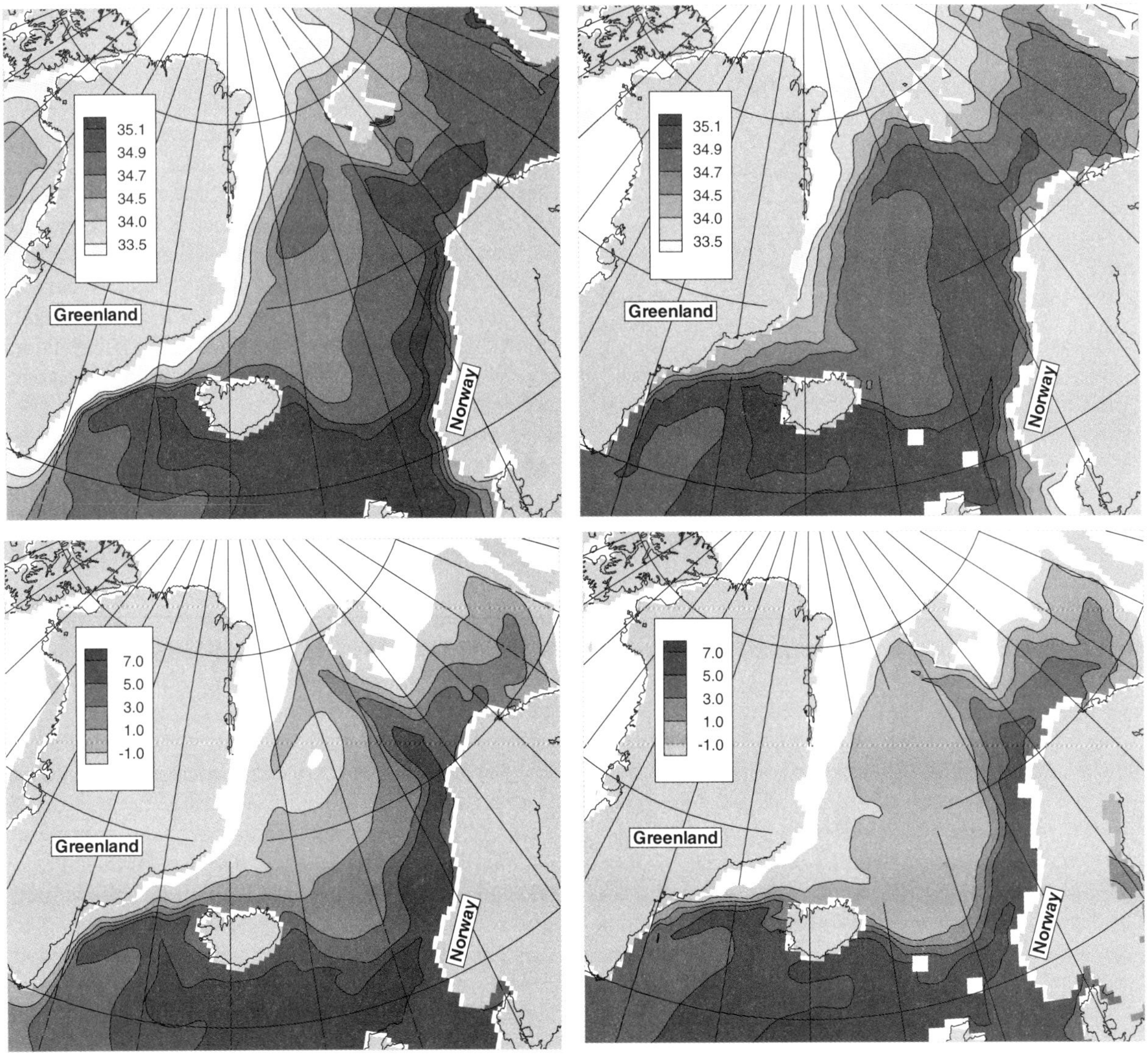

Figure 7. Simulated salinity (psu) (upper panels) and temperature (°C) (lower panels) in March, averaged over the uppermost 150 m of the water column for 1948–2002, as modelled by NAOSIM (left panels) and MICOM (right panels).

eastern parts of the Nordic Seas, and within the Greenland Basin. The history of deep convection in the Nordic Seas and the Labrador Sea, as simulated in NAOSIM, is the topic of *Gerdes et al.* [this volume].

Differences between the two model systems are substantial when interannual variability of convective winter mixing is considered (see Figure 10). Under minimum mixing conditions, NAOSIM produces very shallow convection, which is restricted to the Norwegian Atlantic Current, whereas MICOM shows shallow convection over large parts of the Nordic Seas and maintains relatively deep convection in the Irminger Basin and south of Iceland (not shown). NAOSIM convective mixing depth strongly exceeds the mixing depth in MICOM at maximum mixing in the Greenland Sea (lower panels in Figure 10). In fact, NAOSIM mixes to the sea floor in the Greenland Sea at maximum mixing, whereas MICOM produces less deep but a much more extended and uniform mixing along the eastern and northern rims of the Nordic Seas. A probable reason for this difference is the very weak stratification in NAOSIM compared with MICOM (Figure 8). Dedicated model experiments incorporating passive tracers such as chlorofluorocarbons [*Schlosser et al.*, 1991; *Bönisch et al.*, 1997] or sulphur hexafluoride [*Watson et al.*, 1999; *Olsson et al.*, 2005; *Eldevik et al.*, this issue] are probably required to properly address the mean state and the variability of the simulated mixed layer.

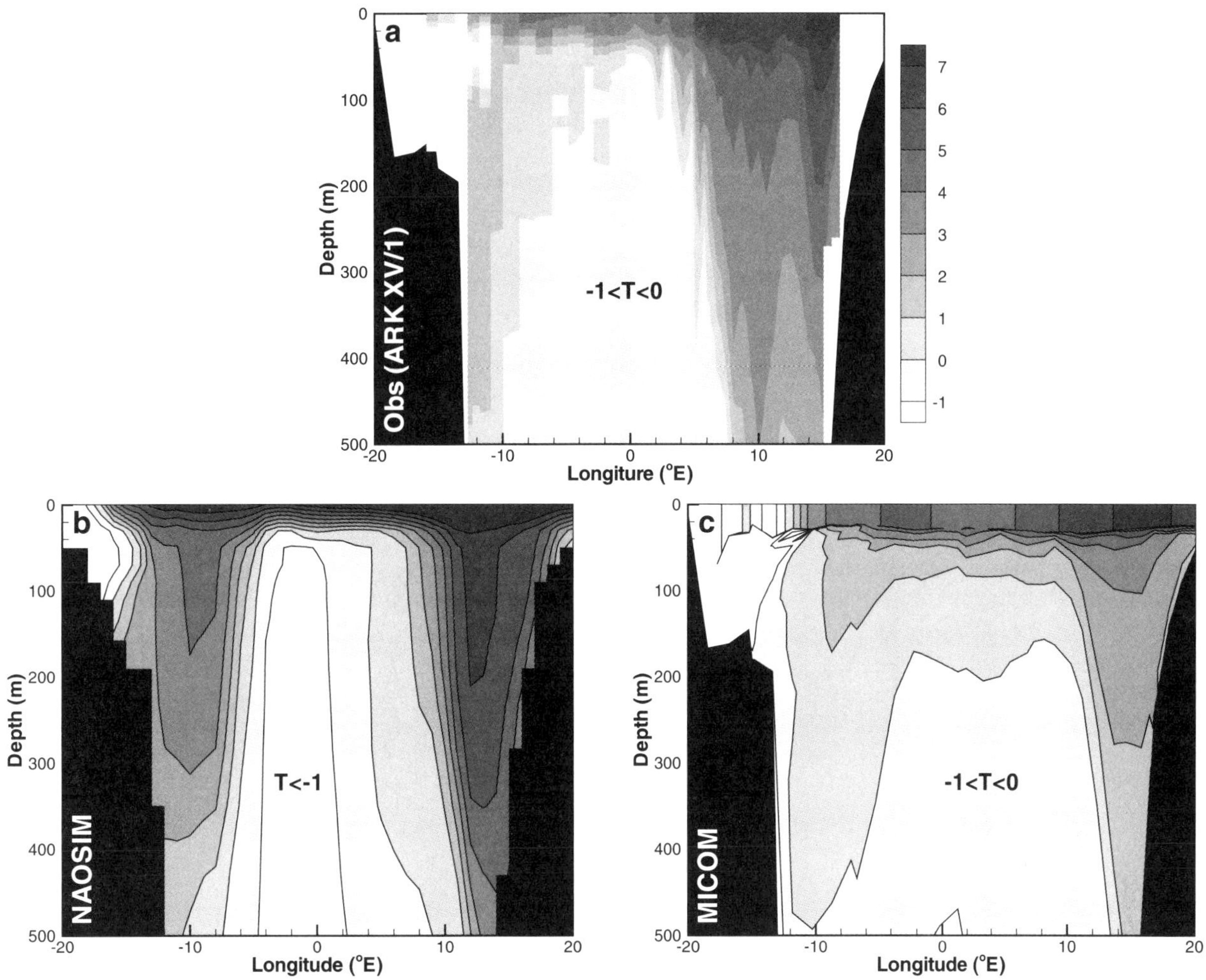

Figure 8. Observed (upper panel) and simulated (lower panels) temperature along 75°N in the Nordic Seas (NAOSIM results to left, MICOM results to right). The observations are from cruise ARK XV/1 with *RV Polarstern* from 23 June to 19 July 1999 [*G. Budeus*, AWI, pers. comm.]. The simulated fields are July monthly means. The same contouring is used for all panels.

4. PARTICULAR CHALLENGES FOR MODELLING THE NORDIC SEAS

4.1. Model-Specific Challenges

In the previous section, particular focus was put on the exchanges of the Nordic Seas with the adjacent ocean basins. We regard proper representation of these exchanges as the foremost challenge for modelling the Nordic Seas as a key region of the global ocean circulation—particularly because processes within the Nordic Seas influence the exchange rates and the properties of the waters exported to the Atlantic and Arctic oceans.

For both the exchanges and the interior processes, proper resolution of the ocean eddies, or the baroclinic Rossby radius, of *O* (10 km) is important. As a rule of thumb; an OGCM will properly describe ocean dynamics on a horizontal scale of about 5 times the grid spacing. Proper model representation of the deformation radius would therefore require an ocean grid spacing of 1–2 km. Such a fine grid mesh for the Nordic Seas and parts of the neighbouring oceans is currently at the limit of what is computationally feasible and certainly beyond the computational resources available for most ocean modelling groups. Models with insufficient resolution—virtually all of the current ocean components of coupled climate models—typically misrepresent the properties of the deep overflows that form the dense constituents of NADW. This can result in the subpolar North Atlantic becoming the main source for Northern Hemisphere dense water in the models, which as *Döscher and Redler* [1997] demonstrated, can imply a false sensitivity of the models to fluctuations in atmospheric forcing and anomalous

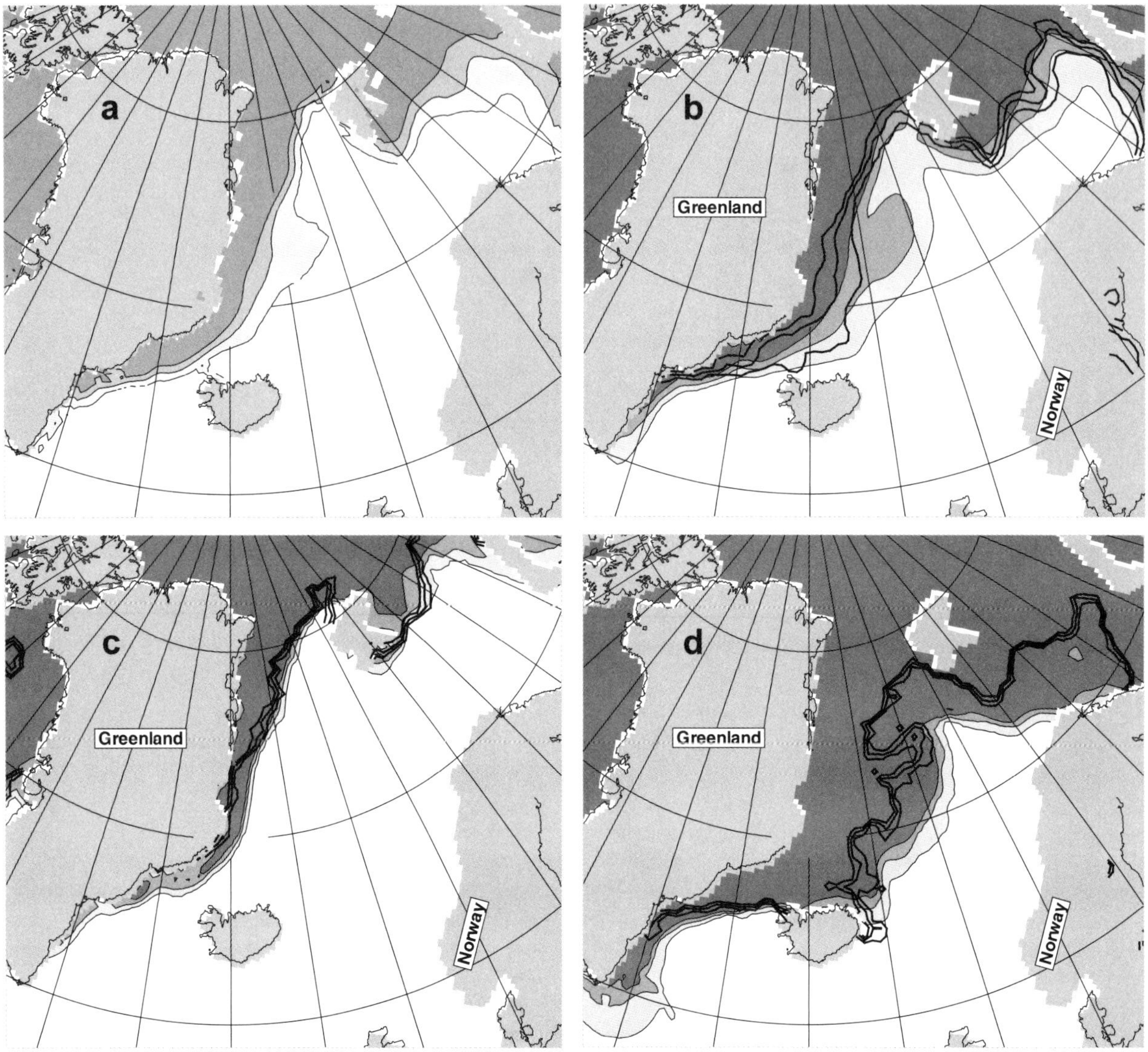

Figure 9. Mean sea-ice concentrations in March for 1978–2002 from observations (a) [*Johannessen et al.*, 1999] and for 1948–2002 from the simulations (b). Simulated minimum (maximum) concentrations are given in panel c (d). Observations and NAOSIM are shown with thin lines and shaded contour levels of 10%, 40%, and 70%, whereas MICOM is shown with thick solid lines and similar contour levels.

freshwater influx to the North Atlantic. The deep convection in the Labrador Sea is much more susceptible to anomalies than the more robust formation of intermediate waters north of GSR, which occurs over long distances and through several processes. Furthermore, the potentially important process of energy transfer from the large-scale potential energy reservoir to the kinetic energy of the large circulation is not captured by the present parameterizations.

Enhanced grid resolution can be obtained by nesting regional, high-resolution models into larger, coarser-resolution models. One variant of this approach is to use regional models that receive boundary conditions from climatology [e.g., *Gerdes et al.*, this issue] or larger-scale models [e.g., *Hátún et al.*, this issue] without feeding back into these "parent" models. Regional OGCMs have the advantage that the spatial grid mesh can be fine, possibly resolving the first baroclinic radius of deformation. The disadvantage with regional model systems is that, in general, water mass fluxes and their properties (i.e., temperature, salinity, and tracer concentrations) have to be prescribed at the lateral boundaries. Technical problems can lead to boundary artefacts propagating into the model domain. For more information about open boundary conditions and nesting of model systems with different resolution, see *Martinsen and Engedahl* [1987]; *Ginis et al.* [1998];

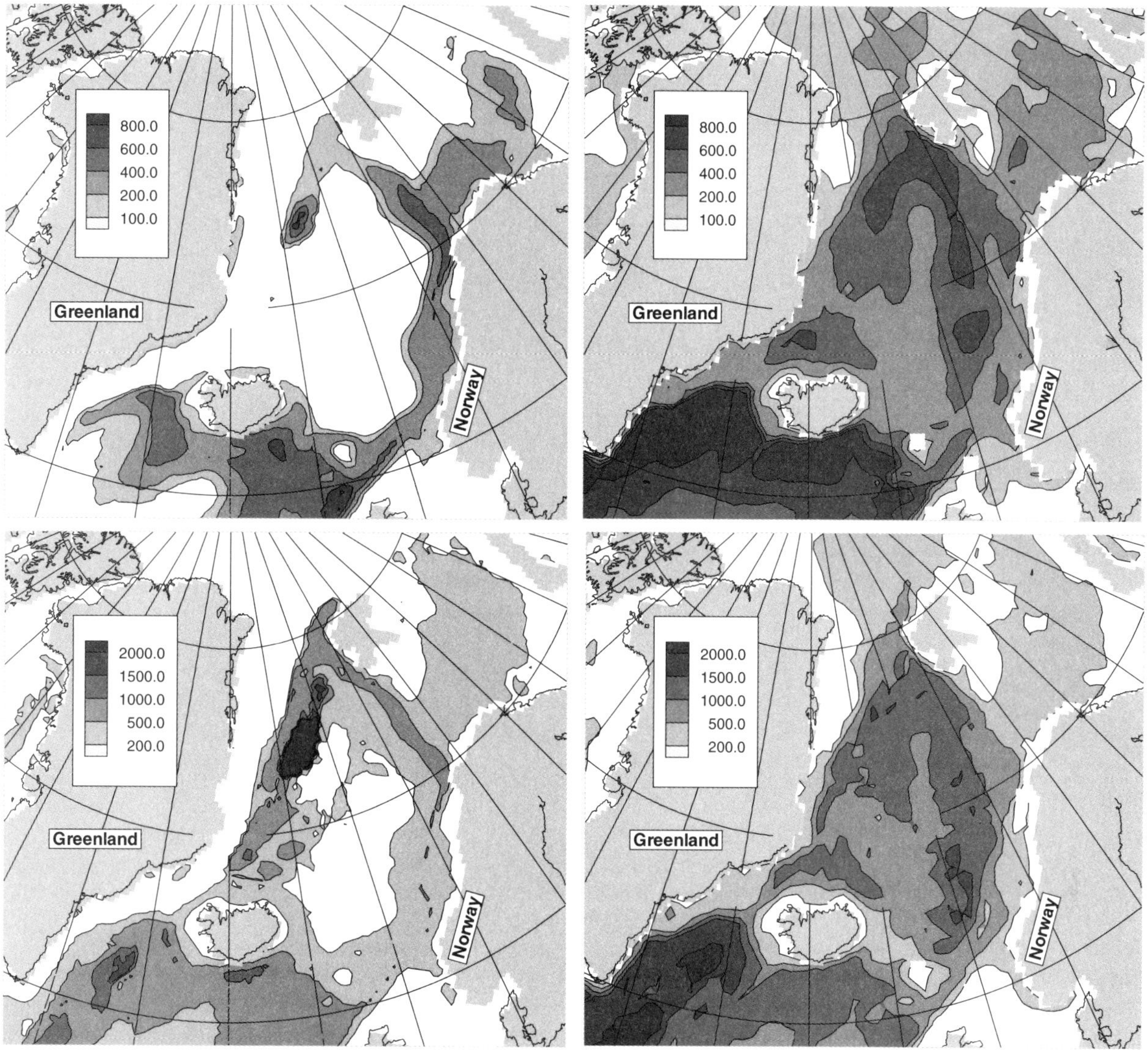

Figure 10. Simulated mean (upper panels) and maximum (lower panels) mixed-layer depth (m) in March for the period 1948–2002 as determined by NAOSIM (left panels) and MICOM (right panels).

Jensen [1998]; *Palma and Matano* [2000]; *Perkins et al.* [1997]; and *Heggelund and Berntsen* [2002].

The overflow of dense water masses across the GSR is associated with excessive mixing in many ocean models [*Gerdes*, 1993; *Roberts and Wood*, 1997]. On the other hand, it has been a problem for isopycnal models to include sufficient entrainment of ambient fluid during the overflow [*Roberts et al.*, 1996]. New parameterizations [*Hallberg*, 2000; *Shi et al.*, 2001] have been developed and implemented to reduce this problem. However, according to *Gerdes* [1993], potential vorticity constraints make mixing or frictional modification of the flow inevitable in models that don't resolve the baroclinic Rossby Radius and when the overflowing water masses experience large changes in layer thickness.

Bottom boundary layer parameterizations in OGCMs are reviewed by *Killworth* [2003], and various schemes have been devised by, among others, *Beckmann and Döscher* [1997] and *Killworth and Edwards* [1999]. These schemes provide a pathway for dense waters down the topographic slopes and avoid excessive entrainment. The schemes have been employed in many large-scale ocean models where they, to some degree, improve overflows and sloping convection from shelf seas at high latitudes. A new approach comes from *Kösters et al.* [2005], who introduce a parameterization based on hydraulic control theory to describe the strength of the Denmark Strait Overflow in a large-scale Atlantic model.

The overflows and other exchanges across the boundaries of the Nordic Seas are strongly linked by continuity. A weaker overflow will thus be associated with a weaker inflow of Atlantic Waters. Weaker inflow of Atlantic Waters will lead to reduced heat release from the ocean to the atmosphere in the Nordic Seas, thus affecting the properties of the return Atlantic Water that is an important contribution to the overflows. The heat release can also affect Arctic sea-ice volume [*Goose et al.*, 2004], although this effect is not included in ocean–sea-ice models forced by prescribed atmospheric fields. More directly, the Atlantic inflow affects the heat supply to the Atlantic layer of the Arctic [*Karcher et al.*, 2003; *Gerdes et al.*, 2003] and the position of the sea-ice edge in the Barents and Greenland seas [*Kauker et al.*, 2003].

The branching of the Atlantic Water into the two Arctic contributaries and the return Atlantic Water as well as the branching of the EGC into the Jan Mayen Current and the East Icelandic Current are heavily influenced by details of the topography. The bathymetry of an OGCM is usually computed as the mean value of the real ocean depth underlying each horizontal grid cell. This implies that bathymetric features like ridges and channels are smoothed. The degree of smoothing is therefore governed by the horizontal length-scale of the bathymetric features compared to the actual grid spacing. In addition, some OGCMs require a smoothed bathymetry to avoid numerical instabilities. This has particularly been the case for terrain-following OGCMs.

Smoothed (or artificially reduced) height of ridges is usually not adjusted in OGCMs, whereas deep channels are, at least in climate modelling, commonly adjusted by artificial widening or deepening of the channels [*Biastoch et al.*, 2003; *Beismann and Barnier*, 2004]. Again, proper resolution of these features is one of the basic requirements for Nordic Seas circulation models. However, horizontal and vertical grid spacing have further, less obvious effects. The energy transfer between baroclinic and barotropic modes seems to be very sensitive to resolution as the much larger energy of barotropic flows in a higher resolution version of NAOSIM indicates [*K. Fieg*, AWI, pers. comm.]. Clearly, the representation of the Joint Effect of Baroclinicity and Relief (JEBAR) [*Sarkisyan and Ivanov*, 1971; *Mertz and Wright*, 1992] in a region of strong water mass boundaries and characterized by deep and narrow channels, steep slopes, and complex ridges is a particular challenge for ocean modelling.

The central Greenland Sea is one of few locations where open-ocean convective mixing takes place [*Marshall and Schott*, 1999]. This process is believed to influence the ventilation of dense, subsurface water masses on climate time scales. Except for the large-scale downward vertical advection proposed by *Budeus et al.* [1998], most vertical mixing processes take place on scales that are not and cannot in the foreseeable future be explicitly resolved by OGCMs. Several parameterization schemes have therefore been proposed to include vertical subgrid scale processes [*Paluskiewicz and Romea*, 1997; *Canuto et al.*, 2004]. However, much more work is needed to properly incorporate a physically consistent description of small-scale vertical mixing processes in climate-type OGCMs. Linked to open-ocean convection is also the treatment of brine waters generated during freezing of sea-ice. Fine-scale modelling has been carried out to describe the dynamics of brine waters released from sea-ice [*Kämpf and Backhaus*, 1999], but incorporation of this effect in OGCMs is typically ignored or included in a highly simplified way.

Finally, the Atlantic inflow into the Arctic is linked to the outflow of much fresher waters near the surface, feeding the EGC. The fresh water carried by the EGC can affect the interior of the Nordic Seas and especially the convection in the central Greenland Sea. Most of the freshwater, however, is carried through the Denmark Strait into the subpolar Atlantic. Proper representation of both the subduction of the northward-flowing saline Atlantic Water under the fresh polar water in the Fram Strait region, the dynamics of the freshwater along the coast of Greenland (and, similarly, along the coast of Norway), and the frontal mixing between the freshwater and the more saline open-ocean surface waters are challenges for OGCMs.

4.2. Observation-Based Evaluation of OGCMs

The only way to proceed from plain comparison of model results and thereby to identify model weaknesses and deficiencies is to actively include observations in the analyses. Figure 6, showing a one-to-one comparison between observed and simulated northward transport through the Faroe–Scotland Channel, provides an example of direct observation-based evaluation of OGCMs. Unfortunately, available observations of the marine climate system are generally scattered in time and space. It is therefore difficult, and in many cases impossible, to adequately address the quality, or realism, of a model system on climate (i.e., multidecadal) time scales, as on the time scale of Figure 5. Nonetheless, the Nordic Seas region is one of the world's oceans with the best observational coverage [*Blindheim and Østerhus*, this issue; *Furevik et al.*, 2002]. Acquisition, quality check, and synthesis of available observations are therefore needed to generate the best possible observation-based background for evaluating the mean state and variability of simulated ocean states. Needless to say, continuous observations of key ocean parameters at key ocean locations are of paramount importance for any evaluation of OGCMs.

For decadal and longer time scales, the use of tracers such as chlorofluorocarbons (CFC-11 and CFC-12) and radiocarbon (^{14}C) has turned out to be useful and cost-effective in assessing the integrated (or net) effect of ventilation of the basin-scale and the world's ocean surface waters [e.g., *Toggweiler et al.*, 1989; *England and Maier-Reimer*, 2001;

Dutay et al., 2002; *Gao et al.*, 2003]. Furthermore, tracers from point sources—e.g.; radioisotopes from the European nuclear reprocessing plants—have been useful to address the interannual transport, mixing, and age properties of, for instance, the Atlantic Water in the Nordic Seas region [*Nies et al.*, 1998; *Karcher et al.*, 2004; *Gao et al.*, 2004, 2005]. On smaller scales of time and space, dedicated tracer experiments have been found to be of great use for assessing small-scale mixing and transport of explicitly tagged water masses. A unique example here is the deliberate release of sulphur hexafluoride in the central Greenland Sea in 1996 [*Watson et al.*, 1999] and the subsequent observation of the spreading of the tracer within and out of the Nordic Seas [*Olsson et al.*, 2005; *Eldevik et al.*, this issue].

5. CONCLUSION

Numerical modelling of the ocean–sea-ice system of the Nordic Seas and adjacent areas has reached a state of maturity similar to that of Atlantic and Arctic modelling efforts. This is the case despite some unique challenges that are posed by the small inherent scale of oceanic motion, the large contrast between inflowing Atlantic and Arctic water masses, and the large influence of fluctuations in freshwater flux. In particular, the mean exchanges into and out of the Nordic Seas have been demonstrated to agree with available observed estimates, and interannual variations to correlate with observations, although the simulated amplitudes tend to be less than the observed ones. This progress is welcome news because it will lead, at some time, to an improved representation of Nordic Seas processes in large-scale ocean–sea-ice models and eventually coupled climate models.

Models have recently been used to put decadal variability and observed trends into a larger-scale perspective, connecting individual observed time series with the large-scale atmospheric forcing fields and the oceanic conditions in adjacent basins. For instance, model experiments have been important tools in identifying fluctuations in the freshwater flux out of the Arctic and their pathways through the Nordic Seas. Similarly, the fate of Atlantic water entering the Nordic Seas in the passages through GSR, the subsequent modification of the water masses in the Nordic Seas, and the branching into different paths entering the Arctic and recirculating within the Nordic Seas have been the subject of analysis of ocean–sea-ice hindcast experiments. These experiments led to identification of multiple Arctic warming events and consideration of the 1990s warming as an outstanding event for at least the last 50 years. Long-term observations at key locations in the Nordic Seas [e.g., *Blindheim and Østerhus*, this issue] have proven invaluable for model evaluation and for indicating dramatic developments in the marine climate system of the region.

Despite this recent progress, many difficulties remain to be overcome and important model improvements realized. As has been demonstrated here, different model system such as NAOSIM and MICOM provide different results under similar atmospheric forcing. These differences pertain to important fluxes through the Nordic Seas and would be of climatic importance if they existed in the real ocean. The cause for the differences are basically unknown and clarification of the cause would require systematic intercomparison efforts as have taken place for the Atlantic (e.g., DYNAMO) and the Arctic (AOMIP) ocean basins. Clearly, some processes—exchanges with the Arctic Ocean through the Fram Strait, interaction of the EGC with the interior of the Greenland Sea, and overflows to the subpolar North Atlantic—require enhanced resolution and perhaps better or new parameterization schemes of small-scale processes. New models with significantly higher horizontal and vertical resolution have recently been implemented or are planned for the near future.

The wealth of existing historical data and the continuous stream of data from observational programs call for a more systematic use of models. This paper has been limited to studies of the present-day climate state. The discussion presented, however, is directly transferable and applicable to studies of past and future climate states as well to studies of the marine biogeochemical cycles. It is therefore foreseen that available and forthcoming OGCMs, even OGCMs tailored for the small Nordic Seas region, will significantly contribute to improved understanding of the variability and stability properties of past, present, and future climate states.

6. APPENDIX

6.1. The Miami Isopycnic Coordinate Ocean Model (MICOM)

The NERSC model used in this study is based on a mixture of versions 2.7 and 2.8 of MICOM [*Bleck et al.*, 1992], fully coupled to a sea-ice module consisting of the *Hibler* [1979] rheology in the implementation of *Harder* [1996] and the thermodynamics of *Drange and Simonsen* [1992].

The model system is identical to that used in *Gao et al.* [2005] to examine the transit time of the poleward-flowing Atlantic Water in the Nordic Seas and a modification of the model used by *Furevik et al.* [2002], *Nilsen et al.* [2003], and *Bentsen et al.* [2004] to address the observed and simulated salinity in the Nordic Seas, the variability of the volume transports across GSR, and the interannual- to decadal-scale variability of the AMOC, respectively. The main modifications to the *Furevik et al.* [2002], *Nilsen et al.* [2003], and *Bentsen et al.* [2004] studies are reduced strengths of the parameterized isopycnal and diapycnal mixing rates, the latter based on the CFC model evaluation study by *Gao et al.* [2003].

The applied model grid configuration is identical to that in *Furevik et al.* [2002], *Nilsen et al.* [2003], and *Gao et al.* [2005], but the horizontal resolution is double that in *Bentsen et al.* [2004].

Specifically, the horizontal grid resolution is about 40 km in the Northern North Atlantic and the Nordic Seas region, and the diffusive velocities (diffusivities divided by the size of the grid cell) for layer-interface diffusion, momentum dissipation, and tracer (temperature and salt) dispersion are 0.015 m s^{-1}, 0.01 m s^{-1} and 0.0025 m s^{-1}, respectively. The diapycnal mixing coefficient K_d (m^2 s^{-1}) is parameterized as $K_d = 5 \times 10^{-8}/N$, where N (s^{-1}) is the Brunt–Väisälä frequency. Consequently, the value of tracer dispersion and diapycnal mixing are factors 2 and 6, respectively, below those used in *Furevik et al.* [2002], *Nilsen et al.* [2003], and *Bentsen et al.* [2004].

The applied forcing is identical to all of the above-mentioned studies: daily atmospheric reanalysis fields from 1948 to the present as provided by the NCAR/NCEP reanalyses project [*Kalnay et al.*, 1996]. The surface forcing is described in *Bentsen et al.* [2004]: Annually repeated heat and freshwater relaxation fluxes with weekly temporal resolution are added to the mixed layer, implying that temperature and salinity anomalies are free to evolve and propagate. Freshwater from rivers is included in the model system. The results presented here are based on integration cycle numbers two and three with NCAR/NCEP forcing, where cycle two is initialised with the full ocean state at the end of cycle one (the spin-up cycle), and cycle three is initiated with the full ocean state at the end of cycle two.

A 20-km version of the model system is available, but output from that model version is not used in this review. See *Hátún et al.* [this issue] for an example of the 20-km version of MICOM.

6.2. The North Atlantic/Arctic Ocean–Sea-Ice Model (NAOSIM)

AWI maintains a hierarchy of coupled sea-ice–ocean models called NAOSIM (North Atlantic/Arctic Ocean–Sea-Ice Model). Models from the NAOSIM hierarchy are described in some detail in *Karcher et al.* [2003] and *Köberle and Gerdes* [2003]. Results for this paper were taken from an experiment with a quarter-degree version of the model described in *Gerdes et al.* [this issue]. The model is forced with daily mean 2-m air temperature, dew point temperature, cloudiness, precipitation, wind speed, and surface wind stress. For the first 50 years of spin-up, a climatological mean seasonal cycle based on the period 1979–1993 with added typical daily variability [OMIP-climatology; *Röske*, 2001] is used. After the spin-up, the forcing consists of daily mean atmospheric data from the NCEP reanalysis for the period 1948–2001 [*Kalnay et al.*, 1996]. Influx of freshwater from rivers is not explicitly included. To account for river runoff and diffuse runoff from the land, as well as to include the effect of flow into the Arctic through the Bering Strait, a restoring flux with an adjustment time scale of 180 days is added to the surface freshwater flux [*Gerdes et al.*, this issue]. The restoring flux is calculated with reference to annual mean climatological surface salinity data, constructed similarly to the initial data. The effect of the restoring flux on the surface salinity for this and other Arctic Ocean models is documented in *Steele et al.* [2001].

Acknowledgments. The European Union DG-XII Climate and Environment Programme has supported the OGCM activities at NERSC and AWI through the projects CONVECTION (contract EVK-CT-2000-00058), DYNAMITE (contract 003903), PREDICATE (EVK2-CT-1999-00020), and NOCES (EVK2-2001-00115). The model development at NERSC has in addition received support from the G. C. Rieber Foundations, the *RegClim, NOClim,* and *ProClim* projects under the Research Council of Norway, and the *Vestnordisk Oceanclima* programme under the Nordic Council of Ministers. The ocean modelling at AWI has also been funded by the German Ministry for Research and Technology under the DEKLIM program (contract 01 LD 0047) and by the Deutsche Forschungsgemeinschaft via the SFB 512. Additional funds were made available through the Arctic Ocean Model Intercomparison Project (AOMIP). Dr. J.E.Ø. Nilsen (NERSC) provided valuable help with some of the figures and tables, and Mads Hvid Ribergaard provided the drifter data in Figure 1. Comments from Torben Schmidt and three anonymous reviewers are also highly acknowledged. This is publication No A 97 from the Bjerknes Centre for Climate Research.

REFERENCES

Aukrust, T. and J.M. Oberhuber, Modeling of the Greenland, Iceland and Norwegian Seas with a coupled sea ice-mixed layer-isopycnal ocean model. *J. Geophys. Res.* 100, 4771-4789, 1995.

Beckmann, A. and R. Döscher, A method for improved representation of dense water spreading over topography in geopotential-coordinate models. *J. Phys. Oceanogr.* 27, 581-591, 1997.

Beismann, J.-O. and B. Barnier, Variability of the meridional overturning circulation of the North Atlantic: sensitivity to overflows of dense water masses. *Ocean Dyn.* 54, 92-106, DOI 10.1007/s10236-003-0088-x, 2004.

Bennett, A.F., *Inverse methods in physical oceanography*. Cambridge University Press, Cambridge, 346 pp, 1992.

Bentsen, M., H. Drange, T. Furevik, and T. Zhou, Simulated variability of the Atlantic meridional overturning circulation, *Clim. Dynam.*, doi: 10.1007/s00382-004-0397-x, 2004.

Biastoch, A., R.H. Käse, and D.H. Stammer, The Sensitivity of the Greenland–Scotland Ridge Overflow to Forcing Changes. *J. Phys. Oceanogr.* 33, 2307-2319, doi: 10.1175/1520-0485(2003)033, 2003.

Bleck, R., An oceanic general circulation model framed in hybrid isopycnic-cartesian coordinates. *Ocean Modelling* 4, 55-88, 2002.

Bleck, R., C. Rooth, D. Hu, and L.T. Smith, Salinity-driven thermocline transients in a wind- and thermohaline-forced isopycnic coordinate model of the North Atlantic. *J. Phys. Oceanogr.* 22, 1486-1505, 1992.

Blumberg, A.F. and G.L. Mellor, A description of a three-dimensional coastal ocean circulation model, in *Three-Dimensional Coastal Ocean Models*, N. Heaps (Ed.), pp. 1-16, American Geophysical Union, Washington, DC, 1987.

Böning, C.W., C. Dieterich, B. Barnier, and Y. Jia, Seasonal cycle of meridional heat transport in the subtropical North Atlantic: a model intercomparison in relation to observations near 25°N. *Prog. Oceanogr.* 48, 231-253, 2001.

Bönisch, G., J. Blindheim, J.L. Bullister, P. Schlosser, and D.W.R. Wallace, Long-term trends of temperature, salinity, density, and transient tracers in the central Greenland Sea. *J. Geophys. Res.* 102, 18553-18571, 1997.

Bryan, K., A numerical model for the study of clirculation of the world oceans. *J. Comput. Phys.* 4, 347-376, 1969.

Budeus, G., W. Schneider, and G. Krause, Winter convective events and bottom water warming in the Greenland Sea. *J. Geophys. Res.* 103, 18513-18527, 1998.

Canuto, V. M., A. Howard, P. Hogan, Y. Cheng, M.S. Dubovikov, and LO.M. Montenegro, Modeling ocean deep convection. *Ocean Modelling* 7, 75-95, 2004.

Collins, M., M. Botzet, A. Carril, H. Drange, A. Jouzeau, M. Latif, O.H. Otterå, H. Pohlmann, A. Sorteberg, R. Sutton, and L. Terray, Interannual to decadal climate predictability: A multi-perfect-model-ensemble study. *J. Climate*, accepted, 2005.

Cox, M.D., *A primitive equation three-dimensional model of the ocean*. Report, GFDL Ocean Group, NOAA, Princeton Univ., Princeton, NJ, 1984.

Curry R., R.R. Dickson, and I. Yashayaev, A Change in the Freshwater Balance of the Atlantic Ocean over the Past four decades. *Nature* 426, 826-829, 2003.

Dickson, B., I. Yashayaev, J. Meincke, B. Turell, S. Dye, and J. Holfort, Rapid freshening of the deep North Atlantic Ocean over the past four decades. *Nature* 416, 832-837, 2002.

Döscher, R. and R. Redler, The relative importance of northern overflow and subpolar deep convection for the North Atlantic Thermohaline circulation. *J. Phys. Oceanogr.* 27, 1894-1902, 1997.

Drange, H and K. Simonsen, *Formulation of air-sea fluxes in the ESOP2 version of MICOM*. Technical Report 125, Nansen Environmental and Remote Sensing Center, Edv. Griegsv. 3A, N-5037 Solheimsviken, Norway, 1996.

Dutay, J.-C., *et al.*, Evaluation of ocean model ventilation with CFC-11: comparison of 13 global ocean models. *Ocean Modelling* 4, 89-120, 2002.

DYNAMO Group (Barnard, S., B. Barnier, A. Beckmann, C.W. Böning, M. Coulibaly, D. de Cuevas, J. Dengg, C. Dieterich, U. Ernst, P. Herrmann, Y. Jia, P.D. Killworth, J. Kröger, M.-M. Lee, C. LeProvost, J.-M. Molines, A. L. New, A. Oschlies, T. Reynaud, L.J. West, and J. Willebrand), *DYNAMO - Dynamics of North Atlantic Models. Simulation and assimilation with high resolution models*. Berichte aus dem Institut für Meereskunde an der Christian-Albrechts-Universität Kiel, 294, 334 pp. 1997.

Engedahl H, B. Ådlandsvik, and E.A. Martinsen, Production of monthly mean climatological archives for the Nordic Seas. *J. Mar. Syst.* 14, 1-26, 1998.

England, M. and E. Maier-Reimer, Using chemical tracers to assess ocean models. *Rev. Geophys.* 39, 29-70, 2001.

Evensen, G. The Ensemble Kalman Filter: theoretical formulation and practical implementation. *Ocean Dynamics* 53, 343-367, DOI 10.1007/s10236-003-0036-9, 2003.

Furevik, T., M. Bentsen, H. Drange, J. Johannessen, and A. Korablev, Temporal and spatial variability of the sea surface salinity in the Nordic Seas. *J. Geophys. Res.* 107, doi:10.1029/2001JC001118, 2002.

Gao Y., H. Drange, M. Bentsen, and O.M. Johannessen, Tracer-derived transient time of the eastern waters in the Nordic Seas. *Tellus*, in press, 2005.

Gao, Y., H. Drange, and M. Bentsen, Effects of diapycnal and isopycnal mixing on the ventilation of CFCs in the North Atlantic in an isopycnic coordinate OGCM. *Tellus* 55B, 837-384, 2003.

Gao, Y., H. Drange, M. Bentsen, and O.M. Johannessen, Simulating the transport of radionuclides in the North Atlantic-Arctic region. *J. Environ. Radioactivity* 71, 1-16, doi:10.1016/S0265-931X(03)00108-5, 2004.

Gaspar, P., Modeling the seasonal cycle of the upper ocean. *J. Phys. Oceanogr.* 18, 161-180, 1988.

Gerdes, R., A primitive equation ocean circulation model using a general vertical coordinate transformation, Part 2 application to an overflow problem. *J. Geophys. Res.* 98, 14703-14726, 1993.

Gerdes, R., M. Karcher, F. Kauker, and U. Schauer. Causes and development of repeated Arctic Ocean warming events. *Geophys. Res. Lett.* 30, doi: 10.1029/2003GL018080, 2003.

Ginis, I., R.A. Richardson, and L.M. Rothstein, Design of a multiply nested primitive equation ocean model. *Mon. Wea. Rev.* 126, 1054-1079, 1998.

Goosse H., R. Gerdes, F. Kauker, and C. Koeberle, Influence of the exchanges between the Atlantic and the Arctic on sea-ice volume variations during the period 1948-1997. *J. Climate* 17, 1294-1305, 2004.

Griffies, S.M., C. Böning, F.O. Bryan, E.P. Chassignet, R. Gerdes, H. Hasumi, A. Hirst, A.-M. Treguier, and D. Webb, Developments in ocean climate modelling. *Ocean Modelling* 2, 123-192, 2000.

Griffies, S.M., M.J. Harrison, R.C. Pacanowski, and A. Rosati, *A technical guide to MOM4*, NOAA/Geophysical Fluid Dynamics Laboratory, Princeton, USA, 337pp, 2004.

Haak, H., J. Jungclaus, U. Mikolajewicz, and M. Latif, Formation and propagation of great salinity anomalies. *Geophys. Res. Lett.* 30, DOI 10.1029/2003GL017065, 2003.

Häkkinen, S., Models and their applications in polar oceanography, in *Polar Oceanography* (Ed. W.O. Smith), Pergamon Press, 1991.

Häkkinen, S., Simulated interannual variability of the Greenland Sea deep water formation and its connection to surface forcing. *J. Geophys. Res.* 100, 4761-4770, 1995.

Häkkinen, S., Variability of the simulated meridional heat transport in the North Atlantic for the period 1951-1993. *J. Geophys. Res.* 104, 10991-11007, 1999.

Häkkinen, S. and G.L. Mellor, Modeling the seasonal variability of a coupled Arctic ice-ocean system. *J. Geophys. Res.* 97, 20,285-20,304, 1992.

Häkkinen, S., Surface salinity variability in the northern North Atlantic during recent decades. *J. Geophys. Res.* 107, 10.1029/2001JC000812, 2002.

Hallberg, R., Time integration of diapycnal diffusion and Richardson number-dependent mixing in isopycnal soordinate ocean models. *Mon. Wea. Rev.* 128, 1402-1419, 2000.

Hansen, B., and S. Østerhus, North Atlantic - Norwegian Sea Exchanges. *Prog. Oceanogr.* 45, 109-208, 2000.

Hansen, B., and S. Østerhus, D. Quadfasel, and W. Turrell, Already the day after tomorrow? *Science* 305, 953-954, DOI: 10.1126/science. 1100085, 2004.

Harder, M., *Dynamik, Rauhigkeit und Alter des Meereises in der Arktis.* PhD thesis Alfred-Wegener-Institut für Polar- und Meeresforschung, Bremerhaven, Germany. 124 pp, 1996.

Heburn, G.W. and C.D. Johnson, Simulations of the mesoscale circulation of the Greenbland-Iceland-Norwegian Seas. *J. Geophys. Res.* 100, 4921-4941, 1995.

Heggelund Y. and J. Berntsen, A method for analysing nesting techniques for the linearized shallow water equations. *Int. J. Numer. Meth. Fl.* 38, 163-185, 2002.

Hibler, III W.D., A Dynamic Thermodynamic Sea Ice Model. *J. Phys. Oceanogr.* 9, 815-846, 1979.

Holland, D.M., L.A. Mysak, and J.M. Oberhuber, Numerical simulation of the mixed-layer circulation in the Arctic Ocean. *J. Geophys. Res.* 101, 1111-1128, 1996.

Hopkins, T.S., The GIN Sea – a synthesis of its physical oceanography and literature review 1972-1985. *Earth-Sci. Rev.* 30, 175-318, 1991.

Ingvaldsen, R.B, L. Asplin, and H. Loeng, The seasonal cycle in the Atlantic transport to the Barents Sea during 1997-2001. *Cont. Shelf Res.* 24, 1015-1032, 2004.

Jacobsen, P.K., M.H. Ribergaard, D. Quadfasel, T. Schmith, and C.W. Hughes, Near surface circulation in the northern North Atlantic as inferred from Lagrangian drifters: variability from the mesoscale to interannual. *J. Geophys. Res.* 108, 3251, 2003.

Jensen, T.G., Open boundary conditions in stratified ocean models. *J. Mar. Sys.* 16, 297-322, 1998.

Johannessen O.M., E.V. Shalina, and M. Miles, Satellite evidence for an Arctic Sea Ice cover in transformation. *Science* 286, 1937-1939, 1999.

Kalnay, E., *et al.*, The NCEP/NCAR 40-year re-analysis project. *Bullet. Amer. Meteorol. Soc.* 77, 437-471, 1996.

Kampf, J., and J.O. Backhaus, Ice-ocean interactions during shallow convection under conditions of steady winds: three dimensional numerical studies. *Deep-Sea Res.* 46(6/7), 1335-1355, 1999.

Karcher, M.J., R. Gerdes, F. Kauker, and C. Köberle, Arctic warming: Evolution and spreading of the 1990s warm event in the Nordic Seas and the Arctic Ocean. *J. Geophys. Res.* 108, 3034, doi:10.1029/2001JC001265, 2003.

Karcher, M.J., S. Gerland, I.H. Harms, M. Iosjpe, GH.E. Heldal, P.J. Kershaw, and M. Sickel, The dispersion of ^{99}Tc in the Nordic Seas and the Arctic Ocean: A comparison of model results and observations. *J. Environm. Radioactivity* 74, 185-198, 2004.

Karcher, M.J. and J.M. Oberhuber, Pathways and modification of the upper and intermediate waters of the Arctic Ocean. *J. Geophys. Res.* 107, 10, 1029/2000JC000530, 2002.

Kauker, F., R. Gerdes, M. Karcher, C. Köberle, and J.L. Lieser, Variability of Arctic and North Atlantic sea ice: A combined analysis of model results and observations from 1978 to 2001. *J. Geophys. Res.* 108, 3182. doi: 10.1029/2002JC001573, 2003.

Killworth, P.D. and N.R. Edwards, A turbulent bottom boundary layer code for use in numerical ocean models. *J. Phys. Oceanogr.* 29, 1221-1238, 1999.

Killworth, P.D., Inclusion of the bottom boundary layer in ocean models, In: Muller, P. (Ed.) *Near-Boundary Processes and Their Parameterization*, Proceedings of the 13th 'Aha Huliko'a Hawaiian Winter Workshop, 21-24 Jan 2003, Honolulu HI, University of Hawaii SOEST, 177-185, 2003.

Köberle, C. and R. Gerdes, Mechanisms determining the variability of Arctic sea ice conditions and export. *J. Climate* 16, 2843-2858, 2003.

Kösters, F., R.H. Käse, A. Schmittner, and P. Herrmann, The effect of Denmark Strait Overflow on the Atlantic meridional overturning circulation. *Geophys. Res. Lett.*, in press, 2005.

Legutke, S., A numerical investigation of the circulation in the Greenland and Norwegian Seas. *J. Phys. Oceanogr.* 21, 118-148, 1991.

Marshall, J. and F. Schott, Open-ocean convection: observations, theory, and models. *Rev. Geophys.* 37, 1-64, 1999.

Martinsen, E.A. and H. Engedahl, Implementation and testing of a lateral boundary scheme as an open boundary condition in a barotropic ocean model. *Coastal Eng.* 11, 603-627, 1987.

Mauritzen, C. and S. Häkkinen, On the relationship between dense water formation and the "Meridional Overturning Cell" in the North Atlantic Ocean. *Deep-Sea Res.* 46, 877-894, 1999.

Mellor, G.L. and S. Häkkinen, A review of coupled ice-ocean models. In: *The Polar Oceans and their Role in Shaping the Global Environment*. Geophysical Monograph, 85, 525pp, 1994.

Mertz, G. and D.G. Wright, Interpretations of the jebar term. *J. Phys. Oceanogr.* 22, 301-305, 1992.

Müller, P. and J. Willebrand, Equations for oceanic motions. In *Landolt-Börnstein, Group V, Oceanography, Volume 3b*, pages 1.14. J. Sündermann (Ed.), Springer Verlag, Berlin, 1989.

New, A.L., S. Barnard, P. Herrmann, and J.M. Molines, On the origin and pathway of the saline inflow to the Nordic Seas: insigths from models. *Prog. Oceanogr* 48, 123-161, 2001.

Nies, H., I.H. Harms, M.J. Karcher, D. Dethleff, C. Bahe, G. Kuhlman, E. Kleine, P. Loewe, J.M. Oberhuber, J.O. Backhaus, D. Matishov, A. Stepanov, and O.F. Vasiliev, Anthropogenic radioactivity in the Nordic Seas and the Arctic Ocean: Results from a joint German project. *Deutsche Hydrographische Zeitung*, (German Journal of Hydrography) 50, 313-343, 1998.

Nilsen, J.E.Ø., Y. Gao, H. Drange, T. Furevik, and M. Bentsen, Simulated North Atlantic-Nordic Seas water mass exchanges in an isopycnic coordinate OGCM. *Geophys. Res. Lett.* 30, doi: 10.1029/2002GL016597, 2003.

Oberhuber, J.M., Simulation of the Atlantic circulation with a coupled sea-ice mixed layer-isopycnal general circulation model, Part I: Model description. *J. Phys. Oceanogr.* 23, 808-829, 1993.

Olsson, K.A., E. Jeansson, L.G. Anderson, B. Hansen, T. Eldevik, R. Kristiansen, M.-J. Messias, T. Johannessen, and A.J. Watson, Intermediate water from the Greenland Sea in the Faroe Bank Channel: spreading of released sulphur hexafluoride. *Deep Sea Res. I* 52, 279-294, 2005.

Østerhus, S., W.R. Turrell, S. Jónsson, and B. Hansen, Measured volume, heat, and salt fluxes from the Atlantic to the Arctic Mediterranean. *Geophys. Res. Lett.* 32, L07603, doi:10.1029/2004GL022188, 2005.

Palma, E.D. and R.P. Matano, On the implementation of open boundary conditions for a general circulation model: the three-dimensional case. *J. Geophys. Res.* 105, 8605-8627, 2000.

Paluskiewicz, T. and R.D. Romea, A one-dimensional model for the parameterization of deep convection in the ocean. *Dyn. Atm. Ocean* 26, 95-130, 1997.

Perkins, A.L., L.F. Smedstad, D.W. Blake, and G.W. Heburn, A new nested boundary condition for a primitive equation ocean model. *J. Geophys. Res.* 102, 3483-3500, 1997.

Proshutinsky, A., M. Steele, J. Zhang, G. Holloway, N. Steiner, S. Häkkinen, D. Holland, R. Gerdes, C. Koeberle, M. Karcher, M. Johnson, W. Maslowski, W. Walczowski, W. Hibler, and J. Wang, 2001. Multinational Effort Studies Differences Among Arctic Ocean Models. *EOS Transactions,* American Geophysical Union, 82, 637, 643, 644.

Roberts, M.J. and R.A. Wood, Topographic Sensitivity Studies with a Bryan–Cox-Type Ocean Model. *J. Phys. Oceanogr.* 27, 823-836, 1997.

Röske, F., An atlas of surface fluxes based on the ECMWF Re-analysis: A climatological dataset to force global ocean general circulation models. *Max-Planck-Institut fur Meteorologie Report* 323, 31 pp, 2001.

Sarkisyan, A.S. and V.F. Ivanov, Joint effect of baroclinicity and borrom relief as an important factor in the dynamics of sea currents. *Izv. Acad. Sci. USSR, Atmos. Oceanic Phys.* 7, 173-178, 1971.

Schauer, U., E. Fahrbach, S. Østerhus, and G. Rohardt, Arctic warming through the Fram Strait - Oceanic heat transport from three years of measurements. *J. Geophys. Res.* 109, C06026, DOI: 10.1029/2003JC001823, 2004.

Schlosser, P., G. Bönisch, M. Rhein, and R. Bayer, Reduction of deepwater formation in the Greenland Sea during the 1980s: Evidence from tracer data. *Science* 251, 1054-1056, 1991.

Semtner, A.J. Jr., Numerical simulation of the Arctic Ocean circulation. *J. Phys. Oceanogr.* 6, 409-425, 1976a.

Semtner, A.J. Jr., A numerical model for the thermodynamic growth of sea ice in numerical investigations of climate. *J. Phys. Oceanogr.* 6, 379-389, 1976b.

Semtner, A.J. Jr., Modeling ocean circulation. *Science* 269, 1379-1385, 1995.

Shi, X.B., L.P. Rød, and B. Hackett, Variability of the Denmark Strait overflow: A numerical study. *J. Geophys. Res.* 106, 22277-22294, 2001.

Simonsen, K. and P.M. Haugan, Heat budgets of the Arctic Mediterranean and sea surface heat flux parametrizations for the Nordic Seas. *J. Geophys. Res.* 101, 6553-6576, 1996.

Stammer, D., C. Wunsch, R. Giering, C. Eckert, P. Haimbach, J. Marotzke, A. Adcroft, C.N. Hill, and J. Marshall, Volume, heat and freshwater of the global ocean circulation 1993-2000, estimated from a general circul;ation model constrained by World Ocean Circulation Experiment (WOCE) data. *J. Geophys. Res.* 108, 3007, doi:10.1029/2001JC001115, 2003a.

Stammer, D., C. Wunsch, R. Giering, C. Eckert, P. Haimbach, J. Marotzke, A. Adcroft, C.N. Hill, and J. Marshall, Global ocean circulation during 1992-1997, estimated from ocean observations and a general circulation model. *J. Geophys. Res.* 108, 3118, doi:10.1029/2001JC000888, 2003b.

Steele, M., W. Ermold, G. Holloway, S. Häkkinen, D.M. Holland, M. Karcher, F. Kauker, W. Maslowski, N. Steiner, and J. Zhang, Adrift in the Beaufort Gyre: A model intercomparison. *Geophys. Res. Lett* 28, 2935-2838, 2001.

Steiner, N., G. Holloway, R. Gerdes, S. Häkkinen, D.M. Holland, M. Karcher, F. Kauker, W. Maslowski, A. Proshutinsky, M. Steele, and J. Zhang, Comparing modeled streamfunction, heat and freshwater content in the Arctic Ocean. *Ocean Modelling* 6, 265-284, doi: doi:10.1016/S1463-5003(03)00013-1, 2004.

Stevens, D.P., A numerical ocean circulation model of the Norwegian and Greenland Seas. *Prog. Oceanog.* 27, 365-402, 1991.

Toggweiler, J., K. Dixon, and K. Bryan, Simulations of radiocarbon in a coarse-resolution world ocean model, 1, Steady state prebomb distributions. *J. Geophys. Res.* 94, 8217-8242, 1989.

Uotila, P., D.M. Holland, M.A. Morales Maqueda, S. Häkkinen, G. Holloway, M. Karcher, F. Kauker, M. Steele, N. Yakovlev, J. Zhang, and A., Proshutinsky, An energy-diagnostics intercomparison of coupled ice-ocean Arctic models. *Ocean Modelling*, doi:10.1016/j.ocemod.2004.11.003, in press, 2005.

Watson, A.J., M.-J. Messias, E. Fogelquist, K.V. Scoy, T. Johannessen, F. Rey, T. Tanhua, F. Carse, K. Simonsen, J. Ledwell, E. Jansen, D. Cooper, J. Kruepke, and E. Guilyardi, Mixing in the Greeland Sea gyre studied by tracer release. *Nature* 401, 901-904, 1999.

Willebrand, J., B. Barnier, C. Boning, C. Dieterich, P.D. Killworth, C. le Provst, Y. Jia, J.-M. Molines, and A.L. New, Ciculation characteristics in three eddy-permitting models of the North Atlantic. *Prog. Oceanogr.* 48, 123-161, 2001.

M. Bentsen, H. Drange, and Y. Gao, Nansen Environmental and Remote Sensing Center and Bjerknes Centre for Climate Research, Thormhølensgt. 47, 5006 Bergen, Norway (mats.bentsen@nersc.no; helge.drange@nersc.no; yongqi.gao@nersc.no)

R. Gerdes, M. Karcher, and F. Kauker, Alfred Wegener Institute for Polar and Marine Research, Bussestrasse 24, 27570 Bremerhaven, Germany (rgerdes@awi-bremerhaven.de; mkarcher@awi-bremerhaven.de; fkauker@awi-bremerhaven.de)

Simulated History of Convection in the Greenland and Labrador Seas, 1948–2001

Rüdiger Gerdes, Jörg Hurka, Michael Karcher, Frank Kauker, and Cornelia Köberle

Alfred Wegener Institute for Polar and Marine Research, Bremerhaven, Germany

Convection in the Greenland and Labrador seas is compared using a sea-ice-ocean model forced with NCEP reanalysis atmospheric data for the period 1948-2001. Model-derived convection rates for the Greenland Sea and the Labrador Sea show good agreement with previous estimates. Composites based on convection indices are used to identify important forcing processes and the relationship to oceanic, atmospheric, and sea-ice fields. Convection in the Labrador Sea is dominated by large-scale atmospheric forcing, especially by the heat fluxes associated with the North Atlantic Oscillation (NAO). On the other hand, we find no robust correlation between Greenland Sea convection and the NAO. In the model, local sea-ice formation, wind direction, and associated sea-ice drift are important parameters affecting convection in the Greenland Sea.

1. INTRODUCTION

Open ocean convection in the Greenland Sea and the Labrador Sea is thought to be a major source of deep water production in the subpolar Atlantic. The intensity of convection at these sites varies on interannual and decadal time scales [e.g. *Dickson et al.*, 1996]. *Dickson et al.* state that the intensity of convection and the associated rate of deep-water formation in the Greenland Sea and Labrador Sea are driven by the atmospheric variability described by the North Atlantic Oscillation (NAO). The negative phase of the NAO, with low air temperatures and large ice extent over the Greenland Sea, favors convective activity in the Greenland Sea, whereas the positive NAO phase, characterized by low air temperatures and strong northwesterly winds over the Labrador Sea, leads to strong convection in the Labrador Sea. For an overview of NAO-related variability in the North Atlantic see *Visbeck et al.* [2003] and references therein.

The temperature evolution of Greenland Sea Deep Water (GSDW) suggests that convective renewal of deep waters in the Greenland Sea occurred at high rates from the late 1950s to the early 1970s and at significantly reduced rates from the mid-1970s onward. Rate estimates based on tracer data indicate an 80-90% reduction of deep-water formation in the Greenland Sea starting at the end of the 1970s [*Schlosser et al.*, 1991; *Rhein*, 1991; *Bönisch and Schlosser*, 1995; *Rhein*, 1996]. *Bönisch and Schlosser* [1995] estimate an annual volume transport rate of 0.47 Sv from convection into depths greater than 1500 m for the period 1965-1979 and a reduced transport rate of 0.1 Sv for the period 1980-1989. Direct observational data collected during the past decade showed convection depths typically shallower than 1500 m, except in 1989, where a depth of 1800 m was inferred from tomography [*Morawitz et al.*, 1996]. Recently, *Gascard et al.* [2002] and *Wadhams et al.* [2002] reported the discovery of long-lived small-scale convective chimneys in the Greenland Sea, with radii of on the order of 10 km and convection depths in excess of 2000 m. However, whether these small-scale chimneys give a significant contribution to the formation of deep water in the Greenland Sea remains an open question.

For the Labrador Sea, the time series of potential temperature and salinity at Ocean Weather Ship *Bravo* indicates that convection was shallow during the 1960s but deepened considerably during the early 1970s, the mid-1980s,

The Nordic Seas: An Integrated Perspective
Geophysical Monograph Series 158

10.1029/158GM15

and especially during the late 1980s and early 1990s. Estimates of the annual formation rate of Labrador Sea Water (LSW) attributable to open ocean convection were given by several authors and range from 2.0 to 12.7 Sv (see Table 2 later). The highest rates are consistently estimated for the early 1990s. LSW can be traced by its low potential vorticity, its relatively low salinity, and its recent ventilation, which is manifest in several tracers. LSW is found throughout the subpolar North Atlantic as well as along the western boundary south to 20°N [*Talley and McCartney*, 1982]. Variability in the LSW formation rate thus has large-scale implications for the North Atlantic. The signal of intensified LSW formation in the late 1980s and early 1990s arrived at Bermuda as a reduction in Upper North Atlantic Deep Water temperatures and salinities in 1996 [*Curry et al.*, 1998]. This was accompanied by an increase in CFC concentrations in the Deep Western Boundary Current at 26°N [*Molinari et al.*, 1998]. Recently, *Pickart et al.* [2003] pointed to the Irminger Sea as another significant source of water in the range of LSW properties.

Here, we analyze the history of open ocean convection in the Greenland and Labrador Seas from the output of a coupled sea-ice-ocean model of the North Atlantic and the Arctic Ocean forced with NCEP/NCAR reanalysis atmospheric data for the period 1948–2001. After a short description of the model setup, we present the temporal development of convection in the Greenland and Labrador Seas over the past 50 years as simulated by the model. We then evaluate the main aspects of the atmospheric forcing and the role of sea-ice in deep convection in both regions.

2. MODEL DESCRIPTION AND EXPERIMENTAL SETUP

Sea-ice and ocean variables were simulated with a coupled ocean-sea-ice model of the North Atlantic/Arctic Ocean-Sea Ice Model (NAOSIM) hierarchy developed at the Alfred Wegener Institute [*Gerdes et al.*, 2001; *Karcher et al.*, 2003; *Kauker et al.*, 2003, *Köberle and Gerdes*, 2003]. The ocean circulation model derives from the Geophysical Fluid Dynamics Laboratory modular ocean model [MOM-2; *Pacanowski*, 1995]. The model domain encloses the Atlantic north of approximately 50°N, the Nordic Seas, and the Arctic Ocean. At the southern boundary an open boundary condition has been implemented, following *Stevens* [1991], allowing the outflow of tracers and the radiation of waves. At inflow points, temperature and salinity are restored with a time constant of 180 days towards an annual mean climatology [*Levitus and Boyer*, 1994; *Levitus et al.*, 1994]. The baroclinic part of the horizontal velocity is calculated from a simplified momentum balance while the barotropic velocities normal to the boundary are specified from a lower-resolution version of the model that covers the entire North Atlantic [*Köberle and Gerdes*, 2003]. The Bering Strait, the Kattegat, and the Hudson Strait are treated as closed walls.

The model is formulated on a rotated spherical grid with a horizontal resolution of 0.25° and 30 unevenly spaced levels in the vertical. The rotation of the grid shifts the equator to the geographical 30°W meridian and the Pole to 60°E on the geographical equator. The bottom topography is based on the Etopo5 data set of the National Geophysical Data Center. Modifications were made to open two channels in the Canadian Archipelago connecting the Arctic Ocean with Baffin Bay. The time step of the model is 900 s.

For the advection of potential temperature, salinity, and other tracers, a flux-corrected transport scheme [*Zalesak*, 1979; *Gerdes et al.*, 1991] is employed, which is characterized by a low implicit diffusion while avoiding false extrema ("overshooting") in advected quantities. There is no explicit diffusion acting on the tracers. Horizontal friction is implemented as a biharmonic diffusion of momentum with a diffusion coefficient of -0.5×10^{13} m^4s^{-1}. Vertical viscosity is Laplacian with a friction coefficient of 10^{-3} m^2s^{-1}. Quadratic bottom friction is included with a drag coefficient of 1.2×10^{-3}. Convective adjustment is used to remove static instabilities from the water column.

A dynamic thermodynamic sea-ice model with viscous-plastic rheology [*Hibler*, 1979] is coupled to the ocean model. The prognostic variables of the sea-ice model are ice thickness, snow thickness, ice age, and ice concentration. The momentum equation is assumed to be in quasi-stationary balance such that ice drift is calculated diagnostically. Snow and ice thickness are mean quantities over a grid box. The thermodynamic evolution of the ice is described by an energy balance of the ocean mixed layer, following *Parkinson and Washington* [1979]. Freezing and melting are calculated by solving the energy budget equation for a single ice layer with a snow layer. The freezing point of sea water is salinity dependent. The sea-ice model is formulated on the ocean model grid and uses the same time step. Flow of ice out of the domain is allowed at the southern boundary and at the Bering Strait. The models are coupled according to the procedure devised by *Hibler and Bryan* [1987]. The surface heat flux is calculated from standard bulk formulae using prescribed atmospheric data and sea-surface temperature predicted by the ocean model. Further details on the present implementation of the sea-ice model can be found in *Hilmer* [2001].

The initial conditions for potential temperature and salinity were taken from the Arctic Ocean Environmental Working Group (EWG) climatology for winter [*NSIDC*, 1997]. Outside of the EWG climatology domain, the climatology of *Levitus and Boyer* [1994] and *Levitus et al.* [1994] have been used. The model is forced with daily mean air temperature at 2 m, dew point temperature, cloudiness, precipitation, wind speed, and surface wind stress. For the first 50 years of spin-up, a climatological mean seasonal cycle based on the period 1979-1993 with added typical daily variability [OMIP climatology; *Röske*, 2001] is used. After the spin-up, the forcing consists of daily mean atmospheric data from the NCEP reanalysis for the period 1948-2001 [*Kalnay*, 1996]. Fresh water influx from rivers is not explicitly included. To account for river runoff and diffuse runoff from the land, as well as to include the effect of flow into the Arctic through the Bering Strait, a restoring flux with an adjustment time scale of 180 days is added to the surface freshwater flux. The restoring flux is calculated in reference to climatological surface salinity data, constructed similar to the initial data. The effect of the restoring flux on the surface salinity for this and other Arctic Ocean models is documented in *Steele et al.* [2001].

3. SPACE AND TIME DISTRIBUTION OF DEEP CONVECTION

The analysis presented here is based on the monthly averaged fields of the modeled sea-ice and ocean quantities. The convection depth is diagnosed in the model as the depth where the potential density exceeds the surface density by a small amount, $\Delta\sigma_{\Theta} = 0.02$ kg m^{-3}. Convection depth is interpolated between level densities. For the hindcast experiment analyzed here, deep-reaching convection in the Greenland and Labrador Seas occurred in the winter months January, February, March, and April (JMFA).

3.1. Spatial Distribution of Deep-Reaching Convection

Inspection of the spatial distribution of deep-reaching convection for the winter months during the NCEP period 1948-2001 revealed a large variability from year to year as well as on longer time scales. Two examples of extremely deep-reaching mixed layers in the model are shown in Figure 1. During the 1950s and 1960s, the Greenland Sea stands out with extreme depths of convection. Deep convection is localized to the interior of the Greenland Sea. Convective mixing in the remainder of the Nordic Seas is associated with the Norwegian Atlantic Current, but mixed layer depths there are typically less than 1000 m. Compared with that of the Greenland Sea, Labrador Sea deep convection occupies a relatively large area. The spatial structure is often shaped by the ice edge and melt-water intrusions into the interior of the Labrador Sea (Figure 1). Convection in the Irminger Sea seldom reaches below 1000 m deep in the simulation.

3.2. Time History of Convective Activity

Indices for the convectively mixed volume are used to describe the temporal development of convection in the Greenland and Labrador Seas. The indices are constructed from the volume of the convectively mixed water column in regions where the mixed layer depth (as defined above) reaches below 1500 m during the winter (January-April). *Marshall and Schott* [1999] refer to such a region as a deep mixed patch. The deep mixed-patch volume was computed

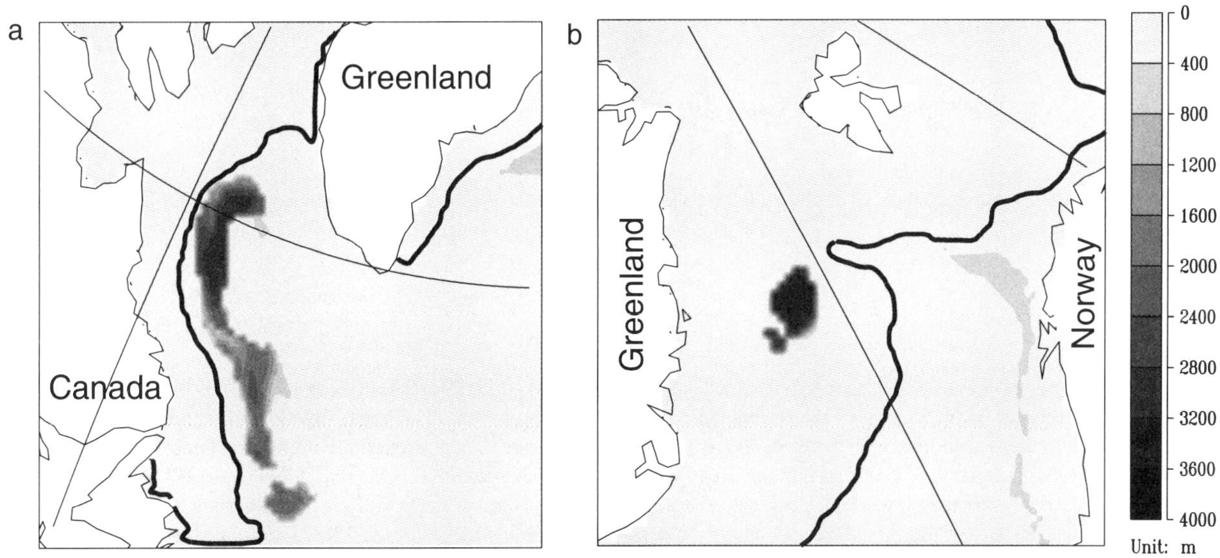

Figure 1. Mixed layer depth in the Labrador Sea in March 1992 (a). Mixed layer depth in the Greenland Sea in February 1963, a year with high convective activity in the Greenland Sea (b). The heavy line indicates the respective position of the ice edge as defined by the monthly mean 15% sea ice concentration line.

from a map of maximal winter convection depth, generated by taking the maximum over the JFMA monthly maps of convection depth. Finally, the volume is converted into an annual rate (Figure 2). Deep-water formation rates derived from deep mixed-patch volumes can overestimate the actual annual transformation rate because one cannot assume that the whole deep-water volume is flushed out each year. Subsequent winter convection thus acts on previously formed deep water [*Marshall and Schott*, 1999]. On the other hand, water can be exchanged and replenished during the time of convection, increasing the amount of water formed compared to the deep mixed-patch volume changes.

During the early 1950s, the 1960s, and the early 1970s, we find high convective activity in the Greenland Sea. From the mid-1970s to the mid-1990s, the model results indicate an almost complete shutdown of deep-reaching convection in the Greenland Sea. Only from the second half of the 1990s onwards do we see a recurrence of convection in the Greenland Sea that goes below 1500 m deep in the model. In the case of the Labrador Sea, deep convection is basically absent before 1970. Deep convective activity in the Labrador Sea clusters around the early 1970s, the early 1980s, and especially during the first half of the 1990s. The timing of deep convection in the model is consistent with what has been inferred about the convective history from observations [e.g., *Dickson et al.*, 1996].

The 1948-2001 average of the convection index for the Greenland Sea amounts to 0.98 Sv. For comparison with the Greenland Sea Deep Water (below 1500 m) formation rates determined by *Bönisch and Schlosser* [1995], we also calculate the volume of water convectively mixed below 1500 m (instead of the whole volume, as is used for the convection index). For the periods 1965-1979 and 1980-1989, we find average yearly rates of 0.69 Sv and 0.06 Sv, respectively—values in good agreement with the rates of 0.47 Sv and 0.1 Sv given by *Bönisch and Schlosser* [1995] for the same periods (Table 1). For the relationship between Greenland Sea convection, water mass spreading in the Nordic Seas, and the overflows into the subpolar North Atlantic, see *Eldevik et al.* (this volume) and Blindheim and Østerhus (this volume).

The 1948-2001 average of the convection index for the Labrador Sea amounts to 3.05 Sv. In Table 2 we compare LSW formation rates from the literature with the convection index from the model results. Note that only the rates of *Clarke and Gascard* [1983] and *Marshall and Schott* [1999] can be directly compared with our model rates since both sets of findings were derived from deep mixed-patch volumes, as our model rates were. Keeping the caveats mentioned above in mind, the model rates overall are in reasonably good agreement with other estimates found in the literature.

Dickson et al. [1996] described an NAO-driven seesaw between Greenland Sea and Labrador Sea convection. Linear

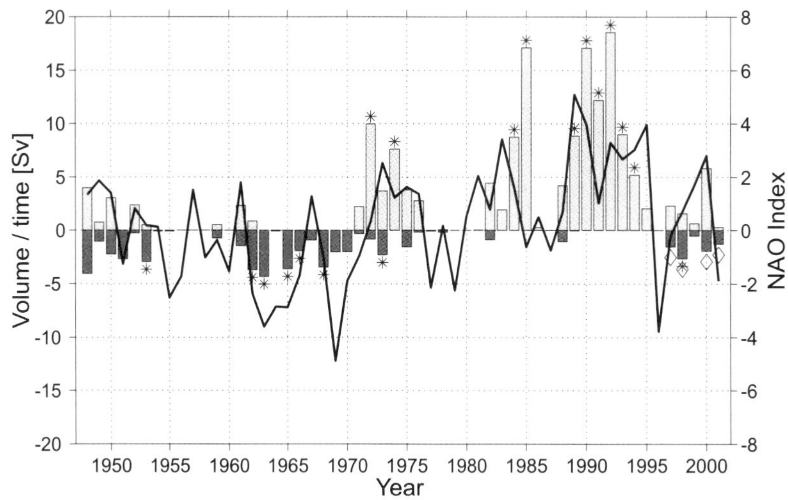

Figure 2. Indices for convective mixing in the Greenland Sea (dark gray bars, plotted as negative values), Labrador Sea (light gray bars), and the NAO Index (solid line; updated from *Hurrel* [1995]). To compute the convection index, we first generate a map of maximum winter convection depth; i.e., in this map the convection depth at a given grid point is the maximum of the JFMA monthly mean convection depth values. For each horizontal grid cell with a maximum convection depth deeper than a certain value, here 1500 m, the horizontal area of the box is multiplied by the convection depth, thus yielding the volume of the mixed water column. Finally, grid cells in the vicinity of the Greenland and Labrador Seas are added and transformed to an annual rate to give the corresponding yearly index values. Stars and diamonds denote those years that are used to compute composites for strong convection in the Greenland Sea and the Labrador Sea, respectively. Composites for weak convection are derived from those years where the respective convection index vanishes.

Table 1. Estimates of Greenland Sea Deep-Water Formation Rates from the Literature Compared With Model-Derived Rates.

	Period	Rate (Sv)	Model Rate[1] (Sv)	Model Rate[2] (Sv)
Bönisch and Schlosser (1995)[3]	1965-1979	0.47 ± 0.14	0.69	1.3
	1980-1989	0.1 ± 0.03	0.06	0.20
Rhein (1991, 1996)[3]	1944-1982 (1948-1982)	0.8 – 1.2	0.62	1.26
	1983-1989	0.1 – 0.38	0.05	0.15
	1989-1993	> 0.14	0.001	0.01
Marowitz *et al.* (1996)[4]	1989	0.12 ± 0.03	0.006	0.04

The following methods were used to derive rate estimates: [1]deep mixed-patch volume below 1500 m, [2]total volume of deep mixed patch, [3]tracer data, [4]tomography.

correlation coefficients between Greenland Sea and Labrador Sea convection indices and the NAO Index amount to $r_G = -0.28$ and $r_L = 0.50$. Furthermore, since we expect a nonlinear dependence of the convective activity on the NAO, we calculated the Spearman rank-ordered correlation coefficients [e.g., *Press et al.*, 1992]—which are better suited to measure a nonlinear correlation—between the convection indices and the NAO and found $r_{sG} = -0.25$ and $r_{sL} = 0.68$. Assuming 54 degrees of freedom, the probability that these correlations are random is 7% for r_{sG} and 10^{-6}% for r_{sL}. The different rank correlations are already apparent from Figure 2. For years with a positive NAO Index, in only four (1957, 1967, 1980, and 1981) did we find no deep convection in the Labrador Sea. Most of these years were probably affected by the influx of anomalous amounts of freshwater at the surface [*Belkin*, 1998]. On the other hand, out of the negative-NAO winters, 10 had no or little deep convection in the Greenland Sea. Thus, Labrador Sea deep convection is much more strongly linked to its phase of the NAO than is the deep convection in the Greenland Sea.

These results indicate that convection in the Labrador Sea is strongly correlated with the NAO, while the anti-correlation for the Greenland Sea convection is considerably weaker (by a factor of 2-3) and not robust. The anti-correlation is mainly due to the period of strong convection and low NAO indices during the 1960s. Deep Greenland Sea convection occurs during positive NAO winters in the 1950s, in the early 1970s, and especially in the late 1990s. Note, however, that atmospheric variability over the central Greenland Sea is not as strongly correlated with the NAO as the East Greenland coast and the Norwegian Sea. Convection in the Labrador Sea is strongly influenced by the large-scale atmospheric conditions, which are dominated by the NAO, while other factors apparently strongly influence the convection in the Greenland Sea. We shall show below that Greenland Sea convection is associated with local processes, in particular the local

Table 2. Estimates of LSW Formation Rates from the Literature Compared with Model-Derived Rates.

	Period	Rate (Sv)	Model Rate (Sv)
Wright (1972)[1]	Annual average (1948-2001)	3.5	3.1
Worthington (1976)[2]	Annual average (1948-2001)	2.0	3.1
Clarke and Gascard (1983)[3]	1976	3.9	2.8
McCartney and Talley (1984)[2]	Annual average (1948-2001)	8.6	3.1
McCartney (1992)[2]	Annual average (1948-2001)	7.0	3.1
Böning *et al.* (1996)[6]	Annual average (1948-2001)	3.5	3.1
Marshall and Schott (1999)[3]	Early 1990s (1990-1992)	12.7	15.9
Marsh (2000)[7]	1980-1997	3.4	6.2
Smethie and Fine (2001)[4]	1970-1990	7.4	4.4
Rhein *et al.* (2002)[4]	1988-1997	4.4–5.6	8.0
	1988-1994	8.1–10.8	10.7
	1995-1997	1.8–2.4	1.5
Khatiwala *et al.* (2002)[5]	1959-1998	2.7	3.7
Böning *et al.* (2003)[8]	1970-1997	4.3	5.1

The following methods were used to derive rate estimates: [1]heat budget, [2]water cooled below 4 °C, [3]observed volume of deep mixed patch, [4]CFC data, [5]surface fluxes (NCEP), [6]MOM-type model forced with a climatology, [7]observed heat and fresh water fluxes, [8]MOM-type model forced with NCEP reanalysis data. The period 1948-2001 denotes the model integration period and is used when estimates based on observations are meant to represent climatological formation rates.

production of sea-ice, which in turn depends on the local wind direction and associated sea-ice drift.

4. COMPOSITES FOR WEAK AND STRONG CONVECTION

To identify the conditions associated with deep convection, we constructed composites of atmospheric quantities, surface buoyancy fluxes, and ice conditions. In this paper we focus on the local surface fluxes and shall only briefly discuss the influence of preconditioning and lateral inflow in the following section. The years marked with stars in Figure 2 are used to construct winter-averaged (JFMA) composites for strong convection. Those years that show no deep convective mixing at all below 1500 m are used for weak convection composites. A second set of composites is constructed from the years marked by the diamonds in Figure 2 to analyze why deep convection in the Greenland Sea sets in again in the late 1990s, even though the large-scale atmospheric conditions then were so different from the previous period of deep convection in the Greenland Sea.

4.1. Sea-Level Pressure

The composites for sea-level pressure (SLP) (Figure 3) confirm the relation of the Labrador Sea convection with the NAO. The pressure pattern for strong Labrador Sea convection shows a strong Icelandic low that extends into the Labrador Sea

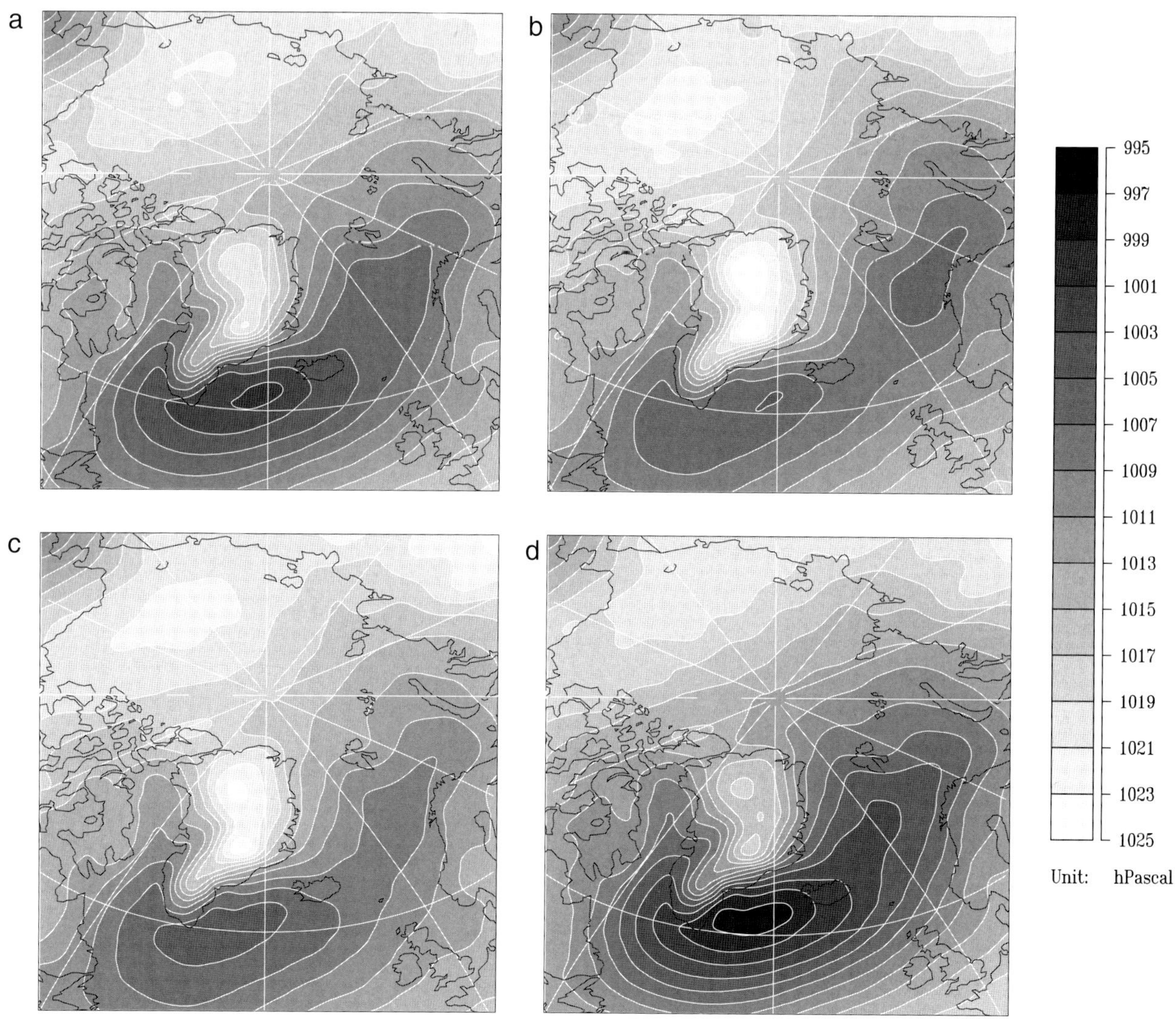

Figure 3. Composites of sea-level pressure. Weak Greenland Sea convection (a), strong Greenland Sea convection (b), weak Labrador Sea convection (c), and strong Labrador Sea convection (d).

and far north over the Nordic Seas into the Barents Sea (Figure 3d). The pressure pattern for weak Greenland Sea convection (Figure 3a) shows a similar but less pronounced pattern. The northerly winds over the Greenland Sea differ between the composites, the advection of cool air extending farther east in the strong convection composite than in the weaker one. Strong convection in the Greenland Sea and weak convection in the Labrador Sea again have similar patterns associated with them, consistent with the notion [*Dickson et al.,* 1996] of a seesaw between convection in the two regions. However, only the differences of the SLP patterns associated with Labrador Sea convection are significant (to 99%, tested with a Student's t-test); the Greenland Sea convection patterns are not significantly different from the mean SLP over the NCEP period.

4.2. Surface Air Temperature Anomalies

Strong Greenland Sea convection is associated with anomalously low air temperatures at 10 m over the Greenland Sea and anomalously high air temperatures over the Labrador Sea (Figure 4). In the weak convection case the situation is reversed, although with smaller signals. Similar air temperature distributions are found in the Labrador convection composites, here with a negative anomaly over the Labrador Sea and a positive anomaly over the Greenland Sea for strong convection. Again, in the low-convection case, the anomalies are less pronounced and show opposite signs. Similar surface air temperature anomalies as for the Labrador Sea convection composites were found by *Deser et al.* [2000] to be

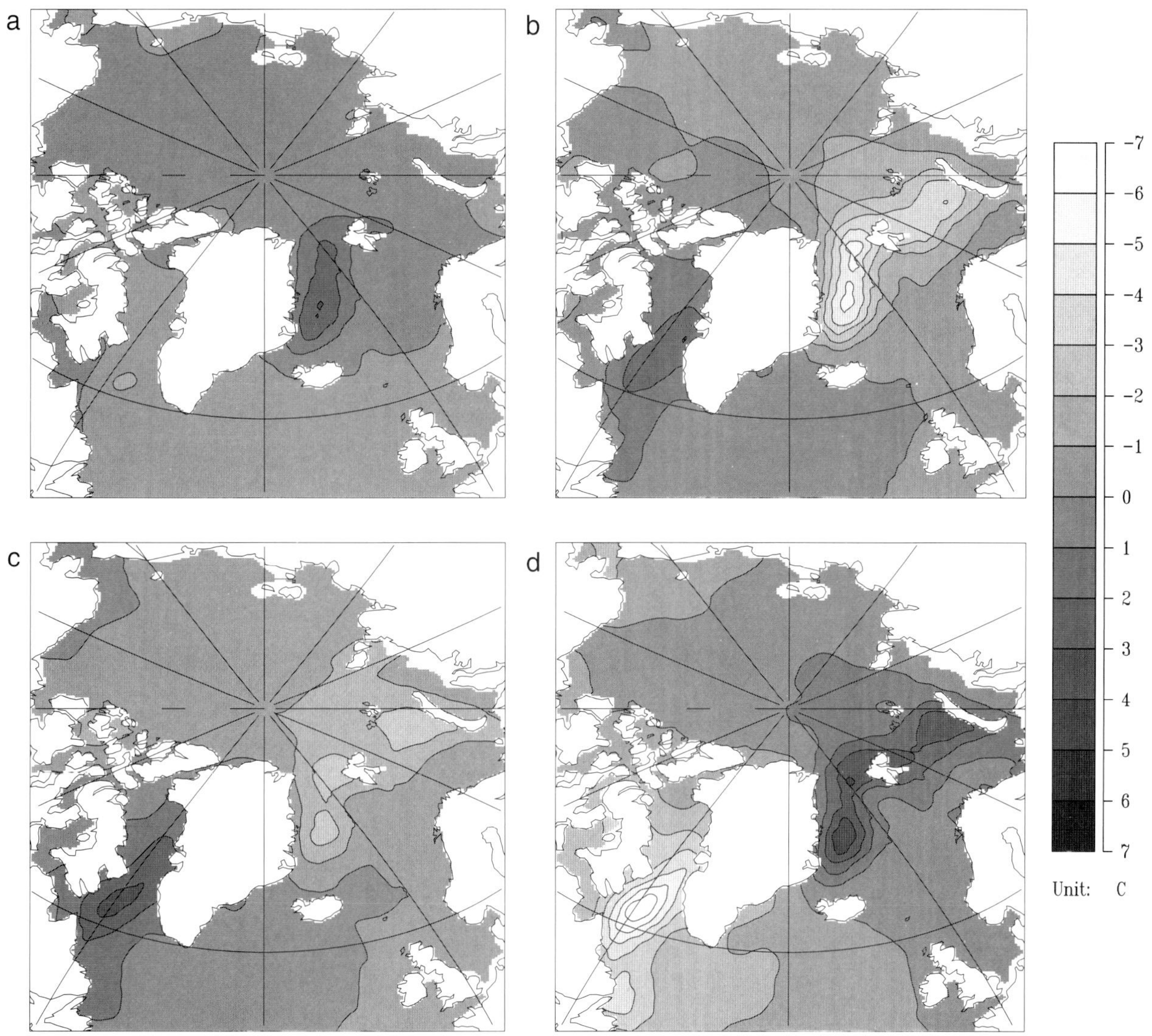

Figure 4. Composites of 10-m air temperature. Weak Greenland Sea convection (a), strong Greenland Sea convection (b), weak Labrador Sea convection (c), and strong Labrador Sea convection (d).

associated with the leading mode of winter sea-ice concentration anomalies for the Northern Hemisphere. This mode is in turn highly correlated with the NAO index time series.

4.3. Sea-Ice Concentration

The dominating feature of the sea-ice distribution in the Greenland Sea area is the Odden, a tongue of sea-ice extending from the East Greenland Current into the Greenland Sea in a northeastward direction. Between the Odden and the dense sea-ice cover of the East Greenland Current, an area of relatively low sea-ice cover, called Nord Bukta, often exists. In the composite for weak Greenland Sea convection, we find sea-ice concentrations below 0.1 in the Nord Bukta area, below 0.5 in the Odden, and over 0.95 over the east Greenland shelf (Figure 5). In the composite for strong convection, the sea-ice cover is larger, the maximum concentrations reaching up to 0.4 in the Nord Bukta and 0.75 in the Odden. Sea-ice cover in the Barents Sea at the same time is relatively small. Likewise, the sea-ice edge in the composite for strong Labrador Sea convection lies farther into the Labrador Sea than does that in the composite for weak convection. The larger sea-ice cover in the case of strong convection may seem surprising because the presence of sea-ice usually isolates the ocean against heat loss to the atmosphere. However,

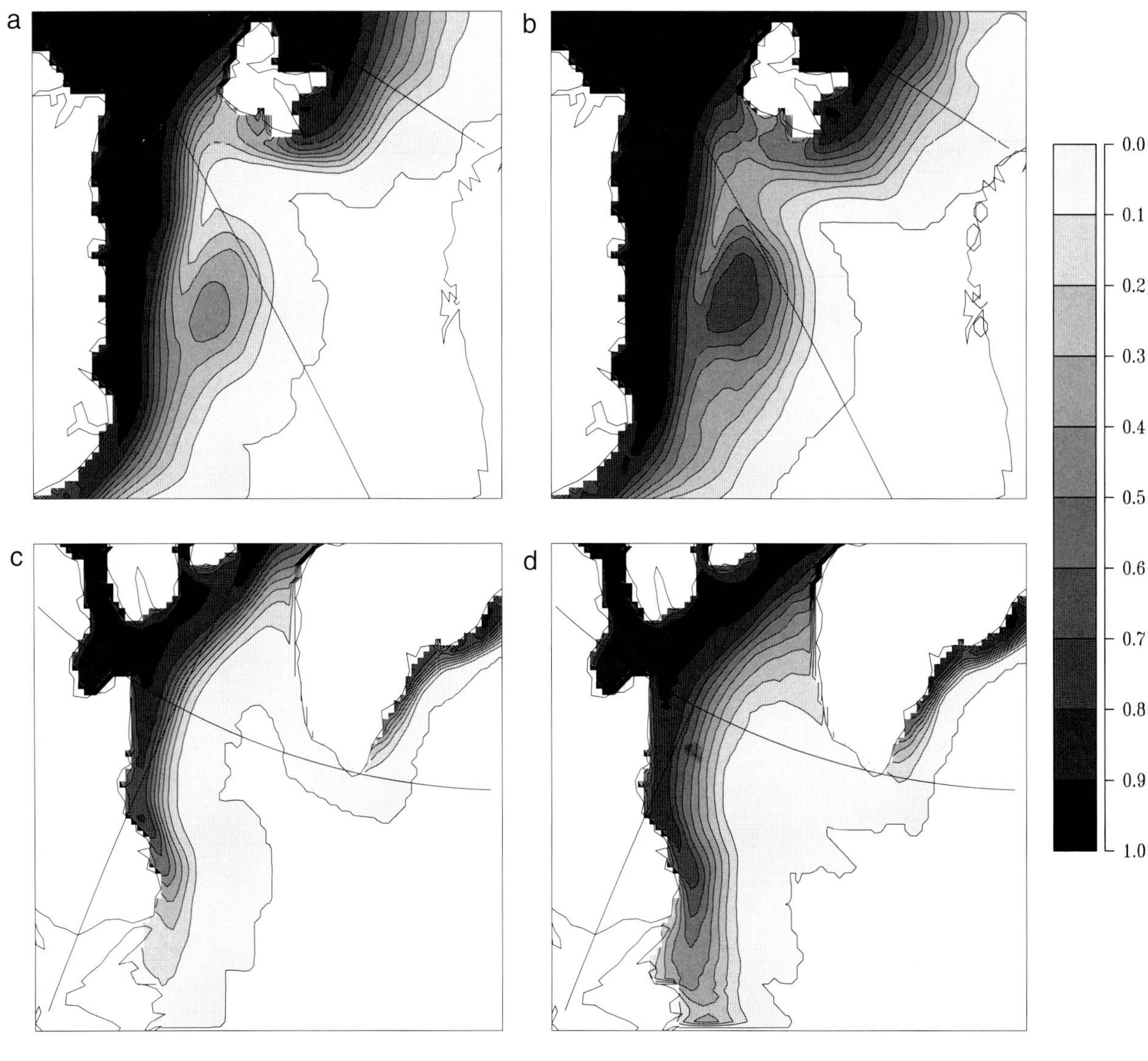

Figure 5. Composites of sea-ice concentration. Weak Greenland Sea convection (a), strong Greenland Sea convection (b), weak Labrador Sea convection (c), and strong Labrador Sea convection (d).

the sea-ice cover is not closed, the concentration is typically less than 0.3, and the heat loss in the area is large—even with the positive sea-ice concentration anomalies of the composite for strong Labrador Sea convection (see below). The combination of low surface air temperature, larger sea-ice concentration, and larger oceanic heat losses is consistent with the observational findings of *Deser et al.* [2000].

4.4. Buoyancy Flux Over the Greenland Sea

Plate 1 shows the surface buoyancy flux composites for weak and strong Greenland Sea convection. In addition to the total buoyancy flux, the contributions due to heat and salt fluxes are displayed,

$$B = \frac{g}{\rho_o}\left[\frac{\alpha}{c_w}Q - \beta\,(P + R - E - G_{th})\,S\right]$$

where Q is the total surface heat flux, G_{th} denotes the thermodynamic growth rate of sea-ice, and α and β are the thermal and saline expansion coefficients, respectively. Surface restoring of salinity is represented by R. The local surface salinity S is used to transform the surface fresh water flux $P + R - E - G_{th}$ into a salt flux. Strong convection in the Greenland Sea (Plate 1d-f) is associated with strong loss of buoyancy in the northern parts of the Greenland Sea. Buoyancy gain in the East Greenland Current remains relatively low, and the area of buoyancy gain adjacent to the Greenland Sea is narrow. During weak convection, the buoyancy loss in the northern Greenland Sea is weaker, and large surrounding areas show buoyancy gain. Heat flux and salt flux both contribute to the negative buoyancy flux in the northern Greenland Sea. There is only a small difference in the buoyancy flux component attributable to heat flux between the composites for weak and strong convection. However, the salt flux contribution in the vicinity of the Greenland Sea differs significantly between the composites (Plate 1b, e). The area of buoyancy loss is always relatively small and restricted to the northern Greenland Sea.

4.5. Thermodynamic Ice Growth and Ice Drift in the Greenland Sea

Plate 2 displays the thermodynamic growth of sea-ice, superimposed by the sea-ice velocity, for weak and strong Greenland Sea convection. In both composites, melting in the West Spitsbergen Current and along the outer edge of the East Greenland Current is large, locally exceeding 1.5 m over the winter months we have chosen for the composites. In the strong convection case, thermodynamic growth of sea-ice in the northern Greenland Sea occurs at higher rates than in the weak convection case. Melting rates are greater in the southern part, consistent with the patterns in the surface buoyancy flux. In fact, Plate 1 and Plate 2 make clear that the difference in the total buoyancy flux over the Greenland Sea between composites for strong and weak convection is caused by the higher thermodynamic growth rate of sea-ice in the strong convection case. In addition to the high ice production in the northern Greenland Sea, the sea-ice velocity indicates a greater transport of ice from the northern to the southern Greenland Sea in the composite for strong convection. This provides the important spatial separation of melting and freezing that enables the sea-ice processes to have a net effect on the ocean surface buoyancy flux.

4.6. Buoyancy Fluxes Over the Labrador Sea

Buoyancy flux composites for the Labrador Sea are given in Plate 3. As in the Greenland Sea, regions of buoyancy gain and buoyancy loss in the Labrador Sea are adjacent to each other, indicating that sea-ice edge processes are important in generating the surface flux pattern. However, unlike the Greenland Sea case, the buoyancy loss in the central Labrador Sea is solely due to heat loss, overcompensating the buoyancy gain caused by the melting of sea-ice and the corresponding positive fresh water flux. The prevailing winds during intense convection in the Labrador Sea push the sea-ice edge into the interior of the Labrador Sea. Although air temperatures are lower than normal, they are well above the freezing point, and the ice transported into the central Labrador Sea melts there. The role of sea-ice in the Labrador Sea is thus opposite that in the Greenland Sea. Extreme buoyancy losses due to enhanced freezing in case of strong Labrador Sea convection occur to the southwest of Baffin Island; however, the water there is always stably stratified due to the fresh water input from the Hudson Strait.

4.7. Wind Stress and Greenland Sea Convection in the Late 1990s

As indicated above, anomalously large net freezing in the northern Greenland Sea and melting in the southern part require a transport of sea-ice from the north to the south. The wind-stress anomalies of the composite for strong Greenland Sea convection reflect strong northerly wind stresses in the interior of the Greenland Sea (Figure 6). These wind-stress anomalies, in turn, lead to large southward sea-ice drift in the Greenland Sea (Plate 2). In the weak convection case, we see a more intense cyclonic atmospheric circulation over the whole Nordic Seas, consistent with the surface pressure field composite (Figure 3). This is associated with a pronounced sea-ice motion within the East Greenland Current along the Greenland coast but smaller sea-ice velocities farther out in the Greenland Sea.

There are only a few years between 1975 and 1995 with deep convection in the Greenland Sea: six years with

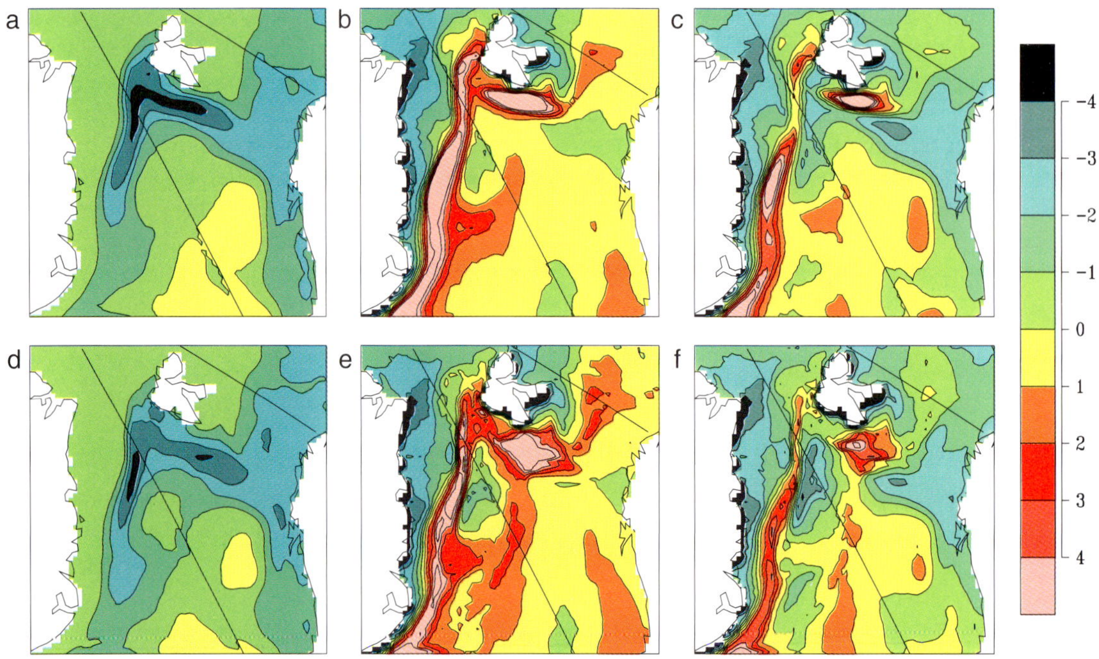

Plate 1. Composites of surface buoyancy flux for Greenland Sea convection in units of 10^{-8} m^2s^{-3}. The composite for weak convection is shown in the top panels (a-c), while the composite for strong Greenland Sea convection is shown below (d-f). Contribution from the surface heat flux (a, d); contribution from the surface salt flux (b, e); total buoyancy flux (c, f).

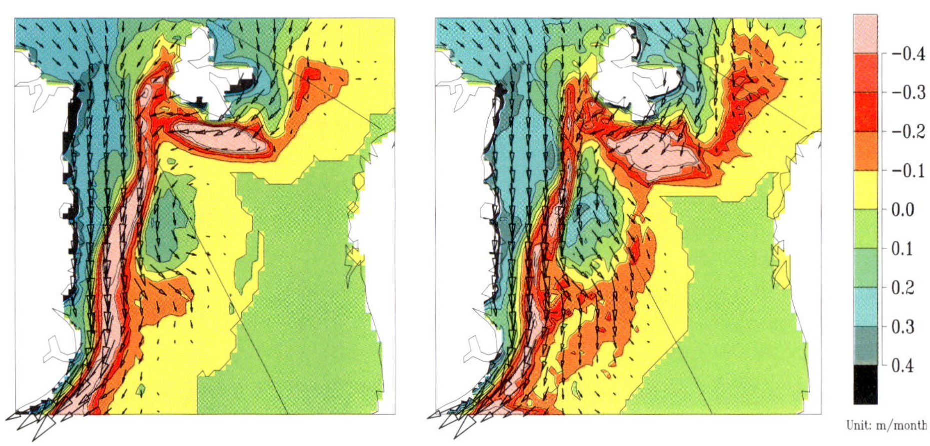

Plate 2. Composites of thermodynamic ice growth in the Greenland Sea overlaid with the corresponding composites for sea-ice velocity. Left: Weak Greenland Sea convection; right: strong convection.

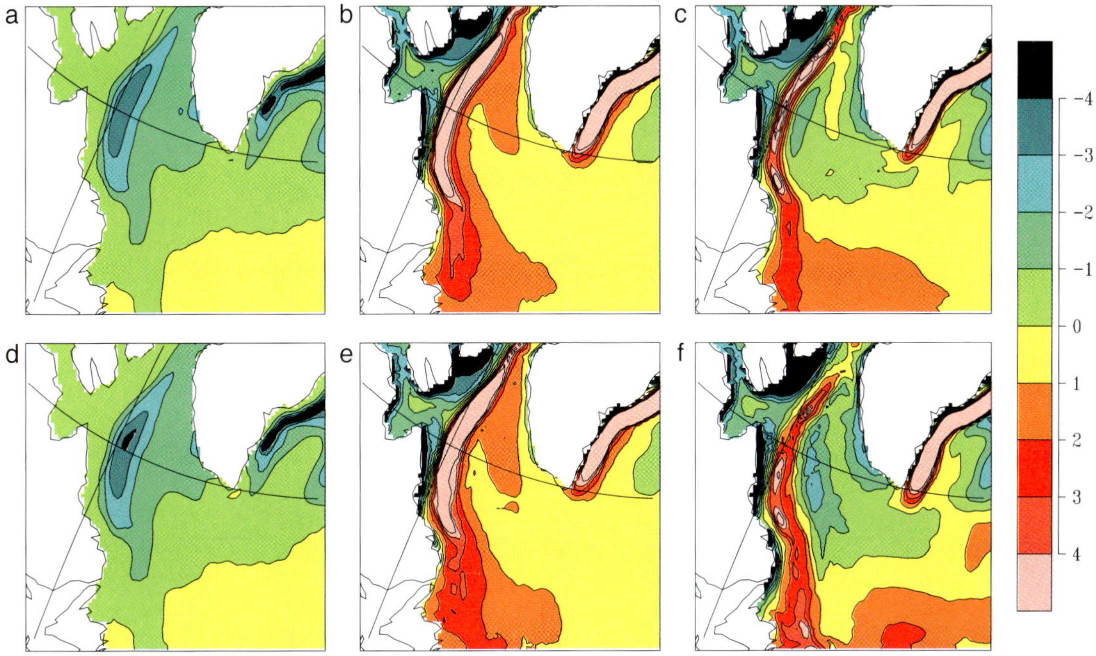

Plate 3. Buoyancy flux composites for the Labrador Sea in units of 10^{-8} m^2s^{-3}. The composite for weak convection is shown in the top panels (a-c), the composite for strong Greenland Sea convection in the lower (d-f). Contribution from the surface heat flux (a, d); contribution from the surface salt flux (b, e); total buoyancy flux (c, f).

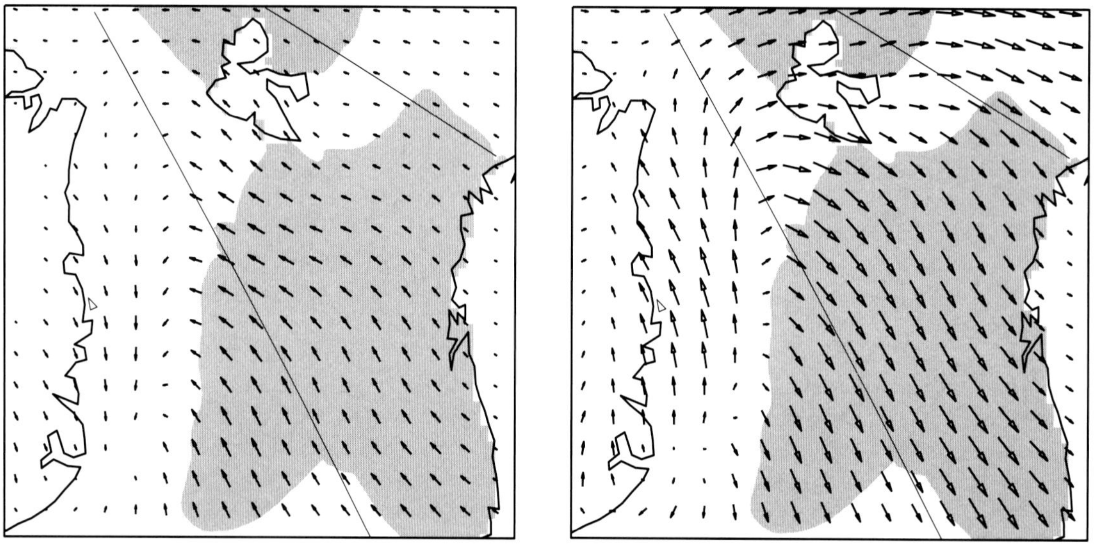

Figure 6. Composites of wind-stress anomalies (with respect to the JFMA average over the complete NCEP period used in the numerical experiment, 1948-2001). Left: The case of weak Greenland Sea convection; right: the case of strong convection. Regions where the composites for weak and strong convection differ at a significance level above 95% are shaded in gray.

convection deeper than 1500 m, with convection deeper than 2000 m in three of these. In those years the affected volume is very small. However, in the second half of the 1990s and in 2000 and 2001, relatively intense convection sets in again in the Greenland Sea. This is especially remarkable because most of the years showed a positive phase of the NAO (Figure 2), an atmospheric condition that has previously been identified with weak convection in the Greenland Sea. These convection events occur at only a few grid points and coincide in time with the observation of convective chimneys having radii on the order of 10 km in the Greenland Sea, where convection penetrates as deep as 2500 m [*Gascard et al.*, 2002; *Wadhams et al.*, 2002]. Given their small spatial scale, the long-lived chimneys observed in the Greenland Sea are probably dynamically different from the deep convective cells in the model. As will be discussed below, we believe that the model convection depends on a certain surface forcing, whereas the observed chimneys might be stable due to their nonlinear dynamics and may need only relatively weak surface fluxes to be recharged each winter [*Budéus et al.*, 2004].

To analyze whether the physical conditions leading to deep Greenland Sea convection in our model in the late 1990s differ from the conditions found in the composites for strong convection discussed above, we constructed a second set of composites from the years marked with diamonds in Figure 2. In the late 1990s wind-stress composite, we find a strong southward wind-stress anomaly over the Fram Strait and the Greenland Sea (Plate 4b). The latter feature is similar to the conditions in the earlier composite for strong Greenland Sea convection (Plate 4a), although the axis of the anomalously southward wind stresses is shifted westward in the late 1990s. The northerly winds over the Greenland Sea are imbedded in a background field very different from that of the composite for strong convection for the earlier years. For example, the 1990s wind-stress composite over the Greenland Sea is dominated by an anomalously low pressure over northern Norway. On the other hand, the anomalies in the earlier composite for strong convection are shaped by a weaker expression of the low-pressure trough that extends from Iceland into the Nordic Seas. Anomalously large net freezing in the northern Greenland Sea and net melting in the southern Greenland Sea, in combination with a high ice drift from north to south, occurs in the late 1990s (Plate 4c, d), as in the early Greenland Sea high-convection composite. Again, it is the brine release associated with this process that causes a high buoyancy loss over the Greenland Sea.

The northerly wind in the Greenland Sea is of great importance for the deep convection in the model. Analyzing winter conditions averaged over pentads from 1950 onward, we find that deep convection never occurred for southerly winds. And only for the pentads 1960-1964 and 1970-1974 do we find deep convection when the wind is not predominantly from northerly directions.

5. DISCUSSION

Variability of convection is determined by the pre-existing density stratification (preconditioning), the surface buoyancy

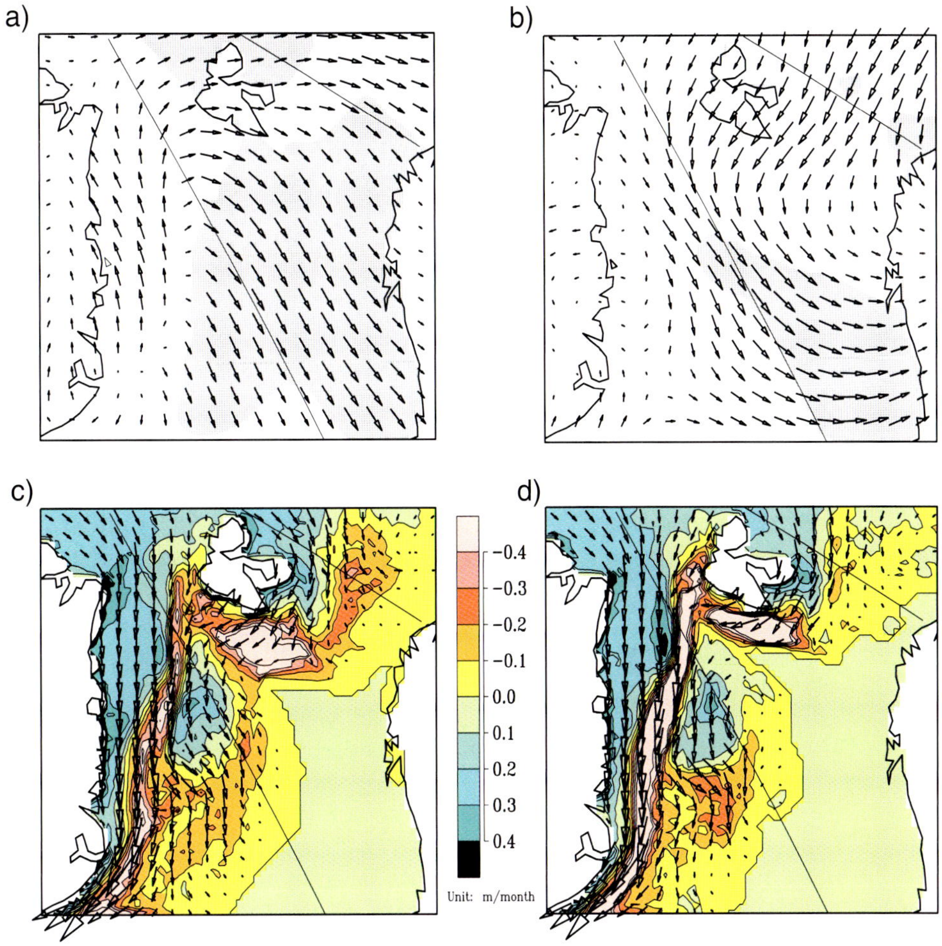

Plate 4. Comparison of the composites for strong Greenland Sea convection (as defined in Figure 2) with the corresponding late-1990s composites. (a) Wind-stress anomolies composite for the strong convection case (same as Figure 6, right); (b) composite for the late 1990s; regions where the composites differ from the composite for weak convection at a significance level above 95% are shaded in gray. (c, d) Corresponding composites for thermodynamic ice growth overlaid by sea-ice velocity composites.

flux, and lateral fluxes of buoyancy. For the example of the Labrador Sea, two extreme cases might be considered. If the convectively mixed water is not involved in lateral exchanges, we expect that the thickness of the LSW layer (the depth of the convectively mixed water in winter) is proportional to the integral of the surface buoyancy fluxes during mixed-layer deepening. Maximum LSW thickness would then be acquired at the end of a positive NAO phase (around 1977, 1985, and 1995), when we assume that the surface buoyancy flux in the Labrador Sea is governed by the NAO. In the second case, as much LSW is carried out of the Labrador Sea during 1 year as is newly formed by convective mixing during the winter. If comparable initial conditions at the beginning of the convective season exist each year, we expect that LSW layer thickness and NAO are in phase (maximum thickness around 1974, 1982, and 1990 or 1994). These extreme cases apparently have not been realized over the last 50 years, neither in the ocean nor in the model.

The LSW layer records the history of deep convection in the Labrador Sea. Generally, decreasing (increasing) temperature of LSW is associated with a thickening (thinning) of the layer. *Curry et al.* [1998] present a time series of the potential temperature at the LSW core that shows increasing values from before 1950 to 1970 and from 1976 to 1981. The temperature decreased from 1970 to 1976 and from 1981 to 1994. These changes correspond with the onset of deep convection in the model in 1971, the period of missing deep convection from 1977 to 1981, and the period of pronounced deep convection from 1982 to 1995 (Figure 2).

Curry et al. [1998] explain the large thickness of the LSW layer in the Labrador Sea in the mid-1970s and the earlier 1990s by the strong heat losses caused by low air temperatures and high winds in the corresponding positive NAO phases. Near-surface lateral fluxes, by eddies or the large-scale flow, tend to restratify the water column. According to *Curry et al.*, the LSW layer thinned in the early 1980s despite positive NAO conditions because of an influx of relatively fresh water [*Belkin et al.*, 1998; *Houghton and Visbeck*, 2002].

Our results confirm the determining role of the surface heat flux-forcing compared with other forcing components found in several studies [*Curry et al.*, 1998; *Houghton and Visbeck*, 2002; *Lazier et al.*, 2002]. On the other hand, the effect of restratification is visible in the delay in the onset of deep convection after the switch from the negative to the positive NAO phase around 1980, in both the model and the observed LSW temperature record. Inspection of the near-surface current field in the model (not shown) suggests that communication between the boundary currents surrounding the Labrador Sea and the interior (where the deep convection occurs) is weak. Time series of winter mean salinity of the upper 300 m averaged over boundary current boxes east and west of southern Greenland exhibit large variability, nearly in phase between these two locations (Figure 7). In comparison, the surface salinity variability in the interior of the Labrador Sea is much smaller and shows little correlation with the boundary current signal. Damping of salinity anomalies due to surface restoring is most effective where advective velocities are small

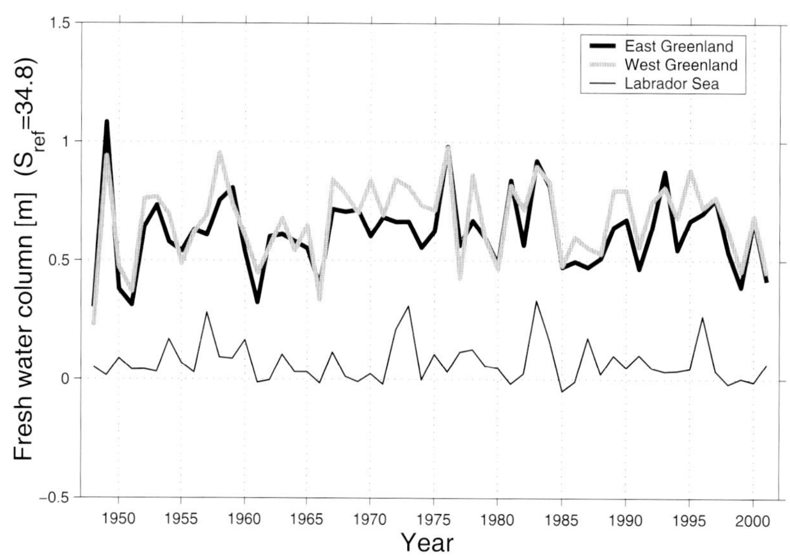

Figure 7. Time series of winter mean upper 300-m freshwater contents (in meters of freshwater column relative to a water column with a constant salinity of 34.8) averaged over boundary current boxes east (thick black) and west (thick gray) of southern Greenland and in the central Labrador Sea (thin black line).

(the restoring time scale is comparable to or smaller than the advective time scale). Thus, anomalies are more likely to be present in the boundary currents than in the interior of the basin. However, the small amplitude of freshwater anomalies in the interior of the Labrador Sea is consistent with the analysis of observed hydrography. *Houghton and Visbeck* [2002] find that the freshwater anomalies in the interior of the Labrador Sea are only one-fifth (measured by volume of fresh water) of those over the Labrador slope. Questioning the traditional cause-and-effect relationship between convection and freshwater anomalies, they hypothesize that salinity fluctuations in the Labrador Sea are caused primarily by changes in the strength of convective mixing. Despite the general reduction of signals in the interior of the Labrador Sea, major freshening events such as the Great Saline Anomaly (somewhat delayed in the model, with its maximum anomaly in 1972 and 1973) are simulated in the interior Labrador Sea for the early 1970s, 1983/1984, and 1996 (Figure 7). The 1983/1984 freshening has a counterpart in the boundary currents (Figure 7). It remains unclear, however, why transport processes between the boundary current and the interior filter out this event in favor of other, comparable signals. The time series of the model for the central Labrador Sea is in good qualitative agreement with the corresponding time series of *Houghton and Visbeck* [2002] (the curve labeled B(r) in their Figure 5), although the anomalies are smaller in the model. The freshening in 1983/1984 is of comparable magnitude to the signal for the Great Saline Anomaly in the model and is apparently able to delay deep convective mixing relative to the positive NAO forcing of the early 1980s.

Another important factor for the existence of deep convection is the preconditioning of the water column in the Labrador and Greenland Seas. From their model results, *Mizoguchi et al.* [2003] state that decadal variations of convective activity in the Labrador Sea are controlled by this preconditioning. However, the subsurface density distribution reacts to many influences, among them convection itself. Therefore, cause and effect are not easily separated. We note that enhanced cyclonic wind stress over the subpolar North Atlantic will reduce the stratification by Ekman suction. Ekman suction raises the isopycnals in the interior of the gyre while mass conservation within the density layers requires the isopycnals to drop in height at the periphery of the gyre. These changes in density distribution reinforce the effect of the surface buoyancy losses during positive NAO phases in the Labrador Sea: Surface buoyancy losses are high and the layer of light fluid at the top is relatively thin such that deep convection is favored.

The situation in the Greenland Sea is more complicated because the position and the strength of the wind stress over the area vary strongly, as indicated in Figure 3. The potential energy distribution (not shown) includes a small-scale dipole, the western part of which is out of phase with the rest of the Nordic Seas. The small spatial scale and the large amplitude of this dipole in the Greenland Sea suggest that the atmospheric forcing is not directly responsible for the potential energy changes. Rather, it is likely that the surface buoyancy fluxes, related to sea-ice freezing and melting, induce the potential energy pattern.

In the Greenland Sea, our model results suggest a connection between the formation of the Odden and deep convection. Furthermore, ice age (not shown) in the Odden is only a few months, indicating that the Odden originates from local sea-ice formation, not from ice advection out of the East Greenland Current. Our results are thus in contradiction to those of *Vinje et al.* [2002], who model freshwater of Arctic origin spreading across the Greenland Sea with strong influence on Greenland Sea convection. The reasons for these different model results are not obvious and can depend on model forcing and model resolution, among other parameters.

The observation of a deep-reaching convective chimney at approximately 75°N, 0°E in 2001 [*Wadhams et al.*, 2002] and in the following years seems to contradict our scenario of a causal relationship between local ice formation and deep convection. After 2000, little or no ice formed in the central Greenland Sea [*Wadhams et al.*, 2002]. On the other hand, the winters of 1997 and 1998, in which the model produces a large volume of convectively mixed water, were characterized by moderate to pronounced positive anomalies in sea-ice cover in the Greenland Sea [*Comiso et al.*, 2001; *Kauker et al.*, 2003]. We believe that the presence of ice indicates new ice formation in the Greenland Sea. The prevailing winds provide for the essential southward transport of sea-ice.

Gascard et al. [2002] describe the trajectories of floats that were deployed in March 1997 at depths of 250 m and 500 m in the Greenland Sea. The trajectories show long-lived anticyclonic eddies with a typical horizontal scale of 5 km. A hydrographic survey along 75°N revealed at 0°E a region of water homogenized over the upper 2000 m of a similar horizontal scale. This suggests that in the winter of 1997 the Greenland Sea was populated with several convective chimneys similar to that observed in 2001. The fact that the 1997 feature was at the exact geographical position of the chimney found in 2001 and had the apparent multiyear life time of the chimney found in 2001 [*Budéus et al.*, 2004] furthermore suggests that even this chimney could have been generated in the winters of 1997 or 1998, when the Greenland Sea was largely ice covered. A mechanism how the chimney could survive so long at its location has been described by *Budéus et al.* [2004].

6. SUMMARY AND CONCLUSIONS

We analyzed open ocean convection in the Greenland Sea and the Labrador Sea from the output of a coupled sea-ice-ocean model forced with NCEP reanalysis atmospheric data over the period 1948-2001. We constructed a time series of the annual convective activity in the Greenland Sea and the Labrador Sea from the volume of ocean regions with convection depths in excess of 1500 m.

Our model-derived convection rates are in good agreement with estimates based on observations in the Greenland Sea (Table 1). Likewise, agreement is good between our model-derived rates for the Labrador Sea and a number of previous rate estimates (Table 2). The 1948-2001 average of our model-derived convection rates, which give an upper limit for the actual deep-water formation rate, is 1.0 Sv for the Greenland Sea, 3.1 Sv for the Labrador Sea.

We find a strong correlation between our Labrador Sea convection index and the NAO index over the simulated period 1948-2001, with a Spearman rank-ordered correlation coefficient of $r_{sL} = 0.68$ and a very high significance level. The convective activity in the Greenland Sea, on the other hand, shows considerably weaker anticorrelation with the NAO, $r_{sG} = -0.25$.

Deep convection in the Labrador Sea is associated with low air temperatures and strong northwesterly winds from the Canadian land mass, a situation typical of strong NAO winters. Surface buoyancy losses over the Labrador Sea are due solely to high oceanic heat loss to the atmosphere, while at the same time freshwater fluxes due to the melting of sea-ice tend to hinder convection. Deep convection in the Labrador Sea occurs when the buoyancy losses due to heat flux overcome the opposing freshwater flux. Occasionally, as in the early 1980s, freshening events in the central Labrador Sea can delay the onset of deep convection that would otherwise be implied by the NAO-related atmospheric forcing. The positive phase of the NAO is also associated with a pronounced doming of isopycnals in the subpolar gyre. This tendency reinforces the effect of the surface buoyancy fluxes and contributes to the robust correlation between NAO and deep Labrador Sea convection.

Convection in the Greenland Sea, on the other hand, seems to be strongly controlled by local processes, particularly sea-ice formation in the northern Greenland Sea and subsequent transport of the newly formed sea-ice into the southern Greenland Sea. Surface buoyancy losses due to brine release associated with the local formation of sea-ice often amount to more than 50% of the total surface buoyancy losses. We find similar local conditions in the Greenland Sea for the period of strong convection before 1975 and the onset of deep convection in the second half of the 1990s—both periods characterized by strong northerly winds over the central Greenland Sea. However, the large-scale conditions over the subpolar North Atlantic, the Nordic Seas, and the Arctic Ocean are vastly different between these two periods, and there is no robust correlation between Greenland Sea convection and the NAO.

Acknowledgments. NCEP reanalysis data was provided by the NOAA/CIRES Climate Diagnostics Center, Boulder, Colorado, www.cdc.noaa.gov. We acknowledge the support of the European Union DG-XII Climate and Environment Programme for support under the CONVECTION project (contract EVK-CT-2000-00058). Further funding for this work was provided by the German ministry for research and technology under the DEKLIM program (contract 01 LD 0047) and by the Deutsche Forschungsgemeinschaft via the SFB 512. This work was completed during a visit of R.G. at the Geophysical Fluid Dynamics Laboratory (GFDL), Princeton. R.G. thanks GFDL for the hospitality and UCAR for financial support.

REFERENCES

Belkin, I.M., S. Levitus, J.I. Antonov, S.-A. Malmberg, "Great Salinity Anomalies" in the North Atlantic, *Progress in Oceanography*, 41, 1-68, 1998.

Blindheim, J., S. Østerhus, The Nordic Seas, main oceanographic features (this volume)

Böning, C.W., F.O. Bryan, W.R. Holland, R. Döscher, Deep-water formation and meridional overturning in a high-resolution model of the North Atlantic, *J. Phys. Oceanogr.*, 25, 1142-1164, 1996.

Böning, C.W., M. Rhein, J. Dengg, C. Dorow, Modeling CFC inventories and formation rates of Labrador Sea Water, *Geophys. Res. Lett.*, 30, 1050, 2003.

Bönisch, G., P. Schlosser, Deep water formation and exchange rates in the Greenland/Norwegian Seas and the Eurasian Basin of the Artic Ocean derived from tracer balances, *Prog. Oceanogr.* 35, 29-52, 1995.

Budeus, G., B. Cisewski, S. Ronski, D. Dietrich, M. Weitere, Structure and effects of a long lived vortex in the Greenland Sea, *Geophys. Res. Lett.*, 31, L053404, doi: 10.1029/2003GL017983, 2004.

Burnham, C. W., Importance of volatile constituents, *in Evolution of Igneous Rocks,* edited by H. D. Yoder, pp. 439-482, Princeton University Press, Princeton, N. J., 1979.

Clarke, R.A., J.-C. Gascard, The formation of Labrador Sea Water, I, Large-scale processes, *J.Phys.Oceanogr.*, 13, 1764-1788, 1983.

Comiso, J.C., P. Wadhams, L.T. Pedersen, R.A. Gersten, Seasonal and interannual variability of the Odden ice tongue and a study of environmental effects, *J. Geophys. Res.*, 106(C5), 9093-9116, 2001.

Curry, R.G., M.S. McCartney, T.M. Joyce, Oceanic transport of subpolar climate signals to mid-depth subtropical waters, *Nature*, 391, 575-577, 1998.

Deser, C., J.E. Walsh, M.S. Timlin, Arctic sea ice variability in the context of recent atmospheric circulation trends, *J.Climate*, 13, 617-633, 2000.

Dickson, R.R., J. Lazier, J. Meinke, P. Rheins, J. Swift, Long term coordinated changes in the convective activity of the North Atlantic, *Prog. Oceanogr.* 38, 241-295, 1996.

Eldevik, T., F. Straneo, A.B. Sandø, T. Furevik, Pathways and export of Greenland Sea Water (this volume).

Gascard, J.-C., A.J. Watson, M.-J. Messias, K.A. Olsson, T. Johannessen, K. Simonsen, Long-lived vortices as a mode of deep ventilation in the Greenland Sea, *Nature*, 416, 525-527, 2002.

Gerdes, R., C. Köberle, J. Willebrand, The influence of numerical advection schemes on the results of ocean general circulation models, *Clim. Dyn.*, 5, 211-226, 1991.

Gerdes, R., M. Karcher, F. Kauker, C. Köberle, Predicting the spread of radioactive substances from the Kursk, *EOS*, Transactions, American Geophysical Union, 82, 253, 256-257, 2001.

Hibler, W.D., A dynamic thermodynamic sea ice model. *J. Phys. Oceanogr.*, 9, 815-846, 1979.

Hibler, W.D. III, K. Bryan, A diagnostic ice-ocean model, *J.Phys.Oceanogr.*, 17, 987-1015, 1987.

Hilmer, M., A model study of Arctic sea ice variability, *Berichte aus dem Institut für Meereskunde*, 320, Institut für Meereskunde Kiel, Düsternbrooker Weg 20, 24105 Kiel, Germany, 2001.

Hilmer, M., M. Harder, P. Lemke, Sea ice transport: A highly variable link between Arctic and North Atlantic, *Geophys.Res.Lett.*, 25, 3359-3362, 1998.

Houghton, R.W., M.H. Visbeck, Quasi-decadal salinity fluctuations in the Labrador Sea, *J.Phys.Oceanogr.*, 32, 687-701, 2002.

Hurrell, J.W., Decadal trends in the North Atlantic oscillation: Regional temperatures and precipitation, *Science*, 269, 676-679, 1995.

Kalnay, E., M. Kanamitsu, R. Kistler, W. Collins, D. Deaven, L. Gandin, M. Iredell, S. Saha, G. White, J. Woollen, Y. Zhu, A. Leetmaa, R. Reynolds, M. Chelliah, W. Ebisuzaki, W. Higgins, J. Janowiak, K.C. Mo, C. Ropelewski, J. Wang, R. Jenne, D. Joseph, The NCEP/NCAR 40-Year Reanalysis Project, *Bulletin of the American Meteorological Society*, 77, 437-495, 1996.

Karcher, M., R. Gerdes, F. Kauker, C. Köberle, Arctic warming-evolution and spreading of the 1990s warm event in the Nordic Seas and the Arctic Ocean, *J.Geophys.Res.*, 108 (C2), 3034, doi: 10.1029/2001JC001265, 2003.

Kauker, F., R. Gerdes, M. Karcher, C. Köberle, J.L. Lieser, Variability of Northern Hemisphere sea ice: A combined analysis of model results and observations from 1978 to 2001, *J.Geophys.Res.*, 108 (C6), 3182, doi:10.1029/2002JC001573, 2003.

Khatiwala, S., P. Schlosser, M. Visbeck, Rates and mechanisms of water mass transformation in the Labrador Sea as inferred from tracer observations, *J.Phys.Oceanogr.*, 32, 666-686, 2002

Köberle, C., R. Gerdes, Mechanisms determining variability of Arctic ice conditions and export, *J.Climate*, 16, 2843-2858, 2003.

Lazier, J., R. Hendry, A. Clarke, I. Yashayaev, P. Rhines, Convection and restratification in the Labrador Sea, 1990-2000, *Deep-Sea Research I*, 49, 1819-1835, 2002.

Levitus, S., T. Boyer, World Ocean Atlas 1994, vol. 4, Temperature, *NOAA Atlas NESDIS 4*, Natl. Occanic and Atmos. Admin., Silver Spring, Md., 1994.

Levitus, S., R. Burgett, T. Boyer, World Ocean Atlas 1994, vol. 3, Salinity, *NOAA Atlas NESDIS 3*, Natl. Oceanic and Atmos. Admin., Silver Spring, Md., 1994.

Marsh, R., Recent variability of the North Atlantic thermohaline circulation inferred from surface heat and fresh water fluxes, *J.Climate*, 13, 3239-3260, 2000.

Marshall, J., F. Schott, Open-ocean convection: Observations, theory, and models, *Reviews of Geophysics*, 37, 1-64, 1999.

McCartney, M.S., Recirculating components to the deep boundary current of the northern North Atlantic, *Progress in Oceanography*, 29, 283-383, 1992.

McCartney, M.S., L.D. Talley, Warm-to-cold water conversion in the northern North Atlantic Ocean, *J.Phys.Oceanogr.*, 14, 922-935, 1984.

Mizoguchi, K., S.L. Morey, J. Zavla-Hidalgo, N. Suginohara, S. Häkkinen, J.J. O'Brien, Convective activity in the Labrador Sea: Preconditioning associated with decadal variability in subsurface ocean stratification, *J. Geophys. Res.*, 108(C10), 3330, doi:10.1029/2002JC001753, 2003.

Molinari, R.L., R.A. Fine, W.D. Wilson, R.G. Curry, J. Abell, M.S. McCartney, The arrival of recently formed Labrador Sea Water in the Deep Western Boundary Current at 26.5°N, *Geophys.Res.Lett.*, 25, 2249-2252, 1998.

Morawitz, W.M.L., P.J. Sutton, P.F. Worcester, B.D. Cornuelle, J.F. Lynch, R. Pawlowicz, Three-dimensional observations of a deep convective chimney in the Greenland Sea during winter 1988/89, *J. Phys. Oceanogr.*, 26, 2316-2343, 1996.

NSIDC, Environmental Working Group, *Joint U.S. Russian Atlas of the Arctic Ocean, Oceanography Atlas for the Winter Period*, CD-Rom, NSIDC/CIRES, University of Colorado, Boulder, 1997.

Pacanowski, R.C., MOM 2 Documentation, user's guide and reference manual, *GFDL Ocean Group Tech. Rep.* No.3, Geophysical Fluid Dynamics Laboratory, Princeton University, Princeton, NJ, 1995.

Parkinson, C., W. Washington, A large-scale numerical model of sea ice, *J. Geophys. Res.*, 84, 311-337, 1979.

Pickart, R.S., M.A. Spall, M.H. Ribergaard, G.W.K. Moore, R.F. Milliff, Deep convection in the Irminger Sea forced by the Greenland tip jet, *Nature*, 424, 152-156, 2003.

Press, W.H., S.A. Teukolsky, W.T. Vetterling, B.P. Fleming, *Numerical Recipes*, Cambridge University Press, New York, 2nd edition, 1992.

Rhein, M., Ventilation rates of the Greenland and Norwegian Seas derived from distribtions of chlorofluoromethanes F11 and F12, *Deep-Sea Res.*, 38, 485-503, 1991.

Rhein, M., Convection in the Greenland Sea, 1982-1993, *J.Geophys.Res.*, 101, 18183-18192, 1996.

Rhein, M., J. Fischer, W.M. Smethie, D. Smythe-Wright, R.F. Weiss, C. Mertens, D.-H. Min, U. Fleischmann, A. Putzka, Labrador Sea Water: Pathways, CFC inventory, and formation rates, *J.Phys.Oceanogr.*, 32, 648-665, 2002.

Röske, F., An atlas of surface fluxes based on the ECMWF reanalysis—a climatological dataset to force global ocean general circulation models, *MPI-Report*, 323, Max-Planck-Institut für Meteorologie, Hamburg, Germany, 2001.

Schlosser, P., G. Bönisch, M. Rhein, R. Bayer, Reduction of deep water formation in the Greenland Sea during the 1980s: evidence from tracer data, *Science*, 251, 1054-1056, 1991.

Smethie, W.M., R.A. Fine, Rates of North Atlantic Deep Water formation calculated from chlorofluorocarbon inventories, *Deep-Sea Research*, I, 48, 189-215, 2001.

Steele, M.W., W. Ermold, S. Häkkinen, D. Holland, G. Holloway, M. Karcher, F. Kauker, W. Maslowski, N. Steiner, J. Zhang, Adrift in the Beaufort Gyre: A model intercomparison, *Geophys. Res. Lett.*, 28, 2935-2938, 2001.

Stevens, D.P., The open boundary condition in the United Kingdom Fine-Resolution Antarctic Model, *J. Phys.Oceanogr.*, 21, 1494-1499, 1991.

Talley, L.D., M.S.McCartney, Distribution and circulation of Labrador Sea Water, *J.Phys.Oceanogr.*, 12, 1189-1205, 1982.

Vinje, T., T.B. Løyning, I. Polyakov, Effects of melting and freezing in the Greenland Sea, *Geophys.Res.Lett.*, 29(23), 2129, doi:10.1029/2002GL015326, 2002.

Visbeck, M., E. Chassignet, R. Curry, T. Delworth, B. Dickson, G. Krahmann, The ocean's response to North Atlantic Oscillation variability, in *The North Atlantic Oscillation*, edited by J. Hurrell, Y. Kushnir, G. Ottersen, M. Visbeck, AGU monograph, Washington, 113-146, 2003.

Wadhams, P., J. Holfort, E. Hansen, J.P. Wilkinson, A deep convective chimney in the winter Greenland Sea, *Geophys. Res. Lett.*, 29, 10.1029/2001GL014306, 2002.

Worthington, L.V., *On the North Atlantic Circulation*, 110pp., John Hopkins Univ. Press, Baltimore, 1976.

Wright, W.R., Northern sources of energy for the deep Atlantic, *Deep Sea Res.*, 19, 865-877, 1972.

Zalesak, S.T., Fully multidimensional flux-corrected transport algorithms for fluids, *J. Comput. Phys.*, 31, 335-362, 1979.

R. Gerdes, J. Hurka, M. Karcher, F. Kauker, and C. Köberle, Alfred-Wegener-Institut, Bussestr. 24, 27570 Bremerhaven, Germany. (rgerdes@awi-bremerhaven.de)

Seasonal to Decadal Temperature Variations in the Faroe–Shetland Inflow Waters

Hjalmar Hátún

Faroese Fisheries Laboratory, Tórshavn, Faroe Islands

Anne Britt Sandø, Helge Drange[1,2,3], and Mats Bentsen[2]

Nansen Environmental and Remote Sensing Center, Bergen, Norway

A 53-year hindcast simulation with a regional version of the Nansen Center version of the Miami Isopycnic Coordinate Ocean Model (MICOM) was conducted to explore the nature of the observed seasonal to interannual variations in the temperature of the poleward flowing Atlantic Water (AW) crossing the Iceland–Scotland Ridge (ISR). The simulated long-term temperature variations closely resembled observations south of the ridge (Rockall Trough), north of the ridge (Svinøy section), and in the Faroe–Shetland Channel (FSC). The simulated temperature on the Faroe Shelf was also compared to daily temperature observations from Mykines, revealing realistic long-term temperature variations, seasonal variations, and a seasonal modulation. The simulated time series in the FSC indicates that the phase and amplitude of the annual temperature cycle of the AW have varied by almost 1 month and 0.15°C between the 1960s and 2001, illustrating the difficulty in unambiguously removing the seasonal cycle from sparsely sampled time series. We argue that the simulated time series can be used to complement the observed time series in periods with sparse sampling. Specifically, the observation-based cold anomaly in the late 1960s and the warming in the early 1980s should be treated with caution. Finally, the analysis indicates that it is not advisable to survey the hydrographic section less than four times a year if reliable decadal-scale temperature variations are of interest.

1. INTRODUCTION

The southeastern part of the Nordic Seas is among the longest [*Helland-Hansen and Nansen*, 1909] and most frequently [*Hansen and Østerhus*, 2000] sampled regions of the World Ocean. It is also, together with the Barents Sea, a region of high biological production [*Anderson et al.*, 2000; *Sakshaug et al.*, 1994]. Mounting evidence shows that the biological production is closely related to the actual state of the marine climate in the region, particularly the flow of warm, saline-, and nutrient-rich Atlantic Water (AW) across the Greenland–Scotland Ridge (GSR)

[1]Geophysical Department, University of Bergen, Bergen, Norway.
[2]Bjerknes Centre for Climate Research, Bergen, Norway.
[3]Nansen-Zhu International Research Centre, Beijing, China.

The Nordic Seas: An Integrated Perspective
Geophysical Monograph Series 158

10.1029/158GM16

[*Beaugrand and Reid*, 2003]. Furthermore, because the inflow of AW keeps the Barents Sea and the waters south of Svalbard ice-free throughout the year, it is consequently of key importance for the local climate in western Scandinavia and northwestern Russia. The salt transported into the Nordic Seas plays a fundamental role in the formation of intermediate waters in the region and is a key component in the formation of abyss water that occasionally takes place in the Greenland Sea. The thermodynamic transformation of AW that takes place north of the GSR is thus also linked to the Atlantic thermohaline circulation [*Dickson and Brown*, 1994].

The flux of AW entering the Nordic Seas between Iceland and the Faroes, and between the Faroes and Scotland, is about equally distributed and amounts to about 3.5 Sv (1 Sv = 10^6 m^3s^{-1}) for each [*Hansen et al.*, 2003; *Turrell et al.*, 2003]. A third branch of AW enters the Nordic Seas through the Denmark Strait. This branch is rather weak with an inflow of less than 1 Sv [*Kristmannsson*, 1998]. The hydrography of the AW entering the Nordic Seas between the Faroes and Scotland has been routinely observed by the Faroese Fisheries Laboratory and the Marine Laboratory in Aberdeen throughout most of the 20th century. An annual average of 3.6 hydrography cruises were undertaken during the second half of the 20th century, yielding a truly unique source of information about the seasonal to decadal variability of the properties of the inflowing AW.

The Faroes–Scotland time series have been used in a series of studies, including tracing hydrographic anomalies that propagate through the Nordic Seas [*Belkin et al.*, 1998; *Furevik*, 2001]. Although frequently sampled, in some periods only one or two observations per year were made. This, combined with a highly variable seasonal cycle in the Faroe–Shetland Channel (FSC) [*Hansen et al.*, 1994], can lead to difficulties in uniquely identifying the seasonal cycle and the interannual temperature anomalies in the series. The consequence of this is that spurious temperature and salinity anomalies can be deduced from the data [*Reverdin et al.*, 1994]. Here we present a hindcast simulation with a regional version of the Miami Isopycnic Coordinate Ocean Model (MICOM) [*Bleck et al.*, 1992] to (1) evaluate one aspect of the model performance, (2) assess whether the observed time series can complement the simulation for the period 1948-2001, and (3) provide guidance for the future sampling strategy of the Faroes–Scotland transect.

The paper starts with a description of the applied model system (Section 2). Then simulated and observed long-term temperature variations from three standard hydrographical sections crossing current branches of the northward-flowing AW are analyzed. In Section 4, the daily temperature record from Mykines is analyzed and used as background for a seasonality and aliasing analysis of the observed and simulated temperature time series from the FSC. In Section 5, some possible applications of a general circulation model (GCM) are illustrated, followed by a discussion and conclusions in Section 6.

2. MODEL DESCRIPTION

The model system adopted in this study consists of the global Nansen Center version of MICOM [*Bentsen et al.*, 2004; *Furevik et al.*, 2002; *Nilsen et al.*, 2003], and a regional version of the same model system covering the Atlantic Ocean between 30°N and 78°N. The global version of the model used in this study has a horizontal resolution of about 40 km over most of the North Atlantic Ocean. The grid configuration of the regional model is identical to the global model, but with doubled horizontal resolution, e.g., with about 20-km grid spacing in the region of the Iceland–Scotland Ridge (ISR).

The nesting approach applies a boundary relaxation scheme towards the outer (i.e., global) solution. This results in a so-called one-way nesting, where the boundary conditions of the regional model are relaxed towards the output from the global model. For the slowly varying baroclinic velocity, temperature, salinity, and layer interface variables, this is an appropriate way to include the boundary conditions. For the barotropic variables, the relaxation approach requires careful tuning to avoid reflection of waves at the open boundaries. The barotropic boundary conditions can be computed exactly while taking into consideration the waves propagating both into and out of the regional model (for details, see ftp://micom.rsmas.miami.edu/bleck/open_bdy.tex). The regional model reads the global fields once a week and interpolates in time to specify the relaxation boundary conditions at each time step.

In the vertical, both model versions have 26 layers, of which the uppermost mixed layer has temporally and spatially varying density and the 25 layers below have constant density. Daily mean NCAR/NCEP reanalyses [*Kalnay et al.*, 1996] of fresh water, heat, and momentum fluxes are used to force the system by applying the scheme of *Bentsen and Drange* [2000]: If the model sea surface state is equal to the NCEP/NCAR sea surface state, the turbulent fluxes of momentum and heat are taken from the reanalysis data. If the sea surface state differs between the ocean model and the reanalysis data, the model fluxes will be modified according to the *Fairall et al.* [1996] scheme.

In the regional model, the mixed layer temperature and salinity fields are linearly relaxed towards the monthly mean climatological values of *Levitus et al.* [1994] and *Levitus and Boyer* [1994], respectively. The e-folding relaxation time scale is set to 30 days for a 50-m-thick mixed layer. Below 50 m, the relaxation is reduced linearly with depth. The applied relaxation to climatology is rather weak, allowing for seasonal to interannual variations in the simulated mixed-layer properties, as is demonstrated below.

2.1. General Remarks on the Applied Model System

The applied horizontal grid resolution of about 20 km implies that the model is far from eddy-resolving in the region of interest but is sufficient to pick up the main topographic steering-effect on the currents and similar to the distance between current meters and the standard hydrographic stations along the standard measurement sections in the region. The model grid will seldom coincide exactly with the actual positions of standard hydrographical stations, but this does not affect the results crucially, at least not when the mean value over a certain region is considered.

The simulated hydrography in the ISR region is, on average, too fresh by about 0.1 psu [*Furevik et al.,* 2002] and too cold by 0.44°C (see below). These deviations may, however, not influence the simulated seasonal to decadal-scale variability in salinity and temperature. Furthermore, the simulated volume transports over the ISR, calculated as the net amount of water passing the openings from surface to bottom between 1948 and 2001, have been compared to transport values from the literature [*Nilsen et al.,* 2003]. These simulated values are found to be reasonable although the branch north and northwest of the Faroes is too weak, possibly as a result of the non-eddy-resolving model resolution.

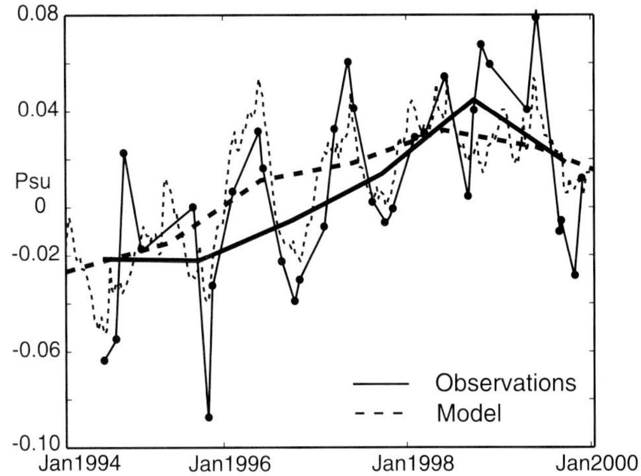

Figure 1. Horizontally averaged salinity deviations at 50-m depths along the Fair Isle–Munken section. The thicker lines show annual averages, and the individual measurements are indicated with the black dots.

3. SIMULATED LONG-TERM HYDROGRAPHY IN THE FSC

High-quality CTD data are available from the Fair Isle–Munken section between the Faroes and Scotland (Plate 1) since 1994. The salinity, horizontally integrated at 50-m depths from the Shetland Shelf halfway towards the Faroes, is compared to the simulated salinity treated in the same way (Figure 1). The model simulates the observed seasonality and the steep salinity increase during the 1990s. However, when the observed and simulated time series are considered for the Shetland Shelf region, the simulated salinity variability is much lower than the observed variability. This indicates that the simulated salinity is too diffusive or lacks details, but that the integrated values are realistic. The simulated temperature, however, showed realistic variability both when considering a limited area on the Shetland Shelf and when looking at the integrated values. This observation is, among other processes, an indication of the weak atmosphere–ocean coupling for salinity (through evaporation minus precipitation and runoff) compared to temperature (through heat fluxes).

The observations from the Fair Isle–Munken section have been merged with observations from the Nolsoy–Flugga section further north in the FSC (Plate 1) to produce a century-long hydrography time series [*Turrell et al.,* 1993]. The averaged temperature field along the Fair Isle–Munken section is shown in the inset in Plate 2b. The relatively warm poleward-flowing Continental Shelf Current (CSC) is evident as the dark red colors on the Shetland Slope, and the cold overflow water flowing in the direction of the equator is seen as the dark blue colors (extending from a depth of 500-600 m to the bottom).

The subregion, shown with the white frame in the inset, covers the observations that are made inshore from the 300-m-depth contour on the Shetland side of the channel. The hydrographic series provided by the Marine Laboratory in Aberdeen is calculated as a spatial average over this region and thus provides a good description of the hydrographical properties of the inflowing AW. A proper comparison of the observed and simulated salinity requires spatially integrated values from an area wider than the white frame, and since the observed salinity is of less quality than temperature [*Turrell et al.,* 1993], we focus on temperature in this study. When the simulated temperature is spatially averaged over the white frame area and compared with the observations, the temporal averages are 9.73°C and 9.29°C, respectively. The temperature anomalies shown in Plate 2a are made by subtracting the average values from the actual time series.

In Plate 2b, the observed data points in Plate 2a have been binned into years and averaged; similarly, the modeled data are picked out at the measurement times and averaged in the same manner. In addition, a 3-year running average has been applied for clarity. The presented curves are raw comparisons between the observed and simulated temperature anomalies

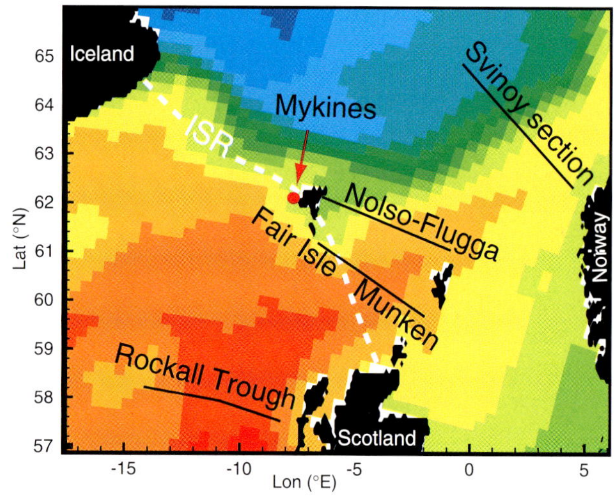

Plate 1. The temperatures T ($T < 3°C$ for the bluish colors; $T > 9°C$ for the pure red color) averaged over the mixed layer in the region of the ISR for the years 1996-2001. The standard hydrographical sections used in the analyses are indicated with the black lines; the coastal station at Mykines is shown with the red dot.

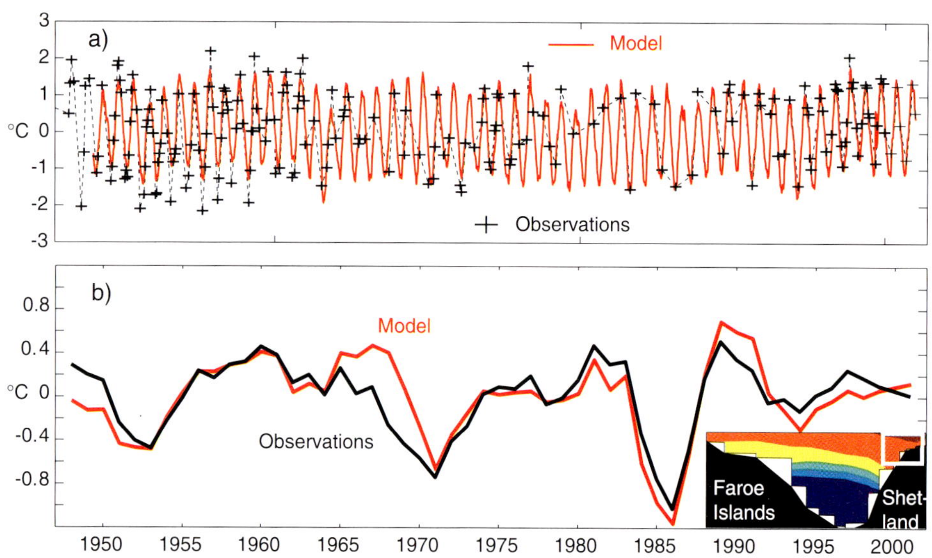

Plate 2. Temperature anomalies in the FSC. Inset: Observed and simulated time series are from the region shown within the white rectangle (see text for details). (a) The observations (plusses) and the weekly resolved simulated time series (solid line); (b) 3-year running averages (see text for averaging details).

and should therefore not be interpreted as the true temperature variation from 1948 to present.

The correlation between the 195 raw temperature observations and the corresponding simulated temperature is 0.93 (before low-passing), although the correlation between annual averages drops to 0.63. The very good raw data correlation is mostly due to a correctly simulated annual cycle, but the annual averages are still significantly related. A comparison of annual averages is, however, not strictly meaningful since there is no straightforward way to remove the seasonal cycle before averaging, as will be discussed in Section 4.2. Nevertheless, this illustrates that the model is able to simulate the observed variability in the FSC. To further support this finding, we examined the variability of the observed and simulated hydrographic conditions at the Rockall Trough and the Svinøy section (see Plate 1).

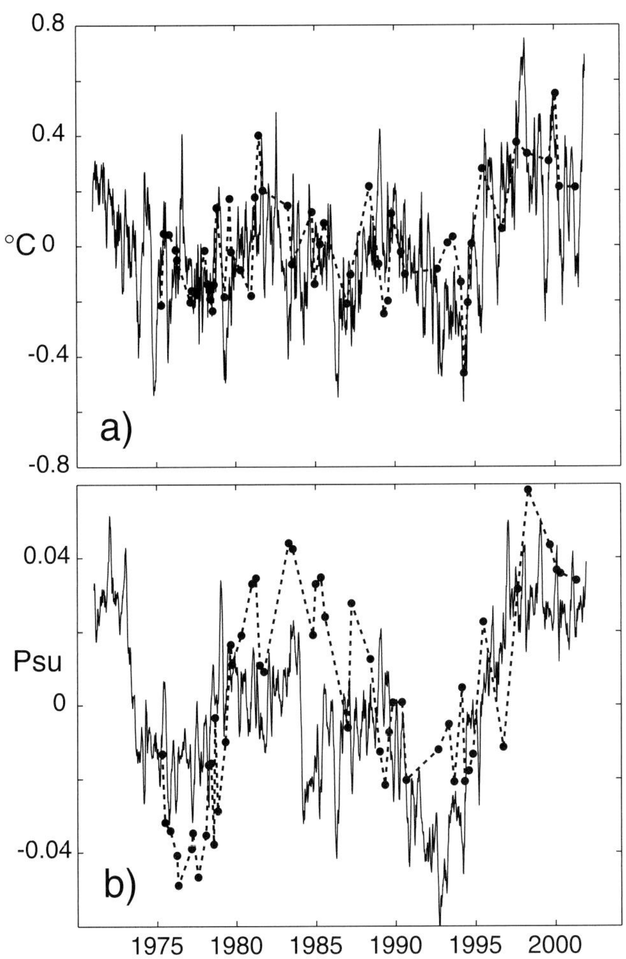

Figure 2. The hydrography along the full Rockall Trough line averaged from surface to 800 m deep: (a) de-seasoned temperature anomalies (observations: dots, simulations: curve) and (b) de-seasoned salinity anomalies.

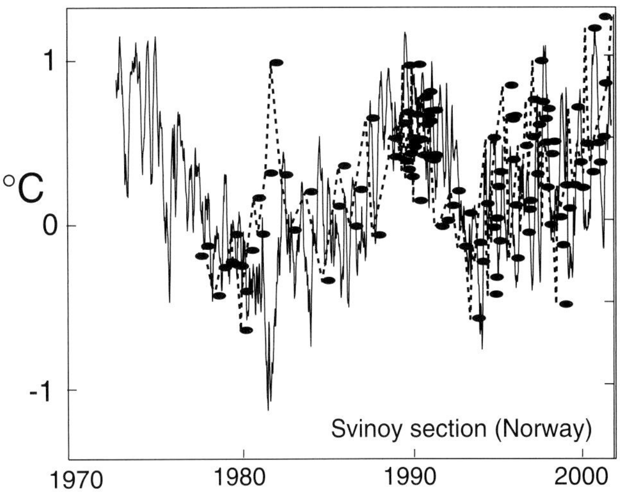

Figure 3. Temperature (de-seasoned anomalies) from the Svinøy section averaged over an area embodying the northward-flowing Atlantic water.

The Rockall Trough represents one pathway by which warm North Atlantic upper water reaches the FSC. Temperature and salinity time series from a standard hydrographic section (the '*Ellet* section'; see Plate 1) that crosses the northern Rockall Trough are available from ICES (www.ices.dk). Figure 2 shows the horizontally averaged, de-seasoned observed anomalies over the uppermost 800 m [*Holliday et al.*, 2000] and the simulated time series processed in a similar way. The temporal averages of the presented simulated and observed time series are 8.97°C and 9.21°C, respectively.

Furthermore, most of the AW inflow that passes through the discussed area in the FSC continues as the Shetland Current [*Hansen and Østerhus*, 2000] to the Svinøy section on the Norwegian Shelf (Plate 1). Temperature measurements from high-quality CTD data along this section (supplied by the Institute of Marine Research, Bergen, Norway) have been spatially averaged over the region representing the AW inflow. Figure 3 shows the de-seasoned observed and simulated time series. The temporal averages of the simulated and observed time series presented are both 7.4°C.

In conclusion, the comparisons from the Rockall Trough and the Svinøy section show that the simulated long-term hydrographic conditions of the northward flowing AW are, indeed, realistic.

4. SEASONALITY AND ALIASING

4.1. Coastal Temperature at Mykines

Daily sea surface temperature (SST) measurements were made at Mykines (Plate 1) during the period 1 January 1914 to 18 September 1969. The observation site is open to the Atlantic, the region has strong tidal currents (implying vigorous mixing), there are no river outlets close to the site, and the series represents Faroe Shelf condition within 0.2°C [*Hansen and Meincke*, 1984].

The correlation between the raw Mykines series and the simulated temperature anomalies on the Faroe Shelf is 0.96 when monthly averages are compared and 0.75 when annual averages are considered. This shows that both the seasonal variations and interannual temperature variations on the Faroe Plateau are realistically simulated in an average sense. Observations and simulations are available concurrently from 1948 to 1969 and the number of samples each month during this period is shown as a stacked histogram in Plate 3b. Only the years 1949, 1950, and 1969 had entire months missing, and on average there are 24 samples each month. With such comprehensive data-coverage, annual averages are simply found by averaging every data point within each year, and by using the average seasonal cycle when isolated months are missing. By doing this, a fairly good correspondence is found between the two time series (Plate 3a, inset). The temporal averages of the presented simulated and observed time series are 8.40°C and 8.02°C, respectively.

The temperature variations are clearly dominated by the seasonal cycle, and aliasing would be a severe problem if only few samplings were available (see later). In analyzing time series with a few observations each year, as is the common case for oceanic records, one must remove the seasonal signal before averaging in order to remove possible bias caused by the few observations. Since the long-term variation is often small compared to the seasonal signal, care must be applied to the seasonal filtering. The fact that the seasonal amplitude and the timing of the annual temperature maximum in the ISR area vary broadly [*Hansen et al.,* 1994] complicates filtering of the annual cycle further. Therefore, filtering the seasonal cycle using an averaged cycle could, for a year with an anomalous seasonal variation, bring in an error to the filtered series. Even though this error is relatively small compared to the seasonal amplitude itself, it can be severe considering the slowly changing long-term averages.

The Mykines series is suitable for evaluating the simulated seasonal cycle. The seasonal cycle is symmetric and well described by a simple trigonometric function of the form $A \cdot \cos[2\pi t/(365.25 - p)]$, where A (°C) is amplitude, t (day-number) is time, and p (rad) is phase. A combination of various trigonometric functions, allowing for nonsymmetric seasonal cycles, did not fit the actual seasonality significantly better than the simple cosine.

Seasonal filtering is performed by fitting the cosine to a time-window of the series and by subtracting the best-fit cosine from the year in the middle of the window. By letting the time-window slide through the full series (Plate 3a, inset)

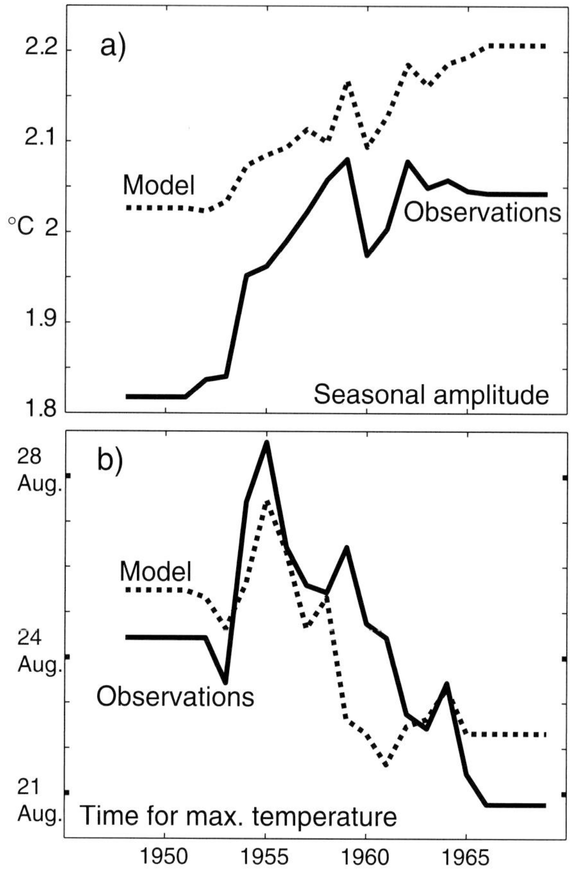

Figure 4. Change in the seasonal temperature variation on the Faroe Shelf: (a) the observed (solid line) and simulated (dashed line) strength in the seasonality, and (b) the time, for each year, when the temperature maximum occurs.

one obtains the de-seasoned (filtered) series shown in Plate 3a (note difference in scales). The amplitude and the phase from this analysis are found to be time-dependent parameters. The narrower the window is, the better any sudden changes in the seasonality are resolved, but a narrow window includes fewer values for the cosine-fitting, thus bringing in an error caused by imperfect fitting. A width of 8 years minimizes the errors caused by imperfect fitting and by the varying seasonality.

Values of the seasonal amplitude and phase, obtained from both observations and the model, are displayed in Figures 4a and 4b. As shown, the simulated seasonal amplitude is larger than that in reality by about 0.2°C, and the changes in the two time series co-vary in time. The model obtains an increase in seasonality by nearly 0.2°C from 1948 to 1969, while the observations show an increase of about 0.25°C. The time for maximum temperature differs by less than 2 days between model and data and, as both time series indicate, the maximum temperature is seen 3-4 days earlier in the late ‘60s than in the late ‘40s. So, in addition to the long-term temperature

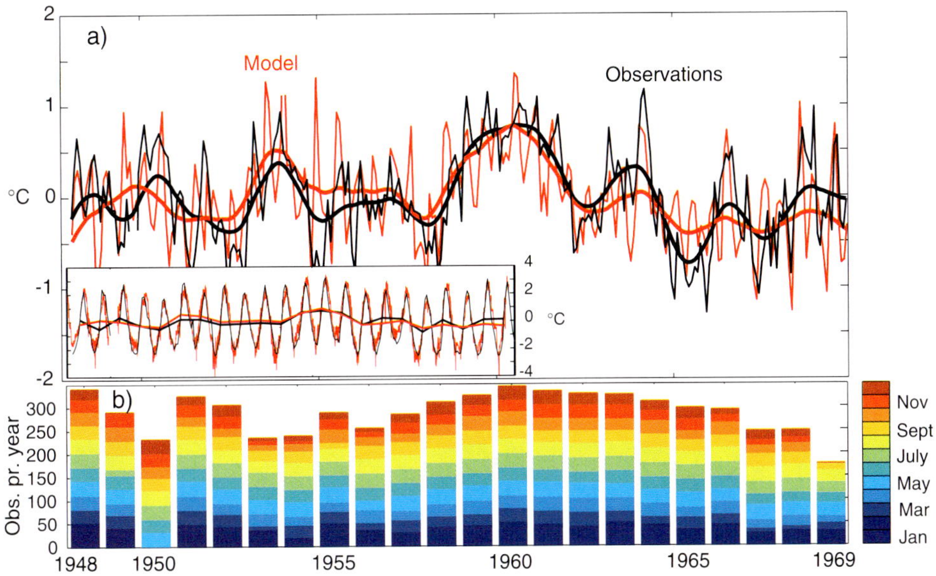

Plate 3. (a) The coastal temperature (with monthly averages and seasonal cycle removed) at Mykines (black) and the simulated temperature (with seasonal cycle removed) on the Faroe Shelf (red). A 12-month running average is shown with the thicker lines. The inset shows the corresponding time series before de-seasoning. (b) The sampling intensity during the period 1948-1969.

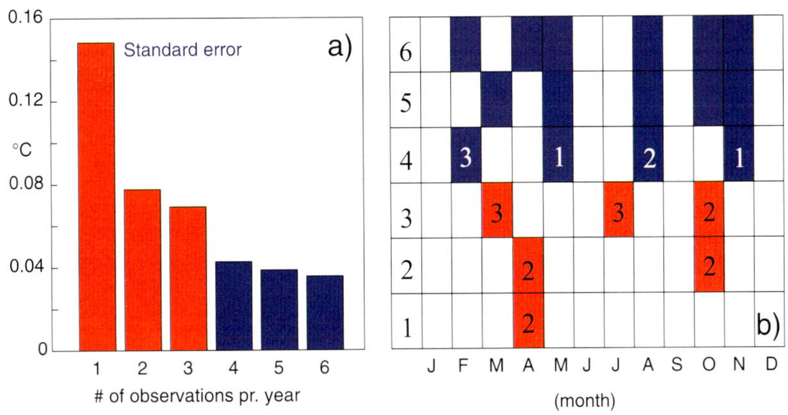

Plate 4. (a) The error introduced by removing the seasonal variation as a function of available data points each year. Red bars show the error when fewer than four observations are taken each year. (b) Optimal month (January to December) and week (indicated numbers) of observation, for 1 to 6 observations per year (left column). Months without numbers indicate midmonth sampling.

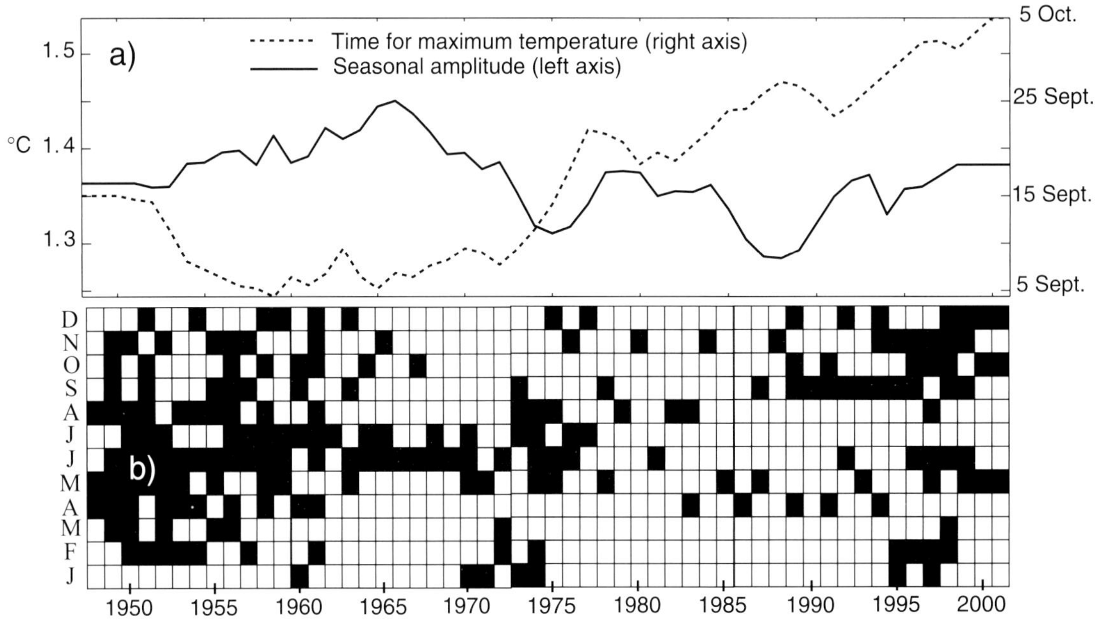

Figure 5. (a) The simulated change in the seasonal temperature variation in the FSC. The unbroken curve (left *y*-axis) shows the amplitude of the seasonality, and the broken curve (right *y*-axis) shows when, for each year, the temperature maximum occurs. (b) Monthly overview of the measurement strategy used in the FSC.

variation and the average seasonal cycle, the model also predicts the seasonal modulation on the Faroe Plateau in a realistic way.

In the densely sampled Mykines series, the 'true' annual averages and the 'true' seasonal cycle are known, and the need for a seasonal cycle with variable amplitude and phase can thus be tested. The temperature series are subsampled in different ways (monthly or more randomly distributed) and the seasonal variations and the annual averages of these subsampled series are found by using both a window-based, and full-series-based de-seasoning method. By comparing the results to the 'true' values, an error estimate can be obtained for each method. The window-based method is found to perform better than the method that assumes an average seasonal cycle.

4.2. The Faroe–Shetland Channel

The observed FSC series in Plate 2a has a character similar to the Mykines series with a dominating seasonal variation and a weaker decadal-scale variation. But, in contrast to the Mykines series, the sampling frequency is scattered throughout the measurement period (Figure 5b), and aliasing will obviously be a severe problem.

No reliable information on a changing seasonality (amplitude and phase) can be obtained from available observations, but the temperature seasonality in the FSC is known to be very variable [*Hansen et al.*, 1994]. This can even be seen by inspecting Plate 2a, although the prominent decrease in the early 1960s could be due to a changed measurement strategy after 1960 [*Turrell et al.*, 1993]. The error brought in by removing the observed seasonal cycle cannot therefore be quantified with confidence, and a comparison of the model data with the true decadal-scale variations is thus not meaningful in itself. A model data benchmarking is therefore limited to the point-wise comparison in Plate 2.

An analysis shows that the modeled and observed seasonal cycles in the FSC are symmetrical and can be modeled with a simple cosine, as was the case for the Mykines series. Fitting a cosine to the observed and simulated time series for the periods 1948 to 2001 gives the same amplitudes (1.32°C), phases differ by 3 days only, and maximum temperature occurs in mid-September. The seasonality is thus, in an average sense, simulated in a correct way.

5. POSSIBLE MODEL APPLICATIONS

5.1. Model-Based De-Seasoned Error in the FSC

The correlation coefficient between the time series from Mykines and FSC is 0.75, and the seasonal amplitude, the phase, and the long-term variations of the Mykines series are well described by the model. It is therefore reasonable to assume that the model should explain some of the seasonal modulation in the FSC as well. With this assumption, one can use the model to quantify the error brought in when the observed seasonality in the FSC series is removed.

The window-based de-seasoning method applied to the complete model time series from the FSC gives the values of the seasonal amplitude and phase as shown in Figure 5a. The figure shows that the simulated temperature amplitude varies more than 0.15 °C during the integration period, and the maximum temperature is found nearly 1 month later in 2001 than in the 1960s. Note the changes around 1965, and the inverse correlatation between amplitudes and the phases.

The main causes of errors when filtering the annual cycle are linked to the observational frequency, the timing of the observations (Figure 5b), or changes in the annual temperature cycle (Figure 5a). The 'true' annually averaged series is then based on the full (weekly averaged) simulated time series. A subsampled model time series with the measurement strategy used in Figure 5b is generated; the seasonal variation based on the subsampled series is found and subtracted (thereby deliberately bringing in errors), and the annual averages are calculated. During this procedure, the full-length time series (1948 to 2001) is used in the cosine fitting. The difference between the 'true' annual average and the one obtained after subsampling gives an estimate on the error brought in (see gray band in Plate 5).

5.2. Recommended Observation Strategy

As a curiosity, we examined the possibility of using the model as a guide for future monitoring. For subsampled model time series generated with one to six values each year, the seasonal variation based on each subsampled series is found and subtracted, and the annual averages are calculated. This has been done for all combinations in each subsampled series, and the standard deviation from the 'true' annual average has been calculated. The standard error found from the optimal way of placing the values throughout the year is provided in Plate 4a. The gain in accuracy is particularly large when up to four values per year are measured. The optimal measurement strategy when using four data-points each year, as based on 1948-2001 as the training period, is to obtain data the third week in February, the first week in May, the second week in August, and the first week in November (Plate 4b), that is, about 1 month prior to the seasonal peaks and the zero-crossings of the annual temperature cycle. This is in accordance with the sampling rate of CTD data employed in the FSC and north of the Faroe Islands from the early '90s to the present [*Hansen et al.*, 2003; *Turrell et al.*, 2003]. These monitoring programs should thus be maintained and not reduced. The standard de-seasoning error produced by the optimal measurement strategy of four cruises per year for the period 1948-2001 is about 0.04°C (Plate 4a). For comparison, the standard error produced by the measurement strategy used in the FSC over the same period is 0.17°C, more than would be obtained with only one optimally located observation each year (Plate 4). The large discrepancy is mainly attributable to periods with few, nonoptimally timed measurements during periods with sudden changes seasonality. In the late '60s only one to two measurements were made, mainly in June–July; this was a period when the seasonal amplitude decreased sharply (Figure 5). The checkerboard in Plate 4b shows that it is not wise to monitor in June–July because this is near the zero-crossings of the annual temperature cycle.

5.3. Homogenization of the Observed Temperature Time Series

If differences in the model–data comparison (Plate 2) are small compared to the long-term variations and to the error brought in by filtering the seasonality, then the simulated time series would have the potential to assist the observations during periods with sparse sampling. This is generally the case, except from the periods in the late 1960s and from 1989 to 1991 (Plate 2b).

In Plate 5, the full time-series model (weekly averaged values) is compared with the observed series, which has been de-seasoned with a cosine fitted to the full-length series. Only anomalies are shown, with the averaging as in Plate 2b. The error from the process of removing the annual cycle, as described above, is illustrated by the gray band.

The discussed uncertain period in the late '60s shows up as a period with a broad gray band and large deviation between simulations and observations. The '80s were surveyed once to twice a year, which is reflected in a model–data deviation and a rather broad error band. Only one cruise was made in 1990, but the previous and the subsequent year had a reasonable sampling (Figure 5b) so the error band was narrow, but still the model and the data disagree during this period. The observed series lie outside the error band throughout the 1990s, although good data are available for this period. The dashed line shows the result when analyzing the high-quality CTD observations from 1994 to 2001 separately. This shows that either a 'contamination' is introduced when analyzing and making anomalies of the observed, full-length time series as a whole, or the high-quality CTD observations reduce the mismatch between the observed and simulated time series.

Sources for errors are many and unclear, and the foundation for taking the analysis this far and for trusting the model more than observations are, admittedly, subjective. However, the presented analyses indicate the potential of using GCMs to interpolate for unevenly sampled ocean time series.

6. DISCUSSION AND CONCLUSIONS

The model compares well to the observed temperature variations on the Shetland side of the FSC at the irregular times

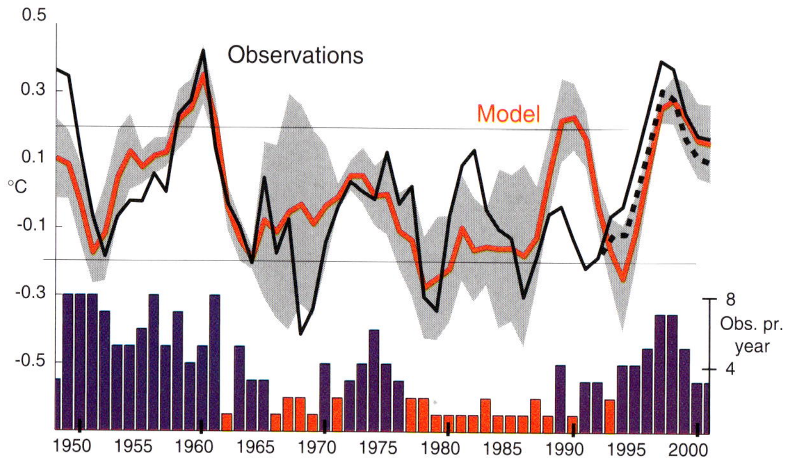

Plate 5. Observed temperature anomalies after removing the seasonal variation (3-year running averages and de-trended), (solid black line) and the weekly resolved simulated time series (red line). The gray zone illustrates the error introduced by de-seasoning as found by comparing the subsampled and the full simulated time series. Histogram: number of measurements made each year (years with fewer than four observations are shown in red). The dashed line shows the observations when only the CTD observations from 1994 to 2001 are considered.

when data are available (Plate 2a,b). Support for the good model–data correspondence is found at the Rockall section ('*Ellet* line') and at the Svinøy section (Figures 2 and 3). Furthermore, the applied model captures most of the seasonality, the changes in the seasonality, and the decadal temperature variations on the Faroe Shelf (Plate 3 and Figure 4).

Since the observed FSC time series is inhomogeneously sampled and with more energy on the seasonal cycle than on lower frequency variations, aliasing is a problem [*Reverdin et al.*, 1994]. *Reverdin et al.* [1994] found that the use of available surface data can critically reduce the aliasing problem in the discussed FSC series. With support from the daily SST measurements from the Faroe Shelf and the fact that the model is able to simulate the observed surface temperatures on the Shetland Shelf practically perfectly (not shown), it should therefore be possible to further pursue the approach used by *Reverdin et al.* [1994].

The variable seasonal cycle in the FSC [*Hansen et al.*, 1994] enlarges the aliasing problem. Applying a time window-based de-seasoning procedure to the modeled series indicates that the seasonal modulation in amplitude and phase vary broadly throughout the integration period 1949-2001.

From telegraph cable measurements, *Hansen et al.* [1994] found that the seasonal temperature variation in the FSC almost turned off around 1910, which coincided with a temperature and salinity minima. They suggested that the changed balance between the Atlantic flow and the East Icelandic Current led to a disruption or reversion of the "normal" seasonal variation of the flow systems. The strongest model-predicted modulation of the temperature seasonality in the FSC (Figure 5a) happened from 1965 to 1977, which coincides with the Great Salinity Anomaly (GSA) of the 1970s [*Dickson et al.*, 1988]. The explanation for the changes in seasonality during the 1970s GSA is likely the same as the explanation for the 1910s GSA [*Hansen* and *Kristiansen*, 1994] and indicates that the simulated seasonal modulation is physical and not just a numerical artifact.

By assuming that the simulated time series is exact, we can estimate the de-seasoning errors introduced by using a subsampled temperature time series. This analysis indicates it is inadvisable to survey the hydrographic section fewer than four times a year if reliable decadal-scale temperature variations are of interest.

This last point is clear when comparing de-seasoned and averaged observations and the simulated low-frequency temperature variations (cf. Plates 2 and 5). The observed and the simulated time series follow each other closely, and the error band from the de-seasoning is narrow, when four or more samples are obtained per year. For periods with poor sampling (red bars in Plate 5), the error band widens and the two time series diverge. This result calls for caution in interpreting the seemingly cold anomaly in the late 1960s and the seemingly warm anomaly in the early 1980s in Plate 5 as real anomalies.

The Mykines series shows no cold anomaly in the late 1960s and thereby supports the simulated time series. However, a cold anomaly is prominent north of Iceland [*Malmberg and Valdimarsson*, 2003] and at 400 m deep at the Ocean Weather Station "Mike" (OWSM) [*Østerhus* and *Gammelsrød*, 1999] in the late 1960s, indicating that a cold anomaly might have been an isolated feature in the Nordic Seas. The warm anomaly in the early 1980s is evident in the observed Rockall series and in the observed Svinøy series (Figures 2 and 3). The model, in contrast, represents the anomaly at the Rockall section but not at the Svinøy section. This anomaly is therefore probably real, but the model is seemingly not able to properly pass it through the FSC. The reason for the model–data misfit during the years 1989-1991 is unclear. The observed Rockall series shows no warm anomaly during this period, but both model and observations show this anomaly very clearly at the Svinøy section.

It should be mentioned that a potential problem in using the observational temperature series presented here is the mix of the Fair Isle–Munken and the Nolsoy–Flugga sections. Analyses of the simulated temperature series from the two sections support fusion of the two lines, although the short overlap between the two observed temperature series [*Ellet and Turrell*, 1992] makes the fusion statistically weak.

Finally, an ultimate description of the ocean dynamics and thermodynamics will be based on data assimilation systems where observations and simulated fields are merged in an optimal way. Unfortunately, the sparse temporal–spatial sampling of the ocean, in combination with the small length scales found in the region, makes such an approach difficult. It is therefore encouraging that hindcast simulations like the one presented here show realistic behavior in several aspects, illustrating the potential for hindcast simulations to expand observed time series and to evaluate the highly observational-demanding and complex coupled atmosphere–sea-ice–ocean climate models.

Acknowledgments. The authors are grateful to Bogi Hansen, the Faroese Fisheries Laboratory, and Bill Turrell, the Marine Laboratory in Aberdeen, for providing the observed time series used in the work. The study has been supported by the Nordic Council of Ministers program *Vestnordisk Oceanklima*; the Research Council of Norway through RegClim, ProClim, and the Program of Supercomputing; and the European Union DG-XII Climate and Environment Programme through the project NOCES (EVK2-2001-00115). This is contribution no. A47 from the Bjerknes Centre for Climate Research.

REFERENCES

Anderson, L.G., H. Drange, M. Chierichi, A. Fransson, T. Johannessen, I. Skjelvan and F. Rey, Seasonal and annual variability in the upper Greenland Sea based on measurements and a box model, *Tellus*, 52B, 1013-1020, 2000.

Beaugrand, G. and P.C. Reid, Long-term changes in phytoplankton, zooplankton, and salmon linked to climate, *Glob. Change Biol.*, 9, 801-817, 2003.

Belkin, I.M., S. Levitus, J. Antonov and S.Aa. Malmberg, "Great Salinity Anomalies" in the North Atlantic, *Progress in Oceanography*, 41, 1-68, 1998.

Bentsen, M. and H. Drange, Parameterizing surface fluxes in ocean models using the NCEP/NCAR reanalysis data. In: RegClim General Technical Report No. 4, pp 149-158. Norwegian Institute for Air Research, Kjeller, Norway, 2000.

Bentsen, M., H. Drange, T. Furevik and T. Zhou, Simulated variability of the Atlantic meridional overturning circulation, *Clim. Dynam., In Press,* 2004.

Bleck, R., C. Rooth, C. Hu and L.T. Smith, Salinity-driven thermohaline transients in a wind- and thermohaline-forced isopycnic coordinate model of the North Atlantic, *Journal of Physical Oceanography*, 22, 1486-1515, 1992.

Dickson, R.R. and J. Brown, The production of North Atlantic Deep Water: Sources, rates, and pathways, *Journal of Geophysical Research*, 99, 12319-12341, 1994.

Dickson, R.R., J. Meincke, S.Aa. Malmberg and A.J. Lee, The "Great Salinity Anomaly" in the Northern North Atlantic 1968-1982, *Progress in Oceanography*, 20, 103-151, 1988.

Ellet, D.J. and W.R. Turrell, Increased salinity levels in the NE Atlantic, *ICES CM 1992/C:20*, 12 pp, 1992.

Fairall, C.W., E.F. Bradley, D. P. Rogers, J.B. Edson, and G.S. Young, Bulk parameterization of air–sea fluxes for Tropical Ocean–Global Atmosphere Coupled-Ocean Atmosphere Response Experiment, J. Geophys. Res., 101, 3747-3764, 1996.

Furevik, T., Annual and interannual variability of Atlantic water temperatures in the Norwegian and Barents Seas: 1980-1996, *Deep Sea Research*, 48, 383-404, 2001.

Furevik, T., M. Bentsen, H. Drange, J.A. Johannessen, and A. Korablev, Temporal and spatial variability of the sea surface salinity in the Nordic Seas, *Journal of Geophysical Research*, 107, 2002.

Hansen, B., H.P. Joensen and V.E. Michelsen, Bottom temperature between Iceland and Shetland 1906-1962 measured in telegraph cables, *ICES CM 1994/S:5*, 14 pp, 1994.

Hansen, B. and R. Kristiansen, Long-term changes in the Atlantic water flowing past the Faroe Islands, *ICES CM 1994/S:4*, 16 pp, 1994.

Hansen, B. and J. Meincke, Long-term coastal sea surface temperature observations at the Faroe Islands, *Rapports et Procès-Verbaux des Réuinons du Conseil International pour l'Exploration de la Mer*, 185, 162-169, 1984.

Hansen, B. and S. Østerhus, North Atlantic-Nordic Seas Exchanges, *Progress in Oceanography*, 45, 109-208, 2000.

Hansen, B., S. Østerhus, H. Hátún, R. Kristiansen and K.M.H. Larsen, The Iceland-Faroe inflow of Atlantic water to the Nordic Seas, *Progress in Oceanography*, 54, 443-474, 2003.

Helland-Hansen, B. and F. Nansen, The Norwegian Sea, *Fiskeridirektoratets Skrifter Serie Havundersøkelser*, 2, 1-390, 1909.

Holliday, N.P., R.T. Pollard, J.F. Read and H. Leach, Water mass properties and fluxes in the Rockall Trough, 1975-1998, *Deep Sea Research*, 47, 1303-1332, 2000.

Kalnay, E., M. Kanamitsu, R. Kistler, W. Collins, D. Deaven, L. Gandin, M. Iredell, S. Saha, G. White, J. Woolen, Y. Zhu, M. Chelliah, W. Ebisuzaki, W. Higgins, J. Janowiak, K.C. Mo, C. Ropelewski, J. Wang, A. Leetma, R. Reynolds, R. Jenne, and D. Joseph, The NCEP/NCAR 40-year reanalysis project. *Bullet. Amer. Meteorol. Soc.*, 77, 437-471, 1996.

Kristmannsson, S.S, Flow of Atlantic Water into the Northern Icelandic Shelf area, 1985-1989, *ICES Cooperative Research Report*, 225, 124-135, 1998.

Levitus, S. and T.P. Boyer, World Ocean Atlas 1994 Volume 4: Temperature. NOAA Atlas NESDIS 4, 1994.

Levitus, S., R. Burgett and T.P. Boyer, World Ocean Atlas 1994 Volume 3: Salinity. NOAA Atlas NESDIS 3, 1994.

Malmberg, S. Aa. and H. Valdimarsson, Hydrographic conditions in Icelandic waters, 1990-1999, 219, 1-453, 2003.

Nilsen, J.E., Y. Gao, H. Drange, T. Furevik and M. Bentsen, Simulated North Atlantic-Nordic Seas water mass exchanges in an isopycnic coordinate OGCM, *Geophysical Research Letters*, 30(10), 1536, doi:10.1029/2002GL016597, 2003.

Østerhus, S. and T. Gammelsrød, The abyss of the Nordic Seas is warming, *Journal of Climate*, 12, 3297-3304, 1999.

Reverdin, G., D. Cayan, H.D. Dooley, D.J. Ellet, S. Levitus, Y.D. Penhoat and A. Dessier, Surface salinity of the North Atlantic: Can we reconstruct its fluctuations over the last one hundred years?, *Progress in Oceanography*, 33, 303-346, 1994.

Sakshaug, E., B. Bjørge, B. Gulliksen, H. Loeng and F. Mehlum, Structure, biomass distribution, and energetics of the pelagic ecosystem in the Barents Sea: A synopsis, *Polar Biol.*, 14, 405-411, 1994.

Turrell, W.R., E. Devonshire, R. Payne and G. Slesser, Analysis of the historic time-series obtained in the Faroe-Shetland Channel, *ICES CM 1993/C:*29, 21 pp, 1993.

Turrell, W.R., B. Hansen, S. Hughes and S. Østerhus, Hydrographic variability during the decade of the 1990s in the Northeast Atlantic and southern Norwegian Sea, *Marine Science Symposia*, 219, 111-120, 2003.

M. Bentsen, H. Drange, and A. Sandøe, Nansen Environmental and Remote Sensing Centre, Edv. Griegsv. 3A, 5059 Bergen, Norway. (mats.bentsen@nersc.no; helge.drange@nersc.no; annebrit@nersc.no)

H. Hátún, Faroese Fisheries Laboratory, Nóatún 1, FO-110, Faroe Islands. (hjalmarh@frs.fo)

Convective Chimneys and Plumes in the Northern Greenland Sea

Ola M. Johannessen,[1,2,3] Kjetil Lygre[1], and Tor Eldevik[1,4]

In the Boreas Basin in the northern Greenland Sea, convection has been observed to be localized as narrow structures—chimneys—in which the water masses may be homogeneous from the surface through great depths. An experiment was carried out in the area by R/V *Håkon Mosby* of the University of Bergen with an aim to detect and investigate chimneys, their structure, and the inherent convective plumes. A chimney was detected at 77.5°N, 0.5°E, by towing an undulating conductivity–temperature–depth (CTD) profiler (Seasoar). Complementary deep CTD stations taken as close as 2.5 km apart, revealed that the chimney was about 30 km in the zonal and 15 km in the meridional directions. The mixed-layer depth inside the chimney was about 500 m, indicating that the chimney was of an upper ocean type. Deep convection was improbable because of the modest cooling of a culminating winter. The Seasoar tows within the chimney revealed weak horizontal temperature, salinity, and density anomalies of 1-5 km extent, i.e. of the order of the internal Rossby deformation radius. Similar scales were characteristic of a nonhydrostatic ocean model set up accordingly. The CTD sections also revealed enhanced sea surface density within cyclonic rim eddies with a diameter of the order of 5 km. Wavelet spectra of the horizontal temperature variability inside the core of the chimney revealed increased power on the 100-m scale, which is of the same magnitude as a plume scale determined from theoretical studies. This is the first time such a plume scale has been directly derived from high-resolution (~10 cm) horizontal CTD (Seasoar) measurements alone.

INTRODUCTION

During the past few decades the understanding of the links between mesoscale ($O(1–10\text{ km})$) dynamics and the global circulation through deep water formation has been a key area of ocean science. A central phenomenon is the *chimney* process [MEDOC-group, 1970], in which active deep convection takes place within a spatially limited area. Its mechanisms have been reviewed in detail by Marshall and Schott [1999], who further claimed that little *net* vertical mass flux takes place inside a chimney and therefore recommended that the term "mixed patch" should be used instead. In the present article, however, we will adhere to the original terminology in accord with the MEDOC-group [1970].

Although the detailed mechanisms of the chimney process remain unknown, chimneys have been observed during wintertime in the weakly stratified waters of the Mediterranean, Labrador, and Weddell seas [*Gascard*, 1991], as well as the Greenland and Boreas basins [*Johannessen et al.*, 1991]. In synthesizing all available observations on chimneys until then, *Gascard* [1991] aimed at a more precise definition of their characteristics. There is now a general agreement that a chimney goes through three main stages. First, there is

[1]Nansen Environmental and Remote Sensing Center, Bergen, Norway.
[2]Geophysical Institute, University of Bergen, Bergen, Norway.
[3]Mohn-Sverdrup Center for Global Ocean Studies and Operational Oceanography, Bergen, Norway.
[4]Bjerknes Centre for Climate Research, Bergen, Norway.

The Nordic Seas: An Integrated Perspective
Geophysical Monograph Series 158

10.1029/158GM17

a preconditioning phase in which the intermediate water is brought to the surface [*Killworth*, 1979]. The triggering mechanism may be baroclinic instability of the mean flow, a topographically induced perturbation, or ice edge upwelling [*Johannessen et al.*, 1983]. Second, there has to be strong forcing (cooling) at the surface in order to convert the intermediate water to deep water by buoyancy loss. A substantial part of the cooled, dense water overturns as plumes with a horizontal scale of the order of 100 m, before rapid mixing of the properties of the plumes is forming a homogeneous mixed patch, tens of kilometers in horizontal extent [*Marshall and Schott*, 1999]. The final stage, decay, involves baroclinic instability of the rim current of the chimney [*Send and Marshall*, 1995], manifested as rim eddies with diameters of the order of the internal Rossby radius of deformation. Numerical model experiments by *Send and Marshall* [1995] showed that the rim features are organized as several smaller cyclones, or *cones*, propagating away from the chimney and thus dispersing properties away from the generation area.

There has recently, however, been a shift in focus from the gyre-scale mixed patch of *Marshall and Schott* [1999] back to the mesoscale chimney and its associated dynamics, e.g., *Gascard et al.*, [2002]. The chimney concept of *Killworth* [1979] is consistent with the recent emphasis that it is the ocean "preconditioning" (as opposed to anomalies in the atmospheric forcing) that mainly localizes convection [*Legg and Marshall*, 1993; *Straneo and Kawase*, 1999]. Also, for the last decade or so there has been no sign of gyre-scale overturning in the Greenland Sea [*Watson et al.*, 1999; *Alekseev et al.*, 2001, 2003]. Ventilated water has instead been associated with pronounced mesoscale eddies of about 5-km radius [*Gascard et al.*, 2002; *Wadhams et al.*, 2002]. In agreement with this scenario such eddies were generated in the numerical model ocean of *Eldevik and Drange* [2002]. This is also consistent with Killworth's model [*Killworth*, 1979], in which chimneys were postulated to be preconditioned by baroclinic instability of the Greenland Sea Gyre rim current. A result of the instability is dense-core, weakly stratified cyclonic eddies. When the gyre region is exposed to wintertime cooling, these anomalies will be prone to convective overturning.

Through several investigations during the Marginal Ice Zone Experiments (MIZEX) campaign, a number of upper ocean chimneys or chimney-like structures were observed and described. During the MIZEX-87 programme in the Boreas Basin [MIZEX-group, 1989; *Sandven et al.*, 1991] structures were typically 10-20 km in diameter, extended from the surface down to about 600 m, and consisted of lower Arctic Intermediate Water (lAIW), which was relatively warm (1.0°C), saline ($S = 34.96$), and dense ($\sigma_\theta = 28.02$) compared to the surrounding water. Two years later during the Seasonal Ice Zone Experiments 89 program [SIZEX group, 1989; *Johannessen et al.*, 1991], a remarkably well-defined chimney containing Deep Water ($T = -0.41$°C, $S = 34.91$) was observed outside the ice edge. It was 20 km in diameter with uniform potential density in the upper 500 m, below which the density increased by only 0.02 sigma-units near bottom at 2800 m. (Hereafter we refer to this as *Chimney89*.)

In order to increase our understanding of chimneys in relation to convective overturning and to characterize the submesoscale and the scale of the plumes within the chimneys, an investigation was carried out in the Boreas Basin in the northern Greenland Sea during February 1994.

In the next section we present the methods and experimental approach. In the third section the observational data are presented together with the results of a nonhydrostatic ocean model. The results are discussed and summarized in the final section.

METHODS AND EXPERIMENTAL LAYOUT

The aim of the R/V *Håkon Mosby* ESOP/CARDEEP[5] experiment in the Boreas Basin during 9-21 February 1994 was to detect chimneys and subsequently describe their structure, scales, and dynamics, including plumes within the chimneys. Whereas previous cruises in the Boreas Basin were more concentrated close to the ice edge [MIZEX-group, 1989; *Johannessen et al.*, 1991], the central part of the basin was the focus of the present investigation. The cruise track is shown in Figure 1. During the initial period of the experiment, the basin was crisscrossed with a towed undulating CTD sensor platform (Seasoar) profiling the upper 200-250 m of the water column, sampling typically every 2.5 km in the horizontal to detect chimneys; these features were further investigated by detailed Seasoar tracks and densely spaced deep CTD casts, to obtain optimal coverage both horizontally and vertically. This is the first time the horizontal structure inside a chimney has been investigated with such high horizontal resolution.

The definition of the detailed experiment centered around 77.5°N, 0.5°E was based on the discovery of a relatively homogeneous, high-density area, subsequently defined as a chimney. It was further investigated first by crisscrossing the site with the Seasoar in a star-patterned grid, followed by a towing of the Seasoar at a fixed depth of 20 m (Figure 2) along 6 zonal legs about 30 km long separated by approximately 4 km, effectively sampling every 10 cm in the horizontal. Hence, the postulated horizontal plume scale of ~100 m would be well resolved. Eventually, 40 CTD casts were taken to 1000 m or near the bottom, with station separation as low as 2.5 km in

[5]European Sub-Polar Ocean Programme/Carbon and Deepwater Formation.

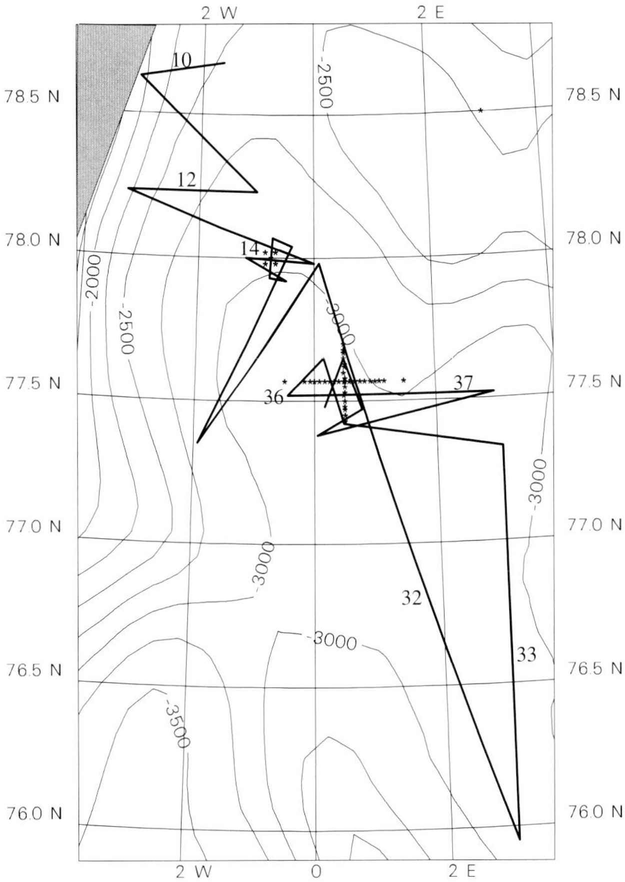

Figure 1. Ship track (lines) with leg numbers referenced in the text, plotted for those tracks on which the Seasoar was vertically profiling the mixed layer down to 250 m. Locations of the deep CTD stations (asterisks) and the ETOPO5 bathymetry (labeled contours) are also shown. The ice edge is indicated in the upper left corner in gray shading.

order to resolve the mesoscale structure of the chimney in the upper part of the ocean. One N–S section was taken along 0.5°E from 77.43°N to 77.7°N, and an E–W section was taken along 77.57°N from 0.5°W to 1.5°E (Figure 2).

The same Neil Brown Mark III CTD sonde was used both in the Seasoar and deep CTD surveys in order to obtain a consistent data set. Salinity has been computed according to PSS78 [UNESCO, 1983], and about 5 salinity samples were taken on each deep station for calibration purposes. An overall r.m.s. difference between the calibrated CTD and salinometer analysis of ~5×10^{-3} was found, sufficiently accurate for this investigation, as our study is focused in the upper 1000 m. Temperature accuracy is about 2×10^{-3} °C. Temperatures θ are hereafter given as referred to surface pressure, which is also the case for the density variable σ_θ, if not otherwise stated. In situ temperature is denoted by T. Except for the fixed-depth tows, all CTD data are averaged over 2 dBar.

A ship-mounted Aanderaa weather station provided standard meteorological parameters from which surface heat flux has been estimated. Furthermore, Synthetic Aperture Radar (SAR) images from the European Space Agency ERS-1 satellite facilitated the determination of characteristic length scales and shapes of the mesoscale structures in the investigation area. A hull-mounted Acoustic Doppler Current Profiler was logging the water velocity. Unfortunately, the quality of navigation information was not good enough to obtain absolute velocities; therefore, these data have been disregarded in the following analysis.

DATA

Before describing the chimney we first provide a brief characterisation of the ambient upper ocean water masses and the mesoscale features captured by the Seasoar profiling, as well as the deep hydrographic structure. We also estimate the turbulent heat flux (the dominant part of the total) to 200 Wm^{-2}. This is based on typical figures for the observed wind speed, 10 m s^{-1}, atmospheric temperature, –10°C, and relative humidity, 75%, logged by the weather station in the mast of the ship, and sea surface temperature of about 0°C. Due to the difficulties in correcting for the effect of the ship on the air flow, we have not considered the details of the temporal evolution of the heat flux.

For the Seasoar we present, for brevity, only surface temperature data; the salinity plots are reported in Johannessen and Lygre [1996], from which the referenced salinity values are taken. From the tracks (Plate 1) it is evident that a major part of the surface water is relatively warm ($T > 0.6$°C) and saline ($S > 34.9$). We will denote this water mass Arctic Surface Water (ASW), even if it is slightly too saline to fit the classical definition of *Carmack* [1990]. In the eastern and northern parts of the area covered by the Seasoar tows there are occurrences of still warmer ($T > 1.5$°C) and saltier ($S > 34.96$) water which apparently is recirculated from the Atlantic domain [*Johannessen et al.*, 1987a,b]; we denote this Atlantic Water (AW). Substantially colder and fresher water masses are found at the boundary of the survey area, north of 78°N and west of 3°W, i.e., Polar Water defining the Polar Front. Similar water masses occur south of 76.3°N in association with the steep southeastward directed topographic ridge—the Greenland Fracture Zone—apparently guiding admixtures of water from the Polar domain into the basin. The Seasoar, limited to span the surface layer, only casually captured the intermediate water masses below the mixed layer (ML).

Characteristic of the region is the mesoscale variability (Plate 1), including cold and less saline water as well as warm and saline water with several patches on the 5-10-km horizontal scale with AW signatures. The chimney is centered at

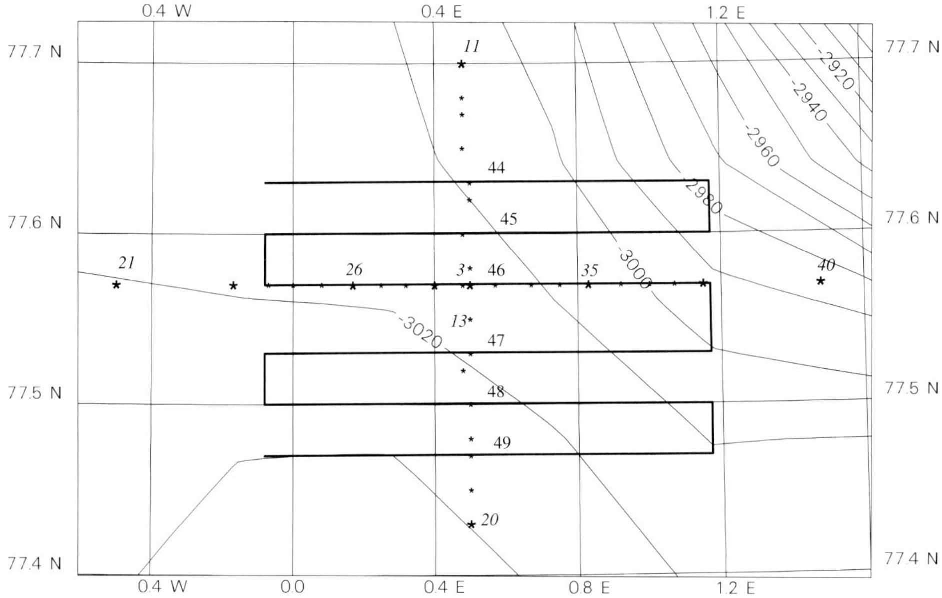

Figure 2. Blow-up of the survey area in the *Chimney94* region. Deep CTD stations are marked with asterisks; smaller symbols indicate 1000-m casts, larger symbols denote casts to near bottom. Station numbers in italics are referenced in other figures. The tracks where the Seasoar was towed at a nominal depth of 20 m are shown with lines; track numbers are in normal types. Coarse bathymetry (ETOPO5) is shown as labeled contours.

about 77.55°N, 0.5°E, roughly bounded by the 0.7-0.8°C isotherms, i.e. belonging to the ASW and surrounded by AW eddy-like intrusions. The size of the horizontal features scales well with the Rossby internal deformation radius R_d, estimated by taking numbers from the vertical density profile in Figure 3 as

$$R_d = \frac{N}{f} H \simeq 3 \text{ km}$$

in which N is the buoyancy frequency, f the local Coriolis parameter, and H a vertical scale here equated with the ML depth (500 m). In computing N, the density gradient is computed from σ_1 at 1000 m and at the surface, using UNESCO [1981]. Even if full two-dimensional coverage is not granted for these patches, they are interpreted as eddies, since they are of the order of R_d and eddies are known to be abundant in the area [*Johannessen et al.*, 1987b, 1995].

Providing further detail in the vertical, the deep CTD stations generally reveal that the ML depth is about 500 m (Figure 3) with $\theta \simeq 0.6$°C, $S \simeq 34.93$, below which the temperature and salinity abruptly change to $\theta \simeq -0.2$°C, $S \simeq 34.89$, with a minor density increase. The following 150 m is the layer of largest static stability, in which the density changes 0.015 units over 150 m.

Between the ML base and about 1000 m depth, the salinity is fairly constant (34.89), while a temperature gradient is stabilising with $\theta \simeq -0.7$°C at 1000 m. The temperature then decreases somewhat more slowly while salinity increases to about 34.91 at 1400 m. Even if a temperature minimum is not observed as in the Greenland Gyre [*Budéus*, 1998], this water mass fits approximately to the definition of upper Arctic Intermediate Water (uAIW). Similarly, the underlying water mass is interpreted as lAIW, although no clear property maxima are found around 1000-1500 m. At 3000 m the temperature is down to −1.15°C, with salinity about 34.91, i.e. well-defined Greenland Sea Deep Water. There is also a very weak indication of a salinity maximum around 2300 m, indicating a weak influence of Arctic Ocean Deep Water [*Aagaard et al.*, 1991].

The upper and intermediate waters do not fit precisely into a common water-mass classification [*Carmack*, 1990], which may be no surprise, as the warm, saline influence in the Boreas Basin comes more directly from the Atlantic and Norwegian Sea waters, compared to the Greenland Basin. In particular, the surface water, which was arbitrarily defined as ASW, is close to lAIW characteristics. However, since lAIW is located deeper and apparently isolated from the surface, it is awkward to categorize the surface water as AIW. It can rather be interpreted as a derivative of AW and uAIW.

Chimney94

When inspecting the vertical deep CTD sections taken through the chimney, it is apparent that it can be defined by

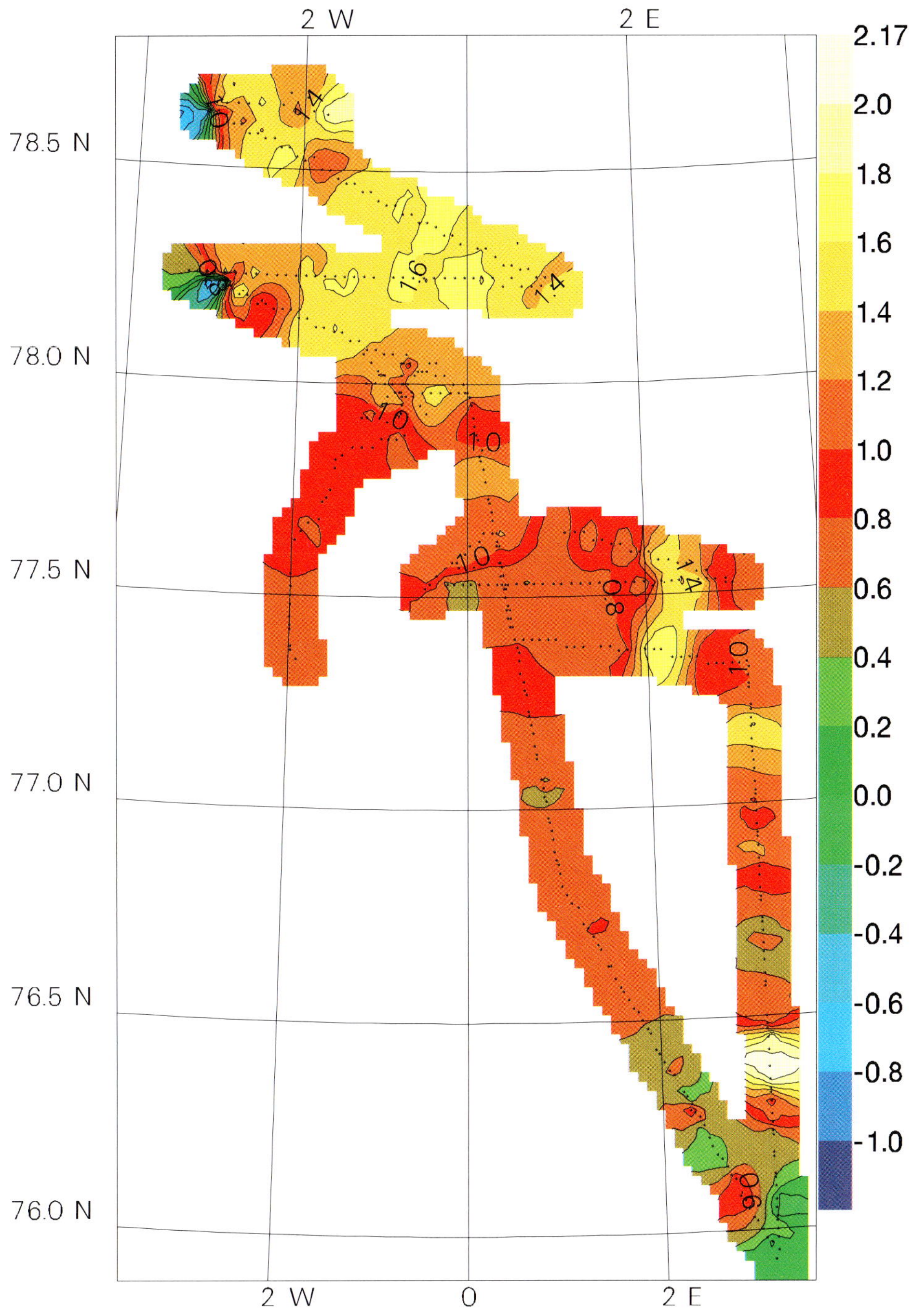

Plate 1. Horizontal plot of near-surface (within 5 m) temperature obtained with the Seasoar on 9-16 February 1994. Temperature scale (°C) to the right; contour interval is 0.2°C. Black dots indicate location of individual profiles, approximately 2.5 km apart.

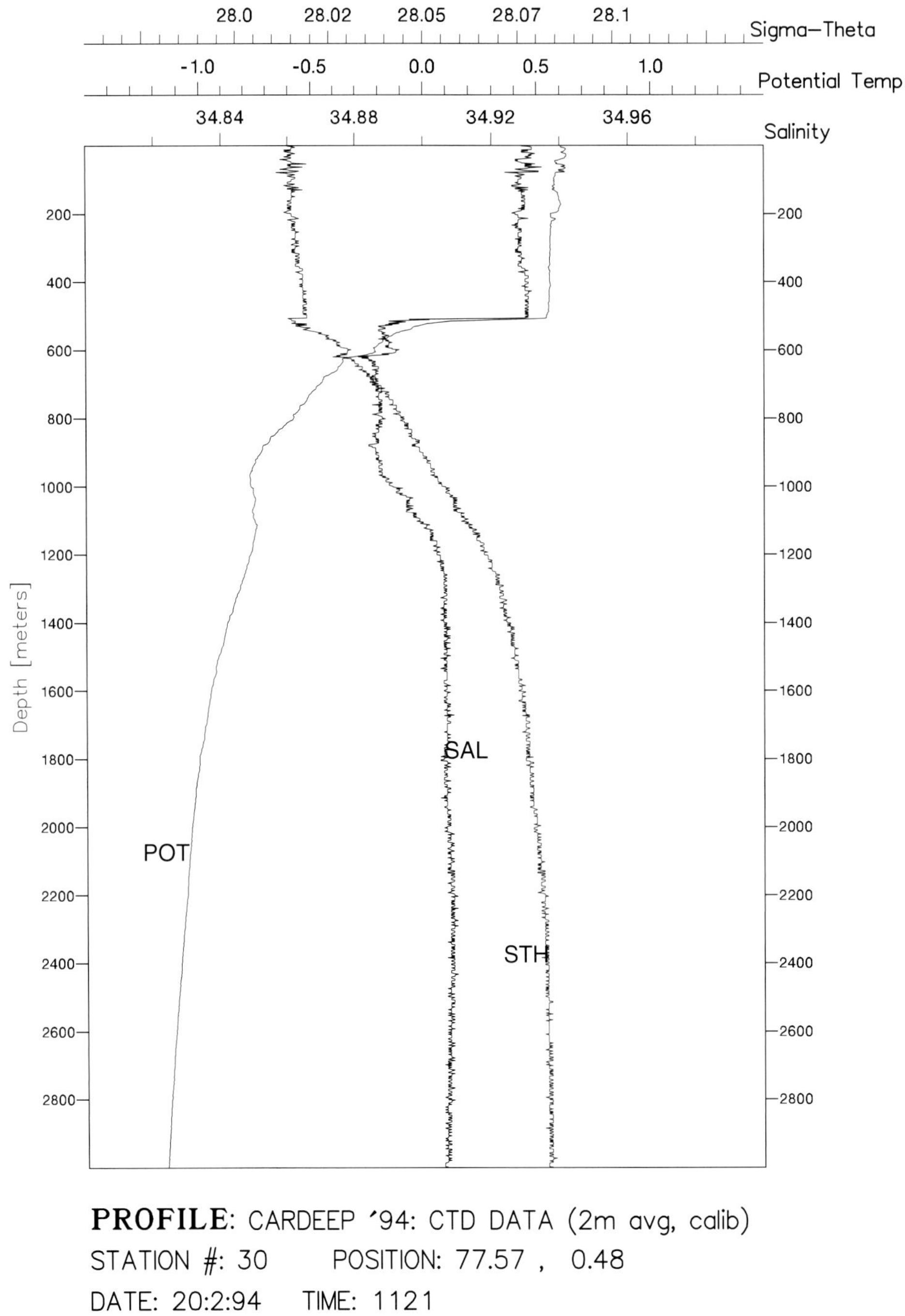

Figure 3. Profiles at station 30 in the center of the chimney. (POT, potential temperature; SAL, salinity; STH, σ_θ).

the fairly homogeneous water mass of the ML, in which salinity is about 34.93, temperature 0.5-0.7°C, and $\sigma_\theta <\simeq 28.02$ sigma-units, with a well-defined boundary towards a warm, saline water mass to the north (Figure 4) and west (Figure 5). The core of the chimney has a north–south and east–west extent of approximately 15 and 30 km, respectively, using the 0.7°C isotherm as a criterion (Figures 4a and 5a). In the southern and eastern parts, the horizontal property gradients are weak, implying a more diffuse transition towards the exterior.

There is also a slight variability within the core, in that the coldest and freshest water appears close to the northern rim (Figures 4a,b between 6 and 10 km). This also corresponds to the densest water ($\sigma_\theta \simeq 28.02$; Figure 4c) with no detectable change over the 500-m-deep ML, consistent with cooling

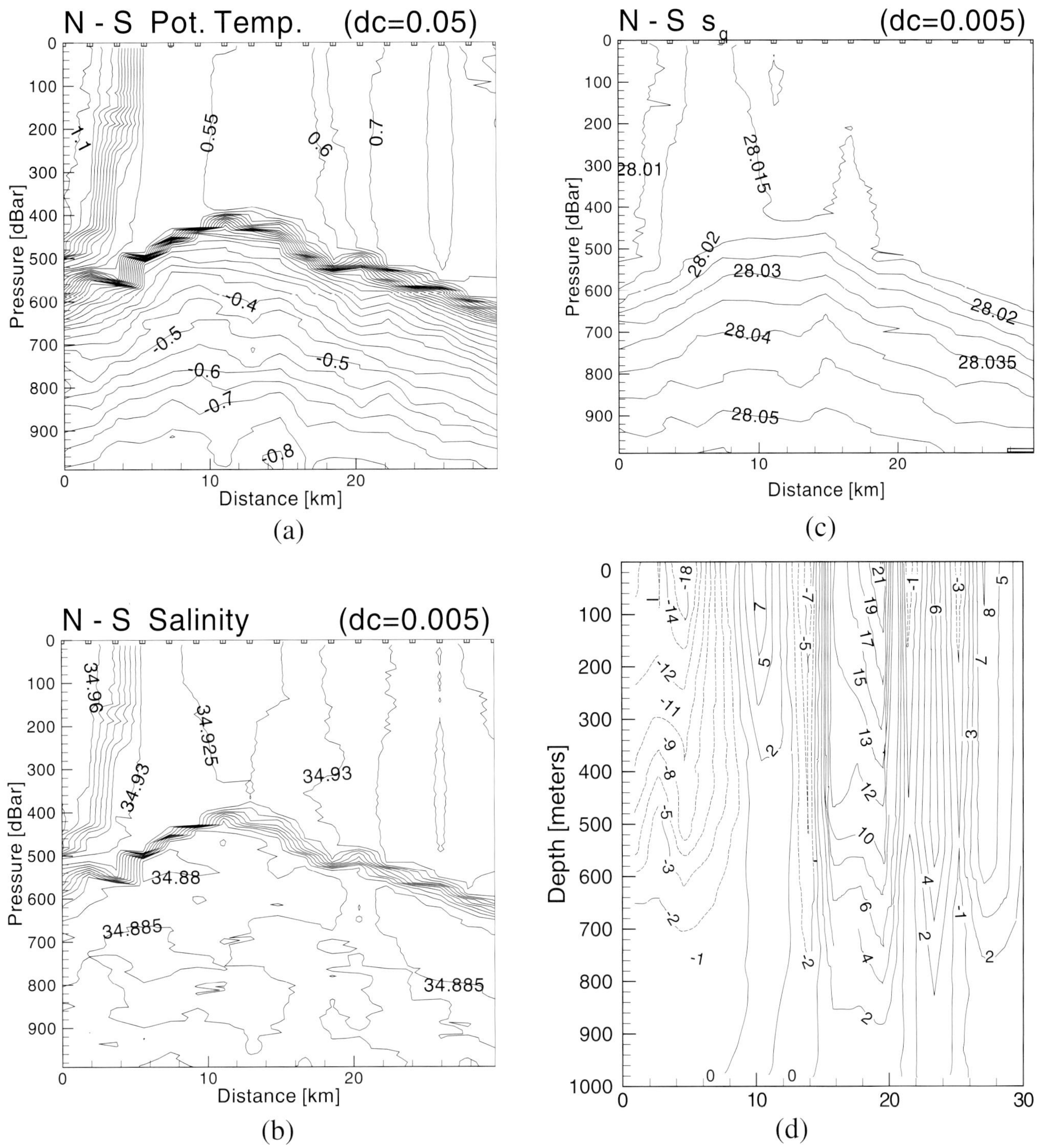

Figure 4. North to south CTD section through the core of chimney of (a) potential temperature, (b) salinity, (c) σ_θ and (d) geostrophic speed [cm s^{-1}] relative to 1000 dBar. Positive current is directed eastward, into the plane of the page. The features between 6 and 10 km and between 10 and 15 km are described in the text.

and buoyancy loss from the surface. Slightly to the south there is a tendency for a weak salinity stratification ($\Delta S \simeq 0.01$ over the uppermost 250 m [Figure 4b], between 10 and 15 km), implying a corresponding density stratification (Figure 4c), whereas all other stations further south have changes in ML density less than 0.01 sigma-unit.

The pycnocline (Figure 4c) shows a well-defined dome shape reaching 500 m deep at the center and deepening

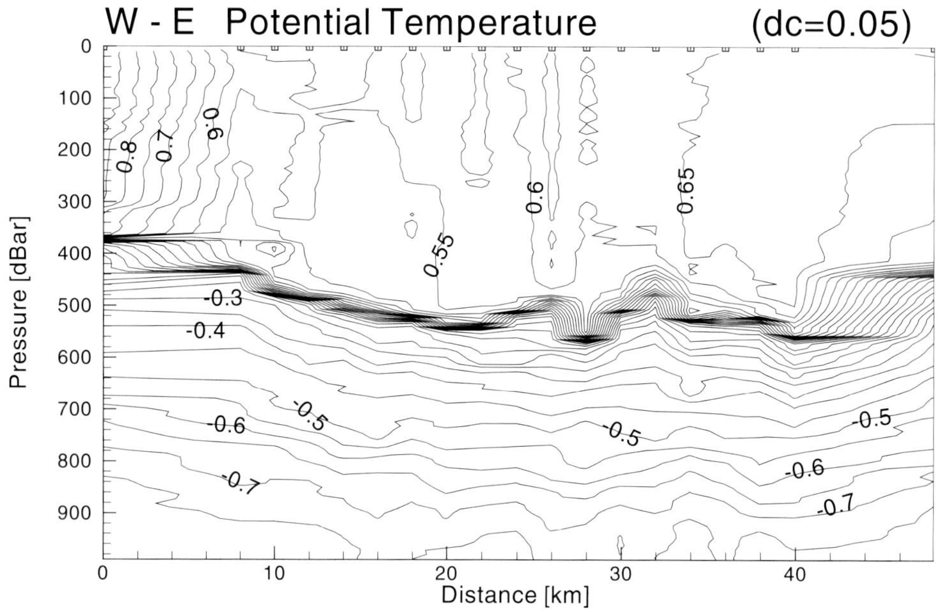

(a)

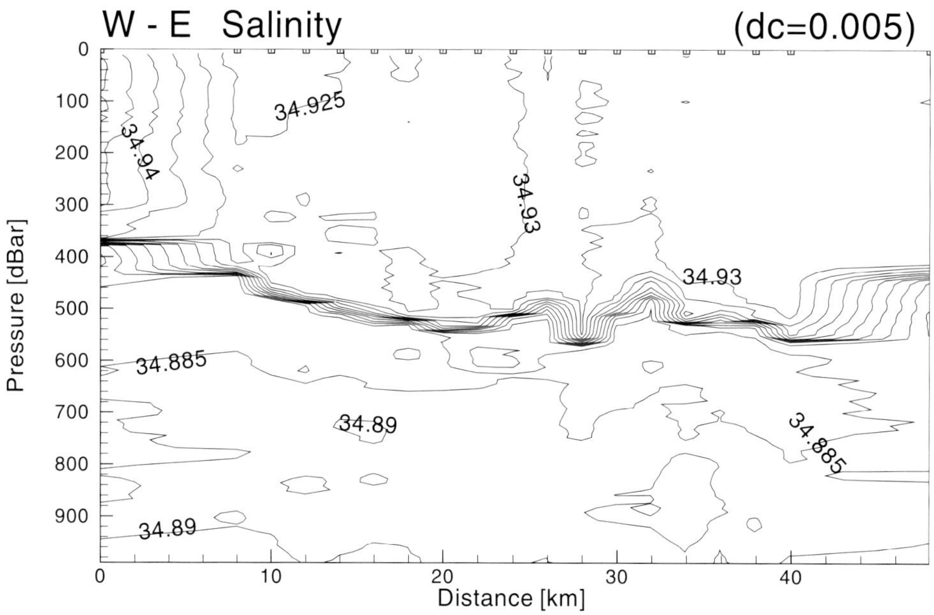

(b)

Figure 5. West to east CTD section through the core of chimney of (a) potential temperature, (b) salinity, (c) σ_θ, and (d) geostrophic speed [cm s^{-1}] relative to 1000 dBar. Positive current is directed northward, into the plane of the page. The features between 8 and 15 km are described in the text.

to about 600 m at the rim. The implied geostrophic speed (Figure 4d) shows a westward surface intensified jet extending over about 10 km near the northern rim reaching about 18 cm s^{-1}, with a corresponding eastward current covering the following 10 km, i.e. spanning the chimney scale. This well-defined cyclonic shear zone is consistent with a rim current, although there are insufficient data to discern this from a background current. The horizontal shear of the geostrophic current is approximately 2×10^{-5} s^{-1}, which is only one order of magnitude smaller than *f*. Suffice it to say that the chosen level of no motion is somewhat arbitrary but in agreement with other studies in the area, e.g., *Johannessen* [1987a].

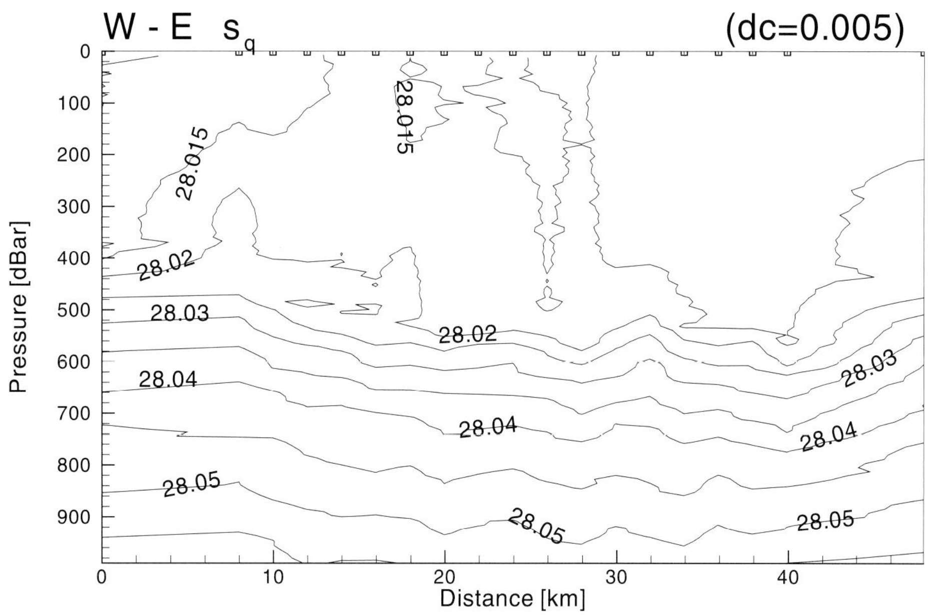

(c)

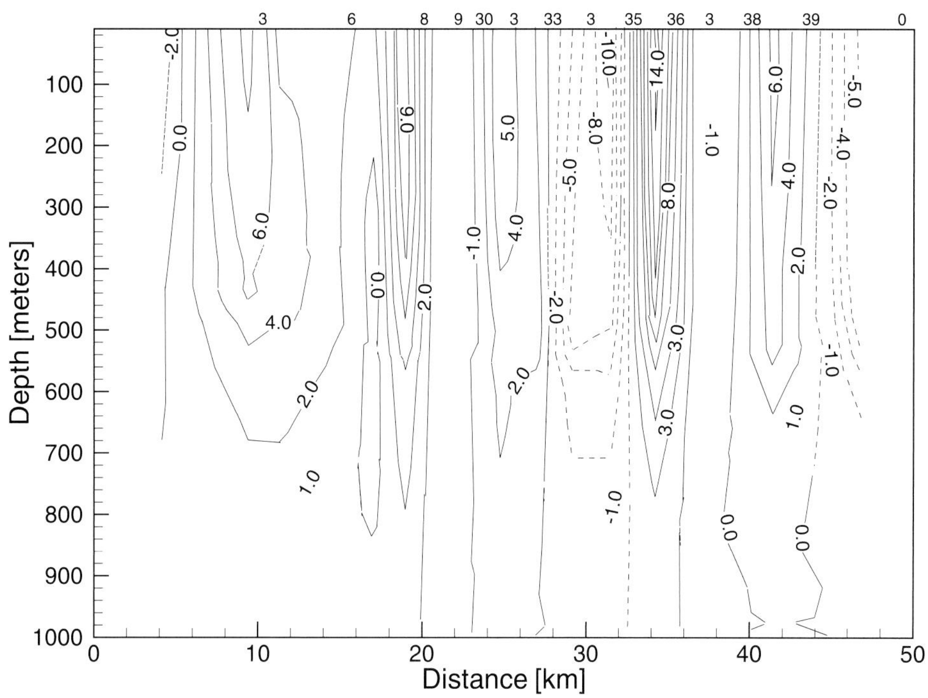

(d)

Figure 5. (Continued).

Elevating it to 500 m would reduce the horizontal shear by roughly 50%. Furthermore, the same authors showed that the barotropic velocities can be considerable, especially over steep bottom features. However, as we are focusing on upper ocean features on the 10-km scale and smaller in an area of modest topographic variability, the geostrophic current should be fairly representative.

As the scale of the current system (20 km) is much larger than R_d (= 3 km), the condition for baroclinic instability is met [*Pedlosky*, 1979]. For the horizontal structure of the geostrophic current, variability on the kilometer scale should be interpreted with caution, since the separation is comparable to R_d, not to mention the nonsynopticity. It is not obvious if there is an inflection point in the chimney-scale horizontal velocity profile, the criterion for barotropic instability [*Pedlosky*, 1979]. Baroclinic instability thus seems to be the most likely generation mechanism for the rim eddies, in agreement with *Send and Marshall* [1995].

As mentioned above, also on the E–W section the core water of the chimney is slightly colder (by less than 0.05°C) and fresher near the rim than in the center (Figures 5a,b, between 8 and 15 km from the west). Again, the density displays a maximum in the same area (Figure 5c, at about 15 km). The relatively fresh water near the surface (between 8 and 15 km) suggests enhanced contact with the fresh pycnocline layer by slightly enhanced convective mixing perhaps in combination with the elevated pycnocline near the rim.

The pycnocline (Figure 5c) is generally more leveled out in comparison to the N–S section, or even has a slight bowl shape, most markedly on the sides. The relatively steep inclination of the pycnocline to the west gives rise to a maximum northward geostrophic speed of about 6 cm s^{-1} in Figure 5d. Besides from somewhat lower speed, there was no well defined cyclonic circulations like that seen on the N–S section.

Mesoscale Structure

The most detailed synoptic map of the near-surface structure of the chimney is that obtained from the 20-m-deep tows. For a better view of its rim we first inspect the last crisscrossings with the undulating Seasoar (Plate 2). Again defining the core as enclosed by the 0.7°C contour, the core apparently extends about 15 km in the meridional and 30 km in the zonal directions, as also seen on the CTD sections. The region to the east of 1°E is apparently very patchy, with eddy-like structures of less than 10 km extent, both cold/fresh and warm/saline anomalies. In the lower right corner in Plates 2a,b resides the warm core and saline anticyclonic feature previously discussed in relation to Plate 1. Its vertical structure appears clearly on Seasoar Section 37 (Figure 1) along 77.52°N (Figure 6; descends 627-634). It is about 15 km in diameter and shaped as a lens, as is characteristic for upper ocean anticyclones. Warm, saline AW is also found in the northwestern part of the chimney, bounded by a front (Plates 2a,b), again consistent with the previous information from the deep sections. Similarly, the minimum of temperature and salinity just inside the chimney core is also evident.

It is also revealed that the temporal variability is significant. When comparing Plates 2a,b from several days of profiling with the final 6 h (Plates 2c,d), it is seen that synopticity cannot be granted, e.g., the northwestern front has moved northwards. The final Seasoar crisscrossing reinforced the previous overall view of the area with respect to temperature (compare Plates 2a and c), although the spatial range was somewhat smaller. Temporal variability is also evident on closer inspection of CTD stations 1, 12, and 30, i.e. those taken at the same location (77.57°N, 0.5°E) separated by time intervals of 13.5 and 29 h, respectively (not shown). In between stations 1 and 12 practically no change has taken place, whereas between stations 12 and 30 the ML properties are virtually unchanged, with the pycnocline depressed by approximately 60 m. Although the latter stations are 500 m apart, the apparent synoptic views presented should be interpreted with caution.

A unique mapping of the internal horizontal structure of the chimney is provided by the Seasoar as it was towed in approximately 20 m depth, sampling approximately every 10 cm, along six E–W tracks 4 km apart (Figure 2). A contour plot of in situ, *T*, *S*, and σ_t is presented in Plate 3, in which the high wavenumber variability has been filtered to emphasize the mesoscale structures. (A consecutive median filter of width 101 was applied, yielding a resolution of about 10 m.) Again, the general structure agrees with the previous observations. The relatively cold, fresh structures inside the core close to the northwestern rim stand out even clearer. Plate 3c shows that the core σ_t typically exceeds 28.016, with slightly higher values in the southwestern part ($\sigma_t \sim 28.022$).

The fine-scale tows also provide insight that was not possible from the 2.5-km-resolution tows. E.g., the warm and saline anomalies of ~5-km scale seen at 1°E, 77.5° and 77.56°N in Plate 3 could be only marginally resolved during the previous profiling (Plate 2). Alternatively, the anomalies can have propagated from the east, with an implied speed of 5 km over a day or so. Third, the warm, saline feature, about 5 km wide, appearing at the southern boundary at about 0.4°E (Plate 3, upper and middle panels), is also missing in the previous data (Plates 2a,b). Although no definite conclusion can be drawn on the time scale and potential propagation on the observed structures on the 5-km scale, it is apparent that these features are characteristic of the dynamics of the rim of the chimney.

Submesoscale Structure

For a detailed view of the horizontal *T–S* structure the individual tracks from the constant depth survey are plotted in Figure 7. Considerable variability on the 1-5-km scale is evident, as is an apparent decorrelation between neighbouring tracks 4 km apart. The gradients are occasionally sharp, i.e., of the order 0.2°C over less than a kilometer (Figure 7c), and the northwestern rim is superposed by anomalies of the order 0.1°C on a few kilometre scale. In the other tracks it is also seen that temperature anomalies of the order 0.1°C

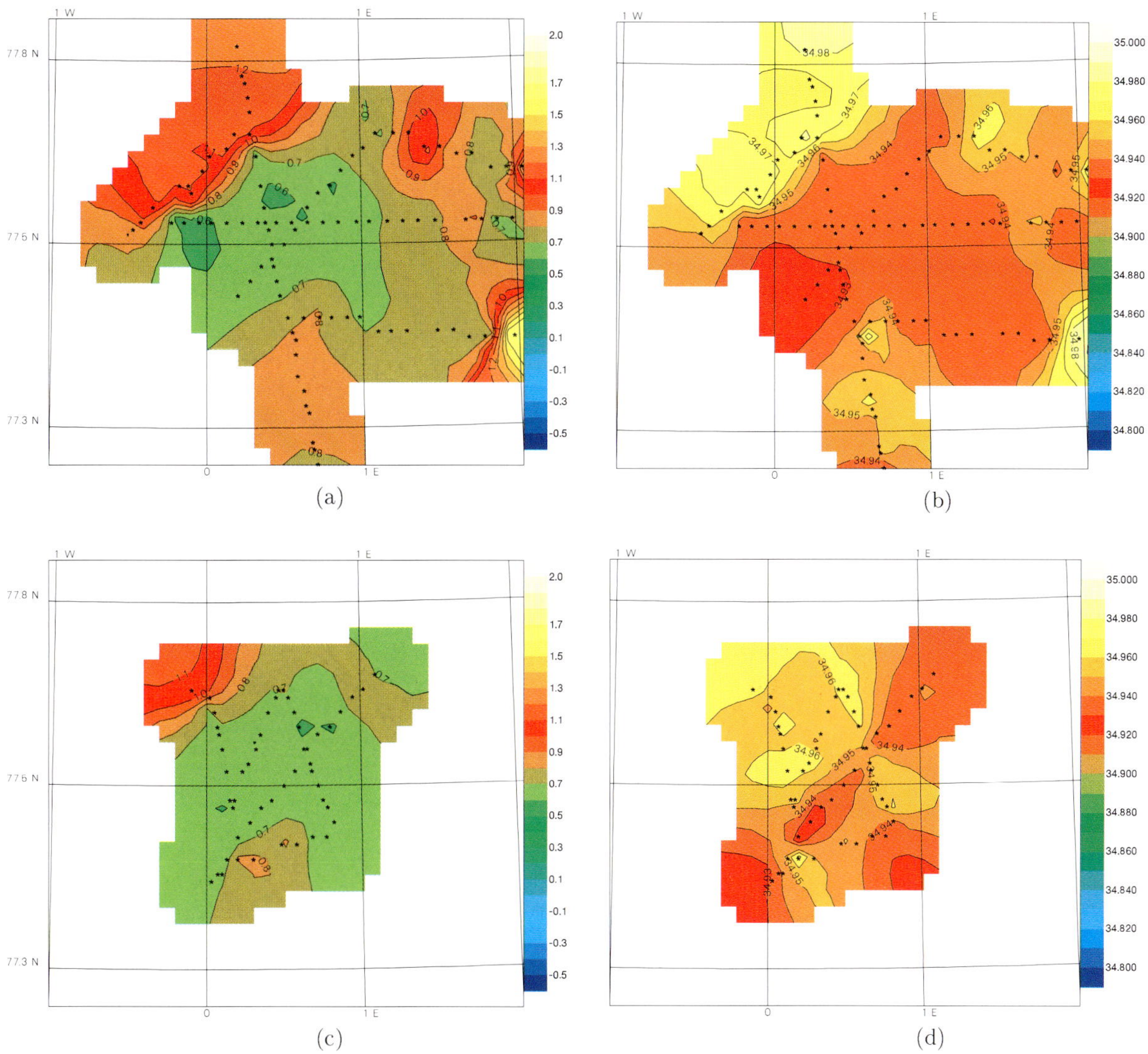

Plate 2. Blow-up of the *Chimney94* area, showing (a,c) temperature and (b,d) salinity. Maps in (a) and (b) were obtained over 3.5 days; those in (c) and (d) were derived from the final 6 h of the survey, immediately following the period for (a) and (b). Contour interval is 0.1°C for temperature and 0.01 for salinity.

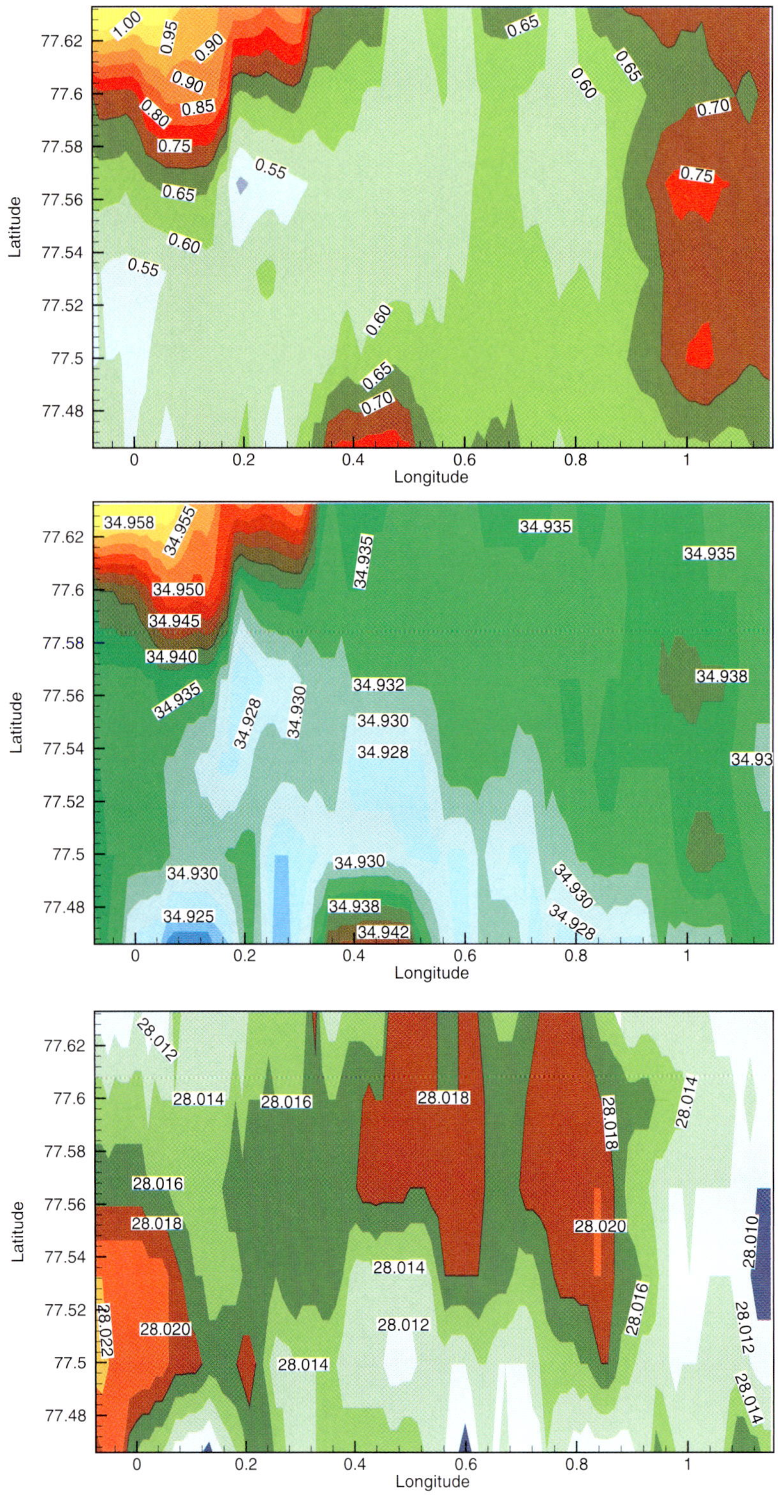

Plate 3. Horizontal plot of the fine-scale tows. Temperature (upper), salinity (middle), σ_θ (lower). Data have been smoothed with a 101-point median filter prior to plotting.

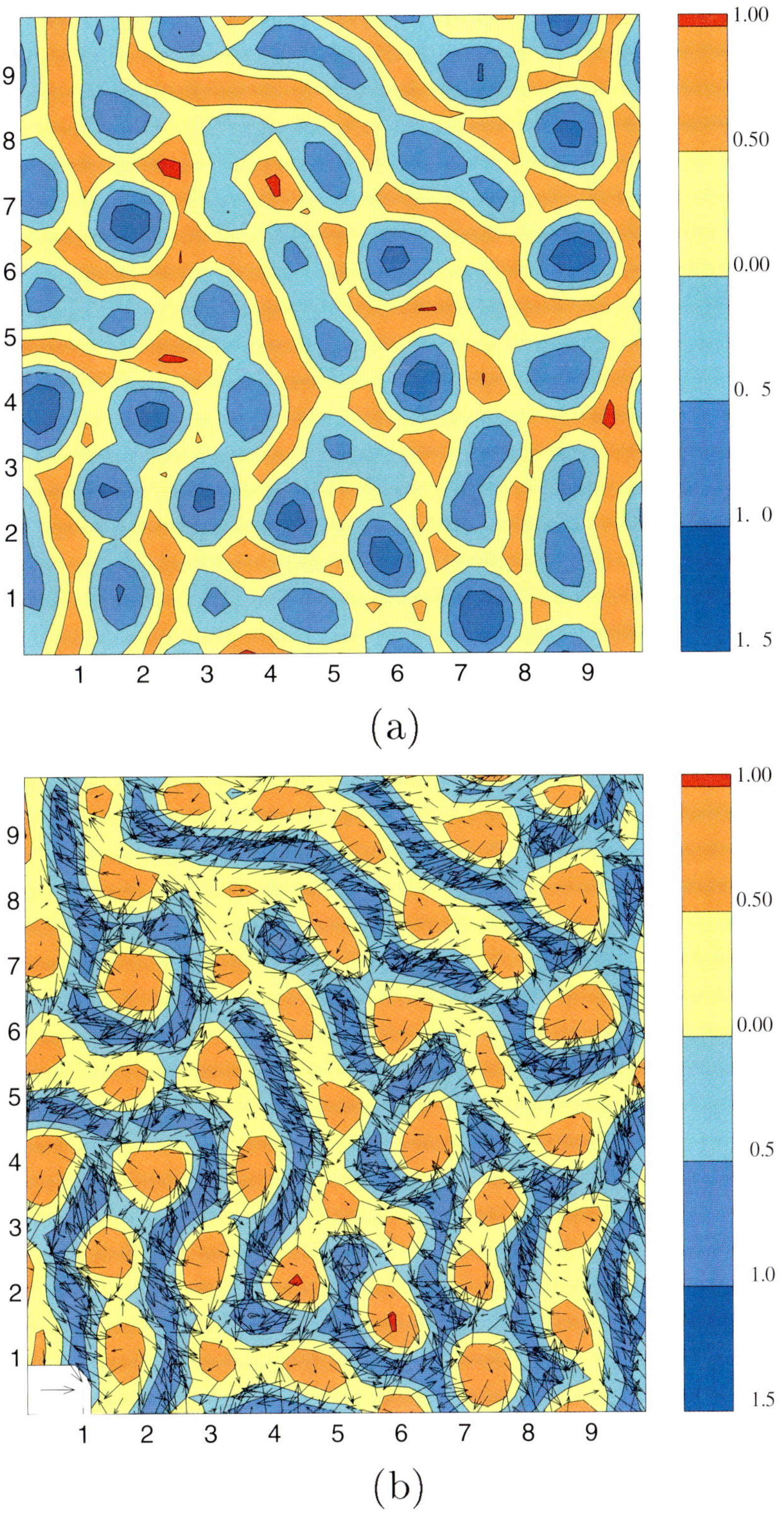

Plate 4. Model state in 20-m depth at $t = 30$ days for (a) density anomaly relative to 28.0405 kg m^{-3} and (b) horizontal divergence superimposed by velocity vectors (reference arrow 2 cm s^{-1}. Units are 10^{-3} kg m^{-3} in (a) and 10^{-4} s^{-1} in (b). Coordinates are in km.

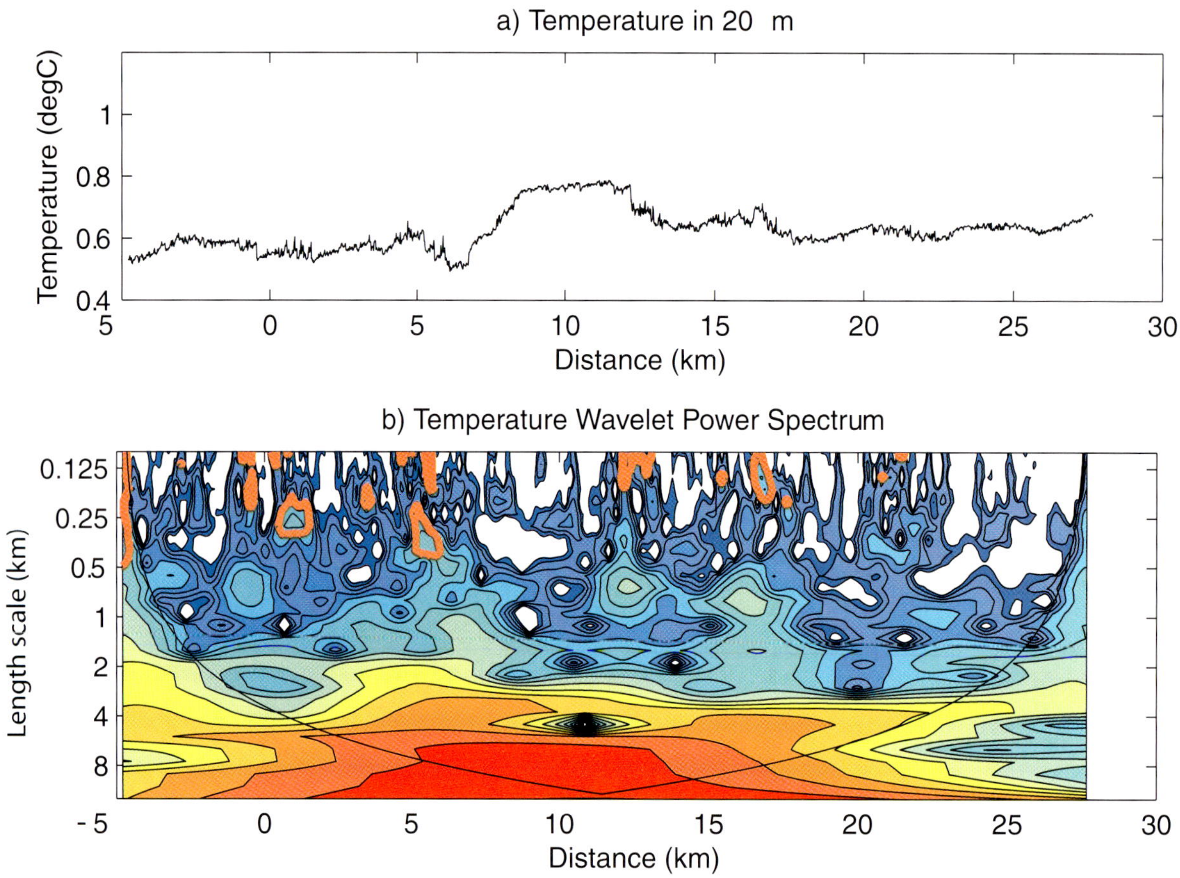

Plate 5. (a) Temperature profile along section 49 (°C) and (b) wavelet power spectrum of the same section as function of distance (km) eastward from the 0° meridian and scale length (km). The color scale indicates relative variation of power with increasing levels from dark blue via green and yellow through red. Power is doubled for each color level. Thick red contours embrace regions where variability is significant at the 5% level. The four peaks between 0 and 2 km in (a), with corresponding maximum in power in (b), are described in the text.

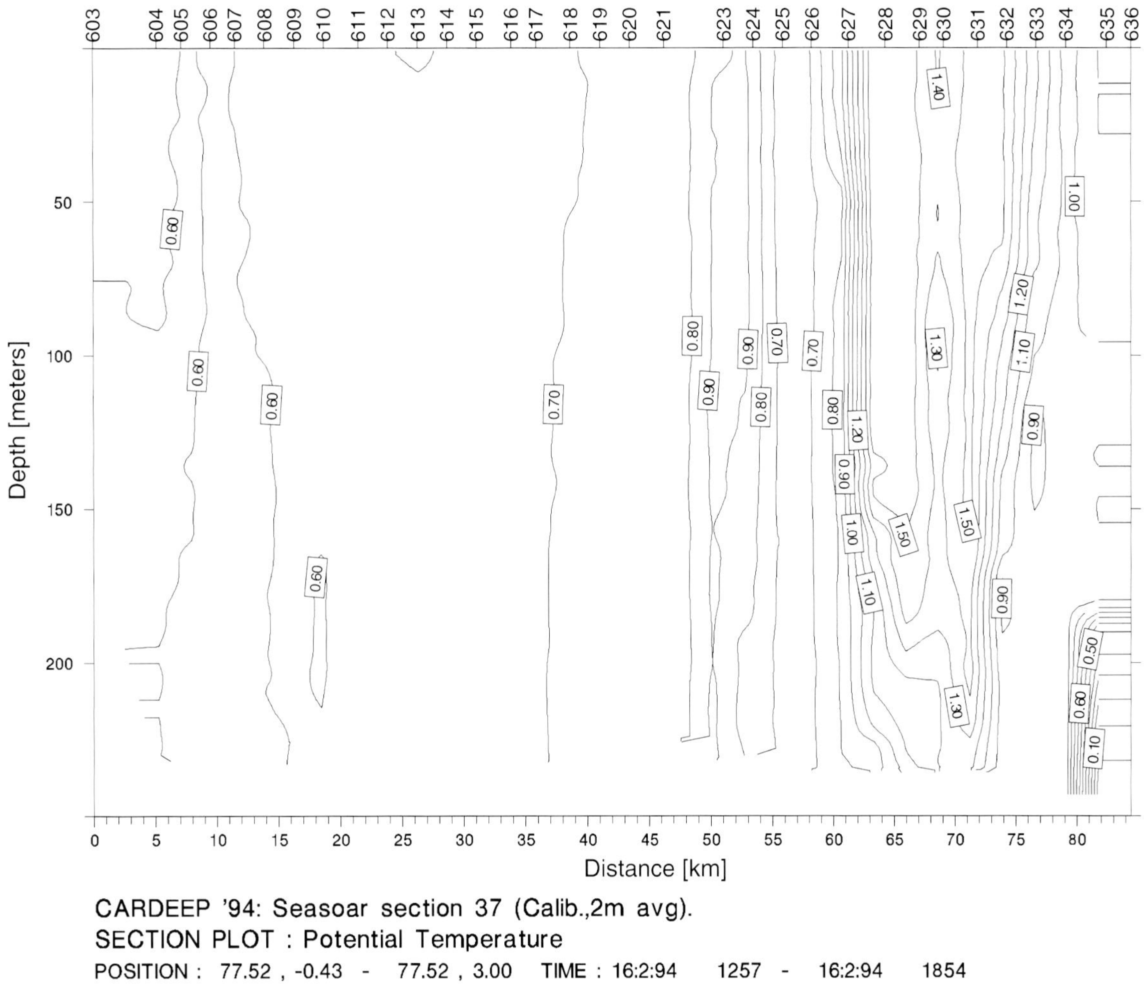

Figure 6. Seasoar Section 37, potential temperature. Contour interval is 0.1°C. The section is running W–E on Plate 2 at 77.52°N, extending further to the east, out of the panel.

over ~1 km is a common feature, i.e., comparable to the cross core temperature range revealed from the deep CTD sections (Figures 4a and 5a). A *T–S* scatter diagram (not shown) including the six tracks reveals an approximate linear fit between temperature and salinity with a slope of $\Delta T/\Delta S \simeq$ 20°C psu^{-1}, a relation that is also apparent from the individual profiles. Thus, the salinity variability closely follows the temperature, consistent with linear mixing of the AW and the uAIW.

Numerical Experiment

To complement our understanding of the observed submesoscale variability, a numerical ocean model was set up to mimic the chimney interior. The model used is the MITgcm [*Marshall et al.*, 1997a,b] in its nonhydrostatic mode. The idealized model domain is 10 km × 10 km, representing the core of the chimney. It is doubly periodic in the horizontal and has a uniform depth of 3000 m, representative of the region. Initially, the fluid is motionless with small-amplitude random horizontal perturbations added to the stratification. The initial stratification is given by the potential density of Figure 3, related to the potential temperature through a linear equation of state, with a thermal expansion coefficient, $\alpha = 0.5 \times 10^{-4}$ (°C)$^{-1}$. The hydrography was smoothed to remove the small-scale noise due to static instabilities. The numerical grid has a uniform horizontal resolution of 250 m and 32 vertical levels with spacing increasing smoothly from 20 m in the upper ocean to 200 m toward the seafloor. The internal radius of deformation (about 3 km) is thus well resolved. The same constant values were used for the Laplacian viscosity and diffusivity. The horizontal was set to 5 m^{2}s^{-1} and the vertical to 0.02 m^{2}s^{-1}, values similar to those used by other authors, e.g., *Haine and Williams* [2002].

The heat flux from the ocean to the atmosphere was estimated from the field observations to be about 200 Wm^{-2}. The ML cooling from this is quite weak, as noted in the previous section. A value of 500 Wm^{-2} has previously been associated with convective chimneys and surface heat fluxes in the area [*Häkkinen*, 1988; *Häkkinen and Cavalieri*, 1989; *Johannessen et al.*, 1991]. The latter was therefore used to initiate convection in the numerical model, producing quite vigorous convection and some deepening of the mixed layer. After the first 20 days, the heat flux was smoothly reduced to the observed 200 Wm^{-2} for the remaining 30 days of the simulation.

The evolution of the vertical stratification is depicted in Figure 8. The horizontally averaged density is plotted for every second day. It shows that most of the ML deepening takes place during the first 20 days of relatively strong cooling. In the horizontal, distinct dense/light (cold/warm) anomalies of subduction/upwelling of 1-km scale emerge after about 1 day. (A power spectrum, not shown, revealed peaks at 0.7 and 1.4 km.) Although the deepening of the ML is quite marginal after day 20, the convective anomalies remain. Snapshots from day 30 of the density anomalies and the horizontal velocity and divergence at 20 m deep are shown in Plates 4a,b. The sinking regions are more intense than the upwelling, consistent with what has been inferred from observations and theory in general, cf. *Marshall and Schott* [1999].

The corresponding temperature difference between the cold and warm anomalies is typically 0.01-0.04°C, and the horizontal speed is 1-4 cm s^{-1}. The spatial scale and temperature anomalies are quite consistent with the observed submesoscale variability (e.g., Figures 7a, e, k). Thus the present observations and numerical experiment suggest that there is a preference for convection to be organized at a kilometer scale, intermediate of the plume (see below) and the chimney scales. Similar patterns are also seen in the large-eddy simulations of *Garwood et al.* [1994], but the agreement may be partly coincidental. Although they are concerned with the high latitudes, their focus is thermobaric convection and the temperature anomalies that emerge are smaller than those reported herein.

The Plume Scale

The observed variability on the scale of a few hundred metres is also prominent, as can be most clearly seen in the temperature, ranging from about 0.02°C through 0.05°C (Figure 7). Hereafter we refer to this as the *plume scale*, since during active convection, plumes are expected to contribute to variability on this scale, as will be discussed below. Over all, the corresponding log-log plots of the power spectra for each section (not shown) revealed a linear decrease of energy with wavenumber, i.e. a "classical" turbulent cascade. However, on smaller scales than the kilometer scale there are marked deviations from this general pattern, especially demonstrated on Section 49, whose temperature profile is repeated in Plate 5a. Four distinct warm anomalies about 300 m wide, i.e. on the plume scale, are seen in the southwestern part of the chimney,

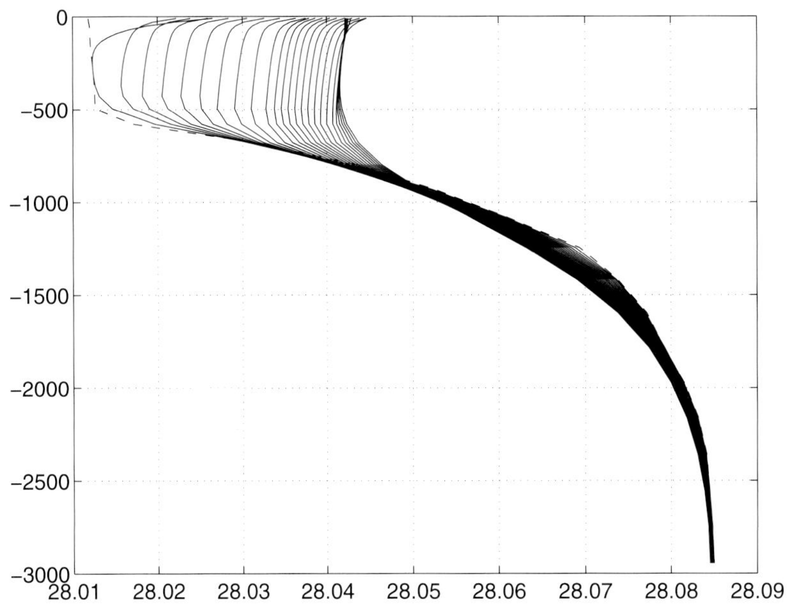

Figure 8. Modeled development of density [sigma-units] versus depth [m] subject to a heat flux of 500 Wm^{-2} for the initial 20 days and 200 Wm^{-2} for the remaining 30. A curve is plotted every 2 days, starting with the dashed curve.

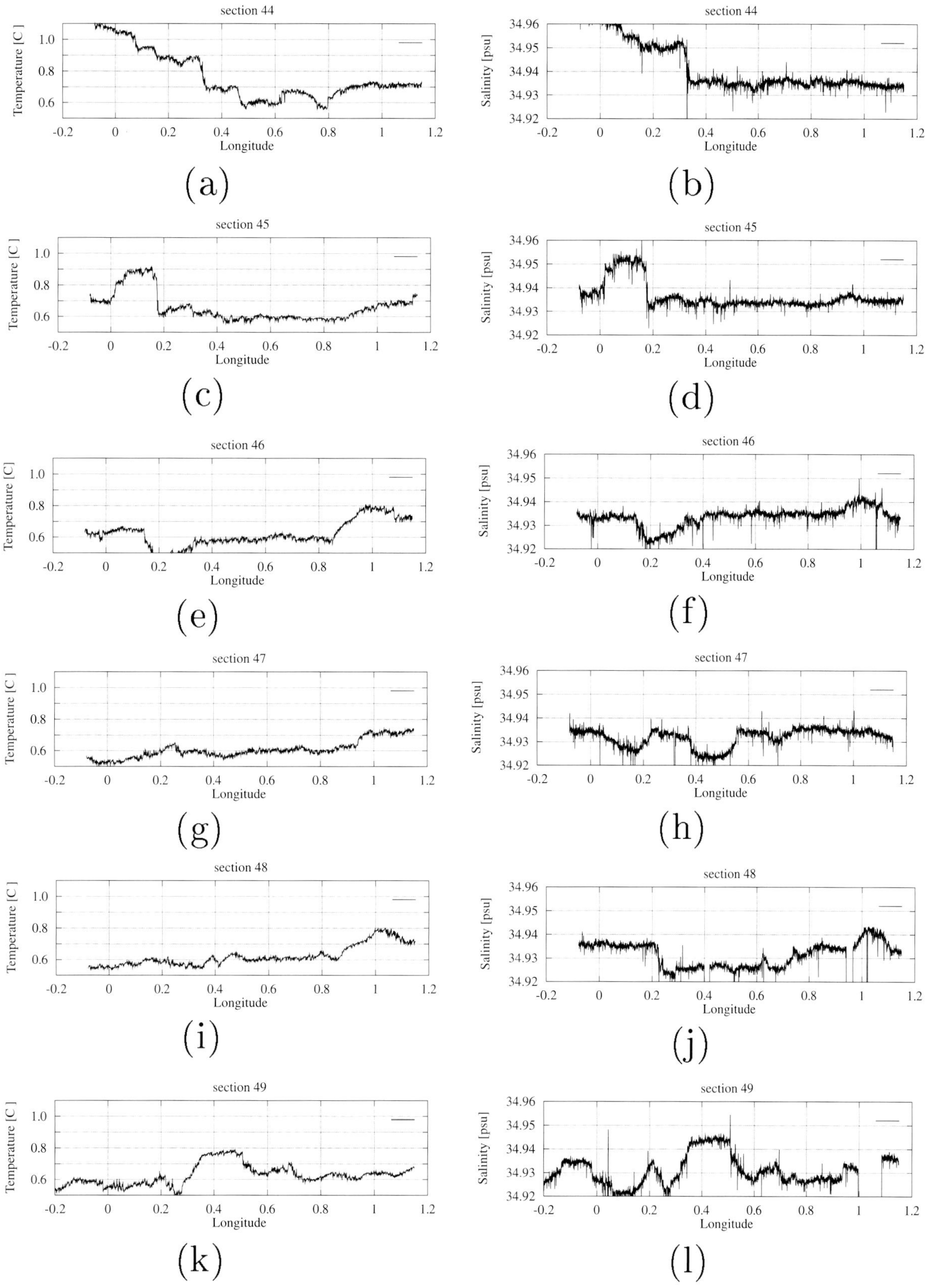

Figure 7. Along-track distribution of in situ temperature (left) and salinity (right) from the legs where the Seasoar was towed in ~20 m. The data have been reduced from a horizontal resolution of ~10 cm to ~10 m by taking consecutive medians.

i.e. between 0 and 2 km each with peaks of ~0.05°C above the background, i.e., comparable to the typical cross ML temperature variation of ~0.1°C. These may represent vertical motion. In a convective and otherwise cold environment, such warm anomalies are consistent with ascending warmer water masses compensating the sinking plumes.

Further quantification of the variability is provided by a wavelet power spectrum [*Kumar and Foufoula-Georgiou*, 1997], shown in Plate 5b, in which the power is plotted as a function of distance along the track and length scale, again for Section 49 as an example. The signature of the aforementioned peaks clearly stands out both with elevated values on the plume scale and falling inside the region where variability is significantly different from noise, at the 5% level, denoted by the thick red contours in Plate 5b. A similarly significant region is seen to the east of 5 km, whereas the small regions on scales smaller than about 150 m are suspicious as the strength of the variability approaches the detection limit. Whether there exists a region of minimum power at about 500-m scale, indicative of a separation of scales on either side, is more speculative.

In order to check if there is any systematic variability of the energy on the plume scale, we examine the energy distribution on the band 150-400 m for the different profiles. We hereafter denote this quantity *Var*. Figure 9 shows there are significant peaks in *Var* on several profiles. In comparison with the raw profiles (Figure 7), these peaks are caused by impulse-like steps in the raw profiles, inducing energy on all scales. The marked exception is the first peak from the left on Section 49 (Figure 9), the previously described region, displaying marked locally increased levels of *Var*. In order to shed more light on this anomalous structure, we first quantify the effect of abrupt steps by computing $|\nabla T|$, evaluated over 1 km. The correlation of *Var* with, respectively, T, σ, and $|\nabla T|$ is shown in Table 1. Not surprisingly, high correlation of *Var* and $|\nabla T|$ is found on the northernmost Sections 44 and 45. Correlation between *Var* and σ is also relatively high on these sections, apparently due to the rim front, which is dominated by salinity. On Section 47 there is virtually no correlation between any of the quantities, probably because it is representative for the chimney interior, with minimal influence from the variability at the rim. The anomalous Section 49 shows a negative correlation between *Var* and T, whereas the other quantities show a positive or no correlation. Repeating the analysis for the westernmost portion of Section 49 by excluding the regions of the second *Var* peak and eastwards retains the high negative correlation between *Var* and T, but enhances the negative correlation between *Var* and σ, implying stronger influence of salinity. Furthermore, the correlation between *Var* and $|\nabla T|$ shows anomalous negative and high correlation, i.e. high variability for a more homogeneous background, which is contrary to the effect of abrupt property steps on the rim.

Taken together with the previous information from the deep CTD sections, we observe that the enhanced plume activity is consistent with increased mixing with the slightly cooler and fresher waters from beneath the ML. Whether the surface cooling is slightly stronger or the stratification slightly weaker than the background cannot be determined without further data. Probably it is a combination of both.

The plume scale appearing on Section 49 agrees well with a plume scale l determined from external parameters (*Marshall and Schott* [1999], Eq. 13a):

$$l = \left(\frac{B}{f^3}\right)^{1/2}$$

Using the turbulent heat flux ($Q = 200$ Wm^{-2}) to derive the buoyancy forcing B based on a representative value for the thermal expansion coefficient, $\alpha = 0.5 \times 10^{-4}(°C)^{-1}$, as for the ocean model, yields

$$B = 0.3 \times 10^{-7}\ \mathrm{m^2 s^{-3}}$$

and

$$l \simeq 100\ \mathrm{m}$$

Finally, an aspect of our observations perhaps linked to the microscale is worth mentioning. The sharp separation of the warm, saline ML water and the cold, fresh water beneath (Figure 3) is striking. The approximate property steps are $\Delta S = 0.05$ and $\Delta\theta = 0.75$°C. This stratification is opposite of what has been observed earlier [*Johannessen et al.*, 1991] and can potentially allow upward thermobaric convection.

DISCUSSION

Our observations of *Chimney94* indicate shallow convection over 400-600 m maintained by approximately 200 Wm^{-2} cooling at the surface. It can easily be shown that even if this flux works exclusively to cool the ML, it will take more than 3 months to erode the thermocline. In other words, deep-reaching convection is very unlikely during the remaining winter. Normally the heat flux in this area exceeds 500 Wm^{-2} [*Häkkinen*, 1988; *Häkkinen and Cavalieri*, 1989]. This often occurs in connection with cold Polar air outbreaks over warm water closer to the ice edge [*Buckley et al.*, 1979]. For such a flux the same deepening is obtained in only about 35 days. The 1994 winter also appeared to be warmer than normal for this area, implying less intense convection.

Despite the modest convective activity, a main finding is the evident horizontal variability on the submeso– and plume scales. In the chimney interior there are both in the observations (Figure 7) and in the numerical simulations (Plate 4) clear signatures of submesoscale anomalies. This suggests there are three distinct dynamical horizontal scales present:

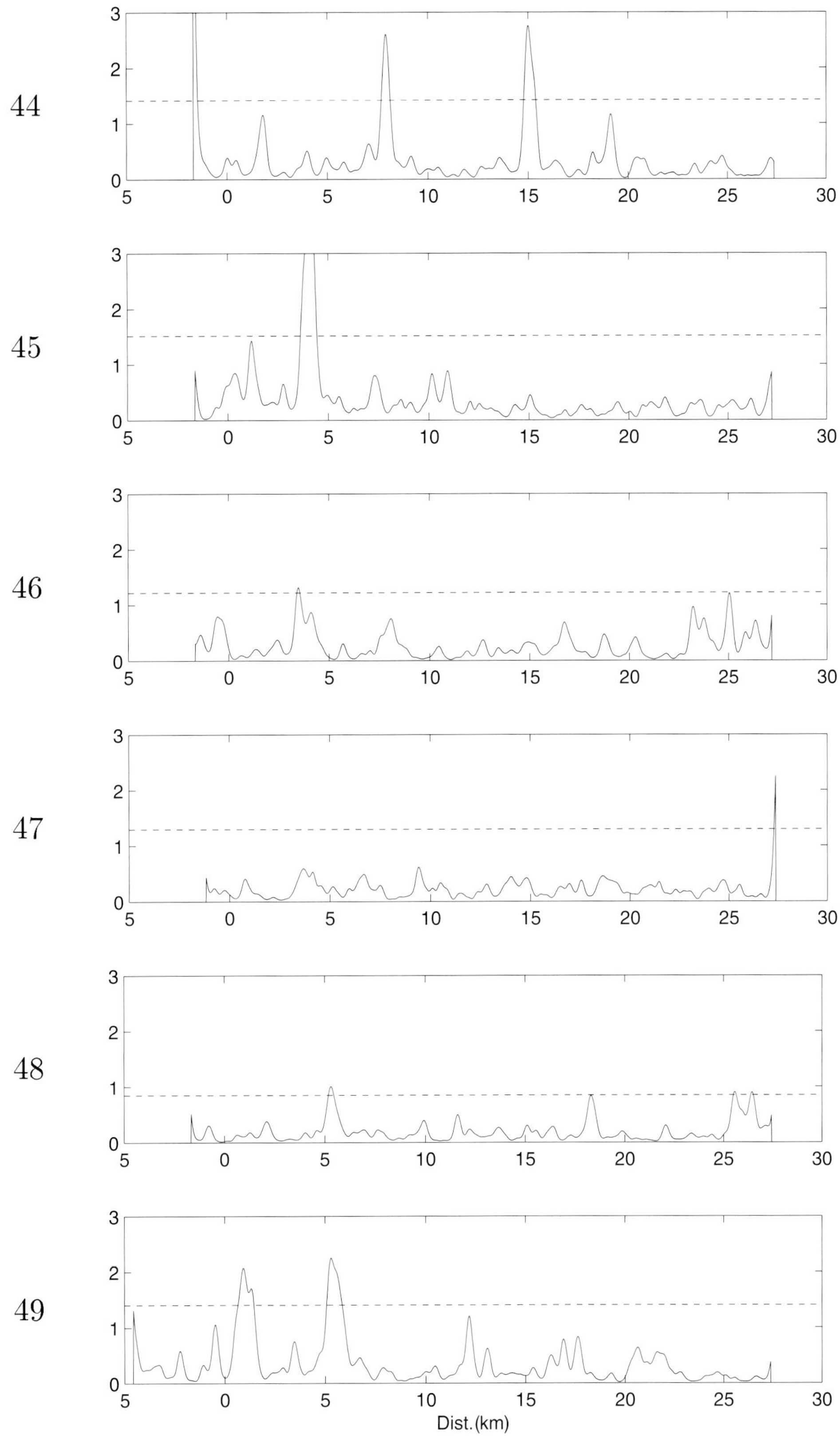

Figure 9. Along-track distribution of temperature variance (10^{-4} (°C)^2) on the 150-400-m scales from the fine-scale tows. The 95% confidence level is indicated with dashed lines. To improve comparison, the ordinate has the same range for each plot.

Table 1. Correlation Between Various Parameters for the Fine-Scale Tows. *Var* denotes the temperature variance on the 150-400-m scales, T is temperature, σ density, and $|\nabla T|$ the absolute value of the temperature gradient computed over 1 km. Section 49 west denotes the portion west of 5 km in Plate 5.

Section	*Var* vs. T	*Var* vs. σ	*Var* vs. $|\nabla T|$
44	0.08	0.12	0.51
45	0.24	0.33	0.81
46	0.21	0.06	0.27
47	0.05	0.00	−0.01
48	0.20	0.08	−0.02
49	−0.30	−0.06	0.15
49 west	−0.26	−0.18	−0.17

the chimney (or meso), the submeso–, and the plume scales. This is in line with the ordering described by Marshall and Schott [1999] for the Labrador and Greenland seas, but with the important difference that they lack the intermediate scale of the present study, the submesoscale.

The observed concentration of energy on the 100-m horizontal scale, the plume scale, coincides with the width of vertically convecting plumes in agreement with several numerical studies and theoretical considerations [*Thorkildsen and Haugan*, 1999; *Marshall and Schott*, 1999]. Furthermore, remote sensing observations in the Greenland Sea by *Carsey and Roach* [1994] also point to a plume scale of 100 m, with a separation between plumes of about 300 m. We thus conclude that our observations show plume activity. More investigations are needed to clarify the important interactions between the various spatial scales.

Also notable is the irregular shape of the rim of the chimney, indicating substantial activity by meanders or eddies, eroding the chimney from the sides, i.e., a restratification process (Plate 2). On the other hand, in the north, the θ–S structure suggests slightly enhanced mixing with the underlying water, where there is also maximum cyclonic shear and the pycnocline is elevated (Figure 4). In this case a cyclone of the destabilized rim current thus moderately assists convection. Taking our observations, modeling, and previous work together, the state of the chimney—the ML depth and the horizontal extent of the chimney—reflects a dynamical balance in which the cooling is the process supporting convection and the erosion by the warm intrusions from the Atlantic domain is the destructive process. With a stronger cooling during the winter one should expect deeper reaching convection and an expansion of the convective region.

In contrast to *Chimney94*, *Chimney89* (also observed in the Boreas Basin by *Johannessen et al.* [1991]) was more well defined. It was by itself colder as it was located closer to the ice edge and exposed to colder air masses typical of the winter of 1989. Perhaps more important is its deep structure with high salinities (34.92), indicating deep water in the core and surrounded by relatively saline lAIW. In the same area deep water was also observed at the surface in several O(10-km) chimney-like structures. Upper ocean chimneys observed 2 years earlier in 1987, discussed by *Johannessen et al.*, [1991], consisted of lAIW but lacked the deep-water core as for the 1989 observation. The more marked presence of deep water in the chimney structures in 1989 relative to 1987 may indicate a gradual renewal of the deep-water mass between these years.

The observations from 1987 and 1989 suggest that some preconditioning by lifting the lAIW to the surface is necessary for a deep-reaching chimney to form. Such a mechanism may be Ekman pumping or a cyclonic eddy, as discussed elsewhere [*Johannessen et al.*, 1991; *Killworth,* 1979]. However, the very small bottom-to-surface density difference, reported to be 0.02-0.03 sigma-units by *Johannessen et al.*, [1991], is probably also a reflection of the preexisting structure of the deep and intermediate waters [*Budéus*, 1998]. In contrast, in 1994 the intermediate waters (500-1000 m) consist of a much fresher uAIW ($S \sim 34.88$) and the bottom-to-surface density difference is 0.07 sigma-units (Figure 3), making penetration by surface convection unlikely, as discussed previously. The inverted hydrography compared to the earlier observation, with a slight capping by warm, saline AW, may also be due to the more eastern location of *Chimney94*.

According to *Gascard* [1990] 20 chimneys can account for a major part of intermediate and deep-water formation in the world ocean. In order to quantify the integral effect of the chimneys in the Greenland Sea, one has to estimate their number and lifetimes in addition to their vertical and horizontal structure. A key issue is the role of baroclinic eddies associated with the rim of the chimney. Is their major role in the postconvective process [*Marshall and Schott*, 1999], or do they also play an active role in the convection process [*Killworth*, 1979; *Straneo and Kawase*, 1999; *Eldevik and Drange*, 2002; *Gascard et al.*, 2002; *Wadhams et al.*, 2002]? Without being conclusive, our results definitely add more complexity to the picture: Apart from the rich submesoscale and finer internal structure inside the chimney described above, there are fronts and large (15-km) anticyclones in the ambient water, obscuring the separation of chimney core and exterior. Structures at the rim of the chimney entangle T–S signatures of water masses from the exterior (warm, saline) and the chimney core (cold, fresh).

Strikingly, the densest near-surface water occurred on the rim of the chimney, with the lightest, warmer, saltier water located only 2.5 km away. The dense water anomaly also coincided with a strong cyclonic shear, a maximum elevation of the pycnocline, and a slightly larger cooling of 0.05°C.

The latter is probably because the surface forcing can operate slightly more effectively over a relatively thinner ML. Again, the modest forcing is taken as the reason why an unambiguous convection process was not observed with well-defined rim eddies in a more symmetric pattern.

Finally, we note the special stratification of the Boreas Basin, distingushing it from the Greenland Basin. In the former, the Atlantic derivative situated at the surface lies directly on top of the fresher and colder intermediate water, whereas in the Greenland Basin the stratification is "normal" with usually a low-salinity water at the surface [*Johannessen and Lygre*, 1996]. In a warm, saline over cold, fresh stratification, a parcel of intermediate water entrained into the convecting ML will become lighter than its surroundings and thus convect freely due to thermobaricity. Such "upward" convection is possible only in this atypical stratification and may serve as a significant mechanism for assisting convection in cases when the surface forcing is not powerful. It is striking that our observations show enhanced plume activity inside a cold and fresh submesoscale anomaly. A closer inspection of the role of the microscale processes is beyond the scope of the present study.

CONCLUSION

For the first time, a chimney has been investigated with a horizontal sampling interval down to 10 cm. *Chimney94* in the Boreas Basin in late winter 1994 was about 30 km in the zonal and 15 km in the meridional directions, the convection reaching typically 500 m. Within the core of the chimney the variability is notable on the submesoscale $O(1\text{ km})$ and the plume scale $O(100\text{ m})$. The latter scale—for the first time observed in this experiment—indicates active plume convection in agreement with theoretical considerations. A nonhydrostatic ocean model corroborates that convection can be organized on the submesoscale even under moderate forcing.

Strong activity on the rim of the chimney indicates that it is vulnerable to destabilisation, inhibiting its further development. Ambient anticyclones are not clearly discriminable from those associated with the rim process. Indications of augmented cooling inside rim cyclones is a sign of active, although weak, convection competing against the break up process. Finally, the special stratification with warm, saline water over cold, fresh water is shown to be a potentially important ingredient in the vertical exchange process.

NOMENCLATURE

B	buoyancy flux
H	thickness of the mixed layer
N	buoyancy frequency
R_d	Rossby radius of deformation
S	Salinity
T	in situ temperature
f	Coriolis parameter
l	length scale
α	thermal expansion coefficient of sea water
σ_θ	potential density anomaly
θ	potential temperature
ASW	Arctic Surface Water
AW	Atlantic Water
CTD	conductivity–temperature–depth
MIZEX	Marginal Ice Zone Experiment
ML	mixed-layer
SAR	Synthetic Aperture Radar
SIZEX	Seasonal Ice Zone Experiment
lAIW	lower Arctic Intermediate Water
uAIW	upper Arctic Intermediate Water

Acknowledgments. This EU ESOP project has been funded by the Norwegian Research Council as contribution to contract MAST2-CT93-0057 under the MAST-2 programme, the Nordic Council of Ministries, and Office of Naval Research. University of Bergen contributed with R/V *Håkon Mosby*. We thank the captain, crew, and scientific staff onboard R/V *Håkon Mosby* for splendid cooperation. Wavelet software was provided by C. Torrence and G. Compo and is available at URL: http://paos.colorado.edu/research/wavelets/ T.E. was supported by grants from the EU 5th Framework Programme project TRACTOR (contract EVK2-CT-2000-00080) and the Norwegian Research Council project ProClim (155923/700). Comments from two anonymous reviewers were helpful in improving the manuscript.

REFERENCES

Aagaard, K., E. Fahrbach, J. Meincke, and J.H. Swift, Saline outflow from the Arctic Ocean: its contribution to the deep waters of the Greenland, Norwegian and Iceland seas, *J. Geophys. Res.*, 96(C11), 20433-20441, 1991.

Alekseev, G., O.M. Johannessen, A. Korablev, V. Ivanov, and D. Kovalevsky, Interannual variability of water masses in the Greenland Sea and adjacent areas, *Polar Research*, 20, 201-208, The Sverdrup Symposium special issue, 2001.

Alekseev, G.V., O.M. Johannessen, A.A. Korablev, and A.Y. Proshutinsky, Arctic Ocean and sea ice, in *Arctic Environment Variability in the Context of Global Change*, edited by Bobylev, L.P., K.Y. Kondratyev, and O.M. Johannessen, pp. 107-202, Springer–Praxis books in environmental sciences, 2003.

Buckley, J.R., T. Gammelsrød, J.A. Johannessen, O.M. Johannessen, and L.P. Røed, Upwelling: Oceanic structure at the edge of the Arctic ice pack in winter, *Science*, 203, 165-167, 1979.

Budéus, G., Winter convective events and bottom water warming in the Greenland Sea, *J. Geophys. Res.*, 103(C9), 18513-18527, 1998.

Carmack, E., Large-scale physical oceanography of polar oceans, in *Polar oceanography. Part A physical science*, edited by Smith, W.O., pp. 171-222, Academic Press, 1990.

Carsey, F.D., and T. Roach, Oceanic convection in the Greenland Sea Odden region as interpreted in satellite data, in *Nansen Centennial Volume: The role of the Polar oceans in shaping the global climate*, edited by Johannessen, O.M., R. Muench, and J. Overland, pp. 211-222, American Geophysical Union Monograph, 1994.

Eldevik, T., and H. Drange, Resolved and parameterized convective eddies in the Greenland Sea, in *Norwegian Ocean Climate Project, Technical Report No. 3*, edited by Sætre Hjøllo, S., pp. 41-44, Geophysical Institute, University of Bergen, Bergen, Norway, 2002.

Garwood, R.W., S.M. Isakari, and P.C. Gallacher, Thermobaric convection, in *Nansen Centennial Volume: The role of the Polar oceans in shaping the global climate*, edited by Johannessen, O.M., R. Muench, and J. Overland, pp. 199-210, American Geophysical Union Monograph, 1994.

Gascard, J.C., Deep Convection and Deep Water Formation: Progress and Directions, *Eos, Transactions*, 71(49), 1837-1839, 1990.

Gascard, J.C., Open ocean convection and deep water formation revisited in the Mediterranean, Labrador, Greenland and Weddell Seas, in *Deep convection and deep water formation in the oceans*, edited by Chu, P.C., and J.C. Gascard, pp. 157-182, Elsevier, 1991.

Gascard, J.-C., A.J. Watson, M.J. Messias, K.A. Olsson, T. Johannessen, and K. Simonsen, Long-lived vortices as a mode of deep ventilation in the Greenland Sea, *Nature*, 416, 525-527, 2002.

Haine, T.W.N., and P. Williams, The role of non-hydrostatic dynamics in controlling development of a surface ocean front, *Ocean Modell.*, 4, 121-135, 2002.

Häkkinen, S., A note on chimney formation in ice edge regions, *J. Geophys. Res.*, 93(C7), 8279-8282, 1988.

Häkkinen, S., and D.J. Cavalieri, A study of surface heat fluxes in the Greenland, Norwegian, and Barents Seas, *J. Geophys. Res.*, 94(C5), 6145-6157, 1989.

Johannessen, J.A., O.M. Johannessen, E. Svendsen, R. Shuchman, T. Manley, W.J. Campbell, E.G. Josberger, S. Sandven, J.C. Gascard, T. Olaussen, K. Davidson, and J. VanLeer, Mesoscale eddies in the Fram Strait marginal ice zone during the 1983 and 1984 Marginal Ice Zone Experiments, *J. Geophys. Res.*, 92(C7), 6754-6772, 1987a.

Johannessen, O.M., V. Jensen, T. Lerøen, and K. Lygre, Data report on the Greenland Sea convective chimney experiment 1994, Special report 36, Nansen Environmental and Remote Sensing Center, 1995.

Johannessen, O.M., J.A. Johannessen,, J. Morison, B. Farrelly, and E. Svendsen, Oceanographic conditions in the marginal ice zone north of Svalbard in early fall 1979 with an emphasis on mesoscale processes, *J. Geophys. Res.*, 88, 2755-2769, 1983.

Johannessen, O.M., J.A. Johannessen, E. Svendsen, R. Shuchman, W.J. Campbell, and E.G. Josberger, Ice-edge eddies in the Fram Strait marginal ice zone, *Science*, 236, 427-429, 1987b.

Johannessen, O.M., and K. Lygre, Observations of convective chimneys, in *European Subpolar Ocean Programme, Sea Ice - Ocean Interactions*, edited by Wadhams, P., J.P. Wilkinson, and S.C.S. Wells, pp. 262-294, Commission of the European Communities, 1996.

Johannessen, O.M., S. Sandven, and J.A. Johannessen, Eddy-related winter convection in the Boreas Basin, in *Deep convection and deep water formation in the oceans*, edited by Chu, P.C., and J.C. Gascard, pp. 87-105, Elsevier, 1991.

Killworth, P.D., On "Chimney" formations in the ocean, *J. Phys. Oceanogr.*, 9, 531-554, 1979.

Kumar, P., and E. Foufoula-Georgiou, Wavelet analysis for geophysical applications, *Rev. Geophysics*, 35, 385-412, 1997.

Legg, S., and J. Marshall, A heton model of the spreading phase of open-ocean deep convection, *J. Phys. Oceanogr.*, 23, 1040-1056, 1993.

Marshall, J., A. Adcroft, C. Hill, L. Perelman, and C. Heisey, A finite volume, incompressible Navier Stokes model for studies of the ocean on parallel computers, *J. Geophys. Res.*, 102(C3), 5753-5766, 1997a.

Marshall, J., C. Hill, L. Perelman, and A. Adcroft, Hydrostatic, quasi hydrostatic, and non-hydrostatic ocean modeling, *J. Geophys. Res.*, 102(C3), 5733-5752, 1997b.

Marshall, J., and F. Schott, Open-ocean convection: observations, theory and models, *Rev. Geophysics*, 37, 1-64, 1999.

MEDOC-group, Observation of formation of deep water in the Mediterranean Sea, *Nature*, 227, 1037-1040, 1970.

MIZEX-group, The winter Marginal Ice Zone Program in the Fram Strait/Greenland Sea, *EOS, Trans. Amer. Geophys. Union*, 70, 545-555, 1989.

Pedlosky, J., *Geophysical Fluid Dynamics*, Springer, New York, 1979.

Sandven, S., O.M. Johannessen, and J.A. Johannessen, Mesoscale eddies and chimneys in the marginal ice zone, *J. Mar. Syst.*, 2, 195-208, 1991.

Send, U., and J. Marshall, Integral effects of deep convection, *J. Phys. Oceanogr.*, 25, 855-872, 1995.

SIZEX-group, SISEX 89 Experiment Report, Technical Report 23, Nansen Remote Sensing Center, 1989.

Straneo, F., and M. Kawase, Comparisons of localized convection due to localized forcing and to preconditioning, *J. Phys. Oceanogr.*, 29, 55-68, 1999.

Thorkildsen, F., and P.M. Haugan, Modeling of deep water renewal through cold convective plumes, *Deep-Sea Res. II*, 46, 1357-1383, 1999.

UNESCO, The Practical Salinity scale 1978 and the International Equation of State of Seawater 1980, *Techn. Pap. in Mar. Sci.*, 36, UNESCO, 1981.

UNESCO, Algorithms for computation of fundamental properties of seawater, *Techn. Pap. in Mar. Sci.* 44, UNESCO, 1983.

Wadhams, P., J. Holfort, E. Hansen, and J.P. Wilkinson, A deep convective chimney in the winter Greenland Sea, *Geophys. Res. Lett.*, 29, DOI 10.1029/2001GL014306, 2002.

Watson, A.J., M.J. Messias, E. Fogelquist, K.A. Van Scoy, T. Johannessen, K.I.C. Oliver, D.P. Stevens, F. Rey, T. Tanhua, K.A. Olsson, F. Carse, K. Simonsen, J.R. Ledwell, E. Jansen, D.E. Cooper, J.A. Kruepke, and E. Guilyardi, Mixing and convection in the Greenland Sea from a tracer-release experiment, *Nature*, 401, 902-904, 1999.

T. Eldevik, O.M. Johannessen, and K. Lygre, Nansen Environmental and Remote Sensing Center, Thormøhlensgate 47, N-5006 Bergen, Norway. (tor.eldevik@nersc.no; ola.johannessen@nersc.no; kjetil.lygre@nersc.no)

Northern High-Latitude Climate Changes During the Holocene as Simulated by Circulation Models

G. Lohmann

Alfred Wegener Institute for Polar and Marine Research, Bremerhaven, Germany

S.J. Lorenz

Max-Planck-Institute for Meteorology, Model and Data Group, Hamburg, Germany

M. Prange

DFG Research Center Ocean Margins, University of Bremen, Germany

To simulate the evolution of Holocene climate, forcing factors for the northern high-latitude climate are examined using different numerical models. A global coupled atmosphere–ocean circulation model driven by astronomical forcing over the last 7,000 years shows a long-term sea-surface temperature decrease in the Nordic Seas region associated with changes in seasonal insolation. A continuous cooling in the northeastern Atlantic is accompanied by a persistent warming in the Labrador Sea from the mid- to late Holocene. This temperature pattern can be attributed to a progressive weakening of the Icelandic Low and altered winds over the Nordic Seas induced by changes in insolation. In the early Holocene, important forcing for the Nordic Seas was probably caused by melting land ice masses and "deepening" of the Bering Strait. The effects of meltwater discharge and Bering Strait inflow are studied in a regional model of the Arctic and North Atlantic Oceans. It is suggested that a gradual increase in the flux of Pacific water through the Bering Strait during the early Holocene slowly affected the polar climate by melting ice and causing ocean circulation changes in the Nordic Seas. This high-latitude response to sea level change is shown to be different from the signature obtained by a freshwater release linked to the final outburst drainage of Laurentide lakes.

1. INTRODUCTION

A number of different processes affected the climatic evolution of the Nordic Seas during the past 10,000 years (the Holocene). Since this time span comprises half a precessional cycle [*Milankovitch*, 1941], astronomical forcing is an important factor that has to be considered when studying Holocene climate variability. Other factors are linked to the melting of land ice. During the early Holocene, remnants of glacial ice sheets were still present in northeastern North America and Scandinavia [e.g., *Peltier*, 1994]. The melting of these ice sheets influenced the oceanic circulation not only by freshening the sea water (i.e. lowering sea-water density) but also by inducing eustatic and isostatic sea-level changes. In particular, a slowly increasing sea level allowed the reopening of the Bering Strait 12-13 kyBP (1,000 years before present). The Bering land-bridge inundation was

The Nordic Seas: An Integrated Perspective
Geophysical Monograph Series 158

10.1029/158GM18

completed 5-6 kyBP [*Elias et al.*, 1996]. The influence of post-glacial meltwater on ocean circulation was not always gradual. One particular abrupt event was presumably caused by a catastrophic release of freshwater into the Labrador Sea from the final outburst drainage of Laurentide lakes at about 8.2 kyBP, resulting in a weakening of the thermohaline circulation (THC) and a reduction in its northward heat transport [*Barber et al.*, 1999].

Here, we present new modeling efforts designed to elucidate the particular roles of different parameters in forcing the Holocene climatic evolution of northern high latitudes. Two fully coupled climate models are employed to analyze the effects of orbital forcing and freshwater release from North America into the Labrador Sea. In addition, a regional ocean–sea-ice model is used to study the roles of reduced Bering Strait throughflow and enhanced freshwater input to the Arctic Ocean during the early Holocene.

Our model studies represent a first attempt to assess the relative importance of different forcing factors for the Holocene evolution of the Nordic Seas. They may help to interpret paleoceanographic records from that region.

2. METHODOLOGY

2.1. Global Climate Model: Varying Orbital Parameters

For the simulation of mid- to late Holocene climate changes owing to orbital forcing, we apply the coupled atmosphere-ocean general circulation model ECHO-G [*Legutke and Voss*, 1999]. The atmospheric part of this model is the circulation model ECHAM4 [*Roeckner et al.*, 1996] with T30 horizontal resolution (approximately 3.8° × 3.8°) and 19 vertical levels. The ocean model includes a dynamic thermodynamic sea-ice model with snow cover and has a resolution of approximately 2.8° × 2.8°. The model consists of 20 irregularly spaced vertical levels with 10 levels covering the upper 300 m. The atmospheric model ECHAM4 has been modified with respect to the standard version to account for sub-grid scale partial ice cover [*Grötzner et al.*, 1996], which is also considered in the ocean model component.

2.1.1. Orbital forcing. The ECHO-G model has been adapted to account for the influence of the annual distribution of solar radiation due to the slowly varying orbital parameters: the eccentricity of the Earth's orbit, the angle between the vernal equinox and the perihelion on the orbit, and the obliquity [*Milankovitch*, 1941]. The calculation follows *Berger* [1978] in providing the incoming solar radiation at the outer boundary of the atmosphere as a function of latitude and time.

The maximum deviation from present insolation was between 13.5 and 8.5 kyBP at the Northern Hemisphere summer. This is due to both a larger tilt of the Earth's rotation axis and the phase in the precession cycle, with the perihelion centered on the boreal summer solstice in the early Holocene. The insolation pattern in the Southern Hemisphere is much like that of the Northern Hemisphere, with a 10-ky phase shift (one half of the precessional cycle). At the winter solstice, a lack of insolation during the early to mid-Holocene compared to today is therefore south of the equator, which is mainly affected by the precession cycle, since the distance to the Sun is then at maximum in boreal winter. At 4.5 kyBP, the insolation at the top of the atmosphere roughly reached the present insolation distribution.

2.1.2. Experimental setup. For the control run, we utilize constant greenhouse gas concentrations typical for the preindustrial era (280 ppm CO_2, 700 ppb CH_4, and 265 ppb N_2O) and modern irradiance. This experiment has been integrated over 3,000 years of model simulation into a preindustrial climate state.

Computer resources for running a complex model like ECHO-G over the time period of the Holocene are very demanding. Therefore, the time scale of the orbital forcing has been shortened by an acceleration factor of 10, which implies that the response of the system to insolation forcing is completed after centuries. In effect, this procedure decouples any slowly responding portions of the system (e.g., deep-water properties) from the feedback structure.

The insolation trends of the last 7,000 years are represented in 700 model years. For the simulation of the Holocene climate we perform two experiments that differ in their initial conditions. For a detailed description of the procedure we refer to *Lorenz and Lohmann* [2004], where the effect of the acceleration factor on Holocene climate trends is evaluated. It is found that (within the model) the magnitude of orbitally forced Holocene trends is largely independent of the chosen acceleration factor.

2.2. Regional Model: Sea Level and Melting Arctic Ice Sheets

To examine the effects of Arctic meltwater discharge and Bering Strait throughflow on high-latitude ocean dynamics, we utilize the Coupled Ocean–Sea-ice Model with Open Surface (COSMOS) covering the Arctic Ocean, the Nordic Seas, and the Atlantic Ocean north of approximately 20 °S [*Prange*, 2003; *Prange and Lohmann*, 2003; *Prange and Gerdes*, 2004]. The model is formulated on a rotated grid to avoid the singularity of geographical coordinates at the pole (Figure 1) and has a horizontal resolution of about 100 km and 19 nonequidistant vertical levels. The ocean–sea-ice model is forced by atmospheric fields from a validated 15-year (1979-1993) period of the European Center for Medium-Range Weather Forecasts [*Prange*, 2003].

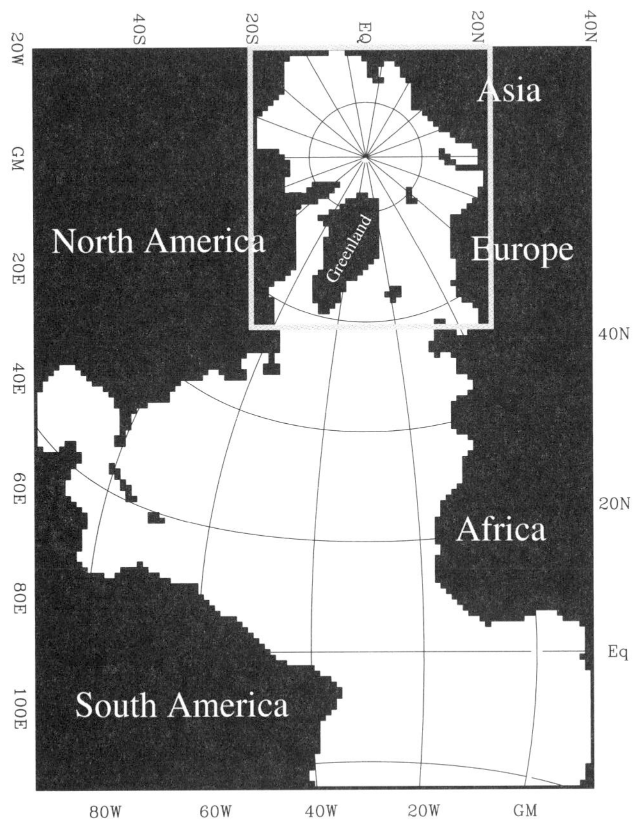

Figure 1. Domain of the regional model COSMOS. Both the geographical and the model grid coordinates are displayed. The frame marks the area that is shown in Figures 8-11.

In addition to atmospheric forcing, the ocean–sea-ice system is forced by river runoff and Bering Strait inflow. The river water inflow is implemented in the model as volume fluxes with zero salinity. Fourteen Arctic rivers are implemented as well as some additional ungauged runoff from the Arctic coastlines. For the Atlantic portion of the model domain, the eight largest rivers are included as well as the freshwater supply from Hudson Bay and the Baltic Sea, as described in *Prange* [2003]. Two experiments are performed, the setups of which differ in freshwater discharge and Pacific water inflow into the Arctic Mediterranean. These forcings refer to the present (control run) and the early Holocene (~10 kyBP).

2.2.1. Control run. A climatology for monthly discharge of the largest Arctic rivers and runoff from the Norwegian coast is applied in the model. Monthly varying inflow of Pacific water through the Bering Strait is implemented based on direct measurements [*Prange*, 2003]. Salinity of this inflow varies between 31.5 psu in September/October and 33.5 psu in March/April [*Roach et al.*, 1995]. The annual mean volume flux amounts to 0.8 Sv. Taking a reference salinity of 35 psu (the salinity of the Atlantic water component in the Arctic Ocean), the annual freshwater flux into the Arctic Ocean due to the Bering Strait inflow is 1,800 km^3. Aside from being a source term in the Arctic Ocean freshwater budget, the Bering Strait inflow is associated with a heat supply during the summer months. The temperature rises by up to 4 °C in September, while winter temperatures (December-May) are at freezing for the salinity [*Roach et al.*, 1995]. In COSMOS, the inflow of Pacific water through the Bering Strait is implemented by the open surface, i.e. the barotropic flow is taken into account [*Prange and Gerdes*, 2004]. This approach is justified by the strait's shallow depth (~50 m) relative to the thickness of the topmost model grid boxes (20 m).

2.2.2. Experiment 10 kyBP. At the beginning of the Holocene, remnants of glacial ice sheets were present in northeastern North America and Scandinavia [e.g., *Peltier*, 1994]. Numerical reconstructions suggest that the waning of the Laurentide ice sheet during the last stages of the deglaciation gave rise to a considerable flux of meltwater into the Arctic Ocean [*Licciardi et al.*, 1999]. Part of this water entered the ocean via the Mackenzie River basin, another part came from the northern islands of the Canadian Archipelago. *Licciardi et al.* [1999] calculated freshwater fluxes derived from meltwater and precipitation runoff from North America. For the early Holocene (~10 kyBP) they found a total freshwater input to the Arctic Ocean of about 1,250 km^3 yr^{-1}. Melting of the Fennoscandian ice sheet remnants provided freshwater to the Norwegian Coastal Current.

Beside the freshwater forcing from melting ice sheets, there is geological evidence for increased river discharge into the Arctic Mediterranean. *Sidorchuk et al.* [1999] reconstructed the paleohydrology of the Vychegda River from the grain size of channel deposits, paleochannel morphology, and paleolandscape features. The Vychegda River is a major tributary of the Northern Dvina, which flows into the Barents Sea. During the early Holocene, the river flow was slightly enhanced. The distribution of freshwater alginite (chlorococcalean algae) in marine sediment cores from the Eurasian continental margin indicates that the freshwater input to the Kara Sea was also enhanced around 10 kyBP [*Boucsein*, 2000].

A quantification of the increased freshwater inputs during the early Holocene is difficult and subject to considerable uncertainty. As an "educated guess", we presume runoff from the Norwegian coast as well as into the Barents and Kara Seas to be 25% higher than today. Extreme runoff, amounting to three times the modern one, is assumed from North American and northern Greenland coasts. Table 1 summarizes the freshwater forcing in our experiments.

Concerning the freshwater budget of the Arctic Sea, the enhanced runoff is partly compensated by a lesser inflow of

Table 1. Freshwater Input From Continents to the Arctic Ocean for the Present-Day Control Run and Experiment 10 kyBP. Units are $km^3\ yr^{-1}$.

Region	Present-day	10 kyBP
Norwegian coast	380	475
Barents Sea	452	565
Kara Sea	1,310	1,638
Laptev Sea	797	797
East Siberian Sea	195	195
North American/ North Greenland coasts	405	1,215
Total	**3,539**	**4,885**

low-saline Pacific water. The lower sea level during the early Holocene restricted the flow through the Bering Strait [*Elias et al.*, 1996]. We assume the 10-kyBP Bering Strait inflow was half of the modern level.

2.3. Global Circulation Model: Release of Freshwater

To analyse the climatic effect of a catastrophic release of freshwater into the Labrador Sea, the coupled atmosphere–ocean circulation model ECHAM3/LSG [*Voss et al.*, 1998] has been employed. The ocean model has a free surface and includes a thermodynamic sea-ice model. The coupled model has a horizontal resolution of about 5.6°.

In the freshwater experiment, a large freshwater anomaly in the northern North Atlantic is released in equal parts to two grid points on the Canadian coast of the Labrador Sea. The experimental setup of this scenario is described elsewhere [*Schiller et al.*, 1997; *Lohmann*, 2003]. The freshwater input to the Labrador Sea forces a transient shutdown of the THC, reducing the northward heat transport in the Atlantic Ocean. The circulation recovers completely after about 600 years, when the freshwater input was switched off [*Schiller et al.*, 1997]. For our analyses, the model output of years 140-150 is taken. At this time, the overturning circulation was about 50% of its present-day value of 18 Sv ($1\ Sv = 1 \times 10^6\ m^3\ s^{-1}$).

3. RESULTS

3.1. ECHO-G: Orbitally Forced Climate Variations

The transient simulations performed with the ECHO-G coupled model, forced by Earth's orbital parameters, exhibit coherent surface temperature trends during the mid- to late Holocene (Figure 2). Regional temperature trends in the North Atlantic realm indicate both positive and negative values: A continuous cooling in the northeastern Atlantic is accompanied by a continuous warming in the region around the Labrador Sea and large areas in the subtropics. The orbitally induced signal of reduced boreal summer insolation in northern mid- and high latitudes (Figure 3) is reflected by a surface temperature drop of up to 2 degrees Kelvin (K) at high latitudes during the last 7,000 years. The most pronounced temperature trends occur over the continents. Enhanced warming during the Holocene occurs in the arid subtropical continents, from northern Africa via western Asia to the Indian subcontinent [*Lorenz and Lohmann*, 2004]. The most distinct cooling takes place over continental and sea-ice–covered northern high latitudes. We find that the trend is robust against the choice of ensemble members (not shown), indicating that the difference in the interannual variability of both experiments has no significant effect on the amplitude and distribution of the regional trends.

Motivated by the Palaeoclimate Modeling Intercomparison Project (PMIP) [*Joussaume and Taylor*, 1995], we take a snapshot of the mid-Holocene time slice at 6 kyBP (sometimes referred to as the Holocene climate optimum). PMIP has fostered a systematic evaluation of climate models, besides others, under conditions of the mid-Holocene. The time slice 6 kyBP was chosen to test the near-equilibrium response of climate models to orbital forcing with CO_2 at preindustrial concentration and present ice sheet distribution. Furthermore, we evaluate the time slices at 4 and 2 kyBP.

Our analyses are based on averages of 60 simulation years out of the set of two 700-year-long transient Holocene experiments, centered at 6, 4, and 0 kyBP. The largest orbitally forced temperature difference between the time slices occurs in the Arctic in October. The simulated near-surface temperature of the 6 kyBP climate for October is shown in Figure 4a, whereas Figure 4b displays temperature anomalies from the latest pre-industrial climate. Sea-ice thickness and its anomalies are also shown in Figure 4. A uniform warming of 3 to 6 K can be detected over the entire Arctic Ocean (Figure 4b), accompanied by a decrease of the Arctic sea-ice thickness of 40 to 80 cm (Figure 4d). The maximum anomalies are located near the New Siberian Islands and Franz Joseph Land. In these regions, the temperature anomaly exceeds 6 K and the sea-ice reduction amounts to more than 80 cm in the same areas. Note the reduction in sea-ice extent in the mid-Holocene simulation in Hudson Bay, the Greenland Sea, and the Barents and Bering Seas, compared with that at 0 kyBP (dark shaded area with sea-ice compactness of more than 20% for 6 kyBP in Figure 4c, and for 0 kyBP in Figure 4d).

The temperature change indicates a strong nonlinear signal in the model response to the radiative forcing: The upper ocean stores the heat of the boreal summer insolation, i. e. $30\ W\ m^{-2}$ more energy input than today in the Arctic from mid-June to the end of July (Figure 3a). The warmer sea-surface

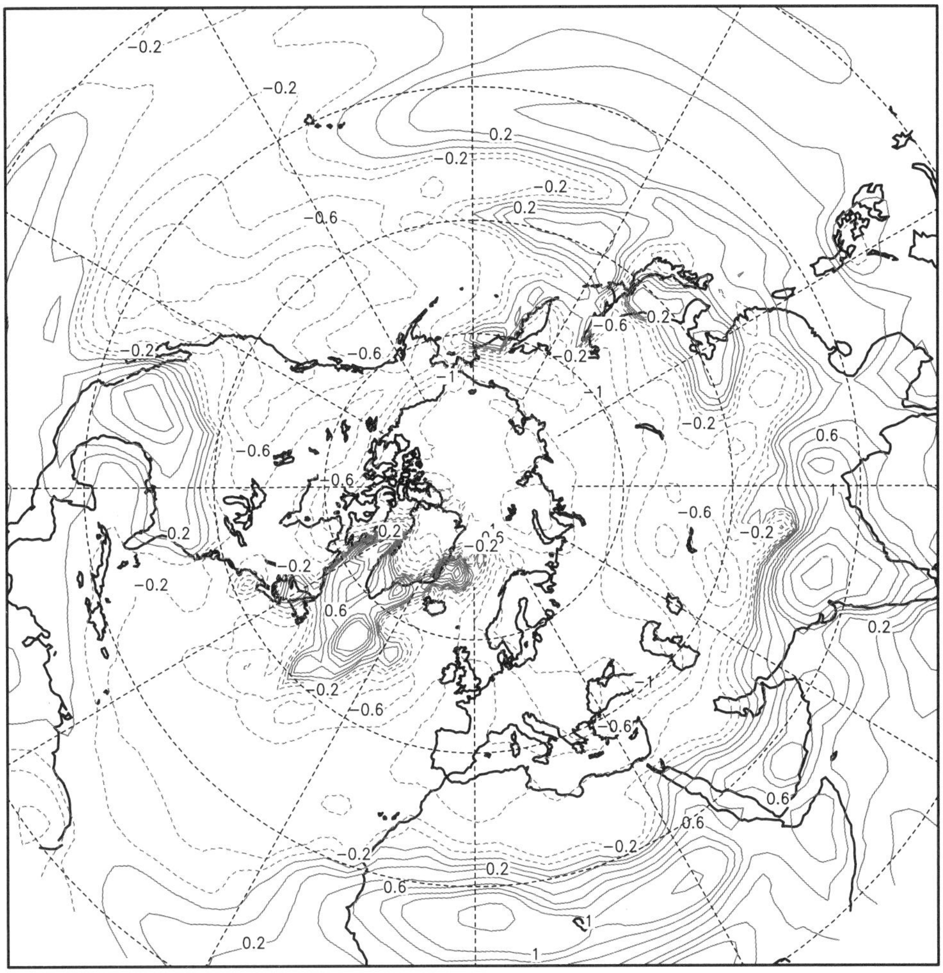

Figure 2. ECHO-G: Mean surface temperature trend from the mid-Holocene into the preindustrial era (7 to 0 kyBP) of the ensemble simulations using an acceleration factor of 10 (700 simulation years). The surface temperature is taken over open water, sea-ice, ground, and snow areas, respectively. Units are K (7 ky)$^{-1}$. Contour lines are at ±0.1, *0.2*, 0.4, *0.6*, 0.8, *1.0*, 1.5, *2.0*, 2.5, where the italicized numbers mark the labeled contour lines.

temperature (SST) during the summer season with high-level solar energy input lengthens the ice-free season and reduces the average sea-ice thickness in October, although the radiation anomaly has already turned its sign. The sea-ice/upper ocean introduces therefore a seasonally delayed response to the Milankovitch forcing. The situation is qualitatively similar for 4 kyBP (Figure 5) and 2 kyBP (not shown).

The model also simulates modified surface winds during the mid-Holocene boreal winter in the Arctic region (Figure 6). We find enhanced southward winds in the western part of the Greenland Sea and over the Denmark Strait. Furthermore, there is intensified cyclonic circulation in the Norwegian Sea. This is consistent with a stronger Icelandic Low during 6 kyBP relative to the preindustrial climate. For 4 kyBP (Figure 7), the enhanced southward winds remain, but the cyclonic circulation in the Norwegian Sea is reduced and shifted to the east. We have evaluated a North Atlantic Oscillation (NAO) Index by taking the pressure difference (in hPa) at sea level between the regions (50 °W to 50 °E, 60 °N to 80 °N) and (60 °W to 60 °E, 15 °N to 50 °N) as in *Felis et al.* [2004]. The modeled NAO trend is 1.5 hPa for 6,000 years, analogous to an NAO Index of 13.5, 13.0, 12.5, and 12.0 hPa for the time slices at 6, 4, 2, and 0 kyBP, respectively.

The wind effects change sea-ice dynamics and, in particular, enhance southward sea-ice transport along the eastern coast of Greenland during the mid-Holocene, consistent with findings of *Tremblay et al.* [1997]. This causes increased sea-ice concentration and a temperature drop southeast of Greenland. Along with the positive phase

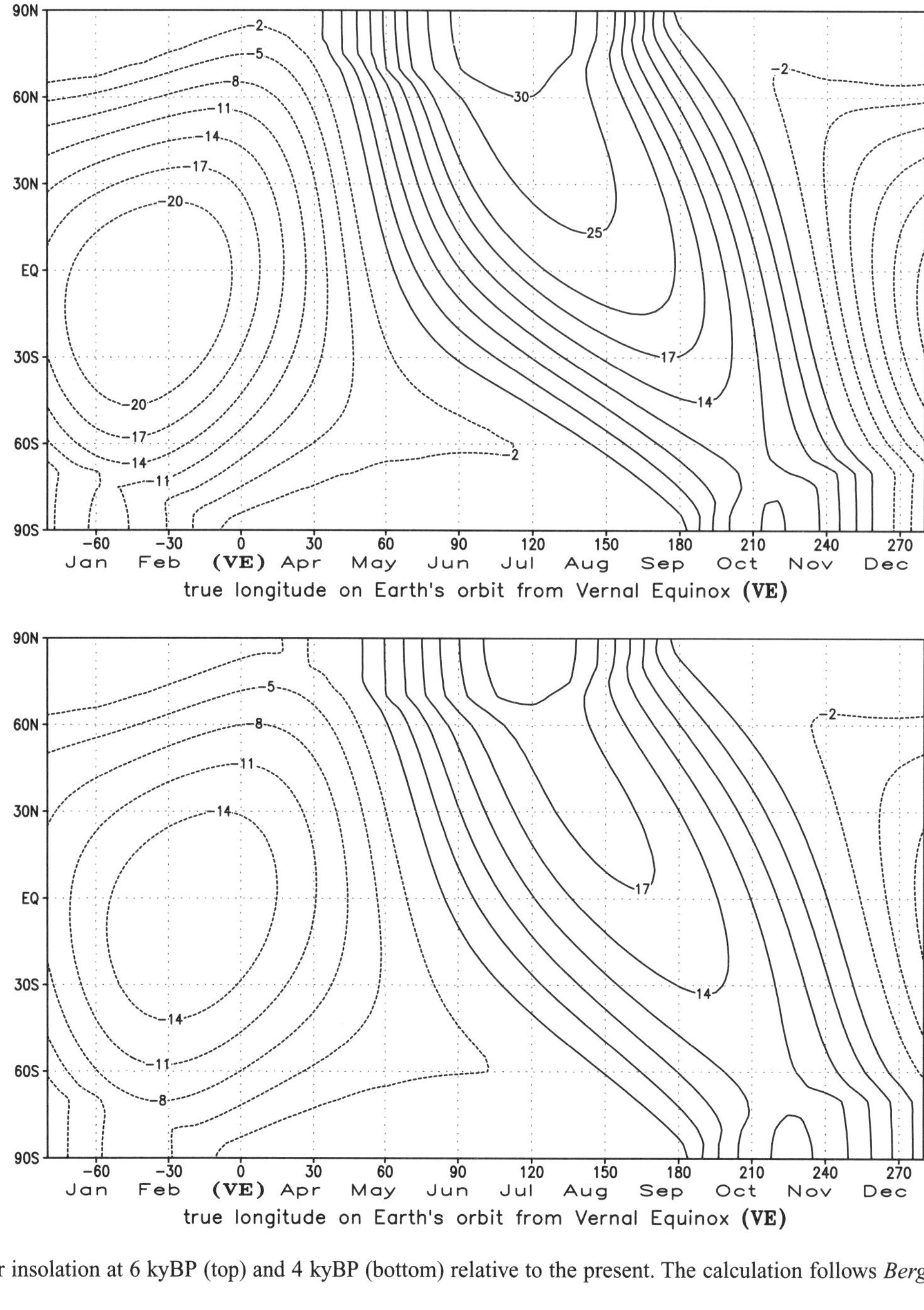

Figure 3. Solar insolation at 6 kyBP (top) and 4 kyBP (bottom) relative to the present. The calculation follows *Berger* (1978). Units are W m^{-2}.

of the NAO, a mid-Holocene temperature rise in Europe can be observed in the model results (Figure 2).

Between 4 and 2 kyBP, we find a warm phase widespread into the northern mid- to high latitudes, which is mainly located over the North American and Eurasian continents [*Lorenz and Lohmann*, 2004]. For 2 kyBP (not shown), the southward wind anomalies are absent and only an increased eastward component of the surface winds is detected. There is a strong temperature decrease in northern mid-latitudes between 7 and 4 kyBP for the boreal winter season and the annual mean [*Lorenz and Lohmann*, 2004].

3.2. COSMOS: Bering Strait and Meltwater Effects

The control run and the experiment 10 kyBP are started from the same spin-up run described by *Prange* [2003].

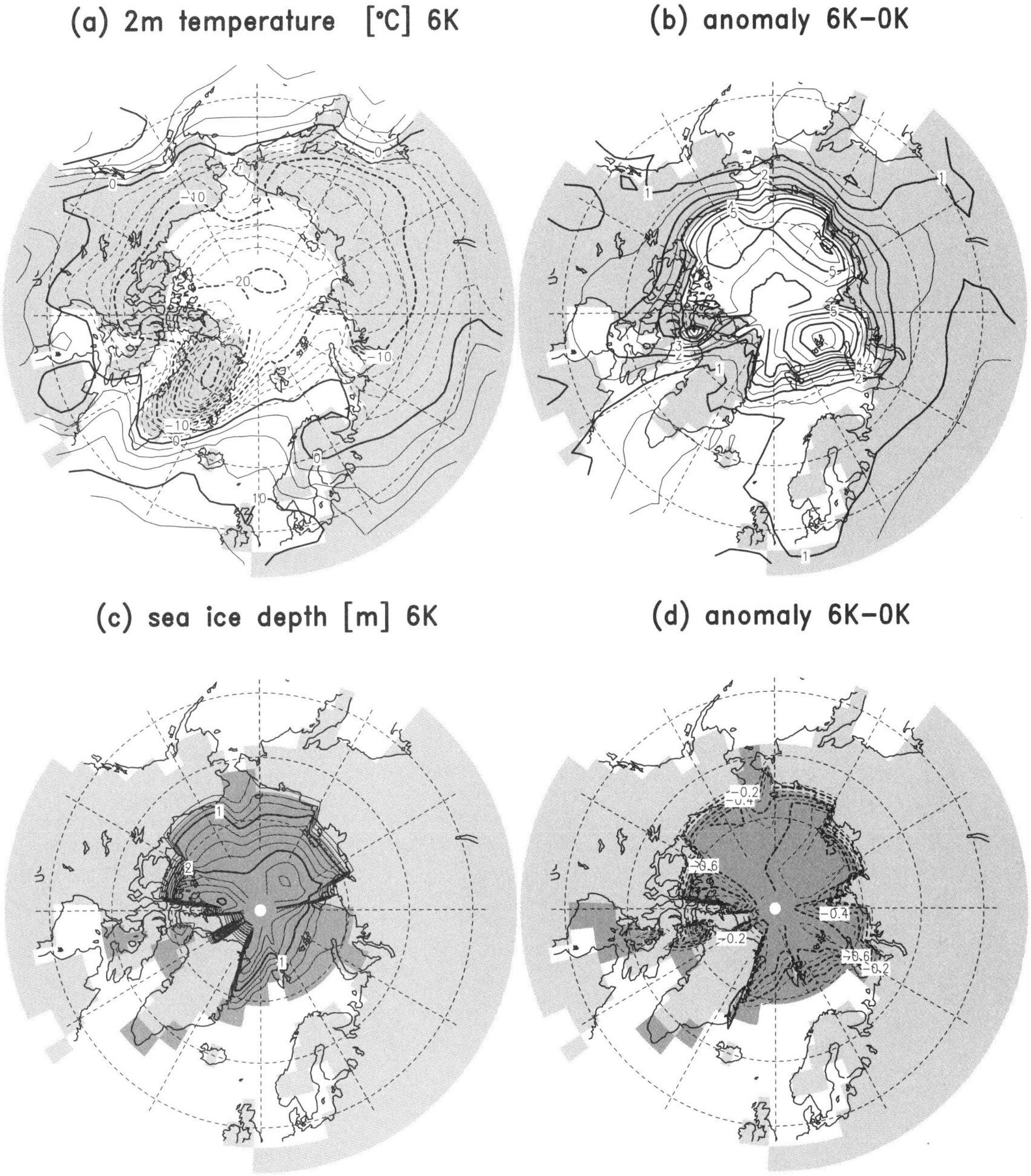

Figure 4. ECHO-G: Surface air temperature (a), sea-ice thickness and compactness (c) in October for the mid-Holocene climate at 6 kyBP; surface air temperature anomaly (b) of the mid-Holocene from the preindustrial climate and sea-ice thickness anomaly and sea-ice compactness (d). Areas where sea-ice compactness exceeds 20% are darkly shaded for 6 kyBP in (c) and for 0 kyBP in (d) (no anomaly). Light shading indicates continents in the T30 resolution of the atmospheric submodel ECHAM. Displayed is the mean of the two-700 year Holocene simulations for October, where 6 kyBP is a mean of 60 model years between 5.7 and 6.3 kyBP, and 0 kyBP is a similar average around the preindustrial climate. The contour intervals for temperature (sea-ice thickness) are 2.5 K (0.25 m) and 0.5 K (0.1 m) for anomalies, respectively.

For each experiment, the model is integrated for 60 years. In the following, we show annual mean fields that apply to the last year of the integration period (multi-year averaging is not necessary, since internal interannual variability proves to be negligible in the modeled Arctic Ocean).

3.2.1. Control experiment. The present-day (model-derived) ocean circulation averaged over the top 80 m in the polar and subpolar seas is presented in Figure 8. The top 80 m are represented by the three topmost levels of the model grid and comprise the surface mixed layer with the upper

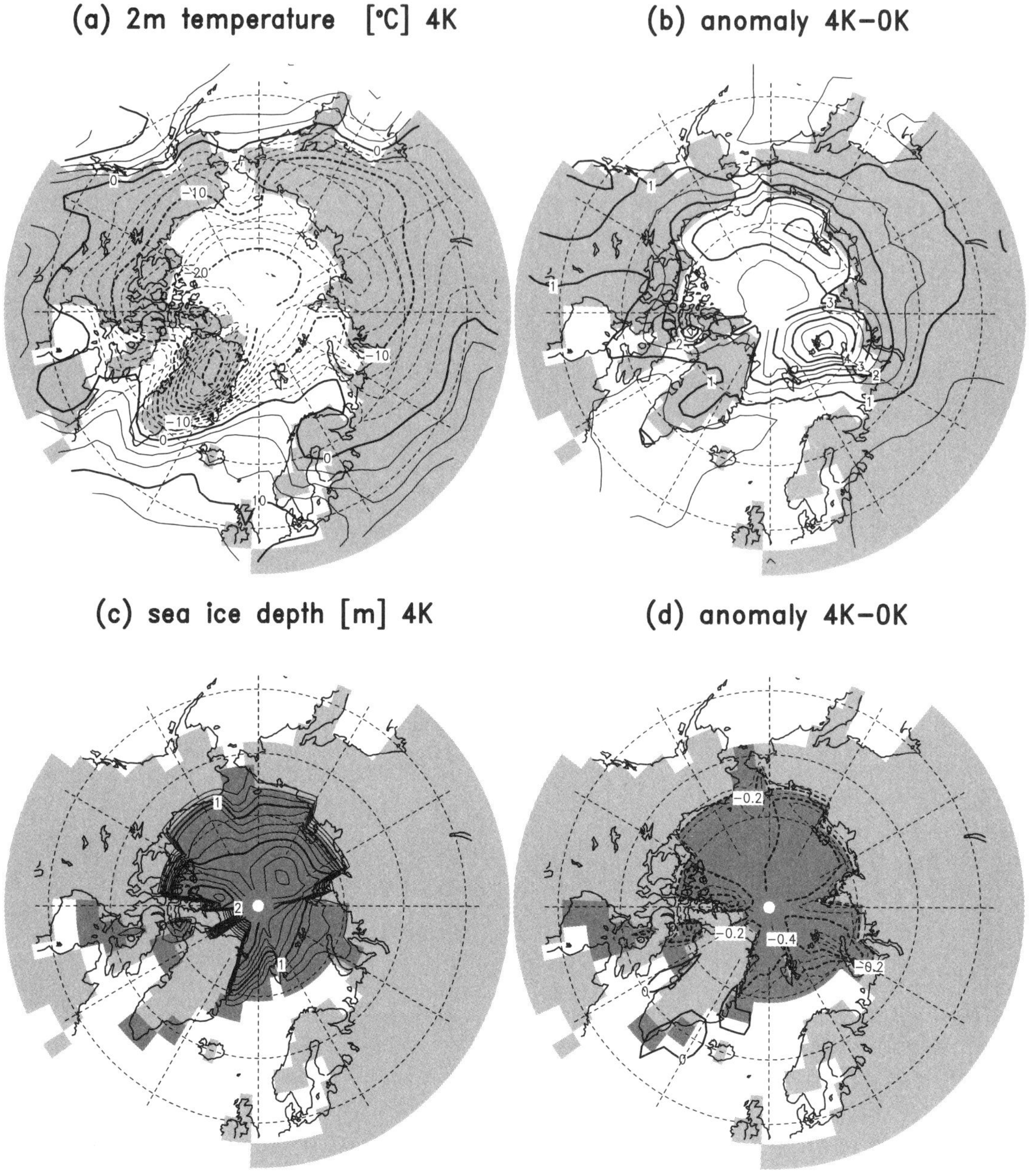

Figure 5. Same as Figure 4, but for 4 kyBP.

part of the cold halocline in the Arctic Ocean. A strong cyclonic gyre dominates the Nordic Seas, consisting of the East Greenland Current (EGC) in the west, and the Norwegian Atlantic Current (NAC) in the east. Atlantic water enters the Barents Sea, bringing some heat into the Arctic Ocean. This current constitutes the southern branch of an overall cyclonic flow pattern in the eastern Arctic Ocean. The Canadian Basin in the western Arctic is dominated by the anticyclonic Beaufort Gyre. The Transpolar Drift (TPD) carries polar waters towards the outlets of the Arctic Ocean, namely, the Fram Strait and the Nares Strait (Canadian Archipelago).

3.2.2. Experiment 10 kyBP. Figure 9 shows differences in upper ocean velocity between experiment 10 kyBP and the present-day control run. Over the Canadian Basin and the East Siberian Sea, the anticyclonic pattern is intensified. A cyclonic gyre north of Greenland and changes in the Eurasian Basin are associated with a modified path of the TPD. The reduced Bering Strait inflow is

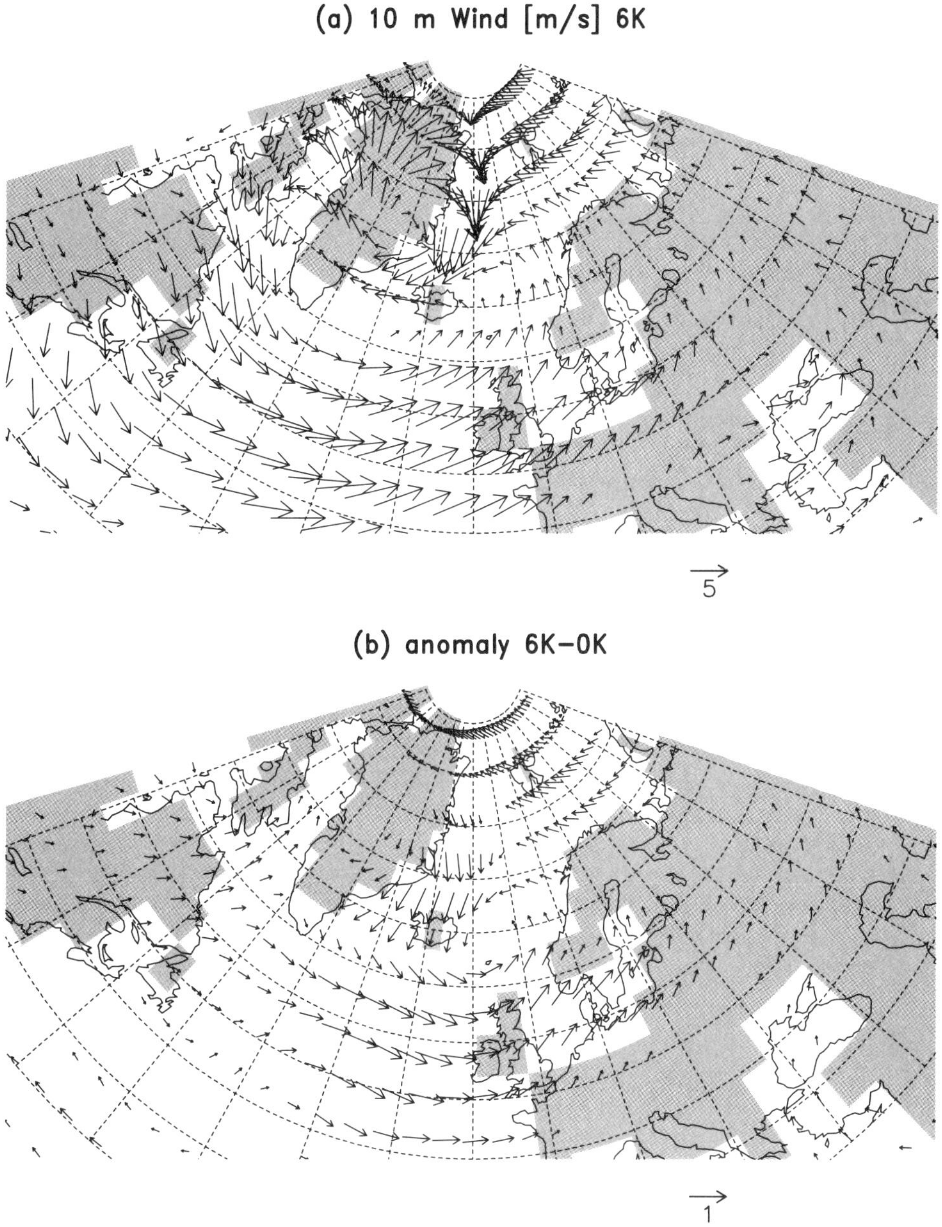

Figure 6. ECHO-G: (a) Surface wind during the boreal winter season (January to March) for the mid-Holocene climate at 6 kyBP. Vectors with a magnitude less than 1 ms^{-1} are suppressed. (b) Wind anomaly of the mid-Holocene from the preindustrial climate during the boreal winter season. Anomaly vectors less than 0.2 ms^{-1} are suppressed. As in Figure 4, the surface winds represent ensemble means of 60 model years averages.

compensated by a weaker outflow through the Canadian Archipelago and an increased inflow via the Barents Sea, maintaining a steady Arctic Ocean mass balance. In the center of the Nordic Seas, an anomalous cyclonic flow pattern is found.

Figure 10 presents annual mean differences between experiment 10 kyBP and the control run. The decreased heat input due to a reduced Bering Strait inflow in Experiment 10 kyBP causes sea-ice to be thicker by up to 1 m in the vicinity of the strait. This thick sea-ice is advected into the central Arctic. In conjunction with lowered heat fluxes from the ocean (not shown) owing to a stronger salinity stratification, this results in a thicker ice cover over wide areas of the Arctic Ocean and in the Greenland Sea. The drift of sea-ice is coupled with ocean currents through ocean–sea-ice stresses. Therefore, differences in sea-ice motion between

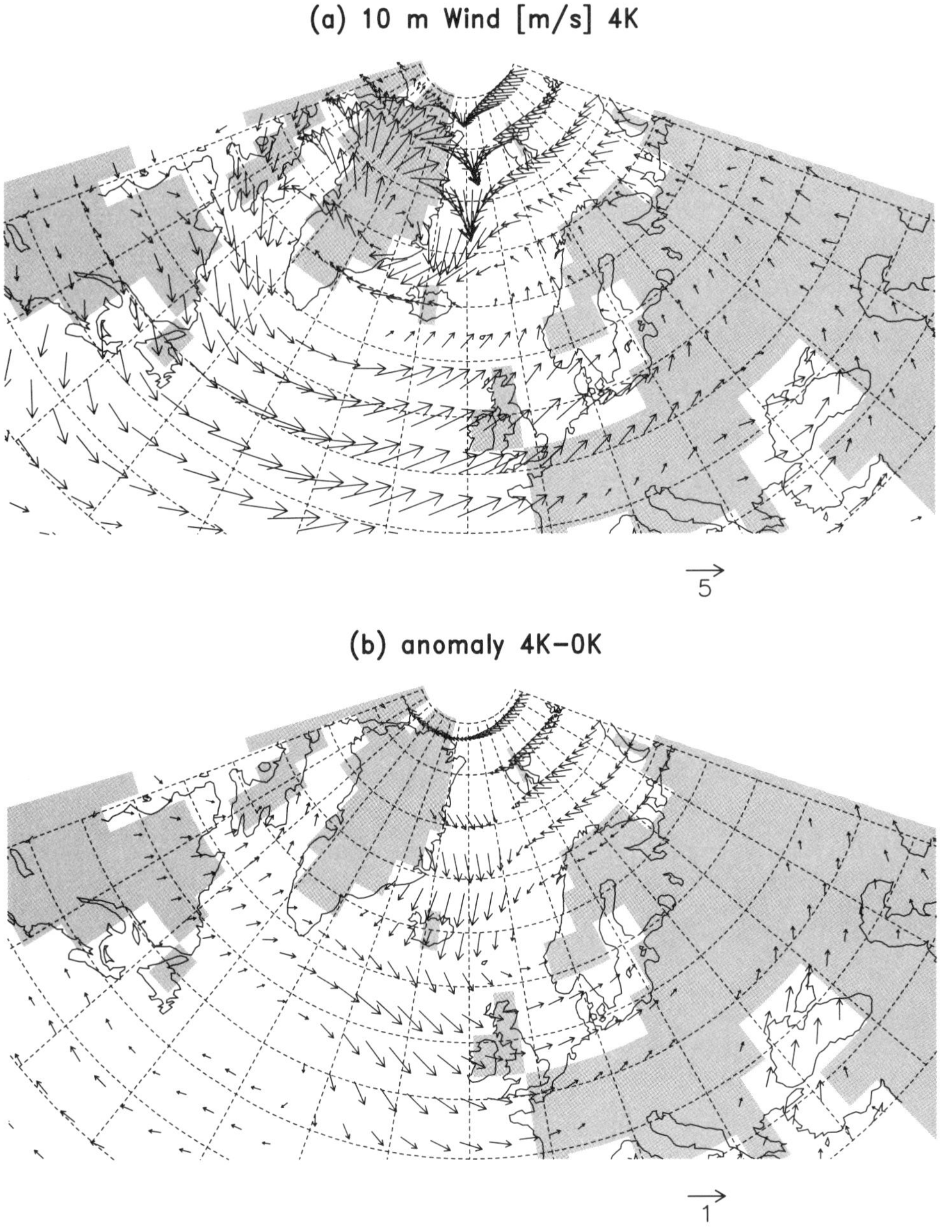

Figure 7. Same as Figure 6, but for 4 kyBP.

experiment 10 kyBP and the control run are almost identical to differences in the upper ocean circulation (see Figure 9).

Changes in the circulation are associated with changes in the field of upper ocean salinity. The meltwater from the Canadian coast is advected by the Beaufort Gyre into the central Arctic. Since the Bering Strait inflow is reduced, part of this freshwater can readily enter the Chukchi and East Siberian Seas. Lowered surface salinities in the Kara, Barents, and Norwegian Seas are directly caused by enhanced local runoff (see Table 1). The eye-catching high-salinity tongue north of Greenland (Figure 11a) results from the reduced inflow of low-saline Pacific water and a changed circulation pattern (Figure 9). Note also the low salinity in the EGC.

Differences in mean potential temperature between experiment 10 kyBP and the control run are presented in Figure 11b for the depth range 80-250 m. Largest values naturally appear in the "high-temperature" North Atlantic. But also in the Arctic Ocean we find differences up to 1 °C. In the central Arctic, these differences can be explained by changes in salinity (i.e., density) stratification rather than

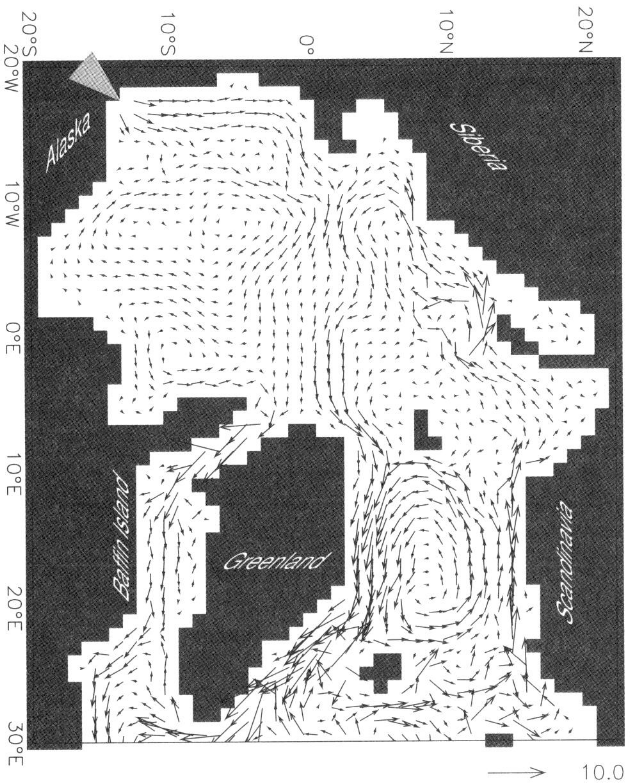

Figure 8. COSMOS: Annual mean upper ocean velocity (averaged over 0-80 m) in the present-day control run. Units are cm s^{-1}. Labels refer to the rotated model grid. The grey arrow marks the position of the Bering Strait.

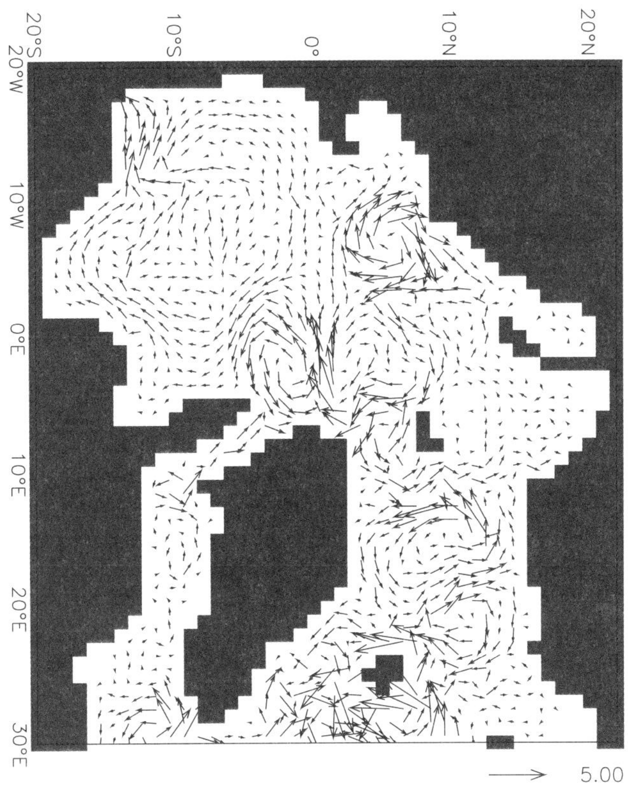

Figure 9. COSMOS: Difference in mean upper ocean velocity (averaged over 0-80 m) between experiment 10 kyBP and the present-day control run. Units are cm s^{-1}.

by horizontal advection. Low surface salinities stabilize the water column, thereby inhibiting convective mixing of cold surface water with warmer water from deeper halocline layers.

In the central Nordic Seas we find overall cooler conditions in experiment 10 kyBP compared with those in the control run. Along the Scandinavian coast and in the EGC, as well as northeast of Iceland, subsurface temperatures are higher.

3.3. ECHAM3/LSG: Freshwater Release

Caused by the freshwater release to the North Atlantic, the THC slows to 9 Sv and the northern North Atlantic cools considerably (Figure 12), where the strongest cooling is over the Nordic Seas caused by advanced sea-ice cover. Figure 13a shows surface winds during the freshwater release for the boreal winter season. We find enhanced southward winds in the Greenland Sea, over the Denmark Strait, and south of Greenland (Figure 13b) that favor the transport of polar water masses into the North Atlantic.

4. DISCUSSION

The impact of the distribution of incident solar radiation on the Holocene climate is investigated by using a coupled atmosphere–ocean–sea-ice model. Justified by the much longer time scale of the astronomical forcing than that of the dynamical feedback processes in the atmosphere-ocean system, we accelerate the time scale of the orbitally varying solar radiation by a factor of 10. This enables the simulation of the mid- to late Holocene climatic evolution in northern high latitudes with a complex circulation model.

This method is limited to the investigation of long-term effects on the atmosphere–ocean–sea-ice system caused by the astronomical forcing. We find no significant changes in the THC in the simulations [*Lorenz and Lohmann*, 2004], nor is our model designed to resolve deep ocean circulation changes.

Based on model simulations that are forced solely by insolation, the surface temperature trend shows a spatial pattern consistent with a weakening of the Icelandic Low from the mid- to late Holocene. The modeled temperature and wind patterns reflect a strong response to a gradual insolation

Figure 10. COSMOS: Difference in mean sea-ice thickness between experiment 10 kyBP and the present-day control run. Units are m.

forcing (Figure 3). A strong Icelandic Low (positive NAO phase) is accompanied by a relatively mild winter over northern Europe. The information about the winter circulation is stored in the subsurface while the summer signal is destroyed in the following winter. Therefore, the annual mean SST response to changes in the atmospheric circulation is almost identical to the winter signal in the NAO. It is therefore not astonishing that the cold season signal is preserved in the reconstructed SSTs. Indeed, alkenone SST reconstructions in the northeastern Atlantic show strong negative SST trends for 6 kyBP relative to the preindustrial climate, 4.4 K at 75 °N,14 °E [*Marchal et al.*, 2002], 1.79 K at 67 °N,8 °E [*Calvo et al.*, 2002], and 3.26 K at 58 °N,9 °E [*Emeis et al.*, 2003], whereas a core further east, at 59 °N,26 °W [*Marchal et al.*, 2002], shows a moderate cooling of 0.62 K within 6,000 years only. These data are consistent with a recent analysis of alkenone SST reconstructions, which shows that negative SST trends in the northeastern Atlantic and western Mediterranean, along with positive SST trends in the eastern Mediterranean, northern Red Sea, and western subtropical Atlantic, are connected with a continuous weakening of the NAO pattern from the early to late Holocene [*Rimbu et al.*, 2003]. The northern hemisphere extratropical response is linked to Rossby wave propagation from low latitudes during the boreal winter. This relationship is also consistent with a global set of collected marine proxy temperature data

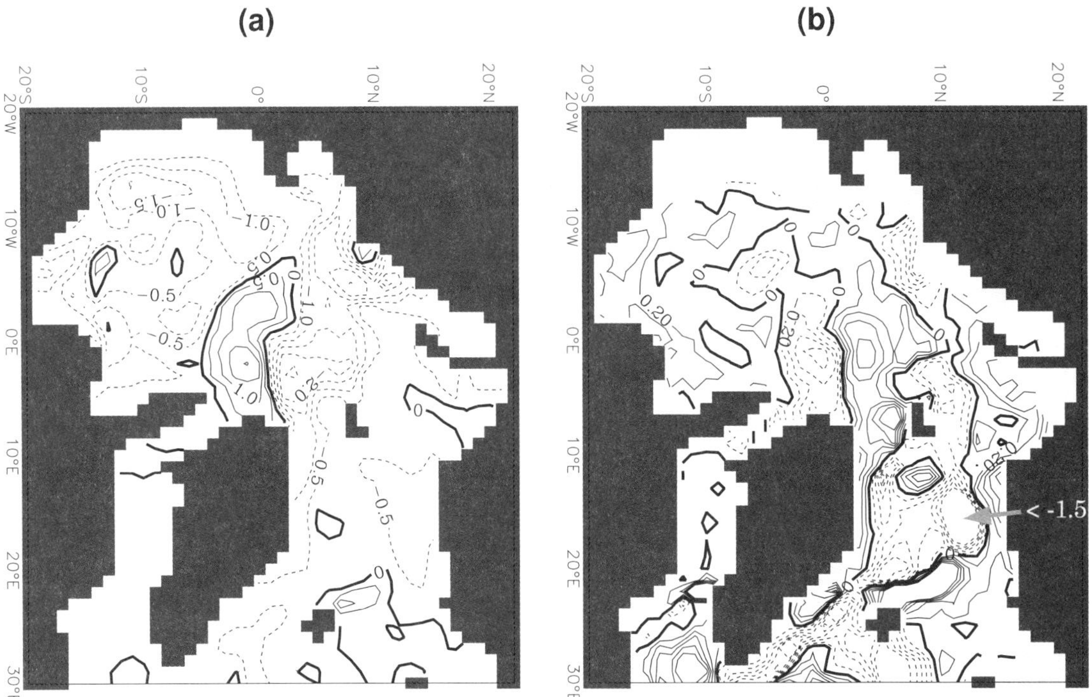

Figure 11. COSMOS: Difference in mean hydrographic fields between experiment 10 kyBP and the present-day control run (i.e., 10 kyBP minus present-day). (a) Salinity averaged over 0-80 m. (b) Potential temperature averaged over 80-250 m. Contour interval is 0.2 °C.

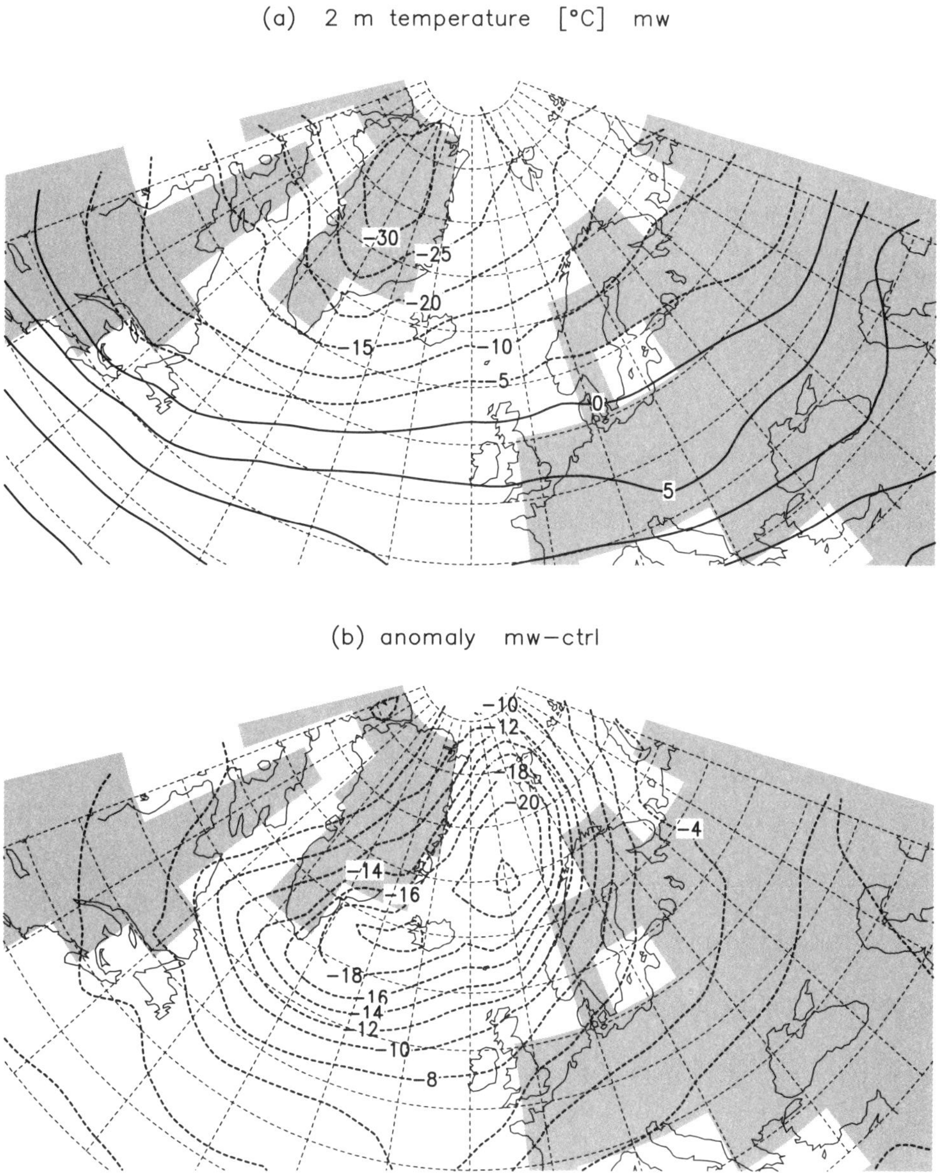

Figure 12. ECHAM3/LSG: a) Near-surface air temperature during the freshwater release for the boreal winter season (January to March). (b) Anomaly from the control integration.

[*Emeis et al.*, 2000; *Marchal et al.*, 2002; *Rohling et al.*, 2002]. A coherent picture of neoglaciation since 7 kyBP is revealed by the model and proxy data: Opposite trends of warming and cooling occur in the tropics and mid-latitudes [*Rimbu et al.*, 2004; *Lorenz et al.*, 2005]. The pronounced warming at 6 kyBP in the Arctic Ocean as seen in the experiments is supported by proxy data [*Texier et al.*, 1997; *Andreev et al.*, 2001] and a northward shift of the Arctic tree lines [*Tarasov et al.*, 1998]. Driftwood records reveal increased Fram Strait outflow by around 6 kyBP [*Dyke et al.*, 1997; *Tremblay et al.*, 1997]. *Prange and Lohmann* [2003] pointed out the importance of freshwater input by Arctic rivers on the Arctic upper ocean and sea-ice circulation. Changes in wind direction may also affect the path of the TPD and ice export from the Arctic which is led to further investigations.

Furthermore, the effect of a freshwater release into the Labrador Sea is shown for present-day background conditions mimicking a perturbation induced by the final outburst drainage of Laurentide lakes at about 8.2 kyBP [*Alley et al.*, 1997]. Although a direct comparison with the simulations for the effect of Earth's orbital parameters (section 3.1) is not possible due to another model and model resolution, it

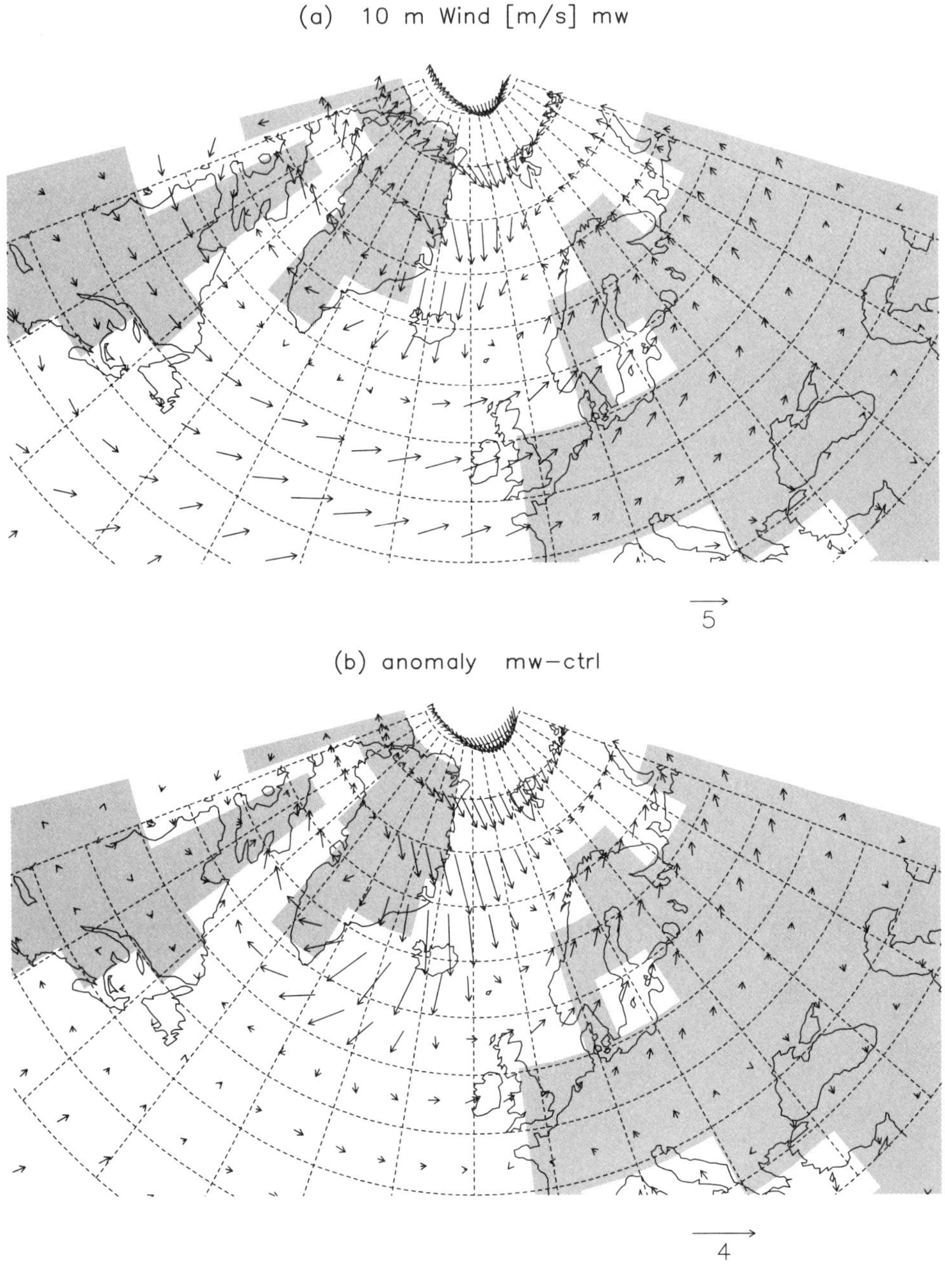

Figure 13. ECHAM3/LSG: (a) Surface wind during the freshwater release for the boreal winter season (January to March). (b) Wind anomaly from the control integration. Units are m s^{-1}.

can be seen that the sudden freshwater input (Figure 13b) has a stronger effect on climate than the gradual Milankovitch forcing for 6 kyBP (cf. Figure 6b).

5. CONCLUDING REMARKS

To avoid the necessity of prescribed SST, coupled atmosphere–ocean simulations of the mid-Holocene climate [*Texier et al.*, 1997; *Hewitt and Mitchell*, 1998; *Voss and Mikolajewicz*, 2001; *Kitoh and Murakami*, 2002] have recently been performed. These experiments simulate near-equilibrium states of the mid-Holocene climate (6 kyBP) or perform a series of time slices stepping through the Holocene [*Liu et al.*, 2003]. *Liu et al.* [2003] applied the time slice concept and also found a cooling at low latitudes and a warming at higher latitudes during the mid-Holocene and attributed this to direct radiation changes. Our method can also be used to obtain the climate at certain time intervals at optimal computing costs [*Lorenz and Lohmann*, 2004; *Felis et al.*, 2004]. In our experiments, we detect changed winds during the mid- to late Holocene in the Nordic Sea region associated with the strength and position of the Icelandic Low and related westward winds. Our study indicates changes in the strength of the Icelandic Low with a significant trend on multi-millenial time scales (1.5 hPa for 6,000 years). Induced by anthropogenic greenhouse gases in the atmosphere, we found with the same model an increase for the modeled index of about 1.7 hPa within the last hundred years [*Felis et al.*, 2004], indicating that the effect of the increased greenhouse gases in the atmosphere on the strength of the Icelandic Low is of similar magnitude than the effect linked to the orbital parameters.

We find a strong warming in the Barents Sea for 6 kyBP to 2 kyBP relative to the preindustrial climate. Interestingly, the anomalous southward component of surface winds in the Nordic Seas at 6 kyBP and 4 kyBP is absent for the 2 kyBP climate. Such nonlinear variations in wind field affect the Holocene hydrography in the Nordic Seas as will be discussed in forthcoming studies. In our experimental setup, we have neglected interactions with changes in vegetation cover which may be of particular importance for the snow cover over land [*Ganopolski et al.*, 1998] and for permafrost [*Velichko et al.*, 1995]. Future model configurations will elaborate vegetation–climate feedbacks and permafrost effects in our framework for Holocene climate modeling.

To simulate the climate of the early Holocene, freshwater fluxes from melting ice sheets and topographic changes have to be taken into account. Our regional model shows that these processes may have an important influence on high-latitude ocean and sea-ice dynamics. In our experiments, we find a substantial effect on the strength and size of the Beaufort Gyre, which is corroborated by geological evidence. On the basis of the heavy-mineral distributions in Arctic Ocean sediment cores, *Behrends* [1999] reconstructed paleo–sea-ice drift patterns. The data suggest that the Beaufort Gyre expanded during the early Holocene.

An important implication of the gradual Bering Strait deepening in the course of the early Holocene is the increasing heat flux into the Arctic Ocean. Our regional model shows that the inflowing heat exerts a strong influence on polar sea-ice. An intensified warm Bering Strait inflow causes a decline in sea-ice coverage in the Chukchi and East Siberian Seas. Where sea-ice is replaced by open water, the surface albedo substantially decreases. It is therefore likely that the gradually increasing influx of Pacific water during the early Holocene slowly affected the polar climate, raising local air temperatures.

Acknowledgments. We appreciate the constructive comments by Wolfgang Berger. This work was supported by Bundesministerium für Bildung und Forschung through DEKLIM, and by Deutsche Forschungsgemeinschaft through DFG Research Centre Ocean Margins at Bremen University (No. RCOM0140).

REFERENCES

Alley, R.B., T. Sowers, P.A. Mayewski, M. Stuiver, K.C. Taylor, P.U. Clark, Holocene climatic instability: A prominent, widespread event 8200 yr ago, *Geology*, 25, 483-486, 1997.

Andreev, A.A., V.A. Klimanov, and L.D. Sulerzhitsky, Vegetation and climate history of the Yana River lowland, Russia, during the last 6400yr, *Quat. Res.*, 20, 259-266, 2001.

Barber, D.C., A. Dyke, C. Hillaire-Marcel, A.E. Jennings, J.T. Andrews, M.W. Kerwin, G. Bilodeau, R. McNeely, J. Southon, M.D. Morehead, J.-M. Gagnon, Forcing of the cold event of 8,200 years ago by catastrophic drainage of Laurentide lakes. *Nature*, 400, 344-348, 1999.

Behrends, M., Rekonstruktion von Meereisdrift und terrigenem Sedimenteintrag im Spätquartär: Schwermineralassoziationen in Sedimenten des Laptev-See-Kontinentalrandes und des zentralen Arktischen Ozeans. *Rep. Polar Res.*, 310, Alfred-Wegener-Institute for Polar and Marine Research, Bremerhaven, 167 pp., 1999.

Berger, A.L., Long-term variations of daily insolation and Quaternary climatic changes, *J. Atmos. Sci.*, 35, 2362-2367, 1987.

Boucsein, B., Organic carbon in Late Quaternary sediments: Responses to paleoenvironmental changes in the Laptev and Kara Seas (Arctic Ocean). *Rep. Polar Res.*, 365, Alfred-Wegener-Institute for Polar and Marine Research, Bremerhaven, 103 pp., 2000

Calvo, E., J. Grimalt, and E. Jansen, High resolution U_{37}^{K} sea surface temperature reconstruction in the Norwegian Sea during the Holocene, *Quat. Sci. Rev.*, 21, 1385-1394, 2002.

Dyke, A.S., J. England, E. Reimnitz, and H. Jette, Changes in driftwood delivery to the Canadian Arctic Archipelago: the hypothesis of postglacial oscillations of the Transpolar Drift. *Arctic*, 50, 1-16, 1997.

Elias, S.A., S.K. Short, C.H. Nelson, and H.H. Birks, Life and times of the Bering Land Bridge. *Nature*, 382, 61-63, 1996.

Emeis, K.-C., U. Struck, H.-M. Schulz, R. Rosenberg, S. Bernasconi, H. Erlenkeuser, T. Sakamoto, and F. Martinez-Ruiz, Temperature and salinity variations of mediterranean sea surface waters over the last 16,000 years from records of planktonic stable oxygen isotopes and alkenone unsaturation ratios, *Palaeogeogr. Palaeoclimatol. Palaeoecol.*, 158, 259-280, 2000.

Felis, T., G. Lohmann, H. Kuhnert, S.J. Lorenz, D. Scholz, J. Pätzold, S.A. Al-Rousan, and S.M. Al-Moghrabi, Increased seasonality in Middle Eastern temperatures during the last interglacial period. *Nature*, 429, 164-168, 2004.

Ganopolski, A., C. Kubatzki, M. Claussen, V. Brovkin, and V. Petoukhov, The influence of vegetation-atmosphere-ocean interaction on climate during the mid-Holocene, *Science*, 280, 1916-1919, 1998.

Grötzner, A., R. Sausen, and M. Claussen, The impact of sub-grid scale sea-ice inhomogeneities on the performance of the atmospheric general circulation model ECHAM3, *Climate Dyn.*, 12, 477-496, 1996.

Hewitt, C.D. and J.F.B. Mitchell, A fully coupled GCM simulation of the climate of the mid-Holocene, *Geophys. Res. Lett.*, 25, 361-364, 1998.

Joussaume, S. and K.E. Taylor, Status of the Paleoclimate Modeling Intercomparison Project (PMIP), in *Proceedings of the First*

International AMIP Scientific Conference, vol. 92, pp. 425-430, WCRP, 1995.

Kim, J.-H., N. Rimbu, S.J. Lorenz, G. Lohmann, S.-I. Nam, S. Schouten, C. Rühlemann, and R. R. Schneider, North Pacific and North Atlantic sea-surface temperature variability during the Holocene. *Quat. Sci. Rev.*, 23, 2141-2154, 2004.

Kitoh, A. and S. Murakami, Tropical Pacific climate at the mid-Holocene and the Last Glacial Maximum simulated by a coupled atmosphere-ocean general circulation model, *Paleoceanography*, 17, doi:10.1029/2001PA000,724, 2002.

Legutke, S. and R. Voss, The Hamburg atmosphere-ocean coupled circulation model ECHO-G, *Technical Report* 18, Deutsches Klimarechenzentrum, Hamburg, Germany, 1999.

Licciardi, J.M., J.T. Teller, and P.U. Clark, Freshwater routing by the Laurentide Ice Sheet during the last deglaciation. In: *Mechanisms of global climate change at millennial time scales*. Edited by P.U. Clark, R.S. Webb, L.D. Keigwin, American Geophysical Union, Geophys. monograph series, No. 112, 177-201, 1999.

Liu, Z., E. Brady, and J. Lynch-Stieglitz, Global ocean response to orbital forcing in the Holocene, *Paleoceanography*, 18, 1041, doi:10.1029/2002PA000,819, 2003.

Lohmann, G., Atmospheric and oceanic freshwater transport during weak Atlantic overturning circulation, *Tellus*, 55 *A*, 438-449, 2003.

Lorenz, S., and G. Lohmann, Acceleration technique for Milankovitch type forcing in a coupled atmosphere-ocean circulation model: method and application for the Holocene. *Clim. Dyn*., 23, 727-743. DOI: 10.1007/s00382-004-0469-y, 2004.

Lorenz, S.J., J.-H. Kim, N. Rimbu, R.R. Schneider, and G. Lohmann, Orbitally driven insolation forcing on Holocene climate trends: evidence from alkenone data and climate modeling, *Paleoceanogr*., 2005, submitted.

Marchal, O., *et al*., Apparent long-term cooling of the sea surface in the northeast Atlantic and Mediterranean during the Holocene, *Quat. Sci. Rev*., 21, 455-483, 2002.

Milankovitch, M., *Kanon der Erdbestrahlung und seine Anwendung auf das Eiszeitenproblem*, 133, Royal Serb. Acad. Spec. Publ., Belgrad, 1941.

Peltier, W.R., Ice Age paleotopography. *Science*, 265, 195-201, 1994.

Prange, M., Einfluss arktischer Süßwasserquellen auf die Zirkulation im Nordmeer und im Nordatlantik in einem prognostischen Ozean-Meereis-Modell. *Report Polar Marine Res*., 468, Alfred-Wegener-Institute for Polar and Marine Research, Bremerhaven, 220 pp., 2003.

Prange, M. and G. Lohmann, Effects of mid-Holocene river runoff on the Arctic ocean-sea ice system: a numerical study. *The Holocene*, 13 (3), 335-342, 2003.

Prange, M. and R. Gerdes, The role of surface freshwater flux boundary conditions in Arctic ocean modelling. *Ocean Modelling*, 2005 (accepted).

Rimbu, N., G. Lohmann, J.-H. Kim, H.W. Arz, and R.R. Schneider, Arctic/North Atlantic Oscillation signature in Holocene sea surface temperature trends as obtained from alkenone data, *Geophys. Res. Lett*., 30, 1280-1283, 2003.

Rimbu, N., G. Lohmann, S.J. Lorenz, J.-H. Kim, and R. Schneider, Holocene climate variability as derived from alkenone sea surface temperature reconstructions and coupled ocean-atmosphere model experiments. *Climate Dyn*., 23, 215-227, 2004.

Roach, A.T., K. Aagaard, C.H. Pease, S.A. Salo, T. Weingartner, V. Pavlov, and M. Kulakov, Direct measurements of transport and water properties through the Bering Strait. *J. Geophys. Res*., 100, 18443-18457, 1995.

Roeckner, E., *et al*., The atmospheric general circulation model ECHAM-4: Model description and simulation of the present-day climate, *Report* 218, Max-Planck-Institut für Meteorologie, 1996.

Rohling, E., P. Mayewski, R. Abu-Zied, J. Casford, and A. Hayes, Holocene atmosphere-ocean interactions: records from Greenland and the Aegean Sea, *Climate Dynamics*, 18, 573-592, 2002.

Schiller, A., U. Mikolajewicz, and R. Voss, The stability of the thermohaline circulation in a coupled ocean-atmosphere general circulation model. *Climate Dyn*., 13, 325-347, 1997.

Sidorchuk, A.Y., O.K. Borisova, N.N. Kovalyukh, A.V. Panin, A.V. Chernov, Paleogidrologiya nizhney Vychegdy v pozdnelednikov'ye i v golotsene (Paleohydrology of the lower Vychegda in the late glacial period and Holocene). Vestnik Moskovskogo Universiteta, Seriya 5: *Geografiya*, 1999/5, 35-42, 1999.

Tarasov, P., *et al*., Present-day and mid-holocene biomes reconstructed from pollen and plant macrofossil data from the former Soviet Union and Mongolia, *Journal of Biogeography*, 25, 1029-1053, 1998.

Texier, D., N. de Noblet, S.P. Harrison, A. Haxeltine, D. Jolly, S. Joussaume, F. Laarif, I.C. Prentice, and P. Tarasov, Quantifying the role of biosphere-atmosphere feedbacks in climate change: coupled model simulations for 6000 years BP and comparison with palaeodata for northern Eurasia and northern Africa, *Climate Dyn*., 13, 865-882, 1997.

Tremblay, L.-B., L. Mysak, and A. Dyke, Evidence from driftwood records for century-to-millennial scale variations of the high latitude atmospheric circulation during the Holocene, *Geophys. Res. Lett*., 24, 2027-2031, 1997.

Velichko, A.A., O.K. Borisova, E.M. Zelikson, and V.P. Nechayev, Permafrost and vegetation response to global warming in North Eurasia. In: *Biotic feedbacks in the global climatic system*. Edited by G.M. Woodwell, F.T. Mackenzie, Oxford Univ. Press, New York, 134-156, 1995.

Voss, R., R. Sausen, and U. Cubasch, Periodically-synchronously coupled integrations with the atmosphere-ocean general circulation model ECHAM3/LSG. *Climate Dyn*., 14, 249-266, 1998.

Voss, R. and U. Mikolajewicz, The climate of 6000 years BP in near-equilibrium simulations with a coupled AOGCM, *Geophys. Res. Lett*., 28, 2213-2216, 2001.

G. Lohmann, Alfred Wegener Institute for Polar and Marine Research, Bussestrasse 24, D-27570 Bremerhaven, Germany. (lohmann@awi-bremerhaven.de)

S.J. Lorenz, Max-Planck-Institute for Meteorology, Model & Data Group, Bundesstr. 55, 20146 Hamburg, Germany. (lorenz@dkrz.de)

M. Prange, University of Bremen, Research Center Ocean Margins, P.O. 330 440, 28334 Bremen, Germany. (mprange@palmod.uni-bremen.de)

Holocene Climate Variability in the Northern North Atlantic Region: A Review of Terrestrial and Marine Evidence

Atle Nesje[1], Eystein Jansen[1], H. John B. Birks[2], Anne Elisabeth Bjune[2], Jostein Bakke[3], Carin Andersson, Svein Olaf Dahl[3], Dorthe Klitgaard Kristensen[1], Stein-Erik Lauritzen[1], Øyvind Lie, Bjørg Risebrobakken[1], and John-Inge Svendsen[1]

Bjerknes Centre for Climate Research, Bergen, Norway

The Holocene epoch, which followed the last major pulse of glaciation (the Younger Dryas) at the end of the last glaciation, encompasses a period before there was any substantial anthropogenic forcing of climate. A synthesis of climatic development during the Holocene (ca. 11,500 cal. yr BP to the present) is based on pollen-based quantitative temperature reconstructions, tree-line variations, chironomids, tree-ring records, speleothem data, glacier variations, and marine records (stable isotopes, species abundance, lithological changes) from the Nordic Seas. The Holocene has been regarded as a period of relatively stable climate, but recent evidence suggests there have been several significant millennial-scale climate fluctuations (larger than the post mid-19th century warming trend) throughout the Holocene. A general climate warming in the first part of the Holocene was punctuated by a few, abrupt climate reversals, centred at 10,000, 9,700, and 8,200 cal. yr BP. The data suggest there was a period of relatively warm conditions in the first half of the Holocene, in many areas warmer than in the 20th century, after which temperatures generally declined. The temperature decline was punctuated by centennial-scale warmer and colder periods with the most recent cold episode (~AD 1550-1925), including the "Little Ice Age", being one of the coldest of the entire Holocene. The kind of data presented here can be used for detecting mechanisms and forcing factors behind the reconstructed climate variations and to study leads and lags in the Earth's climate system. A comparison between the climate-forcing factors and the Holocene records presented here suggests that the early Holocene thermal maximum observed in most records from the Nordic Seas region was caused by increased summer solar insolation to the northern Hemisphere. The suborbital millennial to decennial climate variability observed in the Holocene climate reconstructions was most probably a combined effect of solar activity changes, periods of increased explosive volcanism, and internal feedback mechanisms in the Earth's climate system. Internal modes of the climate system variability (for example, the North Atlantic Oscillation) were most likely responsible for some of the observed climate variability in winter. In addition to

The Nordic Seas: An Integrated Perspective
Geophysical Monograph Series 158

10.1029/158GM19

Also at the Departments of [1]Earth Science, [2]Biology, and [3]Geography, University of Bergen, Norway.

external climate-forcing factors, meltwater from the retreating northern hemisphere ice sheets and abrupt, catastrophic lake drainage from ice-marginal lakes during the termination of the last Ice Age had a significant, regional climate impact. In the near future, it will be possible to combine high-resolution palaeoclimatic reconstructions with fully coupled ocean–atmosphere climate models in order to test the complex interactions between the different forcing factors to help explain the primary forcing factors and the accompanying feedbacks behind the observed Holocene climate development in the North Atlantic region.

1. INTRODUCTION

A major obstacle in producing reliable predictions of global climate change and its environmental impacts is a lack of data on time scales longer than the instrumental records. Archives of past natural climate variability can provide relevant information over longer time scales. Furthermore, a better understanding of past variability can be used to improve the performance of climate models and to test model sensitivity, and palaeoclimatic data may serve as validation data. Model experiments can also be used to investigate the causes and dynamics of past climate variability.

Research on recent climate variability is leaning increasingly towards the view that greenhouse gas–forcing is becoming the dominant, though not the only, process driving the warming trend over the last 60 years or so [e.g., IPCC, 2001]. At the same time, detailed analyses of ice-core records from glacial terminations show how closely atmospheric greenhouse gas concentrations track rapid warming, probably by acting as a major feedback mechanism during the transition to interglacial conditions. These and other lines of evidence from studies of the present-day atmosphere reinforce the view that future climate change will be driven by complex interactions between the effects of increased concentrations of greenhouse gases and the effects of natural climate-forcing factors (e.g., solar variability and volcanic aerosols as well as internal factors in the climate system). There is therefore a need to improve our documentation and understanding of natural climate variability for time periods beyond those for which instrumental records exist.

Knowledge of past climate variability on these time scales should be obtained from both long-term instrumental records and well-calibrated proxy data derived from natural archives. In ideal circumstances such archives provide accurate records of climate history and can be dated with annual to decadal precision on a calendar year time scale. Ideally such records can also be intercorrelated through time and extended to the past to include the full range of climate variability of relevance to climate predictions. Commonly, no single climate archive, however, possesses such properties, and climate reconstructions consequently need to combine information from different sources, within and between geographical regions (Figure 1 [e.g., *Alverson et al.*, 2003]).

The climate in the late 20th through early 21st centuries has been characterised by almost universal warming, the Northern Hemisphere experiencing temperatures significantly higher than in the mid-19th century, when regular instrumental measurements started. Instrumental data provide, however, only a limited temporal perspective on present climate. We may therefore pose the question: How unusual is the last century when compared with the longer-term records? A Holocene perspective encompasses the period before large-scale emissions of greenhouse gases to the global atmosphere by human activities and before major changes in land-surface conditions. Holocene climate variations in the North Atlantic region (Plate 1) have been reconstructed from a number of proxies (see Table 1 for the most widely used proxies for climate reconstruction in the North Atlantic region and relevant literature).

In this chapter it is not possible to give a complete overview of the growing literature on Holocene climate variations in the northern North Atlantic region (Plate 1). We have therefore presented a review of selected records illustrating Holocene climate variability as reconstructed from a range of different proxies. Before discussing these records, we outline the theoretical basis and justification for quantitative climate reconstructions, exemplified by pollenstratigraphical data.

2. METHODS AND APPROACHES

We have compiled records of Holocene palaeoenvironmental change from the North Atlantic region, with a focus on Norway and the Norwegian Sea. Several methods have been used to reconstruct Holocene climate variations. Each individual proxy indicator has a characteristic sensitivity and response time to a change in a given factor in the climate system. Therefore, different proxies from the same record may suggest different magnitudes and timings of climate change. The records presented here are based on terrestrial archives and proxies, such as variations in pollen, tree lines, chironomids, tree rings, speleothems, and marine proxies, such as marine faunal composition, sediment characteristics, and stable isotopes ($\partial^{18}O$) from planktonic and benthic foraminifera.

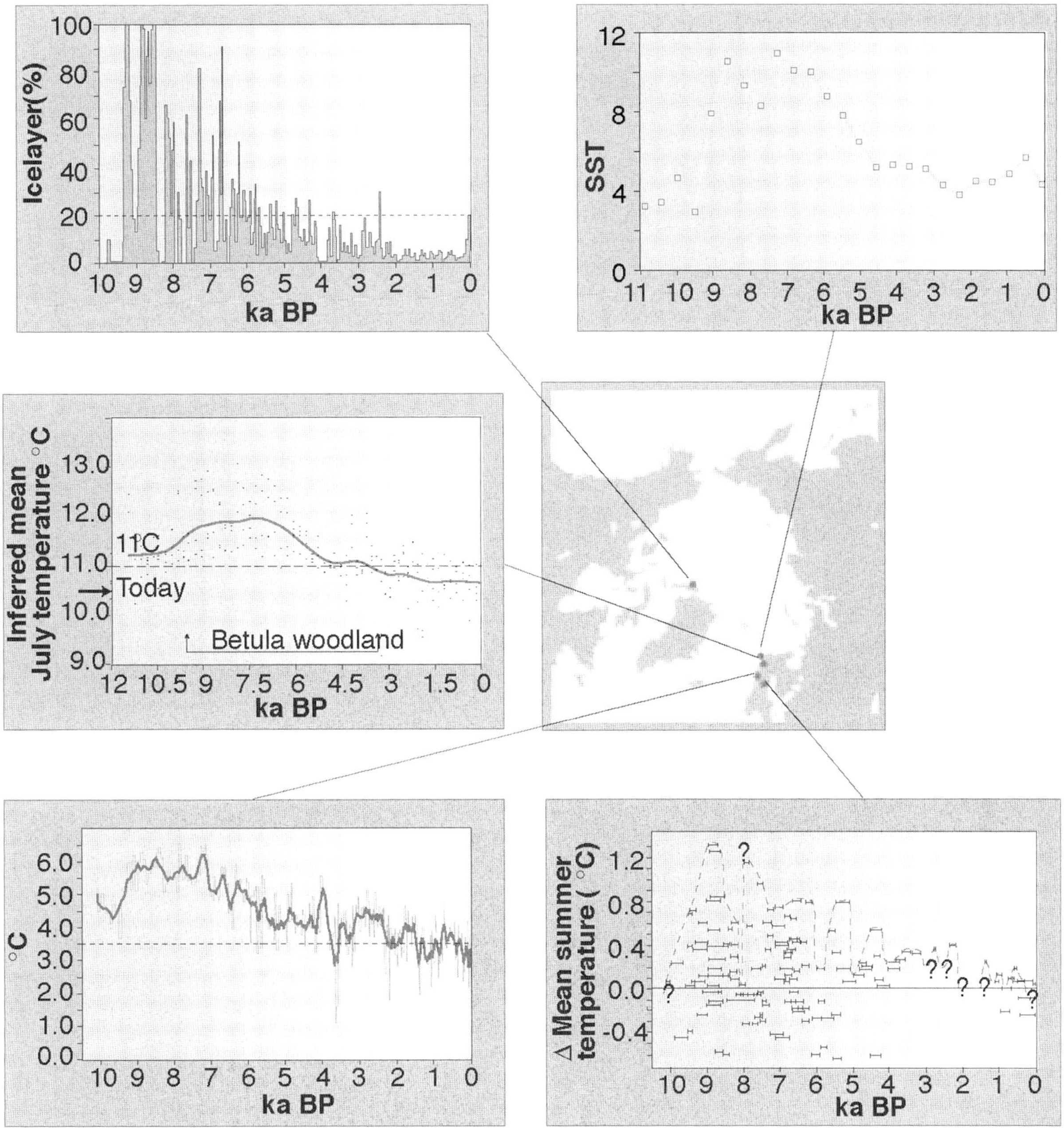

Figure 1. Records showing Holocene temperature changes. Top left: Holocene melt record from the Agassiz Ice Cap [*Koerner and Fisher*, 1990]. Top right: Diatom-inferred sea-surface temperature (SST) in the Norwegian Sea [*Koc et al.*, 1993]. Middle left: Pollen-inferred mean July temperature from northern Norway [*Birks*, 2003]. Bottom left: Summer temperatures from oxygen isotopes in a speleothem from northern Norway [*Lauritzen*, 1996]. Bottom right: Mean summer temperature from the upper limit of pine trees in the Scandes mountains, central Sweden [*Dahl and Nesje*, 1996]. [Source: *Alverson et al.*, 2003.]

Because some of the continental inland sites may have had a significant local effect from the retreating/downwasting ice sheet soon after the deglaciation, it may be difficult to separate the influence of local climate change from regional climate development. Lakes commonly exhibit different sensitivity to climate change and most proxies relate to summer temperature. Most quantitative estimates are based on microfossil assemblages, as converted to temperature and precipitation by using transfer functions based on modern calibration data sets (see below).

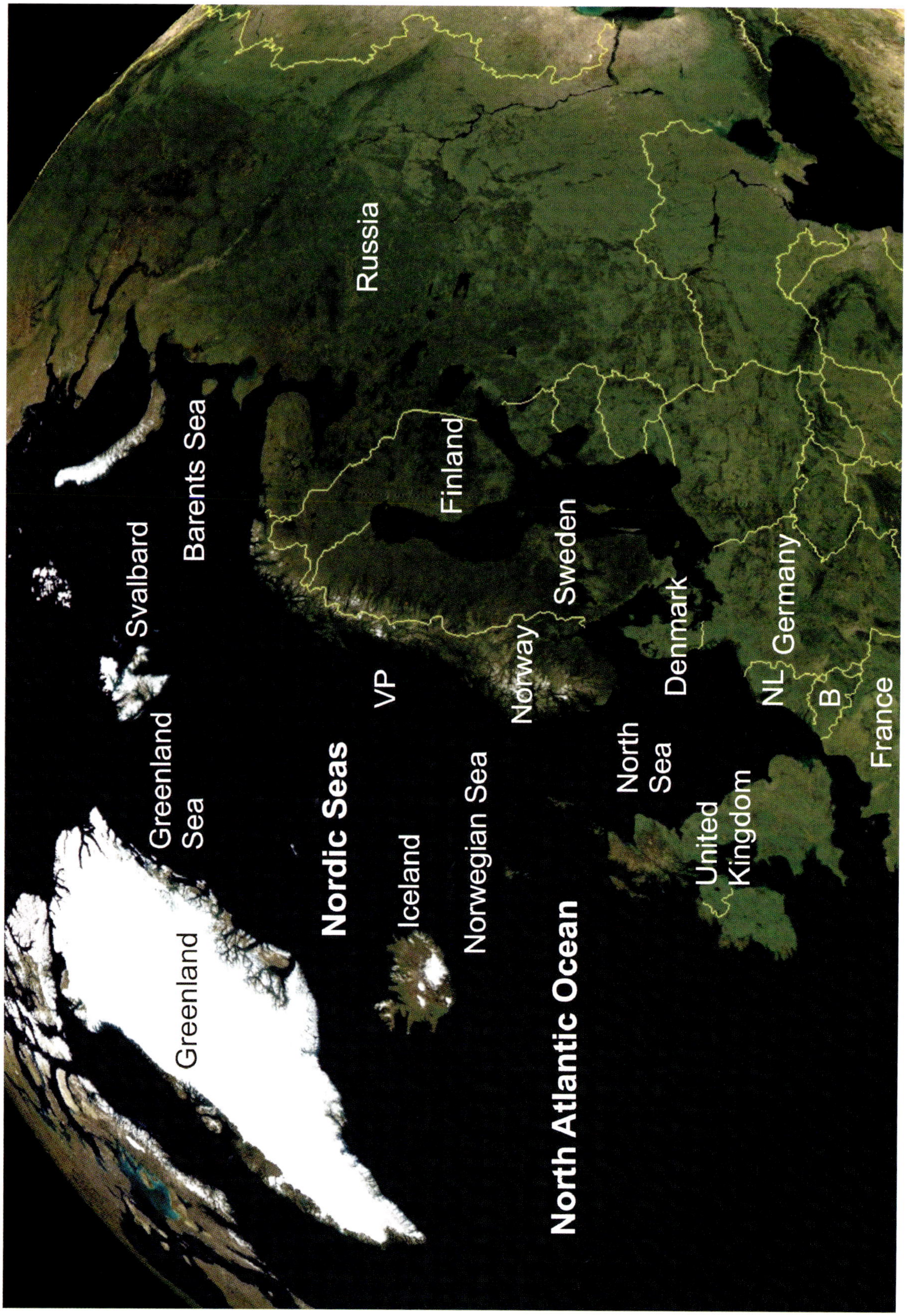

Plate 1. The North Atlantic region and adjacent land areas (from World Wind NASA).

Table 1. The Most Commonly Used Proxies for Holocene Climate Reconstructions in the North Atlantic Region With the Most Relevant References.

Proxy	References
Ice cores	*Dansgaard et al.*, 1993; *Grootes et al.*, 1993; *O'Brian et al.*, 1995; *Stuiver et al.*, 1995; *Alley et al.*, 1997, *Dahl-Jensen et al.*, 1998; *Johnsen et al.*, 2001; *Fisher and Koerner*, 2003
Tree-line fluctuations	*Kullman*, 1981, 1995; *Eronen et al.*, 1993; *Berglund et al.*, 1996; *Karlén and Kuylenstierna*, 1996; *Körner*, 1998; *Barnekow and Sandgren*, 2001
Glacier variations	*Karlén*, 1976, 1988; *Nesje et al.*, 1991, 1994, 1995a, 2000a, 2001; *Dahl and Nesje*, 1992, 1994, 1996; *Karlén and Matthews*, 1992; *Matthews and Karlén*, 1992; *Leemann and Niessen*, 1994; *Karlén et al.*, 1995; *Snowball and Sandgren*, 1996; *Svendsen and Mangerud*, 1997; *Matthews et al.*, 2000; *Snyder et al.*, 2000; *Dahl et al.*, 2002; *Seierstad et al.*, 2002; *Winkler et al.*, 2003; *Shakesby et al.*, 2004; *Lie et al.*, 2004; *Bakke et al.*, 2005a,b; *Matthews*, 2005
Pollen, plant macrofossils, chronomids, and diatoms	*Birks*, 1991; *Lotter et al.*, 1997; *K.E. Barber et al.*, 1999; *Barnekow*, 2000; *Birks and Amman*, 2000; *Birks et al.*, 2000; *Barnett et al.*, 2001; *Rosén et al.*, 2001; *Seppä and Birks*, 2001, 2002; *Bigler et al.*, 2002; *Huntley*, 2002; *Korhola et al.*, 2002; *Seppä et al.*, 2002a, 2002b; *Bigler and Hall*, 2003; *Brooks*, 2003; *Davis et al.*, 2003; *Hannon et al.*, 2003; *Heiri et al.*, 2003; *Heikkilä and Seppä*, 2003; *Bjune et al.*, 2004; *Hammarlund et al.*, 2004; *Larocque and Hall*, 2004; *Bjune*, 2005; *Velle et al.*, 2005a,b
Tree rings	*Briffa et al.*, 1992, 1998, 2002; *Schweingruber and Briffa*, 1996; *Kalela-Brundin*, 1999; *Kirschhefer*, 2001; *Grudd et al.*, 2002; *Helama et al.*, 2002, 2004; *Thun*, 2002
Oxygen isotope records from lake sediments	*Grafenstein et al.*, 1998; *Shemesh et al.*, 2001; *Hammarlund et al.*, 2002, 2003
Physical sediment parameters, such as loss-on-ignition and mineral magnetic properties, in lake sediments	*Thompson et al.*, 1975; *Thompson and Oldfield*, 1986; *Sandgren and Risberg*, 1990; *Snowball and Thompson*, 1990; *Snowball*, 1993; *Sohlenius*, 1996; *Snowball et al.*, 1999; *Stockhausen and Zolitschka*, 1999; *Willemse and Törnquist*, 1999; *Nesje and Dahl*, 2001; *Battarbee et al.*, 2001, 2002; *Kaplan et al.*, 2002; *Lotter and Birks*, 2003; *Rubensdotter and Rosquist*, 2003; *Nesje et al.*, 2004
Peatlands	*Chambers and Charman*, 2004
Speleothems	*Lauritzen*, 1996; *Lauritzen and Lundberg*, 1999
Marine proxies (e.g., stable isotopes, faunal abundance variations, physical sediment parameters)	*Weliky et al.*, 1983; *Lehman et al.*, 1991; *Koc et al.*, 1993; *Haflidason et al.*, 1995; *Bond et al.*, 1997, 2001; *Fronval and Jansen*, 1997; *Klitgaard-Kristensen et al.*, 1998, 2001; *Barber, D.C. et al.*, 1999; *Bianchi and McCave*, 1999; *Grøsfjeld et al.*, 1999; *Jansen and Koc*, 2000; *Hald et al.*, 2001; *Mikalsen et al.*, 2001; *Andrews et al.*, 2002a,b; *Birks and Koc*, 2002; *Geirsdottir et al.*, 2002; *Husum and Hald*, 2002; *Koc and Jansen*, 2002; *Andersson et al.*, 2003; *Andrews, J.T. and Cartee-Schoolfield, S.*, 2003; *Andrews and Helgadottir*, 2003; *Andrews et al.*, 2003a,b; *Oppo et al.*, 2003; *Risebrobakken et al.*, 2003; *Sarnthein et al.*, 2003; *Andrews and Giraudeau*, 2003; *Andersen et al.*, 2004; *Castaneda et al.*, 2004; *Giraudeau et al.*, 2004; *Hald et al.*, 2004; *Husum and Hald*, 2004; *Knudsen et al.*, 2004; *Moros et al.*, 2004; *Solignac et al.*, 2004
Marine molluscs	*Salvigsen et al.*, 1992; *Hjort et al.*, 1995; *Salvigsen*, 2002
Historical/instrumental evidence	*Grove*, 1988, 2004; *Pfister et al.*, 1999; *Jones*, 2001; *Luterbacher et al.*, 2001, 2004; *O'Sullivan et al.*, 2002; *Nesje and Dahl*, 2003; *Nordli et al.*, 2003

Uncertainties in the reconstructions of the spatio-temporal distribution of Holocene climate variations may be related to chronology. Most records have time scales based on radiocarbon dates. The reliability of the age/depth models will therefore vary from site to site as a result of several factors [*Björck and Wohlfarth*, 2001; *Telford et al.*, 2004].

3. QUANTITATIVE RECONSTRUCTIONS OF PAST CLIMATE VARIABLES

Instrumental climate records are not long enough to capture the full range of climate variability. Therefore, Holocene (and longer) climate records must be quantified in terms of physical climate parameters. A major challenge in terrestrial Holocene palaeoclimatology is to derive reliable quantitative reconstructions of past climatic variables (e.g., mean July temperature) from fossil assemblages. Calls for such quantitative reconstructions are increasing, largely in response to the need for quantitative reconstructions of past climate as input to or as validation of climate simulations by earth system models, and may provide a basis for the comparison of terrestrial and marine "proxy" climate records. Transfer functions have been constructed primarily for the calibration and quantification of biological data sets such as pollen. The approach has also been used to calibrate physical data sets, for example, stable isotope ratios [*Lauritzen and Lundberg*, 1999] and glacier mass balance [*Dahl and Nesje*, 1996]. The approach adopted and developed in Bergen involves the use of mathematical transfer functions to transform assemblages of fossil pollen preserved in Holocene lake sediments into quantitative climatic reconstructions. Pollen and spores of terrestrial plants (trees, shrubs, dwarf shrubs, herbs, and ferns) are well preserved in almost all lake sediments in Norway; the particles are present in large amounts, lakes are widespread in Norway, and pollen analysis is a relatively simple, repeatable, and rapid technique compared with, for example, the analysis of plant macrofossils or chironomid head capsules in lake sediments. The basic idea of quantitative climate reconstructions involves transfer or calibration functions and is summarised in *Birks* ([2003]; see also [*ter Braak and Juggins*, 1993; *Birks*, 1995, 1998]).

In the work on reconstructing Holocene climate in Norway from fossil pollen assemblages, we use a modern pollen–climate calibration data set made up of 191 samples from lakes throughout Norway (164 lakes) and northernmost Sweden (27 lakes) [*H.J.B. Birks, S.M. Peglar, and A. Odland*, unpublished]. The data set contains 152 pollen and spore types and covers a range of 7.7-16.4 °C for mean July air temperature. It has a root mean square error of prediction (RMSEP) of 1.03 °C (based on leave-one-out cross-validation) and a maximum bias of 3 °C (at the lowest part of the July temperature gradient). The set has been used for quantitative reconstructions of past climate in northern Sweden [e.g., *Bigler et al.*, 2002; *Hammarlund et al.*, 2002], and in western Norway (see below). It has been combined with 113 samples from Finland that were collected, prepared, and analysed in a way identical to the Norwegian data set to produce a Fennoscandian data set of 304 samples with a RMSEP of 0.99 °C and a maximum bias of 3.6 °C [*Seppä and Birks*, 2001]. This combined data set has been used to reconstruct Holocene climates in northernmost Finland [*Seppä and Birks*, 2001, 2002] and northeast Norway [*Seppä et al.*, 2002a].

4. HOLOCENE PALAEOENVIRONMENTAL RECONSTRUCTIONS FROM TERRESTRIAL SITES

4.1. Pollen Analysis

Quantitative reconstructions of mean July temperature from pollen have been made for 16 sites along the western fringe of Norway from 58 °N to 70 °N (Table 2). The sites cover the last 8,000-10,000 years of the Holocene. A few sites that lie outside of the Younger Dryas ice-sheet cover for the entire 11,500 years of the Holocene. All the reconstructions have been made by using the 191-sample pollen–climate "calibration set" for Norway and northern Sweden (see above) and a two-component weighted averaging partial least squares regression (WA-PLS) model. The reconstructions (Figure 2) are plotted along a south–north gradient on a calibrated age scale (cal. yr BP). The age scales for each site are derived from age–depth models based on a nonparametric regression within the framework of generalised additive models [*Heegaard et al.*, 2004; *Seppä and Birks*, 2001]. Each site age–depth model is based on 6-10 accelerator mass spectrometry (AMS) ^{14}C dates calibrated by using the INTCAL98 calibration data set and method A in the CALIB 4.3 program [*Stuiver and Reimer*, 1993]. The reconstructed mean July values are expressed as deviations from the present-day July temperature as reconstructed from the surface (0-1 cm) sediment through use of the modern pollen–climate data set, and the individual deviations are plotted as circles. The reconstructed values have not been adjusted for isostatic land-uplift.

There is considerable variation in reconstructed values from sample to sample, here and in other related studies [e.g., *Seppä and Birks*, 2001, 2002]. This variation is assumed to reflect the inherent variability in pollen stratigraphical data that may result from several factors including year-to-year variations in pollen production, the vagaries of pollen dispersal and sedimentation, and the inherent statistical uncertainties in percentage pollen counts. In an attempt to highlight the major long-term trends or "signal" in the reconstructions, a locally weighted regression smoothing (LOWESS) (span = 0.2) [*Cleveland*, 1993] has been fitted to each time series. The sample resolution is about 200-250 years, so fine-scale centennial changes cannot be detected. High-resolution (e.g., every 50 years) data are required to detect any such fine-scale changes [e.g., *Seppä and Birks*, 2002; *Heikkilä and Seppä*, 2003].

Nearly all the reconstructions suggest that mean July temperatures were warmer than today with positive deviations for much of the early and mid-Holocene. In general, the highest deviations occurred between 5,500 and 8,000 cal. yr BP at sites in the south (Dalane–Trettetjønn), whereas further north

Table 2. Details of the Sixteen Sites Used in the Reconstruction of Holocene Mean July Temperature in Norway.

	Locality	Altitude, m	Latitude and longitude	Present-day July temperature, °C	Vegetation zone (*Moen*, 1999)
1	Dalane	40	58°14′N, 8°01′E	14.9	Nemoral
2	Grostjørn	180	58°32′N, 7°44′E	15.2	Southern Middle Boreal
3	Øygardtjørn	665	59°37′N, 7°59′E	12.2	Middle Boreal
4	Isbenttjønn	787	59°46′N, 7°26′E	10.5	North Boreal
5	Vestre Øykjamyrtjønn	570	59°48′N, 6°01′E	11.0	Boreonemoral, oceanic
6	Holebudalen	1144	59°50′N, 6°59′E	8.2	Low Alpine
7	Trettetjønn	800	60°43′N, 7°00′E	10.7	Low Alpine
8	Brurskardtjørn	1309	61°25′N, 8°40′E	8.7	Mid Alpine
9	Tiåvatnet	372	63°02′N, 9°25′E	11.3	Middle Boreal, oceanic
10	Svartvatnet	183	63°21′N, 8°51′E	12.1	Southern Boreal, oceanic
11	Storsandvatnet	106	63°27′N, 8°27′E	12.6	Coastal
12	Vouskkujavri	348	68°20′N, 13°06′E	11.0	North Boreal
13	Gammelheimenvatnet	290	68°28′N, 17°45′E	12.8	Mid/Northern Boreal
14	Bjørnfjelltjørn	510	68°26′N, 18°04′E	10.5	Low Alpine
15	Dalmutladdo	355	69°10′N, 20°43′E	11.5	North Boreal
16	Ifjord	317	70°26′N, 27°38′E	10.0	Low Alpine

the highest deviations occurred between 6,000 and 10,000 cal. yr BP. The overall July temperature changes in the Holocene were not large, with positive deviations of 1-2 °C in the north and 1-3 °C in southern and central Norway. The major decrease in July temperature and the attainment of values close to those of today begins almost everywhere at about 4,000-4,500 cal. yr BP, often with a further decrease at about 2,000 cal. yr BP.

There is much consistency between the July temperature deviations from different sites, but there are also some apparent anomalies. The reconstructed deviations at Holebudalen, on the southern edge of Hardangervidda, may be distorted because the uppermost sample has, for unknown reasons, a reconstructed temperature about 2 °C warmer than any of the underlying samples from the last 1,000 years. By expressing temperature deviations relative to the reconstructed value for the uppermost sample only, the magnitude of Holocene temperature change at Holebudalen is considerably reduced. The only site where all the reconstructed values have high positive deviations is Brurskardtjønn in eastern Jotunheimen. This site, like Holebudalen, is above the tree line of today. It is possible that the vegetation in the catchment of Brurskardtjønn and hence the pollen assemblage in the uppermost sediment have been influenced by recent increases in summer grazing, as plants favoured by upland summer grazing (heaths, grasses) have low "optima" for mean July temperature in the modern calibration data set. This would result in an unrealistically reconstructed temperature for today and hence misleading high temperature deviations for the whole Holocene.

The temperature reconstructions from these two low-alpine sites highlight the problems of reconstructing temperatures from pollen-stratigraphical data at high altitudes. When the sites had woodland in their catchments during the early and mid-Holocene (as evidenced by plant macrofossils and pine stomata), their relevant pollen source areas would have been smaller than when the catchments lacked woodland and supported open low-alpine vegetation. They also highlight the problem of presenting reconstructions as deviations from today's temperature value as reconstructed from the pollen assemblage in the uppermost sediment sample. Using the mean of all the reconstructed values for the last 200 years [e.g., *Heikkilä and Seppä*, 2003] or 250 years as the "modern" value for calculating deviations from the recent may be a more robust procedure than expressing all reconstructions relative to the uppermost value only. Despite the considerable advances that have been made in techniques for quantitatively reconstructing past climate from fossil assemblages during the last decade, there is a need for further work on presenting climatic reconstruction time series and for detecting consistent trends and patterns within such time series.

4.2. Tree Lines

Elevational variations in tree limits, based on megafossil wood remains of *Pinus sylvestris* L. (Scots pine), have been reconstructed from several Scandinavian mountains. Two of the most extensive chronologies, one from the southern Scandes mountains (*Kullman* [1981] and references therein; variations in pine tree limit corrected for land uplift

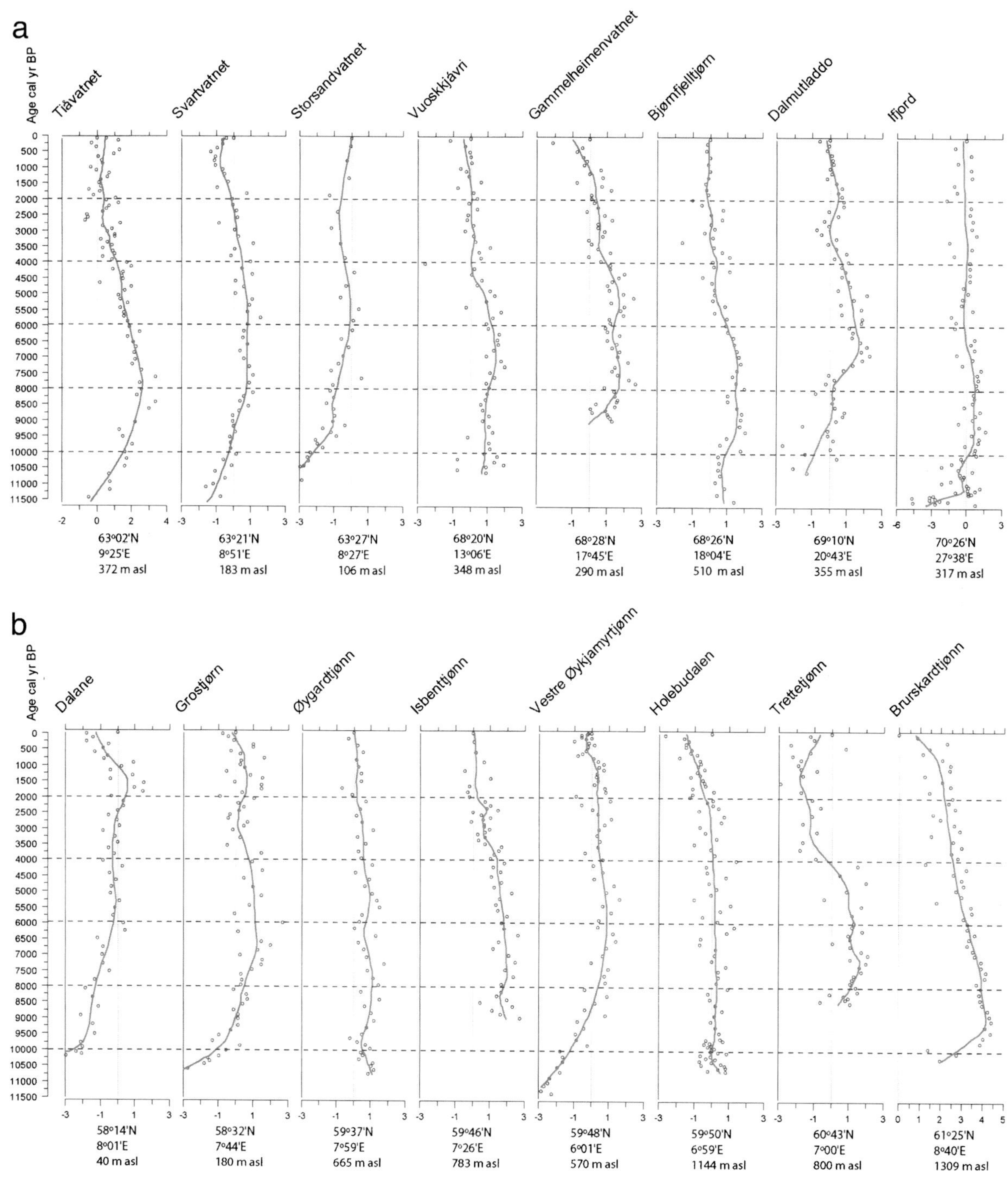

Figure 2. Deviations (°C) in mean July temperatures during the Holocene at 16 sites from southern (upper panel) and northern (lower panel) Norway. The temperature deviations are shown as individual points and are plotted on a common age scale (calibrated years BP). A LOESS smoother (thick line) has been fitted to each time series. The deviations are relative to the reconstructed mean July value for the uppermost sediment sample. The latitude, longitude, and altitude of each site are also given. All the deviations are plotted on an identical horizontal scale except at Ifjord. The pollen-stratigraphical data on which this diagram is based are the work of Sylvia M. Peglar (sites 1-4, 6, 9-14), Anne E. Bjune (sites 5, 7, 8), *Bjune et al.* [2004] (site 15), and Heikki Seppä (site 16) [see *Seppä et al.*, 2002].

and converted to summer temperature variations by *Dahl and Nesje* [1996]; Figure 3A) and the other from northern Sweden [*Karlén and Kuylenstierna*, 1996] (Figure 3B). On the basis of a close modern tree limit/climate relationship, the altitudinal tree limit variations are thought to reflect quite sensitively the secular to millennial-scale summer temperature variations. Temporal gaps in the record are difficult to interpret; they may reflect declines in the tree limit or an absence of preservation due, for example, to dry conditions.

4.3. Chironomids

A chironomid-inferred summer (mean July) temperature reconstruction from Lake Holebudalen, southern Norway, is shown in Figure 3C [*Brooks*, 2003]. The lake, at an altitude of 1,050 m, is located above the present-day tree line. The chironomid-inferred mean July air temperature variations were estimated by using a modern Norwegian calibration set of 153 lakes [*Brooks and Birks*, unpublished data). The reconstructed mean July temperature curve [*Brooks*, 2003] shown by upper and lower (±~1 °C) trend lines] shows a Holocene thermal optimum in the early part of the Holocene and a significant temperature drop around 8,200 cal. yr BP. In addition, mean July temperatures fell significantly around 6,400, 4,300, and 2,200 cal. yr BP, the lowest mean July Holocene temperatures ranging between 2,100 and 1,000 cal. yr BP.

4.4. Tree Rings

As with the instrumental data, tree ring–based records display strong coherence across wide areas of Scandinavia and tree-growth response to summer temperature variations have been demonstrated in many studies. Tree-ring widths from 880 living, dry dead, and subfossil northern Swedish pines have been assembled into a continuous and precisely dated chronology covering the period 5407 BC-AD 1997 [*Grudd et al.*, 2002]. Biological trends in the data were removed to emphasise year-to-year variability. Regional curve standardisation was applied to emphasise variability on time scales from decades to centuries. The 7,400-year-long chronology does not, however, express the full range of millennial-scale temperature changes in northern Sweden, as indicated from other summer temperature proxies (tree lines, pollen). A strong association with mean summer temperature (June-August) enabled the construction of a long-term, centennial-scale temperature reconstruction for the last 7,400 years (Figure 3D).

July temperatures have been reconstructed for the last ~7,500 years by using the long ring-width chronology of Scots pine from northern Finland [*Helama et al.*, 2002]. The chronology (Figure 3E) was constructed by using regional curve standardisation, allowing for long-term (low-frequency) variability to be extracted from this annually resolved record of 1,087 samples from living trees and subfossil pine trees. The warmest 30-yr periods in the northern Finland tree-ring record were AD 531-560 (1,419-1,390 cal. yr before AD 1950), 1190-1161 BC (3,140-3,111 cal. yr before AD 1950), and AD 1541-1570 (409-380 cal. yr before AD 1950), whereas the coldest were 5240-5211 BC (7,190-7,161 cal. yr before AD 1950), 5150-5121 BC (7,100-7,071 cal. yr before AD 1950), and 3710-3681 BC (5660-5631 cal. yr before AD 1950).

Along the coast of northern Norway, tree-ring chronologies, reflecting July-August temperatures, from Scots pine were constructed at Forfjorddalen, Vesterålen (AD 1358-1992), Stonglandseidet, Senja (AD 1548-1994), and Vikran close to Tromsø (AD 1700-1992) [*Kirchhefer*, 2001]. Growth variations were consistent between the three sites for AD 1700-1910, but the amount of low-frequency variability decreased towards the most oceanic site (Forfjorddalen). An extended period of high temperatures occurred around AD 1475-1540. The 17th century, the coldest period of the "Little Ice Age", experienced three cycles of summer temperatures, with minima around AD 1605, 1640, and 1680. The temperature reconstructions show secular trends similar to those observed east of the Scandes but differing in the magnitude and timing of the extremes.

4.5. Speleothems

Lauritzen and Lundberg [1999] presented a speleothem delta function (SDF) that provides a transfer function between the $\partial^{18}O$ signal of speleothem calcite and ground surface temperature. The function is based on physical principles, relating $\partial^{18}O$ of the calcite to thermodynamic fractionation, and to the drip water function, which in turn relates $\partial^{18}O$ of drip waters to that of the local precipitation and thus to the modification of source water in relationship to the geographical position of the site. The SDF was calibrated against reliable and well-dated palaeotemperature points. The end product was a reconstruction of absolute (probable mean annual) temperatures (Figure 3F).

4.6. Holocene Temperature Records From Northern Scandinavia

Holocene temperature records from northern Scandinavia (Figure 4) have been compiled and corrected for Holocene land uplift by *Bakke et al.* [2005b]. The different records, obtained from a region with relatively homogeneous climate, show some similar features (e.g., early Holocene thermal maximum and several millennial-scale temperature fluctuations). Some records, however, deviate significantly in the timing and amplitude of Holocene summer temperature fluctuations.

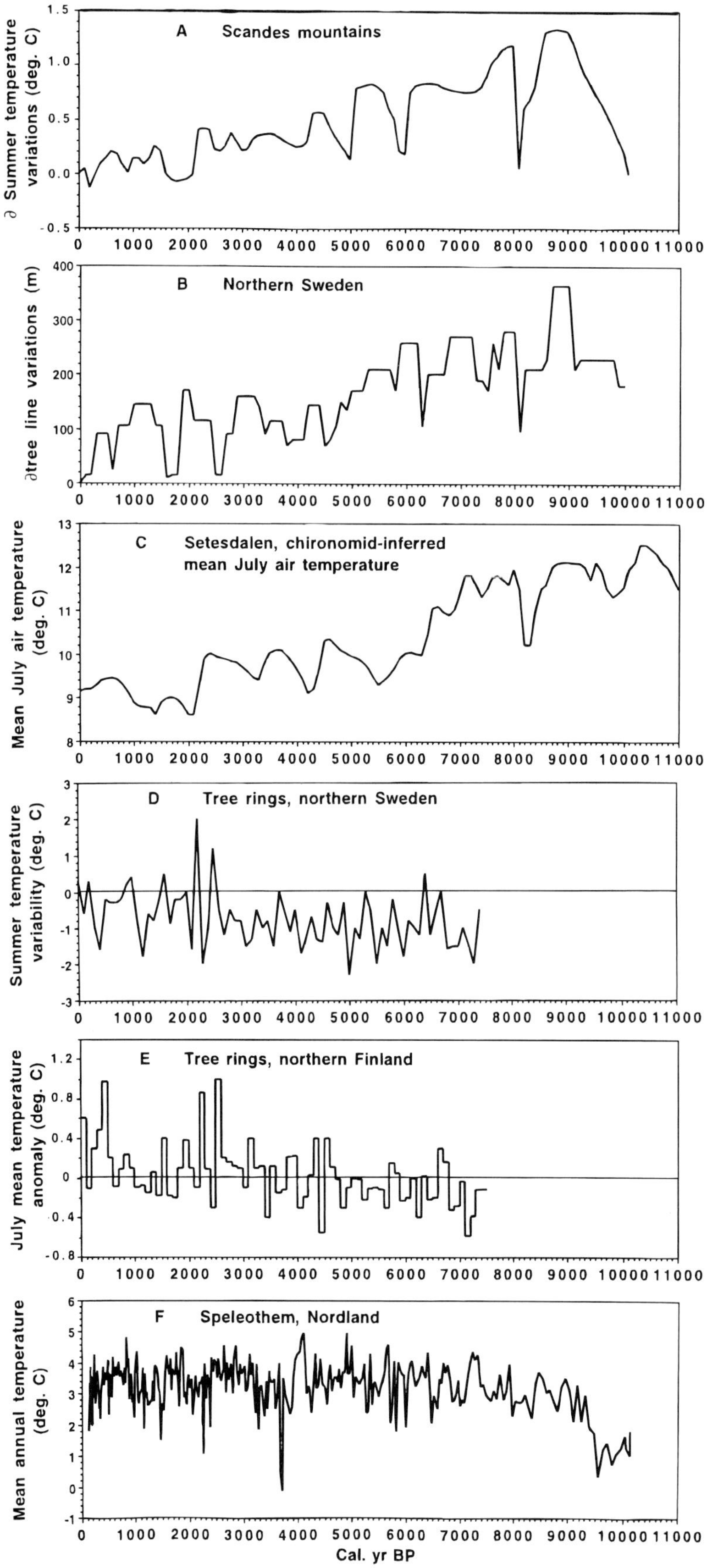

Figure 3. (A) Holocene summer temperature in the southern Scandes Mountains [*Dahl and Nesje*, 1996] mainly based on variations in the pine tree line reconstructed by *Kullman* [1981]. (B) Pine tree limit variations in northern Sweden relative to the modern pine tree line [*Karlén and Kuylenstierna*, 1996]. (C) Chironomid-inferred mean July air temperature variations from the Holebudalen in Setesdalen, southern Norway. Not shown are the error lines of ±1.1 °C with respect to the central line given in the original paper [adapted from *Brooks*, 2003]. (D) Low-frequency (decadal to century) summer temperature variability for the last 7,400 years in northern Sweden, expressed as anomalies from the period 1869-1997 [adapted from *Grudd et al.*, 2002]. (E) Reconstructed July mean temperature anomalies (°C) in Finnish Lapland [adapted from *Helama et al.*, 2002], presented as nonoverlapping 100-year means of century-scale midsummer temperatures. (F) Mean annual temperature record (smoothed) reconstructed from a speleothem (SG93) in Nordland, northern Norway [adapted from *Lauritzen and Lundberg*, 1999].

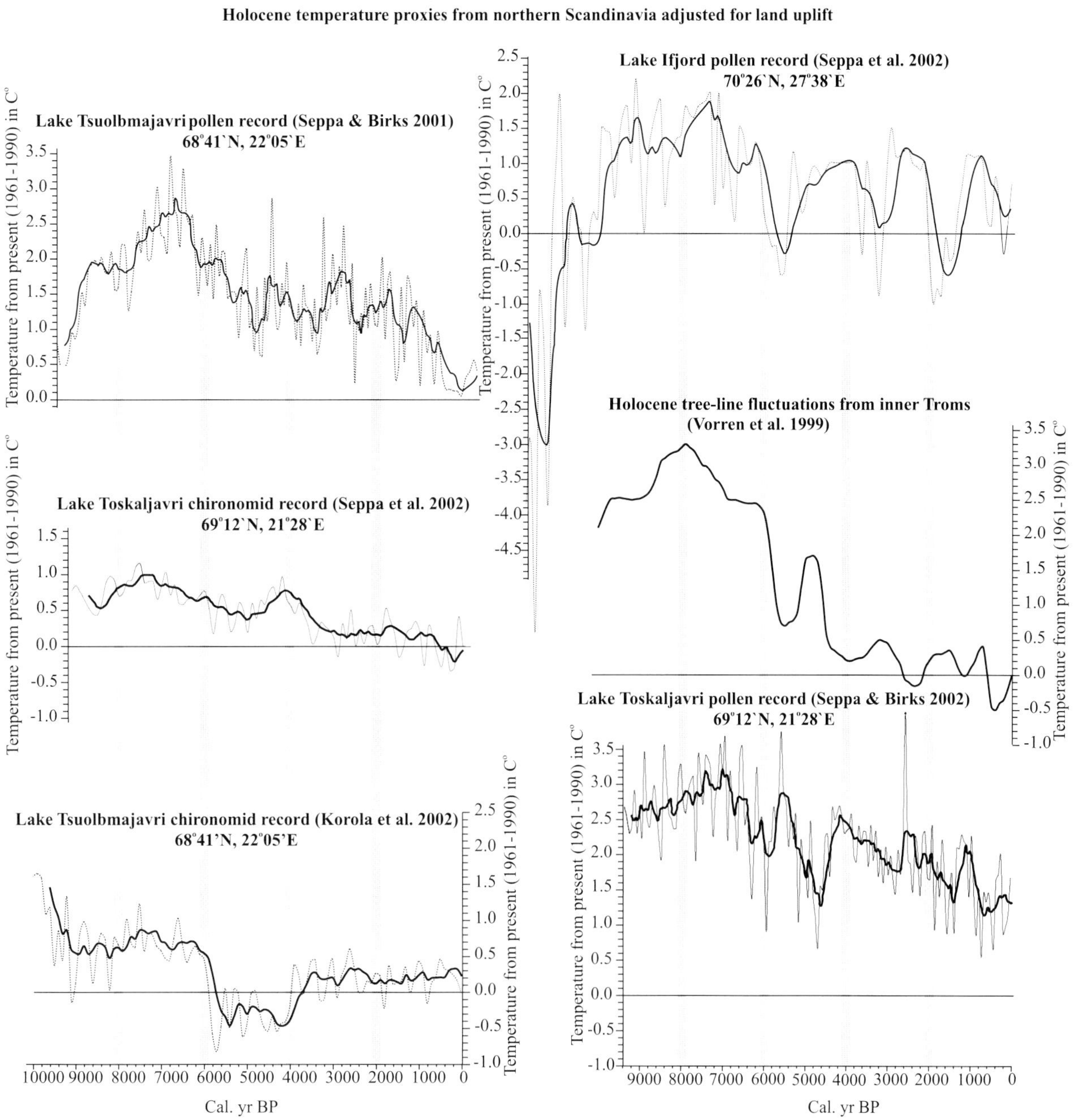

Figure 4. Holocene summer temperature reconstructions from different sites in northern Scandinavia. (A) Lake Tsuolbmajavri pollen record (68°41′N, 22°05′E) [*Seppä and Birks*, 2001], (B) Lake Toskaljavri chronomid record (69°12′N, 21°28′E) [*Seppä et al.*, 2002b], (C) Lake Tsuolbmajavri chrinomid record (68°41′N, 22°05′E) [*Korhola et al.*, 2002], (D) Lake Ifjord pollen record (70°26′N, 27°38′E) [*Seppä et al.*, 2002b], (E) Holocene tree-line fluctuations in Troms, northern Norway [*Alm*, 1993; *Vorren et al.*, 1999], and (F) Lake Toskaljavri pollen record (69°12′N, 21°28′E) [*Seppä and Birks*, 2002]. The records have been compiled and adjusted for land uplift by *Bakke et al.* [2005b].

4.7. Holocene Palaeo-Oceanographic and Climate Variations

High-resolution stable isotope and foraminifera records from cores MD95-2011 and JM97-948 from the Vøring Plateau, eastern Norwegian Sea [*Risebrobakken et al.*, 2003], provide evidence of both relatively large- and small-scale climate variability throughout the Holocene (Figure 5A). During the mid-Holocene a situation possibly driven by more prevailing stronger westerlies increased the eastward

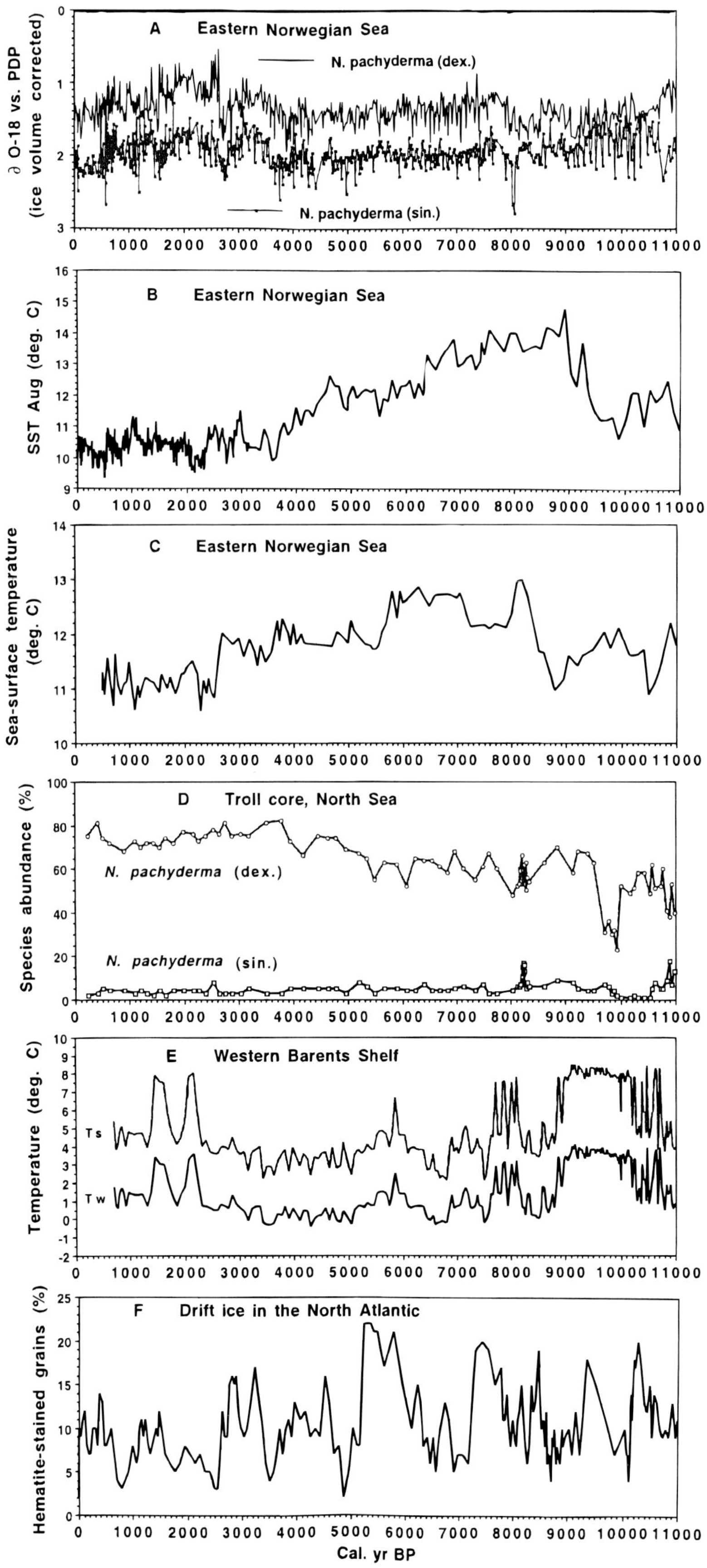

Figure 5. (A) Stable oxygen isotopes (ice-volume corrected) in *N. pachyderma* (dex.) and *N. pachyderma* (sin.) from cores MD95-2011 and JM97-948 obtained from the eastern Norwegian Sea (Vøring Plateau) [*Risebrobakken et al.*, 2003]. (B) Diatom-inferred SST (August) variations in the eastern Norwegian Sea reconstructed from cores MD95-2011 and JM97-948 in the eastern Norwegian Sea (Vøring Plateau) [*Jansen and Koc*, 2000; *Birks and Koc*, 2002; *Koc and Jansen*, 2002; *Andersen et al.*, 2004]. (C) SST variations in the eastern Norwegian Sea reconstructed from alkenone index (U^{K}_{37}) from core MD95-2011 from the Norwegian Sea [*Calvo et al.*, 2002]. (D) Relative frequency of the planktonic foraminiferal species *N. pachyderma* (sin.) to the right-coiled warm-water species of *N. pachyderma* (dex.) in the Troll core, North Sea [adapted from *Klitgaard-Kristensen et al.*, 2001]. (E) SIMMAX-based SST records for summer (Ts) and winter (Tw) for core GIK23258-2 in the northeastern North Atlantic at the western Barents Sea shelf at 75 °N [adapted from *Sarnthein et al.*, 2003]. (F) Variations in the occurrence of hematite-stained grains as an indicator of drift sea-ice in the northern Atlantic [*Bond et al.*, 2001].

influence of Arctic intermediate and near-surface waters. During the late Holocene, a relaxation of the atmospheric-forcing resulted in an increased influence of Atlantic water and higher amplitude centennial-scale variability. There is evidence of consistent multi-decadal variability in the Holocene record. The major Holocene climate changes show, however, no obvious relation to changing solar irradiance (no early Holocene thermal maximum), and spectral analysis reveals no consistent signature for any persistent periodic behaviour of Holocene climate at millennial or centennial timescales.

The same cores (JM97-948 and MD95-2011) from the Vøring Plateau were analysed for fossil diatoms [*Jansen and Koc*, 2000; *Birks and Koc*, 2002; *Koc and Jansen*, 2002; *Andersen*, 2003; *Andersen et al.*, 2004]. Results from diatom assemblages and reconstructed sea-surface temperatures (SSTs) (Figure 5B) indicate a division of the Holocene into three periods: The Holocene climate optimum (9,500-6,500 cal. yr BP), the Holocene transition period (6,500-3,000 cal. yr BP), and the cool late-Holocene period (3,000-70 cal. yr BP). The general temperature development during the Holocene seems to reflect the decreasing insolation on the Northern Hemisphere, but regional differences occur regarding both timing and magnitude of SST changes [*Andersen et al.*, 2004]. The cooling trend of summer SSTs in the early- to mid-Holocene is also reflected in terrestrial systems in Norway and Iceland. Periods of glacial advance can be correlated with cool oscillations in the SST reconstructions. By comparison with records of SSTs from other sites in the Norwegian Sea, spatial and temporal changes in patterns of ocean water masses are reconstructed, to reveal a complex system of feedbacks and influences on the climate of the North Atlantic and Norway.

Calvo et al. [2002] studied the C_{37} alkenone distribution from the same MD95-2011 core in order to test the applicability of the U^{K}_{37} and $U^{K'}_{37}$ indices in the Norwegian Sea. U^{K}_{37} was found most appropriate to reconstruct Holocene SST variations. This proxy suggests that the first part of the Holocene was punctuated by some cold periods, two of the most significant occurring around 10,500 and 8,800 cal. yr BP (Figure 5C). The warmest SSTs were recorded during the first half of the Holocene (8,500-5,500 cal. yr BP). Significant SST drops, on the other hand, occurred approximately 5,700 and 2,700 cal. yr BP.

The region around Iceland is highly sensitive climatically because of its location with respect to atmospheric and oceanographic fronts. The deglaciation of the western part of Iceland was completed by 10,200 cal. yr BP [*Castaneda et al.*, 2004]. The 8,200 cal. yr BP event is present only as a minor isotopic event, and seems not to have had much of a cooling effect on the bottom waters of the northwest Iceland shelf. The Holocene maximum warmth, attributed to a stronger North Icelandic Irminger Current, occurred between ca. 7,800 and 6,200 cal. yr BP. Over the past 4,500 cal. yr BP a general cooling trend has occurred on the North Iceland Shelf. Superimposed on this overall cooling trend are a number of oscillations between periods when relatively warmer and cooler waters occupied the shelf. Relatively cooler waters were present at 4,200-4,000 cal. yr BP, 3,200-2,900 cal. yr BP, 2,500-2,350 cal. yr BP, and 600-200 cal. yr BP, whereas relatively warmer waters were present on the shelf at 3,750-3,450 cal. yr BP, 2,800-2,600 cal. yr BP, and 1,700-1000 cal. yr BP [*Castaneda et al.*, op.cit.].

Late Holocene (the last 3,000 years) sea-surface layer conditions on the Vøring Plateau in the eastern Norwegian Sea are have been reconstructed from planktonic stable isotopes and planktonic foraminiferal assemblage changes [*Andersson et al.*, 2003]. Strong covariance between the planktonic stable oxygen isotope record and abundance changes of *N. pachyderma* (sin.) show that major changes in surface ocean conditions are reflected both in the geochemical composition of the foraminiferal tests and the composition of the foraminifera fauna. During the so-called "Medieval Warm Period", surface ocean conditions were variable, especially marked changes in the SST. The warmest SSTs during this period occurred between 800 and 550 cal. yr BP. The lowest SSTs of the late-Holocene record occurred at 2,750, 1,550, 400, and 100 cal. yr BP. The cooling events at approximately 2,750 and 1,550 cal. yr BP appear to correlate with increases in the ice-rafted debris in the North Atlantic. The onset of the Little Ice Age cooling trend seems to have occurred around 700-600 cal. yr BP, the lowest temperatures in this period occurring at 400 and 100 cal. yr BP.

A record of summer sea-bottom temperatures and SSTs in the North Sea (Norwegian Channel) Troll core during the last 18,000 calendar years has been reconstructed from the content of benthic and planktonic foraminiferal faunas and oxygen isotope ratios ([*Klitgaard-Kristensen et al.*, 2001] (Figure 5D); see also *Birks et al.*, this volume). The early deglaciation (18,000-14,400 cal. yr BP) reflected SSTs below 4 °C and polar environmental conditions accompanied by increased ice rafting and sea-ice cover. The Bølling–Allerød period showed SSTs as warm as today (~9 °C), whereas bottom waters remained cold (0-1 °C) and had normal salinity. The warm Bølling–Allerød period was interrupted twice, at 13,900-13,600 and 13,000-12,800 cal. yr BP, when arctic environmental conditions prevailed and ice rafting increased. The Younger Dryas (12,700-11,500 cal. yr BP) shows variable environmental conditions between arctic/polar water masses, sea-ice cover, and increased ice rafting. Superimposed on the Holocene long-term changes are short-lived, small-amplitude oscillations in the ocean circulation. Before 8,000 cal. yr BP the amplitudes of these short-lived oscillations appear larger than in the rest of the Holocene. This amplification in the early part of the Holocene may have resulted from sudden fresh-water outbursts during deglaciation of the Fennoscandian and Laurentide ice sheets [*Nesje et al.*, 2004].

At the western continental margin of the Barents Sea (75 °N), hemipelagic sediments provided a record of Holocene climate change with a time resolution of 10-70 years [*Sarnthein et al.*, 2003]. Planktonic foraminifera revealed an early Holocene thermal optimum, at 10,700-7,700 cal. yr BP, with summer SSTs of ~8 °C and a much enhanced West Spitsbergen Current (Figure 5E). There was, however, a short cooling between 8,800 and 8,200 cal. yr BP. In the middle and late-Holocene summer, SST dropped to 2.5°-5.0 °C, indicative of reduced Atlantic heat advection, except for two short warm episodes around 2,200 and 1,600 cal. yr BP.

Bond et al. [2001] suggested, on the basis of IRD records, that surface winds and surface ocean hydrography in the subpolar North Atlantic seem to have been influenced by variations in solar activity during the Holocene (Figure 5F). The evidence stems from a close correlation between inferred changes in production rates of the cosmogenic nucleides ^{14}C and ^{10}Be and centennial- to millennial-scale changes in proxies of drift ice measured in deep-sea sediment cores.

Radiocarbon dates of *Mytilus edulis* and *Modiolus modiolus* shells from Svalbard [*Salvigsen et al.*, 1992; *Hjort et al.*, 1995; *Salvigsen*, 2002] indicate that *Mytilus edulis* appeared on Svalbard around 11,000 cal. yr BP (Figure 6). Its maximum extent seems to have been from ~9,700 to 5,600 cal. yr BP, indicating a summer thermal maximum during that interval. *Modiolus modiolus* is dated to have appeared at approximately 9,400 cal. yr BP. *Mytilus edulis* apparently disappeared from the Svalbard waters at approximately 4,500 cal. yr BP,

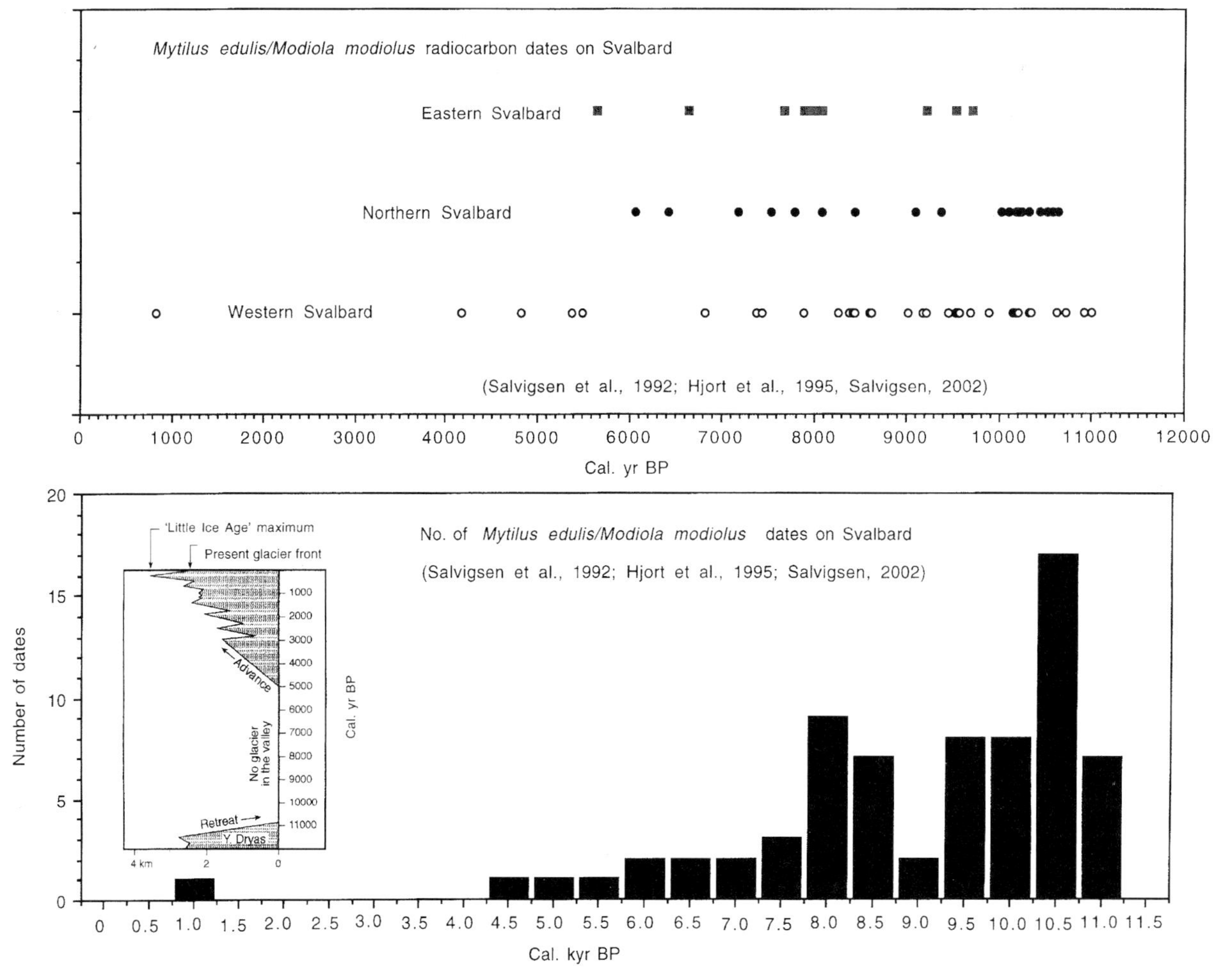

Figure 6. Upper panel: Radiocarbon dates of *Mytilus edulis* and *Modolus modiolus* (2 dates) on Svalbard [data from *Salvigsen et al.*, 1992; *Hjort et al.*, 1995; *Salvigsen*, 2002]. Lower panel: Frequency of shell dates during the Holocene, based on the data in the upper panel. The distribution of dates may to some degree reflect preferential sampling [Chr. Hjort, pers. comm.]. The insert figure shows Holocene glacier variations as reconstructed from Linnévatnet [*Svendsen and Mangerud*, 1997].

marking the end of the Holocene thermal maximum. However, one radiocarbon date of *Mytilus edulis* from around 1,000 cal. yr BP may indicate warmer waters around Svalbard during the Medieval Epoch. Independent evidence of temperatures 1-2 °C higher than at present in Svalbard in the early Holocene comes from plant macrofossil analysis of lake sediments from Skardtjørna, western Spitsbergen [*Birks*, 1991]. The Skardtjørna sequence shows evidence of gradual mid-Holocene climatic cooling between 4,000 and 2,500 cal. yr BP, in accordance with the mollusc record.

4.8. Holocene Glacier Variations

The reconstruction of Holocene glacier variations in Scandinavia was pioneered by Wibjörn Karlén [see review by *Karlén*, 1988]. On Svalbard, Holocene glacial and climatic variations have been discussed by *Svendsen and Mangerud* [1997]. They utilized sediment cores obtained from the proglacial lake Linnévatnet and from the sea floor in Billefjorden, western Spitsbergen [*Svendsen and Mangerud*, 1997]. A low content of glacially derived coal, an absence of regular laminations, and a slow sedimentation rate in the lower part of the lacustrine sediments in Linnévatnet suggest that the glaciers that had existed within the catchment area during the last Ice Age melted away quickly as a result of the Younger Dryas–Holocene warming (Figure 6). During the subsequent 6,000 years there were no glaciers in the valley, which implies that the estimated mean equilibrium-line altitude in this area was more than 100-150 m higher than at present. This would mean 1.5-2.5 °C warmer summers during the early Holocene (10,000-4,400 cal. yr BP) if winter precipitation was similar to that in the present. Increasing coal content, incipient laminations, and rising sedimentation rates indicate that the glacier Linnébreen started to form 4,000-5,000 years ago, probably as a response to late-Holocene cooling. The Holocene maximum extent of the glacier occurred during the Little Ice Age, starting with a pronounced ice advance in the 13th or 14th century. There seems to have been a first short-lived culmination around AD 1400 and a second, longer lasting glacial maximum after AD 1500. The maximum glacial extent during the Little Ice Age has been constrained from sediment cores in front of a present day ice-free cirque near the southwestern end of Linnévatnet [*Snyder et al.*, 2000]. The sediment stratigraphy in these cores indicates that the glacier started to form in this particular cirque some 600 years ago (early 15th century) and that it disappeared again in the 20th century. A well-dated core from Billefjorden, which is assumed to reflect glacial fluctuations in the inner part of the Isfjorden area, indicates a development similar to that in the valley Linnédalen. No glacial signal was recorded during the early and mid-Holocene. The present tidewater glaciers started to form 3,000-4,000 years ago, with glacial maxima around 2,400-2,500 cal. yr BP and during the Little Ice Age. The glacier expansion during the late-Holocene probably was mainly a response to a general cooling trend, but changes in winter precipitation may have been a contributing factor.

The Holocene glacial history of several glaciers in southern Norway, ranging from the maritime (Nordre Folgefonna) to continental (Snøhetta, Dovre) climate regime has been reconstructed (Figure 7). The early part of the Holocene was characterized by retreating glaciers. However, this retreat was punctuated by some significant glacier readvances. The studied glaciers had completely melted at least once during the mid-Holocene. Millennial-scale glacier variability during the Holocene was most pronounced during the last 4,000-5,000 cal. yr. New morphological and stratigraphical evidence has been found about the age and nature of the earliest Holocene glacial advance in Norway, the so-called Erdalen event, first described from the Jostedalsbreen area [e.g., *Nesje et al.*, 1991]. In front of present outlet glaciers from Jostedalsbreen, Hardangerjøkulen, and Folgefonna and glaciers in Jotunheimen, sets of pre-Little Ice Age terminal moraines can be morphostratigraphically correlated with the Erdalen event. Lithostratigraphy in the basal part of a bed-rock basin in Jostedalen provides evidence for a two-phase Erdalen event associated with two glacier readvances of Nigardsbreen [*Dahl et al.*, 2002]. Radiocarbon dates suggest that the first event took place at approximately 10,000 cal. yr BP, whereas the second occurred close to 9,700 cal. yr BP.

High-resolution (0.5-1-cm intervals) loss-on-ignition (LOI) analyses of cores from proglacial and nonglacial lakes revealed a specific pattern, characterised by a two-peaked reduction of LOI in the lower part of the cores [*Nesje and Dahl*, 2001]. In non–glacier-fed lakes, LOI reductions may reflect a decline in lake productivity and hence a decrease in summer temperature (see above), whereas reductions in the LOI in proglacial lakes mainly reflect increased glacier activity in the catchment. AMS ^{14}C dates on bulk fine detritus gyttja and terrestrial plant macrofossils in lake sediments and peat sections suggest that this event, termed the Finse event [*Dahl and Nesje*, 1994, 1996], occurred between ~8,500 and 8,000 cal. yr BP. This event correlates with the "8.2 ka event" recorded in the Greenland ice cores [e.g., *Alley et al.*, 1997], in European lake sediments [*Karlén*, 1976, 1988; *Karlén et al.*, 1995; *Grafenstein et al.*, 1998; *Snowball et al.*, 1999; *Willemse and Törnquist*, 1999; *Matthews et al.*, 2000; *Nesje et al.*, 2000a, 2001], and in North Atlantic marine records [*Bond et al.*, 1997; *Klitgaard-Kristensen et al.*, 1998]. The Finse event, or 8,200 BP event, has been attributed to the breaching of Laurentide ice-sheet dams impounding glacial lakes Agassiz and Ojibway, which led to perturbations of North Atlantic thermohaline circulation [*D.C. Barber et al.*, 1999].

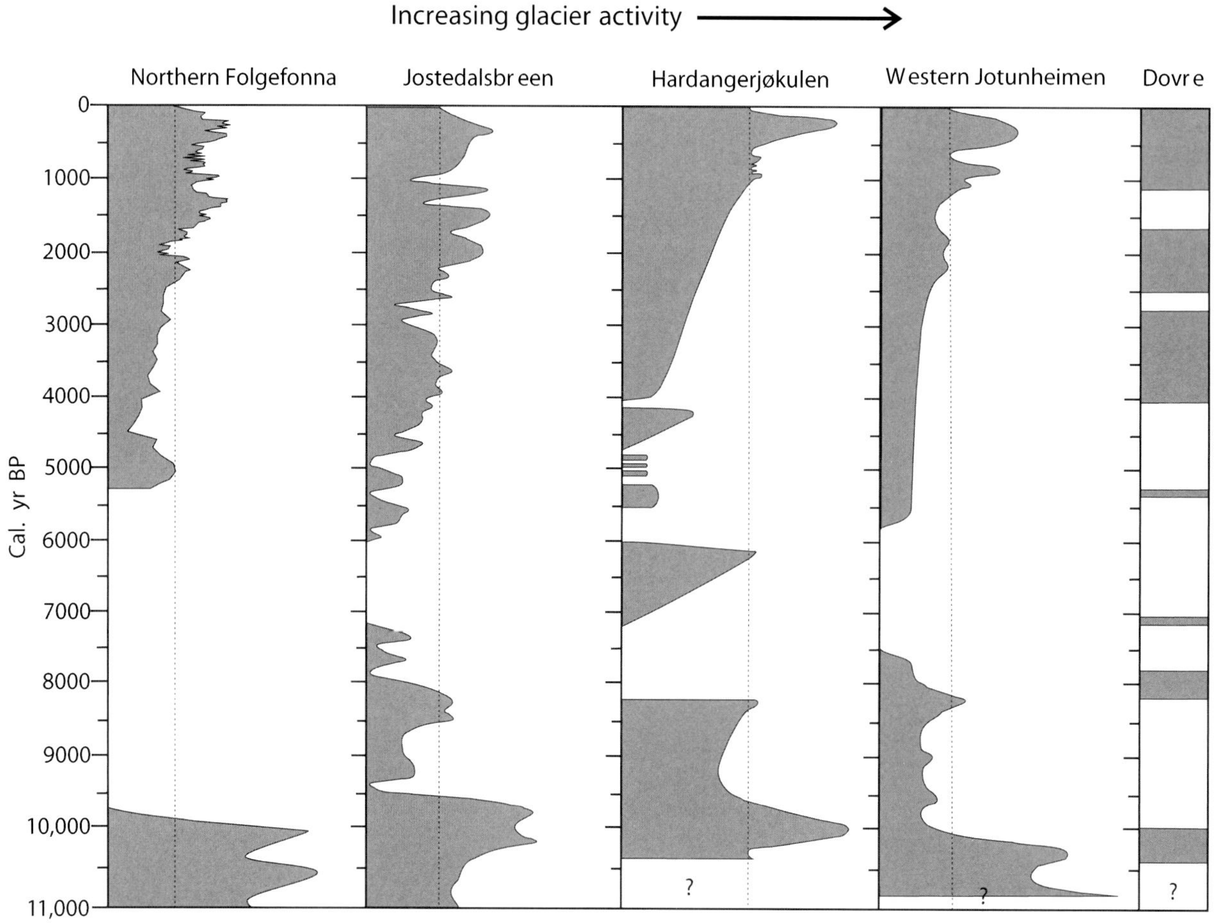

Figure 7. Holocene variations of five glaciers located in maritime (left) and continental (right) regions in southern Norway: Folgefonna [*Bakke et al.*, 2005a], Jostedalsbreen [*Nesje et al.*, 2001], Hardangerjøkulen [*Dahl and Nesje*, 1994, 1996], western Jotunheimen [*Matthews et al.*, 2000], and Dovre (periods with glaciers present and melted indicated; *Dahl et al.*, in prep.]. Vertical dotted lines indicate present position of glacier front.

From investigations so far, most, if not all, glaciers in southern Norway may have melted completely at least once during the early to mid-Holocene (ca. 8,000-4,000 cal. yr BP). Most glaciers seem to have started to reform/advance between 6,000 and 4,000 cal. yr BP. Most glaciers experienced their maximum post-Erdalen event position during the Little Ice Age. However, the timing of the maximum Little Ice Age advance in different parts of southern Norway varies considerably, ranging from the early 18th century to the 1940s. Reconstructed millennial-scale Holocene glacier variations and decadal-scale Little Ice Age glacier variations in southern Norway do not show a consistent pattern. Therefore, regional differences between climate perturbations and glacier response appear to be influenced by differences in the relative importance of summer temperature and winter precipitation operating on different time scales [*Nesje et al.*, 1995; *Nesje and Dahl*, 2003; *Winkler et al.*, 2003].

4.8.1. Norwegian glaciers, modern climate, and the North Atlantic Oscillation (NAO). Glacier mass balance and subsequent length variations are mainly related to both changes in summer temperature and winter precipitation. The world's longest, continuous glacier mass–balance record is from Storglaciären in northern Sweden (Figure 8A). In Scandinavia (including Svalbard), the net mass balance on the maritime glaciers in western Norway is best correlated with the winter balance, whereas the net mass balance on continental glaciers in southern Norway and on the Svalbard glaciers is best correlated with summer balance [e.g., *Nesje et al.*, 1995b, 2000a] (Figure 8B). This is also reflected in cumulative glacier–mass balance records; the maritime glaciers in Scandinavia mainly experienced a positive trend since the early 1960s, whereas the more continental (including Svalbard) glaciers experienced a negative mass–balance trend over the same period (Figure 8C).

In Europe, a high correlation between decadal variations in the NAO and glacier mass balance has been demonstrated [*Nesje et al.*, 2000b; *Reichert et al.*, 2001; *Six et al.*, 2001], the dominant factor being the strong relationship between winter precipitation and the NAO. The NAO winter index is highly correlated with the mass balance from maritime glaciers in southern Norway (Figure 8D). A positive NAO phase means enhanced winter precipitation for maritime glaciers in Scandinavia and reduced winter precipitation for glaciers in the European Alps [*Reichert et al.*, 2001; *Six et al.*, 2001]. This internal climate system mechanism explains the observed strong positive mass balance on maritime glaciers in western Scandinavia, associated with NAO-related changes in precipitation patterns since the 1970s, and partly explains the strong negative mass balance of glaciers in the European Alps. However, in the European Alps, negative mass balance is also related to increased summer temperatures. Some authors suggest that the recent trend in NAO may potentially be related to global warming [e.g., *Hurrell et al.*, 2003].

4.8.2. Reconstruction of Holocene variations in winter precipitation. The equilibrium-line altitude on a glacier is mainly controlled by precipitation as snow during the accumulation-season and by summer temperature during the ablation season. It has been demonstrated that there is a close exponential relationship between mean ablation-season temperature t (1 May-30 September) and winter accumulation A (1 October-30 April) at the equilibrium-line altitude of modern Norwegian glaciers [Liestøl in Sissons, 1979; Sutherland, 1984], which is expressed by the regression equation:

$$A = 0.915\ e^{0.339t}\ (r^2 = 0.989,\ P < 0.0001)$$

where A is in metres water equivalent and t is in °C [*Ballantyne*, 1989].

Based on the regression equation above, present and past mean winter precipitation (A) at the equilibrium-line altitude of a steady-state glacier can be quantified when mean ablation-season temperature (t) is known (see *Dahl and Nesje* [1996] and *Nesje and Dahl* [2000] for further details and examples of how to calculate winter precipitation by combining glacial equilibrium-line altitude and summer temperature data). Variations in winter precipitation at other elevations can be calculated by using a precipitation-elevation gradient. In southern Norway this gradient has been calculated to be ca. 8% precipitation change per 100 m [*Haakensen*, 1989; *Dahl and Nesje*, 1992; *Laumann and Reeh*, 1993].

To reconstruct variations in Holocene winter precipitation at Hardangerjøkulen and Jostedalsbreen, we used Holocene equilibrium-line altitude fluctuations (adjusted for land uplift) inferred from terrestrial and lacustrine sites, and Holocene summer (July) temperature variations at Finse, central southern Norway, reconstructed from chironomids in lake sediments [*Velle*, 1998]. The Holocene winter precipitation variations (percentage variations) (Figure 9) are calculated on the basis of the modern relationship between ablation-season temperature and accumulation-season precipitation normal values (1961-90) at adjacent meteorological stations. The reconstructed winter precipitation curves for the Hardangerjøkulen and Jostedalsbreen areas show significant, millennial-scale variations. The Holocene winter-precipitation curves may reflect fluctuations between periods with prevailing mild and wet winter conditions (positive NAO index weather mode) and periods with prevailing cold and dry winters (negative NAO index weather mode) and thus may indicate large-scale Holocene variability in the atmospheric winter circulation over northwest Europe.

5. CLIMATE DURING THE LAST MILLENNIUM

The circum-North Atlantic region has a number of long instrumental and documentary climate records, and a range of terrestrial and marine palaeoclimatic information, some of which are shown in Figures 10 and 11. In Fennoscandia there is a number of long, homogenous instrumental climate records (see review by [*Jones*, 2001]). Several similarities can be discerned by comparing the records. A positive temperature trend over the last 250 years and a general warming trend during most of the 20th century are clearly observed. None of the records indicate decadal or multi-decadal changes during the last 1,000 years that exceed the mean temperature during the 20th century. There was, however, warm climate across eastern Canada and Greenland during the late 13th, early 15th, and late 19th centuries, in accordance with the warm climate indicated by west European documentary data and the Scandinavian tree-ring record. The middle 14th century was also relatively cold on both sides of the Atlantic. While the west Greenland data indicate cold conditions during most of the late 16th and 17th centuries (in phase with west Europe), the central and northern Greenland data show cool conditions continuing into the 18th century [*Alverson et al.*, 2003].

Figure 11 shows examples of different climate proxies and temperature reconstructions for the last millennium from the northern North Atlantic region: stable oxygen isotopes (ice-volume corrected) in *N. pachyderma* (dex.) and *N. pachyderma* (sin.) from cores MD95-2011 and JM97-948 obtained from the eastern Norwegian Sea (Vøring Plateau) [*Risebrobakken et al.*, 2003]; diatom-inferred SST (in August) variations in the eastern Norwegian Sea reconstructed from cores MD95-2011 and JM97-948 in the eastern Norwegian Sea (Vøring

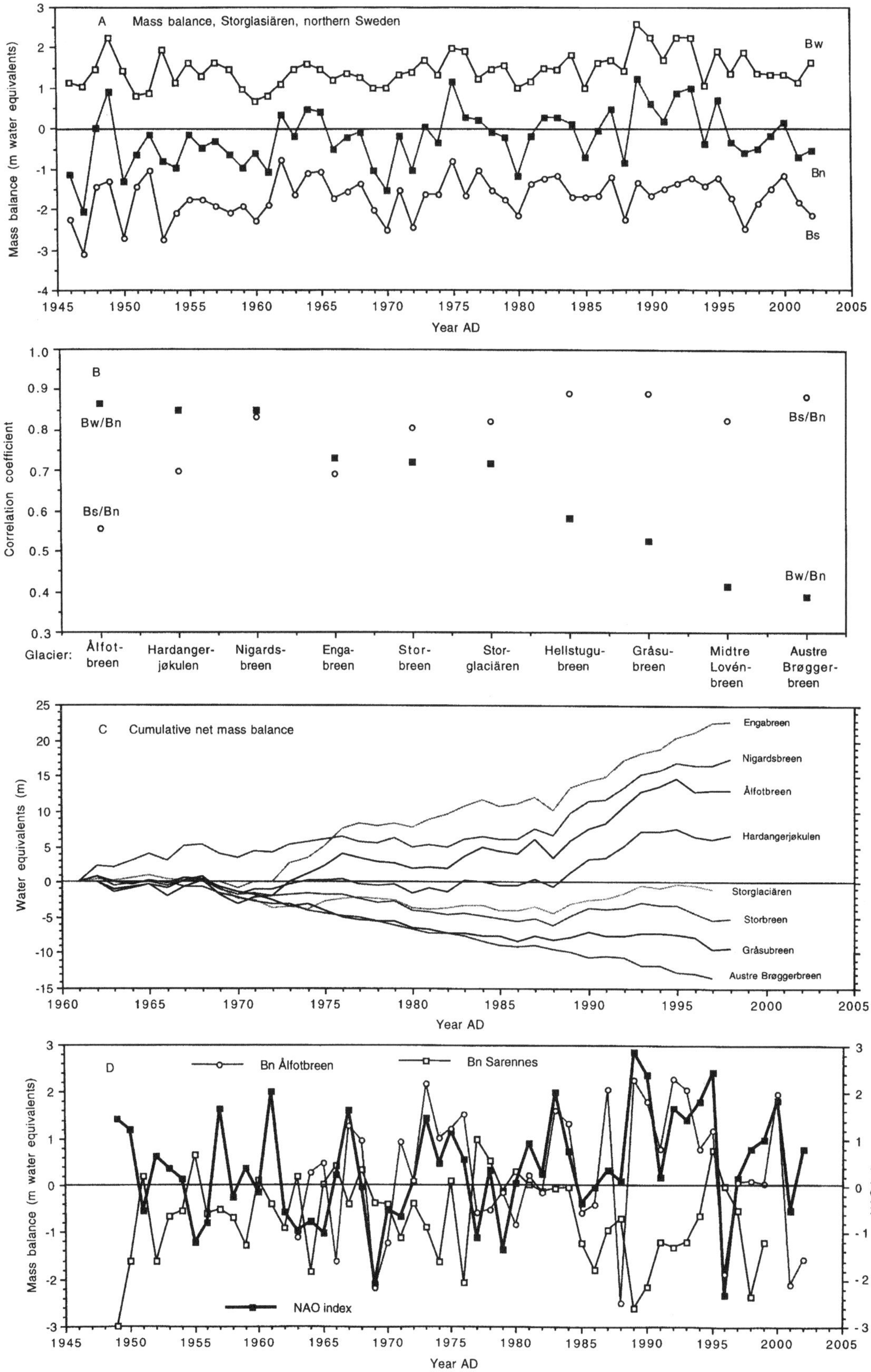
A Mass balance, Storglasiären, northern Sweden
Bw
Bn
Bs
Mass balance (m water equivalents)
Year AD
B
Bw/Bn
Bs/Bn
Correlation coefficient
Glacier:
Ålfot-breen
Hardanger-jøkulen
Nigards-breen
Enga-breen
Stor-breen
Stor-glaciären
Hellstugu-breen
Gråsu-breen
Midtre Lovén-breen
Austre Brøgger-breen
C Cumulative net mass balance
Engabreen
Nigardsbreen
Ålfotbreen
Hardangerjøkulen
Storglaciären
Storbreen
Gråsubreen
Austre Brøggerbreen
Water equivalents (m)
Year AD
D
Bn Ålfotbreen
Bn Sarennes
NAO index
Mass balance (m water equivalents)
NAO index
Year AD

Figure 8. (A) Annual glacier mass balance (Bw, winter balance; Bs, summer balance; Bn, net balance) on Storglaciären, northern Sweden. This is the longest glacier mass-balance record in the world [data: *Holmlund et al.*, 1996], with later updates [data: Peter Jansson and Per Holmlund, Univ. of Stockholm, Sweden]. (B) Correlation between net mass balance (Bn) and winter balance (Bw), and between net mass balance and summer balance (Bs) on ten Scandinavian (including Svalbard) glaciers. (C) Cumulative net mass-balance records of glaciers in Norway (Engabreen, Nigardsbreen, Ålfotbreen, Hardangerjøkulen, Storbreen, and Gråsubreen; data: Norwegian Water Resources and Energy Directorate [NVE]), Sweden [Storglaciären; data: *Holmlund et al.*, 1996, with later updates by Per Jonasson and Per Holmlund, Univ. of Stockholm), and Spitsbergen [Austre Brøggerbreen; data: Norwegian Polar Institute]. (D) Annual net mass balance (Bn) on Ålfotbreen, a maritime glacier in western Norway [data: NVE], on the Sarennes glacier in the French Alps [data: World Glacier Monitoring Service], and the NAO index [*Jones et al.*, 1997, with later updates].

Plateau) [*Jansen and Koc*, 2000; *Birks and Koc*, 2002; *Koc and Jansen*, 2002; *Andersen*, 2003; *Andersen et al.*, 2004]; a tree-ring width series from Trøndelag [*Thun*, 2002]; LOI records from two non–glacier-fed lakes in southern Norway (data from øvre Bukkebergtjørna provided by A. Nesje and from Trettetjørn by Larsen, in prep., and *Bjune* [2005]); mean annual temperature reconstructed from a speleothem in Nordland, northern Norway [*Lauritzen and Lundberg*, 1999]; and LOI values (low values indicate high glacier activity) from the proglacial lake Hafslovatnet draining the southeast part of Jostedalsbreen (data provided by E. Sønstegaard and A. Nesje).

Figure 12 shows reconstructed mean spring and summer temperatures in Trøndelag AD 1701-2003 (top panel) [*Nordli*, 2004], in western Norway AD 1734-2002 (middle panel) [*Nordli et al.*, 2002, 2003], and in southeastern Norway AD 1749-2000 (lower panel) [*Nordli*, 2001]. The time evolution of spring and summer temperature in the three regions was reconstructed based on the first day of grain harvest as obtained from farmers' diaries and, since the mid-19th century, on instrumental data. In addition, for the western Norway series, dated moraines in front of Nigardsbreen (Jostedalsbreen) and Storbreen (Jotunheimen) were used to complement the temperature reconstruction for short periods with incomplete historical data.

In the Trøndelag series, a significant, linear temperature trend for the entire series of +0.6 °C (+0.2 °C per century) was detected. During the 18th century, the temperature trend was negative (–0.4 °C), whereas the temperature trends were positive during the 19th and 20th centuries (+0.3 °C and +0.5 °C, respectively). In the western Norwegian temperature series, the springs of 1889 and 1735 were the warmest in the 19th and 18th (after 1734) centuries, respectively. The extraordinarily cold spring/summer of 1802 was the coldest in the entire series. The summer of 1839 was also extremely cold. In the 18th century, 1784 seems to have been the coldest summer. No significant temperature trend was detected in the western Norwegian temperature series, neither during the 18th century (subsequent to 1734), nor in the 20th century. An abrupt temperature rise, however, was recorded in the 1920s. In the southeastern Norwegian temperature series, a marked temperature trend was detected, amounting to

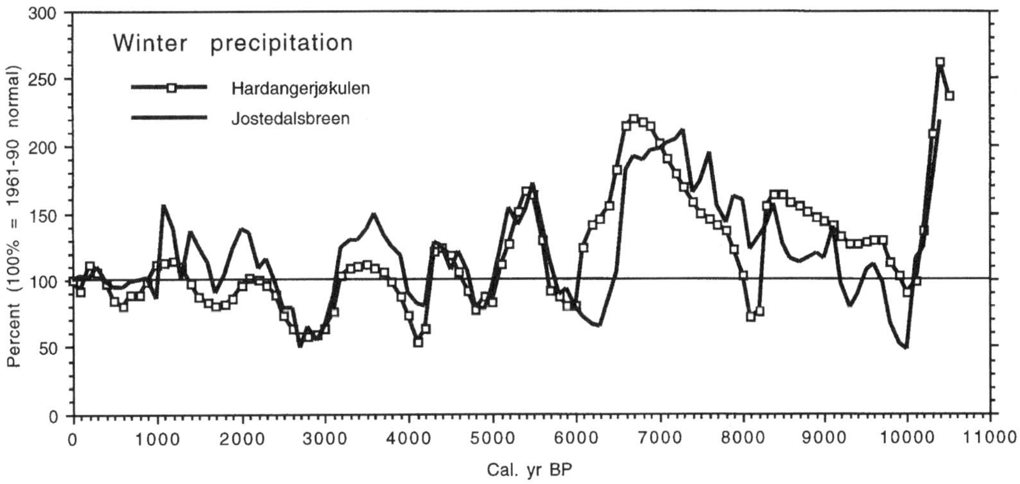

Figure 9. Winter precipitation variations (in percent of the 1961-90 normal) during the Holocene in the Hardangerjøkulen [modified from *Dahl and Nesje*, 1996] and Jostedalsbreen [*Nesje et al.*, 2001] regions.

+1.4 °C. The high temperatures were mainly concentrated within the 20th century, when it was estimated as +0.9 °C. In the 19th century, the temperature trend was only +0.5 °C. In the first half of the 18th century, no significant temperature trend was detected. In the series, 1796-1805 was the coldest decade, whereas the warmest was the period 1988-1997.

6. HOLOCENE CLIMATE-FORCING FACTORS

The influence of orbital variations (changes in precession, obliquity, and eccentricity) played a major role in the long-term Holocene climate variability. The precessional forcing culminated 12,000-10,000 cal. yr BP [*Berger and Loutre*, 1991], and the overall solar insolation in the Northern Hemisphere summer (June) has declined by ~10% at 60 °N over the last 10,000 years (Figure 13A). The effect of such a change in insolation probably explains the overall temperature decline seen in the records from the early Holocene to the present.

By assuming that solar activity variations affect cosmogenic isotope production in the upper atmosphere, the record of ^{14}C and ^{10}Be from tree rings and ice cores, respectively, can provide a long-term index of solar activity [*Beer et al.*, 1996; *Bard et al.*, 2000; *Rind*, 2002]. *Lean et al.* [1992], using long-term observations of the sunspot cycle together with inferences based on the variability of sun-like stars, estimated that there was a change in the solar irradiance from the late 17th/early 18th century solar minimum (Maunder Minimum) to the late 20th century by ~0.24%. Analysis of Northern Hemisphere mean annual temperatures since AD 1600 and solar variations showed that the two records were highly correlated [*Lean et al.*, 1995; *Crowley and Kim*, 1996; *Mann et al.*, 1998]. Based largely on comparisons of proxy records with the ^{14}C anomaly series (Figure 13B), several Holocene palaeoclimatic studies have suggested that solar forcing has been responsible for reconstructed climate changes (e.g., *Bradley* [2003] and references therein). For example, *Bond et al.* [2001] argued that changes in the abundance of hematite-stained grains (ice-rafted debris) in North Atlantic marine sediments vary with the same frequency as ^{14}C

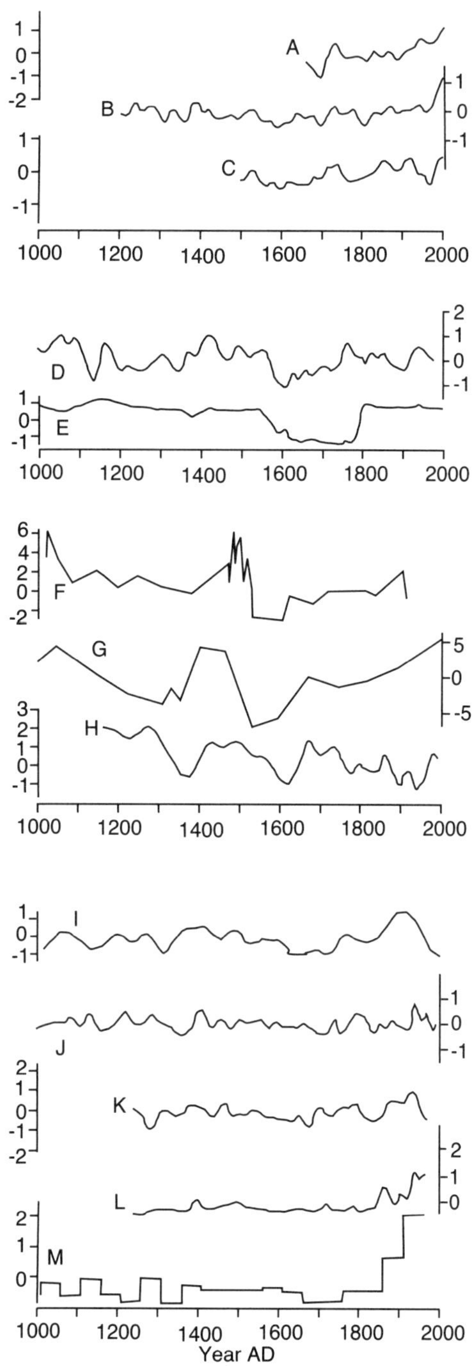

Figure 10. Climate records for the last millennium in the North Atlantic region: (A) Central England mean annual temperatures [*Manley*, 1974]. (B) Annual temperatures for the Benelux countries [*van Engelen et al.*, 2001]. (C) Reconstructed winter NAO indices [*Luterbacher et al.*, 2001]. (D) Warm season temperatures reconstructed from tree-ring data [*Briffa et al.*, 1992]. (E) Moisture index based on bog flora [*K.E. Barber et al.*, 2000]. (F) SST from oxygen isotope composition in foraminifera [*Keigwin*, 1996]. (G) Index of the speed of the deep current flow; low values indicate a reduction in the North Atlantic deep water production [*Bianchi and McCave*, 1999]. (H) Foraminiferal abundance in the Cariaco Basin, indicative of trade wind intensity and possible changes in temperature in the North Atlantic [*Black et al.*, 1999]. (I) Oxygen isotope data from the North GRIP ice core [*Johnsen et al.*, 2001]. (J) Ice-core–derived oxygen isotopes from the GISP2 and GRIP sites [*Johnsen et al.*, 2001]. (K) Several west Greenland ice-core oxygen isotope series [*Fisher et al.*, 1995], (L) High-resolution melt-layer data in an ice core from the Agassiz Ice Cap [*Koerner and Fisher*, 1990; *Fisher et al.*, 1995], (M) A lower-resolution eastern Canadian ice-core melt record [*Alverson et al.*, 2003].

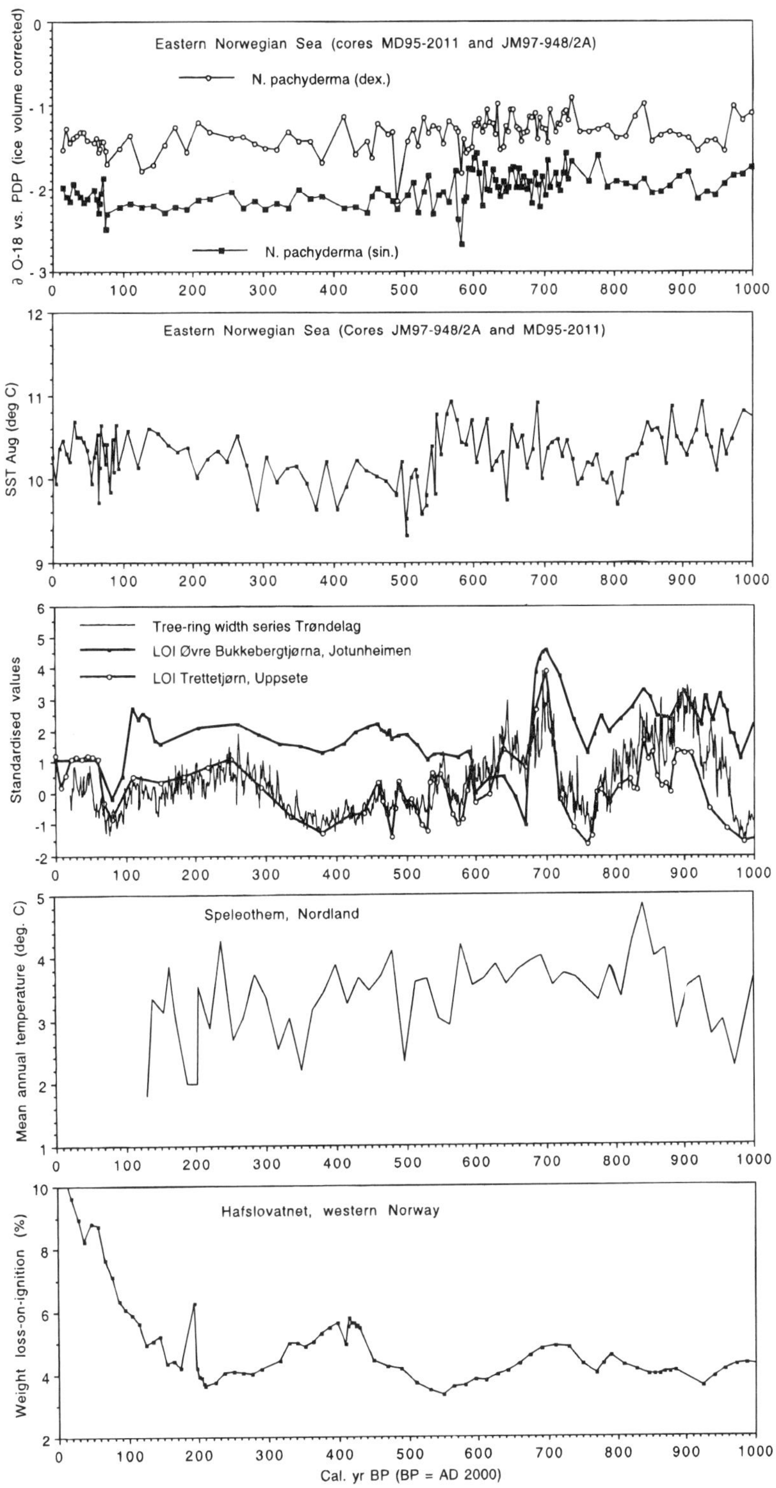

Figure 11. Examples of different climate proxies and temperature reconstructions for the last millennium in the northern North Atlantic region: Stable oxygen isotopes (ice-volume corrected) in *N. pachyderma* (dex.) and *N. pachyderma* (sin.) from cores MD95-2011 and JM97-948 obtained from the eastern Norwegian Sea (Vøring Plateau) [*Risebrobakken et al.*, 2003]. Diatom-inferred SST (August) variations in the eastern Norwegian Sea reconstructed from cores MD95-2011 and JM97-948 in the eastern Norwegian Sea (Vøring Plateau) [*Jansen and Koc*, 2000; *Birks and Koc*, 2002; *Koc and Jansen*, 2002; *Andersen et al.*, 2004]. A tree-ring width series from Trøndelag [*Thun*, 2002] and loss-on-ignition (LOI) records from two non–glacier-fed lakes in southern Norway [data from Øvre Bukkebergtjørna provided by A. Nesje and from Trettetjørn by Larsen, in prep., and *Bjune* [2005]]. Mean annual temperature reconstructed from a speleothem in Nordland [*Lauritzen and Lundberg*, 1999]. LOI (low values indicate high glacier activity) from the proglacial lake Hafslovatnet draining the southeast part of Jostedalsbreen [unpublished data provided by E. Sønstegaard and A. Nesje].

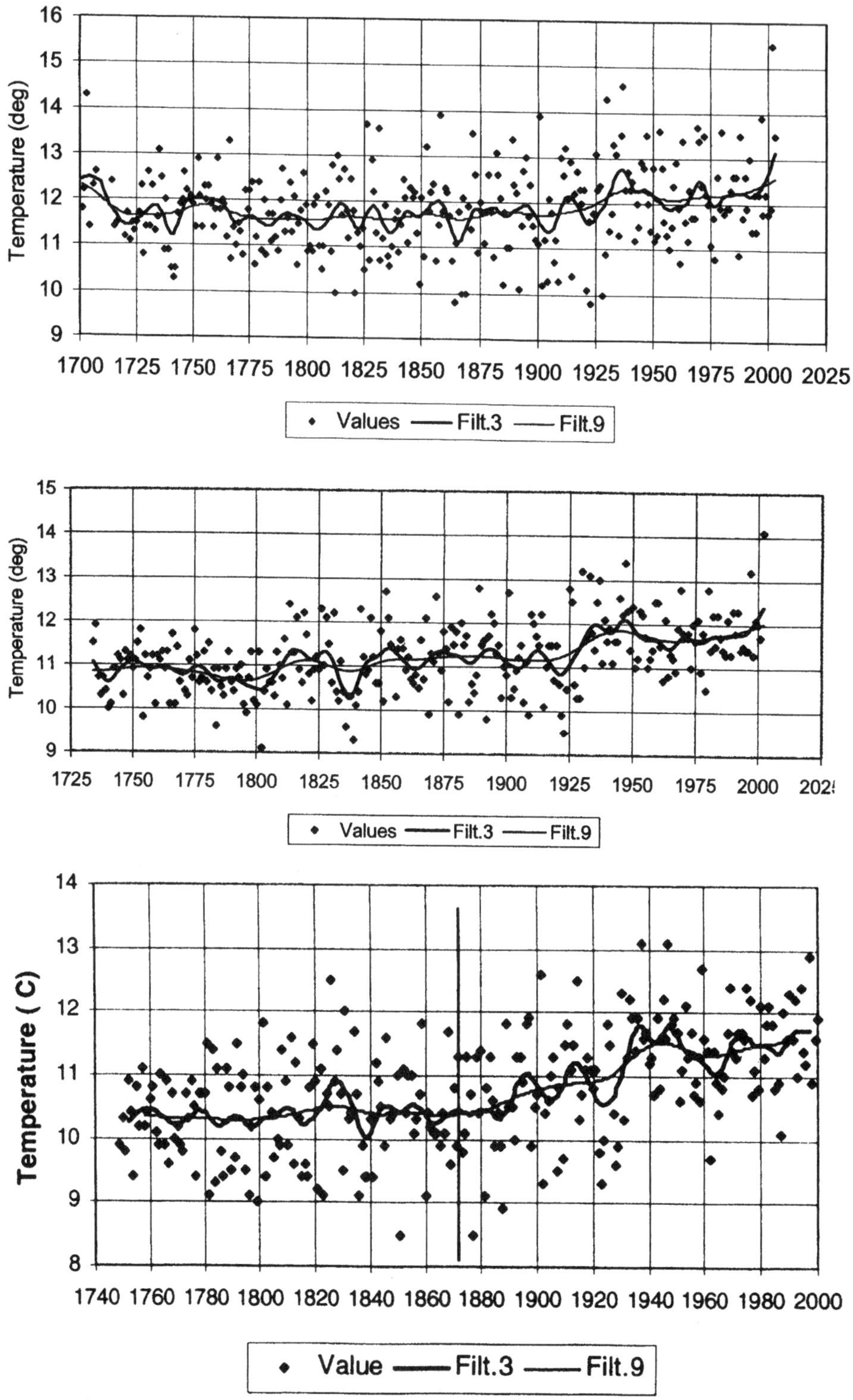

Figure 12. Reconstructed mean spring and summer temperatures in Trøndelag AD 1701-2003 (top panel) [*Nordli*, 2004], in western Norway AD 1734-2002 (middle panel) [*Nordli et al.* 2002, 2003], and in southeastern Norway AD 1749-2000 (lower panel) [*Nordli*, 2001]. Dots represent individual years. The values are filtered by a Gaussian low-pass filter with standard deviations of 3 (Filt.3) and 9 (Filt.9) years in their distributions.

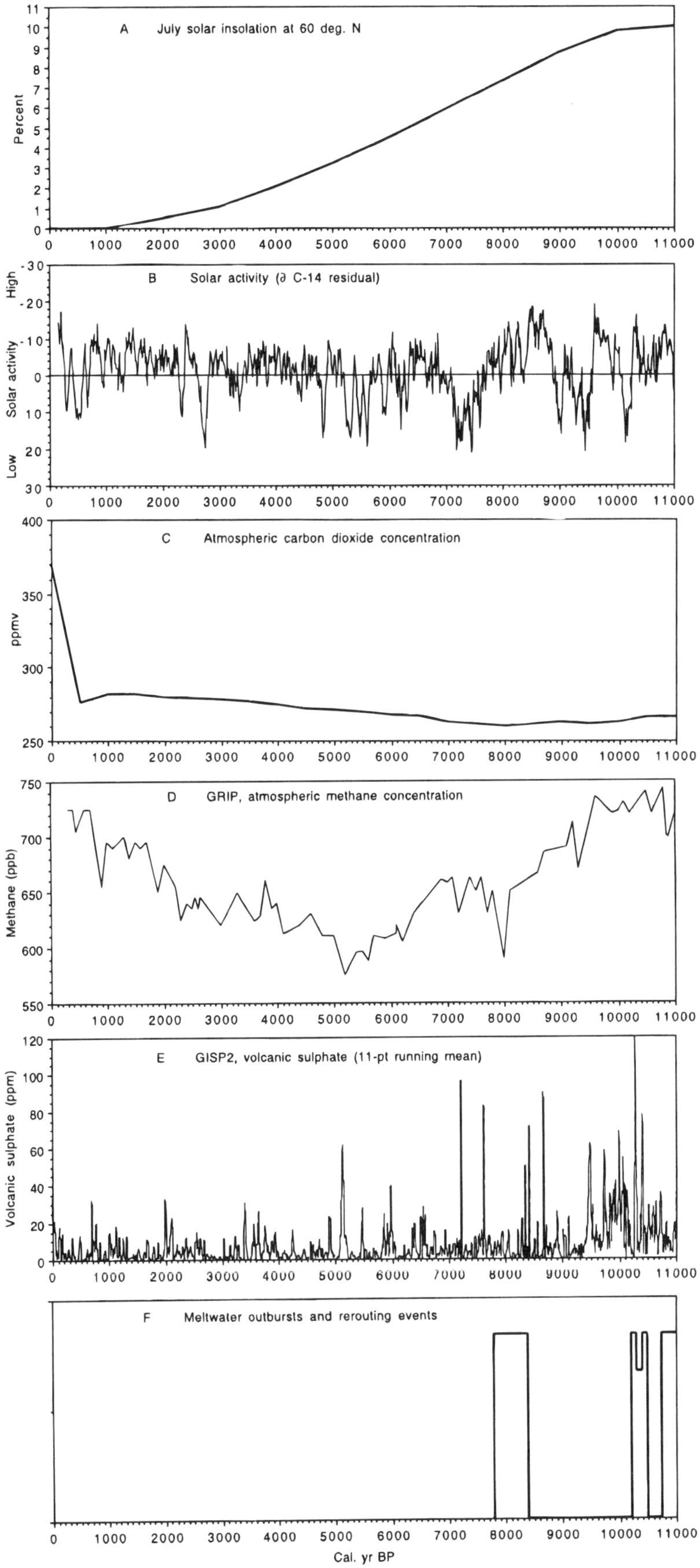

Figure 13. Possible Holocene climate-forcing factors: (A) Insolation anomaly (in percent) at the top of the atmosphere (departure from AD 1950 values) for mid-June (adapted from *Berger* and *Loutre* [1996]). (B) Holocene variations in solar activity as inferred from residual ^{14}C variations (adapted from *Stuiver et al.* [1997]). (C) Holocene variations in atmospheric carbon dioxide concentration in ice cores from Byrd Station, Taylor Dome, and Siple Station [adapted from *Neftel et al.*, 1988; *Staffelbach et al.*, 1991; *Marchal et al.*, 1999; *Indermühle et al.*, 1999] and from the Mauna Loa Observatory, Hawaii [Keeling and Whorf, continuously updated; http://cdiac.esd.ornl.gov/ftp/maunaloa-co2/maunaloa.co2). (D) Holocene variations in atmospheric methane (CH_4) concentration from the Greenland GRIP ice core (adapted from *Blunier et al.* [1995]). (E) The volcanic sulphate record (11-pt running mean) from the Greenland GISP2 ice core (adapted from *Zielinski et al.* [1994] and *Zielinski and Mershon* [1997]). (F) Early Holocene meltwater outburst and rerouting phases from ice-dammed lakes along the margin of the Laurentide ice sheet [e.g., *Clark et al.*, 2002].

anomalies (about 1,450-1,500 years) and *Björck et al.* [2001] suggested that solar forcing was likely to explain a climatic event at 10,300 cal. yr BP recorded in the Nordic Seas.

In principle, air trapped in bubbles of polar ice has the same composition as the atmosphere at the time of ice formation. Based on ice-core records, an increase in atmospheric CO_2 since the beginning of industrialisation was documented [*Neftel et al.*, 1985; *Pearman et al.*, 1986] in Antarctic ice cores (Figure 13C). Atmospheric CO_2 concentrations were also significantly lower during the glacial period than in the preindustrial Holocene [*Neftel et al.*, 1982; *Barnola et al.*, 1987]. A high-resolution ice-core record of atmospheric CO_2 concentration from Taylor Dome shows that the global carbon cycle has not been in steady state during the Holocene epoch [*Indermühle et al.*, 1999], and that changes in terrestrial biomass and SST were the main causes for the observed millennial-scale changes of atmospheric CO_2 concentrations.

The atmospheric concentration of another greenhouse gas, methane (CH_4), during the Holocene has been measured and dated by counting annual layers in the Greenland GRIP ice core [Figure 13D; Blunier et al., 1995]. Cyclic variations in carbon dioxide and methane driven by earth-orbital changes during the last 350,000 years should predict decreases of these gases during the Holocene. However, the CO_2 and CH_4 trends started anomalous increases 8,000 and 5,000 years ago, respectively (Figures 13C, D). *Ruddiman* [2003] suggested that anthropogenic changes due to agricultural expansion in Eurasia, the start of forest clearance 8,000 years ago, and rice irrigation by 5,000 years ago were the main causes for this anomalous development. Radiative forcing by changes in atmospheric trace-gas concentrations was minor during the early Holocene. Carbon dioxide attained near preindustrial levels about 11,000 cal. yr BP and remained nearly constant during the early Holocene. As the climate warmed, in contrast, the water vapour content of the atmosphere probably increased [e.g., *Foley et al.*, 1994] and the heat and moisture flux from the tropics to the Arctic was probably enhanced (positive feedback to warming).

Explosive volcanic eruptions may have short-term cooling effects on hemispheric or global temperatures [e.g., *Robock*, 2000], as shown from instrumental temperature records. Energy-balance and general circulation model studies suggest that explosive volcanism has influenced natural climate variability on a hemispheric and global scale over the recent centuries [e.g., *Meehl et al.*, 2004]. The volcanic aerosol reduces the energy receipt at the surface and, together with circulation interactions that may occur, may lead to lower temperatures a few years after a significant explosive volcanic eruption. Individual explosive eruptions, however, have only a negligible climate effect. If volcanic eruptions in the past happened to occur in clusters, the cumulative effect of the eruptions may have persisted longer, causing decadal to multi-decadal climate change. The GISP2 record indicates that some periods during the Holocene had more frequent volcanic activity than others [*Zielinski et al.*, 1994], especially in the first half of the Holocene (Figure 13E).

Climatic feedbacks of radiative forcing during the early Holocene were probably spatially variable because of the inferred reduction of snow and ice cover and because the vegetation cover was different from the present. These factors influenced the distribution of energy absorbed during the summer and the energy and water balance at the Earth's surface. Retreated and melted glaciers and reductions in sea-ice cover were of particular significance for amplification of warming because vegetation and open sea have much lower albedo and higher heat capacity than ice [e.g., *Kaufmann et al.*, 2004].

A comparison between the climate forcing factors and the Holocene records presented here suggests that the early Holocene thermal maximum observed in most records from the Nordic Seas region was caused by increased summer solar insolation to the northern Hemisphere. The suborbital millennial to decennial climate variability observed in the Holocene climate reconstructions was most probably a combined effect of solar activity changes, periods of increased explosive volcanism, and internal feedback mechanisms in the Earth's climate system. Internal modes of the climate system variability (e.g., the NAO) [*Hurrell et al.*, 2003] may also have been responsible for some of the observed climate variability, especially in winter. In addition to external climate-forcing factors, abrupt, catastrophic lake drainage from ice-marginal lakes during the termination of the last Ice Age (Figure 13F) had a significant, regional climate impact [*D.C. Barber et al.*, 1999; *Nesje et al.*, 2004 and references therein]. The best example of this is perhaps the ~"8.2 ka"/"Finse event" detected in many palaeoclimatic archives that apparently resulted from sudden glacial lake drainage at the margin of the Laurentide ice sheet [e.g., *Clarke et al.*, 2003].

Most palaeoclimatic studies that discuss potential climate-forcing factors have simply made correlations by curve-matching (in the time domain) or by finding spectral peaks that correspond to similar frequencies (in the frequency domain) of one or more potential forcing factors [e.g., *Bond et al.*, 2001]. By the use of realistic climate models (coupled ocean–atmosphere models, preferentially also incorporating hydrological and vegetation feedbacks), it should be possible to combine high-resolution palaeoclimatic reconstructions with climate models to test the complex interactions between the different forcing factors to help explain the primary forcing factors and the accompanying feedbacks behind the observed Holocene climate development.

7. DATA SYNTHESIS AND FUTURE PERSPECTIVES

Holocene marine and terrestrial records presented here show a high degree of consistency (Figures 2-7), indicating strong links between the atmospheric and marine mean climate state in the northern North Atlantic region throughout the Holocene. The Holocene climate records reconstructed from marine and terrestrial proxies indicate that the Holocene climate, at least in this region, has not been as stable as previously suggested from the GRIP and GISP-2 Greenland ice-core records. In particular, the early Holocene (more than ~8,000 cal. yr BP) was punctuated by several significant and abrupt climatic deteriorations (Figures 3-5). Three cooling events were superimposed on the general warming trend that took place between the termination of the Younger Dryas and approximately 8,000 cal. yr BP: the Preboreal Oscillation at 11,300-11,150 [*Björck et al.*, 1997], the Erdalen event 10,300-9700 cal. yr BP [*Nesje et al.*, 1991; *Björck et al.*, 2001; *Dahl et al.*, 2002], and the "8.2 kyr" event [*Alley et al.*, 1997], termed the Finse event in southern Norway [*Dahl and Nesje*, 1994, 1996; *Nesje et al.*, 2001]. In western Scandinavia, a distinctly cool but oceanic summer climate in the early Holocene (e.g., Figure 2) possibly resulted from more persistent westerly airflow that was replaced about 8,000 cal. yr BP by a more meridional flow pattern and by the development of more frequent blocking anticyclonic summer conditions with a more southerly and southeasterly air flow. The inferred low summer temperatures in the early Holocene in western Scandinavia contrasts with the predicted early-Holocene orbitally induced higher radiation than at present. It is possible that the inferred low summer temperatures may be partly due to the influence of remnants of the northern hemisphere ice sheets, and that until about 8000 cal. yr BP the effect of higher summer solar radiation was dampened by the cooling effect of enhanced Atlantic air flow and discharges of meltwater from the diminishing northern hemisphere ice sheets. However, summer temperature reconstructions from more inland continental sites in Scandinavia (Figure 3A-C), restricted glacier coverage, and the presence of thermophilous molluscs around Svalbard (Figure 6) may indicate a very early Holocene summer season thermal maximum in these areas.

Pollen-based reconstructions of summer (mean temperature of the warmest month) anomalies over northern Europe as a whole during the early Holocene show values close to modern temperatures [*Davis et al.*, 2003]. The temperatures rose to a maximum around 6,000 cal. yr BP. At the maximum, the summer temperatures reached +1.5 °C in northwest Europe, before decreasing to modern values. After the 8.2 ka/Finse event, a warm (1.5-2 °C warmer than at present) period occurred between approximately 7600 and 5500 cal. yr BP. The warmest interval seems to have been between 7,600 and 6,300 cal. yr BP, probably associated with a distinct northward shift of the polar vortex or a weaker meridional circulation and prevailing anticyclonic summer conditions. This period was characterised by thermophilous trees and marine species, northward spread of temperature sensitive plant and marine species, small or totally melted glaciers, and low lake levels in small closed basins. Winter (mean temperature of the coldest month) temperatures were significantly greater than late Holocene values in the early part of the Holocene. Subsequent to a rapid warming at 11,500 cal. yr BP, winter temperatures in northwest Europe showed a steady rise to modern-day values. Early Holocene summer temperatures similar to the present-day have been reconstructed in northern Scandinavia [*Seppä and Birks*, 2001; *Rosén et al.*, 2001]. At the coast of western Norway (Kråkenes), a quantitative multi-proxy study indicates that summer temperatures at the start of the Holocene rose quickly to modern values [*Ammann et al.*, 2000; *Birks et al.*, 2000; *Birks and Amman*, 2000]. The early Holocene pine tree line in the Scandes mountains and in northern Scandinavia was close to the modern line (Figure 3A,B). There is strong, multi-proxy evidence for a mid-Holocene thermal maximum in Scandinavia. Tree lines reached their maximum altitude up to 300 m higher than at present [*Eronen and Zetterberg*, 1996; *Barnekow and Sandgren*, 2001]. Glaciers were much reduced in size and periodically absent [*Karlén*, 1988; *Dahl and Nesje*, 1994, 1996; *Nesje et al.*, 2000, 2001; *Seierstad et al.*, 2002]. Quantitative summer temperature reconstructions suggest that summer temperatures were up to 2.0 °C higher than at present (Figure 2) [see also *Barnekow*, 2000; *Barnett et al.*, 2001; *Seppä and Birks*, 2001, 2002; *Rosén et al.*, 2001]. Pollen-based reconstructions of summer (mean temperature of the warmest month) anomalies over the whole of northern Europe indicate a mid-Holocene thermal optimum around 6,000 cal. yr BP [*Davis et al.*, 2003]. Other marine and terrestrial proxies also indicate a mid-Holocene thermal maximum in and around the Nordic Seas, mainly falling between 6,000 and 8,000 cal. yr BP [*Dahl and Nesje*, 1994, 1996; *Seppä and Birks*, 2001; *Korhola et al.*, 2000; *Barnekow*, 2000; *Barnekow and Sandgren*, 2001; *Birks and Koc*, 2002; *Calvo et al.*, 2002; *Seppä et al.*, 2002a,b; *Nesje et al.*, 2000, 2001; *Seierstad et al.*, 2002; *Brooks*, 2003].

Subsequent to the early-mid Holocene thermal maximum, most studies indicate several mid- to late-Holocene coolings and glacier readvance phases in Scandinavia (Figures 2-7) [*Karlén*, 1988; *Dahl and Nesje*, 1994, 1996; *Barnekow*, 2000; *Rosén et al.*, 2001; *Seppä and Birks*, 2001, 2002; *Barnett et al.*, 2001; *Nesje et al.*, 2001; *Korhola et al.*, 2002; *Seierstad et al.*, 2002; *Davis et al.*, 2003], on Svalbard [*Svendsen and Mangerud*, 1997; *Snyder et al.*, 2000] and in the Norwegian Sea [*Birks and Koc*, 2002; *Calvo et al.*, 2002]. Increased interannual variability in the Finnish tree-ring record from approximately 5,800 cal. yr BP indicates increased climatic instability. The most prominent of these occurred around 6,000, 5,000, 4,000-3,800, 2,500, and 1,500 cal. yr BP and at the beginning of the Little Ice Age. The late-Holocene decreasing temperature trend seen in many climate records from the North Atlantic region most probably reflects a regional response to lower summer insolation. Several short-term climate oscillations, however, occurred during the last 4,000-5,000 years. The dominant anti-cyclonic pattern that prevailed during the early to mid-Holocene was probably replaced by a more variable and moister climate. During the Medieval Warm Period (approximately AD 700-1300), temperatures were in the order of 0.5-0.8 °C higher than at present. The subsequent Little Ice Age, was characterised not only by several cold winters and summers, but also by periods of high winter precipitation (causing glacier advance) and extreme weather events (floods and avalanches) [e.g., Grove, 1988; Nesje and Dahl, 2003]. Historical records from northern Europe and Greenland document the profound effects of both the Medieval Warm Period and the Little Ice Age on human society [see reviews by *Grove*, 1988, 2002].

In the future, a larger number of well-dated, high-resolution terrestrial and marine quantitative climate reconstructions are needed for us to be able to better reconstruct Holocene variations in atmospheric and oceanic climate and circulation changes along east–west and north–south gradients in the North Atlantic region. Of particular importance is to define climatic anomalies and the scales and patterns of subdecadal to millennial-scale climate variability. Correlations with records of similar time resolution in lower latitudes are required to understand possible teleconnections and causal mechanisms (solar activity, volcanic eruptions, meltwater spikes, variations in the thermohaline circulation). For this, much improved chronology and better age–depth models are required. The combined Holocene climate reconstructions based on terrestrial and marine proxies provide data for climate modelling experiments of special interest in the early Holocene, when the magnitude of the Scandinavian ice sheet, the meltwater input into the North Atlantic, and the westerly oceanic flow varied significantly, and the climate system was being forced by higher summer insolation.

8. SUMMARY AND CONCLUSIONS

Marine and terrestrial Holocene records from the northeastern North Atlantic region presented here show a highly coherent pattern, indicating strong links between the atmospheric and marine climate variability in this region throughout the Holocene.

The Holocene climate records indicate that the Holocene climate in the North Atlantic region was not stable and the early Holocene, in particular, was punctuated by several significant and abrupt climatic deteriorations.

A cool, oceanic summer climate in western Scandinavia during the early Holocene was probably the result of enhanced westerly airflow, which was replaced about 8,000 cal. yr BP by a more meridional atmospheric flow pattern and by the development of predominantly blocking anticyclonic summer conditions and a more southerly and southeasterly air flow.

Inferred high summer temperatures reconstructed from inland continental sites in Scandinavia, restricted glacier coverage, and the presence of thermophilous molluscs around Svalbard indicate an early Holocene thermal maximum in these areas.

Pollen-based reconstructions of summer temperature anomalies over northern Europe as a whole during the early Holocene show values close to modern temperatures.

In the early part of the Holocene, winter temperatures and winter precipitation were higher than late-Holocene values.

There is strong, multi-proxy evidence for an early to mid-Holocene thermal maximum in the northern North Atlantic region.

Subsequent to the early to mid-Holocene thermal maxima, most records indicate several late-Holocene periods of low summer temperatures and glacier readvances in Scandinavia (including Svalbard) and variable and generally falling temperatures in the Nordic Seas. The general temperature decline was punctuated by centennial-scale warmer (for example, the Medieval Warm Period) and colder periods, the Little Ice Age being one of the coldest of the entire Holocene.

Recently obtained climate reconstructions suggest that a variety of forcing mechanisms, including at least solar activity, ocean circulation, catastrophic outbursts of meltwater from ice sheets, atmospheric concentrations of greenhouse gases and aerosols emitted during volcanic eruptions, have combined to produce the Holocene climate variability in the North Atlantic region. The kind of data presented here may be used to detect mechanisms and forcing factors behind the reconstructed climate variations and to study leads and lags in the Earth's climate system.

Acknowledgments. We express our gratitude to the referees Svante Björck, Christian Hjort and John A. Matthews, whose comments and suggestions helped significantly in improving and clarifying the manuscript.

REFERENCES

Alley, R.B., P.A. Mayewski, T. Sowers, M. Stuiver, K.C. Taylor, and P.U. Clark, Holocene climate instability: A prominent, widespread event 8200 yr ago. *Geology*, 25, 483-486, 1997.

Alm, T. and Øvre Æråsvatnet – palynostratigraphy of a 22,000 to 10,000 BP lacustrine record on Andøya, northern Norway. *Boreas*, 22, 177-188, 1993.

Alverson, K.D., R.S. Bradley, and T.F. Pedersen, *Paleoclimate, Global Change and the Future*. Global Change – The IGBP Series. 221 pp, 2003.

Ammann, B., H.J.B. Birks, S.J. Brooks, U. Eicher, U. von Grafenstein, W. Hofmann, G. Lemdahl, J. Schwander, K. Tobolski, and L. Wick, Quantification of biotic responses to rapid climatic changes around the Younger Dryas: a synthesis. *Palaeogeography, Palaeoclimatology, Palaeo-ecology*, 159, 313-347, 2000.

Andersen, C., *Surface ocean climate development and heat flux variability in the Nordic Seas and the subpolar North Atlantic during the Holocene*. Dr. Scientiarum Thesis, Department of Earth Science, University of Bergen, Norway and Norwegian Polar Institute, Tromsø, Norway, 2003.

Andersen, C., N. Koç, A. Jennings, and T. Andrews, Nonuniform response of the major surface currents of the Nordic Seas to insolation forcing: implications for the Holocene climate variability. *Paleoceanography*, 19, PA2003, doi: 10.1029/2002PA000873, 2004.

Andersson, C., B. Risebrobakken, E. Jansen, and S.O. Dahl, Late Holocene surface ocean conditions of the Norwegian Sea (Vøring Plateau). *Paleoceanography*, 18, 1044, doi: 10.1029/2001PA000654, 2003.

Andrews, J.T. and S. Cartee-Schoolfield, Late Quaternary lithofacies, provenance, and depositional environments (similar to 12-30 cal ka), north and south of the Denmark Strait. *Marine Geology*, 199, 65-82, 2003.

Andrews, J.T., A. Geirsdottir, J. Hardardottir, S. Principato, K. Grønvold, G.B. Kristjansdottir, G. Helgadottir, J. Drexler, and A. Sveinbjornsdottir, Distribution, sediment magnetism and geochemistry of the Saksunarvatn (10,180 +/– 60 cal. yr BP) tephra in marine, lake, and terrestrial sediments, northwest Iceland. *Journal of Quaternary Science*, 17, 731-745, 2002a.

Andrews, J.T. and J. Giraudeau, Multiproxy records showing significant Holocene environmental variability: the inner N. Iceland shelf (Húnaflói). *Quaternary Science Reviews*, 22, 175-193, 2003.

Andrews, J.T., J. Hardardottir, A. Geirsdottir, and G. Helgadottir, Late Quaternary ice extent and glacial history from the Djupall trough, off Vestfirdir peninsula, north-west Iceland: a stacked 36 cal. Ky environmental record. *Polar Research*, 21, 211-226, 2002b.

Andrews, J.T., J. Hardardottir, J.S. Stoner, M.E. Mann, G.B. Kristjansdottir, and N. Koc, Decadal to millennial-scale periodicities in North Iceland shelf sediments over the last 12000 cal yr: long-term North Atlantic oceanographic variability and solar forcing. *Earth and Planetary Science Letters*, 210, 453-465, 2003a.

Andrews, J.T., R. Hardardo, J. Hardardottir, G.B. Kristjansdottir, K. Grønvold, and J.S. Stoner, A high-resolution Holocene sediment record from Hunafloaall, N Iceland margin: century- to millennial-scale variability since the Vedde tephra. *The Holocene*, 13, 625-638, 2003b.

Andrews, J.T. and G. Helgadottir, Late Quaternary ice cap extent and deglaciation, Hunafloaall, Northwest Iceland: Evidence from marine cores. *Arctic Antarctic and Alpine Research*, 35, 218-232, 2003.

Andrews, J.T., R. Kihl, G.B. Kristjansdottir, L.M. Smith, G. Helgadottir, A. Geirsdottir, and A.E. Jennings, Holocene sediment properties of the east Greenland and Iceland continental shelves bordering Denmark Strait (64-68 degrees N), North Atlantic. Sedimentology 49, 5-24. 68 degrees N), North Atlantic. *Sedimentology*, 49, 5-24, 2002c.

Bakke, J., S.O. Dahl, and A. Nesje, Lateglacial and early Holocene palaeoclimatic reconstruction based on glacier fluctuations and equilibrium-line altitudes at northern Folgefonna, Hardanger, western Norway. *Journal of Quaternary Science*, 20, 179-198, 2005a.

Bakke, J., S.O. Dahl, Ø. Paasche, R. Løvlie, and A. Nesje, Glacier fluctuations, equilibrium-line altitudes and palaeoclimate in Lyngen, northern Norway during the Lateglacial and Holocene. *The Holocene*, 15, 387-409, 2005b.

Ballantyne, C.K., The Loch Lomond readvance on the Island of Skye, Scotland: glacier reconstruction and paleoclimatic implications. *Journal of Quaternary Science*, 4, 95-108, 1989.

Barber, D.C., A. Dyke, C. Hillaire-Marcel, A.E. Jennings, J.T. Andrews, M.W. Kerwin, G. Bilodeau, R. McNeely, J. Southon, M.D. Morehead, and J.-M. Gagnon, Forcing of the cold event of 8,200 years ago by catastrophic drainage of Laurentide lakes. *Nature*, 400, 344-348, 1999.

Barber, K.E., R.W. Battarbee, S. Brooks, G. Eglinton, E.Y. Haworth, F. Oldfield, A.C. Stevenson, R. Thompson, P.G. Appleby, W.N. Austin, N.G. Cameron, K.J. Ficken, P. Golding, D.D. Harkness, J. Holmes, R. Hutchinson, J.P. Lishman, D. Maddy, L.C.V. Pinder, N. Rose, and R. Stoneman, Proxy records of climate change in the UK over the last two millennia: documented change and sedimentary records from lakes and bogs. *Journal of the Geological Society of London*, 156, 369-380, 1999.

Bard, E., G. Raisbeck, F. Yiou, and J. Jouzel, Solar irradiance during the last 120years based on cosmogenic nuclides. *Tellus*, 52B, 985-992, 2000.

Barnekow, L., Holocene regional and local vegetation history and lake-level changes in the Torneträsk area, northern Sweden. *Journal of Paleolimnology*, 23, 399-420, 2000.

Barnekow, L. and P. Sandgren, Palaeoclimate and tree-line changes during the Holocene based on pollen and plant macrofossil records from six lakes at different altitudes in northern Sweden. *Review of Palaeobotany and Palynology*, 117, 109-118, 2001.

Barnett, C.T., L. Dumayne-Peaty, and J.A. Matthews, Holocene climatic change and tree-line response in Leirdalen, central Jotunheimen, south central Norway. *Review of Palaeobotany and Palynology*, 117, 119-137, 2001.

Barnola, J.-M., D. Raynaud, Y.S. Korotkevich, and C. Lorius, Vostok ice core provides 160,000 year record of atmospheric CO_2. *Nature*, 329, 408-414, 1987.

Battarbee, R.W., N.G. Cameron, P. Golding, S.J. Brooks, R. Switsur, D. Harkness, P. Appleby, F. Oldfield, R. Thompson, D.T. Monteith, and A. McGovern, Evidence for Holocene climate variability from the sediments of a Scottish remote mountain lake. *Journal of Quaternary Science*, 16, 339-346, 2001.

Battarbee, R.W., J.-A. Grytnes, R. Thompson, P.G. Appleby, J. Catalan, A. Korhola, H.J.B. Birks, E. Heegaard, and A. Lami, Comparing palaeolimnological and instrumental evidence of climate change for remote mountain lakes over the last 200 years. *Journal of Paleolimnology*, 28, 161-179, 2002.

Beer, J., W. Mende, R. Stellmacher, and O.R. White, Intercomparisons of proxies for past solar variability. In: P.D. Jones, R.S. Bradley, and J. Jouzel, (eds.): *Climatic Variations and Forcing mechanisms of the last 2000 years*. Springer Verlag, Berlin, 501-517, 1996.

Berger, A. and M.-F. Loutre, Modeling the climate response to astronomical and CO_2 forcings. Geophys. *Extern. Climat et Envir.*, C.R. Acad. Sci. Paris, 323 (Iia), 1, 1996.

Berglund, B.E., L. Barnekow, D. Hammarlund, P. Sandgren, and I. Snowball, Holocene forest dynamics and climate changes in the Abisko area, northern Sweden – the Sonesson model of vegetation history reconsidered and confirmed. *Ecological Bulletins*, 45, 15-30, 1996.

Bianchi, G.G. and I.N. McCave, Holocene periodicity in North Atlantic climate and deep ocean flow south of Iceland. *Nature*, 397, 515-517, 1999.

Bigler, C. and R.I. Hall, Diatoms as quantitative indicators of July temperature: a validation attempt at century-scale with meteorological data from northern Sweden. *Palaeography, Palaeoclimatology, Palaecology*, 189, 147-160, 2003.

Bigler, C., I. Larocque, S.M. Peglar, H.J.B. Birks, and R.I. Hall, Quantitative multiproxy assessment of long-term patterns of Holocene environmental change from a small lake near Abisko, northern Sweden. *The Holocene*, 12, 481-496, 2002.

Birks, C.J.A. and N. Koc, A high-resolution diatom record of late Quaternary sea-surface temperatures and oceanographic conditions from the eastern Norwegian Sea. *Boreas*, 31, 323-344, 2002.

Birks, H.H., Holocene vegetational history and climatic change in west Spitsbergen – plant macrofossils from Skardtjørna, an Arctic lake. *The Holocene*, 1, 209-218, 1991.

Birks, H.H. and B. Ammann, Two terrestrial records of rapid climatic change during the glacial-Holocene transition (14,000-9000 calendar years B.P.) from Europe. *Proceedings of the National Academy of Science*, 97, 1390-1394, 2000.

Birks, H.H., R.W. Battarbee, and H.J.B. Birks, The development of the aquatic ecosystem at Kråkenes Lake, western Norway, during the late-glacial and early-Holocene – a synthesis. *Journal of Paleolimnology*, 23, 91-114, 2000.

Birks, H.J.B., Quantitative palaeoenvironmental reconstructions. In: D. Maddy, J.S. Brew, (Eds.), *Statistical modelling of Quaternary science data*. Quaternary Research Association, Cambridge, 161-254, 1995.

Birks, H.J.B., Numerical tools in quantitative palaeolimnology – progress, potentialities, and problems. *Journal of Paleolimnology*, 20, 301-332, 1998.

Birks, H.J.B., Quantitative palaeoenvironmental reconstructions from Holocene biological data. In: A. Mackay, R.W. Battarbee, H.J.B. Birks, F. Oldfield, (Eds.), *Global Change in the Holocene*. Arnold, London, 107-122, 2003.

Björck, S., R. Muscheler, B. Kromer, C.S. Andresen, J. Heinemeier, S.J. Johnsen, D. Conley, N. Koc, M. Spurk, and S. Veski, High-resolution analyses of an early Holocene climate event may indicate decreased solar forcing as an important climate trigger. *Geology*, 29, 1107-1110, 2001.

Björck, S., M. Rundgren, Ó. Ingólfson, and S. Funder, The Preboreal oscillation around the Nordic Seas: terrestrial and lacustrine responses. *Journal of Quaternary Science*, 12, 455-465, 1997.

Björck, S. and B. Wohlfarth, ^{14}C chronostratigraphic techniques in paleolimnology. In: *Tracking Environmental Change Using Lake Sediments. Volume 1: Basin Analysis, Coring, and Chronological Techniques*, (eds. W.M. Last, and J.P. Smol). Kluwer Academic Publishers, Dordrecht, The Netherlands, 2001.

Bjune, A.E., Holocene vegetation history and tree-line changes on a north-south transect crossing major climate gradients in southern Norway - evidence from pollen and plant macrofossils in lake sediments. *Review of Palaeobotany and Palynology*, 133, 249-275, 2005.

Bjune, A.E., H.J.B. Birks, and H. Seppä, Holocene vegetation and climate history on a continental – oceanic transect in northern Fennoscandia based on pollen and plant macrofossils. *Boreas*, 33, 211-223, 2004.

Black, D.E., L.C. Peterson, J.T. Overpeck, A. Kaplan, M.N. Evans, and M. Kashgarian, Eight centuries of North Atlantic Ocean atmosphere variability. *Science*, 286, 1709-1713, 1999.

Blunier, T., J. Chappellaz, J. Schwander, J. Stauffer, and D. Raynaud, Variations in atmospheric methane concentrations during the Holocene Epoch. *Nature*, 374, 46, 1995.

Bond, G., B. Kromer, J. Beer, R. Muscheler, M.N. Evans, W. Showers, S. Hoffman, R. Lotti-Bond, I. Hajdas, and G. Bonani, Persistent solar influence on North Atlantic climate during the Holocene. *Science*, 294, 2130-2136, 2001.

Bond, G., W. Showers, M. Chesby, R. Lotti, P. Almasi, P. deMenocal, P. Priore, H. Cullen, I. Hajadas, and G. Bonani, A pervasive millennial-scale cycle in North Atlantic Holocene and glacial climates. *Science*, 278, 1257-1266, 1997.

Bradley, R.S., Climate forcing during the Holocene. In: A. Mackay, R.W. Battarbee, H.J.B. Birks, F. Oldfield, (Eds.), *Global Change in the Holocene*. Arnold, London, 10-19, 2003.

Briffa, K.R., P.D. Jones, T.S. Bartholin, D. Eckstein, F.H. Schweingruber, W. Karlèn, P. Zetterberg, and M. Eronen, Fennoscandian summers from AD 500: temperature changes on short and long timescales. *Climate Dynamics*, 7, 111-119, 1992.

Briffa, K.R., P.D. Jones, F.H. Schweingruber, and T.J. Osborn, Influence of volcanic eruptions on Northern Hemisphere summer temperature over the past 600 years. *Nature*, 393, 450-455, 1998.

Briffa, K.R., T.J. Osborn, F.H. Schweingruber, P.D. Jones, S.G. Shiyatov, and E.A. Vaganov, Tree-ring width and density data around the Northern Hemisphere: Part 1, local and regional climate signals. *The Holocene*, 12, 737-757, 2002.

Brooks, S.J., Chironomid analysis to interpret and quantify Holocene climate change. In: A. Mackay, R.W. Battarbee, H.J.B. Birks, F. Oldfield, (Eds.), *Global Change in the Holocene*. Arnold, London, 328-341, 2003.

Calvo, E., J. Grimalt, and E. Jansen, High resolution U^{K}_{37} sea surface temperature reconstruction in the Norwegian Sea during the Holocene. *Quaternary Science Reviews*, 21, 1385-1394, 2002.

Castaneda, I.S., L.M. Smith, G.L. Kristjansdottir, and J.T. Andrews, Temporal changes in Holocene delta O-18 records from the northwest and central North Iceland Shelf. *Journal of Quaternary Science*, 19, 321-334, 2004.

Chambers, F.M.and D.J. Charman, Holocene environmental change: contributions from the peatland archive. *The Holocene*, 14, 1-6, 2004.

Clark, P.U., N.G. Pisias, T.F. Stocker, and A.J. Weavers, The role of the thermohaline circulation in abrupt climate change. *Nature*, 415, 863-868, 2002.

Clarke, G., D. Leverington, J. Teller, and A. Dyke, Superlakes, megafloods, and abrupt climate change. *Science*, 301, 922-923, 2003.

Cleveland, W.S., Visualizing data. Hobart Press, Summit, 1993.

Crowley, T.J. and K.-Y. Kim, Modeling the temperature response to forced climate change over the last six centuries. *Geophysical Research Letters*, 26, 1901-1904, 1999.

Dahl, S.O. and A. Nesje, Paleoclimatic implications based on equilibrium-line altitude depressions of reconstructed Younger Dryas and Holocene cirque glaciers in inner Nordfjord, western Norway. *Palaeogeography, Palaeoclimatology, Palaeoecology*, 94, 87-97, 1992.

Dahl, S.O. and A. Nesje, Holocene glacier fluctuations at Hardangerjøkulen, central-southern Norway: a high resolution composite chronology from lacustrine and terrestrial deposits. *The Holocene*, 4, 269-277, 1994.

Dahl, S.O. and A. Nesje, A new approach to calculating Holocene winter precipitation by combining glacier equilibrium-line altitudes and pine-tree limits: a case study from Hardangerjøkulen, central southern Norway. *The Holocene*, 6, 381-398, 1996.

Dahl, S.O., A. Nesje, Ø. Lie, K. Fjordheim, and J.A. Matthews, Timing, equilibrium-line altitudes and climatic implications of two early-Holocene glacier readvances during the Erdalen Event at Jostedalsbreen, western Norway. *The Holocene*, 12, 17-25, 2002.

Dahl-Jensen, D., K. Mosegaard, N. Gundestrup, and others, Past temperatures directly from the Greenland ice sheet. *Science*, 282, 268-271, 1998.

Dansgaard, W., S.J. Johnsen, H.B. Clausen, D. Dahl-Jensen, N.S. Gundestrup, C.U. Hammer, C.S. Hvidberg, J.P. Steffensen, A.E. Sveinbjörnsdottir, J. Jouzel, and G. Bond, Evidence for general instability of past climate from a 250-kyr ice-core record. *Nature*, 364, 218-220, 1993.

Davis, B.A.S., S. Brewer, A.C. Stevenson, and J. Guiot, Data contributors, The temperature of Europe during the Holocene reconstructed from pollen data. *Quaternary Science Reviews*, 22, 1701-1716, 2003.

Engelen, A.F.V., J. van Buisman, and F. Ijnsen, A millennium of weather, winds and water in the low countries. In: P.D. Jones, T.D. Davies, A.E.J. Ogilvie, and K.R. Briffa, (Eds.): *History and Climate: Memories of the Future?* Kluwer/Plenum, 101-124, 2001.

Eronen, M. and P. Zetterberg, Climatic changes in Northern Europe since late glacial times, with special reference to dendroclimatological studies in northern Finnish Lapland. *Geophysica*, 2, 35-60, 1996.

Eronen, M., B. Frenzel, and K.-D. Vorren, (Eds.), Oscillations of the Alpine and Polar Tree Limits in the Holocene. *Paleoclimate Research*, 9, 1-234, 1993.

Fisher, D.A. and R.M. Koerner, Holocene ice-core climate history: a multi-variable approach. In: K.D. Alverson, R.S. Bradley, T.F. Pedersen, (Eds.), *Paleoclimate, Global Change and the Future*. Global Change – The IGBP Series, 281-293, 2003.

Fisher, D.A., R.M. Koerner, and N. Reeh, Holocene climatic records from the Agassiz Ice Cap, Ellesmere Island, N.W.T., Canada. *The Holocene*, 5, 19-24, 1995.

Foley, J.A., J.E. Kutzbach, M.T. Coe, and S. Levis, Feedbacks between climate and boreal forests during the Holocene epoch. *Nature*, 371, 52-54, 1994.

Fronval, T., E. Jansen, Eemian and early Weichselian (140-60 ka) paleoceanography and paleoclimate in the Nordic seas with comparisons to Holocene conditions. *Paleoceanography*, 12, 443-462, 1997.

Geirsdottir, A., J.T. Andrews, S. Olafsdottir, G. Helgadottir, and J. Hardardottir, A 36 Ky record of iceberg rafting and sedimentation from north-west Iceland. *Polar Research*, 21, 291-298, 2002.

Giraudeau J., A.E. Jennings, and J.T. Andrews, Timing and mechanisms of surface and intermediate water circulation changes in the Nordic Seas over the last 10,000 cal years: a view from the North Iceland shelf. *Quaternary Science Reviews*, 23, 2127-2139, 2004.

Grafenstein, U., H. von, Erlenkeuser, J. Müller, J. Jouzel, and S. Johnsen, The cold event 8200 years ago documented in oxygen isotope records of precipitation in Europe and Greenland. Climate Dynamics, 14, 73-81, 1998.

Grootes, P.M., M. Stuiver, J.W.C. White, S. Johnsen, and J. Jouzel, Comparison of oxygen isotope records from the GISP2 and GRIP Greenland ice cores. *Nature*, 366, 552-554, 1993.

Grøsfjeld, K., E. Larsen, H.P. Sejrup, A. de Vernal, T. Flatebø, M. Vestbø, H. Haflidason, and I. Aarseth, Dinoflagellate cysts reflecting surface-water conditions in Voldafjorden, western Norway during the last 11,300 years. *Boreas*, 28, 403-415, 1999.

Grove, J.M., *The Little Ice Age*. London: Methuen, 1988.

Grove, J.M., Climatic change in northern Europe over the last two thousand years and its possible influence on human activity. In: G. Wefer, W. Berger, K.-E. Behre, and E. Jansen, (Eds.), *Climate Development and History of the North Atlantic Realm*. Springer-Verlag, Berlin Heidelberg, 313-326, 2002.

Grove, J.M., *Little Ice Ages – Ancient and Modern*. Routledge I and II, 2004.

Grudd, H., K.R. Briffa, W. Karlén, T.S. Bartholin, P.D. Jones, and B. Kromer, A 7400-year tree-ring chronology in northern Swedish Lapland: natural climatic variability expressed on annual to millennial timescales. *The Holocene*, 12, 657-665, 2002.

Haakensen, N., Akkumulasjon på breene i Sør-Norge vinteren 1988-89. *Været* 13, 91-94, 1989.

Haflidason, H., H.P. Sejrup, D. Klitgaard-Kristensen, and S. Johnsen, Coupled response of late glacial climatic shifts of northwest Europe reflected in Greenland ice cores: Evidence from the northern North Sea. *Geology*, 23, 1059-1062, 1995.

Hald, M., T. Dahlgren, T.-E. Olsen, and E. Lebesbye, Late Holocene palaeooceanography in Van Mijenfjorden, Svalbard. *Polar Research*, 20, 23-35, 2001.

Hald, M., H. Ebbesen, M. Forwick, F. Godtliebsen, L. Khomenko, S. Korsun, L. Ringstad Olsen, and T.O. Vorren, Holocene paleoceanography and glacial history of the West Spitsbergen area, Euro-Arctic margin. *Quaternary Science Reviews*, 23, 2075-2088, 2004.

Hammarlund, D., L. Barnekow, H.J.B. Birks, B. Buchardt, and T.W.D. Edwards, Holocene changes in the oxygen-isotope stratigraphy of lacustrine carbonates from northern Sweden. *The Holocene*, 12, 339-351, 2002.

Hammarlund, D., S. Björck, B. Buchardt, C. Israelson, and C.T. Thomsen, Rapid hydrological changes during the Holocene revealed by stable isotope records from lacustrine carbonates from lake Igelsjön, southern Sweden. *Quaternary Science Reviews*, 22, 353-370, 2003.

Hammarlund, D., G. Velle, B.B. Wolfe, T.W.D. Edwards, L. Barnekow, J. Bergman, S. Holmgren, S. Lamme, I. Snowball, B. Wohlfarth, and G. Possnert, Palaeolimnological and sedimentary responses to Holocene forest retreat in the Scandes Mountains, west central Sweden. *The Holocene*, 14, 862-876, 2004.

Hannon, G.E., R.H.W. Bradshaw, and S. Wastegård, Rapid vegetation change during the early Holocene in the Faroe Islands detected in terrestrial and aquatic ecosystems. *Journal of Quaternary Science*, 18, 615-619, 2003.

Heegaard, E., H.J.B. Birks, and R.J. Telford, Relationships between calibrated ages and depth in stratigraphical sequences: an estimation procedure by mixed-effect regression. *The Holocene*, in press, 2005.

Heikkilä, M. and H. Seppä, A 11,000 yr palaeotemperature reconstruction from the southern boreal zone in Finland. *Quaternary Science Reviews*, 22, 541-554, 2003.

Heiri, O., A.F. Lotter, S. Hausmann, and F. Kienast, A chironomid-based Holocene summer air temperature reconstruction from the Swiss Alps. *The Holocene*, 13, 477-484, 2003.

Helama, S., M. Lindholm, M. Timonen, J. Merilänen, and M. Eronen, The supra-long Scots pine tree-ring record for Finnish Lapland: Part 2, interannual to centennial variability in summer temperatures for 7500 years. *The Holocene*, 12, 681-687, 2002.

Helama, S., M. Lindholm, M. Timonen, and M. Eronen, Dendrochronologically dated changes in the limit of pine in northernmost Finland during the past 7.5 millennia. *Boreas*, 33, 250-259, 2004.

Hjort, C., J. Mangerud, L. Adrielsson, S. Bondevik, J.Y. Landvik, and O. Salvigsen, Radiocarbon dated common mussels *Mytilus edulis* from eastern Svalbard and the Holocene marine climatic optimum. *Polar Research*, 14, 239-243, 1995.

Holmlund, P., W. Karlén, and H. Grudd, Fifty years of mass balance and glacier front observations at the Tarfala Research Station. *Geografiska Annaler*, 78A, 105-114, 1996.

Huntley, B., Holocene climatic history of northern Europe as recorded by vegetation changes: Possible influences upon human activity. In: G. Wefer, W. Berger, K.-E. Behre, and E. Jansen, (Eds.), *Climate Development and History of the North Atlantic Realm*. Springer-Verlag, Berlin Heidelberg, 223-232, 2002.

Hurrell, J.W., Y. Kushnir, G. Ottesen, and M. Visbeck, The North Atlantic Oscillation – Climatic Significance and Environmental Impact. *Geophysical Monograph*, 134. 279 pp, 2003.

Husum, K. and M. Hald, Early Holocene cooling events in Malangenfjord and the adjoining shelf, north-east Norwegian Sea. *Polar Research*, 21, 267-274, 2002.

Husum, K. and M. Hald, A continuous marine record 8000-1600 cal. yr BP from the Malangenfjord, north Norway: foraminiferal and isotopic evidence. *The Holocene*, 14, 877-887, 2004.

Indermühle, A., T.F. Stocker, F. Joos, H. Fischer, H.J. Smith, M. Wahlen, B. Deck, D. Mastroianni, J. Tschumi, T. Blunier, R. Meyer, and B. Stauffer, Holocene carbon-cycle dynamics based on CO_2 trapped in ice at Taylor Dome, Antarctica. *Nature*, 398, 121-126, 1999.

IPCC, IPCC *Third Assessment Report: Climate Change 2001*. Geneva Switzerland., 184 pp, 2001.

Jansen, E. and N. Koç, Century to decadal scale records of Norwegian Sea surface temperature variations of the past 2 millennia. *PAGES/CLIVAR Newsletter*, 8(1), 13-14, 2000.

Jennings, A.E., N.J. Weiner, G. Helgadottir, and J.T. Andrews, Modern foraminiferal faunas of the southwestern to northern Iceland shelf: Oceanographic and environmental controls. *Journal of Foraminiferal Research*, 34, 180-207, 2004.

Johnsen, S.J., D. Dahl-Jensen, N. Gundestrup, J.P. Steffensen, H.B. Clausen, H. Miller, V. Masson-Delmotte, and A.E. Sveinbjørnsdottir, Oxygen isotope and paleotemperature records from six Greenland ice-core stations: Camp Century, Dye-3, GRIP, GISP2, Renland and NorthGRIP. *Journal of Quaternary Science*, 16, 299-307, 2001.

Jones, P.D., Early European instrumental records. In: T.D. Davies, and K.R. Briffa, (Eds.), *History and Climate: memories of the Future*. Kluwer Academic/Plenum, New York, 1-8, 2001.

Jones, P.D., T. Jonsson, and D. Wheeler, Extension of the North Atlantic Oscillation using early instrumental pressure observations from Gibraltar and South-west Iceland. *International Journal of Climatology*, 17, 1433-1450, 1997.

Kalela-Brundin, M., Climatic information from tree-rings of *Pinus sylvestris* L. and reconstruction of summer temperatures back to AD 1500 in Femundsmarks, eastern Norway, using partial least squares regression (PLS) analysis. *The Holocene*, 9, 59-77, 1999.

Kaplan, M.R., A.P. Wolfe, and G.H. Miller, Holocene environmental variability in southern Greenland inferred from lake sediments. *Quaternary Research*, 58, 149-159, 2002.

Karlén, W., Lacustrine sediments and tree-limit variations as indicators of Holocene climatic fluctuations in Lapland: Northern Sweden. *Geografiska Annaler*, 58A, 1-34, 1976.

Karlén, W., Scandinavian glacier and climatic fluctuations during the Holocene. *Quaternary Science Reviews*, 7, 199-209, 1988.

Karlén, W., A. Bodin, J. Kuylenstierna, and J.-O. Näslund, Climate of northern Sweden during the Holocene. *Journal of Coastal Research Special Issue* 17: Holocene Cycles: Climate, Sea Levels, and Sedimentation, 49-54, 1995.

Karlén, W. and J. Kuylenstierna, On solar forcing of Holocene climate: evidence from Scandinavia. *The Holocene*, 6, 359-365, 1996.

Karlén, W. and J.A. Matthews, Reconstructing Holocene glacier variations from glacial lake sediments: studies from Nordvestlandet and Jostedalsbreen-Jotunheimen, southern Norway. *Geografiska Annaler*, 74, 327-348, 1992.

Kaufman, D.S., T.A. Ager, N.J. Anderson, P.M. Anderson, J.T. Andrews, P.J. Bartlein, L.B. Brubaker, L.L. Coats, L.C. Cwynar, M.L. Duvall, A.S. Dyke, M.E. Edwards, W.R. Eisner, K. Gajewski, A. Geirsdottir, F.S. Hu, A.E. Jennings, M.R. Kaplan, M.W. Kerwin, A.V. Lozhkin, G.M. McDonald, G.H. Miller, C.J. Mock, W.W. Oswald, B.L. Otto-Bliesner, D.F. Porinchu, K. Rühland, J.P. Smol, E.J. Steig, and B.B. Wolfe, Holocene thermal maximum in the western Arctic (0-180°W). *Quaternary Science Reviews*, 23, 2059-2060, 2004.

Keigwin, L.D., The Little Ice Age and Medieval Warm Period in the Sargasso Sea. *Science*, 274, 1505-1508, 1996.

Kirchhefer, A.J. , Reconstruction of summer temperatures from tree-rings of Scots pine (*Pinus sylvestris* L.) in coastal northern Norway. *The Holocene*, 11, 41-52, 2001.

Klitgaard-Kristensen, D., H.-P. Sejrup, and H. Haflidason, The last 18 kyr fluctuations in Norwegian Sea surface conditions and implications for the magnitude of climatic change: Evidence from the North Sea. *Paleoceanography*, 16, 455-467, 2001.

Klitgaard-Kristensen, D., H.-P. Sejrup, H. Haflidason, S. Johnsen, and M. Spurk, A regional 8200 cal. yr BP cooling event in northwest Europe, induced by final stages of the Laurentide ice-sheet deglaciation? *Journal of Quaternary Science*, 13, 165-169, 1998.

Knudsen, K.L., J. Eirìksson, E. Jansen, H. Jiang, F. Rytter, and E.R. Gudmundsdóttir, Palaeoceanographic changes off North Iceland through the last 1200 years: foraminifera stable isotopes, diatoms and ice rafted debris. *Quaternary Science Reviews*, 23, 2231-246, 2004.

Koç, N. and E. Jansen, Holocene climate evolution of the North Atlantic Ocean and the Nordic Seas - a synthesis of new results. In G. Wefer, W.H. Berger, K.E. Behre, E. Jansen, (Eds.) *Climate and History in the North Atlantic Realm*. Springer-Verlag Berlin Heidelberg, 165-163, 2002.

Koc, N., E. Jansen, and H. Haflidason, Paleoceanographic reconstruction of surface ocean conditions in the Greenland, Iceland and Norwegian Seas through the last 14 ka based on diatoms. *Quaternary Science Reviews*, 12, 115-140, 1993.

Koerner, R.M. and D.A. Fisher, A record of Holocene summer climate from Canadian High Arctic ice core. *Nature*, 343, 630-631, 1990.

Korhola, A., K. Vasko, H.T.T. Toivonen, and H. Olander, Holocene temperature changes in northern Fennoscandia reconstructed from chironomids using Bayesian modelling. *Quaternary Science Reviews*, 21, 1841-1860, 2002.

Kullman, L., Recent tree-limit dynamics of Scots pine (*Pinus sylvestris* L.) in the southern Swedish Scandes. *Wahlenbergia*, 8, 1-67, 1981.

Kullman, L., Holocene tree-limit and climate history from the Scandes Mountains, Sweden. Ecology, 768, 2490-2502, 1995.

Körner, C., A re-assessment of high elevation treeline positions and their explanations. *Oecologia*, 115, 445-459, 1998.

Larocque, I. and R.I. Hall, Holocene temperature estimates and chironomid community composition in the Abisko Valley, northern Sweden. *Quaternary Science Reviews*, 23, 2453-2465, 2004.

Laumann, T. and N. Reeh, Sensitivity to climate change of the mass balance of glaciers in southern Norway. *Journal of Glaciology*, 39, 656-665, 1993.

Lauritzen, S.-E., Calibration of speleothem stable isotopes against historical records: a Holocene temperature curve for north Norway. In: *Climate Change: the Karst Record*. Karst Waters Institute Special Publication 2, Charles Town, West Virginia, 78-80, 1996.

Lauritzen, S.-E. and J. Lundberg, Calibration of the speleothem delta function: an absolute temperature record for the Holocene in northern Norway. *The Holocene*, 9, 659-669, 1999.

Lean, J., J. Beer, and R.S. Bradley, Reconstruction of solar irradiance since 1610: implications for climate change. *Geophysical Research Letters*, 22, 3195-3198, 1995.

Lean, J.L., A. Skumanich, and O.R. White, Estimating the Sun's radiative output during the Maunder Minimum. *Geophysical Research Letters*, 19, 1591-1594, 1992.

Leeman, A. and F. Niessen, Holocene glacial activity and climatic variations in the Swiss Alps: reconstructing a continuous record from proglacial lake sediments. *The Holocene*, 4, 259-268, 1994.

Lehman, S.J., G.A. Jones, L.D. Keigwin, E.S. Andersen, G. Butenko, and S.-R. Østmo, Initiation of Fennoscandian ice-sheet retreat during the last deglaciation. *Nature*, 349, 513-516, 1991.

Lie, Ø., S.O. Dahl, A. Nesje, J.A. Matthews, and S. Sandvold, Holocene fluctuations of a polythermal glacier in high-alpine eastern Jotunheimen, central-southern Norway. *Quaternary Science Reviews*, 23, 1925-1945, 2004.

Lotter, A.F. and H.J.B. Birks, Holocene sediments of Sägistalsee, a small lake at the present-day tree-line in the Swiss Alps. *Journal of Paleolimnology*, 30, 253-260, 2003.

Lotter, A.F., H.J.B. Birks, W. Hofmann, and A. Marchetto, Modern diatom, cladocera, chronomid, and chrysophyte cyst assemblages as quantitative indicators for the reconstruction of past environmental conditions in the Alps. *I. Climate. Journal of Paleolimnology*, 18, 395-420, 1997.

Luterbacher, J., D. Dietrich, E. Xoplaki, M. Grosjean, and H. Wanner, European seasonal and annual temperature variability, trends, and extremes since 1500. *Science*, 303, 1499-1503, 2004.

Luterbacher, J., D. Schmutz, D. Gyalistras, P.D. Jones, T.D. Davies, H. Wanner, and E. Xoplaki, Reconstruction of highly resolved NAO and EU indices back to AD 1500. *Geophysical Research Abstracts*, 2: OA 34, 2001.

Manley, G., Central England temperatures: monthly means 1659 to 1973. *Quaterly Journal of the Royal Meteorological Society*, 100, 389-405, 1974.

Mann, M.E., R.S. Bradley, and M.K. Hughes, Global-scale temperature patterns and climate forcing over the past six centuries. *Nature*, 392, 779-787, 1998.

Marchal, O., T.F. Stocker, F. Joos, A. Indermühle, T. Blunier, and J. Tschumi, Modelling the concentration of atmospheric CO_2 during the Younger Dryas climate event. *Climate Dynamics*, 15, 341-354, 1999.

Matthews, J.A., 'Little Ice Age' glacier variations in Jotunheimen, southern Norway: a study in regionally controlled lichenometric dating of recessional moraines with implications for climate and lichen growth rates. *The Holocene*, 15, 1-19, 2005.

Matthews, J.A. and W. Karlén, Asynchronous Neoglaciation and Holocene climatic change reconstructed from Norwegian glaciolacustrine sedimentary sequences. *Geology*, 20, 991-994, 1992.

Matthews, J.A., S.O. Dahl, A. Nesje, M.S. Berrisford, and C. Andersson, Holocene glacier variations in central Jotunheimen, southern Norway based on distal glaciolacustrine sediment cores. *Quaternary Science Reviews*, 19, 1625-1647, 2000.

Meehl, G.A., W.M. Washington, C.M. Ammann, J.M. Arblaster, T.M.L. Wigley, and C. Tebaldi, Combinations of natural and anthropogenic forcings in the twentieth-century climate. *Journal of Climate*, 17, 3721-3727, 2004.

Mikalsen, G., H.P. Serjup, and I. Aarseth, Late-Holocene changes in ocean circulation and climate: foraminiferal and isotopic evidence from Sulafjord, western Norway. *The Holocene*, 11, 437-466, 2001.

Moen, A., *National Atlas of Norway: Vegetation*. Norwegian Mapping Authority, Hønefoss., 200 pp, 1999.

Moros, M., K. Emeis, B. Risebrobakken, I. Snowball, A. Kuijpers, J. McManus, and E. Jansen, Sea surface temperatures and ice rafting in the Holocene North Atlantic: climate influences on northern Europe and Greenland. *Quaternary Science Reviews*, 23, 2113-2126, 2004.

Neftel, A., E. Moor, H. Oeschger, and B. Stauffer, Evidence from polar ice cores for the increase of atmospheric CO_2 in the past 2 centuries. *Nature*, 315, 45-47, 1985.

Neftel, A., H. Oeschger, J. Schwander, B. Stauffer, and R. Zumbrunn, Ice core sample measurements give atmospheric CO_2 content during the past 40,000 years. *Nature*, 295, 220-223, 1982.

Neftel, A., H. Oeschger, T. Staffelbach, and B. Stauffer, CO_2 record in the Byrd ice core 50,000-5000 years BP. *Nature*, 331, 609-611, 1988.

Nesje, A. and S.O. Dahl, *Glaciers and Environmental Change. Key issues in environmental change*. Arnold, London and New York. 203 pp, 2000.

Nesje, A. and S.O. Dahl, The Greenland 8200 cal. yr BP event detected in loss-on-ignition profiles in Norwegian lacustrine sediment sequences. *Journal of Quaternary Science*, 16, 155-166, 2001.

Nesje, A. and S.O. Dahl, 'The Little Ice Age' – only temperature? *The Holocene*, 13, 139-145, 2003.

Nesje, A., S.O. Dahl, C. Andersson, and J.A. Matthews, The lacustrine sedimentary sequence in Sygneskardvatnet, western Norway: a continuous, high-resolution record of the Jostedalsbreen ice cap during the Holocene. *Quaternary Science Reviews*, 19, 1047-1065, 2000a.

Nesje, A., S.O. Dahl, and J. Bakke, Were abrupt Lateglacial and early-Holocene climatic changes in northwest Europe linked to freshwater outbursts to the North Atlantic and Arctic Oceans? *The Holocene*, 14, 299-310, 2004.

Nesje, A., S.O. Dahl, and Ø. Lie, Holocene millennial-scale summer temperature variability inferred from sediment parameters in a non-glacial mountain lake: Danntjørn, Jotunheimen, central southern Norway. *Quaternary Science Reviews*, 23, 2183-2205, 2004.

Nesje, A., S.O. Dahl, and R. Løvlie, Late Holocene glacier and avalanche activity in the Ålfoten area, western Norway: evidence from a lacustrine sedimentary record. *Norsk Geologisk Tidsskrift*, 75, 120-126, 1995a.

Nesje, A., S.O. Dahl, R. Løvlie, and J. Sulebak, Holocene glacier activity at the southern side of Hardangerjøkulen, central southern Norway; evidence from lacustrine sediments. *The Holocene*, 4, 377-382, 1994.

Nesje, A., T. Johannessen, and H.J.B. Birks, Briksdalsbreen, western Norway: climatic effects on the terminal response of a temperate glacier between AD 1901 and 1994. *The Holocene*, 5, 343-347, 1995b.

Nesje, A., M. Kvamme, N. Rye, and R. Løvlie, Holocene glacial and climate history of the Jostedalsbreen region, western Norway; evidence from lake sediments and terrestrial deposits. *Quaternary Science Reviews*, 10, 87-114, 1991.

Nesje, A., Ø. Lie, and S.O. Dahl, Is the North Atlantic Oscillation reflected in Scandinavian glacier mass balance records? *Journal of Quaternary Science*, 15, 587-601, 2000b.

Nesje, A., J.A. Matthews, S.O. Dahl, M.S. Berrisford, and C. Andersson, Holocene glacier fluctuations of Flatebreen and winter precipitation changes in the Jostedalsbreen region, western Norway, based on glaciolacustrine records. *The Holocene*, 11, 267-280, 2001.

Nordli, P.Ø., *Spring and summer temperatures in south eastern Norway 1749-2000*. DNMI Report Klima, 01/2001, 32 pp, 2001.

Nordli, P.Ø., *Spring and summer temperatures in Trøndelag 1701-2003*. Met.no Report no. 05/2004 Climate. 21 pp, 2004.

Nordli, P.Ø., Ø. Lie, A. Nesje, and S.O. Dahl, Spring-summer temperature reconstruction in western Norway 1734-2003: A data-synthesis approach. International *Journal of Climatology*, 23, 1821-1841, 2003.

O'Brian, S.R., P.A. Mayewski, L.D. Meeker, D.A. Meese, M.S. Twickler, and S.I. Whitlow, Complexity of Holocene climate as reconstructed from a Greenland ice core. *Science*, 270, 1962-1964, 1995.

Oppo, D., J.F. McManus, and J.L. Cullen, Deepwater variability in the Holocene epoch. *Nature*, 422, 277-278, 2003.

O'Sullivan, P.E., R. Moyeed, M.C. Cooper, and M.J. Nicholson, Comparison between instrumental, observational and high resolution proxy sedimentary records of Late Holocene climatic change - a discussion of possibilities. *Quaternary International*, 88, 27-44, 2002.

Pearman, G.I., D. Etheridge, F. Desilva, and P.J. Fraser, Evidence of changing concentrations of atmospheric CO_2, N_2O, and CH_4 from air bubbles in Antarctic ice. *Nature*, 320, 248-250, 1986.

Pfister, C., R. Brázdil, and R. Glaser, (Eds.), *Climatic Variability in Sixteenth Century Europe and its Social Dimension*. Dordrect, Kluwer, 1999.

Reichert, B.K., L. Bengtsson, and J. Oerlemans, Mid latitude forcing mechanisms for glacier mass balance investigated using general circulation models. *Journal of Climate*, 14, 3767-3784, 2001.

Rind, D., The Sun's role in climate variations. *Science*, 296, 673-677, 2002.

Risebrobakken, B., E. Jansen, C. Andersson, E. Mjelde, and K. Hevrøy, A high-resolution study of Holocene paleoclimatic and paleocenographic changes in the Nordic Seas. *Paleoceanography* 18, 1017, doi:10.1029/2002PA000764, 2003.

Robock, A., Volcanic eruptions and climate. *Reviews of Geophysics* 38, 191-219, 2000.

Rosén, P., U. Segerström, L. Eriksson, I. Renberg, and H.J.B. Birks, Holocene climatic change reconstructed from diatoms, chironomids, pollen and near-infrared spectroscopy at an alpine lake (Sjuodjijaure) in northern Sweden. *The Holocene*, 11, 551-562, 2001.

Rubensdotter, L., and G. Rosquist, The effect of geomorphological setting on Holocene lake sediment variability, northern Swedish Lapland. *Journal of Quaternary Science*, 18, 757-767, 2003.

Ruddiman, W.F., The anthropogenic greenhouse era began thousands of years ago. *Climatic Change*, 61, 261-293, 2003.

Salvigsen, O., radiocarbon-dated *Mytilus edulis* and *Modiolus modiolus* from northern Svalbard: climatic implications. *Norsk Geografisk Tidsskrift*, 56, 56-61, 2002.

Salvigsen, O., S.L. Forman, and G.H. Miller, Thermophilous molluscs on Svalbard during the Holocen and their paleoclimatic implications. *Polar Research*, 11, 1-10, 1992.

Sandgren, P. and J. Risberg, Magnetic mineralogy of the lakes in lake Ådran, eastern Sweden, and an interpretation of early Holocene water level changes. *Boreas*, 19, 57-68, 1990.

Sarnthein, M., S. van Kreveld, H. Erlenkeuser, P.M. Grootes, M. Kucera, U. Pflaumann, and M. Schulz, Centennial-to-millennial-scale periodicities of Holocene climate and sediment injections off the western Barents shelf, 75 °N. *Boreas*, 32, 447-461, 2003.

Schweingruber, F.H. and K.R. Briffa, Tree-ring density networks for climate reconstruction. In P.D. Jones, R.S. Bradley, and J. Jouzel, (Eds), *Climatic variations and forcing mechanisms of the last 2000 years*. NATO ASI Series 141, Berlin: Springer, 43-66, 1996.

Seierstad, J., A. Nesje, S.O. Dahl, and J. Riis Simonsen, Holocene glacier fluctuations of Grovabreen and Holocene snow-avalanche activity reconstructed from lake sediments in Grøningstølsvatnet, western Norway. *The Holocene*, 12, 211-222, 2002.

Seppä, H. and H.J.B. Birks, July mean temperature and annual precipitation trends during the Holocene in the Fennoscandian tree-line area: pollen-based climate reconstructions. *The Holocene*, 11, 527-539, 2001.

Seppä, H. and H.J.B. Birks, Holocene climate reconstructions from the Fennoscandian tree-line area based on pollen data from Toskaljarvi. *Quaternary Research*, 57, 191-199, 2002.

Seppä, H., H.H. Birks, and H.J.B. Birks, Rapid climatic changes during the Greenland stadial 1 (Younger Dryas) to early Holocene transition on the Norwegian Barents Sea coast. *Boreas*, 31, 215-225, 2002a.

Seppä, H., M. Nyman, A. Korhola, and J. Weckström, Changes of tree lines and alpine vegetation in relation to post-glacial climate dynamics in northern Fennoscandia based on pollen and chironomid records. *Journal of Quaternary Science*, 17, 287-301, 2002b.

Shakesby, R.A., J.A. Matthews, and S. Winkler, Glacier variations in Breheimen, southern Norway: relative-age dating of Holocene moraine complexes at six high-altitude glaciers. *The Holocene*, 14, 899-910, 2004.

Shemesh, A., G. Rosquist, M. Rietti-Shati, L. Rubensdotter, C. Bigler, R. Yam, and W. Karlén, Holocene climate change in Swedish Lapland inferred from an oxygen isotope record of lacustrine biogenic silica. *The Holocene*, 11, 447-454, 2001.

Sissons, J.B., Palaeoclimatic inferences from former glaciers in Scotland and the Lake District. *Nature*, 278, 518-521, 1979.

Six, D., L. Reynaud, and A. Letréguilly, Bilans de masse des glaciers alpins et scandinaves, leurs relations avec l'oscillation du climat de l'Atlantique nord. *Earth and Planetary Sciences*, 333, 693-698, 2001.

Snowball, I., Mineral magnetic properties of Holocene lake sediments and soils from the Kårsa valley, Lappland, Sweden, and their relevance to palaeomagnetic reconstruction. *Terra Nova*, 5, 258-270, 1993.

Snowball, I. and P. Sandgren, Lake sediment studies of Holocene glacial activity in the Kårsa valley, northern Sweden: contrast in interpretation. *The Holocene*, 6, 367-372, 1996.

Snowball, I. and R. Thompson, A mineral magnetic study of Holocene sedimentation in Lough Catherine, Northern Ireland. *Boreas*, 19, 127-146, 1990.

Snowball, I., P. Sandgren, and G. Petterson, The mineral magnetic properties of an annually laminated Holocene lake sediment sequence in northern Sweden. *The Holocene*, 9, 353-362, 1999.

Snyder, J.A., A. Werner, and G.H. Miller, Holocene cirque glacier activity in western Spitsbergen, Svalbard: sediment redords from proglacial Linnévatnet. *The Holocene*, 10, 555-563, 2000.

Sohlenius, G., Mineral magnetic properties of late Weichselian-Holocene sediments from the northwestern Baltic Proper. *Boreas*, 25, 79-88, 1996.

Solignac, S., A. de Vernal, and C. Hillaire-Marcel, Holocene sea-surface conditions in the North Atlantic – contrasted trends and regimes in the western and eastern sectors (Labrador Sea vs. Iceland Basin). *Quaternary Science Reviews*, 23, 319-334, 2004.

Staffelbach, T., B. Stauffer, A. Sigg, and H. Oeschger, CO_2 measurements from polar ice cores – more data from different sites. *Tellus Series B – Chemical and Physical Meteorology*, 43, 91-96, 1991.

Stockhausen, H. and B. Zolitschka, Environmental changes since 13,000 cal. BP reflected in magnetic and sedimentological properties of sediments from Lake Holzmaar (Germany). *Quaternary Science Reviews*, 18, 913-925, 1999.

Stuiver, M. and P.J. Reimer, Extended ^{14}C data-base and revised CALIB 3.0 ^{14}C age calibration program. *Radiocarbon*, 35, 215-230, 1993.

Stuiver, M., T.F. Braziunas, P.M. Grootes, and G.A. Zielinski, Is there evidence for solar forcing of climate in the GISP2 oxygen isotope record? *Quaternary Research*, 48, 259-266, 1997.

Stuiver, M., P.M. Grootes, and T.F. Braziunas, The GISP2 $\partial^{18}O$ climate record of the past 16,500 years and the role of the sun, ocean, and volcanoes. *Quaternary Research*, 44, 341-354, 1995.

Sutherland, D.G., Modern glacier characteristics as a basis for inferring former climates with particular reference to the Loch Lomond Stadial. *Quaternary Science Reviews*, 3, 291-309, 1984.

Svendsen, J.-I. and J. Mangerud, Holocene glacial and climatic variations on Spitsbergen, Svalbard. *The Holocene*, 7, 45-57, 1997.

Telford, R.J., E. Heegaard, and H.J.B. Birks, All age-depth models are wrong: but how badly? *Quaternary Science Reviews*, 23, 1-5, 2004.

ter Braak, C.J.F. and S. Juggins, Weighted averaging partial least squares regression (WA-PLS): an improved method for reconstructing environmental variables from species assemblages. *Hydrobiologia*, 269/270, 485-502, 1993.

Thompson, R., R.W. Batterbee, P.E. O'Sullivan, and F. Oldfield, Magnetic susceptibility of lake sediments. *Limnology and Oceanography*, 20, 687-698, 1975.

Thomson, R. and F. Oldfield, *Environmental magnetism*. London: George Allen and Unwin, 1986.

Thun, T., *Dendrochronological constructions of Norwegian conifer chronologies providing dating of historical material*. Dr. philos. Thesis, Department of Biology, Faculty of Natural Sciences and

Technology, Norwegian University of Science and Technology (NTNU). 336 pp, 2002.

Velle, G., *A palaeoecological study of chironomids (Insecta: Diptera) with special reference to climate*. Unpublished Cand. scient. thesis, Museum of Zoology, University of Bergen, 1998.

Velle, G., S.J. Brooks, H.J.B. Birks, and E. Willassen, Chronomids as a tool for inferring Holocene climate: an assessment based on six sites in southern Scandinavia. *Quaternary Science Reviews*, (in press), 2005a.

Velle, G., J. Larsen, W. Eide, S.M. Peglar, and H.J.B. Birks, Holocene environmental history and climate of Råtåsjøen, a low alpine lake in south-central Norway. *Journal of Paleolimnology*, 33, 129-153, 2005b.

Vorren, K.-D., C. Jensen, and T. Alm, Klimautviklingen i Troms og Vesterålen de siste 26,000 år. *Ottar*, 227, 29-35, 1999.

Weliky, K., E. Suess, C.A. Ungerer, P.J. Müller, and K. Fischer, Problems with accurate carbon measurements in marine sediments and particulate organic matter in sea water: a new approach. *Limnology and Oceanography*, 28, 1252-1259, 1983.

Willemse, N.W. and T.E. Törnquist, Holocene century-scale temperature variability from West Greenland lake records. *Geology*, 27, 580-584, 1999.

Winkler, S., J.A. Matthews, R.A. Shakesby, and P.Q. Dresser, Glacier variations in Breheimen, southern Norway: dating Little Ice Age moraine sequences at seven low-altitude glaciers. *Journal of Quaternary Science*, 18, 395-413, 2003.

Zielinski, G.A., P.A. Mayewski, and L.D. Meeker, *et al.*, Record of explosive volcanism since 7000 B.C. from the GISP2 Greenland ice core and implications for the volcano-climate system. *Science*, 264, 948-952, 1994.

Zielinski, G.A. and G.R. Mershon, Paleoenvironmental implications of the insolulable microparticle record in the GISP2 (Greenland) ice core during the rapidly changing climate of the Pleistocene-Holocene transition. *Geological Society of America Bulletin*, 109, 547-559, 1997.

C. Andersson and Ø. Lie, Bjerknes Centre for Climate Research, Norway, Allegaten 55, 5007 Bergen, Norway. (carin.andersson@bjerknes.uib.no; oyvind.lie@bjerknes.uib.no)

J. Bakke and S.O. Dahl, Bjerknes Centre for Climate Research and Department of Geography, University of Bergen, Allegaten 55, 5007 Bergen, Norway (jostein.bakke@geog.uib.no; svein.dahl@geog.uib.no)

H.J.B. Birks and A.E. Bjune, Bjerknes Centre for Climate Research and Department of Biology, University of Bergen, Allegaten 41, 5007 Bergen, Norway. (john.birks@bio.uib.no; anne.bjune@bio.uib.no)

E. Jansen, D. Klitgaard Kristensen, S.-E. Lauritzen, A. Nesje, B. Risebrobakken, and J.-I. Svendsen, Bjerknes Centre for Climate Research and Department of Earth Science, University of Bergen, Allegaten 55, 5007 Bergen, Norway. (eystein.jansen@geo.uib.no; dorthe.kristensen@geo.uib.no; stein.lauritzen@geo.uib.no; atle.nesje@geo.uib.no; bjorg.risebrobakken@geo.uib.no; john.svendsen@geo.uib.no)

Extent and Variability of the Meridional Atlantic Circulation in the Eastern Nordic Seas During Marine Isotope Stage 5 and Its Influence on the Inception of the Last Glacial

Bjørg Risebrobakken[1], Trond Dokken, and Eystein Jansen[1]

Bjerknes Centre for Climate Research, Bergen, Norway

Variable climatic and oceanographic conditions characterized the last interglacial at high northern latitudes, probably related to changes in the strength of the Atlantic Meridional Overturning Circulation (AMOC). The magnitudes of these changes are comparable to the Holocene variability, and were thus significantly subdued compared to glacial climate changes. A thermal optimum occurred during the early part of the interglacial, followed by a period of reduced Atlantic inflow to the northernmost Nordic Seas. Subsequently, a new period with increased strength of the AMOC occurred. Significant amounts of Ice-Rafted Debris (IRD) were deposited in the northernmost Nordic Seas before any major change of the global ice volume. This implies an early onset of local ice sheet growth, probably the result of enhanced inflow of Atlantic water to the northernmost Nordic Seas contemporary with a Northern Hemisphere summer insolation minimum. Contrasting sea–land conditions provided large moisture fluxes towards land, giving rise to rapid, early glacial growth. Throughout the glacial part of Marine Isotope Stage (MIS) 5, millennial-scale cold events occurred along the axis of the warm water transport, from the subtropics all the way to the northernmost Nordic Seas. Correlation of IRD events from sites in the Fram Strait, on the Vøring Plateau, and in the North Atlantic provides evidence that the major Northern Hemisphere ice sheets at times responded coherently to the same forcing. The widespread distribution of these events highlights the importance of the oceanic influence on the regional climate system.

1. INTRODUCTION

The drilling of the GISP2 and GRIP ice cores at Greenland initiated several new studies and a debate on the stability of the last interglacial climate. Records obtained from the ice cores indicate large-scale climate variability throughout Marine Isotope Stage (MIS) 5.5, in some contradiction to a more stable expression of MIS 1 [*Dansgaard et al.*, 1993; *GRIP Members*, 1993; *Taylor et al.*, 1993]. Although large variability was found in both GISP2 and GRIP, the cores were not comparable [e.g. *Grootes et al.*, 1993]. Thus, in one or both of the cores the ice had experienced significant stratigraphic deformation. Nevertheless, the discussion on the character of the last interglacial climate has continued. Terrestrial records from Europe [*Aalbersberg and Litt*, 1998; *Caspers et al.*, 2002] and marine records from the North Atlantic [*Adkins et al.*, 1997; *Cortijo et al.*, 1994; *McManus et al.*, 1994; *Oppo et al.*, 1997] indicate a stable MIS 5.5. In contradiction to this, evidence from

[1]Department of Earth Science, University of Bergen, Bergen, Norway.

The Nordic Seas: An Integrated Perspective
Geophysical Monograph Series 158

10.1029/158GM20

the Nordic Seas and northwestern Europe shows relatively well-pronounced cold spells [*Field et al.*, 1994; *Fronval and Jansen*, 1996; *Kristensen et al.*, 1998; *Seidenkrantz et al.*, 1995; *Sejrup et al.*, 1995; *Thouveny et al.*, 1994]. Several studies also argue that the Norwegian Sea and the Iceland Sea were warmer in MIS 5.5 than in MIS 1, while the Greenland Sea and the Fram Strait were characterized by somewhat colder interstadial conditions [*Bauch et al.*, 1999; *Fronval and Jansen*, 1997; *Hebbeln et al.*, 1998; *Köhler and Spielhagen*, 1990].

The onset of a new glaciation following the last interglacial has been considered as a response to a continuous decrease of Northern Hemisphere summer solar insolation, leading to colder atmospheric and surface ocean conditions in the Arctic region, decreasing evaporation, and expansion of sea-ice and snow fields and associated feedbacks. Consequently, the surface water in the Nordic Seas was freshened and the production of Norwegian Sea Deep Water may have decreased as a response to these changes [e.g. *Imbrie et al.*, 1992; *Khodri et al.*, 2001]. In accordance with this, *Cortijo et al.* [1994] reason that cooling of the Norwegian Sea and the North Atlantic preceded growth of the continental ice sheet. Atmospheric general circulation model studies indicate that the initiation of permanent snow cover and the growth of ice sheets is favored by colder sea-surface conditions [*Yoshimori et al.*, 2002]. However, warm interglacial conditions and strong convection prevailed through the inception phase in the subpolar North Atlantic [*Duplessy and Shackleton*, 1985; *Ruddiman and McIntyre*, 1979; *Ruddiman et al.*, 1980]. It has been argued that the moisture necessary to obtain the rapid ice accumulation of the Laurentide and Scandinavian ice sheets was provided by increased storminess resulting from the contrast between these warm water masses and the colder land and ice-covered regions, as the temperature gradient between polar and subpolar areas increased. Several recent papers also argue in favor of an intensified Atlantic Meridional Overturning Circulation (AMOC) during the glacial inception phase, stressing the importance of direct influence of AMOC on the high northern latitude climate during glacial growth [*McManus et al.*, 2002; *Otterå and Drange*, 2004; *Wang and Mysak*, 2002]. The majority of studies that focus on the penultimate glacial inception, however, are based on reconstructions of subpolar areas. Thus, it is of great importance to document the glacial inception history as recorded in the polar region, where feedbacks related to an early change in ice albedo might be important.

Evidence of millennial-scale climate variability as a characteristic feature of the early Weichselian glacial phases is documented from several sites from the North Atlantic [*Chapman and Shackleton*, 1999; *McManus et al.*, 1994; *Oppo et al.*, 2001], subtropical Atlantic [*Lehman et al.*, 2002], the Nordic Seas [*Fronval and Jansen*, 1997], and Greenland ice cores GISP2 and GRIP [*Dansgaard et al.*, 1993; *Grootes et al.*, 1993]. These events are characterized by major input of Ice-Rafted Debris (IRD) to the Nordic Seas and the North Atlantic [*Chapman and Shackleton*, 1999; *McManus et al.*, 1994]. The sea-surface temperature (SST) of the North Atlantic and subtropical Atlantic decreased [*Lehman et al.*, 2002], as did the temperatures on Greenland [*Dansgaard et al.*, 1993; *Grootes et al.*, 1993]. Indications of a reduced overturning circulation associated with these events are also documented [*Hall et al.*, 1998; *Lehman et al.*, 2002]. Several papers argue that the occurrence of millennial-scale cold spells is solely dependent on the crossing of an ice volume threshold [*Chapman and Shackleton*, 1999; *McManus et al.*, 1999; *Schulz et al.*, 1999]. Others find it necessary to introduce additional factors, perhaps through a response in the thermohaline circulation [*Oppo et al.*, 2001]. *Lehman et al.* [2002] argue that a large freshwater influence related to the major IRD input disturbs the overturning circulation and thereby induces the regionally observed cold temperatures. We here present new information concerning the discussion on the last interglacial climate variability and glacial inception at high northern latitudes as well as records that provide new insight into the questions concerning millennial-scale climate variability of MIS 5.4-5.1. Local and regional aspects concerning such rapid climate reorganizations are discussed, emphasizing the influence of the ocean dynamics on the climate system.

2. MATERIAL AND METHODS

Cores MD99-2303 (77°31.18′N, 08°23.98′E; 2277-m water depth) and MD99-2304 (77°37.26′N, 09°56.90′E; 1315-m water depth) were collected from the West Spitsbergen margin. Since there are strong meridional gradients in the Fram Strait, we present results from two nearby cores to be able to assess the existence of such gradients in the past. The core MD95-2010 (66°41.05′N, 04°33.97′E: 1226-m water depth) is from the Vøring Plateau (Figure 1). All cores are sampled at every centimeter throughout the studied interval (MD99-2303: 725-936 m; MD99-2304: 1415-2262 m; MD95-2010: 1074-1254 m).

Stable isotope measurements were performed at the GMS lab at the University of Bergen, using Finnigan MAT 251 and MAT 252 mass spectrometers, both equipped with automatic preparation lines ("Kiel device"). All results are reported in ‰ vs. VPDB, using NBS 19 as reference. Measurements on planktonic foraminifers were performed on left-coiling form of *Neogloboquadrina pachyderma*. Measurements on benthic foraminifers were done on *Oridorsalis umbonatus* (MD99-2303) and on *Cassidulina teretis* and *Cibicides wuellerstorfi* (MD99-2304 and MD95-2010). $\delta^{18}O$ results for

C. wuellerstorfi were adjusted by 0.64‰ [*Shackleton and Opdyke*, 1973] and for *O. umbonatus* by 0.4‰ [*Graham et al.*, 1981] to take account of their departure from isotopic equilibrium. No correction has been performed of the *Cassidulina teretis* $\delta^{18}O$ values, as *C. teretis* calcify close to equilibrium [*Jansen et al.*, 1988]. For MD99-2303 and MD95-2010, measurements were done every centimeter if enough foraminifers were found. For MD99-2304, measurements were performed on one sample every 5 cm from 1415 to 1945 m and one every 1 cm from 1945 to 2262 m, whenever possible.

Minerogenic grains >0.5 mm was counted every 1 cm through the studied intervals of all three cores. These grains are assumed to be ice-rafted. In MD99-2303 and MD99-2304, sedimentary and crystalline grains were counted separately. Percentage of sedimentary rock fragments is presented for samples with a total of more than 100 grains larger then 0.5 mm.

In MD99-2304, planktonic foraminifers were counted at spread intervals in the >150-µm fraction. The polar species *N. pachyderma* (sin) dominates, but minor amounts of the subpolar species *N. pachyderma* (dex) and *Globigerina quinqueloba* occur. Percentages of these species were calculated for samples containing more than 100 planktonic foraminifers. In general, however, the number of planktonic foraminifers in the samples is larger than this limit (78% of the presented counts contain more than 300 individuals). In MD99-2303 we used the same approach for all samples covering MIS 5.5.

3. OCEANOGRAPHY

Presently, the Nordic Seas can be divided into three meridionally separated water masses. The western part is characterized by Polar water (salinity <34.4‰, temperature <0°C [*Fogelquist et al.*, 2003; *Swift*, 1986]). In the eastern part, warm Atlantic water is present: On entering the Norwegian Sea, the Atlantic water is characterized by temperatures typically about 6-9°C and salinities between 35.1‰ and 35.3‰. In the Spitsbergen area, the temperature of the Atlantic water is generally below 5°C and the salinity is often below 35.15‰ [*Blindheim*, 1990]. Polar and Atlantic water masses are separated by Arctic water, a mix of Polar and Atlantic water that is, in general, characterized by a salinity in the range of 34.4-34.9‰ and temperatures >0°C (→ 4°C) [*Swift*, 1986]. The Polar front marks the boundary zone between Polar and Arctic water, and the Arctic front separates Arctic and Atlantic water (Figure 1). MD95-2010 is located on the main axis of the warm Norwegian Atlantic Current (NwAC) that transports Atlantic water masses northwards through the Nordic Seas. At the southwestern Barents Sea boundary, the NwAC bifurcates into the West Spitsbergen Current (WSC), flowing northwards into the Fram Strait, and the North Cape Current, flowing eastward into the Barents Sea. The locations of MD99-2304 and MD99-2303 make it possible to detect changes in the dynamics of the WSC and thus the influence of Atlantic water in the Fram Strait.

4. CHRONOLOGY

The scattered character of the $\delta^{18}O$ records complicates a straightforward correlation of benthic $\delta^{18}O$ to establish the chronology of the cores. Therefore, a stepwise approach was used to obtain good age control. MD95-2010 has the same location as Ocean Drilling Program (ODP) site 644 (66°40′N, 04°34′E; 1227 m depth). *Fronval and Jansen* [1997] established an age model for the MIS 5 interval of ODP 644 primarily by correlating the planktonic $\delta^{18}O$ to the SPECMAP stack [*Martinson et al.*, 1987]. The ODP 644 age model was further supported by a regionally distributed ash horizon in MIS 5.5 and the occurrence of *Pullenia bulloides* in MIS 5.1 [*Haake and Pflaumann*, 1989]. This age model was transferred to MD95-2010 by correlating the planktonic $\delta^{18}O$ records of the two cores. Thereafter, cores MD99-2303 and MD99-2304 were placed on the same age model as MD95-2010 (and ODP 644) by the same approach (Table 1, Figure 2a). The MIS 5.5 ash horizon is not found in the MD99-2303 and MD99-2304, but the increased abundance of *P. bulloides* related to MIS 5.1 is identified in both Fram Strait cores. However, to be able to compare our results from the Nordic Seas with the North Atlantic NEAP18K core [*Chapman and Shackleton*, 1999], a refinement of the age model was necessary. We consider the NEAP18K chronology more reliable as it is based on benthic $\delta^{18}O$ and generally less influenced by local environmental conditions. The northward flowing surface water passing MD95-2010 takes part in the convection processes in the Nordic Seas. Benthic conditions at station NEAP18K are influenced by the south-streaming branch of the deep water formed by these convection processes, with a lag that is negligible in this context. Therefore, we assume a connection between the planktonic $\delta^{13}C$ of MD95-2010 and the benthic $\delta^{13}C$ of NEAP18K and correlate MD95-2010 and NEAP18K on the basis of this assumption, transferring all studied records to the chronology of NEAP18K. The MIS 5.5 plateau is defined by the stable low planktonic $\delta^{18}O$ and the relatively low percentage of *N. pachyderma* (sin) recorded in MD99-2304. The validity of the final chronology is supported by the occurrence of *P. bulloides* in MIS 5.1 in the Fram Strait cores [*Haake and Pflaumann*, 1989] and the high correlation coefficients between the benthic $\delta^{18}O$ record from NEAP18K and all other $\delta^{18}O$ records (Table 2, Figure 2b). The records were correlated by using *AnalySeries* [*Paillard et al.*, 1996]. Uncertainties of MIS 5 age models based on $\delta^{18}O$ correlation are estimated to be ± 2-7 ka [*Martinson et al.*, 1987]. During recent years, an increasing number of absolute dates have provided important improvements to

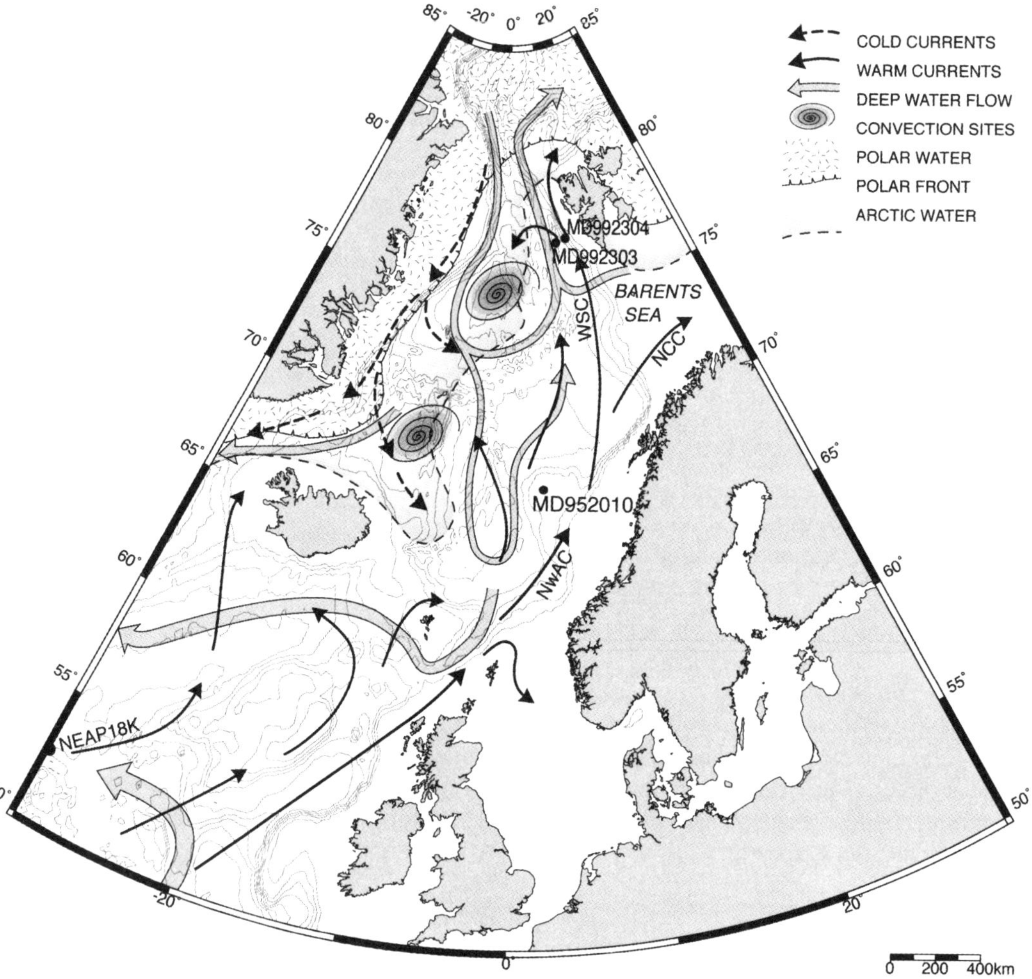

Figure 1. Distribution of different water masses and surface currents in the North Atlantic and the Nordic Seas, together with the location of the studied cores. NwAC: Norwegian Atlantic Current. NCC: North Cape Current. WSC: West Spitsbergen Current.

the chronology of MIS 5 chronologies [*Cutler et al.*, 2003; *Stirling et al.*, 1998]. However, the general uncertainties of absolute age are still in the order of millennia, although the core-to-core uncertainty is probably less.

Table 1. The Correlation Coefficients Results as Given by *Analyseries* [*Paillard et al.*, 1996] After Transferring MD95-2010, MD99-2303, and MD99-2304 to the Same Chronology, Based on the Chronology of ODP Site 644 [*Fronval and Jansen*, 1997].

		ODP 644 δ^{18}Op	MD99-2010 δ^{18}Op	MD99-2303 δ^{18}Op
MD95-2010	δ^{18}Op	0.81	—	—
MD99-2303	δ^{18}Op	0.82	0.78	—
MD99-2304	δ^{18}Op	0.81	0.78	0.75

p = planktonic.

5. RESULTS

In all three cores, the MIS 5 substages are identified in the δ^{18}O records (Figure 3a-c). Superimposed on these large-scale trends, small-scale variability is seen throughout the studied interval. In both the planktonic and the benthic records the main part of this variability is in the range of 0.2-0.5‰. Occasionally, the variability is greater. Gaps in the δ^{18}O records from MD99-2303 and MD99-2304 (indicated by stippled line in Figure 3a,b) are the result of foraminifer-barren sediment samples.

Figure 2. (a) The first step in this approach of developing a common chronology on the discussed cores. The final comparison of planktonic δ^{18}O (*N. pachyderma*) of MD95-2010 (thin black line), MD99-2303 (dark gray line), and MD99-2304 (light gray line) to

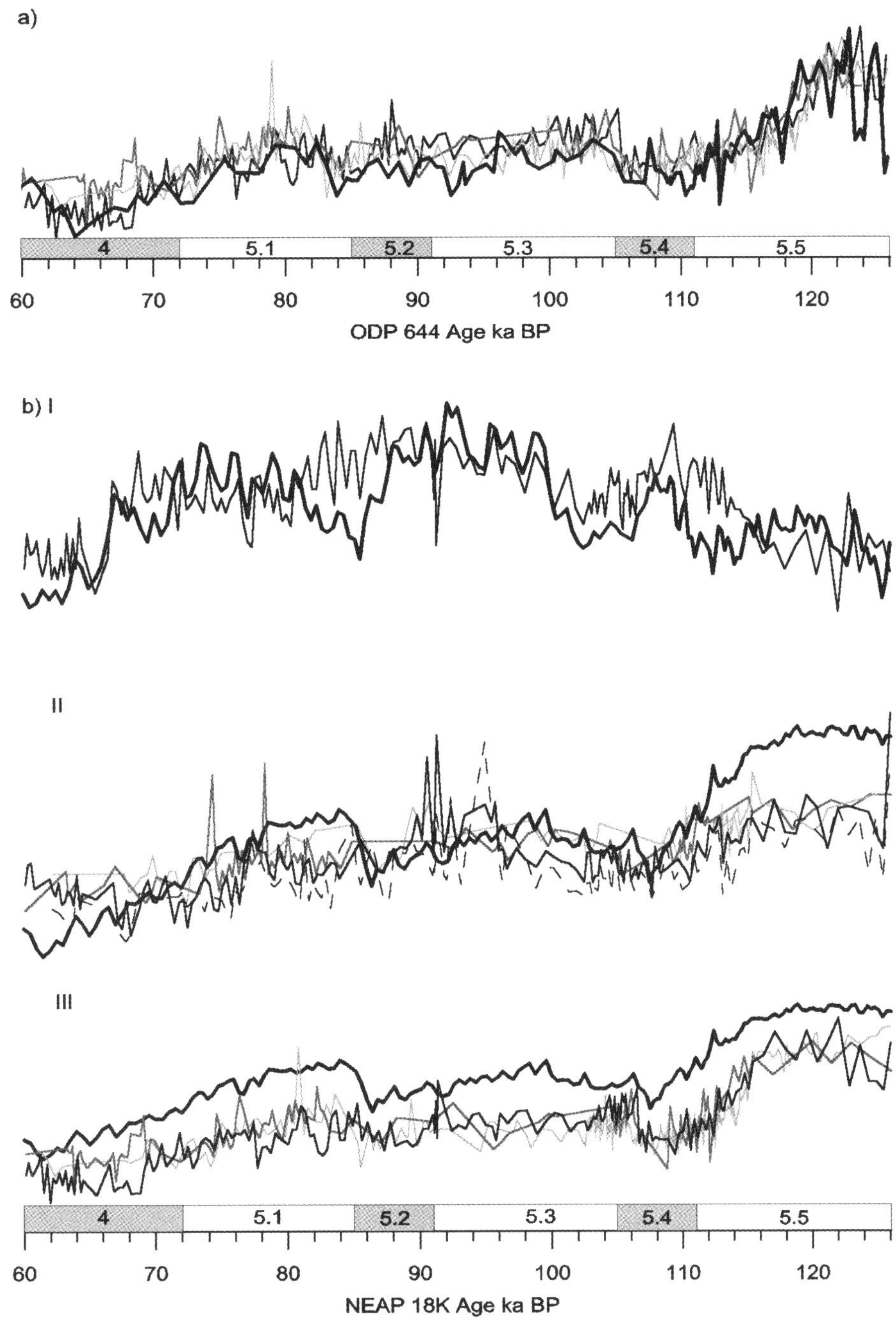

ODP 644 (thick black line), after transferring them to the ODP 644 chronology. (b) The second step in developing a common chronology on the discussed cores. I: Comparison of planktonic $\delta^{13}C$ (*N. pachyderma,* sin) of MD95-2010 (thin black line) and benthic $\delta^{13}C$ (*Cibicides wuellerstorfi*) of NEAP18K (thick black line) after transferring MD95-2010 to the NEAP18K chronology. II: Comparison between benthic $\delta^{18}O$ (*C. wuellerstorfi*) of NEAP18K (thick black line) and benthic $\delta^{18}O$ of MD95-2010 [stippled line (*C. wuellerstorf*) and thin black line (*Cassidulina teretis*)), MD99-2303 (dark gray line) (*O. umbonatus*) and MD99-2304 (light gray line) (*Cibicides wuellerstorfi* and *Cassidulina teretis*) after transferring them to the NEAP18K chronology. The same color code is valid for part III of the figure. III: Comparison between benthic $\delta^{18}O$ (*Cibicides wuellerstorfi*) of NEAP18K and planktonic $\delta^{18}O$ of MD95-2010 (*C. wuellerstorfi* and *Cassidulina teretis*), MD99-2303 (*O. umbonatus*), and MD99-2304 (*Cibicides wuellerstorfi* and *Cassidulina teretis*) after transferring them to the NEAP18K chronology. The records are not necessarily at the same vertical scale. The figure has a purely illustrative purpose and the records are placed on top of each other simply to show the correlating patterns of the different records. The age models are further supported by the results given in Tables 1 and 2.

Table 2. Correlation Coefficients as Given by *Analyseries* [*Paillard et al.*, 1996], After Transferring the NEAP18K Age Model [*Chapman and Shackleton*, 1999] to Cores MD95-2010, MD99-2303, and MD99-2304.

		MD95-2010			MD99-2303			MD99-2304		
		δ^{13}Cp	δ^{18}Ob	δ^{18}Op	δ^{13}Cp	δ^{18}Ob	δ^{18}Op	δ^{13}Cp	δ^{18}Ob[1]	δ^{18}Op
NEAP	δ^{13}Cb	0.65			0.52			0.38		
18K	δ^{18}Ob		0.67[2]	0.9		0.82	0.88		0.83	0.87
	δ^{18}Op			0.73			0.74			0.71

p = planktonic; b = benthic.

[1] The benthic δ^{18}O record from MD99-2304 is based on a spliced benthic record, combining *Cibicides wuellerstorfi* and *Cassidulina teretis* measurements.
[2] This value is the result of correlating δ^{18}O of *Cibicides wuellerstorfi* from NEAP18K with δ^{18}O of *C. wuellerstorfi* from MD95-2010. If correlating δ^{18}O of *Cassidulina teretis* from MD95-2010 with δ^{18}O of *Cibicides wuellerstorfi* from NEAP18K, the value is 0.67.

Both cores MD99-2303 and MD99-2304 document a very early occurrence of IRD after the peak interglacial time, starting at 116 ka BP. This is simultaneous with the first increase in planktonic δ^{18}O at the sites and a major increase in the relative abundance of *N. pachyderma* (sin) in MD99-2304. At high frequency, events characterized by a high content of IRD are seen throughout MIS 5.5. Sedimentary rock fragments dominate most of the high IRD events. Some of the IRD events are, however, characterized by a higher content of crystalline grains (Figure 4b,c). The absolute abundance of IRD in MD95-2010 is less than in the Spitsbergen margin cores. However, some of the same variability and timing of IRD events is seen in all three cores (Figure 4b–d).

Foraminifer data from MD99-2304 show a general dominance of the polar species *N. pachyderma* (sin) (Figure 5). MIS 5.5 is characterized by highly variable conditions, with a rather high, but fluctuating, content of subpolar foraminifers (Figure 6). This is in contrast to the constantly high content of *N. pachyderma* (sin) recorded throughout MIS 5.5 in MD99-2303 (Figure 6). Through MIS 5.4-5.1, some short periods of a somewhat decreased presence of *N. pachyderma* (sin) are documented in MD99-2304, related to intervals of high IRD (Figure 5).

6. DISCUSSION

6.1. The Last Interglacial

The high northern location of cores MD99-2304 and MD99-2303 provides important information on the poleward extension of Atlantic water during MIS 5. The planktonic δ^{18}O record from MD99-2304, the easternmost of the two sites, indicates a climatic optimum in the early part of the interglacial, while more stable and somewhat higher δ^{18}O values characterize the later parts of the warm phase. Assuming that the relative abundance of *N. pachyderma* (sin) is an indicator of temperature, these data suggest a warm early MIS 5.5 (Figure 6b). *Shackleton et al.* [2003] also document slightly lighter δ^{18}O for the early MIS 5.5 plateau. This pattern is comparable with the early Holocene climate optimum [*Birks and Koç*, 2002; *Duplessy et al.*, 2001; *Karlén*, 1988; *Koç Karpuz and Jansen*, 1992], resulting from a summer insolation optimum in the Northern Hemisphere. In general, the foraminifer records indicate variable conditions throughout MIS 5.5 in the Nordic Seas.

In some periods, MD99-2304 records an increased relative content of *N. pachyderma* (dex), which may indicate slightly increased surface temperatures, probably due to enhanced influence of WSC water. The apparent flickering of the percentage of *N. pachyderma* (sin) in the record from 121 to 116 ka BP is related to the noticeable variation in the relative content of *G. quinqueloba* (Figure 6b), a species associated with the Arctic front [*Johannessen et al.*, 1994]. The observed variability may be the result of an alternating location of this front in the Fram Strait region. The Arctic front position may alternate due to a variable influence of the westerlies. Strong westerlies will lead to an enhanced inflow of Atlantic water, yet with restricted lateral extent, the flow of Atlantic water thus being confined in a shelf/slope current. This will increase the eastward influence of Arctic water [*Blindheim et al.*, 2000]. There appears to have been a strong gradient between the two northernmost cores, as the western core, MD99-2303, shows different conditions during the last interglacial. The planktonic δ^{18}O records of the two cores are comparable; however, MD99-2304 shows a clear increase in the content of subpolar foraminifers during the last interglacial while MD99-2303 is dominated by a polar fauna throughout this time interval. The foraminifer production in MIS 5.5 is significantly higher at MD99-2303 than at MD99-2304. Such different results in two nearby cores should be viewed suspiciously, as this may indicate disturbances in the records due to sedimentary or dissolution processes. However, there is no evidence of disturbance or dissolution of the sediment in the cores through this time interval. The δ^{18}O and IRD records of the two cores are comparable,

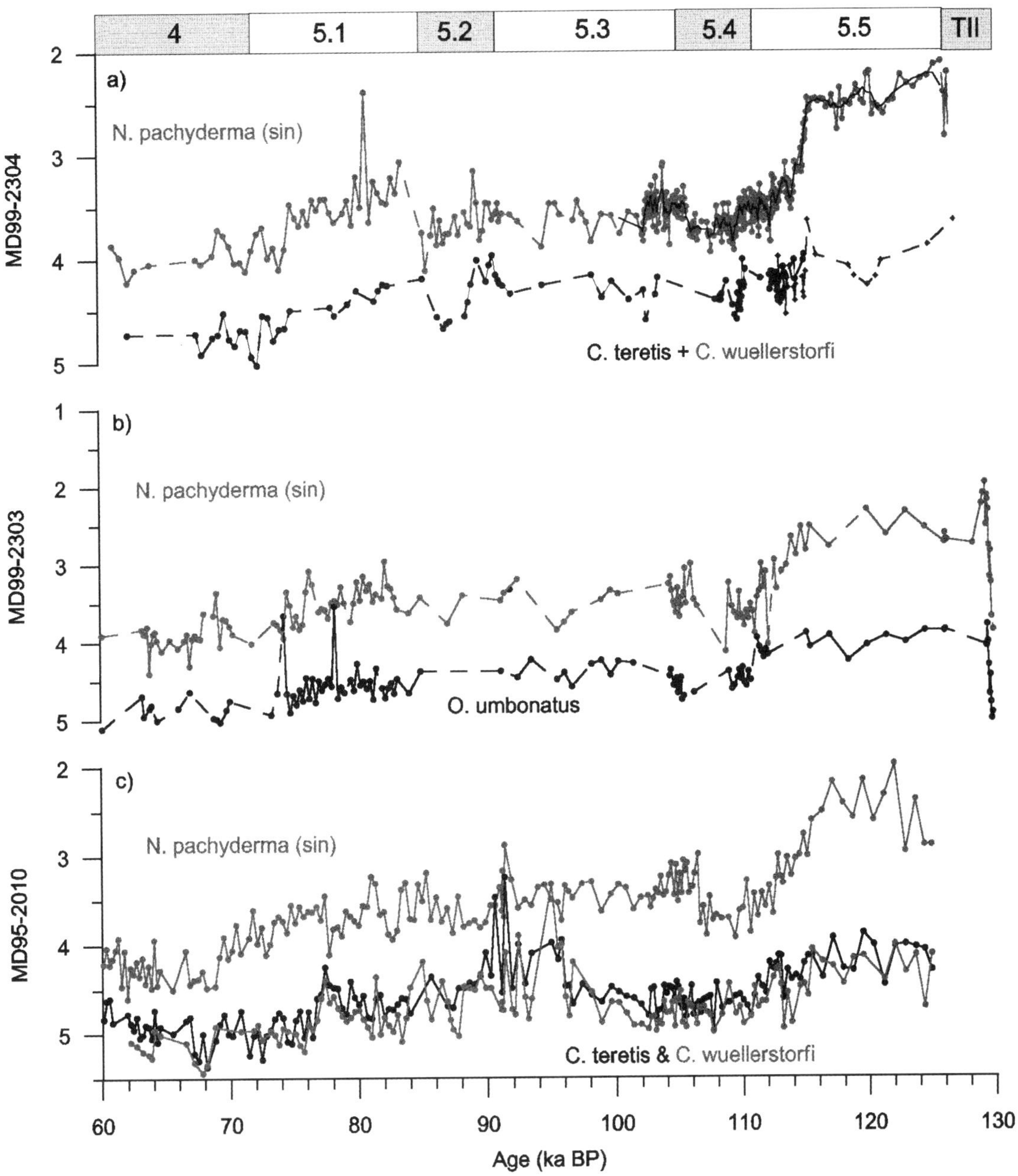

Figure 3. a) Planktonic [*N. pachyderma* (sin)] and benthic [combined *Cassidulina teretis* and *Cibicides wuellerstorfi* (light gray)] $\delta^{18}O$ records from MD99-2304. The thin black line in the *N. pachyderma* (sin) record represents a 5-point running mean of this planktonic $\delta^{18}O$ record through the part of highest resolution. The stippled lines represent intervals where the isotope records are discontinuous due to foraminifer-barren samples. (b) Planktonic [*N. pachyderma* (sin)] and benthic (*O. umbonatus*) $\delta^{18}O$ records from MD99-2303. The stippled lines represent intervals where the isotope records are discontinuous due to foraminifer-barren samples. (c) Planktonic [*N. pachyderma* (sin)] and benthic [*Cassidulina teretis* and *Cibicides wuellerstorfi* (light gray)] $\delta^{18}O$ records from MD95-2010. Oxygen isotope stages are indicated at the top of the uppermost panel.

supporting the interpretation of nondisturbed sediments. MD99-2304 also shows lower planktonic $\delta^{13}C$ values than MD99-2303 (Figure 6), which indicates that the production was lower at that site and that the differing foraminifer content between the two sites is not a result of dissolution at MD99-2304. We thus argue that the observed differences in foraminifer content and assemblage, together with the differences in the planktonic $\delta^{13}C$ records, provide evidence that

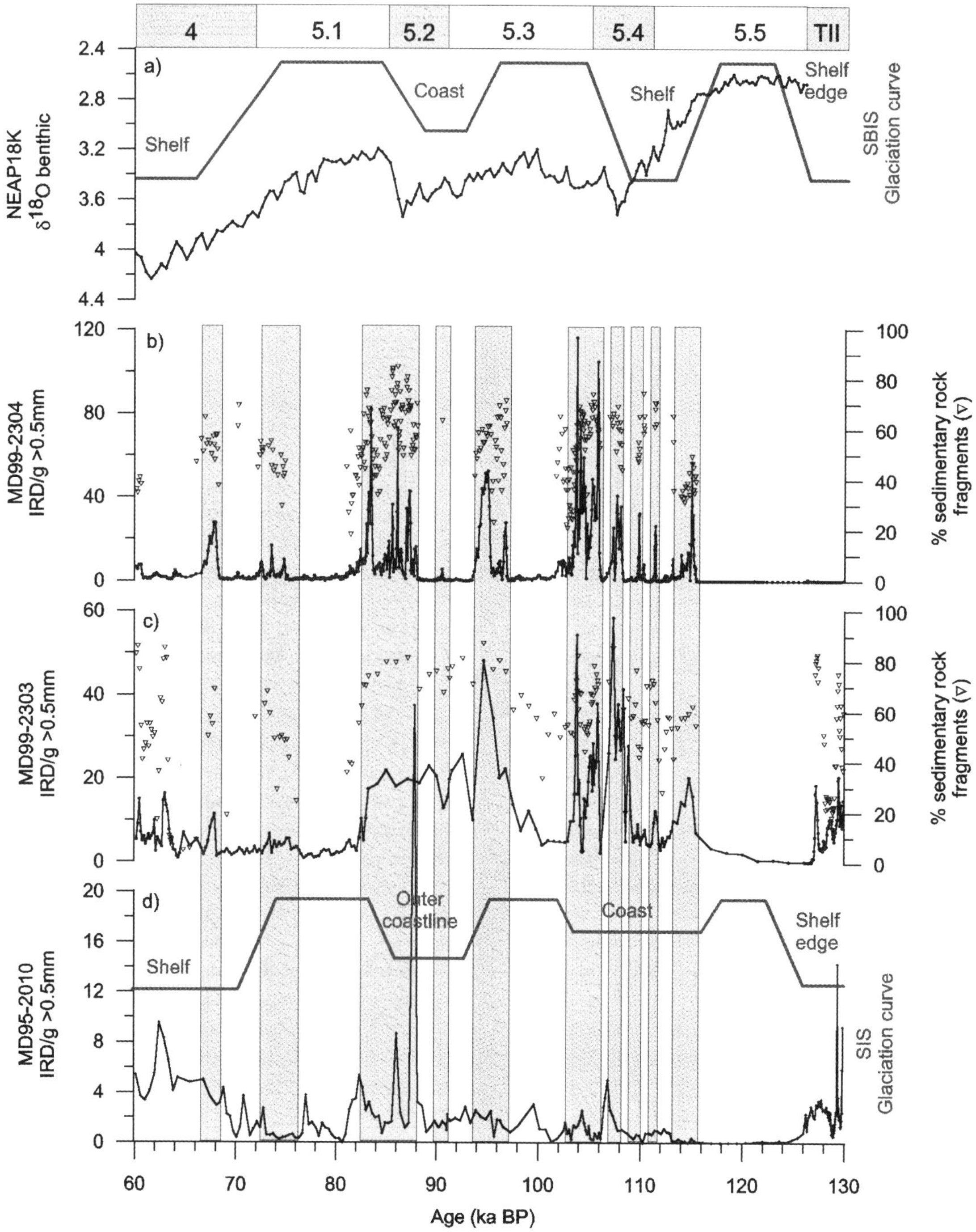

Figure 4. (a) The benthic $\delta^{18}O$ record from NEAP18K [*Chapman and Shackleton*, 1999] plotted together with the reconstructed glaciation curve of the Svalbard–Barents Sea Ice Sheet (SBIS) [*Mangerud et al.*, 1998]. (b) The amount of IRD/g sediment >0.5 mm in MD99-2304 plotted together with the percentage of sedimentary rock fragments (∇) present in the samples. The remaining IRD consists of crystalline grains. (c) The amount of IRD/g sediment >0.5 mm in MD99-2303 plotted together with the percentage of sedimentary rock fragments (∇) present in the samples. The remaining IRD consists of crystalline grains. (d) The amount of IRD/g sediment >0.5 mm in MD95-2010 and the reconstructed glaciation curve of the Scandinavian Ice Sheet (SIS) [*Baumann et al.*, 1995]. Oxygen isotope stages are indicated at the top of the figure. Main IRD events are indicated by gray vertical bars.

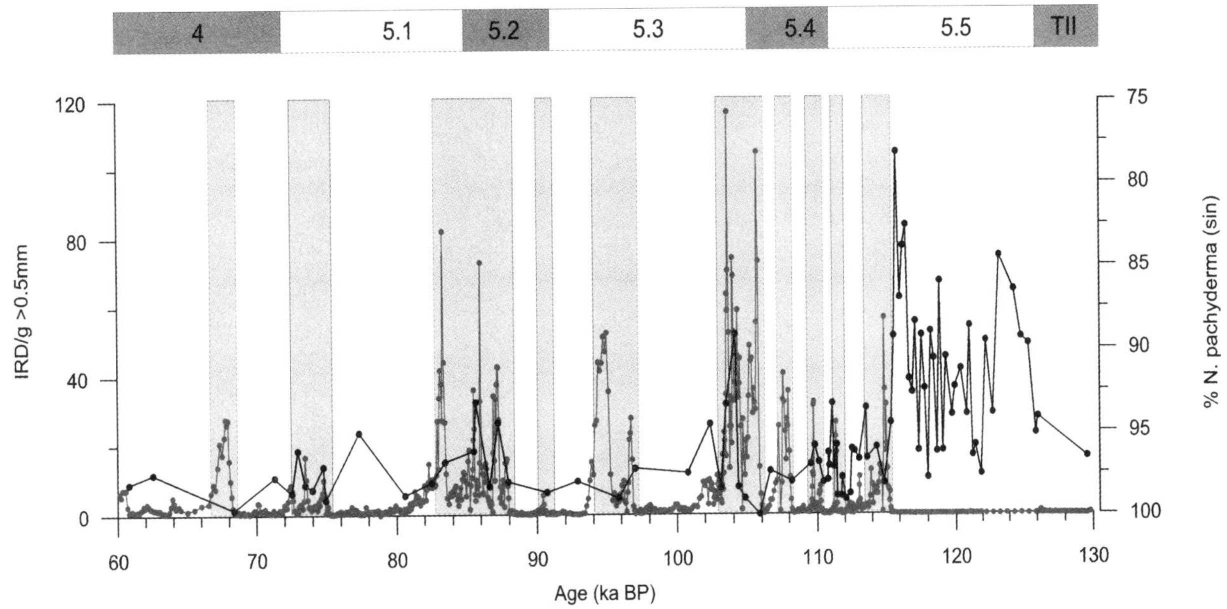

Figure 5. IRD/g sediment >0.5 mm found in MD99-2304 is presented in gray, while the percentage of *N. pachyderma* (sin) is shown as a black line. Gray vertical bars indicate the main IRD events, and oxygen isotope stages are indicated at the top.

the cores were influenced by water masses of contrasting characteristics during MIS 5.5 (Figure 6a,b). We postulate that the two cores were separated by the Arctic front, with a narrow wedge of northward-protruding warmer Atlantic water influencing the eastern core location. The cold, highly productive environment inferred from core MD99-2303 indicates a location underneath Arctic water, close to the Arctic front [*Johannessen et al.*, 1994]. The location of the Arctic front is mainly determined by the bathymetry, because the ocean currents are directed by the topography [*Manley*, 1995; *Saloranta and Svendsen*, 2001]. Results by *Manley* [1995] imply that the location of MD99-2303 today is at the vicinity of the westernmost area under the influence of subsurface Atlantic water, while MD99-2304 is located close to the border between subsurface and surface Atlantic water of the WSC. Our results indicate that during most of MIS 5.5 the Fram Strait was characterized by a narrower, easterly confined position of the warm wedge of northward-flowing Atlantic water than at present, probably the result of the influence of strong westerlies.

Previous studies of the last interglacial conditions in the northernmost Nordic Seas have argued for a more zonal circulation and an increased temperature gradient between the Greenland Sea and the Norwegian Sea, as compared with the meridional pattern seen during the Holocene [*Bauch et al.*, 1999; *Fronval and Jansen*, 1997; *Hebbeln et al.*, 1998; *Köhler and Spielhagen*, 1990]. The interpretations of a zonal circulation pattern are based on sites located further west in the Fram Strait, in the Greenland Sea, or studies of lower resolution. *Hebbeln et al.* [1998] concluded that the northern part of the Nordic Seas remained at interstadial conditions throughout the last interglacial. Our records bring new knowledge to this discussion and indicate that at least the easternmost part of the Fram Strait was characterized by interglacial conditions during MIS 5.5. Evidence for warm-water transport to this area also comes from finding the mollusk *Mytilus edulis* in sediments of Eemian age at Spitsbergen [*Mangerud and Svendsen*, 1992] and is indicated by calcareous nannofossil data from the Fram Strait [*Gard*, 1988]. In general our results support the concept of a non-stable high latitude climate during MIS 5.5 [*Field et al.*, 1994; *Fronval and Jansen*, 1996; *Kristensen et al.*, 1998; *Seidenkrantz et al.*, 1995; *Thouveny et al.*, 1994]. The magnitude of the variability during MIS 5.5 was greatly subdued compared to glacial climate changes; in general the amplitudes are comparable to those of the Holocene [*Bond et al.*, 1997; *Fronval and Jansen*, 1996; and *Risebrobakken et al.*, 2003].

6.2. Glacial Inception

In the present situation, strengthened NwAC is characterized by a restricted lateral extent of warm Atlantic water, as compared to a more widespread but shallower extent at times of a reduced NwAC [*Blindheim et al.*, 2000]. Changes in the strength of the NwAC are primarily the result of changing strength of the westerlies. Our results indicate an intensified AMOC from 117 ka BP. Warm water conditions together

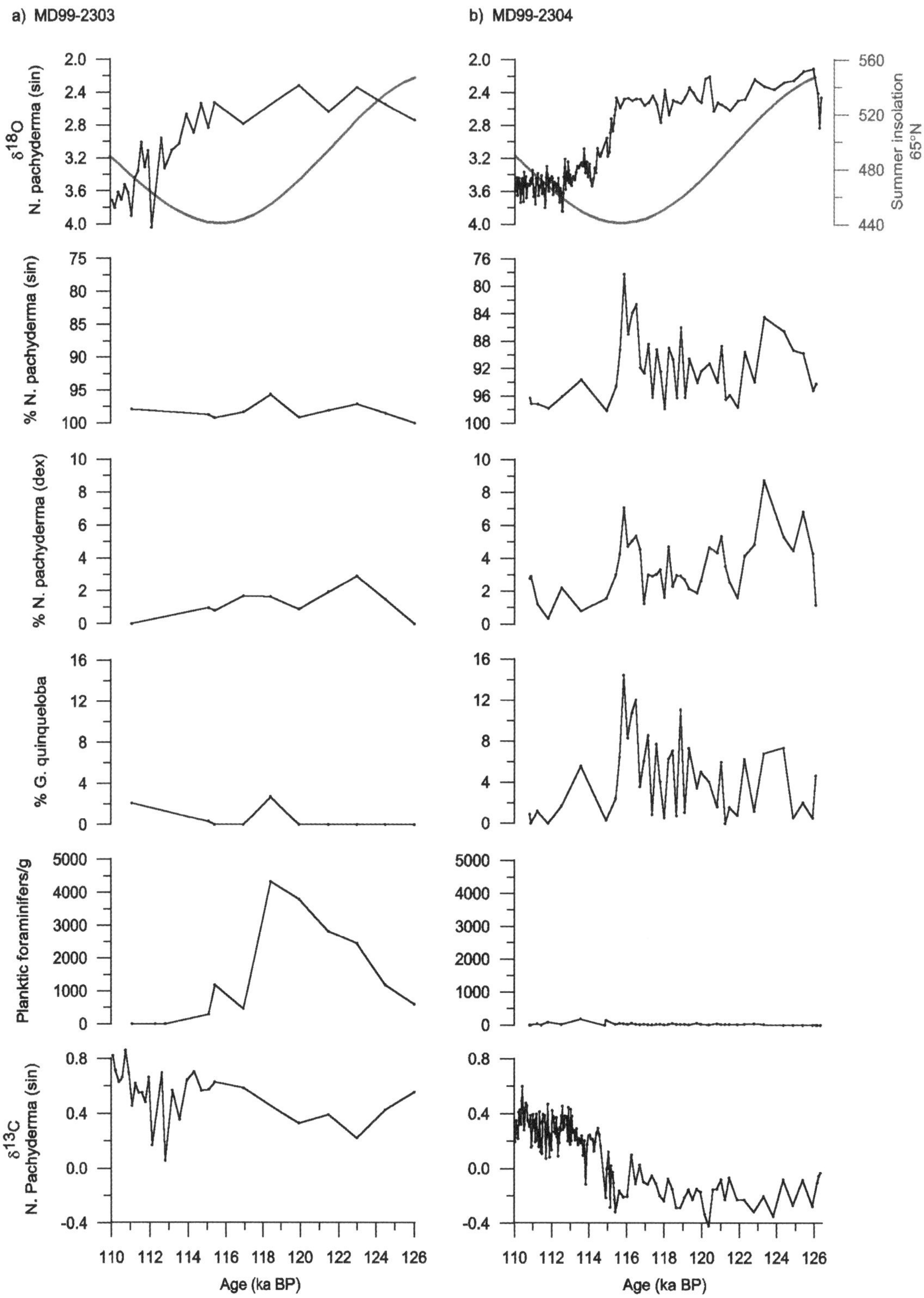

Figure 6. Marine Isotope Stage 5e as recorded in MD99-2304 (a) and MD99-2303 (b): planktonic $\delta^{18}O$, % *N. pachyderma* (sin), % *N. pachyderma* (dex), % *G. quinqueloba*, the amount of planktonic foraminifera/g sediment, and $\delta^{13}C$. Note the major differences seen between most of the properties before 116 ka BP; after this time, conditions were more similar. The summer insolation at 65°N [*Laskar*, 1990] is shown as a thick gray line in both (a) and (b).

with an increased content of the Arctic front indicator *G. quinqueloba* are recorded in MD99-2304, while MD99-2303 was probably bathed by Arctic water. These conditions with a strong AMOC occur at the same time as the summer insolation at 65°N approached its minimum (Figure 6). The enhanced warming characterizing the end of the warm phase of MIS 5.5 at MD99-2304 culminated in a marked cooling, registered by a rapid fall of the relative content of subpolar foraminifers and increased planktonic $\delta^{18}O$ values (Figures 5 and 6a). Immediately after this warm phase, the first major IRD input is recorded in the Fram Strait (Figures 4b,c and 5). This first IRD input coincides with the first MIS 5 global change in ice volume, as indicated by benthic $\delta^{18}O$ from NEAP18K (Figure 4a-c) [*Chapman and Shackleton*, 1999], and with the dated onset of global fall in sea level [*Stirling et al.*, 1998], according to our time scale used in this paper. Thus, glaciers at Spitsbergen grew large enough to reach the coastline and calve before any major global ice sheet growth was observed. The combination of low insolation and strengthened influx of warm water probably favored this rapid growth of the Svalbard–Barents Sea Ice Sheet (SBIS). An increased thermal contrast between sea and land at high latitudes will produce large moisture fluxes towards land [*Wang and Mysak*, 2002], and a steady source of moisture is needed to provide major glacial growth [*Hebbeln et al.*, 1994; *Nesje and Dahl*, 2003]. Strengthening the circulation will also lead to an intensified deflection of Atlantic water into the Barents Sea [*Zhang et al.*, 1998]. If this was the case, the sea–land temperature gradient in the interior of the Barents Sea region may have increased too. Due to a resulting change in the moisture flux, and more favorable conditions for the snow cover to expand due to the low summer insolation, ice sheet growth in northern Russia may have been initiated, similar to the inception of the SBIS. Early initial ice accumulation after the last interglacial was likely important for the growth of large ice sheets in this area during the early Weichselian. In the early Weichselian, large ice crops did exist in the Arctic Russia [*Svendsen et al.*, 1999], while the Scandinavian Ice Sheet (SIS) still had a rather restricted extent [*Baumann et al.*, 1995; and *Sejrup et al.*, 2000]. So far, the hypothesis we present is not supported by direct evidence, and the proposed scenario should be tested further. The maximum extent of the Barents–Kara Sea Ice Sheet is dated to approximately 90 ka BP (MIS 5.2) [*Mangerud et al.*, 2001]. We suggest that warm open water conditions at times of reduced summer insolation initiated the growth of ice on land through the inception phase, while the further development of grounding of ice at the epicontinental Barents and Kara Sea was dependent on the global sea level as well. The possible effect of a strong AMOC late in MIS 5.5 and its potential influence on the initial glacial growth at high northern latitudes was noted by *McManus et al.* [2002]. In an experiment with a global OGCM focusing on the sensitivity of the ocean circulation and sea-ice cover to changes in solar insolation, *Otterå and Drange* [2004] found that the situation during the inception phase favored enhanced AMOC. Terrestrial evidence indicates an early response of the SBIS after the last interglacial [*Mangerud et al.*, 1998], and results from model experiments with a 3-D ice-dynamical model indicate glacial growth in the Barents and Kara Seas region from 118 ka BP [*Bintanja et al.*, 2002]. The model results from *Bintanja et al.* [2002], however, are presently not supported by field data [*Spielhagen et al.*, 2004; *Svendsen et al.*, 2004]. The onset of iceberg calving probably induced increased amounts of fresh water into the Fram Strait. This is supported by correspondingly lower planktonic $\delta^{18}O$ seen in both MD99-2303 and MD99-2304 (Figure 7) and presumably affected the ocean circulation in the area, as well as sea-ice formation, preventing a strong WSC. Evidence for a pronounced change in water mass properties is recorded by MD99-2304 at this time. The differences that occurred between MD99-2303 and MD99-2304 from 126 to 115 ka BP were no longer evident, and Arctic water mass properties characterize the records from both sites (Figure 6a,b).

6.3. Early Weichselian Climate Variability

From 116 ka BP and throughout the remaining part of the studied time interval, high-frequency variability of IRD is recorded in MD99-2303 and MD99-2304. The main loci for

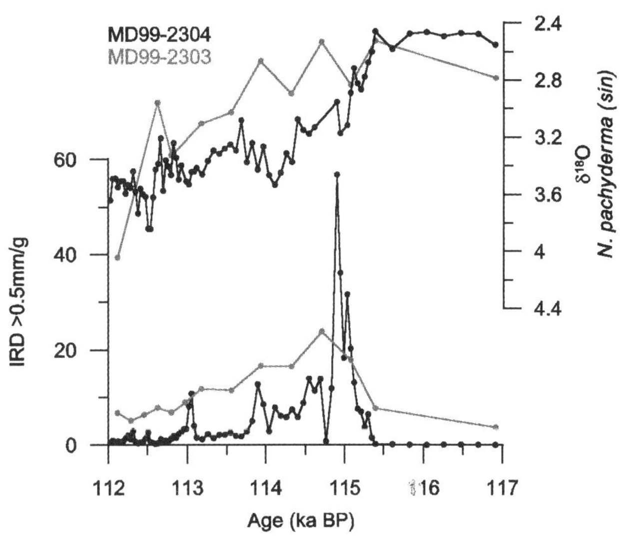

Figure 7. Detail of the relationship between IRD content and planktonic oxygen isotopes through the glacial inception phase. Increased amounts of IRD are followed by somewhat lower $\delta^{18}O$, probably due to increased freshwater influence as icebergs melt.

the SBIS, Svalbard, the West Spitsbergen shelf, and the Barents Sea are essentially sedimentary rocks [*Elverhøi et al.*, 1995]. These are assumed to be the main source areas of IRD found in MD99-2303 and MD99-2304, where sedimentary rock fragments generally dominate. However, a high content of crystalline grains occurs occasionally (Figure 4b,c). The only local crystalline source is a limited exposure along the western part of Spitsbergen [*Elverhøi et al.*, 1995] that covers an area approximately 20-40 km wide [*Winsnes*, 1988]. Periods dominated by crystalline IRD may therefore indicate a calving ice sheet with an extent mainly restricted to Spitsbergen. During MIS 2, IRD from the SIS is found in the Fram Strait [*Hebbeln et al.*, 1994]. Therefore, a more distal source cannot be excluded. However, the generally restricted extent of the SIS during MIS 5 [*Baumann et al.*, 1995; *Sejrup et al.*, 2000] and the low content of IRD found at the Vøring Plateau at this time (Figure 4d) do not support such a scenario. Neither do we see any component that would imply a Scandinavian source, e.g. chalk [*Hebbeln et al.*, 1994]. Chalk sources, however, are restricted to areas south of 59°N, and a northern Scandinavian source of crystalline rock fragments can not be excluded on this basis. The IRD events with a high crystalline content occur during periods of initial ice growth or last decay, if compared with the reconstructed glaciation curve of the SBIS [*Mangerud et al.*, 1998] (Figure 4a-c). This supports the interpretation of these periods as indicative of a restricted ice sheet extent. Dominance of sedimentary rock fragments can be connected to late growth, maximum extent, and decay of the SBIS, as it is reconstructed by *Mangerud et al.* [1998] (Figure 4a–c).

The $\delta^{18}O$ records of MD99-2303 and MD99-2304 show large gaps, due to foraminifer-barren intervals (Figure 3a,b). Most of these intervals correlate to high-IRD events. These events indicate iceberg melting and thus reduced salinity of the surface water. Such conditions are not favorable for foraminifers, and the production of foraminifers may have been reduced due to the changing environment. In addition, during melt water events, the source of deep waters may be from brine water formation [*Dokken and Jansen*, 1999]. During sea-ice freezing virtually no fractionation of oxygen isotopes occurs [*Craig and Gordon*, 1965]. The density will increase due to the salinity enrichment, such that this brine water will sink to the bottom and give rise to extremely light benthic $\delta^{18}O$ values. Brine water is noted today to be corrosive for $CaCO_3$ because of enrichment of CO_2 when it forms on the shelves [*Steinsund and Hald*, 1994]. The foraminifer-barren intervals might therefore be the result of $CaCO_3$ dissolution due to increased brine water influence during the high-IRD periods. Evidence of melt-water is suggested by a strong depletion in the planktonic $\delta^{18}O$. The lack of isotope measurements due to the lack of foraminifers makes it impossible to provide clear evidence that the barren intervals are the result of dissolution due to brine water. However, evidence of both melt water and brine water is observed occasionally in the $\delta^{18}O$ records from MD99-2303, MD99-2304, and MD95-2010, for instance, at approximately 90 ka BP (Figure 3a-c). This supports the assumption that such conditions could have occurred more frequently.

At the Vøring Plateau, much less IRD were deposited throughout MIS 5 than in the Fram Strait (Figure 4d). The main part of the IRD deposited at the Vøring Plateau is assumed to be from the SIS. During the majority of MIS 5, the extent of the SIS was restricted (Figure 4d) [*Baumann et al.*, 1995; *Sejrup et al.*, 2000], preventing major calving and deposition of IRD. Even though the amplitude of the variability differs considerably between the two regions, some of the IRD events recorded at MD95-2010 seem to occur simultaneously with events recorded at MD99-2304 and MD99-2303 (Figure 4b-d). We suggest that the differences seen between the two regions are due to the different configurations of local ice sheets. The similarities, however, suggest that the trigger mechanism behind major iceberg calving may be the same. Slight increases in the proportion of subpolar foraminifers are recorded just before or during some of the IRD events at MD99-2304 (Figure 5), indicating that slight changes in the water mass configuration and oceanic heat transport may work as a destabilizing agent on marine-based ice sheets. Initiation of major calving will eventually increase the global sea level, and an increase in sea level will also have a destabilizing effect on marine-based ice sheets [*Jones and Keigwin*, 1988]. The combined effects of somewhat warmer water masses and sea-level change were probably important for the observed IRD events. The slight warming of water at MD99-2304 may result from an intensification of the WSC, due to increased AMOC. The increased content of subpolar foraminifers are best pronounced in relation to the disintegration of the MIS 5.4 and MIS 5.2 ice sheets (Figures 4a,b and 5). According to *Svendsen et al.* [1999] and *Siegert et al.* [2001],

Figure 8. Amount of IRD/g sediment >0.5 mm in (a) MD99-2304 and (b) MD99-2303. (c) Percentage of IRD >150 μm from NEAP18K [*Chapman and Shackleton*, 1999]. (d) Uk^{37} sea-surface temperature (SST) estimates from MD95-2036 (Bermuda Rise: 33°41.44′N, 57°34.55′W, 4462-m water depth) [*Lehman et al.*, 2002]. The MD95-2036 SST record has been transferred to the same age model as NEAP18K by subtracting 2 ka after and 4 ka before the MIS 5.2 $\delta^{18}O$ maximum, reversing the operation that *Lehman et al.* [2002] performed to transfer NEAP18K to their MD95-2036 age model. IRD events found regionally (C19–C25 as denoted by *Chapman and Shackleton* [1999]) are marked by dark gray bars while light gray bars mark IRD events only found locally in the Fram Strait. Oxygen isotope stages are marked at the top.

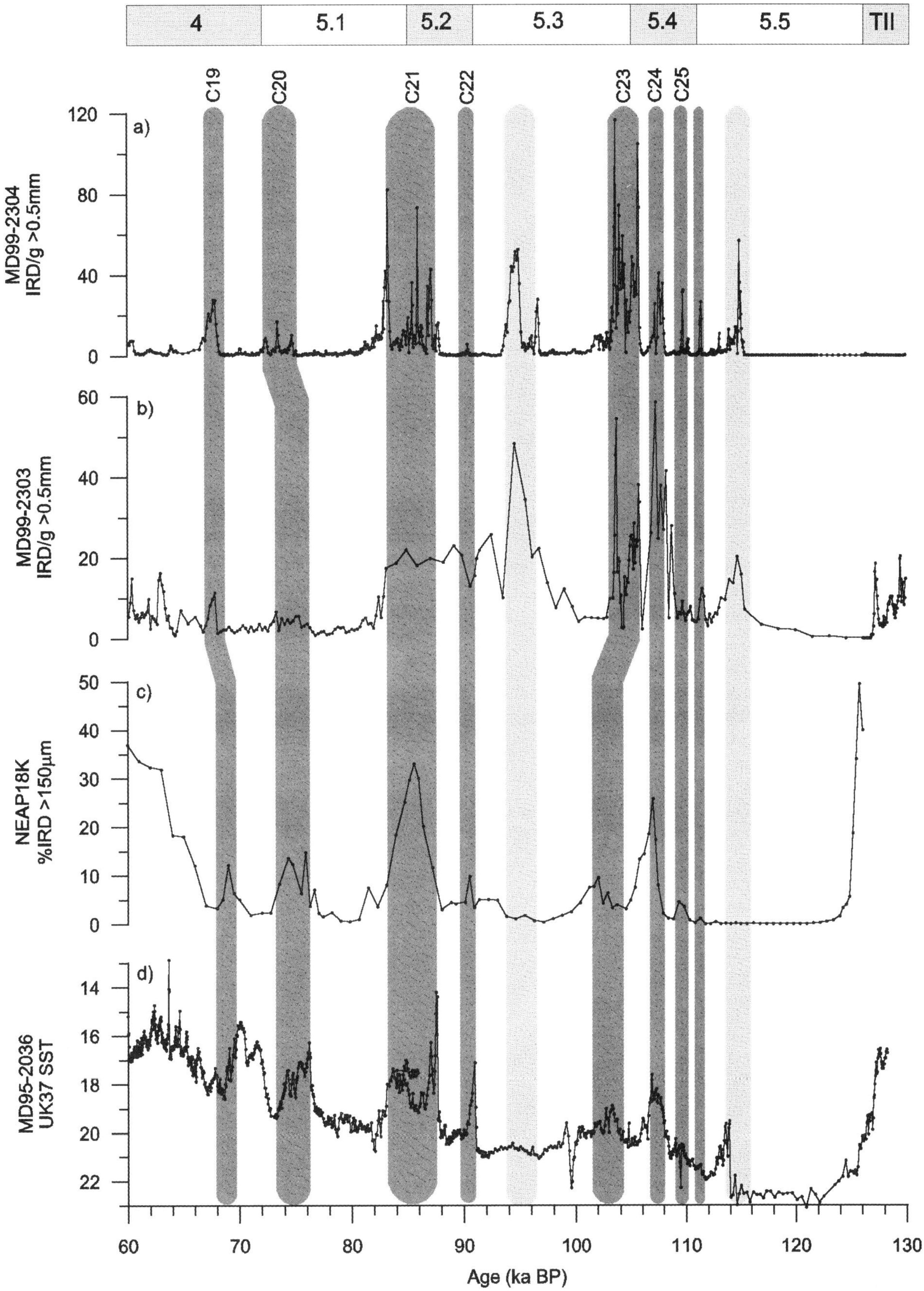
4
5.1
5.2
5.3
5.4
5.5
TII
C19
C20
C21
C22
C23
C24
C25
a)
MD99-2304
IRD/g >0.5mm
b)
MD99-2303
IRD/g >0.5mm
c)
NEAP18K
%IRD >150μm
d)
MD95-2036
UK37 SST
Age (ka BP)

the Barents Sea was covered by the SBIS during the MIS 5.2 glaciation. Model studies have shown that if the influx of Norwegian Atlantic water to the Barents Sea is prevented, the input of warm Atlantic water into the Arctic Ocean through the Fram Strait increases [*Butt et al.*, 2002]. During the maximum position of the MIS 5.2 ice sheet, such a scenario may have occurred and the resulting intensification of the WSC could have triggered the deglaciation. However, no evidence exists for such an ice sheet extent during the MIS 5.4 glaciation [*Mangerud et al.*, 1998]. The above scenario would not be valid for this time, and other processes most likely caused the water mass changes observed.

Most of the IRD events recorded at MD99-2304 and MD99-2303 can be correlated to similar events in the North Atlantic [*Chapman and Shackleton*, 1999] (Figure 8a-c). This indicates a more or less simultaneous response of the major Northern Hemisphere ice sheets throughout MIS 5. The occurrence of contemporaneous IRD events originating from such different sources makes it less likely that they result from internal ice sheet dynamics [*MacAyeal*, 1993]. Rather, several external-forcing mechanisms probably contributed to trigger the calving. The importance of the different mechanisms probably changed through time, depending on the present boundary conditions of the climate system. However, the widespread distribution of these cold spell events (C19–C25 [*Chapman and Shackleton*, 1999; *McManus et al.*, 1994]) along the axis of the Gulf Stream and its subsequent warm northward-flowing currents accentuates the importance of the ocean dynamics and its influence on the climate system. From the results from core MD99-2304, it appears that at times even slight changes in the water mass configuration may destabilize a marine-based ice sheet. *Hall et al.* [1998] showed that a vigorous behavior of the AMOC did characterize at least the glacial inception phase and the period into MIS 5.4. At the periods with major calving of the Northern Hemisphere ice sheets, large amounts of freshwater were introduced to the main areas of deep-water formation, both in the Nordic Seas and in the Labrador Sea. Increased freshwater flux to these sensitive areas most probably perturbed the deep-water formation and weakened the AMOC [e.g. *Clark et al.*, 2002; *Ganopolski and Rahmstorf*, 2001; *Manabe and Stouffer*, 1997]. A reorganization of the deep-water circulation related to the North Atlantic IRD events has been documented [*Chapman and Shackleton*, 1999; *Hall et al.*, 1998; and *Lehman et al.*, 2002], supporting the interpretation of a weakened AMOC as a response of the increased freshwater influx. The regional impact of a weakened AMOC will be large. Generally colder conditions will be induced in the Nordic Seas, the North Atlantic, and surrounding areas [*Manabe and Stouffer*, 1997]. Such coolings are documented by several paleoclimatic studies, e.g. the regionally distributed IRD events are associated with cold atmospheric temperatures at Greenland [*Bender et al.*, 1994] and cold sea-surface conditions in both the subpolar North Atlantic [*McManus et al.*, 1994] and the subtropical North Atlantic [*Lehman et al.*, 2002] (Figure 8a-d).

Two of the main IRD events reconstructed from the Fram Strait cores have no equivalents in the North Atlantic records (Figure 8a–c). The global ice volume at these times was restricted, while locally the SBIS was large enough to initiate pronounced calving during initial growth phases (Figure 4a–c). The event at approximately 95 ka BP shows no indications of regional influence (Figure 8a-d); it is, however, also found in previous studies from the Fram Strait [*Hebbeln and Wefer*, 1997]. The IRD event recorded in the Fram Strait at approximately 116 ka BP can possibly be related to the first cooling event observed in the subtropics after the last interglacial [*Lehman et al.*, 2002] (Figure 8a,b,d). This would imply that even restricted calving might be of regional importance, depending on the prevailing boundary conditions. Several studies have argued that the occurrence of the IRD events is dependent on the crossing of an ice-volume threshold [*Chapman and Shackleton*, 1999; *McManus et al.*, 1999; *Schulz et al.*, 1999]. This may be the case for most large-scale events of overregional importance, while locally the ice sheet configuration of the main source will determine the recorded IRD signature.

7. SUMMARY AND CONCLUSIONS

New high-resolution records from the Fram Strait and the Vøring Plateau provide new knowledge on the climate variability and oceanography of MIS 5. Interglacial conditions are documented to have existed in the Fram Strait during MIS 5.5. However, the area influenced by warm Atlantic water was restricted to the easternmost parts of the strait. Slightly further west, Arctic water masses prevailed throughout the interglacial. The Arctic front was located between the two studied sites. In periods with an increased strength of the WSC, the position of the Arctic front shifted eastwards. This was the case during an early interglacial optimum and during the glacial inception phase. In general, the MIS 5.5 climate seems to have been relatively unstable due to variable AMOC. Nevertheless, the amplitude of this variability was much smaller than glacial climate changes.

Glaciers at Spitsbergen grew large enough to initiate major calving before any large global changes in ice volume occurred. This early glacial inception was initiated by the strong sea–land temperature gradient resulting from the contemporary warm sea-surface and summer insolation minimum. This maintained significant moisture fluxes toward land, concomitant with reduced summer melting, which resulted in rapid glacier growth. We also propose that similar conditions were important for the initiation of glacier growth in high Northern Russia, related to an intensified deflection of Atlantic water into the Barents Sea when the AMOC is strengthened.

The concurrence of millennial-scale cold events during MIS 5.4-5.1, along the axis of the warm water transport from the subtropics to the northernmost Nordic Seas, indicates a common forcing mechanism and emphasizes the importance of ocean dynamics. We suggest that small changes in water mass configuration together with related changes in the global sea level triggered calving of major Northern Hemisphere ice sheets. Large discharges of melt water to the main deep-water formation sites weakened the AMOC and led to regionally colder conditions during these events. Major IRD events are related to growth, maxima, or decay periods of the ice sheet. Overregional important IRD events were probably dependent on the crossing of a global ice-volume threshold, while locally the ice sheet configuration of the main source determined the recorded IRD signal.

Acknowledgments. Dag Inge Blindheim, Svein Tore Gaski, Helge Hellevang, Odd Hansen, and Rune Søraas are thanked for help with sample preparation and laboratory assistance. We would also like to thank Scott Lehman for providing us with data from MD95-2036. The records from NEAP18K were obtained from http://www.ngdc.noaa.gov/paleo/data.html. The reviewers, Robert Spielhagen and Jan Sverre Laberg, are thanked for their comments and suggestions on the manuscript. A third reviewer also provided comments of relevance for the paper. This is publication No. A67 of the Bjerknes Centre for Climate Research.

REFERENCES

Aalbersberg, G., and T. Litt, Multiproxy climate reconstructions for the Eemian and Early Weichselian, *Journal of Quaternary Science*, 13, 367-390, 1998.

Adkins, J.F., E.A. Boyle, L. Keigwin, and E. Cortijo, Variability of the North Atlantic thermohaline circulation during the last interglacial period, *Nature*, 390, 154-156, 1997.

Bauch, H.A., H. Erlenkauser, K. Fahl, R. Spielhagen, M.S. Weinelt, H. Andruleit, and R. Henrich, Evidence for a steeper Eemian than Holocene sea surface temperature gradient between Arctic and sub-Arctic regions, *Palaeogeography, Palaeoclimatology, Palaeoecology*, 145, 95-117, 1999.

Baumann, K.-H., K.S. Lackschewitz, J. Mangerud, R. Spielhagen, T.C.W. Wolf-Welling, R. Henrich, and H. Kassens, Reflection of Scandinavian Ice Sheet Fluctuations in Norwegian Sea Sediment during the Past 150,000 Years, *Quaternary Research*, 43, 185-197, 1995.

Bender, M.,T. Sowers, M.-L. Dickson, J. Orchardo, P. Grootes, P.A. Mayewski, and D.A. Meese, Climate correlations between Greenland and Antarctica during the past 100,000 years, *Nature*, 732, 663-666, 1994.

Bintanja, R., R.S.W. van de Wal, and J. Oerlemans, Global ice volume variations through the last glacial cycle simulated by a 3-D ice-dynamical model, *Quaternary International*, 95-96, 11-23, 2002.

Birks, C.J.A., and N. Koç, A high-resolution diatom record of late-Quaternary sea-surface temperatures and oceanographic conditions from the eastern Norwegian Sea, *Boreas*, 31, 323-344, 2002.

Blindheim, J., Arctic Intermediate Water in the Norwegian Sea, *Deep-Sea Research*, 37, 1475-1489, 1990.

Blindheim, J., V. Borovkov, B. Hansen, S.A. Malmberg, W.R. Turrell, and S. Østerhus, Upper layer cooling and freshening in the Norwegian Sea in relation to atmospheric forcing, *Deep-Sea Research I*, 47, 655-680, 2000.

Bond, G., W. Showers, M. Cheseby, R. Lotti, P. Almasi, P deMenocal, P. Priore, H. Cullen, I. Hajdas, and G. Bonani, A Pervasive Millennial-Scale Cycle in North Atlantic Holocene and Glacial Climates, *Science*, 278, 1257-1266, 1997.

Butt, F.A., H. Drange, A. Elverhøi, O.H. Otterå, and A. Solheim, Modelling Late Cenozoic isostatic elevation changes in the Barents Sea and their implications for oceanic and climatic regimes: preliminary results, *Quaternary Science Reviews*, 21, 1643-1660, 2002.

Caspers, G., J. Merkt, H. Müller, and H. Freund, The Eemian Interglaciation in Northwestern Germany, *Quaternary Research*, 58, 49-52, 2002.

Chapman, M.R., and N.J. Shackleton, Global ice-volume fluctuations, North Atlantic ice-rafting events, and deep-ocean circulation changes between 130 and 70 ka, *Geology*, 27, 795-798, 1999.

Clark, P.U., N.G. Pisias, T. Stocker, and A.J. Weavers, The role of the thermohaline circulation in abrupt climate change, *Nature*, 415, 863-869, 2002.

Cortijo, E., J.C. Duplessy, L. Labeyrie, H. Leclaire, J. Duprat, and T.C.E. van Weering, Eemian cooling in the Norwegian Sea and North Atlantic ocean preceding continental ice-sheet growth, *Nature*, 372, 446-449, 1994.

Craig, H., and L.I. Gordon, Deuterium and Oxygen 18 Variations in the Ocean and the Marine Atmosphere. in *Stable Isotopes in the Oceanographic Studies and paleotemperatures*, edited by Tomgiorgi, E., pp. 9-130, Cons. Naz. Delle Ric. Lab. Di Geol. Nucl., Pisa, 1965.

Cutler, K.B., R.L. Edwards, F.W. Taylor, H. Cheng, J.Adkins, C.D. Gallup, P.M. Cutler, G.S. Burr, and A.L. Bloom, Rapid sea level fall and deep-ocean temperature change since the last interglacial period, *Earth and Planetary Science Letters*, 206, 253-271, 2003.

Dansgaard, W., S.J. Johnsen, H.B. Clausen, D. Dahl-Jensen, N.S. Gundestrup, C.U. Hammer, C.S. Hvidberg, J.P. Steffensen, A.E. Sveinbjörnsdottir, J. Jouzel, and G. Bond, Evidence for general instability of past climate from a 250-kyr ice-core record, *Nature*, 364, 218-220, 1993.

Dokken, T.M., and E. Jansen, Rapid changes in the mechanism of ocean convection during the last glacial period, *Nature*, 401, 458-461, 1999.

Duplessy, J.C., E. Ivanova, I. Murdmaa, M. Paterne, and L. Labeyrie, Holocene paleoceanography of the northern Barents Sea and variations of the northward heat transport by the Atlantic Ocean, *Boreas*, 30, 2-16, 2001.

Duplessy, J.C., and N.J. Shackleton, Response of global deep-water circulation to Earth's climatic change 135,000-107,000 years ago, *Nature*, 316, 500-507, 1985.

Elverhøi, A., E.S. Andersen, T. Dokken, D. Hebbeln, R. Spielhagen, J.I. Svendsen, M. Sørflaten, A. Rørnes, M. Hald, and C.F. Forsberg, The Growth and Decay of the Late Weichselian Ice Sheet in Western Svalbard and Adjecent Areas Based on Provenance Studies of Marine Sediments, *Quaternary Research*, 44, 303-316, 1995.

Field, M.H., B. Huntley, and H.Müller, Eemian climate fluctuations observed in a European pollen record, *Nature*, 371, 779-783, 1994.

Fogelquist, E., J. Blindheim, T. Tanhua, S. Østerhus, E. Buch, and F. Rey, Greenland-Scotland overflow studied by hydro-chemical multivariate analysis, *Deep-Sea Reseach I*, 50, 73-102, 2003.

Fronval, T., and E. Jansen, Rapid changes in ocean circulation and heat flux in the Nordic seas during the last interglacial period, *Nature*, 383, 806-810, 1996.

Fronval, T., and E. Jansen, Eemian and early Weichselian (140-60 ka) paleoceanography and paleoclimate in the Nordic Seas with comparitions to Holocene conditions, *Paleoceanography*, 12, 443-462, 1997.

Ganopolski, A., and S. Rahmstorf, Rapid changes of glacial climate simulated in a coupled climate model, *Nature*, 409, 153-158, 2001.

Gard, G., Late Quaternary calacreous nannofossil biozonation, chronology and palaeo-oceanography in areas North of the Faeroe-Iceland Ridge, *Quaternary Science Reviews*, 7, 65-78, 1988.

Graham, D.W., B.H. Corliss, M.L. Bender, and L.D. Keigwin Jr., Carbon and oxygen isotopic disequilibria of recent deep-sea benthic foraminifera, *Marine Micropaleontology*, 6, 483-497, 1981.

GRIP Members, Climate instability during the last interglacial period recorded in the GRIP ice core, *Nature*, 364, 203-207, 1993.

Grootes, P.M., M. Stuiver, J.W.C. Withe, S. Johnsen, and J. Jouzel, Comparison of oxygen isotope records from the GISP2 and GRIP Greenland ice cores, *Nature*, 366, 552-554, 1993.

Haake, F.W., and U. Pflaumann, Late Pleistocene foraminiferal stratigraphy on the Vøring Plateau, Norwegian Sea, *Boreas*, 18, 343-356, 1989.

Hall, I.R., I.N. McCave, M.R. Chapman, and N.J. Shackleton, Coherent deep flow variation in the Iceland and American basins during the last interglacial, *Earth and Planetary Science Letters*, 164, 15-21, 1998.

Hebbeln, D.,T. Dokken, E.S. Andersen, M. Hald, and A. Elverhøi, Moisture supply for northern ice-sheet growth during the Last Glacial Maximum, *Nature*, 370, 357-360, 1994.

Hebbeln, D., R. Henrich, and K.-H. Baumann, Paleoceanography of the last interglacial/glacial cycle in the polar North Atlantic, *Quaternary Science Reviews*, 17, 125-153, 1998.

Hebbeln, D., and G. Wefer, Late Quaternary paleoceanography in the Fram Strait, *Paleoceanography*, 12, 65-78, 1997.

Imbrie, J., E.A. Boyle, S.C. Clemens, A. Duffy, W.R. Howard, G. Kukla, J. Kutzbach, D.G. Martinson, A. McIntyre, A.C. Mix, B. Molfino, J.J. Morley, L.C. Peterson, N.G. Pisias, W.L. Prell, M.E. Raymo, N.J. Shackleton, and J.R. Toggweiler, On the structure and origin of major glaciation cycles 1. Linear responses to Milankovitch forcing, *Paleoceanography*, 7, 701-738, 1992.

Jansen, E., U. Bleil, R. Henrich, L. Kringstad, and B. Slettemark, Paleoenvironmental changes in the Norwegian Sea and the Northeast Atlantic during the last 2.8 m.y.: Deep Sea Drilling Project/Ocean Drilling Program Sites 610, 642, 643 and 644, *Paleoceanography*, 3, 563-581, 1988.

Johannessen, T., E. Jansen, A. Flatøy, and A.C. Ravelo, The relationship between surface water masses, oceanographic fronts and paleoclimatic proxies in surface sediments of the Greenland, Iceland, Norwegian Seas. in *Carbon Cycling in Glacial Ocean: Constraints on the Ocean's Role in Global Change*, edited by Zahn, R., M. Kominski and L. Labyrie, pp. 61-85, Springer-Verlag, New York, 1994.

Jones, G. A., and L.D. Keigwin, Evidence from Fram Strait (78° N) for early deglaciation, *Nature*, 336, 56-59, 1988.

Karlén, W., Scandinavian glacial and climatic fluctuations during the Holocene, *Quaternary Science Reviews*, 7, 199-209, 1988.

Khodri, M., Y. Leclainche, G. Ramstein, P. Braconnot, O. Marti, and E. Cortijo, Simulating the amplification of orbital forcing by ocean feedbacks in the last glaciation, *Nature*, 410, 570-574, 2001.

Koç Karpuz, N., and E. Jansen, A high-resolution diatom record of the last deglaciation from the SE Norwegian Sea: Documentation of rapid climatic changes, *Paleoceanography*, 7, 499-520, 1992.

Kristensen, P., K.L.Knudsen, H. Lykke-Andersen, E. Nørmark, J.D. Peacock, and A. Sinnott, Interglacial and glacial climate oscillations in a marine shelf sequence from northern Denmark - a multidisciplinary study, *Quaternary Science Reviews*, 17, 813-837, 1998.

Köhler, S.E.I., and R.F. Spielhagen, The enigma of oxygen isotope stage 5 in the central Fram Strait. in *Geological History of the Polar Oceans: Arctic Versus Antarctic*, edited by Bleil, U. and J. Thiede, pp. 489-497, Kluwer Academic Publisher, 1990.

Laskar, J., The chaotic motion of the solar system: A numerical estimate of the chaotic zones, *Icarus*, 88, 266-291, 1990.

Lehman, S.J., J.P. Sachs, A.M. Crotwell, L.D. Keigwin, and E.A. Boyle, Relation of subtropical Atlantic temperature, high-latitude ice rafting, deep water formation, and European climate 130,000-60,000 years ago, *Quaternary Science Reviews*, 21, 1917-1924, 2002.

MacAyeal, D.R., Binge/purge oscillation of the Laurentide Ice Sheet as a cause of the North Atlantic`s Heinrich Events, *Paleoceanography*, 8, 775-784, 1993.

Manabe, S., and R.J. Stouffer, Coupled ocean-atmosphere model response to freshwater input: Comparison to Younger Dryas event, *Paleoceanography*, 12, 321-336, 1997.

Mangerud, J., V.I. Astakhov, A. Murray, and J.I. Svendsen, The chronology of a large ice-dammed lake and the Barents-Kara Ice Sheet advances, Northern Russia, *Global and Planetary Change*, 31, 321-326, 2001.

Mangerud, J., T. Dokken, D. Hebbeln, B. Heggen, Ó. Ingólfsson, J.Y. Landvik, V. Mejdahl, J.I. Svendsen, and T.O. Vorren, Fluctuations of the Svalbard-Barents sea ice sheet during the last 150 000 years, *Quaternary Science Reviews*, 17, 11-42, 1998.

Mangerud, J., and J.I. Svendsen, The last interglacial-glacial period on Spitsbergen, Svalbard, *Quaternary Science Reviews*, 11, 633-664, 1992.

Manley, T.O., Branching of Atlantic Water within the Greenland-Spitsbergen Passage: An estimate of recirculation, *Journal of Geophysical Research*, 100, 20627-20634, 1995.

Martinson, D.G., N.G. Pisias, J.D. Hays, J. Imbrie, T.C. Moore Jr., and N.J. Shackleton, Age Dating and the Orbital Theory of the Ice Ages: Development of a High-Resolution 0 to 300,000-Year Chronostratigraphy, *Quaternary Research*, 27, 1-29, 1987.

McManus, J.F., G.C. Bond, W.S. Broecker, S. Johnsen, L. Labeyrie, and S. Higgins, High-resolution climate records from the North Atlantic during the last interglacial, *Nature*, 371, 326-329, 1994.

McManus, J.F., D.W. Oppo, and J.L. Cullen, A 0.5-Million-Year Record of Millennial-Scale Climate Variability in the North Atlantic, *Science*, 283, 971-975, 1999.

McManus, J.F., D.W. Oppo, L.D. Keigwin, J.L. Cullen, and G.C. Bond, Thermohaline Circulation and Prolonged Interglacial Warmth in the North Atlantic, *Quaternary Research*, 58, 17-21, 2002.

Nesje, A., and S.O. Dahl, The 'Little Ice Age'- only temperature?, *The Holocene*, 13, 139-145, 2003.

Oppo, D.W., M. Horowitz, and S.J. Lehman, Marine core evidence for reduced deep water production during Termination II followed by a relatively stable substage 5e (Eemian), *Paleoceanography*, 12, 51-63, 1997.

Oppo, D.W., L.D. Keigwin, J.F. McManus, and J.L. Culen, Persistent suborbital climate variability in marine isotope stage 5 and Termination II, *Paleoceanography*, 16, 280-292, 2001.

Otterå, O.H., and H.Drange, Effects of solar irradiance forcing on the ocean circulation in the North Atlantic in an isopycnic coordinate OGCM, *Tellus Series A - Dynamic meteorology and oceanography*, 56, 154-166, 2004.

Otterå, O.H., and H. Drange, Effects of solar irradiance forcing on the ocean circulation in the North Atlantic in an isopycnic coordinate OGCM, *Tellus*, in press 2004.

Paillard, D., L. Labeyrie, and P. Yiou, Macintosh program performs time-series analysis, *EOS, Trans. AGU*, TI, 379, 1996.

Risebrobakken, B., E. Jansen, C. Andersson, E. Mjelde, and K. Hevrøy, A high-resolution study of Holocene paleoclimatic and paleoceanographic changes in the Nordic Seas, *Paleoceanography*, 18, 1017, 2003.

Ruddiman, W.F., and A. McIntyre, Warmth of the Subpolar North Atlantic Ocean During Northern Hemisphere Ice-Sheet Growth, *Science*, 204, 173-175, 1979.

Ruddiman, W.F., A. McIntyre, V. Niebler-Hunt, and J.T. Durazzi, Oceanic Evidence for the Mechanism of Rapid Northern Hemisphere Glaciation, *Quaternary Research*, 13, 33-64, 1980.

Saloranta, T.M., and H. Svendsen, Across the Arctic front west of Spitsbergen: high-resolution CTD sections from 1998-2000, *Polar Research*, 20, 177-184, 2001.

Schulz, M., W.H. Berger, M. Sarnthein, and P.M. Grootes, Amplitude variations of 1470-year climate oscillations during the last 100,000 years linked to fluctuations of continental ice mass, *Geophysical Research Letters*, 26, 3385-3388, 1999.

Seidenkrantz, M.-S., P. Kristensen, and K.L. Knudsen, Marine evidence for climatic instability during the last interglacial in shelf records from northwest Europe, *Journal of Quaternary Science*, 10, 77-82, 1995.

Sejrup, H.P., H. Haflidason, D. Klitgaard Kristensen, and S.J. Johnsen, Last interglacial and Holocene climatic development in the Norwegian Sea region: ocean front movements and ice-core data, *Journal of Quaternary Science*, 10, 385-390, 1995.

Sejrup, H.P., E. Larsen, J. Landvik, E.L. King, H. Haflidason, and A. Nesje, Quaternary glaciations in southern Fennoscandia: evidence from southwestern Norway and the northern North Sea region, *Quaternary Science Reviews*, 19, 667-685, 2000.

Shackleton, N.J., and N.D. Opdyke, Oxygen Isotope and Palaeomagnetic Stratigraphy of Equatorial Pacific Core V28-238: Oxygen Isotope Temperatures and Ice Volumes on a 105 Year and 106 Year Scale, Quaternary Research, 3, 39-55, 1973.

Shackleton, N.J., M.F. Sánchez-Goñi, D. Pailler, and Y. Lancelot, Marine Isotope Substage 5e and the Eemian Interglacial, *Global and Planetary Change*, 36, 151-155, 2003.

Siegert, M.J., J.A. Dowdeswell, M. Hald, and J.I. Svendsen, Modelling the Eurasian Ice Sheet through a full (Weichselian) glacial cycle, *Global and Planetary Change*, 31, 367-385, 2001.

Spielhagen, R.F., K.H. Baumann, H. Erlenkauser, N.R. Nowaczyk, N. Nørgaard-Pedersen, C. Vogt, and D. Weiel, Arctic Ocean deep-sea record of Northern Eurasian ice sheet history, *Quaternary Science Reviews*, 23, 1455-1483, 2004.

Steinsund, P.I., and M. Hald, Recent calcium carbonate dissolution in the Barents Sea: Paleoceanographic applications, *Marine Geology*, 117, 303-316, 1994.

Stirling, C.H., T.M. Esat, K. Lambeck, and M.T. McCulloch, Timing and duration of the Last Interglacial: evidence for a restricted interval of widespread coral reef growth, *Earth and Planetary Science Letters*, 160, 745-762, 1998.

Svendsen, J.I., H. Alexanderson, V.I. Astakhov, I. Demidov, J.A. Dowdeswell, S. Funder, G. Gataullin, M. Henriksen, C. Hjort, M. Houmark-Nielsen, H.W. Hubberten, Ó. Ingólfsson, M. Jakobsson, K.H. Kjær, E. Larsen, H.Lokrantz, J.P. Lunkka, A. Lyså, J. Mangerud, A. Matiouchkov, A. Murray, P. Möller, F. Nisessen, O. Nikolskaya, L. Polyak, M. Saarnisto, C. Siegert, M. J. Siegert, R. F. Spielhagen, and R. Stein, Late Quaternary ice sheet history of Northern Eurasia, *Quaternary Science Reviews*, 23, 1229-1271, 2004.

Svendsen, J.I., V.I. Astakhov, D.Y. Bolshiyanov, I. Demidov, J.A. Dowdeswell, V. Gataullin, C. Hjort, H.W. Hubberten, E. Larsen, J. Mangerud, M. Melles, P. Möller, M. Saarnisto, and M.J. Siegert, Maximum extent of the Eurasian ice sheets in the Barents and Kara Sea region during the Weichselian, *Boreas*, 28, 234-242, 1999.

Swift, J.H., The Arctic Waters. in *The Nordic Seas*, edited by Hurdle, B.G., pp. 129-153, Springer-Verlag, 1986.

Taylor, K.C., C.U. Hammer, R.B. Alley, H.B. Clausen, D. Dahl-Jensen, A.J. Gow, N.S. Gundestrup, J. Kipfstuhl, J.C. Moore, and E.D. Waddington, Electrical conductivity measurements from the GISP2 and GRIP Greenland ice cores, *Nature*, 366, 549-552, 1993.

Thouveny, N., J.L. de Beaulieu, E. Bonifay, K.M. Creer, J. Guiot, M. Icole, S. Johnsen, J. Jouzel, M. Reille, T. Williams, and Williamson, Climate variations in Europe over the past 140 kyr deduced from rock magnetism, *Nature*, 371, 503-506, 1994.

Wang, Z., and L.A. Mysak, Simulation of the last glacial inception and rapid ice sheet growth in the McGill Paleoclimate Model, *Geophysical Research Letters*, 29, 2102, 2002.

Winsnes, T.S., Bedrock map of Svalbard and Jan Mayen, Norsk Polarinstitutt, Oslo, 1988.

Yoshimori, M., M.C. Reader, A.J. Weaver, and N.A. McFarlane, On the causes of glacial inception at 116 ka BP, *Climate Dynamics*, 18, 383-402, 2002.

Zhang, J., D.A. Rothrock, and M. Steele, Warming of the Arctic Ocean by a strengthened Atlantic inflow: Model results, *Geophysical Research Letters*, 25, 1745-1748, 1998.

Trond Dokken, Eystein Jansen, and Bjørg Risebrobakken, Bjerknes Centre for Climate Research, University of Bergen, Allégaten 55, 5007 Bergen, Norway. (trond.dokken@bjerknes.uib.no; eystein.jansen@geo.uib.no; bjorg.risebrobakken@geo.uib.no)

Exploratory Comparisons of Quantitative Temperature Estimates Over the Last Deglaciation in Norway and the Norwegian Sea

Hilary H. Birks[1,2], Dorthe Klitgaard Kristensen[3], Trond M. Dokken, and Carin Andersson

Bjerknes Centre for Climate Research, Bergen, Norway

This paper presents an exploratory synthesis of quantitative temperature reconstructions during the last deglaciation in Norway and the Norwegian Sea. The variety of proxy data available permits an overview of climate development and comparisons between land and sea records. Temperature reconstructions from each available proxy were averaged for the time slices "late-glacial interstadial" (Greenland Interstadial 1; ca. 14,000-12,700 cal BP) and "Younger Dryas" (Greenland Stadial 1; 12,700-11,500 cal BP) and at the end of the early Holocene rapid warming phase. When mapped, the data reveal spatial and temporal differentiation. Interstadial terrestrial July temperatures were ca. 10°C in the south to ca. 6°C in the north (gradient of ca. 4°C). Summer sea-surface temperatures were ca. 8°C in the south and ca. 5°C in the north (gradient of 2-3°C). The Younger Dryas July gradient was reduced on land to ca. 2-3°C and to 0°C at sea because southern temperatures cooled more than northern ones. Interstadial and Younger Dryas winter temperatures were similar on land and in the sea. In the early Holocene, the terrestrial July gradient steepened slightly to about 3°C. This is less than today (14°C and 10°C at the south and north Norwegian coasts, respectively) because southern July temperatures then were still 1-2°C lower than at present, but northern temperatures had already reached present values. The marine gradient reestablished but was weaker than the Interstadial gradient at ca. 1-3°C. Winter temperatures rose by at least 5°C on land, but less in the sea. The early Holocene climate was more seasonal than today. These results are compared with other data syntheses and model results.

1. INTRODUCTION

There is an increasing range of quantitative climate reconstructions for the Nordic Seas and the Norwegian region as more modern calibration or training data sets and transfer functions are developed for organisms in relation to environmental variables (Table 1 and references therein). In combination with other types of quantitative climate reconstructions, we can use these results from both marine and terrestrial proxies to compare past regional climates and climate changes over the late-glacial and early Holocene, particularly for temperatures, especially summer (July) temperatures. This is most easily done by a mapped approach, which allows

Also at:
[1]Department of Biology, University of Bergen, Bergen, Norway.
[2]Environmental Change Research Centre, University College London, UK.
[3]Department of Earth Science, University of Bergen, Bergen, Norway.

The Nordic Seas: An Integrated Perspective
Geophysical Monograph Series 158

10.1029/158GM21

Table 1. Sites Shown on Figure 2, the Proxies Used, and the Site References. Temperatures were Reconstructed from all Proxies using Transfer Function Methods, Except for Macrofossils (indicator species and assemblages) and Coleoptera (Mutual Climate Range Method).

Site	Proxy	Reference
I = Ifjord	Pollen	*Seppä et al.* (2002)
H = Jansvatnet, Hammerfest	Pollen, macrofossils	*H.H. Birks & S.M. Peglar* (unpublished data)
D = Dalmutladdo	Pollen, macrofossils	*Bjune et al.* (2004)
A = Austerkjøsen	Pollen	*H.J.B. Birks & S.M. Peglar* (unpublished data)
L = Litlevatnet	Pollen	*H.J.B. Birks & S.M. Peglar* (unpublished data)
S = Svartvatnet	Pollen	*H.J.B. Birks & S.M. Peglar* (unpublished data)
T = Tiåvatnet	Pollen	*H.J.B. Birks & S.M. Peglar* (unpublished data)
G = Godøya	Macrofossils, Coleoptera, pollen	*Birks et al.* (1993)
K = Kråkenes	Macrofossils, pollen, Coleoptera, Chironomidae, Cladocera	*Birks & Ammann* (2000)
BL = Blomøy	Macrofossils	*Birks* (1993)
F = Vestre Øykjamyrtjørn, Folgefonna	Pollen, macrofossils	*Bjune et al.* (2005)
U = Utsira	Macrofossils	*Birks* (1993)
E = Eigebakken	Macrofossils	*van Dinter & Birks* (1996)
Bj = Bjerkreim	Pollen, Chironomidae, Macrofossils	*Aa. Paus, S.J. Brooks & H.H. Birks,* (unpublished data)
Da = Dalane, Setesdal	Pollen	*Eide et al.* (in press)
M = Morttjenn	Pollen	*H.J.B. Birks & S.M. Peglar* (unpublished data)
V = Voldafjord	Dinoflagellates	*Grøsfjeld et al.* (1999), *Sejrup et al.* (2001)
T-88-2	Foraminifera, Diatoms	*Hald et al.* (1996), *Koç et al.* (1996)
T-79-51/2	Foraminifera	*Hald et al.* (1996)
MD 95-2011	Diatoms, Foraminifera	*Birks* (2001), *Birks & Koç* (2001), *Andersson et al.* (2003), *Risebrobakken et al.* (2003), *Dreger* (1999), *T.M. Dokken* (in prep.).
MD 95-2010	Foraminifera	*Dokken & Jansen* (1999)
M 23071	Diatoms	*Koç et al.* (1993)
HM 79-6/4	Diatoms	*Koç Karpuz & Jansen* (1992)
Troll	Foraminifera, Dinoflagellates	*D. Klitgaard Kristiansen et al.* (unpublished data)
MD99-2284	Foraminifera	*T.M. Dokken* (unpublished data)
ENAM 93-21	Foraminifera	*Rasmussen et al.* (1996)

a geographic synthesis and visualisation of patterns in the reconstructions through time and space.

Previous syntheses of the late-glacial terrestrial climate reconstructions have been presented for Norway by *Birks* (1994) and *Birks et al.* (1994) and for the northwest European area by NASP members (*Lowe et al.*, 1994; *Lowe and NASP members*, 1995) and *Coope et al.* (1998). Mapped reconstructions for the Norwegian Sea were presented by *Koç et al.* (1993), who used diatom data, and for the North Atlantic and Nordic Seas by *de Vernal and Hillaire-Marcel* (2000), who used dinoflagellate data. New syntheses for the Last Glacial Maximum, based on foraminifera and other proxy data, are being prepared in the GLAMAP (*Pflaumann et al.*, 2003) and MARGO (unpublished) projects to update the CLIMAP (1981) synthesis. Reconstructions for late-glacial climate parameters from pollen and coleopteran data have been compared to model results by *Renssen and Isarin* (2001) for areas south of 60°N. The synthesis presented here builds on and compares with these previous syntheses, but the most important and novel development is the combination of marine and terrestrial temperature reconstructions, which provides an overview of climate change in the whole Norwegian/Norwegian Sea area over the last deglacial period and the early Holocene.

Proxy data and temporal records are diverse and difficult to compare directly. If they are converted to quantitative climate estimates (temperature) through the use of transfer functions, indicator species or assemblages, or glacial parameters (Table 1), the data are now in a common language and can thus be compared and synthesised (e.g. *Vandenberghe et al.*, 1998). This has several advantages. We can compare marine and terrestrial climates and climate changes. We can map geographical patterns. The reconstructions could be used as input and boundary conditions for climate models or for testing model predictions. Model predictions over areas of steep late-glacial climatic gradients such as are found in the Nordic Seas (e.g. CLIMAP, 1981) and the Norwegian region could also be tested by selecting appropriate new sites. We could thus evaluate the role of different forcing factors and generate or test hypotheses about the processes and causes of climate change over the last deglaciation. Our approach allows us to compare results from different proxies, which could reveal potential anomalies and areas where future research is needed.

In this synthesis, we take a broad-scale exploratory approach with the data currently available to map reconstructed temperatures from marine and terrestrial sites in the Norwegian region through three time-slices during the late-glacial and early Holocene. We look for geographical gradients and compare the magnitude of temperature changes through time on land and in the sea. The deglacial period is well-known for its large climate changes, which are relatively easy to identify in stratigraphic records. A quantitative synthesis of these climate changes over the land and sea in the Norwegian region can be used to further investigate causes of the changes, which include changes in ocean circulation patterns. Information on these rapid climate changes in the past may throw valuable light on possible future changes in the thermohaline circulation related to global warming.

2. METHODS

2.1. Quantitative Climate Reconstructions

There are several ways in which quantitative climate reconstructions can be generated. Here we use reconstructions made by using three general methods.

2.1.1. Transfer functions. These quantitatively relate the present-day occurrence and abundance of organisms in modern surface samples to environmental variables, in our case, climate (*Birks*, 1981; 1995). A unique variety of climate transfer functions is available for the Norwegian region. For terrestrial sites, we have transfer functions between pollen and mean July and mean January temperature and annual precipitation (*Birks et al.*, 2000); between Cladocera and mean July temperature (*Birks et al.*, 2000); and between chironomids and mean July air temperature (*Brooks and Birks*, 2001). Reconstructions from pollen and chironomids are most frequent. For marine sites, we have transfer functions between sea-surface temperature (SST) and salinity and planktic foraminifera, using several different models (e.g. Weighted Averaging Partial Least Squares and the Atlantic core-top database ATL 947; *Pflaumann et al.*, 2003), benthic foraminifera (*Sejrup et al.*, 2004), diatoms (*Birks and Koç*, 2002), and dinoflagellates (*de Vernal et al.*, 1997). There are also transfer functions for some physical-chemical parameters, such as Mg/Ca (*Elderfield and Ganssen*, 2000).

Each transfer function reconstruction has an inherent error range, usually in the region of ±1°C. Therefore, subtle reconstructed temperature changes, such as during the late-glacial Interstadial (Bølling-Allerød), Younger Dryas, and early Holocene, may fall within the error range, and conversely, small reconstructed changes may be a matter of chance and not climatically significant. Therefore, in this survey, we use temperature reconstructions that are averaged over long periods, in an attempt to make sure that these values are not unduly affected by reconstruction uncertainties. Of course, we cannot be certain of this due to the variability in our modern data sets and reconstruction methods and because of factors other than temperature that may be affecting the responses of our proxy organisms.

We have used SST reconstructions from diatom and planktic foraminifera data; some reconstructions from dinoflagellates are also available. The results can be quite different between proxies, for example, between MD95-2011 diatom reconstructions (*Birks and Koç*, 2002) and planktic foraminifera reconstructions (*Andersson et al.*, 2003; *Risebrobakken et al.*, 2003). There are two main reasons for the differences in this example. Firstly, there are ecological differences between the organisms. Diatoms live in the photic zone and can be subjected to large seasonal variations in temperature, as they are directly influenced by insolation in addition to marine currents. The planktic foraminifera live closer to the mixed layer at depths that may range from 10 m to 100 m or even deeper, where temperatures are cooler and less affected by large seasonal changes. These organisms are influenced by the temperature of the water mass, which may differ from that at the surface (*Risebrobakken et al.*, 2003). Secondly, the modern calibration data sets are unsatisfactory in some ways for both proxies, and different reconstructions can result from using different numerical methods on the same data (*Birks*, 2001). All proxies vary for ecological factors in addition to temperature and all transfer function methodologies could be improved.

2.1.2. Indicator species and assemblages. This is essentially an analogue approach, using the estimated or known modern optima and tolerances of species or assemblages (e.g. vegetation types) to reconstruct past conditions. On land, plant macrofossils provide good species and vegetation reconstructions (e.g. *Birks*, 1993, 2003) and Coleoptera distributions are related to climate (*Atkinson et al.*, 1987). In the ocean, grain-size distribution of sediment can give information on current strengths (e.g. *Bianchi and McCave*, 1999), and polar species of particular diatoms and foraminifera (e.g. *Neogloboquadrina pachyderma* var. *sinistral*) can give information on near-surface water temperature.

2.1.3. Glacial proxies. The relationship between equilibrium line altitude (ELA) and ablation season (summer) temperature depends on winter accumulation, and thus the temperature can be reconstructed if the ELA is known and winter precipitation is known or held constant (*Larsen et al.*, 1984). More commonly, summer temperature is independently reconstructed and, with a known ELA, can be used to reconstruct winter precipitation (e.g. *Dahl and Nesje*, 1996; *Larsen and Stalsberg*, 2004; *Bjune et al.*, 2005).

The reconstruction of mean July temperature during the late-glacial and early Holocene at Kråkenes, western Norway (Figure 1) (*Birks and Ammann*, 2000) illustrates the results obtained from a range of multiproxy evidence from the same site. Quantitative transfer functions were used to reconstruct temperatures from each sample for pollen, chironomids, and cladocera; the Mutual Climate Range Method (*Atkinson et al.*, 1987) was used for Coleoptera reconstructions, and an analogue approach was used for the plant macrofossils that was not sample-specific. The patterns of the curves were similar, but the absolute values differed a little among proxies. Various hypotheses can be proposed to explain this, including difficulties with the modern calibration data sets, the different reconstruction methods used, and the responses of organisms to ecosystem factors other than mean July air temperature. The Kråkenes synthesis demonstrates the similarities and discrepancies among transfer functions and other reconstructions and thus the need for refinements in future investigations. However, the temperature reconstructions at Kråkenes and the others used in this paper are the best estimates that can be achieved at the moment with present techniques, existing analogues, and modern calibration data sets.

2.2. Data

The sites with quantitative temperature reconstructions during the late-glacial and early Holocene periods in the Norwegian region are shown in Figure 2. They cover the range 58-72°N, encompassing a north–south temperature gradient. Mean July temperature in Norway is mapped on Figure 2, but there are very sharp temperature, oceanicity, and altitudinal gradients from the coast to inland. Mean July temperature at Lindesnes (southernmost point) is 14°C and at North Cape is 10°C. The winter gradient is hard to assess due to altitudinal and west–east gradients, but areas at the south coast have mean January temperatures of around 1-2.5°C, whereas temperatures at the north coast are around −3.5°C to −5°C. Modern SSTs show a north–south summer gradient of about 7°C at the surface and about 3°C in below-surface temperatures, in addition to a temperature gradient from west to east, depending on the Atlantic and Norwegian currents (Figure 2). Terrestrial sites are situated beyond the limits of the readvance of Younger Dryas ice (Figure 2; see also Figures 4–6) and are concentrated in southern Norway. Few sites are located in the north, and none in the central area. Marine sites are scattered fairly evenly, but they are sparse. Two are located in the coastal zone of fjords. The proxies from each site and the appropriate

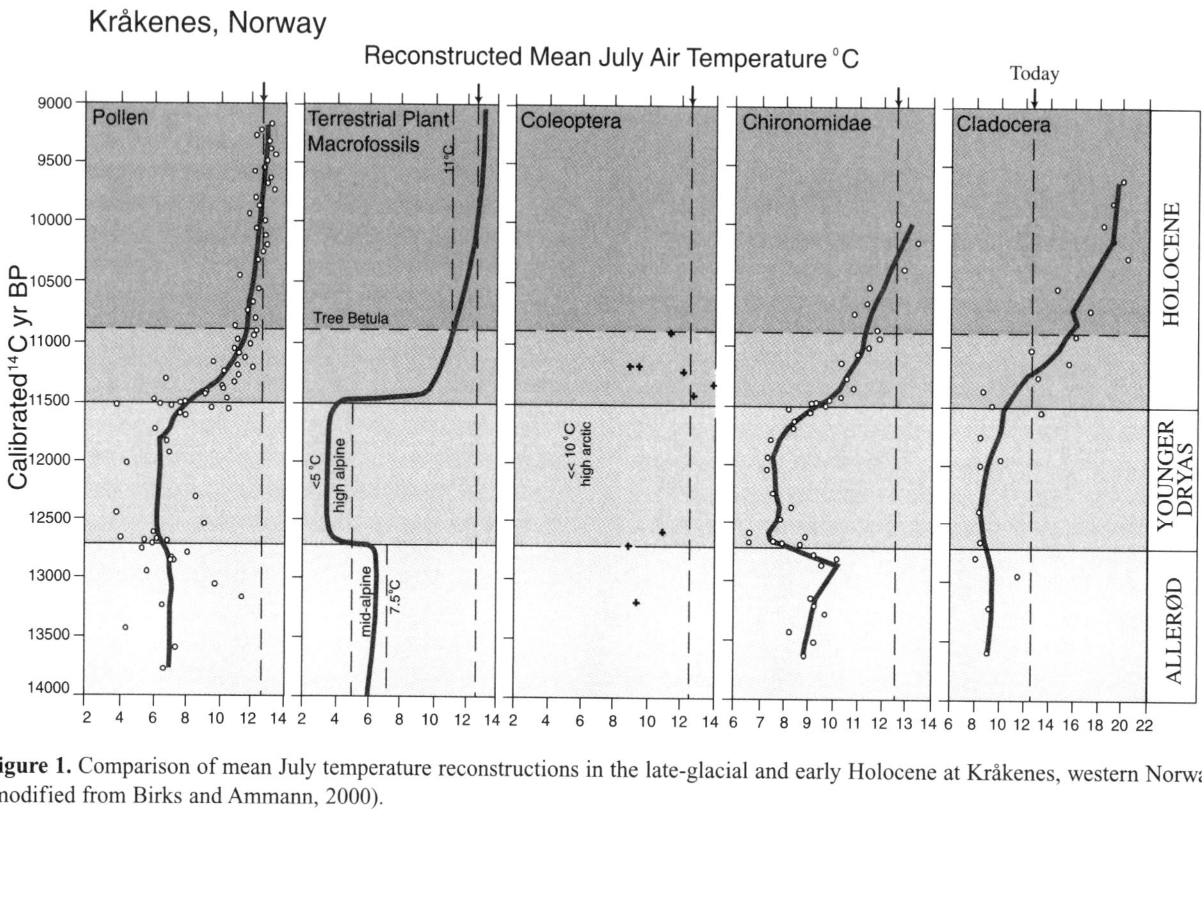

Figure 1. Comparison of mean July temperature reconstructions in the late-glacial and early Holocene at Kråkenes, western Norway (modified from Birks and Ammann, 2000).

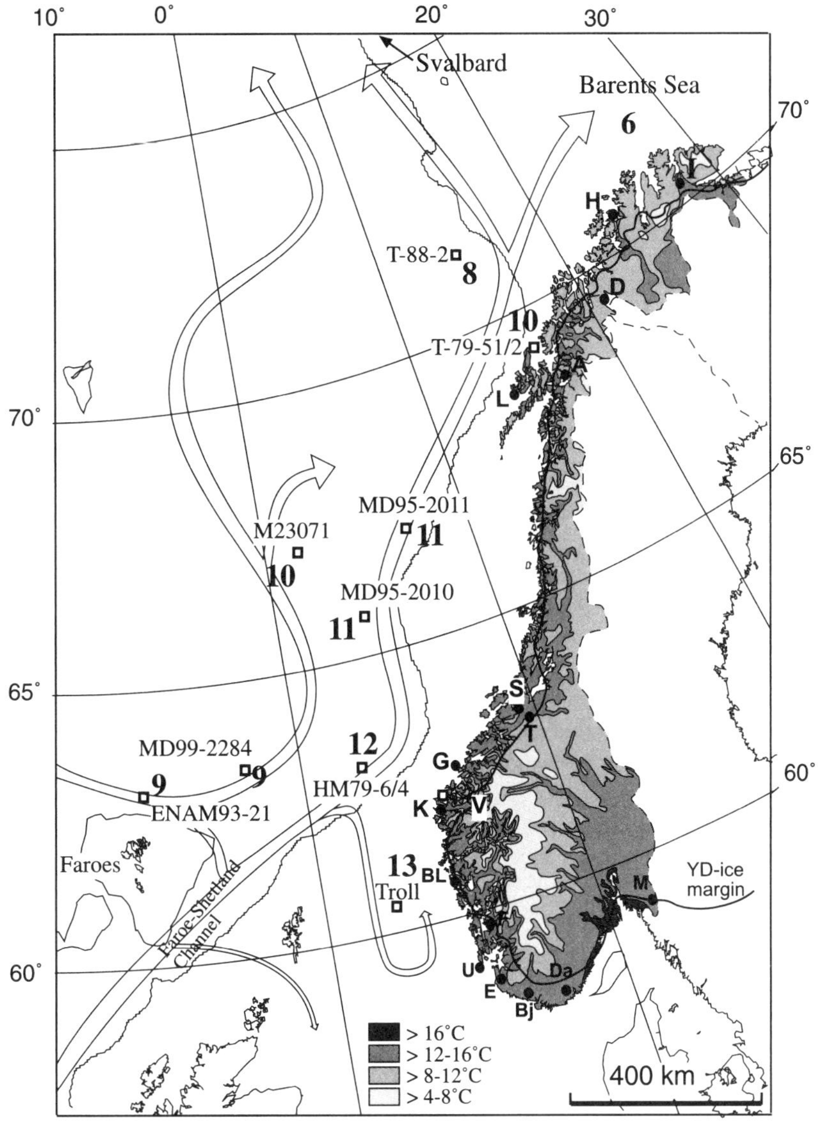

Figure 2. Map of late-glacial and early Holocene sites with quantitative temperature reconstructions in the Norwegian region. Squares are marine sites, circles are terrestrial sites. The courses of the major Atlantic currents at 50-200(500)-m water depth and the 500-m seawater depth contour are shown. The Vøring Plateau lies under MD95-2010 and MD95-2011. The sites and the proxies used to reconstruct temperatures are referenced in Table 1. I = Ifjord, H = Jansvatnet, Hammerfest, D = Dalmutladdo, A = Austerkjøsen, L = Litlevatnet, S = Svartvatnet, T = Tiåvatnet, G = Godøya, K = Kråkenes, BL = Blomøy, F = Vestre Øykjamyrtjørn, Folgefonna, U = Utsira, E = Eigebakken, Bj = Bjerkreim, Da = Dalane, M = Morttjenn, V = Voldafjord.

references or attributions to unpublished data are given in Tablc 1.

2.3. Time Slices and Maps

For this present purpose, three time slices were selected for comparison. The late-glacial Interstadial occurs between initial deglaciation and the Younger Dryas cold period. It is also known as Greenland Interstadial 1 or the Bølling–Allerød interstadial (*Björck et al.*, 1998). The time slice is taken to cover the available Interstadial, from a maximum of 14,000 cal BP to 12,700 cal BP. Records usually cover only the later part, the Allerød, especially on land, due to late ice-melting. If ^{14}C dating is unavailable or uncertain, Interstadial deposits can usually be readily identified by sediment-stratigraphic criteria or by changes in the fossil assemblages. The Younger Dryas is also usually obvious in the records, as sedimentary changes are accompanied by biotic changes indicating cooler temperatures. It is taken to cover 12,700-11,500 cal BP (*Birks et al.*, 2000). In almost all the marine sites and many of the terrestrial sites the Vedde ash layer has been identified. It is dated to 11,980 ice-core years BP in the GRIP ice core (*Grønvold et al.*, 1995) and to ca. 12,000 cal BP at Kråkenes (*Birks et al.*, 2000), giving a stratigraphic time marker within the Younger Dryas. In this synthesis, we use the temperature reached at the end of the Early Holocene steep temperature rise for the maps, not the highest temperature reached in the Holocene. The time at which this occurs varies from site to site, within about 1,000 years. This may be due to differences in the timing of crossing ecological thresholds by the organisms involved, to geographic climate variability, or to uncertainties in dating and low sample resolution. The steep temperature rise is usually clearly distinct from subsequent Holocene temperature rises, which are slower.

The time slices are coarse, but they are relatively easy to delimit. Thus the problems of marine chronology, confused by the marine reservoir effect and slow sedimentation rates, are minimised at this coarse time-scale. The mean July and mean January temperature and SST records were examined through the Interstadial and Younger Dryas, and a mean value was calculated from the estimates for each sample, or by a visual examination of the temperature curves. This procedure ignores fluctuations in the individual records, which can be considerable (e.g. at Troll; *Klitgaard Kristensen et al.*, 2001); together with rounding to whole numbers, the estimates could be about 1°C in error, but the calculation of a standard deviation would give an unjustified impression of precision. Where there is a large variation during the time interval, this is indicated on the maps by a range. This admittedly loses detail, but it also neutralises any fluctuations caused by other factors, such as local environmental factors, problems with transfer functions, taphonomy and preservation, etc. This is an exploratory synthesis, and details can be considered in future analyses.

3. RESULTS

3.1. Interstadial

Figure 3 shows that the reconstructed Bølling–Allerød Interstadial terrestrial mean July temperatures lie between 6°C and 11°C in southern Norway, being warmest in the south. The range for sites is caused by differences in temperatures reconstructed from different proxies or reflects the range of temperatures under which low- or mid-alpine vegetation grows today. These temperatures, derived from several proxies, are all considerably cooler than the present-day and were never warm enough to support the growth of birch forest, in contrast to southern Sweden, Denmark, and Great Britain (*Birks*, 2003). The temperature reconstruction in the north is cooler, at 6-7°C, suggesting a terrestrial south–north temperature gradient of about 3-5°C.

Marine SSTs show a south–north gradient from ca. 11°C (Troll) to ca. 4-6°C (foraminifera reconstructions) or from ca. 8°C to 3-5°C (diatom reconstructions). The two western sites, ENAM93-21 and MD99-2284 (Figure 2), show temperatures cooler than the reconstructions from the eastern site, Troll, at a similar latitude (Figure 2). The Troll foraminifera reconstruction may be comparatively too warm at 11°C. This anomaly can possibly be related to the differences in the sieve-size fraction used for the calibration data set and in the paleo record, leading to a case of non-analogue assemblages. However, the east–west temperature gradient may be real and similar to what it is today (Figure 2) as most of the inflowing Atlantic water followed the Faroe–Shetland channel. Due to coriolis forcing and Ekman drift, this current was pushed against the Shetland margin. The sites of ENAM93-21 and MD99-2284 are situated north and west of the Faroe–Shetland channel (Figure 2), and the fact that the Faroe–Iceland current was weaker than today during the Interstadial may explain why these sites were not experiencing strong warming compared with Troll (*Hansen and Østerhus*, 2000). In addition, the inflow through the Faroe–Shetland channel, instead of penetrating further north, made a turn into the North Sea area, thus creating large east–west and north–south gradients. More data are needed to resolve this detail and establish the temperature values and gradients over the southern Norwegian Sea. The two dinoflagellate reconstructions of 15°C at Voldafjord and Troll are anomalous. They were reconstructed by a modern analogue method and the database used is presently limited in samples or reference sites with similar environments to those found in the North Sea and fjords of western Norway (*de Vernal et al.*, 1997; *Rochon et al.*, 1998, 1999; *Grøsfjeld et al.*, 1999). The record at site M23071 does not extend into the Allerød. The temperatures of around 5°C in the northeast Norwegian Sea (T-88-2) are comparable to the temperatures on nearby land, which are equivalent to those in central Svalbard today.

Mean January temperature reconstructions are sparse (Figure 3). On land, they are consistently as cold as –10°C and to –14°C in the north. Winter SSTs are just above freezing in the south, perhaps a bit warmer at MD95-2011. This small anomaly results from using an enlarged modern diatom training set for reconstruction at this site compared with M23071 and HM79-6/4 (*Birks*, 2001; *Birks and Koç*, 2002). Dinoflagellates give a 2°C reconstruction at Troll and an indication of long-lasting ice cover at Voldafjord. Foraminiferal evidence indicates that the north Norwegian Sea was ice-covered during the winter. This pattern conforms to that reconstructed by *Koç et al.* (1993) from diatom evidence of year-round ice-cover in the Norwegian Sea—apart from a more or less seasonally ice-free corridor along the Norwegian coast, where relatively warm Atlantic water was slowly penetrating northwards, perhaps as a result of coriolis forces in the northern Atlantic (*Berger and Jansen*, 1995).

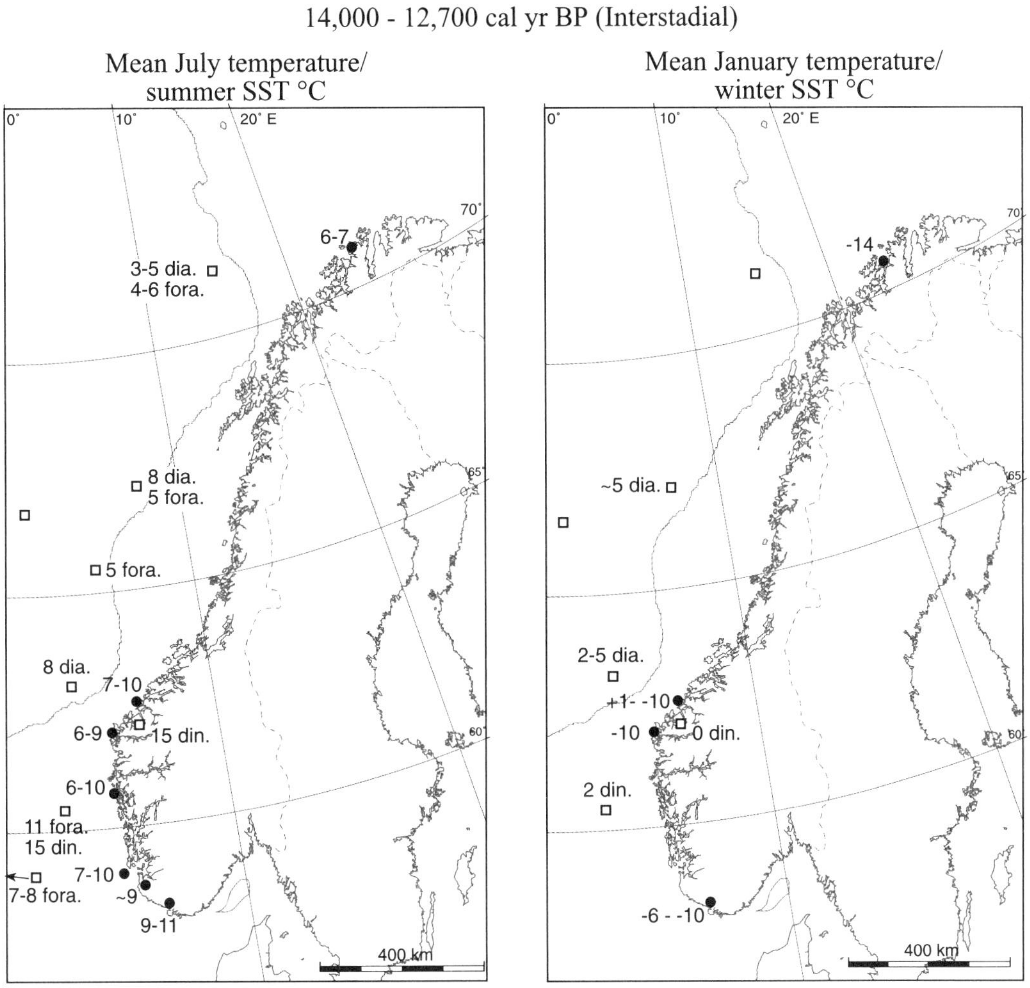

Figure 3. Reconstructed Interstadial (14,000-12,700 cal BP) summer and winter temperatures. Mean July and January temperatures are mapped for terrestrial sites and summer and winter SSTs are mapped for the Norwegian Sea. Figures 3–7: Reconstructions based on diatoms are indicated by "dia.", on foraminifera by "fora.", and on dinoflagellates by "din."

However, the presence of foraminiferal records of interstadial age in the Iceland Sea (*Sejrup et al.*, 1995) suggests that there were, in fact, seasonally ice-free conditions in this area. The absence of a diatom record here (*Koç et al.*, 1993) may be due to factors other than climate (e.g. preservation).

3.2. Younger Dryas

The extent of the Younger Dryas ice-sheet readvance is shown on Figures 2 and 4–6. Terrestrial mean July temperatures in southern Norway reconstructed from biotic proxies are consistent at 8-9°C in the south and are 5-7°C at Kråkenes, where a cirque glacier developed in the catchment. Calculations based on its ELA lowering during the Younger Dryas also reconstruct summer temperatures of 5.5-7.5°C (*Larsen et al.*, 1984). In northern Norway, Younger Dryas summer temperatures are around 6°C, the same as the Interstadial temperatures. There may be some influence of the influx of Atlantic water around the south of Norway and up the west coast, but elsewhere the temperatures were consistently cold, comparable to those in the adjacent sea. The terrestrial south–north July gradient was reduced to about 3°C from 58°N to 60°N and to nothing further north.

Marine summer SSTs from foraminifera are consistent at ca. 4°C, close to the lower limit of temperature reconstructions possible from foraminiferal data; below this temperature, the fauna becomes monospecific (see *Pflaumann et al.*, 1996, 2003). Diatom results are cooler at HM79-6/4, MD23071, and T-88-2 in the north, suggesting that these sites may have been largely ice-covered during the year, but a significantly high abundance of foraminifera indicates the availability of open water during the summer calcification season. Higher diatom reconstructions of ca. 7°C at MD95-2011 may indicate the influence of northwards penetrating Atlantic

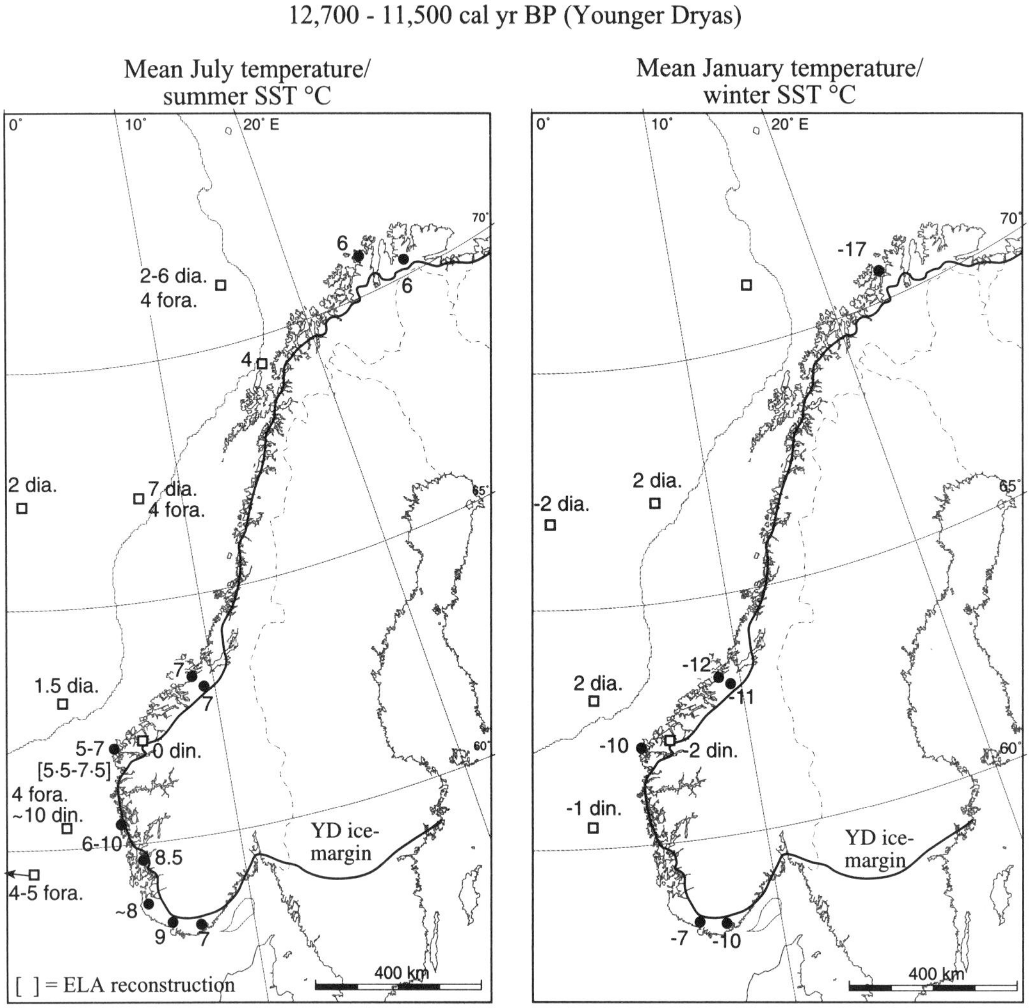

Figure 4. Reconstructed Younger Dryas (12,700-11,500 cal BP) summer and winter temperatures. Mean July and January temperatures are mapped for terrestrial sites and summer SSTs are mapped for the Norwegian Sea. The maximum extent of the Younger Dryas ice readvance in Norway is indicated.

surface water close to the coast. It is likely that relatively high summer insolation (e.g. *Kutzbach and Guetter*, 1986) heated the surface water, giving rise to increased stratification of the upper water column. The relatively high dinoflagellate reconstruction at Troll (10°C) may also suggest the presence of relatively warm water near the sea surface.

The few winter (under the sea-ice) reconstructions of SST (Figure 4) are cold, suggesting extensive sea-ice cover. However, because it is difficult to reconstruct such low temperatures with the calibration data sets available, temperatures may well have been colder. Winter temperatures on land were much colder, reflecting severe conditions along the coast, but were not much different from winter temperatures reconstructed during the Interstadial.

3.3. Interstadial–Younger Dryas Transition

Figure 5 shows the difference in temperatures between the Interstadial and the Younger Dryas. There is an overall decrease in summer temperatures in southern Norway, but only on the order of 1-2°C, even at Kråkenes, where a cirque glacier was initiated. The pollen record from Hammerfest shows no temperature change. The diatom records in the Norwegian Sea reconstruct a drop of about 1°C on the Vøring Plateau and to the north, but considerably more to the south and west, where SSTs were relatively high in the Interstadial but considerably reduced in the Younger Dryas. The increased amount of summer insolation (e.g. *Kutzbach and Guetter*, 1986) provided insufficient heat to counteract the

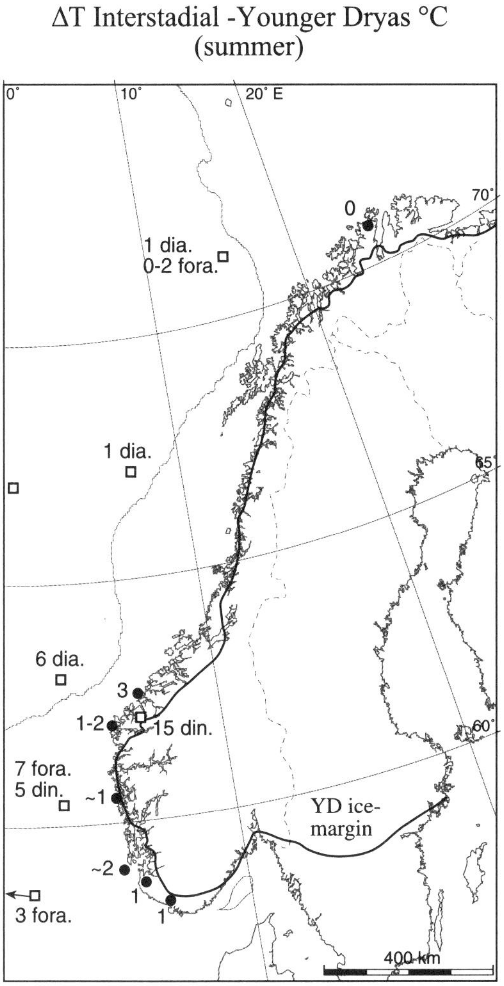

Figure 5. Temperature difference between Interstadial and Younger Dryas summer temperatures in the Norwegian region. The maximum extent of the Younger Dryas ice readvance in Norway is indicated.

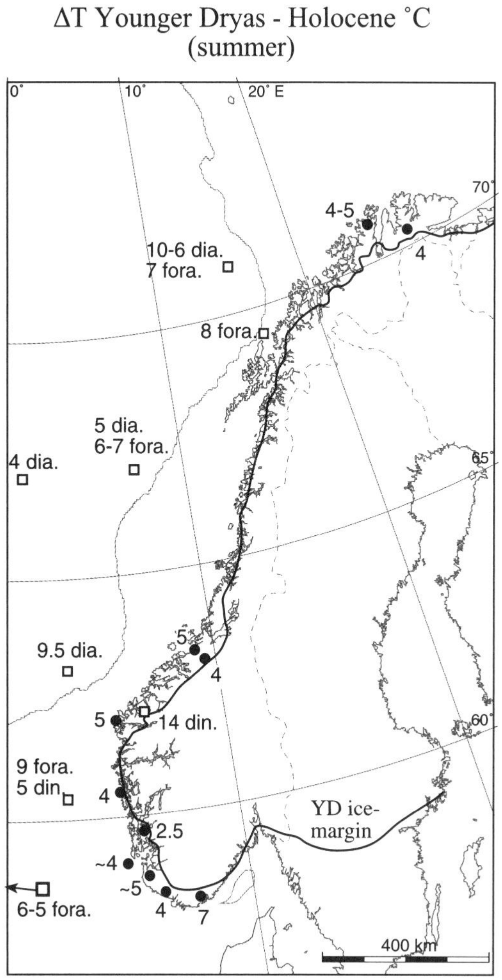

Figure 6. Temperature difference between Younger Dryas and early Holocene summer temperatures in the Norwegian region.

Younger Dryas cooling and could not prevent widespread seasonal sea-ice formation. Although there was substantial cooling in the southern Norwegian Sea, the flow of Atlantic surface waters along the coast still continued towards the Vøring Plateau (see *Koç et al.*, 1993). The dinoflagellate temperature reconstructions show a drop, but the magnitude is probably too large at Voldafjord (see above).

3.4. Early Holocene Temperature Rise

On the land, the steep temperature rise in the early Holocene ended in mean July values up to 14°C in the south, and ~10-11°C in the north. Thus a gradient of about 3°C from south to north was reestablished. However, the July temperatures in southern Norway were generally 1-2°C cooler than they are today, whereas at 67°N they were already the same as they are today, perhaps as a result of the remoteness of the area from the influence of the remnants of the Scandinavian ice sheet (*Seppä et al.*, 2002). The mean July temperature at Iford, north Norway, subsequently rose about 1°C higher by about 9000 cal BP before sinking again to present values (*Seppä et al.*, 2002). A similar pattern was shown in the early Holocene record from Dalmutladdo (*Bjune et al.*, 2005). Summer temperatures on land were similar to those at the sea surface at the same latitude—apart from the southernmost area of Norway, where July temperatures were a little higher.

Summer SSTs reached at the end of the rapid temperature rise ranged around 10-13°C everywhere in both diatom and foraminifera reconstructions, with no detectable south–north gradient. The diatom record at MD23071 to the west shows a continuous smooth temperature rise, 6°C (Figure 7) being reached at a comparable time (ca. 10,000 cal BP) as 10-12°C

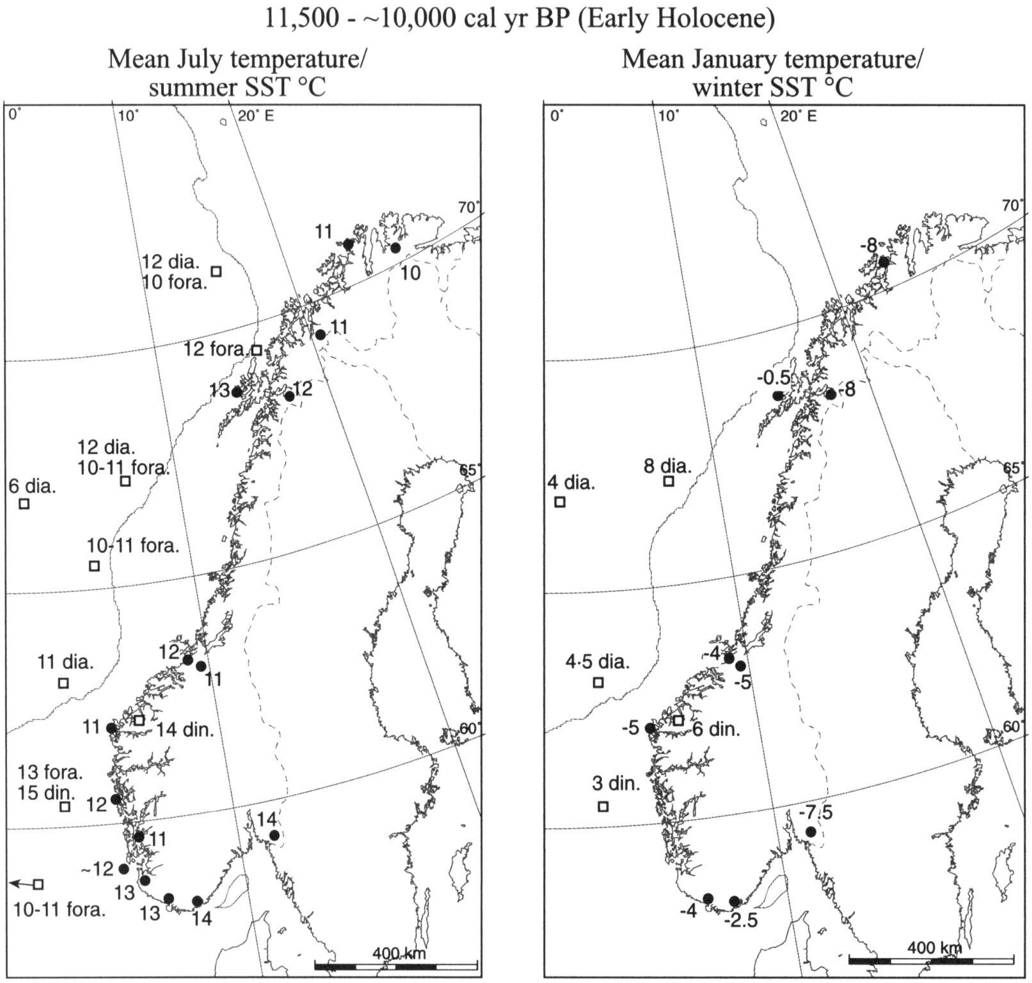

Figure 7. Reconstructed early Holocene winter and summer temperatures reached at the end of the steep temperature rise (11,500–ca. 10,000 cal BP). Mean July and January temperatures are mapped for terrestrial sites and summer and winter SSTs are mapped for the Norwegian Sea.

on the Vøring Plateau. At M23071, 12°C was reached ca. 9500 cal BP, perhaps reflecting the course of the retreat northwestwards of the polar front across this area (see *Koç et al.*, 1993).

Winter SSTs (Figure 7) were about 5-8°C, suggesting cool but ice-free conditions similar to today (Figure 2). Winter temperatures in the sea were influenced by the inflow of Atlantic water, as the Atlantic current became well established in the early Holocene (*Birks and Koç*, 2002). On land in southern Norway, however, winter temperatures were still well below zero and about 5-6°C lower than today's winter temperatures along the coast in the modern oceanic climate. Litlevatnet, on the outer coast of Vesterålen (69°N), is very warm with reconstructed January temperatures of –0.5°C. These temperatures could be influenced by the proximity of the sea, or more likely result from some problems in temperature reconstruction based on pollen data here. Overall, the terrestrial climate was more continental than today with greater seasonality in the early Holocene, but some indications of increasing oceanicity were apparent.

3.5. Younger Dryas–Holocene Transition

The difference in July temperatures between the Younger Dryas and the end of the early Holocene steep temperature rise was about 4-5°C at almost all the terrestrial sites (Figure 6). However, the pattern of change in the sea was different and more complex, with large temperature rises at the southern sites (~9°C), and smaller rises (mostly 5-7°C) north of 65°N. (The large change of 14°C at Voldafjord is a result of the anomalously low Younger Dryas reconstruction (0°C) from dinoflagellates.) Throughout the area, the summer temperature changes were larger in the sea than on land, but winter temperature changes were greater on the land than in the sea. The early Holocene summer temperature values were similar in adjacent land/sea areas (Figure 7), showing that, by this time, the Atlantic circulation was strongly influencing terrestrial conditions (*Seppä et al.*, 2002).

4. DISCUSSION

The synthetic mapping approach, although crude, has revealed an interesting spatial and temporal picture of temperature changes in the Norwegian region. Although several different proxies and reconstruction methods have been used, the results are surprisingly conformable and remarkably consistent.

From the late Interstadial to the Younger Dryas, terrestrial July temperatures decreased by 1-2°C in southern Norway. This rather small drop is probably a result of temperatures in the Interstadial being cool, compared with the rest of northwest Europe. Here, temperatures fell markedly from the warm Interstadial conditions (e.g. *Lowe et al.*, 1994). In southern Sweden and Britain, summer temperatures fell by 3-5°C as reconstructed from Coleoptera (*Coope et al.*, 1998) and model results (*Renssen and Isarin*, 2001). Whereas the British Isles, southern Sweden, and Denmark experienced Interstadial July temperatures around 11-13°C, which supported birch forest development, southern Norway was too cold for tree growth at below 10°C (some coleopteran-based reconstructed temperatures are probably too high for south Norway; *Coope et al.*, 1998), and the Interstadial vegetation was analogous to low- or mid-alpine vegetation in the west Norwegian mountains today (e.g. *Birks*, 1993, 2003), with dwarf shrubs or mountain-herb vegetation.

The Younger Dryas ice readvance was not necessarily only a response to a temperature drop. In the north, the Younger Dryas readvance was much less than in the south (e.g. *Svendsen and Mangerud*, 1997) and the vegetation change was much less marked (*H.H. Birks*, unpublished data), reflected in the fact that summer temperatures, reconstructed from a variety of proxies, hardly changed. A similar pattern is visible in the model results of *Renssen and Isarin* (2001). The enlarged Scandinavian ice sheet influenced the climate pattern in western and southern Norway, reducing temperatures by cold winds. Models simulate strong anticyclonic summer circulation and strengthened westerly winter circulation induced by the strong temperature gradient over the North Atlantic due to the southwards advance of the sea-ice (*Renssen et al.* (1996) and *Isarin et al.* (1997) for the lateglacial and *Siegert and Marsiat* (2001) for the LGM). The enhanced North Atlantic storms picked up moisture from the open sea near the coast (*Koç et al.*, 1993, 1996; *Hebbeln et al.*, 1994; *Sarntheim et al.*, 1995; *Dokken and Hald*, 1996) and brought precipitation from the west and southwest (*Fletcher et al.*, 1993; *Larsen et al.*, 1984). As ice-sheet mass balance depends on both summer melting temperature and winter snow accumulation, it is likely that increased winter precipitation over Scandinavia combined with lowered temperatures and reduced thermohaline circulation all played a role in the Younger Dryas ice-sheet readvance, especially in the west (e.g. *Larsen et al.*, 1984; *Larsen and Stalsberg*, 2004).

During the Interstadial, the Atlantic current was relatively weak and was confined to the eastern part of the Norwegian Sea (*Koç et al.*, 1993). As a result, the summer SST drop into the Younger Dryas was greatest in the southern Norwegian Sea, because Interstadial SSTs were considerably higher in the south. In the north, Interstadial summer SSTs were more or less the same as in the Younger Dryas. Younger Dryas summer SSTs were similar over all the eastern Norwegian Sea, probably near the lower limit of temperature reconstructions.

The July temperature rise into the early Holocene seems to have been generally consistent on land at around 4°C. This was enough to cause the melting of the ice sheet and the withdrawal of the polar and arctic fronts and associated sea ice to the vicinity of Greenland. By ca. 10,000 cal BP terrestrial temperature rises to the south of Norway were much greater, encompassing rises of ca. 5-11°C in south Sweden and 6-10°C in Britain according to coleopteran reconstructions (*Coope et al.*, 1998) and 8-12°C according to model simulations (*Renssen and Isarin*, 2001). The temperature rise in the Norwegian Sea was around 7-9.5°C, tending to be greater in the south, and is compatible with modelled values of 8-10°C (*Renssen and Isarin*, 2001). The Holocene marine temperature gradient was established within ca. 1,000 years after the end of the Younger Dryas, with SSTs at ca. 11-13°C in the south and ca. 10-12°C in the north (Figure 7). Summer SSTs were comparable to mean July air temperatures on land (Figure 7), reflecting the influence of the fully established North Atlantic current on the adjacent land.

Where winter temperature reconstructions are available, they show a strong contrast between land and sea. Land temperatures are reconstructed with values well below zero, but with values about 5°C higher in the early Holocene (Figure 7) than in the Interstadial and the Younger Dryas (Figures 3 and 4). The values of winter temperature modelled by *Renssen and Isarin* (2001) seem to be much colder in the north, but it is difficult to reconstruct temperatures of this magnitude from proxy data. In the sea, winter temperature reconstructions suggest that slow circulation continued during the Younger Dryas (e.g. *Hald et al.*, 1996). Precipitation reconstructions in the early Holocene (e.g. *Dahl and Nesje*, 1996; *Barnekow*, 1999; *Birks et al.*, 2000; *Seppä and Birks*, 2001; *Hammarlund et al.*, 2002) indicate much higher precipitation than today, agreeing with model simulations (*Renssen and Isarin*, 2001).

Exceptions to these consistent patterns appear on the maps, and they highlight difficulties or problems with the interpretation of certain proxy data in terms of temperatures. In particular, the dinoflagellate results differ from those from other proxies. This may be caused by inconsistence in the calibration data set in terms of spatial sample distribution or samples covering all types of environments, the poor quality of marine environmental data, the type of transfer function procedure used, or some combination of these factors (*de Vernal et al.*, 1997; *Grøsfjeld et al.*, 1999). There are also some anomalies among reconstructions based on planktic foraminifera or diatoms. In general, these two proxies show similar trends but differences in values for temperature reconstruction. These differences probably result from the different depth habitats of these lifeforms, but all modern marine calibration data sets suffer from limitations concerning the certain collection of core-top surface samples, the quality of marine environmental data, and the methods of constructing the transfer functions (see e.g. *Birks*, 2001; *Birks and Koç*, 2002) as discussed earlier. Ideally, the variable being reconstructed should directly influence the organisms or be linearly related to factors that control their occurrence (*Birks*, 1995). Where possible, reconstructions made from one proxy should be tested against reconstructions from another proxy or against the occurrence of indicator species or proxies (e.g. *Vandenberghe et al.*, 1998; *Birks and Ammann*, 2000). Such comparisons are valuable to assess the reliability of environmental reconstructions and more studies like that at Kråkenes are needed in the future.

The maps have highlighted connections between land and ocean systems in the Norwegian region. During the late-glacial Interstadial and Younger Dryas, the land was influenced by air-circulation patterns related to the ice sheet and a reduced heat supply due to the albedo of a snow-covered landscape, in spite of the relatively high insolation. The ocean circulation was weak in the Interstadial and Younger Dryas, being effectively restricted to the south Norwegian Sea and to the west Norwegian coast. The whole system was generally cool or cold compared with Europe to the south. However, in the early Holocene, as soon as the polar front retreated and the Atlantic current reached full strength, temperatures on land corresponded much more closely with those in the sea. The ice sheet and sea-ice melted, and a westerly atmospheric circulation pattern was established. However, winter temperatures were still much colder on land than today, reflecting increased seasonality in the early Holocene.

5. FUTURE DIRECTIONS

This synthesis shows that several pathways could be followed in the future. We need to improve transfer function data sets, and to include information about the ecology of the species involved, perhaps using them more frequently in an indicator species or assemblage type of approach in addition to statistically derived transfer functions. It may be that we have reached the limits of the transfer function approach with some proxies. There are particular difficulties with marine proxies because of the limitation of analysing known surface sediments and the quality of marine environmental data. Acquiring known surface marine sediment and measured marine environmental data are both expensive processes. Land-based methods may benefit from future improvements in numerical analysis techniques, such as the use of artificial neural networks, and increased knowledge of the ecological preferences of the organisms. Nevertheless, we consider that the most valuable future activity is to collect good fossil stratigraphic data from new, critically positioned, sites to investigate the nature of the patterns and gradients indicated by the present results. Modelling may help with this approach, in that modelled values are interpolated on a geographical basis. This would allow us to select sites across gradients and ecotones that will test and refine the model results. It would be profitable to extend the model of *Renssen and Isarin* (2001) for northwest Europe to the Norwegian area and to include the marine reconstructions. To facilitate the interactions with modellers, the development of a database is helpful, and a database already developed for some of the results in the NORPAST project is scheduled for further enlargement. With the palaeo data made available, model results can be evaluated and the effects of forcing factors can be explored, such as $p\text{CO}_2$, $p\text{CH}_4$, ice-sheet dimensions, albedo, and wind-blown dust in glacial periods, which are thought to be important drivers of the climate system.

To undertake these various tasks, especially the intensive, time-consuming, and skilful collection of new proxy records, funds to train and support more researchers are needed. The Bjerknes Centre for Climate Research, in combination with the NORPAST, NOClim, and NORPEC projects of the Norwegian Research Council, is making large contributions to Norwegian climate research, where paleo data are considered an important component in the investigations of marine and terrestrial climate processes, in the past, in the present, and in the future.

Acknowledgments. We thank all those whose results we have used, especially unpublished data of H.J.B. Birks, A.E. Bjune, S.J. Brooks, Aa. Paus, and S.M. Peglar. We thank John Birks for making temperature reconstructions from unpublished pollen data at Jansvatnet and Bjerkreim. We thank Heikki Seppä and two anonymous reviewers for valuable comments on the manuscript. We thank Jane Ellingsen and Eva Bjørseth for help with the figures. The work for this synthesis was supported by the Norwegian Research Council Projects NORPAST, NOClim, and NORPEC. This is publication No. A48 of the Bjerknes Centre for Climate Research.

REFERENCES

Andersson, C., B. Risebrobakken, E. Jansen and S.O. Dahl, Late Holocene surface ocean conditions of the Norwegian Sea (Vøring Plateau). *Paleoceanography*, 18(2), 1044, doi:10,1029/2001PA000654, 2003.

Atkinson, T.C., K.R. Briffa and G.R. Coope, Seasonal temperatures in Britain during the past 22,000 years, reconstructed using beetle remains. *Nature*, 325, 587-592, 1987.

Barnekow, L., Holocene tree-line dynamics and inferred climatic changes in the Abisko area, northern Sweden, based on macrofossil and pollen records. *The Holocene*, 9, 253-265, 1999.

Berger, W.H. and E. Jansen, The Younger Dryas episode: ice-collapse and super-fjord heat pump. In: Troelstra, S.R., van Hinte, J.E. & Ganssen, G.M. (Eds.) The Younger Dryas. *Proceedings of the Royal Dutch Academy of Natural Science*, 44, 61-105, 1995.

Bianchi, G.G. and N. McCave, Holocene periodicity in North Atlantic and deep-ocean flow south of Iceland. *Nature*, 397, 515-517, 1999.

Birks, C.J.A., A Younger Dryas–Holocene diatom record of sea-surface temperatures and oceanographic changes from core MD95-2011, Vøring Plateau. MSc Thesis, University of Bergen, 2001.

Birks, C.J.A. and N. Koç, A high-resolution diatom record of late-Quaternary sea-surface temperatures and oceanographic conditions from the eastern Norwegian Sea. *Boreas*, 31, 323-344, 2002.

Birks, H.H., The importance of plant macrofossils in late-glacial climatic reconstructions: an example from western Norway. *Quaternary Science Reviews*, 12, 719-726, 1993.

Birks, H.H., Late-glacial vegetational ecotones and climatic patterns in western Norway. *Vegetation History and Archaeobotany*, 3, 107-119, 1994.

Birks, H.H., The importance of plant macrofossils in the reconstruction of Lateglacial vegetation and climate: examples from Scotland, western Norway and Minnesota, USA. *Quaternary Science Reviews*, 22, 453-473, 2003.

Birks, H.H. and B. Ammann, Two terrestrial records of rapid climate change during the glacial-Holocene transition (14,000-9000 BP) from Europe. *Proceedings National Academy of Sciences USA*, 97, 1390-1394, 2000.

Birks, H.H., R.W. Battarbee and H.J.B. Birks, The development of the aquatic ecosystem at Kråkenes Lake, western Norway, during the late glacial and early Holocene—a synthesis. *Journal of Paleolimnology*, 23, 91-144, 2000.

Birks, H.H., G. Lemdahl, J.I. Svendsen and J.Y. Landvik, The palaeoecology of a late-Allerød peat bed at Godøy, western Norway. *Journal of Quaternary Science*, 8, 147-159, 1993.

Birks, H.H., Aa. Paus, J.I. Svendsen, T. Alm, J. Mangerud and J.Y. Landvik, Late-Weichselian environmental change in Norway, including Svalbard. *Journal of Quaternary Science*, 9, 133-145, 1994.

Birks, H.J.B., The use of pollen analysis in the reconstruction of past climates: a review. In: Wigley, T.M.L., Ingram, M.J. and Farmer, G. (eds.) *Climate and History*. Cambridge University Press, Cambridge. pp. 111-138, 1981.

Birks, H.J.B., Quantitative palaeoenvironmental reconstructions. In: Maddy, D. and Brew, J.S. (eds.) Statistical modelling of Quaternary science data. *Quaternary Research Association Technical Guide 5*, Cambridge. pp. 161-254, 1995.

Björck, S., M.J.C. Walker, L. Cwynar, S.J. Johnsen, K.L. Knudsen, J.J. Lowe, B. Wohlfarth and INTIMATE members, An event stratigraphy for the Last Termination in the North Atlantic region based on the Greenland Ice Core record: a proposal by the INTIMATE group. *Journal of Quaternary Science*, 13, 283-292, 1998.

Bjune, A.E., J. Bakke, A. Nesje and H.J.B. Birks, Holocene mean July temperature and winter precipitation in western Norway inferred from lake sediment proxies. *The Holocene*, 15, 177-189, 2005.

Bjune, A.E., H.J.B. Birks and H. Seppä, Holocene vegetation and climate history on a continental–oceanic transect in northern Fennoscandia based on pollen and plant macrofossils. *Boreas*, 33, 211-223, 2004.

Brooks, S.J. and H.J.B. Birks, Chironomid-inferred air temperatures from Late Glacial and Holocene sites in north-west Europe: progress and problems. *Quaternary Science Reviews*, 20, 1723-1741, 2001.

CLIMAP members, Seasonal reconstructions of the earth's surface at the Last Glacial Maximum. *Geological Society of America, Map and Chart Series MC-36,* 1981.

Coope, G.R., G. Lemdahl, J.J. Lowe and A. Walkling, Temperature gradients in northern Europe during the last glacial-Holocene transition (14-9 ^{14}C yr BP) interpreted from coleopteran assemblages. *Journal of Quaternary Science*, 13, 419-433, 1998.

Dahl, S.O. and A. Nesje, A new approach to calculating Holocene winter precipitation by combining glacier equilibrium-line altitudes and pine-tree limits: a case study from Hardangerjøkulen, central Norway. *The Holocene*, 6, 361-398, 1996.

de Vernal, A., A. Rochon, J.-L. Turon and J. Matthiessen, Organic-walled dinoflagellate cysts: palynological tracers of sea surface-water conditions in middle to high latitude marine environments. *GEOBIOS*, 30, 905-920, 1997.

de Vernal, A. and C. Hillaire-Marcel, Sea-ice cover, sea-surface salinity and halo-/thermocline structure of the north-west North Atlantic: modern versus full-glacial conditions. *Quaternary Science Reviews*, 19, 65-85, 2000.

Dokken, T.M. and M. Hald, Rapid climate shifts during isotope stages 2-4 in the polar North Atlantic. *Geology*, 24, 599-602, 1996.

Dokken, T.M. and E. Jansen, Rapid changes in the mechanism of ocean convection during the last glacial period. *Nature*, 401, 458-461, 1999.

Dreger, D., Decadal- to centennial scale sediment records of ice advance on the Barents shelf and meltwater discharge into the northeastern Norwegian Sea over the last 40 kyr. PhD. thesis, University of Kiel. 63 pp, 1999.

Eide, W., H.H. Birks, N.H. Bigelow, S.M. Peglar and H.J.B. Birks, Holocene tree migrations in the Setesdal valley, southern Norway, reconstructed from macrofossil and pollen evidence. *Vegetation History and Archaeobotany* (in press).

Elderfield, H. and G. Ganssen, Past temperature and δ^{18}O of surface ocean waters inferred from foraminiferal Mg/Ca ratios. *Nature*, 405, 442-445, 2000.

Fletcher, C.H., R.W. Fairbridge, J.J. Moller and A.J. Long, Emergence of the Varanger Peninsula, arctic Norway and climate changes since deglaciation. *The Holocene*, 3, 116-127, 1993.

Grønvold, K., N. Óskarsson, S.J. Johnsen, H.B. Clausen, C. Hammer, G. Bond and E. Bard, Ash layers from Iceland in the Greenland GRIP ice core correlated with oceanic and land sediments. *Earth and Planetary Science Letters*, 135, 149-155, 1995.

Grøsfjeld, K., E. Larsen, H.P. Sejrup, A. de Vernal, T. Flatebø, M. Vestbø, H. Haflidason and I. Aarseth, Dinoflagellate cysts reflecting surface-water conditions in Voldafjorden, western Norway during the last 11,300 years. *Boreas*, 28, 403-415, 1999.

Hald, M., T.M. Dokken and S. Hagen, Palaeoceanography on the European arctic margin during the last deglaciation. In: Andrews, J.T., Austin, W.E.N., Bergsten, H. and Jennings, W.E. (eds.) Late Quaternary Palaeoceanography of the North Atlantic Margins. *Geological Society Special Publication*, 111, 257-287, 1996.

Hammarlund, D., L. Barnekow, H.J.B. Birks, B. Buchardt and T.W.D. Edwards, Holocene changes in atmospheric circulation recorded in the oxygen-isotope stratigraphy of lacustrine carbonates from northern Scandinavia. *The Holocene*, 12, 355-367, 2002.

Hansen, B. and S. Østerhus, North Atlantic–Nordic Seas exchanges. *Progress in Oceanography*, 45, 109-208, 2000.

Hebbeln, D., T. Dokken, E. Andersen, M. Hald and A. Elverhoi, Moisture supply for northern ice sheet growth during the last glacial maximum. *Nature*, 370, 357-360, 1994.

Isarin, R.F.B., H. Renssen and E.A. Koster, Surface wind climate during the Younger Dryas in Europe as inferred from aeolian records and model simulations. *Palaeogeography, Palaeoclimatology, Palaeoecology*, 134, 127-148, 1997.

Klitgaard Kristensen, D., H.P. Sejrup and H. Haflidason, The last 18 kyr fluctuations in Norwegian Sea surface conditions and implications for the magnitude of climatic change: evidence from the North Sea. *Paleoceanography,* 116, 455-467, 2001.

Koç, N., E. Jansen, M. Hald and L. Labeyrie, Late Glacial–Holocene sea surface temperatures and gradients between the North Atlantic and the Norwegian Sea: implications for the Nordic heat pump. In: Andrews, J.T., Austin, W.E.N., Bergsten, H. and Jennings, W.E. (eds.) Late Quaternary Palaeoceanography of the North Atlantic Margins. *Geological Society Special Publication*, 111, 177-185, 1996.

Koç, N., E. Jansen and H. Haflidason, Paleoceanographic reconstructions of the surface ocean conditions in the Greenland, Iceland and Norwegian Seas throughout the last 14 ka, based on diatoms. *Quaternary Science Reviews*, 12, 115-140, 1993.

Koç Karpuz, N. and E. Jansen, A high-resolution diatom record of the last deglaciation from the SE Norwegian Sea: Documentation of rapid climate changes. *Paleoceanography*, 7, 499-520, 1992.

Kutzbach, J.E. and P.J. Guetter, The influence of changing orbital parameters and surface boundary conditions on climate simulations for the past 18,000 years. *Journal of the Atmospheric Sciences*, 43, 1726-1759, 1986.

Larsen, E., F. Eide, O. Longva and J. Mangerud, Allerød-Younger Dryas climatic inferences from cirque glaciers and vegetational development in the Nordfjord area, western Norway. *Arctic and Alpine Research*, 16, 137-160, 1984.

Larsen, E. and M.K. Stalsberg, Younger Dryas glaciolacustrine rhythmites and cirque glacier variations at Kråkenes, western Norway: depositional processes and climate. *Journal of Paleolimnology*, 31, 49-61, 2004.

Lowe, J.J., B. Ammann, H.H. Birks, S. Björck, G.R. Coope, L. Cwynar, J.-L. de Beaulieu, R.J. Mott, D.M. Peteet and M.J.C. Walker, Climatic changes in areas adjacent to the North Atlantic during the last glacial-interglacial transition (14-9 ka BP): a contribution to IGCP-253. *Journal of Quaternary Science*, 9, 185-198, 1994.

Lowe, J.J. and NASP members, Palaeoclimate of the North Atlantic seaboards during the last glacial/interglacial transition. *Quaternary International*, 28, 51-61, 1995.

Pflaumann, U., J. Duprat, C. Pujol and L.D. Labeyrie, SIMMAX: a modern analog technique to deduce Atlantic sea surface temperatures from planktonic foraminifera in deep-sea sediments. *Paleoceanography*, 11, 15-35, 1996.

Pflaumann, U., M. Sarnthein, M. Chapman, M. d'Abreu, B. Funnell, M. Huels, T. Kiefer, M. Maslin, H. Schulz, J. Swallow, S. van Kreveld, M. Vautravers, E. Vogelsang and M. Weinelt, Glacial North Atlantic: Sea-surface conditions reconstructed by GLAMAP 2000. *Paleoceanography*, 18(3), 1065, doi:10.1029/2002PA000774, 2003.

Rasmussen, T., E. Thomsen, L. Labeyrie and T.C.E. van Weering, Circulation changes in the Faroe - Shetland channel correlating with cold events during the last glacial period 59-10 ka BP. *Geology*, 24, 937-940, 1996.

Renssen, H. and R.F.B. Isarin, Two major warming phases of the last deglaciation at ~14.7 and ~11.5 ka cal B.P. in Europe: climate reconstructions and AGCM experiments. *Global and Planetary Change*, 30, 117-153, 2001.

Renssen, H., J. Lautenschlager and C.J.E. Schuurmans, The atmospheric winter circulation during the Younger Dryas stadial in the Atlantic/European sector. *Climate Dynamics*, 12, 813-824, 1996.

Risebrobakken, B., E. Jansen, C. Andersson, E. Mjelde and K. Hevrøy, A high-resolution study of Holocene paleoclimatic and paleoceanographic changes in the Nordic Seas. *Paleoceanography*, 18(1), 1017, doi:10.1029/2002PA000764, 2003.

Rochon, A., A. de Vernal, H.P. Sejrup and H. Haflidason, Palynological evidence of climatic and oceanographic changes in the North Sea during the last deglaciation. *Quaternary Research*, 49, 197-207, 1998.

Rochon, A., A. de Vernal, J.-L. Yuron, J. Matthiessen and M. Head, Distribution of dinoflagellate cysts in surface sediments from the North Atlantic Ocean and adjacent basins and quantitative reconstruction of sea-surface parameters. *American Association of Stratigraphic Palynologists, Contribution Series*, 64 pp, 1999.

Sarnthein, M., E. Jansen, M. Weinelt, M. Arnold, J.C. Duplessy, H. Erlenkeuser, A. Flatøy, G. Johannessen, T. Johannessen, S. Jung, N. Koç, L. Labeyrie, M. Maslin, U. Pflaumann and H. Schulz, Variations in Atlantic surface ocean paleoceanography, 50°-80°N: A time-slice record of the last 30 000 years. *Paleoceanography*, 10(6), 1063-1094, 1995.

Sejrup, H.P., H. Haflidason, D. Kristensen and S. Johnsen, Last interglacial and Holocene climatic development in the Norwegian Sea region: Oceanic front movements and ice core data. *Journal of Quaternary Science,* 10, 385-390, 1995.

Sejrup, H.P., H. Haflidason, T. Flatebø, D. Klitgaard Kristensen, K. Grøsfjeld and E. Larsen, Late-glacial to Holocene environmental changes and climate variability: evidence from Voldafjorden, western Norway. *Journal of Quaternary Science*, 16, 181-198, 2001.

Sejrup, H.P., H.J.B. Birks, D. Klitgaard Kristensen and H. Madsen, Benthonic foraminiferal distributions and quantitative transfer functions for the north-west European continental margin. *Marine Micropaleontology*, 53, 197-226, 2004.

Seppä, H. and H.J.B. Birks, July mean temperature and annual precipitation trends during the Holocene at the Fennoscandian tree-line area: pollen-based climate reconstructions. *The Holocene*, 11, 527-539, 2001.

Seppä, H., H.H. Birks and H.J.B. Birks, Rapid climatic changes during the Greenland stadial 1 (Younger Dryas) to early Holocene transition on the Norwegian Barents Sea coast. *Boreas*, 31, 215-225, 2002.

Siegert, M.J. and I. Marsiat, Numerical reconstruction of LGM climate across the Eurasian Arctic. *Quaternary Science Reviews*, 20, 1595-1605, 2001.

Svendsen, J.I. and J. Mangerud, Holocene glacial and climatic variations on Spitsbergen, Svalbard. *The Holocene*, 7, 45-57, 1997.

Vandenberghe, J., G.R. Coope and K. Kasse, Quantitative reconstructions of palaeoclimates during the last interglacial-glacial in western and central Europe: an introduction. *Journal of Quaternary Science*, 13, 361-366, 1998.

van Dinter, M. and H.H. Birks, Distinguishing fossil *Betula nana* and *B. pubescens* using their wingless fruits: implications for the late-glacial vegetational history of western Norway. *Vegetation History and Archaeobotany*, 5, 229-240, 1996.

Carin Andersson and Trond Dokken, Bjerknes Centre for Climate Research, Bergen, Norway. (carin.andersson@bjerknes.uib.no; trond.dokken@bjerknes.uib.no)

Hilary Birks, Department of Biology, University of Bergen, Bergen, Norway. (hilary.birks@bio.uib.no)

Dorthe Klitgaard Kristensen, Norsk Polarinstitutt, Polarmiljøsenteret, Hjalmar Johanesgt. 14, N-9007, Tromsø, Norway. (dorthe@npolar.no)

Vilhelm Bjerknes' Vision for Scientific Weather Prediction

Sigbjørn Grønås

Department of Geophysics, University of Bergen, Bergen, Norway

With his circulation theorem, Vilhelm Bjerknes extended Kelvin's theorem from barotropic to baroclinic environments. In doing so, he built a bridge between the scientific disciplines of hydrodynamics and thermodynamics. Application of his theorem to circulations in atmosphere and ocean led Bjerknes to publish a paper in 1904 expressing a vision and a programme for scientific weather forecasting. With generous funding from the Carnegie Institution of Washington from 1905, Bjerknes for the rest of his life dedicated his scientific strength to his vision. Needed theories on the dynamics of the atmosphere were given in a number of textbooks that he wrote together with his assistants. With the Bergen School of Meteorology he renewed weather forecasting throughout the world. Bjerknes' work inspired Louis Fry Richardson to his famous experiment to compute a change in the state of the atmosphere. Bjerknes and his assistants also inspired Arnt Eliassen and Ragnar Fjørtoft, who played important roles in developing numerical weather prediction after the Second World War. The name of Vilhelm Bjerknes has now been honoured by the establishment of the Bjerknes Centre for Climate Research in Bergen, Norway.

1. INTRODUCTION

When Vilhelm Bjerknes was a student in the late 19th century, weather forecasting was in its infancy. Many countries already had a national weather service and a network of surface meteorological stations. But weather forecasting was still a descriptive exercise based on experience and intuition (*Bergeron*, 1959). A century earlier *Pierre Simon de Laplace* (1799) had included evolution of weather in the phenomena following the laws of mechanics. But these laws were still poorly understood. This lack of understanding was frustrating to many scientists. The great German scientist Hermann von Helmholtz, quoted a poem by Johann Wolfgang von Goethe (*Helmholtz*, 1884): *"Es regnet wenn es regnen will, es regnet seinen Lauf, und wenn's genug geregnet hat, so hört es wieder auf,"* which expresses the thought that rain comes when it will and stops when there is enough. Von Helmholtz commented

> despite all new knowledge about cause and effect in nature, all new meteorological stations and long records of observations, it does not hit far from the truth. ...among all conditions in nature [weather is] the one that changes most deceptively, that impulsively and uncatchably evade[s] any attempt to be caught under the bounds of the law.

Vilhelm Bjerknes became the one who caught weather forecasting under the laws of the nature and gave birth to scientific weather forecasting. In January 1904 he published a short article in *Meteorologische Zeitschrift*, Wien (Bjerknes 1904a), titled "Das Problem der Wettervorhersage, betrachtet vom Standpunkte der Mechanik und der Physik" (*The Problem of Weather Forecasting as Seen From the Standpoint of Mechanics and Physics*). In this paper Bjerknes introduced hydro- and thermodynamics into meteorology and expressed his vision and programme for future weather forecasting.

The Nordic Seas: An Integrated Perspective
Geophysical Monograph Series 158

10.1029/158GM22

The article begins thus (*Bjerknes*, 1904b: a translation of *Bjerknes,* 1904a by *Mintz*, 1954):

> If it is true, as every scientist believes, that subsequent atmospheric states develop from the preceding ones according to physical law, then it is apparent that the necessary and sufficient conditions for the rational solution of weather forecasting problems are the following:
>
> 1. A sufficiently accurate knowledge of the state of the atmosphere at the initial time.
> 2. A sufficiently accurate knowledge of the laws according to which one state of the atmosphere develops from another.

Bjerknes used the medical expressions *diagnosis* and *prognosis* to describe the two steps. He stated that the diagnosis requires adequate observations to define the three-dimensional state of the atmosphere at a certain time, and that the prognosis must be calculated from known laws of the atmospheric dynamics and physics, i.e. conservation of momentum, mass, and energy.

When Bjerknes wrote his article, observations necessary for a three-dimensional analysis of the atmospheric state did not exist, far less the possibility of performing the calculations needed to estimate future states. Later Bjerknes was involved in measurement campaigns for large parts of Europe to assign dedicated days to atmospheric data collection in order to tackle the first step. At the same time, Bjerknes built up theoretical knowledge to tackle the scientific basis for step 2. His findings inspired Lewis Fry Richardson to make a courageous, but unsuccessful attempt to compute a 6-h forecast from a diagnosis (today often termed analysis) made by Bjerknes and his collaborators in Leipzig, Germany (*Richardson*, 1922).

After World War II, John von Neumann was central in the development of the first electronic computers. He set up a group of scientists to deal with the problem of figuring out the weather with the aid of computers, efforts based on Bjerknes' thinking, and Richardson's numerical experiment. A simplified model of atmospheric dynamics was, for the first time, successfully integrated forward in time to make a 24-h forecast (*Charney et al.*, 1950). The research field was called Numerical Weather Prediction (NWP). In parallel with the development of electronic computers, NWP has since developed tremendously. At the beginning of these developments stand the work and vision of Vilhelm Bjerknes.

2. FROM "ETHEREAL SPACE" TO CIRCULATION THEOREMS

Vilhelm Friman Koren Bjerknes was born 14 March 1862 in Kristiania (now Oslo), Norway. He graduated from the University of Kristiania in 1888, studied mathematics and physics in Paris and Bonn, and obtained his doctoral degree in Kristiania in 1892. He became a professor in mechanics and mathematical physics at Stockholm's Høgskola in 1895; professor in the same subject at the University of Kristiania (1907-1912); professor and head of Geophysics, University of Leipzig (1913-1917); professor in geophysics at Geofysisk Institutt (1917-1926) at Bergen Museum; and then back as professor in mechanics and mathematical physics at the University of Oslo (1926-1932). He died 9 March 1951.

2.1. Ideas of His Father

Bjerknes' interest in hydrodynamics was fostered by his father, Carl Anton Bjerknes, professor in mathematics and physics at the University of Kristiania. In his younger days, Carl Anton had studied hydrodynamics in Göttingen under Pierre Lejeune Drichlet. Much of his work was within the framework of "ethereal" space set by Decartes and Euler. In particular, he sought explanations about "action at distance", the nature of gravitational and electrical forces of attraction. As a student in 1881, Vilhelm Bjerknes, together with his father, demonstrated certain experiments on the motion of solid bodies in fluids, related to "action at distance", at the World Exposition of Electricity in Paris. The exhibition was visited by many of the leading scientists of the world. Their enthusiasm for the experiments resulted in a gold medal to Vilhelm and his father.

After attending lectures by Henri Poincaré in Paris about Maxwell's theory of electromagnetism, Vilhelm Bjerknes wanted to participate directly in experiments with electromagnetic waves, and in 1890 he arrived at Bonn University to work with Professor Heinrich Hertz. In Bonn he was engaged in both experimental and theoretical investigations. He developed a theory of electrical resonance, which even today is familiar to specialists in radio-physics. Hertz died suddenly in 1894. Although Bjerknes continued to work on the physics of electromagnetic waves, and in his older days returned to the idea of building a bridge between hydrodynamics and electromagnetism, his main scientific career was to turn in another direction.

After he obtained his doctoral degree on his work on electromagnetism, he began to teach mechanics in Stockholm in 1893, where he became a full professor in 1895. In Stockholm Vilhelm Bjerknes left the fashionable physics of the time in Bonn and went back to hydrodynamics, guided by ideas of his father. Eventually these studies were ended with two volumes that he wrote in honour of his father, who died in 1903: Vorlesungen über hydrodynamische Fernkräfte nach C.A. Bjerknes' Theorie (*Lectures on Hydrodynamical Forces at a Distance After C.A. Bjerknes' Theory*; Bjerknes 1900, 1902). Vilhelm tried to replace the solid vibrating spheres, studied by his father, with fluid particles of different density.

He then found it necessary to derive the laws of motion for fluids with varying density, i.e. a baroclinic fluid as compared with a barotropic fluid, where density is constant (or is a function of pressure only).

When Vilhelm Bjerknes turned his scientific efforts towards hydrodynamics, loyalty and piety to his father did play a role. *Godske* (1956), in his book about weather prediction, wrote

> This feeling of piety was surely partly a strength of Vilhelm Bjerknes, but maybe also his biggest weakness as a scientist. However, Bjerknes liked to go in his own direction, he had "the courage to do stupidities" (a famous saying by Bjerknes) and never liked to follow the mainstream.

This time his courage was rewarded by his circulation theorems.

2.2. The Circulation Theorems

Well before Vilhelm Bjerknes, Leonard Euler had pioneered the science hydrodynamics. At the end of the century, the science of thermodynamics was developed by scientists such as Helmholtz, Carnot, Kelvin, and Clausius. The summit of hydrodynamics was reached with two famous circulation theorems by Kelvin and Helmholtz, theorems that expressed the same result in two different ways. For ideal barotropic fluids, the theorems expressed that circulation along any closed contour of particles remains constant, thereby making the generation and decay of vortices impossible (Kelvin's theorem; von Helmholtz expressed the same idea as that vorticity is conserved following particles in such a fluid). It was straightforward to conclude that if no vortices were present in the fluid initially, then none would appear later. In this way the theorem was not applicable to the atmosphere, where circulations form and disappear. Only by studying the dynamics of baroclinic motions is it possible to understand real atmospheric circulations. This idea captured Vilhelm Bjerknes' mind for several years.

In 1897 Vilhelm Bjerknes formulated his famous circulation theorem in lectures to his students. That is, along any arbitrary closed contour of particles in a baroclinic fluid, the circulation might change, and the acceleration is equal to the number of solenoids (pipes) that are formed by the intersection of isobaric and isosteric (equal density) surfaces. From a mathematical point of view, the result can be considered a generalisation of the theorem of Kelvin to baroclinic fluids. His findings were published in 1898 (*Bjerknes*, 1898a), and in final form in 1902 (*Bjerknes*, 1902). Recently, *Thorpe et al.* (2003) have given a historical perspective of his work with the theorems.

In an historical essay (contained in *Bjerknes*, 1933, and translated from German by W.H. Berger, Scripps Institution of Oceanography, University of California–San Diegp, La Jolla, California), Vilhelm Bjerknes wrote

> From the beginning, I was greatly surprised that even the extremely simple equation describing velocity eddies was not already known for a long time. That equation is obviously as readily accessible by intuition as it is difficult to avoid analytically when deriving the Helmholtz conservation rule, either in the original form or in the later form provided by Kelvin. For the direct path to the conservation equations proceeds via the general eddy formation equation. Whether one takes notice of it or not depends only on whether one introduces the limiting barotropic condition somewhat later or somewhat earlier.

With his work, Vilhelm Bjerknes crossed the frontier between hydrodynamics and thermodynamics. In a speech he gave in 1912 at the 25th anniversary of the establishment of the Geophysical Department in Leipzig, Bjerknes said (*Bjerknes*, 1938a; translated to English by Lisa Shields, *Bjerknes*, 1938b)

> It would be a hopeless task to approach the more general field of thermodynamics: so all hydrodynamicists thought, and I thought so as well. ...Continuing my father's work, I was led, compellingly, towards the almost inevitable generalisation of the theorems of Helmholtz and Kelvin.

The special accomplishment of Vilhelm Bjerknes was perhaps not that he had generalised the circulation theorem, but that he was able to immediately see the potential of using it for describing circulations taking place in the atmosphere and the ocean, essentially in baroclinic environments. Immediately after his first publication in 1898, Vilhelm Bjerknes wrote a paper in which he demonstrated the possibility of using his theorems on the dynamics of the atmosphere and ocean to explain a wide range of natural phenomena, such as trade winds, breezes, monsoons, and certain ocean currents (*Bjerknes*, 1898b). This paper was the first geophysical paper by Vilhelm Bjerknes.

3. BJERKNES' VISION OF NUMERICAL WEATHER PREDICTION

In Stockholm, Bjerknes got to know many interesting scientists, such as Professor Svante Arrhenius. He became friends with Professor Nils Ekholm, one of the greatest meteorologists at that time. Bjerknes' friendship with Ekholm inspired the application of his theorems to atmosphere and ocean. Two of his students at that time were Vagn Walfrid Ekman, famous for his work on "dead water" (internal waves in the ocean) and wind-driven currents leading to the "Ekman spiral", and Johan W. Sandström, who made the first diagnostic applications of Bjerknes' circulation theorem (*Bjerknes*, 1933).

Working in Stockholm, Bjerknes started to contemplate the fundamental question of how to forecast the weather. He later wrote (*Bjerknes*, 1938b)

> I thought: Perhaps the time is ripe to combine observational meteorology and theoretical thermo-hydrodynamics, which up till then had been developing in isolation. And that is best done, thought I, by boldly and directly addressing the central task of meteorology: that of deducing the future state of the atmosphere from its present state. In other words, the problem of weather forecasting as a mathematical problem had to be stated and, if possible, solved.

Vilhelm Bjerknes' short article in 1904 was a summary and a synthesis of the new dimension of his research, as well as his theoretical and practical programme on scientific weather prediction. He saw what should be done. However, he was uncertain of his own role in solving the problem of weather forecasting. As he later wrote (*Bjerknes*, 1938b),

> I published this programme of meteorological research in 1904 in the journal Meteorologische Zeitschrift—perhaps really more to get rid of the problem than to tackle it myself. I did not consider myself to be the right man for the job. I had no training as a meteorologist, and the problem was a colossal one, for which help was necessary.

When Vilhelm Bjerknes wrote his article, the network of meteorological observations included ground stations only. However, Bjerknes understood the need for upper-air data and wrote (*Bjerknes*, 1904b)

> We can hope...that the time will soon come when either as a daily routine, or for certain designated days, a complete diagnosis of the state of the atmosphere will be available. The first condition for putting forecasting on a rational basis will then be satisfied.

Such designated days, when upper-air observations were made throughout parts of Europe, were organised a little later by the International Commission for Scientific Aeronautics, in which Bjerknes was active for many years. Balloons loaded with meteorological instruments that ascended to more than 10 000 m were released several times daily for certain periods. In the stratosphere the balloons burst, and the instruments that recorded pressure, temperature, and humidity fell to the ground. In his time as director at the geophysical department in Leipzig, Bjerknes and his collaborators analysed such measurements.

In his 1904 article, Bjerknes stated that the prognostic step comprised the solution of the differential equations that express the rate of change of the state of seven dependent variables: pressure, temperature, density, humidity, and the three components of the wind vector. He identified seven independent equations available to compute the change in the state: the three components of the equation of motion (Newton's Second Law): the equation of continuity expressing conservation of mass; the equation of state that ties together pressure, temperature, and density; and two equations on conservation of energy (first and second laws of thermodynamics). Bjerknes' article was of a philosophical kind; he did not express the equations in detail and gave no procedure or algorithm to solve them. *Eliassen* (1999) has pointed out that he made a mistake in including the second law of thermodynamics, when instead he should have specified a continuity equation for water vapour, as was later done by Richardson (*Richardson*, 1922).

Bjerknes realised that the equations from theoretical physics could not readily be solved for real atmospheric states; the computing amount was simply too extensive. He hoped to develop simplified graphical methods but did not succeed except for very simple, idealised flows. Nevertheless, his belief in technical and scientific progress was huge. When in January 1913 he gave his inaugural lecture in Leipzig, he said (*Bjerknes*, 1914a)

> Now that complete observations from an extensive portion of the free air are being published in a regular series, a mighty problem looms before us, and we can no longer disregard it. We must apply the equations of theoretical physics, not to ideal cases only, but to the actual existing atmospheric conditions as they are revealed by modern observations....The problem of accurate pre-calculation that was solved for astronomy centuries ago must now be attacked in all earnest for meteorology.

He realised the complexity of the problem and that a rational forecasting procedure might take more time than the atmosphere itself takes to evolve. He concluded that only if the calculations agreed with facts would a scientific victory be won. Meteorology would then have become an exact science. He himself was convinced that the problems would be overcome and wrote (*Bjerknes*, 1914a)

> It can take years to bore a tunnel through a mountain. Many a worker may not live to see the day of the breakthrough. Nevertheless, that will not prevent others from riding through the tunnel later at the speed of an express train.

4. TOWARDS SCIENTIFIC WEATHER FORECASTING

It was the circulation theorems more than his programme for forecasting that made Vilhelm Bjerknes known and honoured as a scientist in the beginning of the century. The programme expressed in the paper of 1904 influenced efforts internationally to extend the network of observations, in particular to upper air. His idea to compute future states of the atmosphere was ignored, which is understandable because theoretical considerations and any practical application seemed so far away. As already mentioned, also Vilhelm Bjerknes was reluctant to dedicate himself to meteorology. However, a lecture he gave changed his destiny.

4.1. The Grant From Carnegie Institution of Washington

In 1905 Bjerknes accepted an invitation from Colombia University in New York to lecture on the theory of hydrodynamical forces. Following an invitation from the Carnegie Institution of Washington, he presented a lecture about his strategic plan for the solution of the weather forecasting problem (*Jewell*, 1994). Later, Bjerknes wrote (*Bjerknes*, 1938b)

> The programme excited the interest of the Americans, perhaps because of its boldness and colossal dimensions. The Carnegie Institution of Washington, which had then just been set up, decided to give financial support to my work on this programme. This was also done very generously: from 1906 I had, thanks to an annual grant, the possibility of taking on one or more personal scientific assistants. *Thus my fate was sealed.* I entered, as an interloper, into a science strange to me, meteorology. With my Carnegie assistants I immediately began to work out the basics of the forecasting problems, by applying all the available observational material of the day of the upper-air ascents.

If at any time a lecture was a success, it was this. As Harald Ulrik Sverdrup, one of Bjerknes' Carnegie assistants, wrote,

> I do not know any examples of grants of this type, which have yielded strong results and that were used for the good of science in a more effective way.

Backed by this support, Bjerknes systematically developed the methods of analysing the atmospheric states and motions together with various collaborators. The research activity kept the support of the Carnegie Institution until 1948, a few years before Vilhelm died (*Godske et al.*, 1957, p. 624).

Bjerknes' Carnegie assistants—normally very young and sometimes without formal education—made significant scientific contributions. The following were his assistants: Johan W. Sandström, later director of the Swedish weather service; Olav Devik, physicist and legendary executive in the Department of Church and Education; Thor Hesselberg, director of the Norwegian Meteorological Institute from age 29 until he was 70 years old, and general secretary of World's Meteorological Organisation at that time; Harald U. Sverdrup, assistant in Leipzig, professor in meteorology in Bergen after Bjerknes, director for Scripp's Institution of Oceanography, California, director of the Norwegian Institute for Polar Research, and professor at the University of Oslo (Nierenberg at http://www.nap.edu/html/biomems/hsverdrup.html); Svein Rosseland, professor of astrophysics at the University of Oslo (*Elgarøy*, 1995); Jacob Bjerknes, the son of Vilhelm Bjerknes, professor in meteorology in Bergen after Sverdrup and later at UCLA, California (Eliassen at http://www.nap.edu/readingroom/books/biomems/jbjerknes.html); Halvor Solberg, professor in meteorology at the University of Oslo; Tor Bergeron, professor in meteorology at Uppsala University (*Liljequist*, 1980); Carl G. Rossby, professor in meteorology in Chicago and at Stockholm's University (*Phillips*, 1998); Carl L. Godske, professor in meteorology at the University of Bergen; Einar Høiland, professor in applied mathematics at the University of Oslo. Ernst Calwagen, an assistant in Bergen, died in an aircraft accident in 1925 while doing meteorological flight observations.

4.2. Developing Theory

In Stockholm, Vilhelm Bjerknes started to develop theory on the dynamics of the atmosphere as needed steps towards scientific weather forecasting. His first assistant, J. W. Sandström, was a very important collaborator. Entering the university, he had no formal education at all. He was, however, very talented and was considered an "uncut diamond" in science (*Bergeron et al.*, 1962).

Bjerknes' departure from Stockholm in 1907 for a professorship in Kristiania is thought to be connected with the splitting of the union between Norway and Sweden in 1905 (*Godske*, 1956). The extensive work was continued in Norway with his assistants O. Devik and T. Hesselberg. Eventually the work from Stockholm and Kristiania was collected in two volumes, *Dynamic Meteorology and Hydrology* (*Bjerknes et al.*, 1910/1911). The first volume, "Statics", was written with Sandström; the second, "Kinematics", with Devik and Hesselberg. Among other things, the books laid the foundation for weather analyses on pressure levels applied in Leipzig shortly afterwards and inspired Richardson to his marvellous experiment. The idea of scientific weather prediction now became more appealing to scientists.

4.3. From Leipzig to Bergen

Since the days of giants such as von Helmholtz, meteorological research was strong in Germany (*Volkert*, 1999). When Bjerknes in 1912 was called from Kristiania to Leipzig to make his programme come true and to make meteorology an exact science, it therefore represented an international acceptance of Bjerknes' programme. His task was to start the new institute and to work according to lines drawn in his writings, primarily the paper from 1904.

When Bjerknes arrived, 12 eager doctoral students were ready to tackle the theoretical tasks that waited. Bjerknes gave lectures and started research activities, among them analyses of observations from the previously mentioned aerological measurement campaigns. Large-format analysis maps were published. However, when the First World War started, his students became soldiers, and times became very difficult in Germany. Bjerknes was offered a position in Bergen. He accepted and came to Bergen in August 1917 with two of his Carnegie assistants, Jacob Bjerknes and Halvor Solberg.

As Liv Nansen Høyer, daughter of Fridtjof Nansen, a leading Norwegian scientist at the time and an effective diplomat, wrote (*Nansen Høyer*, 1954), Vilhelm Bjerknes' mother, who lived in Oslo, was upset by the living conditions of her son and his Norwegian assistants in Germany. She went to Nansen and asked him to help them leave Germany. The Geophysical Institute in Bergen was at that time newly established, with Professor Helland-Hansen as professor in oceanography. Nansen saw the opportunity to get Bjerknes to come to Bergen, Bergen citizens raised money, and Bjerknes was enticed to return to Norway.

4.4. The Bergen School

In Bergen the main task was directed towards new methods for practical weather prediction, based as far as on possible scientific methods. The strict theoretical research from Leipzig was in this way partly relaxed. *Godske* (1956) writes that the sensible direction for Bjerknes would have been to continue the theoretical approach he started in Leipzig—in a more modest Norwegian way. But Bjerknes realised the importance that forecasting could have for fishing, transport at sea, farming, etc., factors that influenced his choice to accept the offer from Bergen. However, many new questions arose: How should improvements in weather prediction be introduced, and was a typical theoretician the right person for a substantial reorganisation of the weather service?

Bjerknes' efforts were crowned with the Bergen School of Meteorology, the group that discovered the weather fronts as lines of convergence near the surface of the earth and introduced the concepts of air masses, the polar front, and the evolution of cyclones on the polar front, where small disturbances grow to mature extratropical lows (*Bergeron*, 1959; *Friedman*, 1989). Much of this work rested on precursors of frontal theory from the Leipzig period. The research took place as Vilhelm Bjerknes established a weather service, where he managed to increase the number of Norwegian meteorological stations by a factor of 10 (for an account of how this was done, see *Godske*, 1956). The discoveries immediately led to a revision of the methods of weather prediction. In this way the forecasting problem came closer to Bjerknes' dynamical principles, although the prediction step was still far away from the ideas expressed in Bjerknes' programme in 1904. The hope was that concentrating on the diagnostic step—including analyses of fronts in different stages of cyclone developments—would increase insight into the second step and thus improve forecasting.

During this period Bjerknes managed to increase the support from the Carnegie Institution and got more young assistants. The discoveries by the Bergen School and the improvement of the forecasting were due to the work of the youngsters, first of all Halvor Solberg, Jacob Bjerknes, and Tor Bergeron (Ernst Calwagen should also be remembered). *Godske* (1956) described them as follows:

> ...Halvor Solberg, the solid mathematician, Jack Bjerknes, who had the ability to see simplicity in complicated physical systems, and Tor Bergeron, who arrived in 1919, a well of enthusiasm and ideas.

The young scientists wrote about their findings and presented results to colleagues abroad. With the help of "the old one", as they referred to their professor, they also introduced modern weather forecasting to the Norwegian people. Around the three named, more young scientists gathered, among the best known of whom were Carl Gustav Rossby and Svein Rosseland. Scientists and weather forecasters from all over the world, e.g. from Japan (*Fujiwara*, 1949) and Philippines, came to Bergen to learn, and eventually the methods were introduced at weather offices throughout the world. The title of Fujiwara's paper (1949), "The men who had a deep insight into nature—The birth of the Bergen School", indicates the significance of their activity.

As an example of the efforts of the Bergen School to spread their methods internationally, Tor Bergeron gave lectures in Finland (*Palmén*, 1999) and Russia. In Russia he inspired S.P. Chromov to issue a handbook for synoptic weather analysis in Russian in 1934. The book was later translated into German by G. Swoboda (*Chromov*, 1940). In both languages the book was important to meteorologists in eastern Europe and Germany.

Vilhelm Bjerknes inspired and entrusted his young collaborators and gave them freedom and responsibility. He did not draw a single weather map himself but gave all the honour to his young assistants. During the financial boom at the end of the war, it was hard to find workers. Bjerknes later wrote (*Bjerknes*, 1944b)

> We looked in vain for a new stove-boy. I was the one who had free hands to do this work. With my old and slow-working brain I regarded myself as quite unfit to compete with the young meteorologists or assistants in the express-work at the weather

> maps. But I tried to keep the fire going, both spiritually and materially.

All the time Vilhelm Bjerknes gave lectures to his assistants and others, published scientific papers, and toiled to extend his first textbooks from 1910/1911 on dynamical meteorology. In 1933 he published the 797-pp *Physikalische Hydrodynamik mit Anwendungen auf die dynamische Meteorologie*, written by himself, J. Bjerknes, H. Solberg, and T. Bergeron. For a long time this book was the "Bible" to all students in meteorology (*Bjerknes et al.*, 1933). The following year the book was also published in French in Paris (*Bjerknes et al.*, 1934).

4.5. Back to Oslo

When Bjerknes returned to Oslo in 1926, he once more tried to build a bridge between hydrodynamics and electromagnetism. However, meteorology was still strong in his mind. In particular, he was troubled by the weak theoretical foundation for the conceptual cyclone and front models of the Bergen School. Together with Halvor Solberg, now a professor in Oslo, he started to work on these problems. Eventually, the Bergen School on weather forecasting was in Oslo supplemented by theoretical efforts by a new group, led by Bjerknes' last assistant, Einar Høiland. Among the members of that group were Arnt Eliassen and Ragnar Fjørtoft.

5. RICHARDSON'S EXPERIMENT

The Bergen School did not fulfill Bjerknes' programme from 1904, lacking the step to compute future states from the law of physics. However, the British scientist Lewis Fry Richardson made an extremely important attempt on this task.

Richardson was inspired by Bjerknes to compute future states of the atmosphere from an analysis resulting from one of the measuring campaigns mentioned earlier (*Bjerknes*, 1914b). In 1922 he published his remarkable book, *Weather Prediction by Numerical Process* (*Richardson*, 1922). The research had started several years earlier, since the time he took up employment with the British Meteorological Office in 1913 and read about Bjerknes' plan for rational forecasting. In the preface to his book Richardson wrote

> The extensive researches of V. Bjerknes and his School are pervaded by the idea of using the differential equations for all that they are worth. I read his volumes on Static and Kinematics soon after beginning the present study, and they have exercised a considerable influence throughout it.

The writings he referred to are the two volumes of *Bjerknes et al.* (1910/1911).

Richardson made the governing equations suitable for the atmosphere (*Eliassen*, 1955). He was the first to include the quasi-static approximation into the equations. This rests on the assumptions that the vertical accelerations may be ignored in the equation of motion, and that the horizontal component of the earth's angular velocity may be ignored in the expression for the Coriolis force. He replaced the differential equations with numerical methods that gave finite differences on a regular grid, where the computations were made. He reduced the number of points in the grid as much as possible and computed changes in pressure, wind, and temperature at two grid points in the middle of Europe. His results, however, gave a pressure change of 145 hPa in 6 h, a completely unrealistic result. The Irish scientist Peter Lynch has repeated Richardson's computation on a modern computer (*Lynch*, 1999). He obtained exactly the same result and showed that the unrealistic results were due to an imbalance of the field of mass and the field of the wind in the initial conditions. Applying a modern method to balance the initial fields (initialization), he got the reasonable result of 1 hPa change over 6 h.

Richardson was a giant of a scientist and an inspirer in several fields of meteorology. His impressive experiment will stand in history as a masterpiece that after World War II laid the foundation for modern NWP. Richardson met Vilhelm Bjerknes when Richardson was present at the International Aerological Commission's meeting in Bergen in 1921. As a Quaker, Richardson was a great admirer of the peace work by Nansen. After a visit to Richardson's place in Scotland, *Godske* (1956) wrote

> He was one of the great; an unclear, genius person who dug deeper than many magnificent talents. He did not stay long in meteorology. When the Meteorological Office came under the Department of Defence, the Quaker and friend of peace left the service. He wrote works on "the mathematical equations that determine the risk of war"—and ended his days as a teacher in a Scottish school for boys. The big empire did not afford an appropriate position for this strange man, one of the sharpest brains in our time.

A biography of Richardson has been written by *Ashford* (1985). Other articles and essays on Richardson related to his work on NWP are written by *Chapman* (1965), *Platzmann* (1967), *Hunt* (1993), and *Lynch* (1999, 2002).

6. TOWARDS MODERN WEATHER PREDICTION

The development of the computer in Princeton, New Jersey, USA, by von Neumann and his group during the years 1946 to 1952 is well documented (*Goldstine*, 1972; *Macrae,* 1999). One of the leading mathematicians of the century, von Neumann knew about the work of Bjerknes and Richardson.

At the same time he set up a group that started NWP and made the first successful numerical weather forecast. Several authors have reviewed this achievement (e.g. *Platzman*, 1968; *Lynch*, 2002). In historical documentation, Bjerknes' paper from 1904 is always cited as the beginning of NWP, whereas the first results in Princeton came a year or two before Vilhelm Bjerknes died. Bjerknes was probably informed of the achievements. None of his former Carnegie assistants took direct part in the group in Princeton. However, the two Norwegians Arnt Eliassen and Ragnar Fjørtoft were important members of the Princeton group. Both were scientists educated in Oslo in the tradition built by Bjerknes and had been taught by Bjerknes' former assistants: H. Solberg, E. Høiland, and C.L. Godske.

Both Eliassen and Fjørtoft returned to Oslo—Eliassen as professor at the Department of Geophysics at the University of Oslo and Fjørtoft as Director of the Norwegian Meteorological Institute. In several ways they both were active in NWP for many years— Eliassen through his teaching and fundamental research in dynamical meteorology (for which he received the Balzan prize in 1996), and Fjørtoft through operational NWP at the weather service. We have earlier mentioned that Vilhelm Bjerknes believed in graphical methods to integrate from one state of the atmosphere to the next. Indeed, Fjørtoft (1952) developed a graphical method, based on quasi-Lagrangian time integration. Between 1950 and 1960 this method was faster than those using computers.

NWP was developed all over the world, e.g. in Britain, Germany, and Japan, but first of all in USA. A major step towards improved NWP came when the European Centre for Medium Range Weather Forecasts (ECMWF) was established in Reading, England, in 1975. ECMWF became a success and the leading centre for NWP in the world (*Bengtsson*, 1991). The progress has continued until now. Much like the Bergen School, ECMWF has improved weather forecasting through excellent research in an operational environment. Lennart Bengtsson was the first head of research at ECMWF and the second director. He has characterised the methods of NWP as "the methods of brutal force", first of all because of the use of extensive computations by supercomputers. The expression can clearly be traced back to Bjerknes' paper of 1904 and his statement on "tunnels" in his famous talk in Leipzig in 1912.

Some have said that the thoughts of Bjerknes led people to believe that weather, in principle, can be predicted for long time periods only if the initial state is sufficiently accurately known and the computational accuracy sufficiently precise. I have not found any literature that supports such a view. Thoughts in the 1950s on predictability, summarised by *Bergeron* (1959), point to the opposite. Concern is expressed for lack of predictability, in particular when baroclinic instability is involved. I do not think Bjerknes, Richardson, or the scientists in Princeton ever believed in determinism for weather forecasting. Any such belief, if it ever existed, was buried when Edward Lorenz introduced chaos theory (*Lorenz*, 1963). A consequence of the theory limits how far ahead the weather might be predicted. More and better observations, increased computational accuracy with better algorithms and faster computers, and better understanding of the micro-scale physical processes in the atmosphere, have helped increase the practical predictability of NWP. Nevertheless, as Lorenz showed, small errors in the initial state and small errors in the descriptions of the subscale physical relations eventually corrupt the forecast. Today, as a consequence of Lorenz's work, ensemble prediction has been introduced at centres like ECMWF. With this concept, numerous different, but probable, initial states are estimated, and numerical prognoses are made from each of them. In a way, a probabilistic approach is taken, contrary to the customary deterministic approach proposed by Bjerknes.

7. THE HERITAGE OF VILHELM BJERKNES

Vilhelm Bjerknes' main fortune was to transform meteorology and weather forecasting from an activity based on experience to an exact science in line with physics. From 1905, when he was 43 years old, until his death in 1951, Vilhelm Bjerknes dedicated his scientific strength to his programme for scientific weather forecasting. Continuously and systematically he worked to develop and expand theory on dynamic of the atmosphere. Through the Bergen School the methods of weather forecasting came closer to his principles built on physics. His assistants renewed meteorological research and weather forecasting in Scandinavia and throughout the world. Several of Bjerknes' assistants went to USA before World War II (*Newton and Newton*, 1999) and became important in building up meteorological education and research in the USA (*Pettersen*, 1940, 1974). Here, just before Bjerknes' death, the first breakthrough for his vision of scientific weather prediction was accomplished. Today, nearly all weather forecasting follows the principles first set up by Bjerknes.

His assistants have described Vilhelm Bjerknes as a legendary scientist with a wide range of fine qualities (*Bergeron et al.*, 1962). He had visions and believed in them; he had courage to do research in his own way; he owned a natural talent to convince authorities to support his work; his marvellous ability to pick and inspire young scientists could not be overemphasised; he published all his work carefully, not only scientific papers, textbooks and essays for colleagues, but also newspaper articles and speeches for laymen (a bibliography has been set up by his assistants, here referred to as *Bjerknes* [1962]); he showed absolute responsibility for resources entrusted to him and reported carefully to his donors.

Vilhelm Bjerknes was beyond doubt one of the greatest Nordic scientists. The Bergen School might likewise be judged as the most important contribution from our countries to international natural science. The international character of the research should be stressed. Bjerknes was a number of times nominated for the Nobel Prize in physics. Ironically, the central argument of the Swedish Science Academy against him was that his work did not represent an exact science (*Friedman*, 2001).

The story of the life of Vilhelm Bjerknes and his achievements has not yet been properly told. The Bjerknes Centre for Climate Research has recently been established at the University of Bergen. This contributes to honour Bjerknes' name and to renew interest for Bjerknes' work. Perhaps there will be more studies on this determined man, who a hundred years ago set up a realistic programme for modern weather forecasting that very few believed in.

Acknowledgments. The author thanks Professor W.H. Berger, Professor P. Lynch, Mr. A. Persson, and Professor A. Thorpe for their support.

REFERENCES

Ashford, O.M., *Prophet—or Professor? The Life and Work of Lewis Fry Richardson*. Adam Hilger, Bristol and Boston, 1985.

Bengtsson, L., Advances and prospects in numerical weather prediction. *Q.J.R. Meteor. Soc.* 117, 855-902, 1991.

Bergeron, T., Methods in scientific weather analysis and forecasting. An outline in the history of ideas and hints at a program. In *The Rossby Memorial Volume*, The Rockefeller Institute Press in association with Oxford University Press, New York, pp 440-474, 1959.

Bergeron, T., O. Devik, C.L. Godske, Vilhelm Bjerknes, March 14, 1862–April 9, 1951. 1. Family background, education, student of Hertz, by O. Devik. 2. The Stockholm period, by T. Bergeron. 3. The years in Christiania (Oslo and Leipzig), by O. Devik. 4. The Bergen School, by T. Bergeron. 5. Back in Oslo, by C.L. Godske. Issue of *Geophysical Publications* in memory of the 100th anniversary of the birth of V. Bjerknes, pp 7-26, 1962.

Bjerknes, V., Über die Bildung von Zirkulationsbewegungen und Wirbeln in reibungslosen Flüssigheiten. *Skrifter utg. Af Videnskabsselskabet i Christiania*. I. Math-naturv. Klasse, Nr 5, 29 pp, 1898a.

Bjerknes, V., Über einen hydrodynamischen Fundamentalsatz und seine Anwendung besonders auf die Mechanik der Atmosphäre und des Weltmeeres. K. *Svenska vetenskaps-akademiens handlingar*. N.F. 31, Nr 4, 35 pp, 1898b.

Bjerknes, V., *Vorlesungen Über hydrodynamische Fernkräfte nach C.A. Bjerknes's Theorie*. Lpz. 1. Mit vierzig Figuren im Text. 1900 XVI + 388 pp. 2. Mit sechzig Figuren im Text und zwei Tafeln. 1902 XVI + 316 pp, 1900/1902.

Bjerknes, V., Cirkulation relativ zu der Erde. *Meteorological Zeitschrift*, Braunschweig, 19, 97-108, 1902.

Bjerknes, V., Das Problem der Wettervorhersage, betrachtet vom Standpunkte der Mechanik and der Physik. *Meteorologische Zeitschrift*, Wien, 21, 1-7, 1904a.

Bjerknes, V., 1904b. The Problem of Weather Forecasting as a Problem in Mechanics and Physics. Translated by Y Mintz in 1954. In *The Life Cycles of Extratropical Cyclones* (M. Shapiro and S. Grønås, Eds), American Meteorological Society, Boston, 1999.

Bjerknes, V., and different collaborators, *Dynamic Meteorology and Hydrography*. 1. Statics, by V.B and J.W. Sandström, 1910, 146 pp. 2. Kinematics, by V.B., Th. Hesselberg, and O. Devik, 1911, 175 pp. Carnegie Institution of Washington, Publication No 88. 1910/1911.

Bjerknes, V., Meteorology as an exact science. Translation of Die Meteorologie als exakte Wissenschaft. Antrittsvorlesung gehalten am 8. Jan 1913 in der Aula der Universität Leipzig. *Mon. Wea. Rev*., 42, 11-14, 1914a.

Bjerknes, V., *Veröffenlichungen des Geophysikalischen Instituts der Universität Leipzig. Heraugegeben von dessen Direktor V. Bjerknes*. Erste Serie: Synoptische Darstellungen atmosphärischer Zustände. Jahrgang 1910, Heft 3. Zustand der Atmosphäre über Europa am 18., 19., und 20. Mai 1910, Leipzig 1914, 1914b.

Bjerknes, V., Bibliographie mit historischen Erläuterungen. In Bjerknes, V., J. Bjerknes, H. Solberg, T. Bergeron, 1933. *Physikalische Hydrodynamik mit Anwendung auf die dynamische Meteorologie*. Springer Verlag, Berlin, 797 pp, 1933.

Bjerknes, V., Leipzig–Bergen. Festvortrag zur 25-Jahrfeier des Geophysikalischen Instituts der Universität Leipzig. *Zeitschrift für Geophysik*, 14, 49-62, 1938a.

Bjerknes, V., Leipzig–Bergen. Jubilee address on the 25th anniversary of the Geophysical Institute of the University of Leipzig. *Zeitschrift für Geophysik*, 14, 49-62. English translation by Lisa Shields. Available at Met Ëireann, The Irish Meteorological Service, Glasnevin Hill, Dublin, 1938b.

Bjerknes, V., *Hvordan Bergensskolen blev til*. Vervarslinga på Vestlandet 25 år. Festskrift utgitt i anledning av 25-årsjubileet 1. juni 1943. John Grieg Publishing, Bergen, pp. 7-19, 1944a.

Bjerknes, V., 1944b. *How the Bergen School Came into Existence*. English translation of Bjerknes (1944a) by R. Jewell. Available at Geophysical Institute, University of Bergen.

Bjerknes, V., Publications by Vilhelm Bjerknes. In Issue of *Geophysical Publications* in memory of the 100th anniversary of the birth of V. Bjerknes, pp 26-37, 1962.

Bjerknes, V., J. Bjerknes, H. Solberg, T. Bergeron, *Physikalische Hydrodynamik mit Anwendung auf die dynamische Meteorologie*. Springer Verlag, Berlin, 797 pp, 1933.

Bjerknes, V., J. Bjerknes, H. Solberg, T. Bergeron, *Hydrodynamique physique avec applications á la meteorologie dynamique*. Paris, 864 pp, 1934.

Chapman, S., 1965. *Introduction to Dover edition of Weather Prediction by Numerical Process*, pp v-x. (see Richardson 1922).

Charney, J.G., R. Fjørtoft, and J. von Neumann, 1950. Numerical integration of the barotropic vorticity equation. *Tellus*, 2, 237-254.

Chromov, S.P., *Einführung in die synoptische Wetteranalyse*. Unter Mitwirkung von N. Koncek, deutsch bearbeitet von G. Swoboda. Verlag von Julius Springer, Wien, 1940.

Elgarøy, O., Frontiers of Astrophysics. Proceedings of the Rosseland Centenary Symposium, held in Oslo, Norway, June 16-17, 1994, 1995.

Eliassen, A., *Lectures on Physical Weather Prediction*. Given at University of California, Los Angeles. 136 pp. Available at Department of Meteorology, University of Oslo, 1955.

Eliassen, A., Vilhelm Bjerknes' early studies of atmospheric motions and their connection with the cyclone model of the Bergen School. In *The Life Cycles of Extratropical Cyclones*, (M. Shapiro and S. Grønås, Eds) Americal Meteorological Society, Boston, pp 5-13, 1999.

Fjørtoft, R., On a numerical method of integrating the barotropic vorticity equation. *Tellus*, 4,3, 1952.

Friedman, R.M., *Appropriating the Weather: Vilhelm Bjerknes and the Construction of a Modern Meteorology*. Cornell University Press, 251 pp, 1989.

Friedman, R.M., *The Politics of Excellence: Behind the Nobel Price in Science*. Times Book, New York, 379 pp, 2001.

Fujiwara, S., 1949. The men who had a deep insight into nature – The birth of the Norwegian School. *Astronomy and Meteorology*, a Japanese popular magazine, August 1949. English translation at Geophysical Department, University in Bergen.

Godske, C.L, *Hvordan blir været?* J.W Cappelens forlag, Oslo, pp 258, 1956.

Godske, C.L., T. Bergeron, J. Bjerknes, H.C. Bundgaard, *Dynamic Meteorology and Weather Forecasting*. American Meteorological Society, Boston, 800 pp, 1957.

Goldstine, H.H., 1972. *The Computer from Pascal to von Neumann*. Reprinted with new preface, 1993. Princeton University Press.

Helmholtz, H. von, *Wirbelstürme und Gewitter*. Vorträge u. Reden, 2, p. 141, Braunschweig, 1884.

Hunt, J.C.R., A general introduction to the life and work of L.F. Richardson. In *Collected papers of Lewis Fry Richardson*, Vol I (P.G. Drazin Gen. Ed.), Cambridge University Press, pp 1-27, 1993.

Jewell, R., Vilhelm Bjerknes' duty to produce something clear and real in meteorological science. In *The Earth, the heavens and the Carnegie Institution of Washington*, ed. G. Good, 37-46. History of Geophysics, vol. 5 Washington, D.C., American Geophysical Union, 1994.

Laplace, P.S., 1799. *Traité de Mécanique céleste, Tome I*. Paris (English translation by N. Bowditch, Boston, 1829.)

Liljequist, G.H., Tor Bergeron, a biography. *Pure Appl. Geophys.*, 119, 413-442, 1980.

Lorenz. E., Deterministic non-periodic flow. *J. Atmos. Sci*, 20, 130-141, 1963.

Lynch, P., Richardson's marvellous forecast. In The *Life Cycles of Extratropical Cyclones*, (M. Shapiro and S. Grønås, Eds), American Meteorological Society, Boston, pp 61-73, 1999.

Lynch, P., Weather Forecasting – From Woolly Art to Solid Science. In *Meteorology at the Millennium*. Academic Press, pp 106-119, 2002.

Macrae, N., 1999. *John von Neumann*. American Mathematical Society. Originally published by Pantheon Books, 1992.

Nansen Høyer, L., *Nansen og verden*, Cappelen, Oslo, pp 330, 1954.

Newton, C.W., and H.R. Newton, The Bergen School Concepts Come to America. In The *Life Cycles of Extratropical Cyclones*, (M. Shapiro and S. Grønås, Eds), American Meteorological Society, Boston, pp 41-60, 1999.

Palmén, E., Personal recollections about the Bergen School. In The *Life Cycles of Extratropical Cyclones*, (M. Shapiro and S. Grønås, Eds), American Meteorological Society, Boston, pp 75-79, 1999.

Pettersen, S., *Weather analysis and Forecasting*. McGraw-Hill Book Co., 505 pp, 1940.

Pettersen, S., *Kuling fra Nord*. H. Aschehoug, Oslo, 312 pp, 1974.

Phillips, N.A., Cal-Gustaf Rossby: His times, personality, and actions. *Bull. Am. Meteorol. Soc.*, 79, 1097-1112, 1998.

Platzman, G.W., A retrospective view of Richardson's book on weather prediction. *Bull. Am. Meteorol. Soc.*, 49, 496-500, 1967.

Platzman, G.W., The ENIAC computations of 1950 – gateway to numerical weather prediction. *Bull. Am. Meteorol. Soc.*, 60, 302-312, 1968.

Richardson, L.F., 1922. *Weather Prediction by Numerical Process*. Cambridge University Press. (Reprinted by Dover Publications, New York, 1965, with a new introduction by Sydney Chapman.)

Thorpe, A.J., H. Volkert, and M.Z. Ziemianski, The Bjerknes' circulation theorem: an historical perspective. *Bull. Amer. Met. Soc.*, 84, 471-480, 2003.

Volkert, H., Components of the Norwegian Cyclone Model: Observations and Theoretical Ideas in Europe Prior to 1920. In The *Life Cycles of Extratropical Cyclones*, (M. Shapiro and S. Grønås, Eds), American Meteorological Society, Boston, pp 15-28, 1999.

S. Grønås, Department of Geophysics, University of Bergen, Bergen, Norway.